S0-CFL-528

INTEGRALS CONTAINING $\sqrt{a + bu}$

14. $\displaystyle\int u\sqrt{a + bu}\, du = \frac{2}{15b^2}\,(3bu - 2a)(a + bu)^{3/2} + C$

15. $\displaystyle\int u^2\sqrt{a + bu}\, du = \frac{2}{105b^3}\,(15b^2u^2 - 12abu + 8a^2)(a + bu)^{3/2} + C$

16. $\displaystyle\int u^n\sqrt{a + bu}\, du = \frac{2u^n(a + bu)^{3/2}}{b(2n + 3)} - \frac{2an}{b(2n + 3)}\int u^{n-1}\sqrt{a + bu}\, du$

17. $\displaystyle\int \frac{u\, du}{\sqrt{a + bu}} = \frac{2}{3b^2}\,(bu - 2a)\sqrt{a + bu} + C$

18. $\displaystyle\int \frac{u^2\, du}{\sqrt{a + bu}} = \frac{2}{15b^3}\,(3b^2u^2 - 4abu + 8a^2)\sqrt{a + bu} + C$

19. $\displaystyle\int \frac{u^n\, du}{\sqrt{a + bu}} = \frac{2u^n\sqrt{a + bu}}{b(2n + 1)} - \frac{2an}{b(2n + 1)}\int \frac{u^{n-1}\, du}{\sqrt{a + bu}}$

20. $\displaystyle\int \frac{du}{u\sqrt{a + bu}} = \begin{cases} \dfrac{1}{\sqrt{a}}\ln\left|\dfrac{\sqrt{a + bu} - \sqrt{a}}{\sqrt{a + bu} + \sqrt{a}}\right| + C & (a > 0) \\[3mm] \dfrac{2}{\sqrt{-a}}\tan^{-1}\sqrt{\dfrac{a + bu}{-a}} + C & (a < 0) \end{cases}$

21. $\displaystyle\int \frac{du}{u^n\sqrt{a + bu}} = -\frac{\sqrt{a + bu}}{a(n - 1)u^{n-1}} - \frac{b(2n - 3)}{2a(n - 1)}\int \frac{du}{u^{n-1}\sqrt{a + bu}}$

22. $\displaystyle\int \frac{\sqrt{a + bu}\, du}{u} = 2\sqrt{a + bu} + a\int \frac{du}{u\sqrt{a + bu}}$

23. $\displaystyle\int \frac{\sqrt{a + bu}\, du}{u^n} = -\frac{(a + bu)^{3/2}}{a(n - 1)u^{n-1}} - \frac{b(2n - 5)}{2a(n - 1)}\int \frac{\sqrt{a + bu}\, du}{u^{n-1}}$

INTEGRALS CONTAINING $a^2 \pm u^2$ $(a > 0)$

24. $\displaystyle\int \frac{du}{a^2 + u^2} = \frac{1}{a}\tan^{-1}\frac{u}{a} + C$

25. $\displaystyle\int \frac{du}{a^2 - u^2} = \frac{1}{2a}\ln\left|\frac{u + a}{u - a}\right| + C$

26. $\displaystyle\int \frac{du}{u^2 - a^2} = \frac{1}{2a}\ln\left|\frac{u - a}{u + a}\right| + C$

INTEGRALS CONTAINING $\sqrt{u^2 \pm a^2}$ $(a > 0)$

27. $\displaystyle\int \frac{du}{\sqrt{u^2 \pm a^2}} = \ln\left|u + \sqrt{u^2 \pm a^2}\right| + C$

28. $\displaystyle\int \sqrt{u^2 \pm a^2}\, du = \frac{u}{2}\sqrt{u^2 \pm a^2} \pm \frac{a^2}{2}\ln\left|u + \sqrt{u^2 \pm a^2}\right| + C$

29. $\displaystyle\int u\sqrt{u^2 \pm a^2}\, du = \frac{1}{3}\,(u^2 \pm a^2)^{3/2} + C$

30. $\displaystyle\int u^2\sqrt{u^2 \pm a^2}\, du = \frac{u}{8}\,(2u^2 \pm a^2)\sqrt{u^2 \pm a^2} - \frac{a^4}{8}\ln\left|u + \sqrt{u^2 \pm a^2}\right| + C$

31. $\displaystyle\int \frac{\sqrt{u^2 + a^2}\, du}{u} = \sqrt{u^2 + a^2} - a\ln\left|\frac{a + \sqrt{u^2 + a^2}}{u}\right| + C$

32. $\displaystyle\int \frac{\sqrt{u^2 - a^2}\, du}{u} = \sqrt{u^2 - a^2} - a\sec^{-1}\frac{u}{a} + C$

33. $\displaystyle\int \frac{\sqrt{u^2 \pm a^2}\, du}{u^2} = -\frac{\sqrt{u^2 \pm a^2}}{u} + \ln\left|u + \sqrt{u^2 \pm a^2}\right| + C$

34. $\displaystyle\int \frac{u^2\, du}{\sqrt{u^2 \pm a^2}} = \frac{u}{2}\sqrt{u^2 \pm a^2} \mp \frac{a^2}{2}\ln\left|u + \sqrt{u^2 \pm a^2}\right| + C$

35. $\displaystyle\int \frac{du}{u\sqrt{u^2 + a^2}} = -\frac{1}{a}\ln\left|\frac{a + \sqrt{u^2 + a^2}}{u}\right| + C$

36. $\displaystyle\int \frac{du}{u\sqrt{u^2 - a^2}} = \frac{1}{a}\sec^{-1}\frac{u}{a} + C$

37. $\displaystyle\int \frac{du}{u^2\sqrt{u^2 \pm a^2}} = \mp\frac{\sqrt{u^2 \pm a^2}}{a^2u} + C$

38. $\displaystyle\int (u^2 \pm a^2)^{3/2}\, du = \frac{u}{8}\,(2u^2 \pm 5a^2)\sqrt{u^2 \pm a^2} + \frac{3a^4}{8}\ln\left|u + \sqrt{u^2 \pm a^2}\right| + C$

39. $\displaystyle\int \frac{du}{(u^2 \pm a^2)^{3/2}} = \pm\frac{u}{a^2\sqrt{u^2 \pm a^2}} + C$

INTEGRALS CONTAINING $\sqrt{a^2 - u^2}$ $(a > 0)$

40. $\displaystyle\int \frac{du}{\sqrt{a^2 - u^2}} = \sin^{-1}\frac{u}{a} + C$

41. $\displaystyle\int \sqrt{a^2 - u^2}\, du = \frac{u}{2}\sqrt{a^2 - u^2} + \frac{a^2}{2}\sin^{-1}\frac{u}{a} + C$

42. $\displaystyle\int u^2\sqrt{a^2 - u^2}\, du = \frac{u}{8}\,(2u^2 - a^2)\sqrt{a^2 - u^2} + \frac{a^4}{8}\sin^{-1}\frac{u}{a} + C$

43. $\displaystyle\int \frac{\sqrt{a^2 - u^2}\, du}{u} = \sqrt{a^2 - u^2} - a\ln\left|\frac{a + \sqrt{a^2 - u^2}}{u}\right| + C$

44. $\displaystyle\int \frac{\sqrt{a^2 - u^2}\, du}{u^2} = -\frac{\sqrt{a^2 - u^2}}{u} - \sin^{-1}\frac{u}{a} + C$

45. $\displaystyle\int \frac{u^2\, du}{\sqrt{a^2 - u^2}} = -\frac{u}{2}\sqrt{a^2 - u^2} + \frac{a^2}{2}\sin^{-1}\frac{u}{a} + C$

46. $\displaystyle\int \frac{du}{u\sqrt{a^2 - u^2}} = -\frac{1}{a}\ln\left|\frac{a + \sqrt{a^2 - u^2}}{u}\right| + C$

47. $\displaystyle\int \frac{du}{u^2\sqrt{a^2 - u^2}} = -\frac{\sqrt{a^2 - u^2}}{a^2u} + C$

48. $\displaystyle\int (a^2 - u^2)^{3/2}\, du = -\frac{u}{8}\,(2u^2 - 5a^2)\sqrt{a^2 - u^2} + \frac{3a^4}{8}\sin^{-1}\frac{u}{a} + C$

49. $\displaystyle\int \frac{du}{(a^2 - u^2)^{3/2}} = \frac{u}{a^2\sqrt{a^2 - u^2}} + C$

Multivariable
Calculus

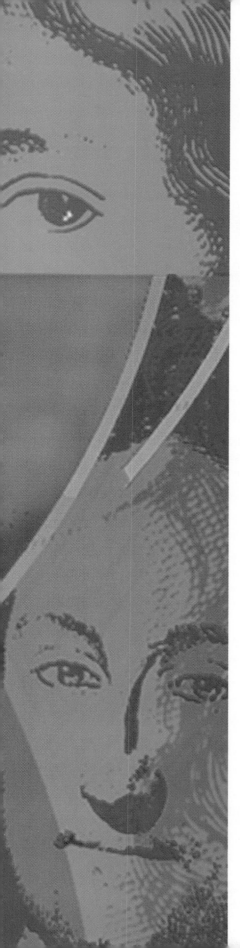

Multivariable
Calculus

FIFTH EDITION

HOWARD ANTON
Drexel University

in collaboration with
ALBERT HERR, Drexel University

JOHN WILEY & SONS, INC.
New York   Chichester   Brisbane   Toronto   Singapore

Mathematics Editor: Barbara Holland
Developmental Editor: Nancy Perry
Marketing Manager: Susan Elbe
Senior Production Editor: Nancy Prinz
Copy Editor: Lilian Brady
Design Supervisor: Madelyn Lesure
Manufacturing Manager: Susan Stetzer
Photo Editor: Hilary Newman
Illustration Coordinator: Sigmund Malinowski
Production, Cover and Text Design: HRS Electronic Text Management, Inc.
Electronic Illustration: Techsetters, Inc.

This book was set in Times Roman by General Graphic Services and printed and bound by Von Hoffmann Press, Inc. The cover was printed by The Lehigh Press, Inc.

Cover photo credits: Maria Gaetana Agnesi: Culver Pictures, Inc. Archimedes and Carl Friedrich Gauss: The Bettmann Archive. Augustin Louis Cauchy, Gottfried Wilhelm Leibniz and Georg Friedrich Bernhard Riemann: Courtesy of The David Eugene Smith Collection, Columbia University. Leonhard Euler and Sir Isaac Newton: Courtesy of The New York Public Library.

Recognizing the importance of preserving what has been written, it is a policy of John Wiley & Sons, Inc. to have books of enduring value published in the United States printed on acid-free paper, and we exert our best efforts to that end.

The paper on this book was manufactured by a mill whose forest management programs include sustained yield harvesting of its timberlands. Sustained yield harvesting principles ensure that the number of trees cut each year does not exceed the amount of new growth.

Copyright © 1995, by Anton Textbooks, Inc.

All rights reserved. Published simultaneously in Canada.

Reproduction or translation of any part of this work beyond that permitted by Sections 107 and 108 of the 1976 United States Copyright Act without the permission of the copyright owner is unlawful. Requests for permission or further information should be addressed to the Permissions Department, John Wiley & Sons, Inc.

Derive is a registered trademark of Soft Warehouse, Inc.
Maple is a registered trademark of Waterloo Maple Software, Inc.
Mathematica is a registered trademark of Wolfram Research, Inc.

Library of Congress Cataloging in Publication Data:

Anton, Howard.
    Multivariable calculus / Howard Anton, in collaboration with Albert Herr.— 5th ed.
        p.   cm.
    Consists of the multivariable portion of Calculus with analytic geometry, 5th ed., by Howard Anton, with added material.
    Includes index.
    ISBN 0-471-13909-2 (pbk. ; alk. paper)
    1. Calculus. I. Herr, Albert. II. Anton, Howard. Calculus with analytic geometry. 5th ed. III. Title.
QA03.A54   1995
515—dc20   95-32954
                                                                     CIP

Printed in the United States of America

10 9 8 7 6 5 4 3 2 1

About Howard Anton

Howard Anton obtained his B.A. from Lehigh University, his M.A. from the University of Illinois, and his Ph.D. from the Polytechnic Institute of Brooklyn, all in mathematics. In the early 1960's he worked for Burroughs Corporation and Avco Corporation at Cape Canaveral, Florida, where he was involved with missile tracking problems for the manned space program. In 1968 he joined the Mathematics Department at Drexel University, where he taught full time until 1983. Since that time he has been an adjunct professor at Drexel and has devoted the majority of his time to textbook writing and activities for mathematical associations. Dr. Anton was President of the EPADEL Section of the Mathematical Association of America (MAA), served on the Board of Governors of that organization, and guided the creation of the Student Chapters of the MAA.

He has published numerous research papers in Functional Analysis, Approximation Theory, and Topology, as well as pedagogical papers on applications of mathematics. He is best known for his textbooks in mathematics, which are among the most widely used in the world. There are currently more than eighty versions of his books including translations into Spanish, Arabic, Portuguese, Italian, Indonesian, French, Japanese, Chinese, and German.

Dr. Anton has an avid interest in computer technology as it relates to mathematical education and publishing. He has developed pedagogical software for teaching linear algebra as well as various software programs for the publishing industry that automate the production of four-color mathematical text and art. For relaxation he enjoys traveling and photography.

About Albert Herr

It is my sad duty to report that Albert J. Herr, my colleague, collaborator, and friend passed away shortly after the first printing of this book appeared. He will be missed greatly by all of those who had the good fortune to work with him.

Al held degrees in Electrical Engineering and Biomedical Engineering from Drexel University. He joined the Department of Mathematics and Computer Science at Drexel in 1964, eventually becoming Assistant Department Head for Undergraduate Programs until his retirement in 1993. Al's career always had a strong focus on the teaching of calculus. In addition to supervising graduate teaching assistants, he coordinated the calculus program at Drexel for many years, wrote the solutions manuals for this text, and assisted in its various revisions. Starting in 1980, he became actively involved in developing software and fostering the use of computers in mathematics education at both the high school and college levels. In 1984 he designed and incorporated the first computer-related materials into the calculus curriculum at Drexel.

In addition to his work in collegiate mathematics, Al actively participated in a variety of programs to stimulate an interest in mathematics and engineering among high school students. He gave numerous invited talks and workshops on the use of computers in mathematics education, and he served as project director of an NSF grant for creating computer graphics mathematics laboratories.

Al received the Lindback Award for excellence in teaching, as well as three outstanding teaching awards from student organizations at Drexel. He also received the Drexel University College of Science Award for dedication and service to students.

To
My Wife Pat
My Children Brian, David, and Lauren
My Mother Shirley

In Memory of
Stephen Girard (1750–1831) Benefactor

PREFACE

ABOUT THIS EDITION

This book consists of the multivariable portion of my text, *Calculus with Analytic Geometry*, 5th Edition (Chapters 14–18), as well as Chapter 19 on second-order differential equations. Also included are: the chapter on infinite series (Chapter 11), the section on first-order differential equations (Section 7.7), the appendix on Cramer's rule (Appendix D), and the appendix on complex numbers (Appendix E).

This book, in combination with the *Brief Edition of Calculus with Analytic Geometry*, 5th Edition, covers all of the material in *Calculus with Analytic Geometry*, 5th Edition.

It is asssumed that the reader has completed a course on single variable calculus, but for reference a brief review of key concepts from single-variable calculus is given at the front of this text.

- **Technology** — Each chapter ends with a set of exercises that are designed to be solved using computer algebra systems or graphing calculators. Many of the exercises involve applications, and almost all of them can be solved in a variety of ways that are limited only by the student's imagination.

- **Revision of Multivariate Calculus** — The multivariate calculus material was completely rewritten, incorporating the concept of a vector field and focusing more on the major applications of vector analysis to physics and engineering.

- **New Material** — Material not included in previous editions was added: Jacobians, parametric representations of surfaces, Kepler's laws, conics in polar coordinates, integrals with respect to arc length, vector fields, and an appendix with some basic material on complex variables that can serve as a reference for engineers and students who need this material for other courses.

- **More Use of Calculator Computations in the Exposition** — We assume in this edition that the student has a numerical calculator available as he or she reads the text, and numerical computations are used more extensively in developing concepts.

OTHER FEATURES

- **Rule of Four** — The term "rule of four" has recently been coined to describe exposition that presents ideas from the symbolic, geometric, computational, and verbal viewpoints. Readers familiar with earlier editions of this text will recognize that this has always been an integral part of my writing style. This style continues in this edition.

- **Rigor** — The challenge of writing a good calculus book is to strike the right balance between rigor and clarity. My goal is to present precise mathematics to the fullest extent possible for the freshman audience, but where clarity and rigor conflict I choose clarity. However, I believe it to be essential that the student understand the difference between a careful proof and an informal argument, so I try to make it clear to the reader when arguments are informal.

- **Historical Notes** — The biographies and historical notes have been a hallmark of this text from its first edition, and new biographies have been added in this edition. All of the biographical material has been distilled from standard sources with the goal of capturing the personalities of the great mathematicians and bringing them to life for the student.

■ **Section Exercises** — Section exercise sets begin with routine problems and progress gradually toward problems of greater difficulty. Exercises that require a calculator are listed at the beginning of the exercise set and marked with the icon $\boxed{C}$. Many exercise sets contain so-called "spiral" problems, which revisit earlier problem types using concepts from the current section.

ABOUT THE TECHNOLOGY EXERCISES

■ The purpose of the technology exercises is to introduce the student to techniques of problem solving using graphing calculators and/or computer algebra systems such as *Mathematica*™ , *Maple*™ , or *Derive*™ . Many of these exercises involve applications of calculus, and most of them can be solved using *either* a graphing calculator or a computer algebra system. Thus, part of the challenge to the student is to develop a problem-solving strategy that is appropriate for the technology that he or she has available.

■ Many of the problems cannot be solved by a blind, unintelligent use of technology; they may require some preliminary hand calculation to put the problem in an appropriate form or some thoughtful analysis to ensure that solutions are not missed when technology is applied.

■ Many problems will raise issues of accuracy, since some students may be able to avoid decimal approximations using a computer algebra system and other students may obtain different levels of decimal accuracy depending on their strategy and technology. This is the opportunity for an instructor to explore issues of error analysis if so inclined. However, it is not essential.

■ The technology exercises are more open-ended than the exercises at the end of each section, making them more like problems that arise in the real world. Instructors can either leave the students on their own or can provide a level of guidance that fits their own teaching philosophy. Some instructors may want to use these exercises for group projects.

SUPPLEMENTS

GRAPHING CALCULATOR SUPPLEMENTS

The following supplement contains a collection of problems that are intended to be solved on a graphing calculator. The problems are not specific to a particular brand of calculator. Also provided is an overview of the types of calculators available and general instructions for calculator use.

- *Discovering Calculus with Graphing Calculators*, *Second Edition*
  ISBN: 0-471-00974-1

The following free supplement provides a brief overview of those aspects of graphing calculators that are relevant to the problems in this text. Topics include: choice of viewing window, roundoff error, techniques for finding roots, and common pitfalls associated with graphing calculators.

- *Graphing Calculator Survival Guide*
  ISBN: 0-471-13172-5

SYMBOLIC ALGEBRA SUPPLEMENTS

The following supplements are collections of problems for the student to solve. Each contains a brief set of instructions for using the software as well as an extensive set of problems utilizing the capabilities of the software. The problems range from very basic to those involving real-world applications.

- *Discovering Calculus with DERIVE™*, *Second Edition*
  Jerry Johnson, *University of Nevada–Reno*
  Benny Evans, *Oklahoma State University*
  ISBN: 0-471-00972-5

- *Discovering Calculus with MAPLE™*
  Kent Harris, *Western Illinois University*
  Robert J. Lopez, *Rose–Hulman Institute of Technology*
  ISBN: 0-471-55156-2

- *Discovering Calculus with MATHEMATICA™*
  Cecilia A. Knoll, Florida Institute of Technology
  Michael D. Shaw, Florida Institute of Technology
  Jerry Johnson, University of Nevada-Reno
  Benny Evans, Oklahoma State University
  ISBN: 0-471-00976-8

CD-ROM VERSION OF CALCULUS FOR IBM COMPATIBLE COMPUTERS

This supplement is an electronic version of Anton's *Calculus,* the *Student's Solutions Manual*, and the *Calculus Companion* on compact disk for use with IBM compatible computers equipped with a CD-ROM drive. All text material and illustrations are stored on disk with an interconnecting network of hyperlinks that allows the student to access related items that do not appear in proximity in the text. A complete keyword glossary and step-by-step discussions of key concepts are also included.

- *CD-ROM Version of Anton Calculus: An Electronic Study Environment*
  Developed by Smart Books. Inc.
  ISBN: 0-471-55803-6

CD-ROM MULTIMEDIA SUPPLEMENT FOR IBM COMPATIBLE COMPUTERS

This highly interactive multimedia CD provides opportunities for students to ask "what if" questions, change parameters, enter their own functions and see the effects of their mathematical decisions in real time. There are 24 multimedia modules accompanied by a laboratory workbook that covers key concepts and spans the entire calculus sequence.

■ *Calculus Connections: A Multimedia Adventure*
Douglas Quinney, University of Keele
Robert Harding, Cambridge University
IntelliPro, Inc.
ISBN: 0-471-13795-2

EARLY TRANSCENDENTAL SUPPLEMENT

This free supplement is designed for those who want an early treatment of exponentials and logarithms. In this short supplement the material in Section 7.2 is broken into smaller self-contained units for easy placement earlier in the text, and a guide for implementing the early transcendental option is provided.

■ *Early Transcendental Supplement to Accompany Anton Calculus 5/E*
ISBN: 0-471-13173-3

LINEAR ALGEBRA SUPPLEMENT

This free supplement is a brief introduction to those aspects of linear algebra that are of immediate concern to the calculus student. The emphasis is on methods rather than proof.

■ *Linear Algebra Supplement to Accompany Anton Calculus / 5E*
ISBN: 0-471-10677-1

STUDENT STUDY RESOURCES

The following supplement is a tutorial, review, and study aid for the student.

■ *The Calculus Companion to Accompany Anton Calculus / 5E*
William H. Barker and James E. Ward, *Bowdoin College*
ISBN: 0-471-10678-x

The following supplement contains detailed solutions to all odd-numbered exercises.

■ *Student's Solutions Manual to Accompany Anton Calculus / 5E*
Albert Herr, *Drexel University*
ISBN: 0-471-10589-9

RESOURCES FOR THE INSTRUCTOR

There is a resource package for the instructor that includes hard copy and electronic test banks and other materials. These can be obtained by writing on your institutional letterhead to Debra Riegert, Senior Marketing Manager, John Wiley & Sons, Inc., 605 Third Avenue, New York, N.Y., 10158-0012.

ACKNOWLEDGMENTS

It has been my good fortune to have the advice and guidance of many talented people, whose knowledge and skills have enhanced this book in many ways. For their valuable help I thank:

REVIEWERS AND CONTRIBUTORS TO EARLIER EDITIONS

Edith Ainsworth, *University of Alabama*
David Armacost, *Amherst College*
Larry Bates, *University of Calgary*
Irl C. Bivens, *Davidson College*
Harry N. Bixler, *Bernard M. Baruch College, CUNY*
Marilyn Blockus, *San Jose State University*
Ray Boersma, *Front Range Community College*
David Bolen, *Virginia Military Institute*
Daniel Bonar, *Denison University*
George W. Booth, *Brooklyn College*
Mark Bridger, *Northeastern University*
John Brothers, *Indiana University*
Robert C. Bueker, *Western Kentucky University*
Robert Bumcrot, *Hofstra University*
James Caristi, *Valparaiso University*
Chris Christensen, *Northern Kentucky University*
Hannah Clavner, *Drexel University*
David Cohen, *University of California, Los Angeles*
Michael Cohen, *Hofstra University*
Robert Conley, *Precision Visuals*
Terrance Cremeans, *Oakland Community College*
Michael Dagg, *Numerical Solutions, Inc.*
Stephen L. Davis, *Davidson College*
A. L. Deal, *Virginia Military Institute*
Charles Denlinger, *Millersville State College*
Dennis DeTurck, *University of Pennsylvania*
Jacqueline Dewar, *Loyola Marymount University*
Irving Drooyan, *Los Angeles Pierce College*
Tom Drouet, *East Los Angeles College*
Ken Dunn, *Dalhousie University*
Hugh B. Easler, *College of William and Mary*
Joseph M. Egar, *Cleveland State University*

Garret J. Etgen, *University of Houston*
James H. Fife, *University of Richmond*
Barbara Flajnik, *Virginia Military Institute*
Daniel Flath, *University of South Alabama*
Nicholas E. Frangos, *Hofstra University*
Katherine Franklin, *Los Angeles Pierce College*
Michael Frantz, *University of La Verne*
Susan L. Friedman, *Bernard M. Baruch College, CUNY*
William R. Fuller, *Purdue University*
G. S. Gill, *Brigham Young University*
Raymond Greenwell, *Hofstra University*
Gary Grimes, *Mt. Hood Community College*
Jane Grossman, *University of Lowell*
Michael Grossman, *University of Lowell*
Douglas W. Hall, *Michigan State University*
Nancy A. Harrington, *University of Lowell*
Kent Harris, *Western Illinois University*
Albert Herr, *Drexel University*
Peter Herron, *Suffolk County Community College*
Konrad J. Heuvers, *Michigan Technological University*
Robert Higgins, *Quantics Corporation*
Louis F. Hoelzle, *Bucks County Community College*
Herbert Kasube, *Bradley University*
Phil Kavanaugh, *Illinois Wesleyan University*
Maureen Kelly, *Northern Essex Community College*
Harvey B. Keynes, *University of Minnesota*
Paul Kumpel, *SUNY, Stony Brook*
Leo Lampone, *Quantics Corporation*
Bruce Landman, *Hofstra University*
Benjamin Levy, *Lexington H.S., Lexington, Mass.*
Phil Locke, *University of Maine, Orono*
John Lucas, *University of Wisconsin–Oshkosh*

Stanley M. Lukawecki, *Clemson University*
Nicholas Macri, *Temple University*
Melvin J. Maron, *University of Louisville*
Thomas McElligott, *University of Lowell*
Judith McKinney, *California State Polytechnic University, Pomona*
Joseph Meier, *Millersville State College*
Ron Moore, *Ryerson Polytechnical Institute*
Barbara Moses, *Bowling Green State University*
David Nash, *VP Research, Autofacts, Inc.*
Richard Nowakowski, *Dalhousie University*
Robert Phillips, *University of South Carolina at Aiken*
Mark A. Pinsky, *Northeastern University*
David Randall, *Oakland Community College*
William H. Richardson, *Wichita State University*
David Sandell, *U.S. Coast Guard Academy*
George Shapiro, *Brooklyn College*
Donald R. Sherbert, *University of Illinois*
Wolfe Snow, *Brooklyn College*
Ian Spatz, *Brooklyn College*
Jean Springer, *Mount Royal College*
Norton Starr, *Amherst College*
Richard B. Thompson, *The University of Arizona*
William F. Trench, *Trinity University*
Walter W. Turner, *Western Michigan University*
Richard C. Vile, *Eastern Michigan University*
Shirley Wakin, *University of New Haven*
James Warner, *Precision Visuals*
Peter Waterman, *Northern Illinois University*
Evelyn Weinstock, *Glassboro State College*
Candice A. Weston, *University of Lowell*
Yihren Wu, *Hofstra University*
Richard Yuskaitis, *Precision Visuals*

DEVELOPMENT TEAM FOR THE FIFTH EDITION

The following survey respondents critiqued the previous edition and recommended many of the changes that found their way into the new edition.

Robert C. Banash, *St. Ambrose University*
George R. Barnes, *University of Louisville*
John P. Beckwith, *Michigan Technological University*
Joan E. Bell, *Northeastern Oklahoma State University*
Barbara Bohannon, *Hofstra University*
Phyllis Boutilier, *Michigan Technological University*
Stephen L. Brown, *Olivet Nazarene University*
Virginia Buchanan, *Hiram College*
Carlos E. Caballero, *Winthrop University*
Stan R. Chadick, *Northwestern State University*
Hongwei Chen, *Christopher Newport University*
Robert D. Cismowski, *San Bernardino Valley College*
David Clydesdale, *Sauk Valley Community College*
Cecil J. Coone, *State Technical Institute at Memphis*
Norman Cornish, *University of Detroit*
William H. Dent, *Maryville College*
Preston Dinkins, *Southern University*
Scott Eckert, *Cuyamaca College*
Judith Elkins, *Sweet Briar College*
Brett Elliott, *Southeastern Oklahoma State University*
Dorothy M. Fitzgerald, *Golden West College*
Ernesto Franco, *California State University–Fresno*
Daniel B. Gallup, *Pasadena City College*
Mahmood Ghamsary, *Long Beach City College*
Michael Gilpin, *Michigan Technological University*
S. B. Gokhale, *Western Illinois University*
Morton Goldberg, *Broome Community College*
Mordechai Goodman, *Rosary College*
Sid Graham, *Michigan Technological University*
Kent Harris, *Western Illinois University*
Jim Hefferson, *St. Michael College*
Warland R. Hersey, *North Shore Community College*
Konrad J. Heuvers, *Michigan Technological University*

Robert Homolka, *Kansas State University–Salina*
John M. Johnson, *George Fox College*
Wells R. Johnson, *Bowdoin College*
Richard Krikorian, *Westchester Community College*
Fat C. Lam, *Gallaudet University*
James F. Lanahan, *University of Detroit–Mercy*
Kuen Hung Lee, *Los Angeles Trade–Technology College*
Marshall J. Leitman, *Case Western Reserve University*
Darryl A. Linde, *Northeastern Oklahoma State University*
Leland E. Long, *Muscatine Community College*
Mauricio Marroquin, *Los Angeles Valley College*
Larry Matthews, *Concordia College*
Phillip McGill, *Illinois Central College*
Aileen Michaels, *Hofstra University*
Janet S. Milton, *Radford University*
Robert Mitchell, *Rowan College of New Jersey*
Marilyn Molloy, *Our Lady of the Lake University*
Kylene Norman, *Clark State Community College*
Roxie Novak, *Radford University*
Donald Passman, *University of Wisconsin*
Walter M. Patterson, *Lander University*
Edward Peifer, *Ulster County Community College*
Richard Remzowski, *Broome Community College*
Guanshen Ren, *College of Saint Scholastica*
Naomi Rose, *Mercer County Community College*
David Ryeburn, *Simon Fraser University*
Ned W. Schillow, *Lehigh County Community College*
Parashu R. Sharma, *Grambling State University*
Howard Sherwood, *University of Central Florida*
Bhagat Singh, *University of Wisconsin Centers*
Martha Sklar, *Los Angeles City College*
John L. Smith, *Rancho Santiago Community College*
Jean Springer, *Mount Royal College*
David Voss, *Western Illinois University*
Bruce F. White, *Lander University*
Gary L. Wood, *Azusa Pacific University*
Michael L. Zwilling, *Mount Union College*

The following people contributed numerous new and imaginative problems to the text:

Loren Argabright, *Drexel University*
Patricia Clark, *Rochester Institute of Technology*

Lawrence Cusick, *California State University–Fresno*
Benny Evans, *Oklahoma State University*
Rebecca Hill, *Rochester Institute of Technology*
Jerry Johnson, *University of Nevada–Reno*
Michael Zeidler, *Milwaukee Area Technical College*

The following people assisted with the critically important job of preparing the answer section, solutions for the *Student's Solutions Manual*, answers to technology exercises, and preparing the index:

Chris Butler, *Case Western Reserve University*
Stephen L. Davis, *Davidson College*
Michael Dagg, *Numerical Solutions, Inc.*
Blaise DeSesa, *Drexel University*

Clyde Dubbs, *New Mexico Institute of Mining and Technology*
Sheldon Dyck, *Waterloo Maple Software*
Diane Hagglund, *Waterloo Maple Software*
Majid Masso, *Brookdale Community College*
Kylene Norman, *Clark State Community College*
Stanley Ocken, *City College—CUNY*
Sharon Ross, *DeKalb College*
Dennis Schneider, *Knox College*
Dan Seth, *Morehead State University*
Shirley Wakin, *University of New Haven*

The following telesession participants provided crucial feedback on the previous edition and suggestions for the revision:

Chris Butler, *Case Western Reserve University*
Clyde Dubbs, *New Mexico Institute of Mining and Technology*
Della Duncan, *California State University–Fresno*

Kaplana Godbole, *Michigan Technological University*
Ed Hoefer, *Rochester Institute of Technology*
David Patterson, *West Texas A&M*
Father Bernard Portz, *Creighton University*
David Rollins, *University of Central Florida*
Jean Springer, *Mount Royal College*

The following people reviewed various stages of the fifth edition for its content and accuracy:

John Bailey, *Clark State Community College*
Irl Bivens, *Davidson College*
Christopher Butler, *Case Western Reserve University*
Clyde Dubbs, *New Mexico Institute of Mining and Technology*
Della Duncan, *California State University–Fresno*
Edwin Hoefer, *Rochester Institute of Technology*

Susan Friedman, *Bernard M. Baruch College, CUNY*
Marc Frantz, *Indiana University–Purdue University at Indianapolis*
Konrad J. Heuvers, *Michigan Technological University*
Majid Masso, *Brookdale Community College*
Kylene Norman, *Clark State Community College*
Stanley Ocken, *City College–CUNY*
Sharon Ross, *DeKalb College*
Dennis Schneider, *Knox College*

The following people provided additional material for tests and other supplements:

Pasquale Condo, *University of Lowell*
Maureen Kelley, *Northern Essex Community College*
Catherine H. Pirri, *Northern Essex Community College*

THE CONTRIBUTIONS OF ALBERT HERR

This revision was prepared in collaboration with Professor Albert Herr of Drexel University, a gifted, award-winning teacher with years of experience in the calculus classroom. Al participated in every aspect of this revision, and many of the improvements in the exposition and the mathematics are due to him. This edition has benefited greatly from Al's background in engineering mathematics and his wonderful skills as a problem creator—many of the technology problems are his, and many improvements in the text material reflect his ideas. I feel most fortunate to have had this opportunity to learn some new mathematics and new ideas about teaching calculus from such a skilled mathematician and talented teacher.

SPECIAL CONTRIBUTIONS

I am indebted to:

- Barbara Holland, my editor, for her perceptive understanding of contemporary calculus issues, her faith in my work, and doing everything that a fine editor should. She is a joy to work with.
- Ann Berlin, Lucille Buonocore, and Nancy Prinz of the Wiley Production Department for working miracles with a tight production schedule. Thank you all once again.
- Lilian Brady, whose eye for detail and aesthetics of typography created hundreds of purses out of sows' ears.
- Sharon Smith for magically juggling a gaggle of supplements that would stagger a sumo wrestler.
- The group at HRS for biting their tongues at just the right moments.
- The illustration group at Techsetters for tolerating my compulsion for detail and believing that I can see a $\frac{1}{10}$ point difference.
- Irl Bivens of Davidson College whose keen sense of sound pedagogy and good mathematics guided us in structuring this revision.
- Benny Evans of Oklahoma State University and Jerry Johnson of the University of Nevada–Reno for providing a wonderfully imaginative set of technology exercises for us to work with.
- My assistant, Dolores Morgan, for keeping the coffee hot, the project on schedule, and the author sane. I can't thank her enough.

HOWARD ANTON

CONTENTS

CONTENTS

SUPPLEMENTARY MATERIAL

INTRODUCTION

Calculus is the mathematical tool used to analyze changes in physical quantities. It was developed in the seventeenth century to study four major classes of scientific and mathematical problems of the time:

■ Find the tangent line to a curve at a point.

■ Find the length of a curve, the area of a region, and the volume of a solid.

■ Find the maximum or minimum value of a quantity — for example, the maximum and minimum distances of a planet from the sun, or the maximum range attainable for a projectile by varying its angle of fire.

■ Given a formula for the distance traveled by a body in any specified amount of time, find the velocity and acceleration of the body at any instant. Conversely, given a formula that specifies the acceleration or velocity at any instant, find the distance traveled by the body in a specified period of time.

These problems were attacked by the greatest minds of the seventeenth century, culminating in the crowning achievement of Gottfried Wilhelm Leibniz and Isaac Newton — the creation of calculus.

Gottfried Wilhelm Leibniz (1646–1716)

This gifted genius was one of the last people to have mastered most major fields of knowledge — an impossible accomplishment in our own era of specialization. He was an expert in law, religion, philosophy, literature, politics, geology, metaphysics, alchemy, history, and mathematics.

Leibniz was born in Leipzig, Germany. His father, a professor of moral philosophy at the University of Leipzig, died when Leibniz was six years old. The precocious boy then gained access to his father's library and began reading voraciously on a wide range of subjects, a habit that he maintained throughout his life. At age 15 he entered the University of Leipzig as a law student and by the age of 20 received a doctorate from the University of Altdorf. Subsequently, Leibniz followed a career in law and international politics, serving as counsel to kings and princes.

During his numerous foreign missions, Leibniz came in contact with outstanding mathematicians and scientists who stimulated his interest in mathematics—most notably, the physicist Christian Huygens. In mathematics Leibniz was self-taught, learning the subject by reading papers and journals. As a result of this fragmented mathematical education, Leibniz often duplicated the results of others, and this ultimately led to a raging conflict as to whether he or Isaac Newton should be regarded as the inventor of calculus. The argument over this question engulfed the scientific circles of England and Europe, with most scientists on the continent supporting Leibniz and those in England supporting Newton. The conflict was unfortunate, and both sides suffered in the end: The continent lost the benefit of Newton's discoveries in astronomy and physics for more than 50 years, and for a long period England became a second-rate country mathematically because its mathematicians were hampered by Newton's inferior calculus notation. It is of interest to note that Newton and Leibniz never went to the lengths of vituperation of their advocates — both were sincere admirers of each other's work. The fact is that both men invented calculus independently; Leibniz invented it 10 years after Newton, in 1685, but he published his results 20 years before Newton.

Leibniz never married. He was moderate in his habits, quick-tempered, but easily appeased, and charitable in his judgment of other people's work. In spite of his great achievements, Leibniz never received the honors showered on Newton, and he spent his final years as a lonely embittered man. At his funeral there was one mourner, his secretary. An eyewitness stated, "He was buried more like a robber than what he really was — an ornament of his country."

Isaac Newton (1642–1727)

Newton was born in the village of Woolsthorpe, England. His father died before he was born and his mother raised him on the family farm. As a youth he showed little evidence of his later brilliance, except for an unusual talent with mechanical devices — he apparently built a working water clock and a toy flour mill powered by a mouse. In 1661 he entered Trinity College in Cambridge with a deficiency in geometry. Fortunately, Newton caught the eye of Isaac Barrow, a gifted mathematician and teacher. Under Barrow's guidance Newton immersed himself in mathematics and science, but he graduated without any special distinction. Because the Plague was spreading rapidly through London, Newton returned to his home in Woolsthorpe and stayed there during the years of 1665 and 1666. In those two momentous years the entire framework of modern science was miraculously created in Newton's mind — he discovered calculus, recognized the underlying principles of planetary motion and gravity, and determined that "white" sunlight was composed of all colors, red to violet. For some reason he kept his discoveries to himself. In 1667 he returned to Cambridge to obtain his Master's degree and upon graduation became a teacher at Trinity. Then in 1669 Newton succeeded his teacher, Isaac Barrow, to the Lucasian chair of mathematics at Trinity, one of the most honored chairs of mathematics in the world. Thereafter, brilliant discoveries flowed from Newton steadily. He formulated the law of gravitation and used it to explain the motion of the moon, the planets, and the tides; he formulated basic theories of light, thermodynamics, and hydrodynamics; and he devised and constructed the first modern reflecting telescope.

Gottfried Leibniz
(Culver Pictures)

Isaac Newton
(Culver Pictures)

Throughout his life Newton was hesitant to publish his major discoveries, revealing them only to a select circle of friends, perhaps because of a fear of criticism or controversy. In 1687, only after intense coaxing by the astronomer, Edmond Halley (Halley's comet), did Newton publish his masterpiece, *Philosophiae Naturalis Principia Mathematica* (The Mathematical Principles of Natural Philosophy). This work is generally considered to be the most important and influential scientific book ever written. In it Newton explained the workings of the solar system and formulated the basic laws of motion which to this day are fundamental in engineering and physics. However, not even the pleas of his friends could convince Newton to publish his discovery of calculus. Only after Leibniz published his results did Newton relent and publish his own work on calculus.

After 35 years as a professor, Newton suffered depression and a nervous breakdown. He gave up research in 1695 to accept a position as warden and later master of the London mint. During the 25 years that he worked at the mint, he did virtually no scientific or mathematical work. He was knighted in 1705 and on his death was buried in Westminster Abbey with all the honors his country could bestow. It is interesting to note that Newton was a learned theologian who viewed the primary value of his work to be its support of the existence of God. Throughout his life he worked passionately to date biblical events by relating them to astronomical phenomena. He was so consumed with this passion that he spent years searching the Book of Daniel for clues to the end of the world and the geography of hell.

Newton described his brilliant accomplishments as follows: "I seem to have been only like a boy playing on the seashore and diverting myself in now and then finding a smoother pebble or prettier shell than ordinary, whilst the great ocean of truth lay all undiscovered before me."

Multivariable
Calculus

The Discoverers
of Calculus

Gottfried Wilhelm Leibniz (1646-1716)

Isaac Newton.

Sir Isaac Newton

PRELIMINARIES

■ REVIEW OF SELECTED CONCEPTS FROM SINGLE-VARIABLE CALCULUS

We assume that the reader has previously studied single-variable calculus, but for convenience we have provided lists of the key single-variable concepts required in each chapter, followed by appropriate extracts from the Brief Edition of this text. The numbering of theorems, figures, and appendices in this review is from the Brief Edition. Most references in the text to material in the Brief Edition can be found in this review. This is not intended to provide a complete summary of required material, and should you need a more in-depth review, refer to the text from which you studied single-variable calculus.

REVIEW TOPICS FOR CHAPTER 14

- Distance between points in the plane
- Midpoint of a line segment
- Equations of circles
- Graph of $Ax^2 + Ay^2 + Dx + Ey + F = 0$
- Parametric equations
- Graphs of conic sections
- Translation of coordinate axes
- Relationship between polar and rectangular coordinates
- Work

1.5.1 THEOREM. *The distance d between two points (x_1, y_1) and (x_2, y_2) in a coordinate plane is given by*

$$d = \sqrt{(x_2 - x_1)^2 + (y_2 - y_1)^2}$$

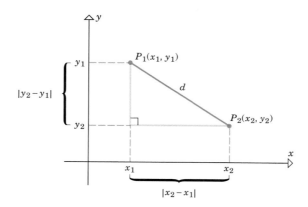

Figure 1.5.1

1.5.2 THEOREM (*The Midpoint Formula*).   *The midpoint of the line segment joining two points* (x_1, y_1) *and* (x_2, y_2) *in a coordinate plane is*

$$\left(\tfrac{1}{2}(x_1 + x_2), \tfrac{1}{2}(y_1 + y_2)\right)$$

Standard Form of the Equation of a Circle

$$(x - x_0)^2 + (y - y_0)^2 = r^2$$

1.5.3 THEOREM.   *An equation of the form*

$$Ax^2 + Ay^2 + Dx + Ey + F = 0$$

where $A \neq 0$, *represents a circle, or a point, or else has no graph.*

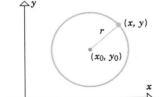

Figure 1.5.4

PARABOLAS

ORIENTATION	DESCRIPTION	STANDARD EQUATION
	• Vertex at the origin. • Parabola opens in the positive x-direction. • Symmetric about the x-axis.	$y^2 = 4px$
	• Vertex at the origin. • Parabola opens in the negative x-direction. • Symmetric about the x-axis.	$y^2 = -4px$
	• Vertex at the origin. • Parabola opens in the positive y-direction. • Symmetric about the y-axis.	$x^2 = 4py$
	• Vertex at the origin. • Parabola opens in the negative y-direction. • Symmetric about the y-axis.	$x^2 = -4py$

ELLIPSES

ORIENTATION	DESCRIPTION	STANDARD EQUATION
	• Foci and major axis on the x-axis. • Minor axis on the y-axis. • Center at the origin. • x-intercepts: $\pm a$. • y-intercepts: $\pm b$. • $a \geq b$	$\dfrac{x^2}{a^2} + \dfrac{y^2}{b^2} = 1$
	• Foci and major axis on the y-axis. • Minor axis on the x-axis. • Center at the origin. • x-intercepts: $\pm b$. • y-intercepts: $\pm a$. • $a \geq b$	$\dfrac{x^2}{b^2} + \dfrac{y^2}{a^2} = 1$

HYPERBOLAS

ORIENTATION	DESCRIPTION	STANDARD EQUATION	ASYMPTOTE EQUATIONS
	• Foci on the x-axis. • Conjugate axis on the y-axis. • Center at the origin.	$\dfrac{x^2}{a^2} - \dfrac{y^2}{b^2} = 1$	$y = \dfrac{b}{a}x$ $y = -\dfrac{b}{a}x$
	• Foci on the y-axis. • Conjugate axis on the x-axis. • Center at the origin.	$\dfrac{y^2}{a^2} - \dfrac{x^2}{b^2} = 1$	$y = \dfrac{a}{b}x$ $y = -\dfrac{a}{b}x$

Translation Equations

$$x' = x - h, \quad y' = y - k \qquad x = x' + h, \quad y = y' + k$$

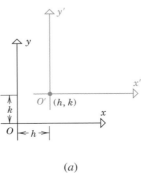

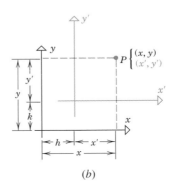

Figure 12.2.8   *(a)*   *(b)*

TRANSLATED CONICS

Parabola with Vertex (h, k) and Axis Parallel to y-axis

$$(x - h)^2 = \pm 4p(y - k)$$

where the $+$ sign occurs if the parabola opens in the positive y-direction and the $-$ sign if it opens in the negative y-direction.

Parabola with Vertex (h, k) and Axis Parallel to x-axis

$$(y - k)^2 = \pm 4p(x - h)$$

where the $+$ sign occurs if the parabola opens in the positive x-direction and the $-$ sign if it opens in the negative x-direction.

Ellipse with Center (h, k) and Major Axis Parallel to x-axis

$$\frac{(x - h)^2}{a^2} + \frac{(y - k)^2}{b^2} = 1 \quad (a \geq b)$$

Ellipse with Center (h, k) and Major Axis Parallel to y-axis

$$\frac{(x - h)^2}{b^2} + \frac{(y - k)^2}{a^2} = 1 \quad (a \geq b)$$

Hyperbola with Center (h, k) and Focal Axis Parallel to x-axis

$$\frac{(x - h)^2}{a^2} - \frac{(y - k)^2}{b^2} = 1$$

Hyperbola with Center (h, k) and Focal Axis Parallel to y-axis

$$\frac{(y - k)^2}{a^2} - \frac{(x - h)^2}{b^2} = 1$$

Relationship between Polar and Rectangular Coordinates

$$x = r \cos \theta$$
$$y = r \sin \theta$$

and

$$r^2 = x^2 + y^2$$
$$\tan \theta = \frac{y}{x}$$

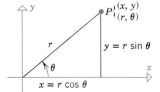

Figure 13.1.5

Also review the concept of work.

REVIEW TOPICS FOR CHAPTER 15

- Functions, domain, and range
- Concept of a limit
- Properties of limits
- Continuity
- Definition of a derivative
- Tangent lines
- Properties of derivatives
- Chain rule
- Indefinite integrals and properties
- Definite integrals and properties
- Arc length
- Velocity, speed, and acceleration
- Conics in polar coordinates

DEFINITION.   A *function* is a rule that assigns to each element in a nonempty set *A* one and only one element in a set *B*. The set *A* is called the ***domain*** of the function, and the set of all values of $f(x)$ is called the ***range***.

*If a function is defined by a formula and there is no domain explicitly stated, then it is understood that the domain consists of all real numbers for which the formula makes sense, and the function has a real value. This is called the **natural domain** of the function.*

LIMIT NOTATION

MATHEMATICAL SITUATION	NOTATION	HOW TO READ THE NOTATION
The value of $f(x)$ approaches the number L_1 as x approaches x_0 from the right side.	$\lim\limits_{x \to x_0^+} f(x) = L_1$	The limit of $f(x)$ as x approaches x_0 from the right is equal to L_1.
The value of $f(x)$ approaches the number L_2 as x approaches x_0 from the left side.	$\lim\limits_{x \to x_0^-} f(x) = L_2$	The limit of $f(x)$ as x approaches x_0 from the left is equal to L_2.
The value of $f(x)$ approaches the number L as x approaches x_0 from either the left or right side; that is, $\lim\limits_{x \to x_0^+} f(x) = \lim\limits_{x \to x_0^-} f(x) = L$	$\lim\limits_{x \to x_0} f(x) = L$	The limit of $f(x)$ as x approaches x_0 is equal to L.

Also, review limits involving

$$\lim_{x \to +\infty} f(x), \qquad \lim_{x \to -\infty} f(x), \qquad \lim_{x \to x_0} f(x) = \pm\infty$$

2.5.1 THEOREM. *Let* $\lim$ *stand for one of the limits* $\lim\limits_{x \to a}$, $\lim\limits_{x \to a^-}$, $\lim\limits_{x \to a^+}$, $\lim\limits_{x \to +\infty}$, *or* $\lim\limits_{x \to -\infty}$. *If* $L_1 = \lim f(x)$ *and* $L_2 = \lim g(x)$ *both exist, then*

(a) $\lim [f(x) + g(x)] = \lim f(x) + \lim g(x) = L_1 + L_2$

(b) $\lim [f(x) - g(x)] = \lim f(x) - \lim g(x) = L_1 - L_2$

(c) $\lim [f(x)g(x)] = \lim f(x) \lim g(x) = L_1 L_2$

(d) $\lim \dfrac{f(x)}{g(x)} = \dfrac{\lim f(x)}{\lim g(x)} = \dfrac{L_1}{L_2}$ *if* $L_2 \neq 0$

(e) $\lim \sqrt[n]{f(x)} = \sqrt[n]{\lim f(x)} = \sqrt[n]{L_1}$ *provided* $L_1 \geq 0$ *if* n *is even.*

2.7.1 DEFINITION. A function f is said to be ***continuous at a point c*** if the following conditions are satisfied:

1. $f(c)$ is defined.

2. $\lim\limits_{x \to c} f(x)$ exists.

3. $\lim\limits_{x \to c} f(x) = f(c)$.

If one or more of the conditions in this definition fails to hold, then f is called ***discontinuous at c*** and c is called a ***point of discontinuity*** of f. If f is continuous at all points of an open interval (a, b), then f is said to be ***continuous on (a, b)***. A function that is continuous on $(-\infty, +\infty)$ is said to be ***continuous everywhere*** or simply ***continuous***.

Also review continuity on a closed interval.

2.7.3 THEOREM.   *If the functions f and g are continuous at c, then*

(a)   *$f + g$ is continuous at c;*
(b)   *$f - g$ is continuous at c;*
(c)   *$f \cdot g$ is continuous at c;*
(d)   *f/g is continuous at c if $g(c) \neq 0$ and is discontinuous at c if $g(c) = 0$.*

2.7.5 THEOREM.   *Let* lim *stand for one of the limits* $\lim_{x \to c}$, $\lim_{x \to c^-}$, $\lim_{x \to c^+}$, $\lim_{x \to +\infty}$, *or* $\lim_{x \to -\infty}$. *If* $\lim g(x) = L$ *and if the function f is continuous at L, then* $\lim f(g(x)) = f(L)$. *That is,* $\lim f(g(x)) = f(\lim g(x))$.

2.7.6 THEOREM.   *If the function g is continuous at the point c and the function f is continuous at the point g(c), then the composition $f \circ g$ is continuous at c.*

3.2.2 DEFINITION.   The function f' defined by the formula

$$f'(x) = \lim_{h \to 0} \frac{f(x + h) - f(x)}{h}$$

is called the ***derivative with respect to x*** of the function f. The domain of f' consists of all x for which the limit exists.

Geometric Interpretation of the Derivative
f' is the function whose value at x is the slope of the tangent line to the graph of f at x.

Rate of Change Interpretation of the Derivative
If $y = f(x)$, then f' is the function whose value at x is the instantaneous rate of change of y with respect to x at the point x.

Derivative Notations for $y = f(x)$

$$f'(x) = \lim_{\Delta x \to 0} \frac{\Delta f}{\Delta x} \qquad \frac{dy}{dx} = \lim_{\Delta x \to 0} \frac{f(x + \Delta x) - f(x)}{\Delta x}$$

$$\frac{dy}{dx} = \lim_{\Delta x \to 0} \frac{\Delta y}{\Delta x} \qquad \frac{dy}{dx} = f'(x) \qquad \frac{d}{dx}[f(x)] = f'(x)$$

$$\left.\frac{dy}{dx}\right|_{x=x_0} = f'(x_0) \qquad \left.\frac{d}{dx}[f(x)]\right|_{x=x_0} = f'(x_0)$$

Tangent Line Equation to $y = f(x)$ at x_0

$$y - y_0 = f'(x_0)(x - x_0)$$

Derivative Formulas

$$\frac{d}{dx}[c] = 0 \qquad \frac{d}{dx}[f(x) + g(x)] = \frac{d}{dx}[f(x)] + \frac{d}{dx}[g(x)]$$

$$\frac{d}{dx}[cf(x)] = c\frac{d}{dx}[f(x)] \qquad \frac{d}{dx}[f(x) - g(x)] = \frac{d}{dx}[f(x)] - \frac{d}{dx}[g(x)]$$

$$\frac{d}{dx}[f(x)g(x)] = f(x)\frac{d}{dx}[g(x)] + g(x)\frac{d}{dx}[f(x)]$$

$$\frac{d}{dx}\left[\frac{f(x)}{g(x)}\right] = \frac{g(x)\dfrac{d}{dx}[f(x)] - f(x)\dfrac{d}{dx}[g(x)]}{[g(x)]^2}$$

3.5.2   THEOREM (**The Chain Rule**).   *If g is differentiable at the point x and f is differentiable at the point $g(x)$, then the composition $f \circ g$ is differentiable at the point x. Moreover, if*

$$y = f(g(x)) \quad and \quad u = g(x)$$

then $y = f(u)$ and

$$\frac{dy}{dx} = \frac{dy}{du} \cdot \frac{du}{dx}$$

4.2.2   THEOREM.   *Let f be a function that is continuous on a closed interval $[a, b]$ and differentiable on the open interval (a, b).*

(a)   *If $f'(x) > 0$ for every value of x in (a, b), then f is increasing on $[a, b]$.*
(b)   *If $f'(x) < 0$ for every value of x in (a, b), then f is decreasing on $[a, b]$.*
(c)   *If $f'(x) = 0$ for every value of x in (a, b), then f is constant on $[a, b]$.*

5.2.1   DEFINITION.   A function F is called an ***antiderivative*** of a function f on a given interval if $F'(x) = f(x)$ for all x in that interval.

5.2.2   THEOREM.   *If $F(x)$ is any antiderivative of $f(x)$ on a given interval, then for any value of C, the function $F(x) + C$ is also an antiderivative of $f(x)$ on that interval; moreover, every antiderivative of $f(x)$ on the interval is expressible in the form $F(x) + C$, where C is a constant.*

The relationship in Theorem 5.2.2 can be expressed using an indefinite integral:

$$\int f(x)\,dx = F(x) + C$$

5.6.3 DEFINITION.   Let the interval $[a, b]$ be divided into n subintervals. Let the kth subinterval have width Δx_k, let x_k^* be any point in the kth subinterval, and let max Δx_k be the largest of the widths of the subintervals. If the function f is defined on $[a, b]$, then f is called **Riemann integrable** on $[a, b]$ or more simply **integrable** on $[a, b]$ if the limit

$$\lim_{\max \Delta x_k \to 0} \sum_{k=1}^{n} f(x_k^*)\,\Delta x_k$$

exists. If f is integrable on $[a, b]$, then we define the **definite integral** of f from a to b by

$$\int_a^b f(x)\,dx = \lim_{\max \Delta x_k \to 0} \sum_{k=1}^{n} f(x_k^*)\,\Delta x_k$$

5.6.9 THEOREM.   *Let f be a function that is defined at all points in the interval $[a, b]$.*
(a) *If f is continuous on $[a, b]$, then f is integrable on $[a, b]$.*
(b) *If f is bounded on $[a, b]$ and has only finitely many points of discontinuity on $[a, b]$, then f is integrable on $[a, b]$.*
(c) *If f is not bounded on $[a, b]$, then f is not integrable on $[a, b]$.*

Area Interpretation of the Definite Integral
If the function f is continuous on $[a, b]$ and $f(x) \geq 0$ for all x in $[a, b]$, then

$$\int_a^b f(x)\,dx$$

is the area under the curve $y = f(x)$ over the interval $[a, b]$. If $f(x)$ has both positive and negative values on $[a, b]$, then this integral represents the difference of two areas: the area under $y = f(x)$ above $[a, b]$ minus the area over $y = f(x)$ below $[a, b]$.

$$\int_a^b f(x)\,dx = (A_I + A_{III}) - A_{II} = \begin{bmatrix} \text{area above} \\ [a, b] \end{bmatrix} - \begin{bmatrix} \text{area below} \\ [a, b] \end{bmatrix}$$

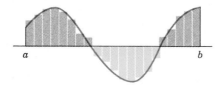
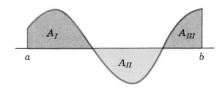

Figure 5.6.8

5.7.1 THEOREM (**The First Fundamental Theorem of Calculus**). *If f is continuous on $[a, b]$ and if F is an antiderivative of f on $[a, b]$, then*

$$\int_a^b f(x)\, dx = F(b) - F(a)$$

5.9.1 THEOREM (**The Second Fundamental Theorem of Calculus**). *Let f be a continuous function on an interval I, and let a be any point in I. If F is defined by*

$$F(x) = \int_a^x f(t)\, dt$$

then $F'(x) = f(x)$ at each point x in the interval I.

It follows from Theorems 5.7.1 and 5.9.1 that

$$\int_a^b f(x)\, dx = \left[\int f(x)\, dx \right]_a^b \qquad \frac{d}{dx}\left[\int_a^x f(t)\, dt \right] = f(x)$$

5.6.5 THEOREM. *If f and g are integrable on $[a, b]$ and if c is a constant, then cf, $f + g$, and $f - g$ are integrable on $[a, b]$ and*

(a) $\displaystyle \int_a^b cf(x)\, dx = c \int_a^b f(x)\, dx$

(b) $\displaystyle \int_a^b [f(x) + g(x)]\, dx = \int_a^b f(x)\, dx + \int_a^b g(x)\, dx$

(c) $\displaystyle \int_a^b [f(x) - g(x)]\, dx = \int_a^b f(x)\, dx - \int_a^b g(x)\, dx$

The equalities in Theorem 5.6.5 also hold for indefinite integrals.

5.6.6 THEOREM. *If f is integrable on a closed interval containing the three points a, b, and c, then*

$$\int_a^b f(x)\, dx = \int_a^c f(x)\, dx + \int_c^b f(x)\, dx$$

no matter how the points are ordered.

Integration by Parts for Indefinite Integrals

$$\int u\, dv = uv - \int v\, du$$

Integration by Parts for Definite Integrals

$$\int_a^b u \, dv = uv \Big]_a^b - \int_a^b v \, du$$

13.4.1 ARC-LENGTH FORMULA FOR PARAMETRIC CURVES.   If no segment of the curve represented by the parametric equations

$$x = x(t), \quad y = y(t) \quad (a \le t \le b)$$

is traced more than once as t increases from a to b, and if dx/dt and dy/dt are continuous functions for $a \le t \le b$, then the arc length L of the curve is given by

$$L = \int_a^b \sqrt{\left(\frac{dx}{dt}\right)^2 + \left(\frac{dy}{dt}\right)^2} \, dt$$

4.10.1 DEFINITION.   If $s(t)$ is the position function of a particle moving on a coordinate line, then the ***instantaneous velocity*** at time t is defined by

$$v(t) = s'(t) = \frac{ds}{dt}$$

and the ***instantaneous acceleration*** at time t is defined by

$$a(t) = v'(t) = \frac{dv}{dt} = s''(t)$$

The ***instantaneous speed*** of a particle moving on a coordinate line is defined to be the absolute value of the velocity. Thus,

$$\begin{bmatrix} \text{instantaneous} \\ \text{speed} \end{bmatrix} = |v(t)| = |s'(t)| = \left|\frac{ds}{dt}\right|$$

Let a fixed point F (the focus) and a fixed line D (the directrix) be given so that the directrix does not contain the focus. If a point P moves in the plane determined by the focus and the directrix so that the distance between P and the focus is proportional to the distance between P and the directrix, then the curve traced out by P is a conic.

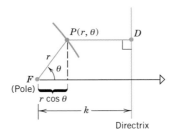

Figure 13.2.17

CONICS IN POLAR COORDINATES

ORIENTATION	EQUATION	CONIC
• Focus at pole. • Directrix perpendicular to polar axis. • Directrix k units to right of pole.	$r = \dfrac{ek}{1 + e \cos \theta}$	
• Focus at pole. • Directrix perpendicular to polar axis. • Directrix k units to left of pole.	$r = \dfrac{ek}{1 - e \cos \theta}$	$0 < e < 1$ Ellipse $e = 1$ Parabola $e > 1$ Hyperbola
• Focus at pole. • Directrix parallel to polar axis. • Directrix k units above the pole.	$r = \dfrac{ek}{1 + e \sin \theta}$	
• Focus at pole. • Directrix parallel to polar axis. • Directrix k units below the pole.	$r = \dfrac{ek}{1 - e \sin \theta}$	

REVIEW TOPICS FOR CHAPTER 16

- Triangle inequality
- $\delta - \epsilon$ definition of a limit
- Relationship between continuity and differentiability
- Mean-value theorem
- Differentials
- Relative and absolute extrema
- First and second derivative tests

1.2.5   THEOREM (*Triangle Inequality*).   *If a and b are any real numbers, then*

$$|a + b| \le |a| + |b|$$

2.6.3   LIMIT DEFINITION.   Let $f(x)$ be defined for all x in some open interval containing the number a, with the possible exception that $f(x)$ may not be defined at a. We shall write

$$\lim_{x \to a} f(x) = L$$

if, given any number $\epsilon > 0$, we can find a number $\delta > 0$ such that

$$|f(x) - L| < \epsilon \quad \text{if } x \text{ satisfies} \quad 0 < |x - a| < \delta$$

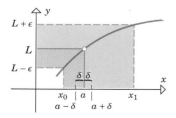

Figure 2.6.2

3.2.3   THEOREM.   *If f is differentiable at a point x_0, then f is also continuous at x_0.*

Theorem 3.2.3 shows that differentiability at a point implies continuity at that point. The converse, however, is false—*a function may be continuous at a point but not differentiable there.* Whenever the graph of a function has a corner at a point, but no break or gap there, we have a point where the function is continuous, but not differentiable.

4.9.2   THEOREM (***Mean-Value Theorem***).   *Let f be differentiable on (a, b) and continuous on $[a, b]$. Then there is at least one point c in (a, b) where*

$$f'(c) = \frac{f(b) - f(a)}{b - a}$$

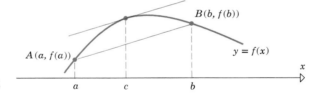

Figure 4.9.4

Increments

   If $y = f(x)$, and x changes from an initial value x_0 to a final value x_1, then there is a corresponding change in the value of y from $y_0 = f(x_0)$ to $y_1 = f(x_1)$. The change in x, denoted by Δx, is called the ***increment in x*** and the change in y, denoted by Δy, is called the ***increment in y***:

$$\Delta x = x_1 - x_0 \qquad \Delta y = y_1 - y_0 = f(x_1) - f(x_0)$$

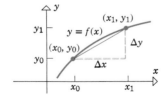

Figure 3.7.1

Differentials

Regard x as fixed and *define dx* to be an independent variable that can be assigned an arbitrary value. If f is differentiable at x, then we *define dy* by the formula

$$dy = f'(x)\,dx$$

It is important to understand the distinction between the increment Δy and the differential dy. To see the difference, let us assign the independent variables dx and Δx the same value, so $dx = \Delta x$. Then Δy represents the change in y that occurs when we start at x and travel *along the curve* $y = f(x)$ until we have moved $\Delta x\,(= dx)$ units in the x-direction, while dy represents the change in y that occurs if we start at x and travel *along the tangent* line until we have moved $dx\,(= \Delta x)$ units in the x-direction (Figure 3.7.4).

For $\Delta x = dx$ near zero, dy is commonly used as an approximation to Δy:

$$\Delta y \approx dy$$

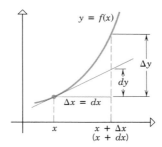

Figure 3.7.4

4.3.1   DEFINITION.   A function f is said to have a ***relative maximum*** at x_0 if $f(x_0) \geq f(x)$ for all x in some open interval containing x_0.

4.3.2   DEFINITION.   A function f is said to have a ***relative minimum*** at x_0 if $f(x_0) \leq f(x)$ for all x in some open interval containing x_0.

4.3.3   DEFINITION.   A function f is said to have a ***relative extremum*** at x_0 if it has either a relative maximum or a relative minimum at x_0.

4.3.4   THEOREM.   *If f has a relative extremum at x_0, then either $f'(x_0) = 0$ or f is not differentiable at x_0.*

4.3.5 DEFINITION.   A *critical point* for a function f is any value of x in the domain of f at which $f'(x) = 0$ or at which f is not differentiable; the critical points where $f'(x) = 0$ are called *stationary points* of f.

4.3.6 THEOREM (*First Derivative Test*).   *Suppose f is continuous at a critical point x_0.*

(a)   *If $f'(x) > 0$ on an open interval extending left from x_0 and $f'(x) < 0$ on an open interval extending right from x_0, then f has a relative maximum at x_0.*

(b)   *If $f'(x) < 0$ on an open interval extending left from x_0 and $f'(x) > 0$ on an open interval extending right from x_0, then f has a relative minimum at x_0.*

(c)   *If $f'(x)$ has the same sign [either $f'(x) > 0$ or $f'(x) < 0$] on an open interval extending left from x_0 and on an open interval extending right from x_0, then f does not have a relative extremum at x_0.*

4.3.7 THEOREM (*Second Derivative Test*).   *Suppose that f is twice differentiable at a stationary point x_0.*

(a)   *If $f''(x_0) > 0$, then f has a relative minimum at x_0.*

(b)   *If $f''(x_0) < 0$, then f has a relative maximum at x_0.*

4.6.1 DEFINITION.   If $f(x_0) \geq f(x)$ for all x in the domain of f, then $f(x_0)$ is called the *absolute maximum value* or simply the *maximum value* of f.

4.6.2 DEFINITION.   If $f(x_0) \leq f(x)$ for all x in the domain of f, then $f(x_0)$ is called the *absolute minimum value* or simply the *minimum value* of f.

4.6.3 DEFINITION.   A number that is either the maximum or the minimum value of a function f is called an *absolute extreme value* or *extreme value* of f. Sometimes the terms *absolute extremum* or *extremum* are also used.

4.6.4   THEOREM (***Extreme-Value Theorem***).   *If a function f is continuous on a closed interval [a, b], then f has both a maximum value and a minimum value on [a, b].*

4.6.5   THEOREM.   *If a function f has an extreme value (either a maximum or a minimum) on an open interval (a, b), then the extreme value occurs at a critical point of f.*

4.6.6   THEOREM.   *Let f be continuous on an interval I and assume that f has exactly one relative extremum on I, say at x_0.*

(*a*)   *If f has a relative minimum at x_0, then $f(x_0)$ is the minimum value of f on the interval I.*

(*b*)   *If f has a relative maximum at x_0, then $f(x_0)$ is the maximum value of f on the interval I.*

REVIEW TOPICS FOR CHAPTER 17

- Graphs in polar coordinates
- Volumes by slicing (cross-sections)
- Surface area

LINES IN POLAR COORDINATES

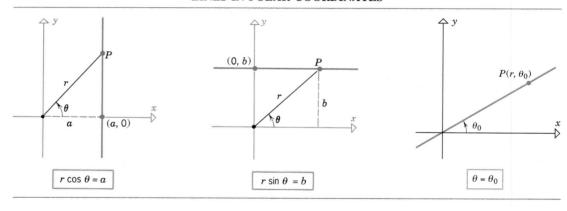

$r \cos \theta = a$ $r \sin \theta = b$ $\theta = \theta_0$

CIRCLES IN POLAR COORDINATES

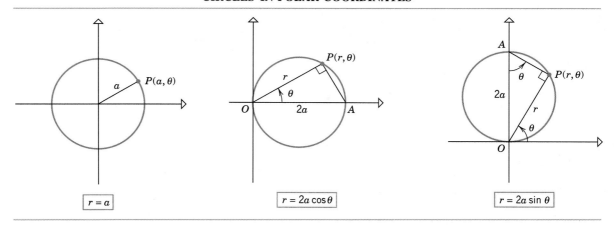

| $r = a$ | $r = 2a \cos \theta$ | $r = 2a \sin \theta$ |

LIMAÇONS AND CARDIOIDS

$$r = a + b \sin \theta, \quad r = a - b \sin \theta$$
$$r = a + b \cos \theta, \quad r = a - b \cos \theta$$

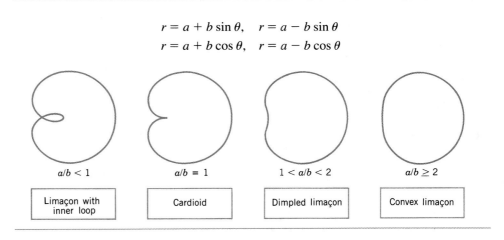

| $a/b < 1$ | $a/b = 1$ | $1 < a/b < 2$ | $a/b \geq 2$ |

| Limaçon with inner loop | Cardioid | Dimpled limaçon | Convex limaçon |

LEMNISCATES

$$r^2 = a^2 \cos 2\theta, \quad r^2 = -a^2 \cos 2\theta$$
$$r^2 = a^2 \sin 2\theta, \quad r^2 = -a^2 \sin 2\theta$$

A lemniscate

ROSE CURVES

$$r = a \sin n\theta$$

$$r = a \cos n\theta$$

A three-petal rose
(n = 3)

A four-petal rose
(n = 2)

6.2.1   VOLUME FORMULA.   Let S be a solid bounded by two parallel planes perpendicular to the x-axis at $x = a$ and $x = b$. If, for each x in $[a, b]$, the cross-sectional area of S perpendicular to the x-axis is $A(x)$, then the volume of the solid is

$$V = \int_a^b A(x)\, dx$$

provided $A(x)$ is integrable.

There is a similar result for cross sections perpendicular to the y-axis.

6.2.2   VOLUME FORMULA.   Let S be a solid bounded by two parallel planes perpendicular to the y-axis at $y = c$ and $y = d$. If, for each y in $[c, d]$, the cross-sectional area of S perpendicular to the y-axis is $A(y)$, then the volume of the solid is

$$V = \int_c^d A(y)\, dy$$

provided $A(y)$ is integrable.

6.5.2   SURFACE AREA FORMULAS.   Let f be a smooth, nonnegative function on $[a, b]$. Then the **surface area S** generated by revolving the portion of the curve $y = f(x)$ between $x = a$ and $x = b$ about the x-axis is

$$S = \int_a^b 2\pi f(x)\sqrt{1 + [f'(x)]^2}\, dx$$

For a curve expressed in the form $x = g(y)$, where g' is continuous on $[c, d]$, and $g(y) \geq 0$ for $c \leq y \leq d$, the surface area S generated by revolving the portion of the curve from $y = c$ to $y = d$ about the y-axis is given by

$$S = \int_c^d 2\pi g(y)\sqrt{1 + [g'(y)]^2}\, dy$$

REVIEW TOPICS FOR CHAPTER 18

- See review material for Chapters 14–17.

REVIEW TOPICS FOR CHAPTER 19

- First-order linear and separable differential equations (see Section 7.7).

Jakob Bernoulli (1654–1705)

14 THREE-DIMENSIONAL SPACE; VECTORS

In this section we shall discuss coordinate systems in three-dimensional space and some basic facts about surfaces in three dimensions.

☐ **RECTANGULAR COORDINATE SYSTEMS**

In the remainder of this text we shall call three-dimensional space *3-space*, two-dimensional space (a plane) *2-space*, and one-dimensional space (a line) *1-space*. Just as points in 2-space can be placed in one-to-one correspondence with pairs of real numbers using two perpendicular coordinate lines, so points in 3-space can be placed in one-to-one correspondence with triples of real numbers by using three mutually perpendicular coordinate lines, called the *x-axis*, the *y-axis*, and the *z-axis*, positioned so that their origins coincide (Figure 14.1.1). The three coordinate axes form a three-dimensional *rectangular coordinate system* (or *Cartesian coordinate system*). The point of intersection of the coordinate axes is called the *origin* of the coordinate system.

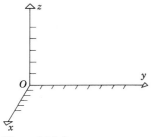

Figure 14.1.1

Rectangular coordinate systems in 3-space fall into two categories: *left-handed* and *right-handed*. A right-handed system has the property that when the fingers of the right hand are cupped so that they curve from the positive *x*-axis toward the positive *y*-axis, the thumb points (roughly) in the direction of the positive *z*-axis (Figure 14.1.2*a*). A system that is not right-handed is called left-handed (Figure 14.1.2*b*). We shall use only right-handed coordinate systems.

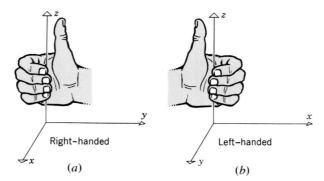

Figure 14.1.2   (*a*)   (*b*)

The coordinate axes, taken in pairs, determine three *coordinate planes*: the *xy-plane*, the *xz-plane*, and the *yz-plane*. To each point *P* in 3-space we can assign a triple of real numbers by passing three planes through *P* parallel to the coordinate planes and letting *a*, *b*, and *c* be the coordinates of the intersections of those planes with the *x*-axis, *y*-axis, and *z*-axis, respectively (Figure 14.1.3). We call *a*, *b*, and *c* the *x-coordinate*, *y-coordinate*, and *z-coordinate* of *P*, respectively, and we denote the point *P* by (a, b, c) or by $P(a, b, c)$. Figure 14.1.4 shows the points $(4, 5, 6)$ and $(-3, 2, -4)$.

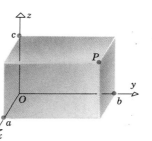

Figure 14.1.3

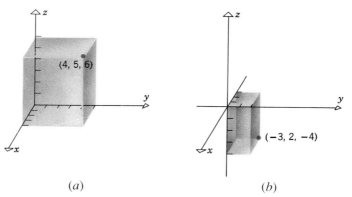

Figure 14.1.4   (*a*)   (*b*)

Just as the coordinate axes in a two-dimensional coordinate system divide 2-space into four quadrants, so the coordinate planes of a three-dimensional coordinate system divide 3-space into eight parts, called *octants* (count them). Those points having three positive coordinates form the *first octant*; the remaining octants have no standard numbering.

The reader should be able to visualize the following results about three-dimensional rectangular coordinate systems:

REGION	DESCRIPTION
xy-plane	Consists of all points of the form $(x, y, 0)$
xz-plane	Consists of all points of the form $(x, 0, z)$
yz-plane	Consists of all points of the form $(0, y, z)$
x-axis	Consists of all points of the form $(x, 0, 0)$
y-axis	Consists of all points of the form $(0, y, 0)$
z-axis	Consists of all points of the form $(0, 0, z)$

□ **DISTANCE, MIDPOINT,
AND SPHERES**

A formula for the distance between two points in 3-space can be obtained by starting with a box with dimensions a, b, and c and considering the right triangle in Figure 14.1.5 that has one side along the diagonal of the base, one side along a vertical edge, and its hypotenuse along a diagonal of the box. By the Theorem of Pythagoras, the side in the base of the box has length $\sqrt{a^2 + b^2}$, and by a second application of the Theorem of Pythagoras the hypotenuse has length

$$d = \sqrt{(\sqrt{a^2 + b^2})^2 + c^2} = \sqrt{a^2 + b^2 + c^2} \tag{1}$$

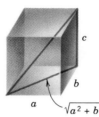

Figure 14.1.5

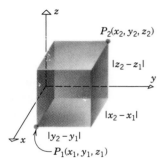

Figure 14.1.6

If we now apply (1) to the box in Figure 14.1.6, which has $P_1(x_1, y_1, z_1)$ and $P_2(x_2, y_2, z_2)$ as diagonal corners and hence sides of length

$$|x_2 - x_1|, \quad |y_2 - y_1|, \quad \text{and} \quad |z_2 - z_1|$$

then we find that the distance d between P_1 and P_2 is

$$d = \sqrt{(x_2 - x_1)^2 + (y_2 - y_1)^2 + (z_2 - z_1)^2} \tag{2}$$

(where we have dropped the absolute value signs, since $|x|^2 = x^2$ for all x.)

Observe that Formula (2) has the same form as the distance formula in 2-space, except for an additional term to account for the z-coordinate. Similarly, we shall prove later that the formula for the midpoint of a line segment in 3-space has the same form as the formula in 2-space, except for an additional coordinate:

$$\begin{bmatrix} \text{The midpoint of the} \\ \text{line segment joining} \\ P_1(x_1, y_1, z_1) \text{ and } P_2(x_2, y_2, z_2) \end{bmatrix} = (\tfrac{1}{2}(x_1 + x_2), \tfrac{1}{2}(y_1 + y_2), \tfrac{1}{2}(z_1 + z_2)) \tag{3}$$

Example 1   Find the distance d between the points $(4, -1, 3)$ and $(2, 3, -1)$, and find the midpoint M of the line segment that joins them.

Solution.   From Formulas (2) and (3)

$$d = \sqrt{(4 - 2)^2 + (-1 - 3)^2 + (3 + 1)^2} = \sqrt{36} = 6$$
$$M = (\tfrac{1}{2}(4 + 2), \tfrac{1}{2}(-1 + 3), \tfrac{1}{2}(3 - 1)) = (3, 1, 1) \quad \blacktriangleleft$$

In 2-space, the set of points satisfying an equation in x and y is usually a *curve* in the xy-plane. Similarly, in 3-space the set of points satisfying an equation in x, y, and z is usually a *surface* in an xyz-coordinate system. The surface is called the **graph** of the equation. For example, Figure 14.1.7 shows a sphere in 3-space with center (x_0, y_0, z_0) and radius r. The sphere consists of all points whose distance from (x_0, y_0, z_0) is r, so from (2) it consists of those points (x, y, z) whose coordinates satisfy

$$\sqrt{(x - x_0)^2 + (y - y_0)^2 + (z - z_0)^2} = r$$

or equivalently,

$$(x - x_0)^2 + (y - y_0)^2 + (z - z_0)^2 = r^2 \tag{4}$$

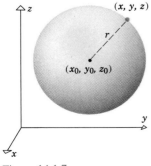

Figure 14.1.7

This is called the ***standard form of the equation of the sphere*** with center (x_0, y_0, z_0) and radius r. Observe that this equation has the same form as the standard form of the equation of a circle in 2-space, except for the additional term to account for the z-coordinate. Some examples are given in the following table.

EQUATION	GRAPH
$(x - 3)^2 + (y - 2)^2 + (z - 1)^2 = 9$	Sphere with center $(3, 2, 1)$ and radius 3
$(x + 1)^2 + y^2 + (z + 4)^2 = 5$	Sphere with center $(-1, 0, -4)$ and radius $\sqrt{5}$
$x^2 + y^2 + z^2 = 1$	Sphere with center $(0, 0, 0)$ and radius 1

If the terms in (4) are squared out and like terms are then collected, then the resulting equation has the form

$$x^2 + y^2 + z^2 + Gx + Hy + Iz + J = 0 \tag{5}$$

As the following example shows, the center and radius of a sphere expressed in this form can be obtained by completing the squares.

Example 2 Find the center and radius of the sphere

$$x^2 + y^2 + z^2 - 2x - 4y + 8z + 17 = 0$$

Solution. We can put the equation in the form of (4) by completing the squares:

$$(x^2 - 2x) + (y^2 - 4y) + (z^2 + 8z) = -17$$

$$(x^2 - 2x + 1) + (y^2 - 4y + 4) + (z^2 + 8z + 16) = -17 + 21$$

$$(x - 1)^2 + (y - 2)^2 + (z + 4)^2 = 4$$

which is the equation of the sphere with center $(1, 2, -4)$ and radius 2. ◄

In general, completing the squares in (5) produces an equation of the form

$$(x - x_0)^2 + (y - y_0)^2 + (z - z_0)^2 = k$$

which represents a sphere of radius $\sqrt{k}$ and center (x_0, y_0, z_0) if $k > 0$, the point (x_0, y_0, z_0) if $k = 0$, or has no graph if $k < 0$ (why?). The following theorem summarizes these observations.

14.1.1 THEOREM. *An equation of the form*

$$x^2 + y^2 + z^2 + Gx + Hy + Iz + J = 0$$

represents either a sphere or a point, or else has no graph.

☐ **CYLINDRICAL SURFACES**

Many important surfaces in 3-space can be generated by translating a plane curve along a line. For example, the surface in Figure 14.1.8 is obtained by translating the curve $y = x^2$ in the xy-plane along a line parallel to the z-axis. The process of translating a curve along a line to generate a surface is called ***extrusion***, and surfaces that are generated by extrusion are called ***cylindrical surfaces***.

A surface in an xyz-system that is obtained by extrusion along a line parallel to a coordinate axis has an equation with only two of the three variables x, y, and z. For example, the surface in Figure 14.1.8 is represented by the equation $y = x^2$ (with no z variable). This can be seen by observing that if (x, y) satisfies the equation $y = x^2$, then for *arbitrary* values of z the points of the form (x, y, z) lie on the surface; they are directly above or below the point $(x, y, 0)$ on the curve $y = x^2$ in the xy-plane (Figure 14.1.9).

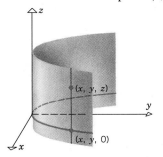

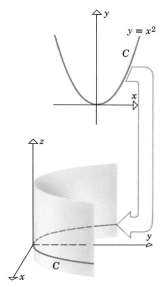

Figure 14.1.9

Figure 14.1.8

In summary, we have the following theorem.

> **14.1.2**   THEOREM.   *An equation containing only two of the three variables x, y, and z represents a cylindrical surface in 3-space. The surface is obtained by extrusion parallel to the axis corresponding to the missing variable.*

REMARK.   Just as $x = a$ can represent a point on the x-axis or a line in the xy-plane parallel to the y-axis, so an equation with two variables, such as $y = x^2$, can represent either a curve in the xy-plane in 2-space or a cylindrical surface in an xyz-coordinate system in 3-space. The appropriate interpretation will usually be clear from the context in which the equation appears.

Example 3   Sketch the graph of $x^2 + z^2 = 1$ in 3-space.

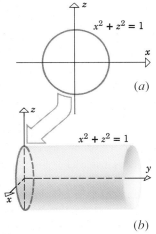

Figure 14.1.10

Solution.   Since y does not appear in this equation, the graph is a cylindrical surface generated by extrusion parallel to the y-axis. It is helpful to begin with a sketch of the equation in 2-space. In the xz-plane the curve $x^2 + z^2 = 1$ is a circle (Figure 14.1.10*a*). Thus, in 3-space this equation represents a right-circular cylinder parallel to the y-axis (Figure 14.1.10*b*).   ◄

Example 4   Sketch the graph of $z = \sin y$ in 3-space.

Solution.   (See Figure 14.1.11.)   ◄

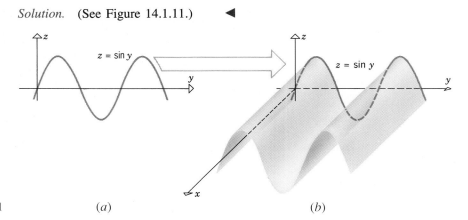

Figure 14.1.11   (*a*)   (*b*)

▶ Exercise Set 14.1

1. Plot the points P and Q in a right-handed coordinate system. Then, find the distance between them and the midpoint of the line segment joining them.

   (a) $P(0, 0, 0)$; $Q(2, 1, 3)$

   (b) $P(5, 2, 3)$; $Q(4, 1, 6)$

   (c) $P(-2, -1, 3)$; $Q(3, 0, 5)$

   (d) $P(-1, -1, -3)$; $Q(4, 3, -2)$.

2. A cube of side 4 has its geometric center at the origin and its faces parallel to the coordinate planes. Sketch the cube and give the coordinates of the vertices.

3. A rectangular parallelepiped has its faces parallel to the coordinate planes and has $(4, 2, -2)$ and $(-6, 1, 1)$ as endpoints of a diagonal. Sketch the parallelepiped and give the coordinates of the vertices.

4. Show that $(4, 5, 2)$, $(1, 7, 3)$, and $(2, 4, 5)$ are vertices of an equilateral triangle.

5. (a) Show that $(2, 1, 6)$, $(4, 7, 9)$, and $(8, 5, -6)$ are the vertices of a right triangle.

   (b) Which vertex is at the 90° angle?

   (c) Find the area of the triangle.

6. Find the distance from the point $(-5, 2, -3)$ to the

   (a) xy-plane   (b) xz-plane   (c) yz-plane

   (d) x-axis   (e) y-axis   (f) z-axis.

7. Show that the distance from a point (x_0, y_0, z_0) to the z-axis is $\sqrt{x_0^2 + y_0^2}$, and find the distances from the point to the x- and y-axes.

In Exercises 8–11, find an equation for the sphere with center C and radius r.

8. $C(0, 0, 0)$; $r = 8$.

9. $C(-2, 4, -1)$; $r = 6$.

10. $C(5, -2, 4)$; $r = \sqrt{7}$.

11. $C(0, 1, 0)$; $r = 3$.

12. In each part find an equation for the sphere with center $(-3, 5, -4)$ and satisfying the given condition.

   (a) Tangent to the xy-plane

   (b) Tangent to the xz-plane

   (c) Tangent to the yz-plane.

13. In each part, find an equation for the sphere with center $(2, -1, -3)$ and satisfying the given condition.

   (a) Tangent to the xy-plane

   (b) Tangent to the xz-plane

   (c) Tangent to the yz-plane.

In Exercises 14–19, find the standard equation of the sphere satisfying the given conditions.

14. Center $(1, 0, -1)$; diameter $= 8$.

15. A diameter has endpoints $(-1, 2, 1)$ and $(0, 2, 3)$.

16. Center $(-1, 3, 2)$ and passing through the origin.

17. Center $(3, -2, 4)$ and passing through $(7, 2, 1)$.

18. Center $(-3, 5, -4)$; tangent to the sphere of radius 1 centered at the origin (two answers).

19. Center $(0, 0, 0)$; tangent to the sphere of radius 1 centered at $(3, -2, 4)$ (two answers).

In Exercises 20–25, describe the surface whose equation is given.

20. $x^2 + y^2 + z^2 - 2x - 6y - 8z + 1 = 0$.

21. $x^2 + y^2 + z^2 + 10x + 4y + 2z - 19 = 0$.

22. $x^2 + y^2 + z^2 - y = 0$.

23. $2x^2 + 2y^2 + 2z^2 - 2x - 3y + 5z - 2 = 0$.

24. $x^2 + y^2 + z^2 + 2x - 2y + 2z + 3 = 0$.

25. $x^2 + y^2 + z^2 - 3x + 4y - 8z + 25 = 0$.

26. Find the largest and smallest distances between the point $P(1, 1, 1)$ and the sphere
$$x^2 + y^2 + z^2 - 2y + 6z - 6 = 0.$$

27. Find the largest and smallest distances between the origin and the sphere $x^2 + y^2 + z^2 + 2x - 2y - 4z - 3 = 0$.

28. Describe the set of all points in 3-space whose coordinates satisfy the inequality $x^2 + y^2 + z^2 - 2x + 8z \le 8$.

29. Describe the set of all points in 3-space whose coordinates satisfy the inequality $y^2 + z^2 + 6y - 4z > 3$.

30. The distance between a point $P(x, y, z)$ and the point $A(1, -2, 0)$ is twice the distance between P and the point $B(0, 1, 1)$. Show that the set of all such points is a sphere, and find the center and radius of the sphere.

31. A bowling ball of radius R is placed inside a box just large enough to hold it, and it is secured for shipping by packing a Styrofoam sphere into each corner of the box. Find the radius of the largest Styrofoam sphere that can be used. [*Hint:* Take the origin of a Cartesian coordinate system at a corner of the box with the coordinate axes along the edges.] (See Figure 14.1.12.)

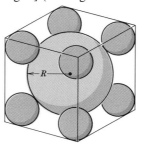

Figure 14.1.12

In Exercises 32–39, sketch the surface whose equation is given.

32. (a) $y = x$ (b) $y = z$ (c) $x = z$.

33. (a) $x^2 + y^2 = 25$ (b) $y^2 + z^2 = 25$

    (c) $x^2 + z^2 = 25$.

34. (a) $y = x^2$ (b) $z = x^2$ (c) $y = z^2$.
35. (a) $y = e^x$ (b) $x = \ln z$ (c) $yz = 1$.
36. (a) $2x + 3y = 6$ (b) $2x + z = 3$.
37. (a) $y = \sin x$ (b) $z = \cos x$.
38. (a) $z = 1 - y^2$ (b) $z = \sqrt{3 - x}$.
39. (a) $4x^2 + 9z^2 = 36$ (b) $y^2 - 4z^2 = 4$.
40. In each part of Figure 14.1.13 find an equation for the right-circular cylinder of radius a shown.
41. Show that for all values of θ and ϕ, the point

$(a \sin \phi \cos \theta,\ a \sin \phi \sin \theta,\ a \cos \phi)$ lies on the sphere $x^2 + y^2 + z^2 = a^2$.

42. Consider the equation

$$x^2 + y^2 + z^2 + Gx + Hy + Iz + J = 0$$

and let $K = G^2 + H^2 + I^2 - 4J$.

(a) Prove that the equation represents a sphere, a point, or no graph, according to whether $K > 0$, $K = 0$, or $K < 0$.

(b) In the case where $K > 0$, find the center and radius of the sphere.

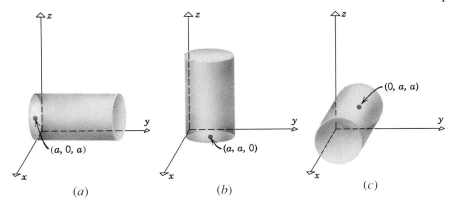

Figure 14.1.13

14.2 VECTORS

*Many physical quantities such as area, length, mass, and temperature are completely described once the magnitude of the quantity is given. Such quantities are called scalars. Other physical quantities, called **vectors**, are not completely determined until both a magnitude and a direction are specified. For example, wind movement is usually described by giving the speed and the direction, say 20 mi/hr northeast. The wind speed and wind direction together form a vector quantity called the wind velocity. Other examples of vectors are force and displacement. In this section we shall develop the basic mathematical properties of vectors.*

□ **VECTORS VIEWED GEOMETRICALLY**

Vectors can be represented geometrically as directed line segments or arrows in two- or three-dimensional space; the direction of the arrow specifies the direction of the vector and the length of the arrow describes its magnitude. The tail of the arrow is called the *initial point* of the vector, and the tip of the arrow the *terminal point*. We shall denote vectors by lowercase boldface type such as **a**, **k**, **v**, **w**, and **x**. When discussing vectors, we shall refer to real numbers as *scalars*. Scalars will be denoted by lowercase italic type such as a, k, v, w, and x.

If, as in Figure 14.2.1a, the initial point of a vector **v** is A and the terminal point is B, we write $\mathbf{v} = \overrightarrow{AB}$. Vectors having the same length and same direction, such as those in Figure 14.2.1b, are called *equivalent*. Since we want a vector to be determined solely by its length and direction, equivalent vectors are regarded as *equal* even though they may be located in different positions. If **v** and **w** are equivalent, we write $\mathbf{v} = \mathbf{w}$.

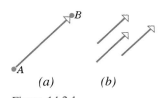

Figure 14.2.1

The vector whose initial and terminal points coincide has length zero. We call this the *zero vector* and denote it by **0**. The zero vector has no natural direction, so we shall agree that it can be assigned any direction that is convenient for a problem at hand.

There are various algebraic operations that are performed on vectors, all of whose definitions originated in physics. We begin with vector addition.

14.2.1 DEFINITION. If **v** and **w** are any two nonzero vectors, then the *sum* **v** + **w** is the vector determined as follows: Position the vector **w** so that its initial point coincides with the terminal point of **v**. The vector **v** + **w** is represented by the arrow from the initial point of **v** to the terminal point of **w** (Figure 14.2.2*a*).

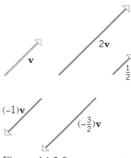

In Figure 14.2.2*b* we have constructed two sums, **v** + **w** (purple arrows) and **w** + **v** (green arrows). It is evident that

$$\mathbf{v} + \mathbf{w} = \mathbf{w} + \mathbf{v}$$

and that the sum coincides with the diagonal of the parallelogram determined by **v** and **w** when these vectors are located so that they have the same initial point.

Definition 14.2.1 does not apply if either of the vectors is zero, but we shall agree that for all vectors **v**

$$\mathbf{0} + \mathbf{v} = \mathbf{v} + \mathbf{0} = \mathbf{v}$$

We now consider another basic operation on vectors.

Figure 14.2.2

14.2.2 DEFINITION. If **v** is a nonzero vector and k is a nonzero real number (scalar), then the *scalar multiple* $k\mathbf{v}$ is defined to be the vector whose length is $|k|$ times the length of **v** and whose direction is the same as that of **v** if $k > 0$ and opposite to that of **v** if $k < 0$. We define $k\mathbf{v} = \mathbf{0}$ if $k = 0$ or $\mathbf{v} = \mathbf{0}$.

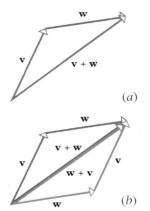

Figure 14.2.3 shows the geometric relationship between a vector **v** and various scalar multiples of it. Observe that the vector $(-1)\mathbf{v}$ has the same length as **v** but is oppositely directed. We call this vector the *negative of* **v** and denote it by

$$(-1)\mathbf{v} = -\mathbf{v}$$

(Figure 14.2.4). In addition, we define $-\mathbf{0} = (-1)\mathbf{0} = \mathbf{0}$. Subtraction of vectors is defined as follows.

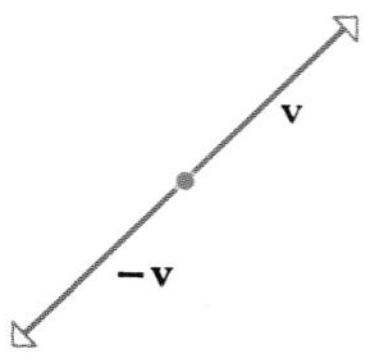

Figure 14.2.3

14.2.3 DEFINITION. If **v** and **w** are any two vectors, then the *difference* **v** − **w** is defined by

$$\mathbf{v} - \mathbf{w} = \mathbf{v} + (-\mathbf{w})$$

Figure 14.2.4

The difference **v** − **w** can be obtained geometrically by the parallelogram method shown in Figure 14.2.5*a*, or more directly, as in Figure 14.2.5*b*, by drawing the vector from the terminal point of **w** to the terminal point of **v**. We leave it for the reader to deduce that

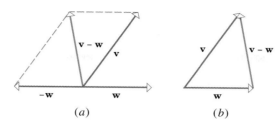

Figure 14.2.5   (a)   (b)

$$\mathbf{v} + (-\mathbf{v}) = \mathbf{v} - \mathbf{v} = \mathbf{0}$$

☐ **VECTORS IN**
  COORDINATE SYSTEMS

Problems involving vectors can often be simplified by introducing a rectangular coordinate system, in which case one must distinguish between vectors in 2-space and vectors in 3-space.

If **v** is a vector in 2-space or 3-space with its initial point at the origin of a rectangular coordinate system (Figure 14.2.6), then the coordinates (v_1, v_2) or (v_1, v_2, v_3) of the terminal point are called the **components** of **v** and we write

$$\mathbf{v} = \langle v_1, v_2 \rangle \quad \text{or} \quad \mathbf{v} = \langle v_1, v_2, v_3 \rangle$$

depending on whether the vector is in 2-space or 3-space.

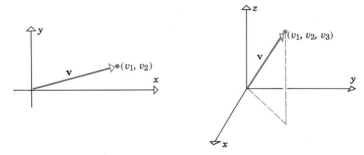

Figure 14.2.6

REMARK.   We emphasize that a vector must be positioned with its initial point at the origin in order for the components to be the coordinates of the terminal point. Later we shall show how to obtain the components of a vector that does not have its initial point at the origin.

If the zero vector is positioned with its initial point at the origin, then the terminal point is also at the origin, so

$$\mathbf{0} = \langle 0, 0 \rangle \quad \text{and} \quad \mathbf{0} = \langle 0, 0, 0 \rangle$$
  2-space   3-space

Since equivalent vectors have the same length and direction, such vectors must have the same terminal point when they are positioned with their initial points at the origin. Hence, equivalent vectors have the same components. Conversely, vectors with the same components have the same length and direction, and hence must be equivalent. Thus, for vectors **v** and **w** in either 2-space or 3-space we have **v** = **w** if and only if corresponding components of **v** and **w** are the same. For example,

$$\langle a, b, c \rangle = \langle 1, -4, 2 \rangle$$

if and only if $a = 1$, $b = -4$, and $c = 2$.

☐ **ARITHMETIC OPERATIONS ON VECTORS**

The following theorem shows how to perform arithmetic operations on vectors using components.

> **14.2.4   THEOREM.**   *If $\mathbf{v} = \langle v_1, v_2 \rangle$ and $\mathbf{w} = \langle w_1, w_2 \rangle$ are vectors in 2-space and k is any scalar, then*
>
> $$\mathbf{v} + \mathbf{w} = \langle v_1 + w_1, v_2 + w_2 \rangle \tag{1a}$$
>
> $$\mathbf{v} - \mathbf{w} = \langle v_1 - w_1, v_2 - w_2 \rangle \tag{1b}$$
>
> $$k\mathbf{v} = \langle kv_1, kv_2 \rangle \tag{1c}$$
>
> *Similarly, if $\mathbf{v} = \langle v_1, v_2, v_3 \rangle$ and $\mathbf{w} = \langle w_1, w_2, w_3 \rangle$ are vectors in 3-space and k is any scalar, then*
>
> $$\mathbf{v} + \mathbf{w} = \langle v_1 + w_1, v_2 + w_2, v_3 + w_3 \rangle \tag{2a}$$
>
> $$\mathbf{v} - \mathbf{w} = \langle v_1 - w_1, v_2 - w_2, v_3 - w_3 \rangle \tag{2b}$$
>
> $$k\mathbf{v} = \langle kv_1, kv_2, kv_3 \rangle \tag{2c}$$

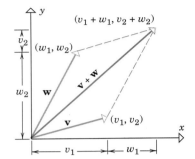

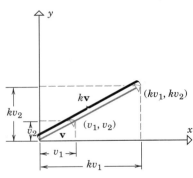

Figure 14.2.7

We shall not prove this theorem. However, results (1a) and (1c) should be evident from Figure 14.2.7. Similar figures in 3-space can be used to motivate (2a) and (2c). Formulas (1b) and (2b) can be obtained from parts (a) and (c) by writing $\mathbf{v} - \mathbf{w} = \mathbf{v} + (-1)\mathbf{w}$.

Example 1   If $\mathbf{v} = \langle -2, 0, 1 \rangle$ and $\mathbf{w} = \langle 3, 5, -4 \rangle$, then

$$\mathbf{v} + \mathbf{w} = \langle -2, 0, 1 \rangle + \langle 3, 5, -4 \rangle = \langle 1, 5, -3 \rangle$$

$$3\mathbf{v} = \langle -6, 0, 3 \rangle$$

$$-\mathbf{w} = \langle -3, -5, 4 \rangle$$

$$\mathbf{w} - 2\mathbf{v} = \langle 3, 5, -4 \rangle - \langle -4, 0, 2 \rangle = \langle 7, 5, -6 \rangle \quad \blacktriangleleft$$

REMARK.   Except for the number of components, there is no difference between arithmetic computations on vectors in 2-space and 3-space.

☐ **VECTORS WITH INITIAL POINT NOT AT THE ORIGIN**

If a vector in 2-space or 3-space is positioned with its initial point at the origin, then the coordinates of the terminal point are the components of the vector. If the vector does not have its initial point at the origin (Figure 14.2.8), then the following theorem shows that the components can be obtained by subtracting the coordinates of the initial point from the coordinates of the terminal point.

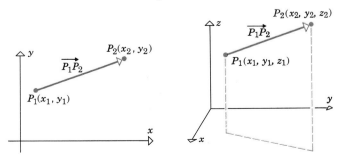

Figure 14.2.8

14.2.5 THEOREM. *If $\overrightarrow{P_1P_2}$ is a vector in 2-space with initial point $P_1(x_1, y_1)$ and terminal point $P_2(x_2, y_2)$, then*

$$\overrightarrow{P_1P_2} = \langle x_2 - x_1, y_2 - y_1 \rangle \tag{3a}$$

Similarly, if $\overrightarrow{P_1P_2}$ is a vector in 3-space with initial point $P_1(x_1, y_1, z_1)$ and terminal point $P_2(x_2, y_2, z_2)$, then

$$\overrightarrow{P_1P_2} = \langle x_2 - x_1, y_2 - y_1, z_2 - z_1 \rangle \tag{3b}$$

Proof. We shall give the proof in 2-space. The proof in 3-space is similar. The vector $\overrightarrow{P_1P_2}$ is the difference of vectors $\overrightarrow{OP_2}$ and $\overrightarrow{OP_1}$ (Figure 14.2.9). Thus,

$$\overrightarrow{P_1P_2} = \overrightarrow{OP_2} - \overrightarrow{OP_1} = \langle x_2, y_2 \rangle - \langle x_1, y_1 \rangle = \langle x_2 - x_1, y_2 - y_1 \rangle \quad \blacksquare$$

Example 2 In 2-space the vector from $P_1(1, 3)$ to $P_2(4, -2)$ is

$$\overrightarrow{P_1P_2} = \langle 4 - 1, -2 - 3 \rangle = \langle 3, -5 \rangle$$

and in 3-space the vector from $A(0, -2, 5)$ to $B(3, 4, -1)$ is

$$\overrightarrow{AB} = \langle 3 - 0, 4 - (-2), -1 - 5 \rangle = \langle 3, 6, -6 \rangle \quad \blacktriangleleft$$

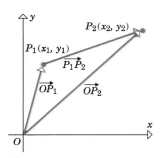

Figure 14.2.9

In the preceding section we stated without proof that the midpoint $M(x, y, z)$ of the line segment joining the points $P_1(x_1, y_1, z_1)$ and $P_2(x_2, y_2, z_2)$ has coordinates

$$x = \tfrac{1}{2}(x_1 + x_2), \quad y = \tfrac{1}{2}(y_1 + y_2), \quad z = \tfrac{1}{2}(z_1 + z_2) \tag{4}$$

To see that this is so we need only observe that

$$\overrightarrow{P_1M} = \tfrac{1}{2}\overrightarrow{P_1P_2}$$

(Figure 14.2.10), so that

$$\langle x - x_1, y - y_1, z - z_1 \rangle = \tfrac{1}{2}\langle x_2 - x_1, y_2 - y_1, z_2 - z_1 \rangle$$

or, on equating components,

$$x - x_1 = \tfrac{1}{2}(x_2 - x_1), \quad y - y_1 = \tfrac{1}{2}(y_2 - y_1), \quad z - z_1 = \tfrac{1}{2}(z_2 - z_1)$$

from which (4) follows.

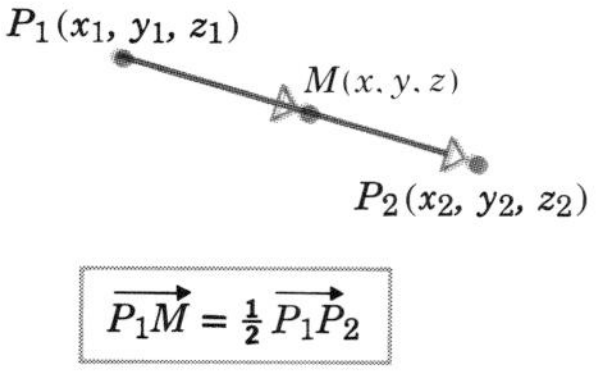

$P_1(x_1, y_1, z_1)$

$M(x, y, z)$

$P_2(x_2, y_2, z_2)$

$$\boxed{\overrightarrow{P_1M} = \tfrac{1}{2}\overrightarrow{P_1P_2}}$$

Figure 14.2.10

☐ **RULES OF VECTOR ARITHMETIC**

The following theorem shows that many of the familiar rules of ordinary arithmetic also hold for vector arithmetic.

14.2.6 THEOREM. *For any vectors $\mathbf{u}$, $\mathbf{v}$, and $\mathbf{w}$ and any scalars k and l, the following relationships hold*:

(*a*) $\mathbf{u} + \mathbf{v} = \mathbf{v} + \mathbf{u}$
(*b*) $(\mathbf{u} + \mathbf{v}) + \mathbf{w} = \mathbf{u} + (\mathbf{v} + \mathbf{w})$
(*c*) $\mathbf{u} + \mathbf{0} = \mathbf{0} + \mathbf{u} = \mathbf{u}$
(*d*) $\mathbf{u} + (-\mathbf{u}) = \mathbf{0}$
(*e*) $k(l\mathbf{u}) = (kl)\mathbf{u}$
(*f*) $k(\mathbf{u} + \mathbf{v}) = k\mathbf{u} + k\mathbf{v}$
(*g*) $(k + l)\mathbf{u} = k\mathbf{u} + l\mathbf{u}$
(*h*) $1\mathbf{u} = \mathbf{u}$

Before discussing the proof, we note that we have developed two approaches to vectors: *geometric*, in which vectors are represented by arrows or directed line segments, and *analytic*, in which vectors are represented by pairs or triples of numbers called components. As a consequence, the results in this theorem can be established either geometrically or analytically. As an illustration, we shall prove part (*b*) both ways. The remaining proofs are left as exercises.

Proof (b) (Analytic in 2-space).   Let $\mathbf{u} = \langle u_1, u_2 \rangle$, $\mathbf{v} = \langle v_1, v_2 \rangle$, and $\mathbf{w} = \langle w_1, w_2 \rangle$. Then

$$
\begin{aligned}
(\mathbf{u} + \mathbf{v}) + \mathbf{w} &= (\langle u_1, u_2 \rangle + \langle v_1, v_2 \rangle) + \langle w_1, w_2 \rangle \\
&= \langle u_1 + v_1, u_2 + v_2 \rangle + \langle w_1, w_2 \rangle \\
&= \langle (u_1 + v_1) + w_1, (u_2 + v_2) + w_2 \rangle \\
&= \langle u_1 + (v_1 + w_1), u_2 + (v_2 + w_2) \rangle \\
&= \langle u_1, u_2 \rangle + \langle v_1 + w_1, v_2 + w_2 \rangle \\
&= \mathbf{u} + (\mathbf{v} + \mathbf{w})
\end{aligned}
$$

Proof (b) (Geometric).   Let $\mathbf{u}$, $\mathbf{v}$, and $\mathbf{w}$ be represented by $\overrightarrow{PQ}$, $\overrightarrow{QR}$, and $\overrightarrow{RS}$ as shown in Figure 14.2.11. Then

$$\mathbf{v} + \mathbf{w} = \overrightarrow{QS} \quad \text{and} \quad \mathbf{u} + (\mathbf{v} + \mathbf{w}) = \overrightarrow{PS}$$

$$\mathbf{u} + \mathbf{v} = \overrightarrow{PR} \quad \text{and} \quad (\mathbf{u} + \mathbf{v}) + \mathbf{w} = \overrightarrow{PS}$$

Therefore,

$$(\mathbf{u} + \mathbf{v}) + \mathbf{w} = \mathbf{u} + (\mathbf{v} + \mathbf{w}) \qquad \blacksquare$$

REMARK.   In light of part (*b*) of this theorem, the symbol $\mathbf{u} + \mathbf{v} + \mathbf{w}$ is unambiguous since the same result is obtained no matter where parentheses are inserted. Moreover, if the vectors $\mathbf{u}$, $\mathbf{v}$, and $\mathbf{w}$ are placed "tip to tail," then the sum $\mathbf{u} + \mathbf{v} + \mathbf{w}$ is the vector from the initial point of $\mathbf{u}$ to the terminal point of $\mathbf{w}$ (Figure 14.2.11).

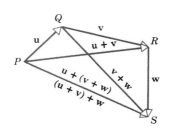

Figure 14.2.11

□ **LENGTH OF A VECTOR**

Geometrically, the ***length*** of a vector $\mathbf{v}$, also called the ***norm*** of $\mathbf{v}$, is the distance between its initial and terminal points. The length (or norm) of $\mathbf{v}$ is denoted by $\|\mathbf{v}\|$. It follows from the distance formulas in 2-space and 3-space that the norm of a vector $\mathbf{v} = \langle v_1, v_2 \rangle$ in 2-space is given by

$$\|\mathbf{v}\| = \sqrt{v_1^2 + v_2^2} \tag{5a}$$

and the norm of a vector $\mathbf{v} = \langle v_1, v_2, v_3 \rangle$ in 3-space is given by

$$\|\mathbf{v}\| = \sqrt{v_1^2 + v_2^2 + v_3^2} \tag{5b}$$

(Figures 14.2.12*a* and 14.2.12*b*).

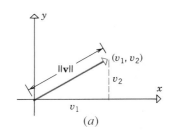

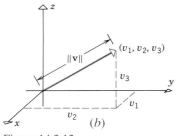

Figure 14.2.12

Example 3   Find the norm of $\mathbf{v} = \langle -2, 3 \rangle$ and $\mathbf{w} = \langle 2, 3, 6 \rangle$.

Solution.   From (5a) and (5b)

$$\|\mathbf{v}\| = \sqrt{(-2)^2 + 3^2} = \sqrt{13}$$

$$\|\mathbf{w}\| = \sqrt{2^2 + 3^2 + 6^2} = \sqrt{49} = 7 \qquad \blacktriangleleft$$

Recall from Definition 14.2.2 that the length of $k\mathbf{v}$ is $|k|$ times the length of $\mathbf{v}$. Expressed as an equation, this statement says that

$$\|k\mathbf{v}\| = |k|\, \|\mathbf{v}\| \tag{6}$$

This formula applies to both vectors in 2-space and 3-space.

☐ **UNIT VECTORS**

Vectors of length 1 are called ***unit vectors***. The unit vectors that run along the positive coordinate axes of a Cartesian coordinate system are especially important: In 2-space or 3-space the unit vectors along the positive x- and y-axes are denoted by **i** and **j**, respectively, and in 3-space the unit vector along the positive z-axis is denoted by **k**. Thus, we have (Figure 14.2.13)

$$\mathbf{i} = \langle 1, 0 \rangle, \qquad \mathbf{j} = \langle 0, 1 \rangle \qquad \boxed{\text{In 2-space}}$$

$$\mathbf{i} = \langle 1, 0, 0 \rangle, \quad \mathbf{j} = \langle 0, 1, 0 \rangle, \quad \mathbf{k} = \langle 0, 0, 1 \rangle \qquad \boxed{\text{In 3-space}}$$

Every vector in 2-space is expressible uniquely in terms of **i** and **j**, and every vector in 3-space is expressible uniquely in terms of **i**, **j**, and **k** as follows:

$$\mathbf{v} = \langle v_1, v_2 \rangle = \langle v_1, 0 \rangle + \langle 0, v_2 \rangle = v_1 \langle 1, 0 \rangle + v_2 \langle 0, 1 \rangle = v_1 \mathbf{i} + v_2 \mathbf{j}$$

$$\mathbf{v} = \langle v_1, v_2, v_3 \rangle = v_1 \langle 1, 0, 0 \rangle + v_2 \langle 0, 1, 0 \rangle + v_3 \langle 0, 0, 1 \rangle = v_1 \mathbf{i} + v_2 \mathbf{j} + v_3 \mathbf{k}$$

REMARK. The notations $\langle v_1, v_2, v_3 \rangle$ and $v_1 \mathbf{i} + v_2 \mathbf{j} + v_3 \mathbf{k}$ are interchangeable, as are $\langle v_1, v_2 \rangle$ and $v_1 \mathbf{i} + v_2 \mathbf{j}$. For example, (5b) can be written as

$$\| v_1 \mathbf{i} + v_2 \mathbf{j} + v_3 \mathbf{k} \| = \sqrt{v_1^2 + v_2^2 + v_3^2}$$

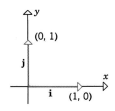

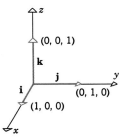

Figure 14.2.13

Example 4

2-SPACE	3-SPACE

2-SPACE

$\langle 2, 3 \rangle = 2\mathbf{i} + 3\mathbf{j}$

$\langle -4, 0 \rangle = -4\mathbf{i} + 0\mathbf{j} = -4\mathbf{i}$

$\langle 0, 0 \rangle = 0\mathbf{i} + 0\mathbf{j} = \mathbf{0}$

$(3\mathbf{i} + 2\mathbf{j}) + (4\mathbf{i} + \mathbf{j}) = 7\mathbf{i} + 3\mathbf{j}$

$5(6\mathbf{i} - 2\mathbf{j}) = 30\mathbf{i} - 10\mathbf{j}$

$\| 2\mathbf{i} - 3\mathbf{j} \| = \sqrt{2^2 + (-3)^2} = \sqrt{13}$

3-SPACE

$\langle 2, -3, 4 \rangle = 2\mathbf{i} - 3\mathbf{j} + 4\mathbf{k}$

$\langle 0, 3, 0 \rangle = 3\mathbf{j}$

$\langle 0, 0, 0 \rangle = 0\mathbf{i} + 0\mathbf{j} + 0\mathbf{k} = \mathbf{0}$

$(3\mathbf{i} + 2\mathbf{j} - \mathbf{k}) - (4\mathbf{i} - \mathbf{j} + 2\mathbf{k}) = -\mathbf{i} + 3\mathbf{j} - 3\mathbf{k}$

$2(\mathbf{i} + \mathbf{j} - \mathbf{k}) + 4(\mathbf{i} - \mathbf{j}) = 6\mathbf{i} - 2\mathbf{j} - 2\mathbf{k}$

$\| \mathbf{i} + 2\mathbf{j} - 3\mathbf{k} \| = \sqrt{1^2 + 2^2 + (-3)^2} = \sqrt{14}$ ◀

If **v** is a nonzero vector, then it follows from (6) with $k = 1/\|\mathbf{v}\|$ that

$$\left\| \frac{1}{\|\mathbf{v}\|} \mathbf{v} \right\| = \left| \frac{1}{\|\mathbf{v}\|} \right| \|\mathbf{v}\| = \frac{1}{\|\mathbf{v}\|} \|\mathbf{v}\| = 1$$

which tells us that multiplying a nonzero vector by the reciprocal of its length produces a unit vector. We call the process of multiplying **v** by $1/\|\mathbf{v}\|$ ***normalizing*** **v**. Thus, by normalizing, a nonzero vector **v** in 2-space or 3-space can be expressed as

$$\mathbf{v} = \|\mathbf{v}\| \left(\frac{1}{\|\mathbf{v}\|} \mathbf{v} \right) \tag{7}$$

which is the length of **v** times a unit vector in the same direction as **v**.

Example 5 The vector $\mathbf{v} = \langle 3, 4 \rangle$ has length $\|\mathbf{v}\| = \sqrt{3^2 + 4^2} = 5$, so that

$$\frac{1}{\|\mathbf{v}\|} \mathbf{v} = \tfrac{1}{5} \langle 3, 4 \rangle = \left\langle \tfrac{3}{5}, \tfrac{4}{5} \right\rangle$$

is a unit vector in the same direction as **v**, and from (7) we can express **v** as

$$\langle 3, 4 \rangle = 5 \left\langle \tfrac{3}{5}, \tfrac{4}{5} \right\rangle$$

which is the length of **v** times a unit vector in the same direction as **v**. ◀

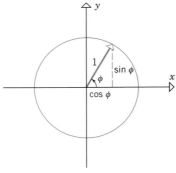

Figure 14.2.14

The components of a unit vector $\mathbf{u}$ in 2-space have a useful geometric interpretation: If ϕ denotes the angle from the positive x-axis to $\mathbf{u}$, then, as suggested by Figure 14.2.14, the x- and y-components of $\mathbf{u}$ are $\cos \phi$ and $\sin \phi$, respectively; that is,

$$\mathbf{u} = \langle \cos \phi, \sin \phi \rangle = (\cos \phi)\mathbf{i} + (\sin \phi)\mathbf{j}$$

Thus, if $\mathbf{v}$ is any nonzero vector in 2-space, then it follows from (7) that

$$\mathbf{v} = \|\mathbf{v}\| \langle \cos \phi, \sin \phi \rangle \qquad \text{or equivalently,} \qquad \mathbf{v} = \|\mathbf{v}\| \cos \phi\, \mathbf{i} + \|\mathbf{v}\| \sin \phi\, \mathbf{j} \qquad (8)$$

where ϕ is the angle from the positive x-axis to $\mathbf{v}$.

Example 6 Find the vector of length 2 that makes an angle of $\pi/4$ with the positive x-axis.

Solution. From (8)

$$\mathbf{v} = 2 \cos \frac{\pi}{4} \mathbf{i} + 2 \sin \frac{\pi}{4} \mathbf{j} = \sqrt{2}\mathbf{i} + \sqrt{2}\mathbf{j} \qquad \blacktriangleleft$$

► **Exercise Set 14.2**

In Exercises 1–4, sketch the vectors with the initial point at the origin.

1. (a) $\langle 2, 5 \rangle$ **2.** (a) $\langle -3, 7 \rangle$
    (b) $\langle -5, -4 \rangle$ (b) $\langle 6, -2 \rangle$
    (c) $\langle 2, 0 \rangle$.   (c) $\langle 0, -8 \rangle$.

3. (a) $\langle 1, -2, 2 \rangle$ **4.** (a) $\langle -1, 3, 2 \rangle$
    (b) $\langle 2, 2, -1 \rangle$.   (b) $\langle 3, 4, 2 \rangle$.

In Exercises 5–8, sketch the vectors with the initial point at the origin.

5. (a) $-5\mathbf{i} + 3\mathbf{j}$ **6.** (a) $4\mathbf{i} + 2\mathbf{j}$
    (b) $3\mathbf{i} - 2\mathbf{j}$ (b) $-2\mathbf{i} - \mathbf{j}$
    (c) $-6\mathbf{j}$.   (c) $4\mathbf{i}$.

7. (a) $-\mathbf{i} + 2\mathbf{j} + 3\mathbf{k}$ **8.** (a) $2\mathbf{j} - \mathbf{k}$
    (b) $2\mathbf{i} + 3\mathbf{j} - \mathbf{k}$.   (b) $\mathbf{i} - \mathbf{j} + 2\mathbf{k}$.

In Exercises 9–12, find the components of the vector $\overrightarrow{P_1P_2}$.

9. (a) $P_1(3, 5)$, $P_2(2, 8)$
    (b) $P_1(7, -2)$, $P_2(0, 0)$
    (c) $P_1(-6, -2)$, $P_2(-4, -1)$
    (d) $P_1(0, 0)$, $P_2(-8, 7)$.

10. (a) $P_1(1, 3)$, $P_2(4, 1)$
    (b) $P_1(6, -4)$, $P_2(0, 0)$
    (c) $P_1(-8, -1)$, $P_2(-3, -2)$
    (d) $P_1(0, 0)$, $P_2(-3, -5)$.

11. (a) $P_1(5, -2, 1)$, $P_2(2, 4, 2)$
    (b) $P_1(-1, 3, 5)$, $P_2(0, 0, 0)$.

12. (a) $P_1(0, 0, 0)$, $P_2(-1, 6, 1)$
    (b) $P_1(4, 1, -3)$, $P_2(9, 1, -3)$.

13. Find the terminal point of $\mathbf{v} = 3\mathbf{i} - 2\mathbf{j}$ if the initial point is $(1, -2)$.

14. Find the terminal point of $\mathbf{v} = \langle 7, 6 \rangle$ if the initial point is $(2, -1)$.

15. Find the initial point of $\mathbf{v} = \langle -2, 4 \rangle$ if the terminal point is $(2, 0)$.

16. Find the terminal point of $\mathbf{v} = \mathbf{i} + 2\mathbf{j} - 3\mathbf{k}$ if the initial point is $(-2, 1, 4)$.

17. Find the initial point of $\mathbf{v} = \langle -3, 1, 2 \rangle$ if the terminal point is $(5, 0, -1)$.

18. Let $\mathbf{u} = \langle 1, 3 \rangle$, $\mathbf{v} = \langle 2, 1 \rangle$, and $\mathbf{w} = \langle 4, -1 \rangle$. Find
    (a) $\mathbf{u} - \mathbf{w}$ (b) $7\mathbf{v} + 3\mathbf{w}$
    (c) $-\mathbf{w} + \mathbf{v}$ (d) $3(\mathbf{u} - 7\mathbf{v})$
    (e) $-3\mathbf{v} - 8\mathbf{w}$ (f) $2\mathbf{v} - (\mathbf{u} + \mathbf{w})$.

19. Let $\mathbf{u} = 2\mathbf{i} + 3\mathbf{j}$, $\mathbf{v} = 4\mathbf{i}$, $\mathbf{w} = -\mathbf{i} - 2\mathbf{j}$. Find
    (a) $\mathbf{w} - \mathbf{v}$ (b) $6\mathbf{u} + 4\mathbf{w}$
    (c) $-\mathbf{v} - 2\mathbf{w}$ (d) $4(3\mathbf{u} + \mathbf{v})$
    (e) $-8(\mathbf{v} + \mathbf{w}) + 2\mathbf{u}$ (f) $3\mathbf{w} - (\mathbf{v} - \mathbf{w})$.

20. Let $\mathbf{u} = \langle 2, -1, 3 \rangle$, $\mathbf{v} = \langle 4, 0, -2 \rangle$, $\mathbf{w} = \langle 1, 1, 3 \rangle$. Find
    (a) $\mathbf{u} - \mathbf{w}$ (b) $7\mathbf{v} + 3\mathbf{w}$
    (c) $-\mathbf{w} + \mathbf{v}$ (d) $3(\mathbf{u} - 7\mathbf{v})$
    (e) $-3\mathbf{v} - 8\mathbf{w}$ (f) $2\mathbf{v} - (\mathbf{u} + \mathbf{w})$.

21. Let $\mathbf{u} = 3\mathbf{i} - \mathbf{k}$, $\mathbf{v} = \mathbf{i} - \mathbf{j} + 2\mathbf{k}$, $\mathbf{w} = 3\mathbf{j}$. Find
    (a) $\mathbf{w} - \mathbf{v}$ (b) $6\mathbf{u} + 4\mathbf{w}$
    (c) $-\mathbf{v} - 2\mathbf{w}$ (d) $4(3\mathbf{u} + \mathbf{v})$
    (e) $-8(\mathbf{v} + \mathbf{w}) + 2\mathbf{u}$ (f) $3\mathbf{w} - (\mathbf{v} - \mathbf{w})$.

In Exercises 22–25, compute the norm of **v**.

22. (a) $\mathbf{v} = \langle 3, 4 \rangle$ (b) $\mathbf{v} = -\mathbf{i} + 7\mathbf{j}$
   (c) $\mathbf{v} = -3\mathbf{j}$.

23. (a) $\mathbf{v} = \langle 1, -1 \rangle$ (b) $\mathbf{v} = \langle 2, 0 \rangle$
   (c) $\mathbf{v} = \sqrt{2}\mathbf{i} - \sqrt{7}\mathbf{j}$.

24. (a) $\mathbf{v} = \mathbf{i} + \mathbf{j} + \mathbf{k}$ (b) $\mathbf{v} = \langle -1, 2, 4 \rangle$.

25. (a) $\mathbf{v} = -3\mathbf{i} + 2\mathbf{j} + \mathbf{k}$ (b) $\mathbf{v} = \langle 0, -3, 0 \rangle$.

26. Let $\mathbf{u} = \langle 1, -3 \rangle$, $\mathbf{v} = \langle 1, 1 \rangle$, and $\mathbf{w} = \langle 2, -4 \rangle$. Find
   (a) $\|\mathbf{u} + \mathbf{v}\|$ (b) $\|\mathbf{u}\| + \|\mathbf{v}\|$
   (c) $\|-2\mathbf{u}\| + 2\|\mathbf{v}\|$ (d) $\|3\mathbf{u} - 5\mathbf{v} + \mathbf{w}\|$.

27. Let $\mathbf{u} = 2\mathbf{i} - 5\mathbf{j}$, $\mathbf{v} = 2\mathbf{i}$, and $\mathbf{w} = 3\mathbf{i} + 4\mathbf{j}$. Find
   (a) $\|\mathbf{v} + \mathbf{w}\|$ (b) $\|\mathbf{v}\| + \|\mathbf{w}\|$
   (c) $\|-3\mathbf{u}\| + 4\|\mathbf{v}\|$ (d) $\|\mathbf{u} - \mathbf{v} - \mathbf{w}\|$
   (e) $\dfrac{1}{\|\mathbf{w}\|}\mathbf{w}$ (f) $\left\|\dfrac{1}{\|\mathbf{w}\|}\mathbf{w}\right\|$.

28. Let $\mathbf{u} = \langle 2, -1, 0 \rangle$ and $\mathbf{v} = \langle 0, 1, -1 \rangle$. Find
   (a) $\|\mathbf{u} + \mathbf{v}\|$ (b) $\|\mathbf{u}\| + \|\mathbf{v}\|$
   (c) $\|3\mathbf{u}\|$ (d) $\|2\mathbf{u} - 3\mathbf{v}\|$.

29. Let $\mathbf{u} = \mathbf{i} - 3\mathbf{j} + 2\mathbf{k}$, $\mathbf{v} = \mathbf{i} + \mathbf{j}$, and $\mathbf{w} = 2\mathbf{i} + 2\mathbf{j} - 4\mathbf{k}$. Find
   (a) $\|\mathbf{u} + \mathbf{v}\|$ (b) $\|\mathbf{u}\| + \|\mathbf{v}\|$
   (c) $\|-2\mathbf{u}\| + 2\|\mathbf{v}\|$ (d) $\|3\mathbf{u} - 5\mathbf{v} + \mathbf{w}\|$
   (e) $\dfrac{1}{\|\mathbf{w}\|}\mathbf{w}$ (f) $\left\|\dfrac{1}{\|\mathbf{w}\|}\mathbf{w}\right\|$.

30. Let $\mathbf{u} = \langle -1, 1 \rangle$, $\mathbf{v} = \langle 0, 1 \rangle$, and $\mathbf{w} = \langle 3, 4 \rangle$. Find the vector $\mathbf{x}$ that satisfies $\mathbf{u} - 2\mathbf{x} = \mathbf{x} - \mathbf{w} + 3\mathbf{v}$.

31. Let $\mathbf{u} = \langle 1, 3 \rangle$, $\mathbf{v} = \langle 2, 1 \rangle$, $\mathbf{w} = \langle 4, -1 \rangle$. Find the vector $\mathbf{x}$ that satisfies $2\mathbf{u} - \mathbf{v} + \mathbf{x} = 7\mathbf{x} + \mathbf{w}$.

32. Find $\mathbf{u}$ and $\mathbf{v}$ if $\mathbf{u} + \mathbf{v} = \langle 2, -3 \rangle$ and $3\mathbf{u} + 2\mathbf{v} = \langle -1, 2 \rangle$.

33. Find $\mathbf{u}$ and $\mathbf{v}$ if $\mathbf{u} + 2\mathbf{v} = 3\mathbf{i} - \mathbf{k}$ and $3\mathbf{u} - \mathbf{v} = \mathbf{i} + \mathbf{j} + \mathbf{k}$.

34. Give a geometric argument to show that if $\mathbf{u}$ and $\mathbf{v}$ are nonzero and $\mathbf{u}$ is not parallel to $\mathbf{v}$, then any vector $\mathbf{w}$ in the plane of $\mathbf{u}$ and $\mathbf{v}$ can be written as $\mathbf{w} = c_1\mathbf{u} + c_2\mathbf{v}$ for a suitable choice of c_1 and c_2.

35. Give a geometric argument to show that if $\mathbf{u}$, $\mathbf{v}$, and $\mathbf{w}$ are not coplanar, then any vector $\mathbf{z}$ can be written as $\mathbf{z} = c_1\mathbf{u} + c_2\mathbf{v} + c_3\mathbf{w}$ for a suitable choice of scalars c_1, c_2, and c_3.

36. Let $\mathbf{u} = 2\mathbf{i} - \mathbf{j}$ and $\mathbf{v} = 4\mathbf{i} + 2\mathbf{j}$. Find scalars c_1 and c_2 such that $c_1\mathbf{u} + c_2\mathbf{v} = -4\mathbf{j}$.

37. Let $\mathbf{u} = \langle 1, -3 \rangle$ and $\mathbf{v} = \langle -2, 6 \rangle$. Show that there do not exist scalars c_1 and c_2 such that $c_1\mathbf{u} + c_2\mathbf{v} = \langle 3, 5 \rangle$.

38. Let $\mathbf{u} = \langle 1, 0, 1 \rangle$, $\mathbf{v} = \langle 3, 2, 0 \rangle$, and $\mathbf{w} = \langle 0, 1, 1 \rangle$. Find scalars c_1, c_2, and c_3 such that
$$c_1\mathbf{u} + c_2\mathbf{v} + c_3\mathbf{w} = \langle -1, 1, 5 \rangle$$

39. Let $\mathbf{u} = \mathbf{i} - \mathbf{j}$, $\mathbf{v} = 3\mathbf{i} + \mathbf{k}$, and $\mathbf{w} = 4\mathbf{i} - \mathbf{j} + \mathbf{k}$. Show that there do not exist scalars c_1, c_2, and c_3 such that
$$c_1\mathbf{u} + c_2\mathbf{v} + c_3\mathbf{w} = 2\mathbf{i} + \mathbf{j} - \mathbf{k}$$

40. Let $\mathbf{v} = 4\mathbf{i} - 3\mathbf{j}$. Find all scalars k such that $\|k\mathbf{v}\| = 3$.

41. Verify parts (b), (e), (f), and (g) of Theorem 14.2.6 for $\mathbf{u} = \langle 1, -3 \rangle$, $\mathbf{v} = \langle 6, 6 \rangle$, $\mathbf{w} = \langle -8, 1 \rangle$, $k = 3$, and $l = 6$.

42. Find a unit vector having the same direction as $-\mathbf{i} + 4\mathbf{j}$.

43. Find a unit vector oppositely directed to $3\mathbf{i} - 4\mathbf{j}$.

44. Find a unit vector having the same direction as $2\mathbf{i} - \mathbf{j} - 2\mathbf{k}$.

45. Find a unit vector oppositely directed to $6\mathbf{i} - 4\mathbf{j} + 2\mathbf{k}$.

46. Find a unit vector having the same direction as the vector from the point $A(-3, 2)$ to the point $B(1, -1)$.

47. Find a unit vector having the same direction as the vector from the point $A(-1, 0, 2)$ to the point $B(3, 1, 1)$.

48. Find a vector having the same direction as the vector $\mathbf{v} = -2\mathbf{i} + 3\mathbf{j}$ but with three times the length of $\mathbf{v}$.

49. Find a vector oppositely directed to $\mathbf{v} = \langle 3, -4 \rangle$ but with half the length of $\mathbf{v}$.

50. Find a vector with the same direction as $\mathbf{v} = \langle 7, 0, -6 \rangle$ but with twice the length of $\mathbf{v}$.

51. Find a vector oppositely directed to $\mathbf{v} = -3\mathbf{i} + 4\mathbf{j} + \mathbf{k}$ but with twice the length of $\mathbf{v}$.

52. Let $\mathbf{r} = \langle x, y \rangle$. Describe the set of points (x, y) for which $\|\mathbf{r}\| = 1$.

53. Let $\mathbf{r}_0 = \langle x_0, y_0 \rangle$ and $\mathbf{r} = \langle x, y \rangle$. Describe the set of all points (x, y) for which $\|\mathbf{r} - \mathbf{r}_0\| = 1$.

54. Let $\mathbf{r}_1 = \langle x_1, y_1 \rangle$, $\mathbf{r}_2 = \langle x_2, y_2 \rangle$, and $\mathbf{r} = \langle x, y \rangle$. Describe the set of all points (x, y) for which $\|\mathbf{r} - \mathbf{r}_1\| + \|\mathbf{r} - \mathbf{r}_2\| = k$, where $k > \|\mathbf{r}_2 - \mathbf{r}_1\|$.

55. Let $\mathbf{r}_0 = \langle x_0, y_0, z_0 \rangle$ and $\mathbf{r} = \langle x, y, z \rangle$. Describe the set of all points (x, y, z) for which
   (a) $\|\mathbf{r}\| = 2$ (b) $\|\mathbf{r} - \mathbf{r}_0\| = 3$
   (c) $\|\mathbf{r} - \mathbf{r}_0\| \leq 1$.

56. Find two unit vectors in 2-space parallel to the line $y = 3x + 2$.

57. (a) Find two unit vectors in 2-space parallel to the line $x + y = 4$.
   (b) Find two unit vectors in 2-space perpendicular to the line in part (a).

58. Let P be the point $(2, 3)$ and Q the point $(7, -4)$. Use vectors to find the point on the line segment joining P and Q that is $\frac{3}{4}$ of the way from P to Q.

59. For the points P and Q in Exercise 58, use vectors to find the point on the line segment joining P and Q that is $\frac{3}{4}$ of the way from Q to P.

60. Find a unit vector in 2-space making an angle of $135°$ with the x-axis.

61. (a) Find a unit vector in 2-space making an angle of $\pi/3$ with the positive x-axis.

(b)   Find a vector of length 4 in 2-space making an angle of $3\pi/4$ with the positive x-axis.

62.   Use vectors to find the length of the diagonal of the parallelogram determined by $\mathbf{i} + \mathbf{j}$ and $\mathbf{i} - 2\mathbf{j}$.

63.   Use vectors to find the fourth vertex of a parallelogram, three of whose vertices are $(0, 0)$, $(1, 3)$, and $(2, 4)$. [*Note:* There is more than one answer.]

64.   Prove: $\|\mathbf{u} + \mathbf{v}\| \le \|\mathbf{u}\| + \|\mathbf{v}\|$ geometrically.

65.   Prove parts (a), (c), and (e) of Theorem 14.2.6 analytically in 2-space.

66.   Prove parts (d), (g), and (h) of Theorem 14.2.6 analytically in 2-space.

67.   Prove part (f) of Theorem 14.2.6 geometrically.

68.   Use vectors to prove that the line segment joining the midpoints of two sides of a triangle is parallel to the third side and half as long.

69.   Use vectors to prove that the midpoints of the sides of a quadrilateral are the vertices of a parallelogram.

70.   Find $\overrightarrow{AB} + \overrightarrow{BC} + \overrightarrow{CA}$, where A, B, and C are any three distinct points.

In Exercises 71–73, let A, B, C, and D be any four points in 3-space.

71.   If M is the midpoint of BC, show that $\overrightarrow{AB} + \overrightarrow{AC} = 2\overrightarrow{AM}$.

72.   If M and N are the midpoints of AC and BD, show that $\overrightarrow{AB} + \overrightarrow{CD} = 2\overrightarrow{MN}$.

73.   If M and N are the midpoints of AC and BD, show that $\overrightarrow{AB} + \overrightarrow{AD} + \overrightarrow{CB} + \overrightarrow{CD} = 4\overrightarrow{MN}$.

74.   Let A and B be distinct points on a straight line L. If a point P different from B is on L, then $\overrightarrow{AP} = t\overrightarrow{PB}$ for some value of the scalar t. Let $\mathbf{a}$, $\mathbf{b}$, and $\mathbf{r}$ be vectors from the origin to the points A, B, and P, respectively. Show that $\mathbf{r} = (\mathbf{a} + t\mathbf{b})/(1 + t)$.

75.   Let $\mathbf{r}_1, \mathbf{r}_2, \dots, \mathbf{r}_n$ be vectors from the origin to points $P_1, P_2, \dots, P_n$, respectively. The *centroid* of points $P_1, P_2, \dots, P_n$ is defined as the point P for which
$$\sum_{k=1}^{n} \overrightarrow{PP_k} = \mathbf{0}.$$
Let $\mathbf{r}$ be the vector from the origin to P. Show that $\mathbf{r} = \dfrac{1}{n}\sum_{k=1}^{n}\mathbf{r}_k$. [*Hint:* Write $\overrightarrow{PP_k}$ as a difference of vectors $\mathbf{r}$ and $\mathbf{r}_k$.]

76.   Let $P_1, P_2, \dots, P_n$ be consecutive vertices of a polygon in 2-space, all of whose interior angles are less than π. Let $\mathbf{r}_k$ be the vector from the origin to the point P_k for $k = 1, 2, \dots, n$. From Exercise 75, the endpoint of the vector
$$\mathbf{r} = \frac{1}{n}\sum_{k=1}^{n}\mathbf{r}_k$$
drawn from the origin is the centroid of the points P_1, $P_2, \dots, P_n$. It can be shown that the centroid is the balance point of the polygon.

(a)   Find the centroid of the triangle with vertices $P_1(1, 1)$, $P_2(3, 3)$, and $P_3(5, 0)$.

(b)   Cut the polygon described in part (a) out of cardboard and show that it balances when the tip of a pencil is placed at the centroid.

77.   Repeat Exercise 76 for the quadrilateral with vertices $P_1(-1, 2)$, $P_2(2, 3)$, $P_3(5, -2)$, and $P_4(0, -1)$.

■ **14.3**   DOT PRODUCT; PROJECTIONS

In this section we shall introduce a type of multiplication of vectors in 2-space and 3-space. We shall also discuss the arithmetic properties of this multiplication and give some of its applications.

☐ **ANGLE BETWEEN VECTORS**

Let $\mathbf{u}$ and $\mathbf{v}$ be two nonzero vectors in 2-space or 3-space, and assume these vectors have been positioned so that their initial points coincide. By the *angle between* $\mathbf{u}$ *and* $\mathbf{v}$, we shall mean the angle θ determined by $\mathbf{u}$ and $\mathbf{v}$ that satisfies $0 \le \theta \le \pi$ (Figure 14.3.1).

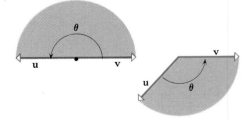

Figure 14.3.1

14.3.1 DEFINITION. If **u** and **v** are vectors in 2-space or 3-space and θ is the angle between **u** and **v**, then the *dot product* or *Euclidean inner product* **u** · **v** is defined by

$$\mathbf{u} \cdot \mathbf{v} = \begin{cases} \|\mathbf{u}\| \|\mathbf{v}\| \cos \theta, & \text{if } \mathbf{u} \neq \mathbf{0} \text{ and } \mathbf{v} \neq \mathbf{0} \\ 0, & \text{if } \mathbf{u} = \mathbf{0} \text{ or } \mathbf{v} = \mathbf{0} \end{cases}$$

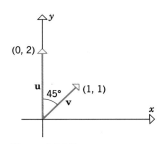

Figure 14.3.2

Example 1 As shown in Figure 14.3.2, the angle between vectors $\mathbf{u} = \langle 0, 2 \rangle$ and $\mathbf{v} = \langle 1, 1 \rangle$ is 45°. Thus,

$$\mathbf{u} \cdot \mathbf{v} = \|\mathbf{u}\| \|\mathbf{v}\| \cos \theta = \sqrt{0^2 + 2^2} \sqrt{1^2 + 1^2} \cos 45° = (2)(\sqrt{2}) \frac{1}{\sqrt{2}} = 2 \quad ◀$$

☐ **FORMULA FOR THE DOT PRODUCT**

For purposes of computation, it is desirable to have a formula that expresses the dot product of two vectors in terms of the components of the vectors. We shall derive such a formula for vectors in 3-space and just state the corresponding formula for vectors in 2-space.

Let $\mathbf{u} = \langle u_1, u_2, u_3 \rangle$ and $\mathbf{v} = \langle v_1, v_2, v_3 \rangle$ be two nonzero vectors. If, as in Figure 14.3.3, θ is the angle between **u** and **v**, then the law of cosines yields

$$\|\overrightarrow{PQ}\|^2 = \|\mathbf{u}\|^2 + \|\mathbf{v}\|^2 - 2\|\mathbf{u}\| \|\mathbf{v}\| \cos \theta \tag{1}$$

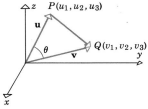

Figure 14.3.3

Since $\overrightarrow{PQ} = \mathbf{v} - \mathbf{u}$, we can rewrite (1) as

$$\|\mathbf{u}\| \|\mathbf{v}\| \cos \theta = \tfrac{1}{2}(\|\mathbf{u}\|^2 + \|\mathbf{v}\|^2 - \|\mathbf{v} - \mathbf{u}\|^2)$$

or

$$\mathbf{u} \cdot \mathbf{v} = \tfrac{1}{2}(\|\mathbf{u}\|^2 + \|\mathbf{v}\|^2 - \|\mathbf{v} - \mathbf{u}\|^2)$$

Substituting

$$\|\mathbf{u}\|^2 = u_1^2 + u_2^2 + u_3^2, \quad \|\mathbf{v}\|^2 = v_1^2 + v_2^2 + v_3^2$$

and

$$\|\mathbf{v} - \mathbf{u}\|^2 = (v_1 - u_1)^2 + (v_2 - u_2)^2 + (v_3 - u_3)^2$$

we obtain, after simplifying,

$$\mathbf{u} \cdot \mathbf{v} = u_1 v_1 + u_2 v_2 + u_3 v_3 \tag{2a}$$

This formula also holds if $\mathbf{u} = \mathbf{0}$ or $\mathbf{v} = \mathbf{0}$. If $\mathbf{u} = \langle u_1, u_2 \rangle$ and $\mathbf{v} = \langle v_1, v_2 \rangle$ are two vectors in 2-space, then the formula corresponding to (2a) is

$$\mathbf{u} \cdot \mathbf{v} = u_1 v_1 + u_2 v_2 \tag{2b}$$

If **u** and **v** are nonzero vectors, the formula in Definition 14.3.1 can be written as

$$\cos \theta = \frac{\mathbf{u} \cdot \mathbf{v}}{\|\mathbf{u}\| \|\mathbf{v}\|} \tag{3}$$

Example 2 Consider the vectors

$$\mathbf{u} = 2\mathbf{i} - \mathbf{j} + \mathbf{k} \quad \text{and} \quad \mathbf{v} = \mathbf{i} + \mathbf{j} + 2\mathbf{k}$$

Find **u** · **v** and determine the angle θ between **u** and **v**.

Solution.

$$\mathbf{u} \cdot \mathbf{v} = u_1 v_1 + u_2 v_2 + u_3 v_3 = (2)(1) + (-1)(1) + (1)(2) = 3$$

For the given vectors, $\|\mathbf{u}\| = \|\mathbf{v}\| = \sqrt{6}$, so that

$$\cos\theta = \frac{3}{\sqrt{6}\sqrt{6}} = \frac{1}{2}$$

Thus, $\theta = 60°$ ◀

Example 3   Find the angle between a diagonal of a cube and one of its edges.

Solution.   Let k be the length of an edge and let us introduce a coordinate system as shown in Figure 14.3.4.

If we let $\mathbf{u}_1 = \langle k, 0, 0 \rangle$, $\mathbf{u}_2 = \langle 0, k, 0 \rangle$, and $\mathbf{u}_3 = \langle 0, 0, k \rangle$, then the vector

$$\mathbf{d} = \langle k, k, k \rangle = \mathbf{u}_1 + \mathbf{u}_2 + \mathbf{u}_3$$

is a diagonal of the cube. The angle θ between $\mathbf{d}$ and the edge $\mathbf{u}_1$ satisfies

$$\cos\theta = \frac{\mathbf{u}_1 \cdot \mathbf{d}}{\|\mathbf{u}_1\|\|\mathbf{d}\|} = \frac{k^2}{(k)(\sqrt{3k^2})} = \frac{1}{\sqrt{3}}$$

The same results hold for $\mathbf{u}_2$ and $\mathbf{u}_3$. Thus, with the help of a calculator

$$\theta = \cos^{-1} 1/\sqrt{3} \approx 54°44'$$ ◀

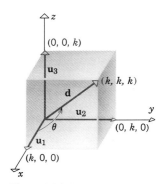

Figure 14.3.4

The sign of the dot product provides useful information about the angle between two vectors.

14.3.2   THEOREM.   *If $\mathbf{u}$ and $\mathbf{v}$ are nonzero vectors in 2-space or 3-space, and if θ is the angle between them, then*

θ is acute	*if and only if $\mathbf{u} \cdot \mathbf{v} > 0$*
θ is obtuse	*if and only if $\mathbf{u} \cdot \mathbf{v} < 0$*
$\theta = \pi/2$	*if and only if $\mathbf{u} \cdot \mathbf{v} = 0$*

Proof.   Since $\mathbf{u}$ and $\mathbf{v}$ are nonzero vectors, $\|\mathbf{u}\| > 0$ and $\|\mathbf{v}\| > 0$. Thus,

$$\mathbf{u} \cdot \mathbf{v} = \|\mathbf{u}\|\|\mathbf{v}\|\cos\theta$$

is positive, negative, or zero according to whether $\cos\theta$ is positive, negative, or zero. Since $0 \le \theta \le \pi$, it follows that θ is acute if and only if $\cos\theta > 0$; θ is obtuse if and only if $\cos\theta < 0$; and $\theta = \pi/2$ if and only if $\cos\theta = 0$. ∎

Example 4   If $\mathbf{u} = \mathbf{i} - 2\mathbf{j} + 3\mathbf{k}$, $\mathbf{v} = -3\mathbf{i} + 4\mathbf{j} + 2\mathbf{k}$, and $\mathbf{w} = 3\mathbf{i} + 6\mathbf{j} + 3\mathbf{k}$, then

$$\mathbf{u} \cdot \mathbf{v} = (1)(-3) + (-2)(4) + (3)(2) = -5$$

$$\mathbf{v} \cdot \mathbf{w} = (-3)(3) + (4)(6) + (2)(3) = 21$$

$$\mathbf{u} \cdot \mathbf{w} = (1)(3) + (-2)(6) + (3)(3) = 0$$

Therefore $\mathbf{u}$ and $\mathbf{v}$ make an obtuse angle, $\mathbf{v}$ and $\mathbf{w}$ make an acute angle, and $\mathbf{u}$ and $\mathbf{w}$ are perpendicular. ◀

☐ **ORTHOGONAL VECTORS**   Perpendicular vectors are also called ***orthogonal*** vectors. In light of Theorem 14.3.2, two nonzero vectors are orthogonal if and only if their dot product is zero. If we agree to consider $\mathbf{u}$ and $\mathbf{v}$ to be perpendicular when either or both of these vectors is $\mathbf{0}$, then we can state without exception that two vectors $\mathbf{u}$ and $\mathbf{v}$ are orthogonal (perpendicular) if and only if $\mathbf{u} \cdot \mathbf{v} = 0$.

☐ **DIRECTION COSINES**

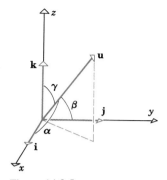

Figure 14.3.5

Of special interest are the angles α, β, and γ that a vector $\mathbf{u}$ in 3-space makes with the vectors $\mathbf{i}$, $\mathbf{j}$, and $\mathbf{k}$ (Figure 14.3.5). These are called the *direction angles* of $\mathbf{u}$. The numbers $\cos\alpha$, $\cos\beta$, and $\cos\gamma$ are called the *direction cosines* of $\mathbf{u}$. Formulas for the direction cosines follow easily from (3).

14.3.3 THEOREM. *The three direction cosines of a nonzero vector* $\mathbf{u} = u_1\mathbf{i} + u_2\mathbf{j} + u_3\mathbf{k}$ *in 3-space are*

$$\cos\alpha = \frac{u_1}{\|\mathbf{u}\|}, \quad \cos\beta = \frac{u_2}{\|\mathbf{u}\|}, \quad \cos\gamma = \frac{u_3}{\|\mathbf{u}\|}$$

Proof. Since $\mathbf{u} \cdot \mathbf{i} = (u_1)(1) + (u_2)(0) + (u_3)(0) = u_1$, it follows that

$$\cos\alpha = \frac{\mathbf{u} \cdot \mathbf{i}}{\|\mathbf{u}\|\,\|\mathbf{i}\|} = \frac{u_1}{\|\mathbf{u}\|}$$

Similarly for $\cos\beta$ and $\cos\gamma$. ∎

The direction cosines of a vector $\mathbf{u} = u_1\mathbf{i} + u_2\mathbf{j} + u_3\mathbf{k}$ can be obtained by simply reading off the components of the unit vector $\mathbf{u}/\|\mathbf{u}\|$ since

$$\frac{\mathbf{u}}{\|\mathbf{u}\|} = \frac{u_1}{\|\mathbf{u}\|}\mathbf{i} + \frac{u_2}{\|\mathbf{u}\|}\mathbf{j} + \frac{u_3}{\|\mathbf{u}\|}\mathbf{k} = (\cos\alpha)\mathbf{i} + (\cos\beta)\mathbf{j} + (\cos\gamma)\mathbf{k}$$

Example 5 Find the direction cosines of the vector $\mathbf{u} = 2\mathbf{i} - 4\mathbf{j} + 4\mathbf{k}$, and approximate the direction angles to the nearest degree.

Solution. $\|\mathbf{u}\| = \sqrt{4 + 16 + 16} = 6$, so that $\mathbf{u}/\|\mathbf{u}\| = \frac{1}{3}\mathbf{i} - \frac{2}{3}\mathbf{j} + \frac{2}{3}\mathbf{k}$. Thus,

$$\cos\alpha = \tfrac{1}{3}, \quad \cos\beta = -\tfrac{2}{3}, \quad \cos\gamma = \tfrac{2}{3}$$

With the help of a calculator that can compute inverse trigonometric functions, one obtains

$$\alpha = \cos^{-1}\left(\tfrac{1}{3}\right) \approx 71°, \quad \beta = \cos^{-1}\left(-\tfrac{2}{3}\right) \approx 132°, \quad \gamma = \cos^{-1}\left(\tfrac{2}{3}\right) \approx 48° \quad ◀$$

☐ **PROPERTIES OF THE DOT PRODUCT**

The following arithmetic properties of the dot product are useful in calculations involving vectors.

14.3.4 THEOREM. *If* $\mathbf{u}$, $\mathbf{v}$, *and* $\mathbf{w}$ *are vectors in 2- or 3-space and* k *is a scalar, then*

(a) $\mathbf{u} \cdot \mathbf{v} = \mathbf{v} \cdot \mathbf{u}$
(b) $\mathbf{u} \cdot (\mathbf{v} + \mathbf{w}) = \mathbf{u} \cdot \mathbf{v} + \mathbf{u} \cdot \mathbf{w}$
(c) $k(\mathbf{u} \cdot \mathbf{v}) = (k\mathbf{u}) \cdot \mathbf{v} = \mathbf{u} \cdot (k\mathbf{v})$
(d) $\mathbf{v} \cdot \mathbf{v} = \|\mathbf{v}\|^2$

We shall prove parts (c) and (d) for vectors in 3-space and omit the remaining proofs.

Proof (c). Let $\mathbf{u} = \langle u_1, u_2, u_3 \rangle$ and $\mathbf{v} = \langle v_1, v_2, v_3 \rangle$; then

$$k(\mathbf{u} \cdot \mathbf{v}) = k(u_1 v_1 + u_2 v_2 + u_3 v_3) = (k u_1)v_1 + (k u_2)v_2 + (k u_3)v_3 = (k\mathbf{u}) \cdot \mathbf{v}$$

Similarly, $k(\mathbf{u} \cdot \mathbf{v}) = \mathbf{u} \cdot (k\mathbf{v})$.

Proof (d). $\mathbf{v} \cdot \mathbf{v} = v_1 v_1 + v_2 v_2 + v_3 v_3 = v_1{}^2 + v_2{}^2 + v_3{}^2 = \|\mathbf{v}\|^2$. ∎

Part (d) of the last theorem is sometimes expressed in the following alternative form:

$$\|\mathbf{v}\| = \sqrt{\mathbf{v} \cdot \mathbf{v}} \tag{4}$$

☐ **ORTHOGONAL PROJECTIONS OF VECTORS**

In many applications it is of interest to "decompose" a vector **u** into a sum of two vectors, one parallel to a specified nonzero vector **b** and the other perpendicular to **b**. If **u** and **b** are positioned so that their initial points coincide at a point Q, we may decompose the vector **u** as follows (Figure 14.3.6): Drop a perpendicular from the tip of **u** to the line through **b** and construct the vector $\mathbf{w}_1$ from Q to the foot of this perpendicular; next, form the difference

$$\mathbf{w}_2 = \mathbf{u} - \mathbf{w}_1$$

As indicated in Figure 14.3.6, the vector $\mathbf{w}_1$ is parallel to **b**, the vector $\mathbf{w}_2$ is perpendicular to **b**, and

$$\mathbf{u} = \mathbf{w}_1 + \mathbf{w}_2 = \mathbf{w}_1 + (\mathbf{u} - \mathbf{w}_1) \tag{5}$$

Figure 14.3.6

The vector $\mathbf{w}_1$ is called the ***vector component of u along b***, and the vector $\mathbf{w}_2$ is called the ***vector component of u orthogonal to b***. The vector component of **u** along **b** is also called the ***orthogonal projection of u on b*** and is denoted by $\text{proj}_\mathbf{b}\,\mathbf{u}$. With this notation the decomposition in (5) can be expressed as

$$\mathbf{u} = \text{proj}_\mathbf{b}\,\mathbf{u} + (\mathbf{u} - \text{proj}_\mathbf{b}\,\mathbf{u})$$

Vector   Vector
component + component
along **b**   orthogonal to **b**

The following theorem gives formulas for calculating $\text{proj}_\mathbf{b}\,\mathbf{u}$ and $\mathbf{u} - \text{proj}_\mathbf{b}\,\mathbf{u}$.

14.3.5   THEOREM.   *If **u** and **b** are vectors in 2-space or 3-space and if* $\mathbf{b} \neq \mathbf{0}$*, then*

$$\text{proj}_\mathbf{b}\,\mathbf{u} = \frac{\mathbf{u} \cdot \mathbf{b}}{\|\mathbf{b}\|^2}\,\mathbf{b} \quad \text{(vector component of u along b)} \tag{6}$$

$$\mathbf{u} - \text{proj}_\mathbf{b}\,\mathbf{u} = \mathbf{u} - \frac{\mathbf{u} \cdot \mathbf{b}}{\|\mathbf{b}\|^2}\,\mathbf{b} \quad \begin{array}{l}\text{(vector component of u}\\ \text{orthogonal to b)}\end{array} \tag{7}$$

Proof.   We shall begin by expressing the vector $\text{proj}_\mathbf{b}\,\mathbf{u}$ as its length times a unit vector in the same direction. If θ is the angle between **u** and **b**, then a unit vector in the same direction as $\text{proj}_\mathbf{b}\,\mathbf{u}$ is $\mathbf{b}/\|\mathbf{b}\|$ in the case where $0 \leq \theta \leq \pi/2$ or $-\mathbf{b}/\|\mathbf{b}\|$ in the case where $\pi/2 < \theta \leq \pi$ (Figure 14.3.7). From this and Theorem 14.3.2 it follows that

$$\text{proj}_\mathbf{b}\,\mathbf{u} = \begin{cases} \|\text{proj}_\mathbf{b}\,\mathbf{u}\|\,(\mathbf{b}/\|\mathbf{b}\|) & \text{if} \quad \mathbf{u} \cdot \mathbf{b} \geq 0 \\ \|\text{proj}_\mathbf{b}\,\mathbf{u}\|\,(-\mathbf{b}/\|\mathbf{b}\|) & \text{if} \quad \mathbf{u} \cdot \mathbf{b} < 0 \end{cases} \tag{8}$$

Again referring to Figure 14.3.7, it follows that

$$\|\text{proj}_\mathbf{b}\,\mathbf{u}\| = \|\mathbf{u}\|\,|\cos\theta| \tag{9}$$

or on multiplying by $\|\mathbf{b}\|/\|\mathbf{b}\|$

$$\|\text{proj}_\mathbf{b}\,\mathbf{u}\| = \frac{\|\mathbf{u}\|\,\|\mathbf{b}\|\,|\cos\theta|}{\|\mathbf{b}\|} = \frac{|\mathbf{u} \cdot \mathbf{b}|}{\|\mathbf{b}\|} \tag{10}$$

Substituting this formula in (8) yields (6) in all cases (verify). Formula (7) follows directly from (6).   ∎

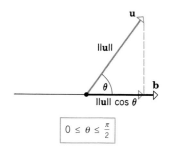

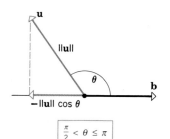

Figure 14.3.7

Example 6   Let $\mathbf{u} = 2\mathbf{i} - \mathbf{j} + 3\mathbf{k}$ and $\mathbf{b} = 4\mathbf{i} - \mathbf{j} + 2\mathbf{k}$. Find the vector component of **u** along **b** and the vector component of **u** orthogonal to **b**.

Solution.

$$\mathbf{u} \cdot \mathbf{b} = (2)(4) + (-1)(-1) + (3)(2) = 15$$

$$\|\mathbf{b}\|^2 = 4^2 + (-1)^2 + 2^2 = 21$$

Thus, the vector component of **u** along **b** is

$$\text{proj}_{\mathbf{b}}\,\mathbf{u} = \frac{\mathbf{u} \cdot \mathbf{b}}{\|\mathbf{b}\|^2}\,\mathbf{b} = \frac{15}{21}(4\mathbf{i} - \mathbf{j} + 2\mathbf{k}) = \frac{20}{7}\mathbf{i} - \frac{5}{7}\mathbf{j} + \frac{10}{7}\mathbf{k}$$

and the vector component of **u** orthogonal to **b** is

$$\mathbf{u} - \text{proj}_{\mathbf{b}}\,\mathbf{u} = (2\mathbf{i} - \mathbf{j} + 3\mathbf{k}) - \left(\frac{20}{7}\mathbf{i} - \frac{5}{7}\mathbf{j} + \frac{10}{7}\mathbf{k}\right) = -\frac{6}{7}\mathbf{i} - \frac{2}{7}\mathbf{j} + \frac{11}{7}\mathbf{k}$$

As a check, the reader may wish to verify that the vectors $\mathbf{u} - \text{proj}_{\mathbf{b}}\,\mathbf{u}$ and **b** are perpendicular by showing that their dot product is zero.   ◀

☐ **WORK**

It follows from Section 6.7 that the work W done by a constant force **F** of magnitude $\|\mathbf{F}\|$ acting in the direction of motion on a particle moving from P to Q on a line is

$$W = (\text{force}) \times (\text{distance}) = \|\mathbf{F}\|\,\|\overrightarrow{PQ}\|$$

If the force **F** is constant, but makes an angle θ with the direction of motion (Figure 14.3.8), then we *define* the work done by **F** to be

$$W = (\|\mathbf{F}\| \cos \theta)\,\|\overrightarrow{PQ}\| = \mathbf{F} \cdot \overrightarrow{PQ} \tag{11}$$

The quantity $\|\mathbf{F}\| \cos \theta$ is the "component" of force in the direction of motion and $\|\overrightarrow{PQ}\|$ is the distance traveled by the particle.

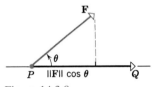

Figure 14.3.8

Example 7   A wagon is pulled horizontally by exerting a continual force of 10 lb on the handle at an angle of 60° with the horizontal. How much work is done in moving the wagon 50 ft?

Solution.   Introduce an xy-coordinate system so that the wagon moves from $P(0, 0)$ to $Q(50, 0)$ along the x-axis (Figure 14.3.9). In this coordinate system

$$\overrightarrow{PQ} = 50\mathbf{i}$$

and

$$\mathbf{F} = (10 \cos 60°)\mathbf{i} + (10 \sin 60°)\mathbf{j} = 5\mathbf{i} + 5\sqrt{3}\mathbf{j}$$

so that the work done is

$$W = \mathbf{F} \cdot \overrightarrow{PQ} = (5\mathbf{i} + 5\sqrt{3}\mathbf{j}) \cdot (50\mathbf{i}) = 250 \text{ (foot-pounds)}$$ ◀

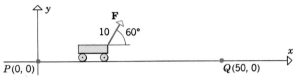

Figure 14.3.9

▶ **Exercise Set 14.3**

1. Find $\mathbf{u} \cdot \mathbf{v}$.

(a) $\mathbf{u} = \mathbf{i} + 2\mathbf{j}, \ \mathbf{v} = 6\mathbf{i} - 8\mathbf{j}$

(b) $\mathbf{u} = \langle -7, -3 \rangle, \ \mathbf{v} = \langle 0, 1 \rangle$

(c) $\mathbf{u} = \mathbf{i} - 3\mathbf{j} + 7\mathbf{k}, \ \mathbf{v} = 8\mathbf{i} - 2\mathbf{j} - 2\mathbf{k}$

(d) $\mathbf{u} = \langle -3, 1, 2 \rangle, \ \mathbf{v} = \langle 4, 2, -5 \rangle.$

2. In each part of Exercise 1, find the cosine of the angle θ between **u** and **v**.

3. Determine whether **u** and **v** make an acute angle, an obtuse angle, or are orthogonal.

(a) $\mathbf{u} = 7\mathbf{i} + 3\mathbf{j} + 5\mathbf{k}, \ \mathbf{v} = -8\mathbf{i} + 4\mathbf{j} + 2\mathbf{k}$

(b) $\mathbf{u} = 6\mathbf{i} + \mathbf{j} + 3\mathbf{k}$, $\mathbf{v} = 4\mathbf{i} - 6\mathbf{k}$

(c) $\mathbf{u} = \langle 1, 1, 1 \rangle$, $\mathbf{v} = \langle -1, 0, 0 \rangle$

(d) $\mathbf{u} = \langle 4, 1, 6 \rangle$, $\mathbf{v} = \langle -3, 0, 2 \rangle$.

4.   Find the orthogonal projection of $\mathbf{u}$ on $\mathbf{a}$.

(a) $\mathbf{u} = 2\mathbf{i} + \mathbf{j}$, $\mathbf{a} = -3\mathbf{i} + 2\mathbf{j}$

(b) $\mathbf{u} = \langle 2, 6 \rangle$, $\mathbf{a} = \langle -9, 3 \rangle$

(c) $\mathbf{u} = -7\mathbf{i} + \mathbf{j} + 3\mathbf{k}$, $\mathbf{a} = 5\mathbf{i} + \mathbf{k}$

(d) $\mathbf{u} = \langle 0, 0, 1 \rangle$, $\mathbf{a} = \langle 8, 3, 4 \rangle$.

5.   In each part of Exercise 4, find the vector component of $\mathbf{u}$ orthogonal to $\mathbf{a}$.

6.   Find $\| \text{proj}_{\mathbf{a}}\, \mathbf{u} \|$.

(a) $\mathbf{u} = 2\mathbf{i} - \mathbf{j}$, $\mathbf{a} = 3\mathbf{i} + 4\mathbf{j}$

(b) $\mathbf{u} = \langle 4, 5 \rangle$, $\mathbf{a} = \langle 1, -2 \rangle$

(c) $\mathbf{u} = 2\mathbf{i} - \mathbf{j} + 3\mathbf{k}$, $\mathbf{a} = \mathbf{i} + 2\mathbf{j} + 2\mathbf{k}$

(d) $\mathbf{u} = \langle 4, -1, 7 \rangle$, $\mathbf{a} = \langle 2, 3, -6 \rangle$.

7.   Verify part (c) of Theorem 14.3.4 for $\mathbf{u} = 6\mathbf{i} - \mathbf{j} + 2\mathbf{k}$, $\mathbf{v} = 2\mathbf{i} + 7\mathbf{j} + 4\mathbf{k}$, and $k = -5$.

8.   Find two vectors in 2-space of norm 1 that are orthogonal to $3\mathbf{i} - 2\mathbf{j}$.

9.   Let $\mathbf{u} = \langle 1, 2 \rangle$, $\mathbf{v} = \langle 4, -2 \rangle$, and $\mathbf{w} = \langle 6, 0 \rangle$. Find

(a) $\mathbf{u} \cdot (7\mathbf{v} + \mathbf{w})$

(b) $\|(\mathbf{u} \cdot \mathbf{w})\mathbf{w}\|$

(c) $\|\mathbf{u}\|(\mathbf{v} \cdot \mathbf{w})$

(d) $(\|\mathbf{u}\|\mathbf{v}) \cdot \mathbf{w}$.

10.   Explain why each of the following expressions makes no sense.

(a) $\mathbf{u} \cdot (\mathbf{v} \cdot \mathbf{w})$

(b) $(\mathbf{u} \cdot \mathbf{v}) + \mathbf{w}$

(c) $\|\mathbf{u} \cdot \mathbf{v}\|$

(d) $k \cdot (\mathbf{u} + \mathbf{v})$.

11.   Use vectors to find the cosines of the interior angles of the triangle with vertices $(-1, 0)$, $(2, -1)$, and $(1, 4)$.

12.   Find two unit vectors in 2-space that make an angle of $45°$ with $4\mathbf{i} + 3\mathbf{j}$.

13.   Show that $A(2, -1, 1)$, $B(3, 2, -1)$, and $C(7, 0, -2)$ are vertices of a right triangle. At which vertex is the right angle?

14.   Find k so that the vector from the point $A(1, -1, 3)$ to the point $B(3, 0, 5)$ is perpendicular to the vector from A to the point $P(k, k, k)$.

15.   Let $\mathbf{r}_0 = \langle x_0, y_0 \rangle$ and $\mathbf{r} = \langle x, y \rangle$. Describe the set of all points (x, y) for which

(a) $\mathbf{r} \cdot \mathbf{r}_0 = 0$

(b) $(\mathbf{r} - \mathbf{r}_0) \cdot \mathbf{r}_0 = 0$

(c) $\mathbf{r} \cdot (\mathbf{r} - \mathbf{r}_0) = 0$.

16.   Suppose that $\mathbf{a} \cdot \mathbf{b} = \mathbf{a} \cdot \mathbf{c}$ and $\mathbf{a} \neq \mathbf{0}$. Does it follow that $\mathbf{b} = \mathbf{c}$? Explain.

17.   Let $\mathbf{a} = k\mathbf{i} + \mathbf{j}$ and $\mathbf{b} = 4\mathbf{i} + 3\mathbf{j}$. Find k so that

(a) $\mathbf{a}$ and $\mathbf{b}$ are orthogonal

(b)   the angle between $\mathbf{a}$ and $\mathbf{b}$ is $\pi/4$

(c)   the angle between $\mathbf{a}$ and $\mathbf{b}$ is $\pi/6$

(d) $\mathbf{a}$ and $\mathbf{b}$ are parallel.

18.   Find the direction cosines of $\mathbf{u}$ and estimate the direction angles to the nearest degree.

(a) $\mathbf{u} = \mathbf{i} + \mathbf{j} - \mathbf{k}$

(b) $\mathbf{u} = 2\mathbf{i} - 2\mathbf{j} + \mathbf{k}$

(c) $\mathbf{u} = 3\mathbf{i} - 2\mathbf{j} - 6\mathbf{k}$

(d) $\mathbf{u} = 3\mathbf{i} - 4\mathbf{k}$.

19.   Prove: The direction cosines of a vector satisfy the equation $\cos^2 \alpha + \cos^2 \beta + \cos^2 \gamma = 1$.

20.   Prove: Two nonzero vectors $\mathbf{u}_1$ and $\mathbf{u}_2$ are perpendicular if and only if their direction cosines satisfy

$$\cos \alpha_1 \cos \alpha_2 + \cos \beta_1 \cos \beta_2 + \cos \gamma_1 \cos \gamma_2 = 0$$

21.   Given the points $A(2, -3)$, $B(5, 1)$, and $P(1, 0)$,

(a)   find $\|\text{proj}_{\overrightarrow{AB}}\, \overrightarrow{AP}\|$

(b)   use the Pythagorean Theorem and the result of part (a) to find the distance from P to the line through A and B.

22.   Follow the directions of Exercise 21 for the points $A(1, 1, 0)$, $B(-2, 3, -4)$, and $P(-3, 1, 2)$.

23.   Follow the directions of Exercise 21 for the points $A(2, 1, -3)$, $B(0, 2, -1)$, and $P(4, 3, 0)$.

24.   A boat travels 100 meters due north while the wind exerts a force of 50 newtons toward the northeast. How much work does the wind do?

25.   Find the work done by a force $\mathbf{F} = -3\mathbf{j}$ (pounds) applied to a point that moves on a line from $(1, 3)$ to $(4, 7)$. Assume that distance is measured in feet.

26.   Let $\mathbf{u}$ and $\mathbf{v}$ determine a parallelogram. Use vectors to prove that the diagonals of the parallelogram are perpendicular if and only if the sides are equal in length.

27.   Let $\mathbf{u}$ and $\mathbf{v}$ determine a parallelogram. Use vectors to prove that the parallelogram is a rectangle if and only if the diagonals are equal in length.

28.   Prove: $\|\mathbf{u} + \mathbf{v}\|^2 + \|\mathbf{u} - \mathbf{v}\|^2 = 2\|\mathbf{u}\|^2 + 2\|\mathbf{v}\|^2$.

29.   Prove: $\mathbf{u} \cdot \mathbf{v} = \frac{1}{4}\|\mathbf{u} + \mathbf{v}\|^2 - \frac{1}{4}\|\mathbf{u} - \mathbf{v}\|^2$.

30.   Find, to the nearest degree, the angle between the diagonal of a cube and a diagonal of one of its faces.

31.   Find, to the nearest degree, the acute angle formed by two diagonals of a cube.

32.   Find, to the nearest degree, the angles that a diagonal of a box with dimensions 10 in. by 15 in. by 25 in. makes with the edges of the box.

33.   Prove: If vectors $\mathbf{v}_1$, $\mathbf{v}_2$, and $\mathbf{v}_3$ are nonzero and mutually perpendicular, then any vector $\mathbf{v}$ can be written as

$$\mathbf{v} = c_1\mathbf{v}_1 + c_2\mathbf{v}_2 + c_3\mathbf{v}_3$$

where $c_i = (\mathbf{v} \cdot \mathbf{v}_i)/\|\mathbf{v}_i\|^2$, $i = 1, 2, 3$.

34.   Show that the three vectors

$$\mathbf{v}_1 = 3\mathbf{i} - \mathbf{j} + 2\mathbf{k}, \quad \mathbf{v}_2 = \mathbf{i} + \mathbf{j} - \mathbf{k}, \quad \mathbf{v}_3 = \mathbf{i} - 5\mathbf{j} - 4\mathbf{k}$$

are mutually perpendicular. Use the result of Exercise 33 to find scalars c_1, c_2, and c_3 so that

$$c_1\mathbf{v}_1 + c_2\mathbf{v}_2 + c_3\mathbf{v}_3 = \mathbf{i} - \mathbf{j} + \mathbf{k}$$

35. Prove: If **v** is orthogonal to $\mathbf{w}_1$ and $\mathbf{w}_2$, then **v** is orthogonal to $k_1\mathbf{w}_1 + k_2\mathbf{w}_2$ for all scalars k_1 and k_2.

36. Let **u** and **v** be nonzero vectors, and let $k = \|\mathbf{u}\|$ and $l = \|\mathbf{v}\|$. Prove that

$$\mathbf{w} = l\mathbf{u} + k\mathbf{v}$$

bisects the angle between **u** and **v**.

■ **14.4** CROSS PRODUCT

In many applications of vectors to problems in geometry, physics, and engineering, it is of interest to construct a vector in 3-space that is perpendicular to two given vectors. In this section we shall introduce a type of vector multiplication that facilitates this construction.

□ **DETERMINANTS**

Determinants are functions that assign numerical values to square arrays of numbers. For example, if a_1, a_2, b_1, and b_2 are real numbers, then a **2 × 2 *determinant*** is defined by

$$\begin{vmatrix} a_1 & a_2 \\ b_1 & b_2 \end{vmatrix} = a_1 b_2 - a_2 b_1$$

For example,

$$\begin{vmatrix} 3 & -2 \\ 4 & 5 \end{vmatrix} = (3)(5) - (-2)(4) = 15 + 8 = 23$$

A **3 × 3 *determinant*** is defined in terms of 2 × 2 determinants by

$$\begin{vmatrix} a_1 & a_2 & a_3 \\ b_1 & b_2 & b_3 \\ c_1 & c_2 & c_3 \end{vmatrix} = a_1 \begin{vmatrix} b_2 & b_3 \\ c_2 & c_3 \end{vmatrix} - a_2 \begin{vmatrix} b_1 & b_3 \\ c_1 & c_3 \end{vmatrix} + a_3 \begin{vmatrix} b_1 & b_2 \\ c_1 & c_2 \end{vmatrix}$$

The right side of this formula is easily remembered by noting that a_1, a_2, and a_3 are the entries in the first "row" of the left side, and the 2 × 2 determinants on the right side arise by deleting the first row and an appropriate column from the left side. The pattern is as follows:

$$\begin{vmatrix} a_1 & a_2 & a_3 \\ b_1 & b_2 & b_3 \\ c_1 & c_2 & c_3 \end{vmatrix} = a_1 \begin{vmatrix} a_1 & a_2 & a_3 \\ b_1 & b_2 & b_3 \\ c_1 & c_2 & c_3 \end{vmatrix} - a_2 \begin{vmatrix} a_1 & a_2 & a_3 \\ b_1 & b_2 & b_3 \\ c_1 & c_2 & c_3 \end{vmatrix} + a_3 \begin{vmatrix} a_1 & a_2 & a_3 \\ b_1 & b_2 & b_3 \\ c_1 & c_2 & c_3 \end{vmatrix}$$

For example,

$$\begin{vmatrix} 3 & -2 & -5 \\ 1 & 4 & -4 \\ 0 & 3 & 2 \end{vmatrix} = 3 \begin{vmatrix} 4 & -4 \\ 3 & 2 \end{vmatrix} - (-2) \begin{vmatrix} 1 & -4 \\ 0 & 2 \end{vmatrix} + (-5) \begin{vmatrix} 1 & 4 \\ 0 & 3 \end{vmatrix}$$

$$= 3(20) + 2(2) - 5(3) = 49$$

Higher-order determinants can be defined as well, but we will not need them in this text.

□ **CROSS PRODUCT**

14.4.1 DEFINITION. If $\mathbf{u} = \langle u_1, u_2, u_3 \rangle$ and $\mathbf{v} = \langle v_1, v_2, v_3 \rangle$ are vectors in 3-space, then the ***cross product*** $\mathbf{u} \times \mathbf{v}$ is the vector defined by

$$\mathbf{u} \times \mathbf{v} = \begin{vmatrix} u_2 & u_3 \\ v_2 & v_3 \end{vmatrix} \mathbf{i} - \begin{vmatrix} u_1 & u_3 \\ v_1 & v_3 \end{vmatrix} \mathbf{j} + \begin{vmatrix} u_1 & u_2 \\ v_1 & v_2 \end{vmatrix} \mathbf{k} \tag{1}$$

Formula (1) can be remembered by writing it in the form

$$\mathbf{u} \times \mathbf{v} = \begin{vmatrix} \mathbf{i} & \mathbf{j} & \mathbf{k} \\ u_1 & u_2 & u_3 \\ v_1 & v_2 & v_3 \end{vmatrix} \tag{2}$$

However, this is just a mnemonic device since the entries in a determinant must be numbers, not vectors.

Example 1   Find $\mathbf{u} \times \mathbf{v}$, where $\mathbf{u} = \langle 1, 2, -2 \rangle$ and $\mathbf{v} = \langle 3, 0, 1 \rangle$.

Solution.

$$\mathbf{u} \times \mathbf{v} = \begin{vmatrix} \mathbf{i} & \mathbf{j} & \mathbf{k} \\ 1 & 2 & -2 \\ 3 & 0 & 1 \end{vmatrix}$$

$$= \begin{vmatrix} 2 & -2 \\ 0 & 1 \end{vmatrix} \cdot \mathbf{i} - \begin{vmatrix} 1 & -2 \\ 3 & 1 \end{vmatrix} \mathbf{j} + \begin{vmatrix} 1 & 2 \\ 3 & 0 \end{vmatrix} \mathbf{k} = 2\mathbf{i} - 7\mathbf{j} - 6\mathbf{k} \quad \blacktriangleleft$$

Observe that the cross product of two vectors is another vector, whereas the dot product of two vectors is a scalar. Moreover, the cross product is defined only for vectors in 3-space, whereas the dot product is defined for vectors in 2-space and 3-space.

The following theorem gives an important relationship between dot product and cross product and also shows that $\mathbf{u} \times \mathbf{v}$ is orthogonal to both $\mathbf{u}$ and $\mathbf{v}$.

14.4.2   THEOREM.   *If* $\mathbf{u}$ *and* $\mathbf{v}$ *are vectors in 3-space, then*

(*a*) $\mathbf{u} \cdot (\mathbf{u} \times \mathbf{v}) = 0$ ($\mathbf{u} \times \mathbf{v}$ *is orthogonal to* $\mathbf{u}$)
(*b*) $\mathbf{v} \cdot (\mathbf{u} \times \mathbf{v}) = 0$ ($\mathbf{u} \times \mathbf{v}$ *is orthogonal to* $\mathbf{v}$)
(*c*) $\|\mathbf{u} \times \mathbf{v}\|^2 = \|\mathbf{u}\|^2 \|\mathbf{v}\|^2 - (\mathbf{u} \cdot \mathbf{v})^2$ (*Lagrange's identity*)

Proof.   Let $\mathbf{u} = \langle u_1, u_2, u_3 \rangle$ and $\mathbf{v} = \langle v_1, v_2, v_3 \rangle$.

(*a*)   By definition

$$\mathbf{u} \times \mathbf{v} = \begin{vmatrix} u_2 & u_3 \\ v_2 & v_3 \end{vmatrix} \mathbf{i} - \begin{vmatrix} u_1 & u_3 \\ v_1 & v_3 \end{vmatrix} \mathbf{j} + \begin{vmatrix} u_1 & u_2 \\ v_1 & v_2 \end{vmatrix} \mathbf{k}$$

which may be rewritten as

$$\mathbf{u} \times \mathbf{v} = \langle u_2 v_3 - u_3 v_2, \ u_3 v_1 - u_1 v_3, \ u_1 v_2 - u_2 v_1 \rangle \tag{3}$$

so that

$$\mathbf{u} \cdot (\mathbf{u} \times \mathbf{v}) = u_1(u_2 v_3 - u_3 v_2) + u_2(u_3 v_1 - u_1 v_3) + u_3(u_1 v_2 - u_2 v_1) = 0$$

(*b*)   Similar to (*a*).

(*c*)   From (3) we obtain

$$\|\mathbf{u} \times \mathbf{v}\|^2 = (u_2 v_3 - u_3 v_2)^2 + (u_3 v_1 - u_1 v_3)^2 + (u_1 v_2 - u_2 v_1)^2 \tag{4}$$

Moreover,

$$\|\mathbf{u}\|^2 \|\mathbf{v}\|^2 - (\mathbf{u} \cdot \mathbf{v})^2 = (u_1{}^2 + u_2{}^2 + u_3{}^2)(v_1{}^2 + v_2{}^2 + v_3{}^2)$$
$$- (u_1 v_1 + u_2 v_2 + u_3 v_3)^2 \tag{5}$$

Lagrange's identity can be established by ''multiplying out'' the right sides of (4) and (5) and verifying their equality.   ∎

Example 2 Let $\mathbf{u} = \langle 1, 2, -2 \rangle$ and $\mathbf{v} = \langle 3, 0, 1 \rangle$. In Example 1 we showed that

$$\mathbf{u} \times \mathbf{v} = \langle 2, -7, -6 \rangle$$

Thus,

$$\mathbf{u} \cdot (\mathbf{u} \times \mathbf{v}) = (1)(2) + (2)(-7) + (-2)(-6) = 0$$

and

$$\mathbf{v} \cdot (\mathbf{u} \times \mathbf{v}) = (3)(2) + (0)(-7) + (1)(-6) = 0$$

so that the vector $\mathbf{u} \times \mathbf{v}$ is orthogonal to both $\mathbf{u}$ and $\mathbf{v}$ as guaranteed by Theorem 14.4.2. ◄

If $\mathbf{u}$ and $\mathbf{v}$ are nonzero vectors in 3-space, then the length of $\mathbf{u} \times \mathbf{v}$ has a useful geometric interpretation. Lagrange's identity, given in Theorem 14.4.2, states that

$$\|\mathbf{u} \times \mathbf{v}\|^2 = \|\mathbf{u}\|^2 \|\mathbf{v}\|^2 - (\mathbf{u} \cdot \mathbf{v})^2$$

If θ denotes the angle between $\mathbf{u}$ and $\mathbf{v}$, then $\mathbf{u} \cdot \mathbf{v} = \|\mathbf{u}\| \|\mathbf{v}\| \cos \theta$, so the preceding equation can be rewritten as

$$\|\mathbf{u} \times \mathbf{v}\|^2 = \|\mathbf{u}\|^2 \|\mathbf{v}\|^2 - \|\mathbf{u}\|^2 \|\mathbf{v}\|^2 \cos^2 \theta$$
$$= \|\mathbf{u}\|^2 \|\mathbf{v}\|^2 (1 - \cos^2 \theta)$$
$$= \|\mathbf{u}\|^2 \|\mathbf{v}\|^2 \sin^2 \theta$$

Since $0 \le \theta \le \pi$, it follows that $\sin \theta \ge 0$, so

$$\|\mathbf{u} \times \mathbf{v}\| = \|\mathbf{u}\| \|\mathbf{v}\| \sin \theta \qquad (6)$$

But $\|\mathbf{v}\| \sin \theta$ is the altitude of the parallelogram determined by $\mathbf{u}$ and $\mathbf{v}$ (Figure 14.4.1). Thus, from (7), the area A of this parallelogram is given by

$$A = (\text{base})(\text{altitude}) = \|\mathbf{u}\| \|\mathbf{v}\| \sin \theta = \|\mathbf{u} \times \mathbf{v}\| \qquad (7)$$

In other words, the length of $\mathbf{u} \times \mathbf{v}$ is numerically equal to the area of the parallelogram determined by $\mathbf{u}$ and $\mathbf{v}$.

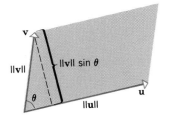

Figure 14.4.1

Example 3 Find the area of the triangle that is determined by the points $P_1(2, 2, 0)$, $P_2(-1, 0, 2)$, and $P_3(0, 4, 3)$.

Solution. The area A of the triangle is half the area of the parallelogram determined by the vectors $\overrightarrow{P_1P_2}$ and $\overrightarrow{P_1P_3}$ (Figure 14.4.2). But $\overrightarrow{P_1P_2} = \langle -3, -2, 2 \rangle$ and $\overrightarrow{P_1P_3} = \langle -2, 2, 3 \rangle$, so

$$\overrightarrow{P_1P_2} \times \overrightarrow{P_1P_3} = \langle -10, 5, -10 \rangle$$

(verify), and consequently

$$A = \tfrac{1}{2} \|\overrightarrow{P_1P_2} \times \overrightarrow{P_1P_3}\| = \tfrac{15}{2} \qquad ◄$$

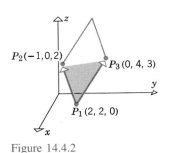

Figure 14.4.2

It follows from (7) that $\mathbf{u} \times \mathbf{v} = \mathbf{0}$ if and only if

$$\mathbf{u} = \mathbf{0}, \quad \text{or} \quad \mathbf{v} = \mathbf{0}, \quad \text{or} \quad \sin \theta = 0$$

In all three cases the vectors $\mathbf{u}$ and $\mathbf{v}$ are parallel. For the first two cases this is true because $\mathbf{0}$ is parallel to every vector, and in the third case, $\sin \theta = 0$ implies that the angle θ between $\mathbf{u}$ and $\mathbf{v}$ is $\theta = 0$ or $\theta = \pi$. In summary, we have the following result.

14.4.3 THEOREM. *If* $\mathbf{u}$ *and* $\mathbf{v}$ *are vectors in 3-space, then* $\mathbf{u} \times \mathbf{v} = \mathbf{0}$ *if and only if* $\mathbf{u}$ *and* $\mathbf{v}$ *are parallel vectors.*

The main arithmetic properties of the cross product are listed in the next theorem.

14.4.4   THEOREM.   *If* **u**, **v**, *and* **w** *are any vectors in 3-space and k is any scalar, then*

(a) $\mathbf{u} \times \mathbf{v} = -(\mathbf{v} \times \mathbf{u})$
(b) $\mathbf{u} \times (\mathbf{v} + \mathbf{w}) = (\mathbf{u} \times \mathbf{v}) + (\mathbf{u} \times \mathbf{w})$
(c) $(\mathbf{u} + \mathbf{v}) \times \mathbf{w} = (\mathbf{u} \times \mathbf{w}) + (\mathbf{v} \times \mathbf{w})$
(d) $k(\mathbf{u} \times \mathbf{v}) = (k\mathbf{u}) \times \mathbf{v} = \mathbf{u} \times (k\mathbf{v})$
(e) $\mathbf{u} \times \mathbf{0} = \mathbf{0} \times \mathbf{u} = \mathbf{0}$
(f) $\mathbf{u} \times \mathbf{u} = \mathbf{0}$

We shall prove (a) and leave the remaining proofs as exercises.

Proof (a).   First note that interchanging the rows of a 2×2 determinant changes the sign of the determinant, since

$$\begin{vmatrix} c & d \\ a & b \end{vmatrix} = bc - ad = -(ad - bc) = -\begin{vmatrix} a & b \\ c & d \end{vmatrix}$$

It follows that interchanging **u** and **v** in (1) interchanges the rows of the three determinants on the right side of (1) and thereby changes the sign of each component in the cross product. Thus, $\mathbf{u} \times \mathbf{v} = -(\mathbf{v} \times \mathbf{u})$. This means that(unlike the dot product)the cross product is not commutative.   ∎

REMARK.   The fact that interchanging the rows of a 2×2 determinant reverses the sign of the determinant is important in its own right. We leave it for the reader to prove the analogous result for 3×3 determinants; that is, interchanging any two rows reverses the sign of the determinant.

Cross products of the unit vectors **i**, **j**, and **k** are of special interest. We obtain, for example,

$$\mathbf{i} \times \mathbf{j} = \begin{vmatrix} \mathbf{i} & \mathbf{j} & \mathbf{k} \\ 1 & 0 & 0 \\ 0 & 1 & 0 \end{vmatrix} = \begin{vmatrix} 0 & 0 \\ 1 & 0 \end{vmatrix} \mathbf{i} - \begin{vmatrix} 1 & 0 \\ 0 & 0 \end{vmatrix} \mathbf{j} + \begin{vmatrix} 1 & 0 \\ 0 & 1 \end{vmatrix} \mathbf{k} = \mathbf{k}$$

The reader should have no trouble obtaining the following list of cross products:

$$\mathbf{i} \times \mathbf{j} = \mathbf{k} \qquad \mathbf{j} \times \mathbf{k} = \mathbf{i} \qquad \mathbf{k} \times \mathbf{i} = \mathbf{j}$$
$$\mathbf{j} \times \mathbf{i} = -\mathbf{k} \qquad \mathbf{k} \times \mathbf{j} = -\mathbf{i} \qquad \mathbf{i} \times \mathbf{k} = -\mathbf{j}$$
$$\mathbf{i} \times \mathbf{i} = \mathbf{0} \qquad \mathbf{j} \times \mathbf{j} = \mathbf{0} \qquad \mathbf{k} \times \mathbf{k} = \mathbf{0}$$

The diagram in Figure 14.4.3 is helpful for remembering these results. In this diagram, the cross product of two consecutive vectors going clockwise is the next vector around, and the cross product of two consecutive vectors going counterclockwise is the negative of the next vector around.

WARNING.   It is *not* true in general that $\mathbf{u} \times (\mathbf{v} \times \mathbf{w}) = (\mathbf{u} \times \mathbf{v}) \times \mathbf{w}$. For example, we have

$$\mathbf{i} \times (\mathbf{j} \times \mathbf{j}) = \mathbf{i} \times \mathbf{0} = \mathbf{0} \quad \text{and} \quad (\mathbf{i} \times \mathbf{j}) \times \mathbf{j} = \mathbf{k} \times \mathbf{j} = -\mathbf{i}$$

so that

$$\mathbf{i} \times (\mathbf{j} \times \mathbf{j}) \neq (\mathbf{i} \times \mathbf{j}) \times \mathbf{j}$$

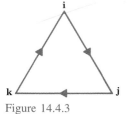

Figure 14.4.3

We know from Theorem 14.4.2 that $\mathbf{u} \times \mathbf{v}$ is orthogonal to both $\mathbf{u}$ and $\mathbf{v}$. It can be shown that if $\mathbf{u}$ and $\mathbf{v}$ are nonzero vectors, then the direction of $\mathbf{u} \times \mathbf{v}$ can be determined using the following "right-hand rule"* (Figure 14.4.4). Let θ be the angle between $\mathbf{u}$ and $\mathbf{v}$, and suppose that $\mathbf{u}$ is rotated through the angle θ until it coincides with $\mathbf{v}$. If the fingers of the right hand are cupped so that they point in the direction of rotation, then the thumb indicates (roughly) the direction of $\mathbf{u} \times \mathbf{v}$. The reader may find it instructive to practice this rule with the products

$$\mathbf{i} \times \mathbf{j} = \mathbf{k}, \qquad \mathbf{j} \times \mathbf{k} = \mathbf{i}, \qquad \mathbf{k} \times \mathbf{i} = \mathbf{j}$$

Equation (7) and the right-hand rule show that both the magnitude and direction of $\mathbf{u} \times \mathbf{v}$ can be determined purely geometrically; that is, $\mathbf{u} \times \mathbf{v}$ does not depend on the coordinate system being used. Thus, we say that the definition of $\mathbf{u} \times \mathbf{v}$ is **coordinate free**. This result is important to physicists and engineers, since it allows them to choose any convenient coordinate system with assurance that the choice will not affect the end result of any vector computations.

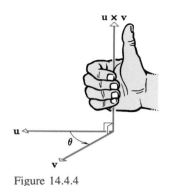

Figure 14.4.4

□ **TRIPLE SCALAR PRODUCTS**

If $\mathbf{a} = \langle a_1, a_2, a_3 \rangle$, $\mathbf{b} = \langle b_1, b_2, b_3 \rangle$, and $\mathbf{c} = \langle c_1, c_2, c_3 \rangle$ are vectors in 3-space, then the number

$$\mathbf{a} \cdot (\mathbf{b} \times \mathbf{c})$$

is called the **triple scalar product** of $\mathbf{a}$, $\mathbf{b}$, and $\mathbf{c}$. The triple scalar product may be conveniently calculated from the formula

$$\mathbf{a} \cdot (\mathbf{b} \times \mathbf{c}) = \begin{vmatrix} a_1 & a_2 & a_3 \\ b_1 & b_2 & b_3 \\ c_1 & c_2 & c_3 \end{vmatrix} \qquad (8)$$

The validity of this formula may be seen by writing

$$\mathbf{a} \cdot (\mathbf{b} \times \mathbf{c}) = \mathbf{a} \cdot \left(\begin{vmatrix} b_2 & b_3 \\ c_2 & c_3 \end{vmatrix} \mathbf{i} - \begin{vmatrix} b_1 & b_3 \\ c_1 & c_3 \end{vmatrix} \mathbf{j} + \begin{vmatrix} b_1 & b_2 \\ c_1 & c_2 \end{vmatrix} \mathbf{k} \right)$$

$$= \begin{vmatrix} b_2 & b_3 \\ c_2 & c_3 \end{vmatrix} a_1 - \begin{vmatrix} b_1 & b_3 \\ c_1 & c_3 \end{vmatrix} a_2 + \begin{vmatrix} b_1 & b_2 \\ c_1 & c_2 \end{vmatrix} a_3$$

$$= \begin{vmatrix} a_1 & a_2 & a_3 \\ b_1 & b_2 & b_3 \\ c_1 & c_2 & c_3 \end{vmatrix}$$

Example 4 Calculate the triple scalar product $\mathbf{a} \cdot (\mathbf{b} \times \mathbf{c})$ of the vectors

$$\mathbf{a} = 3\mathbf{i} - 2\mathbf{j} - 5\mathbf{k}, \qquad \mathbf{b} = \mathbf{i} + 4\mathbf{j} - 4\mathbf{k}, \qquad \mathbf{c} = 3\mathbf{j} + 2\mathbf{k}$$

Solution.

$$\mathbf{a} \cdot (\mathbf{b} \times \mathbf{c}) = \begin{vmatrix} 3 & -2 & -5 \\ 1 & 4 & -4 \\ 0 & 3 & 2 \end{vmatrix}$$

$$= 3 \begin{vmatrix} 4 & -4 \\ 3 & 2 \end{vmatrix} - (-2) \begin{vmatrix} 1 & -4 \\ 0 & 2 \end{vmatrix} + (-5) \begin{vmatrix} 1 & 4 \\ 0 & 3 \end{vmatrix}$$

$$= 60 + 4 - 15 = 49 \qquad ◄$$

---

*Recall that we agreed to consider only right-handed coordinate systems in this text. Had we used left-handed systems instead, a "left-hand rule" would apply here.

REMARK. The symbol $(\mathbf{a} \cdot \mathbf{b}) \times \mathbf{c}$ makes no sense since we cannot form the cross product of a scalar and a vector. Thus, no ambiguity arises if we write $\mathbf{a} \cdot \mathbf{b} \times \mathbf{c}$ rather than $\mathbf{a} \cdot (\mathbf{b} \times \mathbf{c})$.

Recall from the remark following Theorem 14.4.4 that interchanging two rows of a 3×3 determinant reverses the sign of the determinant. It follows from this that

$$\mathbf{a} \cdot (\mathbf{b} \times \mathbf{c}) = \mathbf{c} \cdot (\mathbf{a} \times \mathbf{b}) = \mathbf{b} \cdot (\mathbf{c} \times \mathbf{a}) \tag{9}$$

since the 3×3 determinants that represent these products can be obtained from one another by *two* row interchanges. (Verify.) These relationships may be remembered by moving the vectors $\mathbf{a}$, $\mathbf{b}$, and $\mathbf{c}$ clockwise around the vertices of the triangle in Figure 14.4.5. We also note that from part (*a*) of Theorem 14.3.4, the first equality in (9) can be written as $\mathbf{a} \cdot (\mathbf{b} \times \mathbf{c}) = (\mathbf{a} \times \mathbf{b}) \cdot \mathbf{c}$, which, on dropping parentheses, yields

$$\mathbf{a} \cdot \mathbf{b} \times \mathbf{c} = \mathbf{a} \times \mathbf{b} \cdot \mathbf{c} \tag{10}$$

This shows that interchanging the dot and cross does not change the value of a triple scalar product.

The triple scalar product $\mathbf{a} \cdot (\mathbf{b} \times \mathbf{c})$ has a useful geometric interpretation. If we assume, for the moment, that the vectors $\mathbf{a}$, $\mathbf{b}$, $\mathbf{c}$ do not all lie in the same plane when they are positioned with a common initial point, then the three vectors form adjacent sides of a parallelepiped (Figure 14.4.6). If the parallelogram determined by $\mathbf{b}$ and $\mathbf{c}$ is regarded as the base of the parallelepiped, then the area of the base is $\|\mathbf{b} \times \mathbf{c}\|$, and the height h is the length of the orthogonal projection of $\mathbf{a}$ on $\mathbf{b} \times \mathbf{c}$ (Figure 14.4.6). Therefore, by Formula (10) of Section 14.3 we have

$$h = \|\text{proj}_{\mathbf{b} \times \mathbf{c}} \, \mathbf{a}\| = \frac{|\mathbf{a} \cdot (\mathbf{b} \times \mathbf{c})|}{\|\mathbf{b} \times \mathbf{c}\|}$$

It follows that the volume V of the parallelepiped is

$$V = (\text{area of base}) \cdot \text{height} = \|\mathbf{b} \times \mathbf{c}\| \frac{|\mathbf{a} \cdot (\mathbf{b} \times \mathbf{c})|}{\|\mathbf{b} \times \mathbf{c}\|}$$

or, on simplifying,

$$V = \begin{bmatrix} \text{volume of parallelepiped with} \\ \text{adjacent sides } \mathbf{a}, \mathbf{b}, \text{ and } \mathbf{c} \end{bmatrix} = |\mathbf{a} \cdot (\mathbf{b} \times \mathbf{c})| \tag{11}$$

REMARK. It follows from this formula that

$$\mathbf{a} \cdot (\mathbf{b} \times \mathbf{c}) = \pm V$$

where the $+$ or $-$ results depending on whether $\mathbf{a}$ makes an acute or obtuse angle with $\mathbf{b} \times \mathbf{c}$ (Theorem 14.3.2).

If the vectors $\mathbf{a}$, $\mathbf{b}$, and $\mathbf{c}$ do not lie in a plane when they are positioned with a common initial point, then they determine a parallelepiped of positive volume. Thus, from (11), $\mathbf{a} \cdot (\mathbf{b} \times \mathbf{c}) \neq 0$ if $\mathbf{a}$, $\mathbf{b}$, and $\mathbf{c}$ do not lie in a plane. It follows, therefore, that if $\mathbf{a} \cdot (\mathbf{b} \times \mathbf{c}) = 0$, then $\mathbf{a}$, $\mathbf{b}$, and $\mathbf{c}$ lie in a plane. Conversely, it can be shown that if $\mathbf{a}$, $\mathbf{b}$, and $\mathbf{c}$ lie in a plane, then $\mathbf{a} \cdot (\mathbf{b} \times \mathbf{c}) = 0$. In summary, we have the following result.

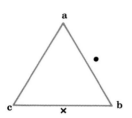

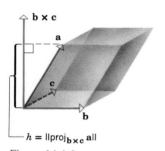

$h = \|\text{proj}_{\mathbf{b} \times \mathbf{c}} \mathbf{a}\|$

Figure 14.4.6

14.4.5 THEOREM. *If the vectors* $\mathbf{a} = \langle a_1, a_2, a_3 \rangle$, $\mathbf{b} = \langle b_1, b_2, b_3 \rangle$, *and* $\mathbf{c} = \langle c_1, c_2, c_3 \rangle$ *have the same initial point, then they lie in a plane if and only if*

$$\mathbf{a} \cdot (\mathbf{b} \times \mathbf{c}) = \begin{vmatrix} a_1 & a_2 & a_3 \\ b_1 & b_2 & b_3 \\ c_1 & c_2 & c_3 \end{vmatrix} = 0$$

▶ Exercise Set 14.4

In Exercises 1–4, find $\mathbf{a} \times \mathbf{b}$.

1. $\mathbf{a} = \langle 1, 2, -3 \rangle$, $\mathbf{b} = \langle -4, 1, 2 \rangle$.
2. $\mathbf{a} = 3\mathbf{i} + 2\mathbf{j} - \mathbf{k}$, $\mathbf{b} = -\mathbf{i} - 3\mathbf{j} + \mathbf{k}$.
3. $\mathbf{a} = \langle 0, 1, -2 \rangle$, $\mathbf{b} = \langle 3, 0, -4 \rangle$.
4. $\mathbf{a} = 4\mathbf{i} + \mathbf{k}$, $\mathbf{b} = 2\mathbf{i} - \mathbf{j}$.
5. Let $\mathbf{u} = \langle 2, -1, 3 \rangle$, $\mathbf{v} = \langle 0, 1, 7 \rangle$, and $\mathbf{w} = \langle 1, 4, 5 \rangle$. Find
   (a) $\mathbf{u} \times (\mathbf{v} \times \mathbf{w})$
   (b) $(\mathbf{u} \times \mathbf{v}) \times \mathbf{w}$
   (c) $\mathbf{u} \times (\mathbf{v} - 2\mathbf{w})$
   (d) $(\mathbf{u} \times \mathbf{v}) - 2\mathbf{w}$
   (e) $(\mathbf{u} \times \mathbf{v}) \times (\mathbf{v} \times \mathbf{w})$
   (f) $(\mathbf{v} \times \mathbf{w}) \times (\mathbf{u} \times \mathbf{v})$.
6. Find a vector orthogonal to both $\mathbf{u}$ and $\mathbf{v}$.
   (a) $\mathbf{u} = -7\mathbf{i} + 3\mathbf{j} + \mathbf{k}$, $\mathbf{v} = 2\mathbf{i} + 4\mathbf{k}$
   (b) $\mathbf{u} = \langle -1, -1, -1 \rangle$, $\mathbf{v} = \langle 2, 0, 2 \rangle$.
7. Verify Theorem 14.4.2 for the vectors $\mathbf{u} = \mathbf{i} - 5\mathbf{j} + 6\mathbf{k}$ and $\mathbf{v} = 2\mathbf{i} + \mathbf{j} + 2\mathbf{k}$.
8. Verify Theorem 14.4.4 for $\mathbf{u} = \langle 2, 0, -1 \rangle$, $\mathbf{v} = \langle 6, 7, 4 \rangle$, $\mathbf{w} = \langle 1, 1, 1 \rangle$, and $k = -3$.
9. Find all unit vectors parallel to the yz-plane that are perpendicular to the vector $3\mathbf{i} - \mathbf{j} + 2\mathbf{k}$.
10. Prove: If θ is the angle between $\mathbf{u}$ and $\mathbf{v}$ and $\mathbf{u} \cdot \mathbf{v} \neq 0$, then $\tan \theta = \|\mathbf{u} \times \mathbf{v}\|/(\mathbf{u} \cdot \mathbf{v})$.
11. Simplify $(\mathbf{u} + \mathbf{v}) \times (\mathbf{u} - \mathbf{v})$.
12. Find the area of the parallelogram determined by the vectors $\mathbf{u}$ and $\mathbf{v}$.
    (a) $\mathbf{u} = \mathbf{i} - \mathbf{j} + 2\mathbf{k}$, $\mathbf{v} = 3\mathbf{j} + \mathbf{k}$
    (b) $\mathbf{u} = 2\mathbf{i} + 3\mathbf{j}$, $\mathbf{v} = -\mathbf{i} + 2\mathbf{j} - 2\mathbf{k}$.
13. Find the area of the triangle having vertices P, Q, and R.
    (a) $P(1, 5, -2)$, $Q(0, 0, 0)$, $R(3, 5, 1)$
    (b) $P(2, 0, -3)$, $Q(1, 4, 5)$, $R(7, 2, 9)$.
14. Use the cross product to find the sine of the angle between the vectors $\mathbf{a} = 2\mathbf{i} + 3\mathbf{j} - 6\mathbf{k}$ and $\mathbf{b} = 2\mathbf{i} + 3\mathbf{j} + 6\mathbf{k}$.
15. (a) Find the area of the triangle having vertices $A(1, 0, 1)$, $B(0, 2, 3)$, and $C(2, 1, 0)$.
    (b) Use the result of part (a) to find the length of the altitude from vertex C to side AB.
16. Show that if $\mathbf{u}$ is a vector from any point on a line to a point P not on the line, and $\mathbf{v}$ is a vector parallel to the line, then the distance between P and the line is given by $\|\mathbf{u} \times \mathbf{v}\|/\|\mathbf{v}\|$.
17. Use the result of Exercise 16 to find the distance between the point P and the line through the points A and B.
    (a) $P(-3, 1, 2)$, $A(1, 1, 0)$, $B(-2, 3, -4)$
    (b) $P(4, 3, 0)$, $A(2, 1, -3)$, $B(0, 2, -1)$.
18. Find all unit vectors in the plane of $\mathbf{u} = 3\mathbf{i} + \mathbf{k}$ and $\mathbf{v} = \mathbf{i} - \mathbf{j} - \mathbf{k}$ that are perpendicular to the vector $\mathbf{w} = \mathbf{i} + 2\mathbf{j}$.
19. What is wrong with the expression $\mathbf{u} \times \mathbf{v} \times \mathbf{w}$?

In Exercises 20–23, find $\mathbf{a} \cdot (\mathbf{b} \times \mathbf{c})$.

20. $\mathbf{a} = \langle 1, -2, 2 \rangle$, $\mathbf{b} = \langle 0, 3, 2 \rangle$, $\mathbf{c} = \langle -4, 1, -3 \rangle$.
21. $\mathbf{a} = 2\mathbf{i} - 3\mathbf{j} + \mathbf{k}$, $\mathbf{b} = 4\mathbf{i} + \mathbf{j} - 3\mathbf{k}$, $\mathbf{c} = \mathbf{j} + 5\mathbf{k}$.
22. $\mathbf{a} = \langle 2, 1, 0 \rangle$, $\mathbf{b} = \langle 1, -3, 1 \rangle$, $\mathbf{c} = \langle 4, 0, 1 \rangle$.
23. $\mathbf{a} = \mathbf{i}$, $\mathbf{b} = \mathbf{i} + \mathbf{j}$, $\mathbf{c} = \mathbf{i} + \mathbf{j} + \mathbf{k}$.
24. Suppose that $\mathbf{u} \cdot (\mathbf{v} \times \mathbf{w}) = 3$. Find
    (a) $\mathbf{u} \cdot (\mathbf{w} \times \mathbf{v})$
    (b) $(\mathbf{v} \times \mathbf{w}) \cdot \mathbf{u}$
    (c) $\mathbf{w} \cdot (\mathbf{u} \times \mathbf{v})$
    (d) $\mathbf{v} \cdot (\mathbf{u} \times \mathbf{w})$
    (e) $(\mathbf{u} \times \mathbf{w}) \cdot \mathbf{v}$
    (f) $\mathbf{v} \cdot (\mathbf{w} \times \mathbf{w})$.
25. Find the volume of the parallelepiped with sides $\mathbf{a}$, $\mathbf{b}$, and $\mathbf{c}$.
    (a) $\mathbf{a} = \langle 2, -6, 2 \rangle$, $\mathbf{b} = \langle 0, 4, -2 \rangle$, $\mathbf{c} = \langle 2, 2, -4 \rangle$
    (b) $\mathbf{a} = 3\mathbf{i} + \mathbf{j} + 2\mathbf{k}$, $\mathbf{b} = 4\mathbf{i} + 5\mathbf{j} + \mathbf{k}$, $\mathbf{c} = \mathbf{i} + 2\mathbf{j} + 4\mathbf{k}$.
26. Determine whether $\mathbf{u}$, $\mathbf{v}$, and $\mathbf{w}$ lie in the same plane.
    (a) $\mathbf{u} = \langle 1, -2, 1 \rangle$, $\mathbf{v} = \langle 3, 0, -2 \rangle$, $\mathbf{w} = \langle 5, -4, 0 \rangle$
    (b) $\mathbf{u} = 5\mathbf{i} - 2\mathbf{j} + \mathbf{k}$, $\mathbf{v} = 4\mathbf{i} - \mathbf{j} + \mathbf{k}$, $\mathbf{w} = \mathbf{i} - \mathbf{j}$
    (c) $\mathbf{u} = \langle 4, -8, 1 \rangle$, $\mathbf{v} = \langle 2, 1, -2 \rangle$, $\mathbf{w} = \langle 3, -4, 12 \rangle$.
27. Consider the parallelepiped with sides
    $$\mathbf{a} = 3\mathbf{i} + 2\mathbf{j} + \mathbf{k}$$
    $$\mathbf{b} = \mathbf{i} + \mathbf{j} + 2\mathbf{k}$$
    $$\mathbf{c} = \mathbf{i} + 3\mathbf{j} + 3\mathbf{k}$$
    (a) Find the volume.
    (b) Find the area of the face determined by $\mathbf{a}$ and $\mathbf{c}$.
    (c) Find the angle between $\mathbf{a}$ and the plane containing the face determined by $\mathbf{b}$ and $\mathbf{c}$.
28. Find a vector $\mathbf{n}$ perpendicular to the plane determined by the points $A(0, -2, 1)$, $B(1, -1, -2)$, and $C(-1, 1, 0)$.
29. Simplify $(\mathbf{u} + k\mathbf{v}) \times \mathbf{v}$.
30. Let $\mathbf{u}$, $\mathbf{v}$, and $\mathbf{w}$ be vectors in 3-space with the same initial point and such that no two are collinear. Prove:
    (a) $\mathbf{u} \times (\mathbf{v} \times \mathbf{w})$ lies in the plane determined by $\mathbf{v}$ and $\mathbf{w}$
    (b) $(\mathbf{u} \times \mathbf{v}) \times \mathbf{w}$ lies in the plane determined by $\mathbf{u}$ and $\mathbf{v}$.
31. Prove parts (*b*) and (*c*) of Theorem 14.4.4.
32. Prove parts (*d*), (*e*), and (*f*) of Theorem 14.4.4.
33. Prove that $\mathbf{x} \times (\mathbf{y} \times \mathbf{z}) = (\mathbf{x} \cdot \mathbf{z})\mathbf{y} - (\mathbf{x} \cdot \mathbf{y})\mathbf{z}$. [*Hint:* First prove the result for the case $\mathbf{z} = \mathbf{i}$, then for $\mathbf{z} = \mathbf{j}$, and then for $\mathbf{z} = \mathbf{k}$. Finally, prove it for an arbitrary vector $\mathbf{z} = z_1\mathbf{i} + z_2\mathbf{j} + z_3\mathbf{k}$.]
34. For the vectors $\mathbf{a} = \mathbf{i} + 3\mathbf{j} - \mathbf{k}$, $\mathbf{b} = \mathbf{i} + \mathbf{j} + 2\mathbf{k}$, and $\mathbf{c} = 3\mathbf{i} - \mathbf{j} + 2\mathbf{k}$, calculate $\mathbf{a} \times (\mathbf{b} \times \mathbf{c})$ using Exercise 33, then check your result by calculating directly.
35. Prove: If $\mathbf{a}$, $\mathbf{b}$, $\mathbf{c}$, and $\mathbf{d}$ lie in the same plane when positioned with a common initial point, then
    $$(\mathbf{a} \times \mathbf{b}) \times (\mathbf{c} \times \mathbf{d}) = \mathbf{0}$$

36. It is a theorem of solid geometry that the volume of a tetrahedron is $\frac{1}{3}$(area of base) $\cdot$ (height). Use this result to prove that the volume of a tetrahedron whose sides are the vectors $\mathbf{a}$, $\mathbf{b}$, and $\mathbf{c}$ is $\frac{1}{6}|\mathbf{a}\cdot(\mathbf{b}\times\mathbf{c})|$.

37. Use the result of Exercise 36 to find the volume of the tetrahedron with vertices P, Q, R, and S.

   (a) $P(-1,2,0)$, $Q(2,1,-3)$, $R(1,0,1)$, $S(3,-2,3)$

   (b) $P(0,0,0)$, $Q(1,2,-1)$, $R(3,4,0)$, $S(-1,-3,4)$.

■ 14.5 PARAMETRIC EQUATIONS OF LINES

In this section we shall discuss parametric equations of lines in 2-space and 3-space. In 3-space, parametric equations of lines are especially important because they are generally the most convenient form for representing such lines algebraically.

☐ **LINES DETERMINED BY A POINT AND A VECTOR**

A line in 2-space or 3-space can be determined uniquely by specifying a point on the line and a nonzero vector parallel to the line (Figure 14.5.1). The following theorem gives parametric equations of the line through a point P_0 and parallel to a nonzero vector $\mathbf{v}$.

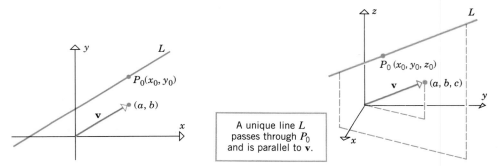

A unique line L passes through P_0 and is parallel to $\mathbf{v}$.

Figure 14.5.1

14.5.1   THEOREM.

(a) *The line in 2-space that passes through the point $P_0(x_0, y_0)$ and is parallel to the nonzero vector $\mathbf{v} = \langle a, b\rangle = a\mathbf{i} + b\mathbf{j}$ has parametric equations*

$$x = x_0 + at, \quad y = y_0 + bt \tag{1a}$$

(b) *The line in 3-space that passes through the point $P_0(x_0, y_0, z_0)$ and is parallel to the nonzero vector $\mathbf{v} = \langle a, b, c\rangle = a\mathbf{i} + b\mathbf{j} + c\mathbf{k}$ has parametric equations*

$$x = x_0 + at, \quad y = y_0 + bt, \quad z = z_0 + ct \tag{1b}$$

We shall prove part (b). The proof of (a) is similar.

Proof (b). If L is the line in 3-space passing through the point $P_0(x_0, y_0, z_0)$ and parallel to the nonzero vector $\mathbf{v} = \langle a, b, c\rangle$, it is clear (Figure 14.5.2) that L consists precisely of those points $P(x, y, z)$ for which the vector $\overrightarrow{P_0P}$ is parallel to $\mathbf{v}$. In other words, the point $P(x, y, z)$ is on L if and only if $\overrightarrow{P_0P}$ is a scalar multiple of $\mathbf{v}$, say

$$\overrightarrow{P_0P} = t\mathbf{v}$$

This equation can be written as

$$\langle x - x_0, y - y_0, z - z_0\rangle = \langle ta, tb, tc\rangle$$

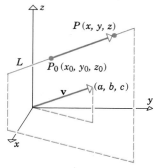

Figure 14.5.2

Equating components yields

$$x - x_0 = ta, \quad y - y_0 = tb, \quad z - z_0 = tc$$

from which (1b) follows.   ∎

REMARK.   Although it is not stated explicitly, it is understood in (1a) and (1b) that $-\infty < t < +\infty$, which reflects the fact that lines extend indefinitely.

Example 1   Find parametric equations of the line

(a)   passing through $(4, 2)$ and parallel to $\mathbf{v} = \langle -1, 5 \rangle$;

(b)   passing through $(1, 2, -3)$ and parallel to $\mathbf{v} = 4\mathbf{i} + 5\mathbf{j} - 7\mathbf{k}$;

(c)   passing through the origin in 3-space and parallel to $\mathbf{v} = \langle 1, 1, 1 \rangle$.

Solution (a).   From (1a) with $x_0 = 4$, $y_0 = 2$, $a = -1$, and $b = 5$ we obtain

$$x = 4 - t, \quad y = 2 + 5t$$

Solution (b).   From (1b) we obtain

$$x = 1 + 4t, \quad y = 2 + 5t, \quad z = -3 - 7t$$

Solution (c).   From (1b) with $x_0 = 0$, $y_0 = 0$, $z_0 = 0$, $a = 1$, $b = 1$, and $c = 1$ we obtain

$$x = t, \quad y = t, \quad z = t \qquad ◀$$

Example 2

(a)   Find parametric equations of the line L passing through the points $P_1(2, 4, -1)$ and $P_2(5, 0, 7)$.

(b)   Where does the line intersect the xy-plane?

Solution (a).   Since the vector $\overrightarrow{P_1P_2} = \langle 3, -4, 8 \rangle$ is parallel to L and $P_1(2, 4, -1)$ lies on L, the line L is given by

$$x = 2 + 3t, \quad y = 4 - 4t, \quad z = -1 + 8t$$

Had we used P_2 as the point on L rather than P_1, we would have obtained the equations

$$x = 5 + 3t, \quad y = -4t, \quad z = 7 + 8t$$

Although these equations look different from those obtained using P_1, the two sets of equations are actually equivalent in that both generate L as t varies from $-\infty$ to $+\infty$.

Solution (b).   From the first parametrization in part (a) the line intersects the xy-plane at the point where $z = -1 + 8t = 0$, that is, when $t = \frac{1}{8}$. Substituting this value of t in the parametric equations for L yields the point of intersection

$$(x, y, z) = \left(\tfrac{19}{8}, \tfrac{7}{2}, 0 \right) \qquad ◀$$

Example 3   Let L_1 and L_2 be the lines

$$L_1: x = 1 + 4t, \quad y = 5 - 4t, \quad z = -1 + 5t$$

$$L_2: x = 2 + 8t, \quad y = 4 - 3t, \quad z = 5 + t$$

(a)   Are the lines parallel?

(b)   Do the lines intersect?

Solution (a).   The line L_1 is parallel to the vector $4\mathbf{i} - 4\mathbf{j} + 5\mathbf{k}$, and the line L_2 is parallel to the vector $8\mathbf{i} - 3\mathbf{j} + \mathbf{k}$. These vectors are not parallel since neither is a scalar multiple of the other. Thus, the lines are not parallel.

Solution (b).   In order for the lines to intersect at some point (x_0, y_0, z_0) these coordinates would have to satisfy the equations of both L_1 and L_2. In other words, there would have to exist values t_1 and t_2 for the parameters such that

$$x_0 = 1 + 4t_1, \quad y_0 = 5 - 4t_1, \quad z_0 = -1 + 5t_1$$

and

$$x_0 = 2 + 8t_2, \quad y_0 = 4 - 3t_2, \quad z_0 = 5 + t_2$$

This leads to three conditions on t_1 and t_2,

$$1 + 4t_1 = 2 + 8t_2$$

$$5 - 4t_1 = 4 - 3t_2 \tag{2}$$

$$-1 + 5t_1 = 5 + t_2$$

We shall try to solve these equations for t_1 and t_2. If we obtain a solution, then the lines intersect; if we find that there is no solution, then the lines do not intersect, since the three conditions cannot be satisfied.

The first two equations in (2) may be solved by adding them together to obtain

$$6 = 6 + 5t_2$$

or $t_2 = 0$. Substituting $t_2 = 0$ in the first equation yields

$$1 + 4t_1 = 2$$

or $t_1 = \frac{1}{4}$. However, the values $t_1 = \frac{1}{4}$, $t_2 = 0$ do not satisfy the third equation in (2), so there is no simultaneous solution to the three equations. Thus, the lines do not intersect.   ◀

Two lines in 3-space that are not parallel and do not intersect (such as those in Example 3) are called *skew* lines. As illustrated in Figure 14.5.3, any two skew lines lie in parallel planes.

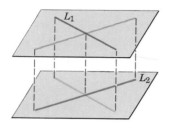

Parallel planes containing skew lines L_1 and L_2 may be determined by translating each line until it intersects the other.

Figure 14.5.3

☐ **LINE SEGMENTS**

Sometimes one is not interested in an entire line, but rather some *segment* of a line. Parametric equations of a line segment can be obtained by finding parametric equations for the entire line, then restricting the parameter appropriately so that only the desired segment is generated as the parameter varies.

Example 4   Find parametric equations for the line segment that joins the points $P_1(2, 4, -1)$ and $P_2(5, 0, 7)$.

Solution.   From Example 2, the line through P_1 and P_2 has parametric equations $x = 2 + 3t$, $y = 4 - 4t$, $z = -1 + 8t$. With these equations, the point P_1 corresponds to $t = 0$ and P_2 to $t = 1$. Thus, the line segment from P_1 to P_2 is given by

$$x = 2 + 3t, \quad y = 4 - 4t, \quad z = -1 + 8t \qquad (0 \le t \le 1) \quad ◀$$

☐ **VECTOR EQUATIONS OF LINES**

We shall now show how vector notation can be used to express the parametric equations of a line in a more compact form. Because two vectors are equal if and only if their components are equal, (1a) and (1b) can be written as

$$\langle x, y \rangle = \langle x_0 + at, y_0 + bt \rangle$$

and
$$\langle x, y, z \rangle = \langle x_0 + at, y_0 + bt, z_0 + ct \rangle$$
or, equivalently, as
$$\langle x, y \rangle = \langle x_0, y_0 \rangle + t\langle a, b \rangle \tag{3a}$$
and
$$\langle x, y, z \rangle = \langle x_0, y_0, z_0 \rangle + t\langle a, b, c \rangle \tag{3b}$$
In 2-space let us introduce the vectors $\mathbf{r}$, $\mathbf{r}_0$, and $\mathbf{v}$ given by
$$\mathbf{r} = \langle x, y \rangle, \quad \mathbf{r}_0 = \langle x_0, y_0 \rangle, \quad \mathbf{v} = \langle a, b \rangle \tag{4a}$$
and in 3-space
$$\mathbf{r} = \langle x, y, z \rangle, \quad \mathbf{r}_0 = \langle x_0, y_0, z_0 \rangle, \quad \mathbf{v} = \langle a, b, c \rangle \tag{4b}$$
Substituting (4a) and (4b) in (3a) and (3b), respectively, yields the equation
$$\mathbf{r} = \mathbf{r}_0 + t\mathbf{v} \tag{5}$$

in both cases. We call this the **vector equation of a line** in 2-space or 3-space. In this equation $\mathbf{v}$ is a nonzero vector parallel to the line and $\mathbf{r}_0$ is a vector whose components are the coordinates of a point on the line.

Figure 14.5.4 illustrates how Equation (5) can be interpreted geometrically: $\mathbf{r}_0$ can be viewed as a vector from the origin to a point P_0 on L, $t\mathbf{v}$ is a scalar multiple of a nonzero vector $\mathbf{v}$ that is parallel to L, and $\mathbf{r} = \mathbf{r}_0 + t\mathbf{v}$ can be interpreted as a vector from the origin to a point on L. As the parameter t varies from $-\infty$ to $+\infty$, the terminal point of $\mathbf{r}$ traces out the line L.

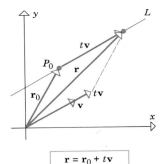

$$\mathbf{r} = \mathbf{r}_0 + t\mathbf{v}$$

Figure 14.5.4

Example 5 The equation
$$\langle x, y, z \rangle = \langle -1, 0, 2 \rangle + t\langle 1, 5, -4 \rangle$$
is of form (5) with
$$\mathbf{r}_0 = \langle -1, 0, 2 \rangle \quad \text{and} \quad \mathbf{v} = \langle 1, 5, -4 \rangle$$
Thus, the equation represents the line in 3-space that passes through the point $(-1, 0, 2)$ and is parallel to the vector $\langle 1, 5, -4 \rangle$. ◄

Example 6 Find a vector equation of the line in 3-space that passes through the points $P_1(2, 4, -1)$ and $P_2(5, 0, 7)$.

Solution. The vector
$$\overrightarrow{P_1P_2} = \langle 3, -4, 8 \rangle$$
is parallel to the line, so it can be used as $\mathbf{v}$ in (5). For $\mathbf{r}_0$ we can use either the vector from the origin to P_1 or the vector from the origin to P_2. Using the former yields
$$\mathbf{r}_0 = \langle 2, 4, -1 \rangle$$
Thus, a vector equation of the line through P_1 and P_2 is
$$\langle x, y, z \rangle = \langle 2, 4, -1 \rangle + t\langle 3, -4, 8 \rangle$$
(Note that equating corresponding components on the two sides of this vector equation yields the parametric equations for the line that were obtained in Example 2.) ◄

▶ Exercise Set 14.5   Ⓒ 53, 54, 57

In Exercises 1–8, find parametric equations for the line through P_1 and P_2.

1. $P_1(3, -2)$, $P_2(5, 1)$.   **2.** $P_1(0, 1)$, $P_2(-3, -4)$.

3. $P_1(4, 1)$, $P_2(4, 3)$.   **4.** $P_1(5, 2)$, $P_2(3, 7)$.

5. $P_1(5, -2, 1)$, $P_2(2, 4, 2)$.

6. $P_1(-1, 3, 5)$, $P_2(-1, 3, 2)$.

7. $P_1(0, 0, 0)$, $P_2(-1, 6, 1)$.

8. $P_1(4, 0, 7)$, $P_2(-1, -1, 2)$.

In Exercises 9–16, find parametric equations for the line segment joining P_1 and P_2.

9. $P_1(3, -2)$, $P_2(5, 1)$.   **10.** $P_1(0, 1)$, $P_2(-3, -4)$.

11. $P_1(4, 1)$, $P_2(4, 3)$.   **12.** $P_1(5, 2)$, $P_2(3, 7)$.

13. $P_1(5, -2, 1)$, $P_2(2, 4, 2)$.

14. $P_1(-1, 3, 5)$, $P_2(-1, 3, 2)$.

15. $P_1(0, 0, 0)$, $P_2(-1, 6, 1)$.

16. $P_1(4, 0, 7)$, $P_2(-1, -1, 2)$.

In Exercises 17–25, find parametric equations for the line.

17. The line through $(-5, 2)$ and parallel to $2\mathbf{i} - 3\mathbf{j}$.

18. The line through $(0, 3)$ and parallel to the line $x = -5 + t$, $y = 1 - 2t$.

19. The line tangent to the circle $x^2 + y^2 = 25$ at the point $(3, -4)$.

20. The line tangent to the parabola $y = x^2$ at the point $(-2, 4)$.

21. The line through $(-1, 2, 4)$ and parallel to $3\mathbf{i} - 4\mathbf{j} + \mathbf{k}$.

22. The line through $(2, -1, 5)$ and parallel to $\langle -1, 2, 7 \rangle$.

23. The line through $(-2, 0, 5)$ and parallel to the line $x = 1 + 2t$, $y = 4 - t$, $z = 6 + 2t$.

24. The line through the origin and parallel to the line $x = t$, $y = -1 + t$, $z = 2$.

25. The line through $(3, 7, 0)$ and parallel to the x-axis.

26. Where does the line $x = 1 + 3t$, $y = 2 - t$ intersect
(a)   the x-axis   (b)   the y-axis?

27. Where does the line $x = 2t$, $y = 3 + 4t$ intersect the parabola $y = x^2$?

28. Where does the line $x = -1 + 2t$, $y = 3 + t$, $z = 4 - t$ intersect
(a)   the xy-plane   (b)   the xz-plane
(c)   the yz-plane?

29. Where does the line $x = -2$, $y = 4 + 2t$, $z = -3 + t$ intersect
(a)   the xy-plane   (b)   the xz-plane
(c)   the yz-plane?

30. Where does the line $x = 2 - t$, $y = 3t$, $z = -1 + 2t$ intersect the plane $2y + 3z = 6$?

31. Where does the line $x = 1 + t$, $y = 3 - t$, $z = 2t$ intersect the cylinder $x^2 + y^2 = 16$?

32. Find parametric equations for the line through (x_0, y_0, z_0) and (x_1, y_1, z_1).

33. Find parametric equations for the line through (x_1, y_1, z_1) and parallel to the line
$$x = x_0 + at, \quad y = y_0 + bt, \quad z = z_0 + ct$$

34. Prove: If a, b, and c are nonzero, then each point on the line $x = x_0 + at$, $y = y_0 + bt$, $z = z_0 + ct$ satisfies
$$\frac{x - x_0}{a} = \frac{y - y_0}{b} = \frac{z - z_0}{c}$$
and conversely, each point (x, y, z) satisfying these equations lies on the line. (These are called the **symmetric equations** of the line.)

35. Show that the lines
$$x = 2 + t, \quad y = 2 + 3t, \quad z = 3 + t$$
and
$$x = 2 + t, \quad y = 3 + 4t, \quad z = 4 + 2t$$
intersect and find the point of intersection.

36. Show that the lines
$$x + 1 = 4t, \quad y - 3 = t, \quad z - 1 = 0$$
and
$$x + 13 = 12t, \quad y - 1 = 6t, \quad z - 2 = 3t$$
intersect and find the point of intersection.

37. Show that the lines
$$x = 1 + 7t, \quad y = 3 + t, \quad z = 5 - 3t$$
and
$$x = 4 - t, \quad y = 6, \quad z = 7 + 2t$$
are skew.

38. Show that the lines
$$x = 2 + 8t, \quad y = 6 - 8t, \quad z = 10t$$
and
$$x = 3 + 8t, \quad y = 5 - 3t, \quad z = 6 + t$$
are skew.

39. Determine whether P_1, P_2, and P_3 lie on the same line.
(a) $P_1(6, 9, 7)$, $P_2(9, 2, 0)$, $P_3(0, -5, -3)$
(b) $P_1(1, 0, 1)$, $P_2(3, -4, -3)$, $P_3(4, -6, -5)$.

40. Find k_1 and k_2 so that the point $(k_1, 1, k_2)$ lies on the line passing through $(0, 2, 3)$ and $(2, 7, 5)$.

41. Find the point on the line segment joining $P_1(1, 4, -3)$ and $P_2(1, 5, -1)$ that is $\frac{2}{3}$ of the way from P_1 to P_2.

42. In each part, determine whether the lines are parallel.
(a) $x = 3 - 2t$, $\quad y = 4 + t$, $\quad z = 6 - t$
and
$x = 5 - 4t$, $\quad y = -2 + 2t$, $\quad z = 7 - 2t$

(b) $x = 5 + 3t, \quad y = 4 - 2t, \quad z = -2 + 3t$
and
$x = -1 + 9t, \quad y = 5 - 6t, \quad z = 3 + 8t.$

43. Show that the equations
$$x = 3 - t, \quad y = 1 + 2t$$
and
$$x = -1 + 3t, \quad y = 9 - 6t$$
represent the same line.

44. Show that the equations
$$x = 1 + 3t, \quad y = -2 + t, \quad z = 2t$$
and
$$x = 4 - 6t, \quad y = -1 - 2t, \quad z = 2 - 4t$$
represent the same line.

45. Find a vector equation of the line in 2-space that passes through the points P_1 and P_2.
(a) $P_1(2, -1), \; P_2(-5, 3)$
(b) $P_1(0, 3), \; P_2(4, 3).$

46. Find a vector equation of the line in 3-space that passes through the points P_1 and P_2.
(a) $P_1(3, -1, 2), \; P_2(0, 1, 1)$
(b) $P_1(2, 4, 1), \; P_2(2, -1, 1).$

47. Describe the line segment represented by the vector equation $\langle x, y \rangle = \langle 1, 0 \rangle + t\langle -2, 3 \rangle, \; 0 \le t \le 2.$

48. Describe the line segment represented by the vector equation $\langle x, y, z \rangle = \langle -2, 1, 4 \rangle + t\langle 3, 0, -1 \rangle, \; 0 \le t \le 3.$

In Exercises 49–52, use the result in Exercise 16, Section 14.4.

49. Find the distance between the point $(-2, 1, 1)$ and the line $x = 3 - t, \, y = t, \, z = 1 + 2t.$

50. Find the distance between the point $(1, 4, -3)$ and the line $x = 2 + t, \, y = -1 - t, \, z = 3t.$

51. Verify that the lines $x = 2 - t, \; y = 2t, \; z = 1 + t$ and $x = 1 + 2t, \, y = 3 - 4t, \, z = 5 - 2t$ are parallel, and find the distance between them.

52. Verify that the lines $x = 2t, \, y = 3 + 4t, \, z = 2 - 6t$ and $x = 1 + 3t, \, y = 6t, \, z = -9t$ are parallel, and find the distance between them.

53. Let L_1 and L_2 be the lines whose parametric equations are
$$x = 1 + 2t, \quad y = 2 - t, \quad z = 4 - 2t$$
and
$$x = 9 + t, \quad y = 5 + 3t, \quad z = -4 - t$$
(a) Show that L_1 and L_2 intersect at the point $(7, -1, -2).$
(b) Find, to the nearest degree, the acute angle between L_1 and L_2 at their intersection.
(c) Find parametric equations for the line that is perpendicular to L_1 and L_2 and passes through their point of intersection.

54. Let L_1 and L_2 be the lines whose parametric equations are
$$x = 4t, \quad y = 1 - 2t, \quad z = 2 + 2t$$
and
$$x = 1 + t, \quad y = 1 - t, \quad z = -1 + 4t$$
(a) Show that L_1 and L_2 intersect at the point $(2, 0, 3).$
(b) Find, to the nearest degree, the acute angle between L_1 and L_2 at their intersection.
(c) Find parametric equations for the line that is perpendicular to L_1 and L_2 and passes through their point of intersection.

55. Find parametric equations for the line that contains the point $(0, 2, 1)$ and intersects the line $x = 2t, \, y = 1 - t, \, z = 2 + t$ at a right angle.

56. Find parametric equations for the line that contains the point $(3, 1, -2)$ and intersects the line $x = -2 + 2t, \, y = 4 + 2t, \, z = 2 + t$ at a right angle.

57. Let $x = 4 - t, \, y = 1 + 2t, \, z = 2 + t$ and $x = t, \, y = 1 + t, \, z = 1 + 2t$ be the straight-line paths of motion of two particles. Suppose that t is the time in seconds and that x, y, and z are measured in centimeters.
(a) How far apart are the particles when $t = 0$?
(b) How close can the particles get?

■ 14.6 PLANES IN 3-SPACE

In this section we shall use vectors to derive equations of planes in 3-space, and we shall use these equations to solve some basic geometric problems.

□ **PLANES PARALLEL TO THE COORDINATE PLANES**

The plane parallel to the xy-plane and passing through the point $(0, 0, k)$ on the z-axis consists of all points (x, y, z) for which $z = k$ (Figure 14.6.1); thus, the equation of this plane is $z = k$. Similarly, $x = k$ represents a plane parallel to the yz-plane and passing through $(k, 0, 0)$, while $y = k$ represents a plane parallel to the xz-plane and passing through $(0, k, 0).$

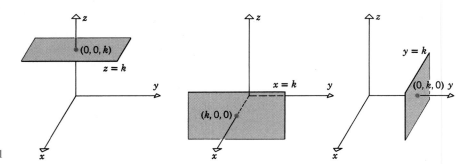

Figure 14.6.1

☐ **PLANES DETERMINED BY A POINT AND A NORMAL VECTOR**

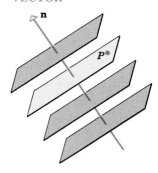

The colored plane is uniquely determined by the point P and the vector $\mathbf{n}$ perpendicular to the plane.

Figure 14.6.2

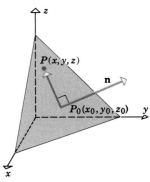

Figure 14.6.3

A plane in 3-space is uniquely determined by specifying a point in the plane and a vector perpendicular to the plane (Figure 14.6.2). A vector perpendicular to a plane is called a *normal* to the plane.

Suppose that we want the equation of the plane passing through the point $P_0(x_0, y_0, z_0)$ and perpendicular to the nonzero vector $\mathbf{n} = \langle a, b, c \rangle$. It is evident from Figure 14.6.3 that the plane consists precisely of those points $P(x, y, z)$ for which the vector $\overrightarrow{P_0P}$ is perpendicular to $\mathbf{n}$; or, phrased as an equation,

$$\mathbf{n} \cdot \overrightarrow{P_0P} = 0 \tag{1}$$

Since $\overrightarrow{P_0P} = \langle x - x_0, y - y_0, z - z_0 \rangle$, (1) can be rewritten as

$$a(x - x_0) + b(y - y_0) + c(z - z_0) = 0 \tag{2}$$

We shall call this the *point-normal form* of the equation of a plane.

Example 1   Find an equation of the plane passing through the point $(3, -1, 7)$ and perpendicular to the vector $\mathbf{n} = \langle 4, 2, -5 \rangle$.

Solution.   From (2), a point-normal form of the equation is

$$4(x - 3) + 2(y + 1) - 5(z - 7) = 0 \quad \blacktriangleleft$$

By multiplying out and collecting terms, (2) can be rewritten in the form

$$ax + by + cz + d = 0 \tag{3}$$

where a, b, c, and d are constants, and a, b, and c are not all zero. To illustrate, the equation in Example 1 can be rewritten as

$$4x + 2y - 5z + 25 = 0$$

As our next theorem shows, every equation of form (3) represents a plane in 3-space.

14.6.1   THEOREM.   *If a, b, c, and d are constants, and a, b, and c are not all zero, then the graph of the equation*

$$ax + by + cz + d = 0$$

is a plane having the vector $\mathbf{n} = \langle a, b, c \rangle$ as a normal.

Proof.   By hypothesis, a, b, and c are not all zero. Assume, for the moment, that $a \neq 0$, and note that the equation $ax + by + cz + d = 0$ can be rewritten in the form $a[x + (d/a)] + by + cz = 0$. But this is a point-normal form of the plane passing through the point $(-d/a, 0, 0)$ and having $\mathbf{n} = \langle a, b, c \rangle$ as a normal.

If $a = 0$, then either $b \neq 0$ or $c \neq 0$. A straightforward modification of the above argument will handle these other cases. ▪

The equation in Theorem 14.6.1 is called the **general form** of the equation of a plane.

Example 2 Determine whether the planes

$$3x - 4y + 5z = 0 \quad \text{and} \quad -6x + 8y - 10z - 4 = 0$$

are parallel.

Solution.   It is clear geometrically that two planes are parallel if and only if their normals are parallel vectors. A normal to the first plane is

$$\mathbf{n}_1 = \langle 3, -4, 5 \rangle$$

and a normal to the second plane is

$$\mathbf{n}_2 = \langle -6, 8, -10 \rangle$$

Since $\mathbf{n}_2$ is a scalar multiple of $\mathbf{n}_1$, the normals are parallel. Thus, the planes are also parallel. ◀

A plane cannot be specified by giving a point on it and *one* vector parallel to it, since there are infinitely many such planes (Figure 14.6.4a). However, as shown in Figure 14.6.4b, a plane is uniquely determined by giving a point in the plane and *two* nonparallel vectors that are parallel to the plane. A plane is also uniquely determined by specifying three noncollinear points in the plane (Figure 14.6.4c).

Example 3   Find an equation of the plane through the points $P_1(1, 2, -1)$, $P_2(2, 3, 1)$, and $P_3(3, -1, 2)$.

Solution.   Since the points P_1, P_2, and P_3 lie in the plane, the vectors $\overrightarrow{P_1P_2} = \langle 1, 1, 2 \rangle$ and $\overrightarrow{P_1P_3} = \langle 2, -3, 3 \rangle$ are parallel to the plane. Therefore,

$$\overrightarrow{P_1P_2} \times \overrightarrow{P_1P_3} = \begin{vmatrix} \mathbf{i} & \mathbf{j} & \mathbf{k} \\ 1 & 1 & 2 \\ 2 & -3 & 3 \end{vmatrix} = 9\mathbf{i} + \mathbf{j} - 5\mathbf{k}$$

is normal to the plane, since it is perpendicular to both $\overrightarrow{P_1P_2}$ and $\overrightarrow{P_1P_3}$. By using this normal and the point $P_1(1, 2, -1)$ in the plane, we obtain the point-normal form

$$9(x - 1) + (y - 2) - 5(z + 1) = 0$$

which may be rewritten as

$$9x + y - 5z - 16 = 0 \quad ◀$$

Example 4   Determine whether the line

$$x = 3 + 8t, \quad y = 4 + 5t, \quad z = -3 - t$$

is parallel to the plane $x - 3y + 5z = 12$.

Solution.   The vector $\mathbf{v} = \langle 8, 5, -1 \rangle$ is parallel to the line and the vector $\mathbf{n} = \langle 1, -3, 5 \rangle$ is normal to the plane. In order for the line and plane to be parallel, the vectors $\mathbf{v}$ and $\mathbf{n}$ must be perpendicular. But this is not so, since the dot product

$$\mathbf{v} \cdot \mathbf{n} = (8)(1) + (5)(-3) + (-1)(5) = -12$$

is nonzero. Thus, the line and plane are not parallel. ◀

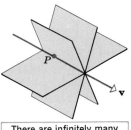

There are infinitely many planes containing P and parallel to **v**.

(a)

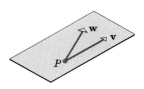

There is a unique plane through P that is parallel to both **v** and **w**.

(b)

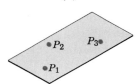

There is a unique plane through three noncollinear points.

(c)

Figure 14.6.4

Example 5   Find the intersection of the line and plane in Example 4.

Solution.   If we let (x_0, y_0, z_0) be the point of intersection, then the coordinates of this point satisfy both the equation of the plane and the parametric equations of the line. Thus,

$$x_0 - 3y_0 + 5z_0 = 12 \tag{4}$$

and for some value of t, say $t = t_0$,

$$x_0 = 3 + 8t_0, \quad y_0 = 4 + 5t_0, \quad z_0 = -3 - t_0 \tag{5}$$

Substituting (5) in (4) yields

$$(3 + 8t_0) - 3(4 + 5t_0) + 5(-3 - t_0) = 12$$

Solving for t_0 yields $t_0 = -3$ and on substituting this value in (5), we obtain

$$(x_0, y_0, z_0) = (-21, -11, 0) \quad \blacktriangleleft$$

Two intersecting planes determine two angles of intersection, an acute angle θ ($0 \leq \theta \leq 90°$) and its supplement $180° - \theta$ (Figure 14.6.5a). If $\mathbf{n}_1$ and $\mathbf{n}_2$ are normals to the planes, then the angle between $\mathbf{n}_1$ and $\mathbf{n}_2$ is θ or $180° - \theta$, depending on the directions of the normals (Figure 14.6.5b). Thus, the angles of intersection of two planes may be determined from the normals.

Example 6   Find the acute angle of intersection between the two planes

$$2x - 4y + 4z = 7 \quad \text{and} \quad 6x + 2y - 3z = 2$$

Solution.   From the given equations, we obtain the normals $\mathbf{n}_1 = \langle 2, -4, 4 \rangle$ and $\mathbf{n}_2 = \langle 6, 2, -3 \rangle$. Since the dot product $\mathbf{n}_1 \cdot \mathbf{n}_2 = -8$ is negative, the normals make an obtuse angle. To obtain normals making an acute angle, we can reverse the direction of either $\mathbf{n}_1$ or $\mathbf{n}_2$. To be specific, let us reverse the direction of $\mathbf{n}_1$. Thus, the acute angle θ between the planes will be the angle between $-\mathbf{n}_1 = \langle -2, 4, -4 \rangle$ and $\mathbf{n}_2 = \langle 6, 2, -3 \rangle$. From (3) of Section 14.3 we obtain

$$\cos \theta = \frac{(-\mathbf{n}_1) \cdot \mathbf{n}_2}{\|-\mathbf{n}_1\| \|\mathbf{n}_2\|} = \frac{8}{\sqrt{36}\sqrt{49}} = \frac{4}{21}$$

With the aid of a calculator one obtains

$$\theta = \cos^{-1}\left(\frac{4}{21}\right) \approx 79° \quad \blacktriangleleft$$

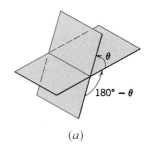

(a)

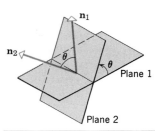

In this figure the angle between $\mathbf{n}_1$ and $\mathbf{n}_2$ is θ. If the direction of one of the normals is reversed, then the angle between the normals would be $180° - \theta$.

(b)

Figure 14.6.5

☐ **DISTANCE PROBLEMS INVOLVING PLANES**

We conclude this section by discussing three basic "distance problems" in 3-space:

- Find the distance between a point and a plane.
- Find the distance between two parallel planes.
- Find the distance between two skew lines.

The three problems are related. If we can find the distance between a point and a plane, then we can find the distance between parallel planes by computing the distance between one of the planes and an arbitrary point P_0 in the other plane (Figure 14.6.6a). Moreover, we can find the distance between two skew lines by computing the distance between parallel planes containing them (Figure 14.6.6b).

14.6.2   THEOREM.   *The distance D between a point $P_0(x_0, y_0, z_0)$ and the plane $ax + by + cz + d = 0$ is*

$$D = \frac{|ax_0 + by_0 + cz_0 + d|}{\sqrt{a^2 + b^2 + c^2}} \tag{6}$$

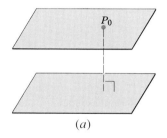

(a)

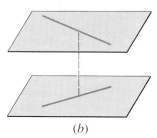

(b)

Figure 14.6.6

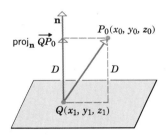

Figure 14.6.7

Proof. Let $Q(x_1, y_1, z_1)$ be any point in the plane, and position the normal $\mathbf{n} = \langle a, b, c \rangle$ so that its initial point is at Q. As illustrated in Figure 14.6.7, the distance D is equal to the length of the orthogonal projection of $\overrightarrow{QP_0}$ on $\mathbf{n}$. Thus, from (10) of Section 14.3,

$$D = \|\text{proj}_{\mathbf{n}} \overrightarrow{QP_0}\| = \frac{|\overrightarrow{QP_0} \cdot \mathbf{n}|}{\|\mathbf{n}\|}$$

But

$$\overrightarrow{QP_0} = \langle x_0 - x_1, y_0 - y_1, z_0 - z_1 \rangle$$

$$\overrightarrow{QP_0} \cdot \mathbf{n} = a(x_0 - x_1) + b(y_0 - y_1) + c(z_0 - z_1)$$

$$\|\mathbf{n}\| = \sqrt{a^2 + b^2 + c^2}$$

Thus,

$$D = \frac{|a(x_0 - x_1) + b(y_0 - y_1) + c(z_0 - z_1)|}{\sqrt{a^2 + b^2 + c^2}} \tag{7}$$

Since the point $Q(x_1, y_1, z_1)$ lies in the plane, its coordinates satisfy the equation of the plane, so that

$$ax_1 + by_1 + cz_1 + d = 0$$

or

$$d = -ax_1 - by_1 - cz_1$$

Substituting this expression in (7) yields (6). ∎

REMARK.   See Exercise 31 for an analog of Formula (6) in 2-space.

Example 7   Find the distance D between the point $(1, -4, -3)$ and the plane

$$2x - 3y + 6z = -1$$

Solution. To apply (6), we first rewrite the equation of the plane in the form

$$2x - 3y + 6z + 1 = 0$$

Then

$$D = \frac{|(2)(1) + (-3)(-4) + 6(-3) + 1|}{\sqrt{2^2 + (-3)^2 + 6^2}} = \frac{|-3|}{7} = \frac{3}{7} \blacktriangleleft$$

Example 8   The planes

$$x + 2y - 2z = 3 \quad \text{and} \quad 2x + 4y - 4z = 7$$

are parallel since their normals, $\langle 1, 2, -2 \rangle$ and $\langle 2, 4, -4 \rangle$, are parallel vectors. Find the distance between these planes.

Solution. To find the distance D between the planes, we may select an arbitrary point in one of the planes and compute its distance to the other plane. By setting $y = z = 0$ in the equation $x + 2y - 2z = 3$, we obtain the point $P_0(3, 0, 0)$ in this plane. From (6), the distance from P_0 to the plane $2x + 4y - 4z = 7$ is

$$D = \frac{|(2)(3) + 4(0) + (-4)(0) - 7|}{\sqrt{2^2 + 4^2 + (-4)^2}} = \frac{1}{6} \blacktriangleleft$$

Example 9   It was shown in Example 3 of Section 14.5 that the lines

$$L_1: x = 1 + 4t, \quad y = 5 - 4t, \quad z = -1 + 5t$$

and

$$L_2: \quad x = 2 + 8t, \quad y = 4 - 3t, \quad z = 5 + t$$

are skew. Find the distance between them.

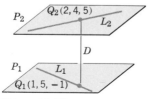

Figure 14.6.8

Solution. Let P_1 and P_2 denote parallel planes containing L_1 and L_2, respectively (Figure 14.6.8). To find the distance D between L_1 and L_2, we shall calculate the distance from a point in P_1 to the plane P_2. Setting $t = 0$ in the equations of L_1 yields the point $Q_1(1, 5, -1)$ in plane P_1. (The value $t = 0$ was chosen for simplicity. Any value of t will suffice.) Next we shall obtain an equation for plane P_2.

The vector $\mathbf{u}_1 = \langle 4, -4, 5 \rangle$ is parallel to line L_1, and therefore also parallel to planes P_1 and P_2. Similarly, $\mathbf{u}_2 = \langle 8, -3, 1 \rangle$ is parallel to L_2 and hence parallel to P_1 and P_2. Therefore, the cross product

$$\mathbf{n} = \mathbf{u}_1 \times \mathbf{u}_2 = \begin{vmatrix} \mathbf{i} & \mathbf{j} & \mathbf{k} \\ 4 & -4 & 5 \\ 8 & -3 & 1 \end{vmatrix} = 11\mathbf{i} + 36\mathbf{j} + 20\mathbf{k}$$

is normal to both P_1 and P_2. Using this normal and the point $Q_2(2, 4, 5)$ found by setting $t = 0$ in the equations of L_2, we obtain an equation for P_2:

$$11(x - 2) + 36(y - 4) + 20(z - 5) = 0$$

or

$$11x + 36y + 20z - 266 = 0$$

The distance between $Q_1(1, 5, -1)$ and this plane is

$$D = \frac{|(11)(1) + (36)(5) + (20)(-1) - 266|}{\sqrt{11^2 + 36^2 + 20^2}} = \frac{95}{\sqrt{1817}}$$

This is also the distance between L_1 and L_2. ◀

▶ **Exercise Set 14.6** ☐ 11

In Exercises 1–4, find an equation of the plane passing through P and having $\mathbf{n}$ as a normal.

1. $P(2, 6, 1)$; $\mathbf{n} = \langle 1, 4, 2 \rangle$.

2. $P(-1, -1, 2)$; $\mathbf{n} = \langle -1, 7, 6 \rangle$.

3. $P(1, 0, 0)$; $\mathbf{n} = \langle 0, 0, 1 \rangle$.

4. $P(0, 0, 0)$; $\mathbf{n} = \langle 2, -3, -4 \rangle$.

5. Find an equation of the plane passing through the given points.

    (a) $(-2, 1, 1)$, $(0, 2, 3)$, and $(1, 0, -1)$

    (b) $(3, 2, 1)$, $(2, 1, -1)$, and $(-1, 3, 2)$.

6. Determine whether the planes are parallel.

    (a) $3x - 2y + z = 4$ and $6x - 4y + 3z = 7$

    (b) $2x - 8y - 6z - 2 = 0$ and $-x + 4y + 3z - 5 = 0$

    (c) $y = 4x - 2z + 3$ and $x = \frac{1}{4}y + \frac{1}{2}z$.

7. Determine whether the line and plane are parallel.

    (a) $x = 4 + 2t$, $y = -t$, $z = -1 - 4t$; $3x + 2y + z - 7 = 0$

    (b) $x = t$, $y = 2t$, $z = 3t$; $x - y + 2z = 5$.

8. Determine whether the planes are perpendicular.

    (a) $x - y + 3z - 2 = 0$, $2x + z = 1$

    (b) $3x - 2y + z = 1$, $4x + 5y - 2z = 4$.

9. Determine whether the line and plane are perpendicular.

    (a) $x = -1 + 2t$, $y = 4 + t$, $z = 1 - t$; $4x + 2y - 2z = 7$

    (b) $x = 3 - t$, $y = 2 + t$, $z = 1 - 3t$; $2x + 2y - 5 = 0$.

10. Find the point of intersection of the line and plane.

    (a) $x = t$, $y = t$, $z = t$; $3x - 2y + z - 5 = 0$

    (b) $x = 1 + t$, $y = -1 + 3t$, $z = 2 + 4t$; $x - y + 4z = 7$

    (c) $x = 2 - t$, $y = 3 + t$, $z = t$; $2x + y + z = 1$.

11. Find the acute angle of intersection of the planes (to the nearest degree).

    (a) $x = 0$ and $2x - y + z - 4 = 0$

    (b) $x + 2y - 2z = 5$ and $6x - 3y + 2z = 8$.

12. Find an equation of the plane through $(-1, 4, -3)$ and perpendicular to the line $x - 2 = t$, $y + 3 = 2t$, $z = -t$.

13. Find an equation of
   (a) the xy-plane   (b) the xz-plane
   (c) the yz-plane.

14. Find an equation of the plane that contains the point (x_0, y_0, z_0) and is
   (a) parallel to the xy-plane
   (b) parallel to the yz-plane
   (c) parallel to the xz-plane.

15. Find an equation of the plane through the origin that is parallel to the plane $4x - 2y + 7z + 12 = 0$.

16. Find an equation of the plane containing the line $x = -2 + 3t$, $y = 4 + 2t$, $z = 3 - t$ and perpendicular to the plane $x - 2y + z = 5$.

17. Find an equation of the plane through $(-1, 4, 2)$ and containing the line of intersection of the planes $4x - y + z - 2 = 0$ and $2x + y - 2z - 3 = 0$.

18. Show that the points $(1, 0, -1)$, $(0, 2, 3)$, $(-2, 1, 1)$, and $(4, 2, 3)$ lie in the same plane.

19. Find parametric equations of the line through $(5, 0, -2)$ that is parallel to the planes $x - 4y + 2z = 0$ and $2x + 3y - z + 1 = 0$.

20. Find an equation of the plane through $(-1, 2, -5)$ and perpendicular to the planes $2x - y + z = 1$ and $x + y - 2z = 3$.

21. Find an equation of the plane through $(1, 2, -1)$ and perpendicular to the line of intersection of the planes $2x + y + z = 2$ and $x + 2y + z = 3$.

22. Find a plane through the points $P_1(-2, 1, 4)$, $P_2(1, 0, 3)$ and perpendicular to the plane $4x - y + 3z = 2$.

23. Show that the lines

$$x = -2 + t, \quad y = 3 + 2t, \quad z = 4 - t$$

and

$$x = 3 - t, \quad y = 4 - 2t, \quad z = t$$

are parallel and find an equation of the plane they determine.

24. Find an equation of the plane that contains the point $(2, 0, 3)$ and the line $x = -1 + t$, $y = t$, $z = -4 + 2t$.

25. Find an equation of the plane, each of whose points is equidistant from $(2, -1, 1)$ and $(3, 1, 5)$.

26. Find an equation of the plane containing the line $x = 3t$, $y = 1 + t$, $z = 2t$ and parallel to the intersection of the planes $2x - y + z = 0$ and $y + z + 1 = 0$.

27. Show that the line $x = 0$, $y = t$, $z = t$
   (a) lies in the plane $6x + 4y - 4z = 0$
   (b) is parallel to and below the plane $5x - 3y + 3z = 1$
   (c) is parallel to and above the plane $6x + 2y - 2z = 3$.

28. Show that the lines
$$x + 1 = 4t, \quad y - 3 = t, \quad z - 1 = 0$$
and
$$x + 13 = 12t, \quad y - 1 = 6t, \quad z - 2 = 3t$$
intersect and find an equation of the plane they determine.

29. Find parametric equations of the line of intersection of the planes
   (a) $-2x + 3y + 7z + 2 = 0$ and
       $x + 2y - 3z + 5 = 0$
   (b) $3x - 5y + 2z = 0$ and $z = 0$.

30. Show that the plane whose intercepts with the coordinate axes are $x = a$, $y = b$, and $z = c$ has equation

$$\frac{x}{a} + \frac{y}{b} + \frac{z}{c} = 1$$

provided a, b, and c are nonzero.

31. (a) Prove that the distance D between a point $P_0(x_0, y_0)$ and the line $ax + by + c = 0$ is given by

$$D = \frac{|ax_0 + by_0 + c|}{\sqrt{a^2 + b^2}}$$

[Hint: First establish that the vector $\mathbf{n} = a\mathbf{i} + b\mathbf{j}$ is normal to the line by showing that for any distinct points $P_1(x_1, y_1)$ and $P_2(x_2, y_2)$ on the line, the vector $\mathbf{n}$ is perpendicular to $\overrightarrow{P_1P_2}$.]

   (b) Use the formula in part (a) to find the distance between the point $P(1, -2)$ and the line $3x + 4y + 7 = 0$.

32. Use the formula in part (a) of Exercise 31 to find the distance between the point $P(-3, 5)$ and the line $y = -2x + 1$.

In Exercises 33–35, find the distance between the point and the plane.

33. $(1, -2, 3)$; $2x - 2y + z = 4$.

34. $(0, 1, 5)$; $3x + 6y - 2z - 5 = 0$.

35. $(7, 2, -1)$; $20x - 4y - 5z = 0$.

In Exercises 36–38, find the distance between the given parallel planes.

36. $2x - 3y + 4z = 7$, $4x - 6y + 8z = 3$.

37. $-2x + y + z = 0$, $6x - 3y - 3z - 5 = 0$.

38. $x + y + z = 1$, $x + y + z = -1$.

In Exercises 39–41, find the distance between the given skew lines.

39. $x = 1 + 7t$, $y = 3 + t$, $z = 5 - 3t$; $x = 4 - t$, $y = 6$, $z = 7 + 2t$.

40. $x = 3 - t$, $y = 4 + 4t$, $z = 1 + 2t$; $x = t$, $y = 3$, $z = 2t$.

41. $x = 2 + 4t$, $y = 6 - 4t$, $z = 5t$; $x = 3 + 8t$, $y = 5 - 3t$, $z = 6 + t$.

42. Show that the line $x = -1 + t$, $y = 3 + 2t$, $z = -t$ and the plane $2x - 2y - 2z + 3 = 0$ are parallel, and find the distance between them.

43. Prove: The planes
$$a_1x + b_1y + c_1z = d_1$$
and
$$a_2x + b_2y + c_2z = d_2$$
are perpendicular if and only if $a_1a_2 + b_1b_2 + c_1c_2 = 0$.

44. Let $\mathbf{r}_0 = \langle x_0, y_0, z_0 \rangle$ and $\mathbf{r} = \langle x, y, z \rangle$. Describe the set of all points (x, y, z) for which

(a) $\mathbf{r} \cdot \mathbf{r}_0 = 0$ (b) $(\mathbf{r} - \mathbf{r}_0) \cdot \mathbf{r}_0 = 0$.

45. Find an equation of the sphere with its center at $(2, 1, -3)$ and tangent to the plane $x - 3y + 2z = 4$.

46. Find where the line that passes through the point $(3, 1, 0)$ and is normal to the plane $2x + y - z = 0$ intersects the plane.

■ **14.7** QUADRIC SURFACES

In this section we shall study an important class of surfaces that are the three-dimensional analogs of the conic sections.

☐ MESH PLOTS OF SURFACES

In 2-space the general shape of a curve can be obtained by plotting points. However, for surfaces in 3-space, point-plotting is not generally helpful since too many points are needed to obtain even a crude picture of the surface. It is better to build up the shape of the surface using curves obtained by cutting the surface with planes parallel to the coordinate planes. This is called a ***mesh plot*** of the surface. For example, Figure 14.7.1 shows a mesh plot of the surface $z = x^3 - 3xy^2$. (This surface is called a "monkey saddle" because a monkey sitting astride the x-axis has a place for its two legs and tail.)

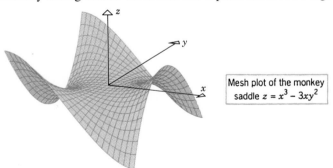

Mesh plot of the monkey saddle $z = x^3 - 3xy^2$

Figure 14.7.1

☐ TRACES OF SURFACES

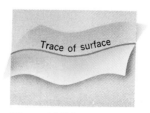

Trace of surface

Figure 14.7.2

The curve of intersection of a surface with a plane is called the ***trace*** of the surface in the plane (Figure 14.7.2). Mesh plots are usually built up from traces in planes parallel to the coordinate planes. Equations for such traces can be obtained by substituting the equation of the plane into the equation of the surface. For example, the trace of the monkey saddle of Figure 14.7.1 in the plane $x = 3$ is obtained by substituting $x = 3$ into

$$z = x^3 - 3xy^2$$

which yields

$$z - 27 = -9y^2 \quad (x = 3) \tag{1}$$

This is a parabola with vertex at the point $(x, y, z) = (3, 0, 27)$ opening in the negative z-direction (why?). Observe that this result is consistent with Figure 14.7.1.

REMARK. In (1) we explicitly noted the restriction $x = 3$ in parentheses. This is necessary because the equation by itself does not convey the information that the trace lies in the plane $x = 3$.

THE QUADRIC SURFACES

Earlier in the text, we saw that the graph in two dimensions of a second-degree equation in x and y,

$$Ax^2 + Bxy + Cy^2 + Dx + Ey + F = 0$$

is a conic section (possibly degenerate). In three dimensions, the graph of a second-degree equation in x, y, and z,

$$Ax^2 + By^2 + Cz^2 + Dxy + Exz + Fyz + Gx + Hy + Iz + J = 0$$

is called a **quadric surface** or a **quadric**. The most important quadric surfaces, shown in Table 14.7.1, are the **ellipsoids**, **hyperboloids of one and two sheets**, **elliptic cones**, **elliptic paraboloids**, and **hyperbolic paraboloids**.

The simplest equations for the quadric surfaces result when the surfaces are positioned in certain "standard positions" relative to the coordinate axes. Table 14.7.1 illustrates

Table 14.7.1

SURFACE	EQUATION	SURFACE	EQUATION
ELLIPSOID	$\dfrac{x^2}{a^2} + \dfrac{y^2}{b^2} + \dfrac{z^2}{c^2} = 1$ The traces in the coordinate planes are ellipses, as are the traces in planes parallel to the coordinate planes.	**ELLIPTIC CONE**	$z^2 = \dfrac{x^2}{a^2} + \dfrac{y^2}{b^2}$ The trace in the xy-plane is a point (the origin), and the traces in planes parallel to the xy-plane are ellipses. The traces in the yz- and xz-planes are pairs of lines intersecting at the origin. The traces in planes parallel to these are hyperbolas.
HYPERBOLOID OF ONE SHEET	$\dfrac{x^2}{a^2} + \dfrac{y^2}{b^2} - \dfrac{z^2}{c^2} = 1$ The trace in the xy-plane is an ellipse, as are the traces in planes parallel to the xy-plane. The traces in the yz-plane and xz-plane are hyperbolas, as are the traces in planes parallel to these.	**ELLIPTIC PARABOLOID**	$z = \dfrac{x^2}{a^2} + \dfrac{y^2}{b^2}$ The trace in the xy-plane is a point (the origin), and the traces in planes parallel to and above the xy-plane are ellipses. The traces in the yz- and xz-planes are parabolas, as are the traces in planes parallel to these.
HYPERBOLOID OF TWO SHEETS	$\dfrac{x^2}{a^2} + \dfrac{y^2}{b^2} - \dfrac{z^2}{c^2} = -1$ There is no trace in the xy-plane. In planes parallel to the xy-plane, which intersect the surface, the traces are ellipses. In the yz- and xz-planes, the traces are hyperbolas, as are the traces in planes parallel to these.	**HYPERBOLIC PARABOLOID**	$z = \dfrac{y^2}{b^2} - \dfrac{x^2}{a^2}$ The trace in the xy-plane is a pair of lines intersecting at the origin. The traces in planes parallel to the xy-plane are hyperbolas. The hyperbolas above the xy-plane open in the y-direction, and those below in the x-direction. The traces in the yz- and xz-planes are parabolas, as are the traces in planes parallel to these.

some typical standard positions and the equations that result. The table also describes the traces of the quadric surfaces in planes parallel to the coordinate planes. As we shall see, such traces are important in studying properties of quadric surfaces. The constants a, b, and c that appear in the equations in the table are all assumed to be positive.

Example 1   To illustrate how the traces described in Table 14.7.1 were obtained, we shall consider the case of the elliptic cone

$$z^2 = \frac{x^2}{a^2} + \frac{y^2}{b^2} \tag{2}$$

The other cases are similar. The trace of (2) in the plane $z = 0$ (the xy-plane) is

$$\frac{x^2}{a^2} + \frac{y^2}{b^2} = 0 \qquad (z = 0)$$

which implies that $x = 0$, $y = 0$, $z = 0$. Thus, the trace is the single point $(0, 0, 0)$.

For $k \neq 0$, the trace of (2) in the plane $z = k$ is (after simplification)

$$\frac{x^2}{(ak)^2} + \frac{y^2}{(bk)^2} = 1 \qquad (z = k)$$

which is an ellipse. As $|k|$ increases, so do $(ak)^2$ and $(bk)^2$. Thus, the dimensions of these ellipses increase as the planes containing them recede from the xy-plane. This is consistent with the picture of the elliptic cone in Table 14.7.1.

The trace of (2) in the xz-plane is

$$z^2 = \frac{x^2}{a^2} \qquad (y = 0)$$

which is equivalent to the pair of equations

$$z = \frac{x}{a} \quad \text{and} \quad z = -\frac{x}{a} \qquad (y = 0)$$

These are the equations of two lines in the xz-plane that intersect at the origin. Similarly, the trace of (2) in the yz-plane is the pair of intersecting lines

$$z = \frac{y}{b} \quad \text{and} \quad z = -\frac{y}{b} \qquad (x = 0)$$

For $k \neq 0$, the trace of (2) in the plane $y = k$, which is parallel to the xz-plane, is (after simplification)

$$\frac{z^2}{k^2/b^2} - \frac{x^2}{a^2k^2/b^2} = 1 \qquad (y = k)$$

This is a hyperbola in the plane $y = k$ opening along a line parallel to the z-axis. Similarly, for $k \neq 0$ the trace in the plane $x = k$, which is parallel to the yz-plane, is (after simplification)

$$\frac{z^2}{k^2/a^2} - \frac{y^2}{b^2k^2/a^2} = 1 \qquad (x = k)$$

This is a hyperbola in the plane $x = k$ opening along a line parallel to the z-axis.   ◄

REMARK.   An elliptic cone in which the elliptical cross sections are circles is called a **circular cone**; similarly, elliptic paraboloids with circular cross sections are called **circular paraboloids**.

☐ **TECHNIQUES FOR GRAPHING QUADRIC SURFACES**

Accurate graphs of quadric surfaces are best left to computers. However, the techniques that follow can be used to obtain rough sketches of these surfaces that are useful for many purposes.

ELLIPSOIDS   A sketch of the ellipsoid

$$\frac{x^2}{a^2} + \frac{y^2}{b^2} + \frac{z^2}{c^2} = 1 \quad (a > 0, b > 0, c > 0) \tag{3}$$

can be obtained by first plotting the intersections with the coordinate axes, then sketching the elliptical traces in the coordinate planes, and then sketching the surface itself using the traces as a guide.

Example 2   Sketch the graph of the ellipsoid

$$\frac{x^2}{4} + \frac{y^2}{16} + \frac{z^2}{9} = 1 \tag{4}$$

Solution.   The x-intercepts, obtained by setting $y = 0$ and $z = 0$ in (4), are $x = \pm 2$. Similarly, the y-intercepts are $y = \pm 4$, and the z-intercepts are $z = \pm 3$. From these intercepts we obtain the elliptical traces and the ellipsoid sketched in Figure 14.7.3.   ◄

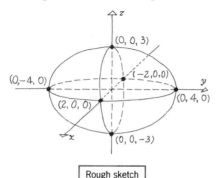

Rough sketch

Figure 14.7.3

HYPERBOLOIDS OF ONE SHEET   A sketch of the hyperboloid of one sheet

$$\frac{x^2}{a^2} + \frac{y^2}{b^2} - \frac{z^2}{c^2} = 1 \quad (a > 0, b > 0, c > 0) \tag{5}$$

can be obtained by first sketching the elliptical trace in the xy-plane, then the elliptical traces in the planes $z = c$ and $z = -c$, and then the hyperbolic curves that join the endpoints of the axes of these ellipses.

Example 3   Sketch the graph of the hyperboloid of one sheet

$$x^2 + y^2 - \frac{z^2}{4} = 1 \tag{6}$$

Solution.   The trace in the xy-plane, obtained by setting $z = 0$ in (6), is

$$x^2 + y^2 = 1 \quad (z = 0)$$

which is a circle of radius 1 centered on the z-axis. The traces in the planes $z = 2$ and $z = -2$, obtained by setting $z = \pm 2$ in (6), are given by

$$x^2 + y^2 = 2 \quad (z = \pm 2)$$

which are circles of radius $\sqrt{2}$ centered on the z-axis. Joining these circles by the hyperbolic traces in the vertical coordinate planes yields Figure 14.7.4.   ◄

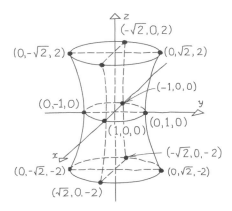

Figure 14.7.4

Rough sketch

HYPERBOLOIDS OF TWO SHEETS   A sketch of the hyperboloid of two sheets

$$\frac{x^2}{a^2} + \frac{y^2}{b^2} - \frac{z^2}{c^2} = -1 \quad (a > 0, b > 0, c > 0) \tag{7}$$

can be obtained by first plotting the intersections with the z-axis, then sketching the elliptical traces in the planes $z = 2c$ and $z = -2c$, and then sketching the hyperbolic traces that connect the z-axis intersections and the endpoints of the axes of the ellipses. (It is not essential to use the planes $z = \pm 2c$; any pair of horizontal planes above $z = c$ and below $z = -c$ would do just as well.)

Example 4   Sketch the graph of the hyperboloid of two sheets

$$x^2 + \frac{y^2}{4} - z^2 = -1 \tag{8}$$

Solution.   The z-intercepts, obtained by setting $x = 0$ and $y = 0$ in (8), are $z = \pm 1$. The traces in the planes $z = 2$ and $z = -2$, obtained by setting $z = \pm 2$ in (8), are given by

$$\frac{x^2}{3} + \frac{y^2}{12} = 1 \quad (z = \pm 2)$$

Sketching these ellipses and the hyperbolic traces in the vertical coordinate planes yields Figure 14.7.5.   ◀

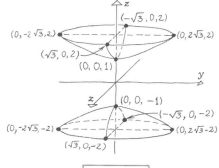

Figure 14.7.5

Rough sketch

ELLIPTIC CONES   A sketch of the elliptic cone

$$z^2 = \frac{x^2}{a^2} + \frac{y^2}{b^2} \quad (a > 0, b > 0) \tag{9}$$

can be obtained by first sketching the elliptical traces in the planes $z = \pm 1$, then sketching the linear traces that connect the endpoints of the axes of the ellipses.

Example 5 Sketch the graph of the elliptic cone

$$z^2 = x^2 + \frac{y^2}{4} \tag{10}$$

Solution. The traces of (10) in the planes $z = \pm 1$ are given by

$$x^2 + \frac{y^2}{4} = 1 \qquad (z = \pm 1)$$

Sketching these ellipses and the linear traces in the vertical coordinate planes yields the graph in Figure 14.7.6.   ◄

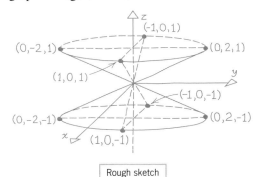

Figure 14.7.6

ELLIPTIC PARABOLOIDS   A sketch of the elliptic paraboloid

$$z = \frac{x^2}{a^2} + \frac{y^2}{b^2} \quad (a > 0, b > 0) \tag{11}$$

can be obtained by first sketching the elliptical trace in the plane $z = 1$, then sketching the parabolic traces (in the vertical coordinate planes) whose vertices are at the origin and pass through the endpoints of the axes of the ellipse.

Example 6 Sketch the graph of the elliptic paraboloid

$$z = \frac{x^2}{4} + \frac{y^2}{9} \tag{12}$$

Solution. The trace of (12) in the plane $z = 1$ is

$$\frac{x^2}{4} + \frac{y^2}{9} = 1 \qquad (z = 1)$$

Sketching this ellipse and the parabolic traces in the vertical coordinate planes yields the graph in Figure 14.7.7.   ◄

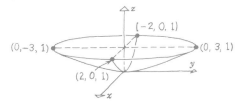

Figure 14.7.7

HYPERBOLIC PARABOLOIDS   The graph of a hyperbolic paraboloid is difficult to draw, but fortunately a rough sketch showing the orientation of the surface relative to the coordinate axes suffices for most purposes. The orientation of the hyperbolic paraboloid

$$z = \frac{y^2}{b^2} - \frac{x^2}{a^2} \quad (a > 0, b > 0) \tag{13}$$

can be obtained by first sketching the two parabolic traces that pass through the origin; one of these results from setting $x = 0$ in (13) and the other from setting $y = 0$. After the parabolic traces are drawn, sketch the hyperbolic traces in the planes $z = \pm 1$ with their proper orientation. Finally, sketch in any missing edges.

Example 7   Sketch the graph of the hyperbolic paraboloid

$$z = \frac{y^2}{4} - \frac{x^2}{9} \tag{14}$$

Solution.   Setting $x = 0$ in (14) yields

$$z = \frac{y^2}{4} \quad (x = 0)$$

which is a parabola in the yz-plane with vertex at the origin and opening in the positive z-direction (why?), and setting $y = 0$ yields

$$z = -\frac{x^2}{9} \quad (y = 0)$$

which is a parabola in the xz-plane with vertex at the origin and opening in the negative z-direction.

The trace in the plane $z = 1$ is

$$\frac{y^2}{4} - \frac{x^2}{9} = 1 \quad (z = 1)$$

which is a hyperbola that opens along a line parallel to the y-axis (verify), and the trace in the plane $z = -1$ is

$$\frac{x^2}{9} - \frac{y^2}{4} = 1 \quad (z = -1)$$

which is a hyperbola that opens along a line parallel to the x-axis. Combining all of the above information leads to the sketch in Figure 14.7.8.   ◄

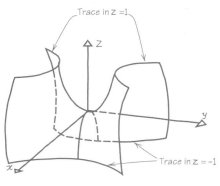

Figure 14.7.8

Rough sketch

REMARK.   The hyperbolic paraboloid (13) has an interesting behavior near the origin. The trace in the xz-plane has a relative maximum at the origin, and the trace in the yz-plane has a relative minimum there (see Figure 14.7.8, for example). On this surface the origin is commonly described as a *saddle point* or a *minimax point*.

☐ **REFLECTIONS OF QUADRIC SURFACES**

In our earlier study of curves in the xy-plane, we observed that interchanging the variables x and y in an equation has the effect of reflecting the graph of the equation about the line $y = x$. Analogous results occur in xyz-coordinate systems. For example, interchanging the variables x and z in the equation of a surface has the geometric effect of reflecting that surface symmetrically about the plane $x = z$. Thus, the equation

$$x^2 = \frac{z^2}{a^2} + \frac{y^2}{b^2}$$

represents an elliptic cone opening along the x-axis rather than the z-axis as shown in Table 14.7.1.

We also note that replacing z by $-z$ in the equation of the elliptic paraboloid in Table 14.7.1 has the geometric effect of reflecting the surface symmetrically about the xy-plane, thereby producing an elliptic paraboloid that opens down. The equation for this surface can be written as

$$z = -\left(\frac{x^2}{a^2} + \frac{y^2}{b^2}\right) \tag{15}$$

☐ **TRANSLATION OF AXES IN 3-SPACE**

Let an $x'y'z'$-coordinate system be obtained by translating an xyz-coordinate system so that the $x'y'z'$-origin is at the point whose xyz-coordinates are $(x, y, z) = (h, k, l)$ (Figure 14.7.9). It can be shown that the $x'y'z'$-coordinates and xyz-coordinates of a point P are related by

$$x' = x - h, \quad y' = y - k, \quad z' = z - l \tag{16}$$

[Compare this to (6) in Section 12.2.]

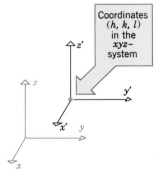

Figure 14.7.9

Example 8   Sketch the surface $z = 1 - x^2 - y^2$.

Solution.   Rewrite the equation in the form

$$z - 1 = -(x^2 + y^2) \tag{17}$$

If we translate the coordinate axes so that the new origin is at the point $(h, k, l) = (0, 0, 1)$, then the translation equations (16) are

$$x' = x, \quad y' = y, \quad z' = z - 1$$

so that in $x'y'z'$-coordinates (17) becomes

$$z' = -(x'^2 + y'^2)$$

which is of form (15) with $a = b = 1$. Thus, the surface is a circular paraboloid, opening down, with vertex $(0, 0, 1)$ in xyz-coordinates (Figure 14.7.10). ◀

Example 9 Sketch the surface

$$4x^2 + 4y^2 + z^2 + 8y - 4z = -4$$

Solution. Completing the squares yields

$$4x^2 + 4(y + 1)^2 + (z - 2)^2 = -4 + 4 + 4$$

or

$$x^2 + (y + 1)^2 + \frac{(z - 2)^2}{4} = 1 \tag{18}$$

If we translate the coordinate axes so that the new origin is $(h, k, l) = (0, -1, 2)$, then the translation equations (16) are

$$x' = x, \quad y' = y + 1, \quad z' = z - 2$$

so that in $x'y'z'$-coordinates (18) becomes

$$x'^2 + y'^2 + \frac{z'^2}{4} = 1$$

which is of form (3) with $a = 1$, $b = 1$, and $c = 2$. Thus, the surface is the ellipsoid shown in Figure 14.7.11. ◀

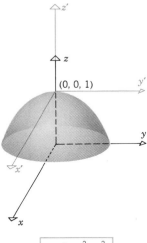

$$z = 1 - x^2 - y^2$$

Figure 14.7.10

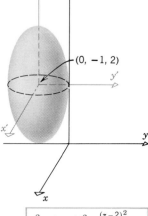

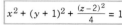

$$x^2 + (y + 1)^2 + \frac{(z-2)^2}{4} = 1$$

Figure 14.7.11

▶ **Exercise Set 14.7**

In Exercises 1–6, find an equation for the trace of the surface in the given plane and identify the trace.

1. $4x^2 + y^2 + z^2 = 4$:

    (a) $z = 0$ (b) $x = 1/2$ (c) $y = 1$.

2. $9x^2 - y^2 + 4z^2 = 9$:

    (a) $z = 0$ (b) $x = 2$ (c) $y = 4$.

3. $9x^2 - y^2 - z^2 = 16$:

    (a) $y = 0$ (b) $x = 2$ (c) $z = 2$.

4. $x^2 + 4y^2 - 9z^2 = 0$:

    (a) $x = 0$ (b) $y = 1$ (c) $z = 1$.

5. $z = 9x^2 + 4y^2$:

   (a) $x = 0$ (b) $y = 2$ (c) $z = 4$.

6. $z = x^2 - 4y^2$:

   (a) $y = 0$ (b) $x = 1$ (c) $z = 4$.

In Exercises 7–28, name and sketch the quadric surface.

7. $x^2 + y^2/4 + z^2/9 = 1$.   **8.** $x^2 + 4y^2 + 9z^2 = 36$.

9. $4x^2 + y^2 + 4z^2 = 16$.   **10.** $4x^2 + 4y^2 + z^2 = 9$.

11. $x^2/4 + y^2/9 - z^2/16 = 1$.

12. $4x^2 - y^2 + 4z^2 = 16$.

13. $2y^2 - x^2 + 2z^2 = 8$.   **14.** $x^2 + y^2 - z^2 = 9$.

15. $y^2 - 2x^2 - 2z^2 = 1$.   **16.** $x^2 - 3y^2 - 3z^2 = 9$.

17. $9z^2 - 4y^2 - 9x^2 = 36$.   **18.** $y^2 - 4x^2 - z^2 = 4$.

19. $4z^2 = x^2 + 4y^2$.   **20.** $y^2 = x^2 + z^2$.

21. $x^2 - 3y^2 - 3z^2 = 0$.   **22.** $9x^2 + 4y^2 - 36z^2 = 0$.

23. $y = x^2 + z^2$.   **24.** $z - 3x^2 - 3y^2 = 0$.

25. $4z = x^2 + 2y^2$.   **26.** $x - y^2 - 4z^2 = 0$.

27. $z = x^2/4 - y^2/9$.   **28.** $z = y^2 - x^2$.

29. The following equations represent quadric surfaces with orientations different from those in Table 14.7.1. Name and sketch the surface.

   (a) $\dfrac{z^2}{c^2} - \dfrac{y^2}{b^2} + \dfrac{x^2}{a^2} = 1$ (b) $\dfrac{x^2}{a^2} - \dfrac{y^2}{b^2} - \dfrac{z^2}{c^2} = 1$

   (c) $x = \dfrac{y^2}{b^2} + \dfrac{z^2}{c^2}$ (d) $x^2 = \dfrac{y^2}{b^2} + \dfrac{z^2}{c^2}$

   (e) $y = \dfrac{z^2}{c^2} - \dfrac{x^2}{a^2}$ (f) $y = -\left(\dfrac{x^2}{a^2} + \dfrac{z^2}{c^2}\right)$.

In Exercises 30–33, sketch the graph of the equation.

30. $z = \sqrt{1 - x^2 - y^2}$.   **31.** $z = \sqrt{x^2 + y^2}$.

32. $z = \sqrt{1 + x^2 + y^2}$.   **33.** $z = \sqrt{x^2 + y^2 - 1}$.

In Exercises 34–39, name and sketch the surface.

34. $\dfrac{(x-1)^2}{4} + \dfrac{(y-2)^2}{9} + \dfrac{(z-4)^2}{16} = 1$.

35. $z = (x + 2)^2 + (y - 3)^2 - 9$.

36. $4x^2 - y^2 + 16(z - 2)^2 = 100$.

37. $9x^2 + y^2 + 4z^2 - 18x + 2y + 16z = 10$.

38. $z^2 = 4x^2 + y^2 + 8x - 2y + 4z$.

39. $z = 4 - x^2 - y^2 - 2y$.

40. Obtain the results in Table 14.7.1 for the ellipsoid $x^2/a^2 + y^2/b^2 + z^2/c^2 = 1$.

41. Obtain the results in Table 14.7.1 for the hyperboloid of one sheet $x^2/a^2 + y^2/b^2 - z^2/c^2 = 1$.

42. Obtain the results in Table 14.7.1 for the hyperboloid of two sheets $x^2/a^2 + y^2/b^2 - z^2/c^2 = -1$.

43. Obtain the results in Table 14.7.1 for the elliptic paraboloid $z = x^2/a^2 + y^2/b^2$.

44. Obtain the results in Table 14.7.1 for the hyperbolic paraboloid $z = y^2/b^2 - x^2/a^2$.

In Exercises 45–52, find an equation of the projection onto the xy-plane of the curve of intersection of the surfaces. Identify the curve. [*Hint:* The values of z on both surfaces are the same along the curve of intersection.]

45. The paraboloids $z = x^2 + y^2$ and $z = 4 - x^2 - y^2$.

46. The paraboloids $z = x^2 + y^2$ and $z = 1 - 4x^2 - y^2$.

47. The paraboloid $z = x^2 + y^2$ and the plane $z = 2x$.

48. The paraboloid $z = 4 - x^2 - y^2$ and the parabolic cylinder $z = y^2$.

49. The cone $z^2 = x^2 + y^2$ and the plane $z = y + 1$.

50. The cone $z^2 = x^2 + y^2$ and the portion of the parabolic cylinder given by $z = 2\sqrt{y}$.

51. The ellipsoid $x^2 + y^2 + 4z^2 = 5$ and the portion of the parabolic cylinder given by $z = \sqrt{x}$.

52. The ellipsoid $x^2 + 2y^2 + z^2 = 2$ and the plane $z = x$.

53. For the elliptic paraboloid

$$z = \frac{x^2}{9} + \frac{y^2}{4}$$

   (a) find the focus and vertex of the (parabolic) trace in the plane $x = k$

   (b) find the foci and the endpoints of the major and minor axes of the (elliptic) trace in the plane $z = k$.

54. Use the method of slicing to find the volume of the ellipsoid

$$\frac{x^2}{a^2} + \frac{y^2}{b^2} + \frac{z^2}{c^2} = 1$$

   [*Hint:* The area of the ellipse $x^2/a^2 + y^2/b^2 = 1$ is πab.]

55. Find an equation of the surface consisting of the points $P(x, y, z)$ for which the distance between P and the plane $z = -1$ is equal to the distance between P and the point $(0, 0, 1)$. Identify the surface.

56. Find an equation of the surface consisting of the points $P(x, y, z)$ for which the distance between P and the plane $z = -1$ is twice the distance between P and the point $(0, 0, 1)$. Identify the surface.

57. (a) Show that the lines

$$x = 3 + t, \quad y = 2 + t, \quad z = 5 + 2t$$

   and

$$x = 3 + t, \quad y = 2 - t, \quad z = 5 + 10t$$

   both lie completely on the hyperbolic paraboloid $z = x^2 - y^2$.

(b) Let $P_0(x_0, y_0, z_0)$ be any point on the surface $z = x^2 - y^2$. Show that it is always possible to find two lines with equations of the form $x = x_0 + t$, $y = y_0 + at$, $z = z_0 + bt$ that pass through P_0 and lie completely on the surface.

58. (a) Show that the lines $x = 2 + \frac{3}{5}t, y = 1 + \frac{4}{5}t, z = 2 + t$ and $x = 2 + t$, $y = 1$, $z = 2 + t$ both lie completely on the hyperboloid $x^2 + y^2 - z^2 = 1$.

(b) Let $P_0(x_0, y_0, z_0)$ be any point on the surface $x^2 + y^2 - z^2 = 1$. Show that it is always possible to find two lines with equations of the form $x = x_0 + at$, $y = y_0 + bt$, $z = z_0 + t$ that pass through P_0 and lie completely on the surface.

14.8 CYLINDRICAL AND SPHERICAL COORDINATES

In this section we shall discuss two new types of coordinate systems in three dimensions that are extremely important in applied problems. These coordinate systems produce simpler equations than rectangular coordinate systems for surfaces with various kinds of symmetries.

☐ **CYLINDRICAL AND SPHERICAL COORDINATES**

To have a useful coordinate system in three dimensions, each point in space must be associated with a triple of real numbers (the coordinates of the point), and each triple of real numbers must determine a unique point. Figure 14.8.1 illustrates three possible ways for doing this. The rectangular coordinates (x, y, z) of a point P are shown in part (a) of the figure, the **cylindrical coordinates** (r, θ, z) of P are shown in part (b), and the **spherical coordinates** (ρ, θ, ϕ) of P are shown in part (c).

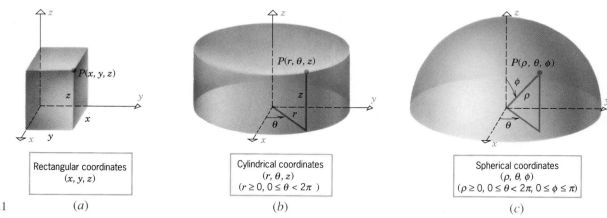

Rectangular coordinates
(x, y, z)

Cylindrical coordinates
(r, θ, z)
$(r \geq 0, 0 \leq \theta < 2\pi)$

Spherical coordinates
(ρ, θ, ϕ)
$(\rho \geq 0, 0 \leq \theta < 2\pi, 0 \leq \phi \leq \pi)$

Figure 14.8.1   (a)   (b)   (c)

The restrictions on the values of the cylindrical and spherical coordinates noted in Figure 14.8.1 are fairly standard; they ensure that each point not on the z-axis has a unique set of coordinates.

☐ **CONSTANT SURFACES**

In rectangular coordinates the surfaces represented by equations of the form

$$x = x_0, \quad y = y_0, \quad \text{and} \quad z = z_0$$

where x_0, y_0, and z_0 are constants, are planes parallel to the yz-plane, xz-plane, and xy-plane, respectively (Figure 14.8.2a). In cylindrical coordinates the surfaces represented by equations of the form

$$r = r_0, \quad \theta = \theta_0, \quad \text{and} \quad z = z_0$$

where r_0, θ_0, and z_0 are constants, are shown in Figure 14.8.2b.

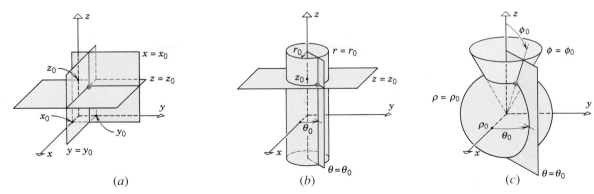

Figure 14.8.2   (a)   (b)   (c)

- The surface $r = r_0$ is a right-circular cylinder of radius r_0 centered on the z-axis. At each point (r, θ, z) on this cylinder, r has the value r_0, but θ and z are unrestricted except for our blanket assumption that $0 \leq \theta < 2\pi$.
- The surface $\theta = \theta_0$ is a half-plane attached along the z-axis and making an angle θ_0 with the positive x-axis. At each point (r, θ, z) on this surface, θ has the value θ_0, but r and z are unrestricted except for our blanket assumption that $r \geq 0$.
- The surface $z = z_0$ is a horizontal plane. At each point (r, θ, z) on this plane, z has the value z_0, but r and θ are unrestricted except for our blanket assumptions.

In spherical coordinates the surfaces represented by equations of the form

$$\rho = \rho_0, \quad \theta = \theta_0, \quad \text{and} \quad \phi = \phi_0$$

where ρ_0, θ_0, and ϕ_0 are constants, are shown in Figure 14.8.2c.

- The surface $\rho = \rho_0$ consists of all points whose distance ρ from the origin is ρ_0. Assuming ρ_0 to be nonnegative, this is a sphere of radius ρ_0 centered at the origin.
- As in cylindrical coordinates, the surface $\theta = \theta_0$ is a half-plane attached along the z-axis, making an angle of θ_0 with the positive x-axis.
- The surface $\phi = \phi_0$ consists of all points from which a line segment to the origin makes an angle of ϕ_0 with the positive z-axis. Depending on whether $0 < \phi_0 < \pi/2$ or $\pi/2 < \phi_0 < \pi$, this will be a cone opening up or opening down. (If $\phi_0 = \pi/2$, then the cone is flat and the surface is the xy-plane.)

☐ CONVERTING
COORDINATES

Frequently, the coordinates of a point are known in one type of coordinate system and it is of interest to find the coordinates in one of the other types. Table 14.8.1 lists the formulas for making such coordinate conversions.

Table 14.8.1

CONVERSION	FORMULAS
$(r, \theta, z) \rightarrow (x, y, z)$	$x = r\cos\theta, \quad y = r\sin\theta, \quad z = z$
$(x, y, z) \rightarrow (r, \theta, z)$	$r = \sqrt{x^2 + y^2}, \quad \tan\theta = y/x, \quad z = z$
$(\rho, \theta, \phi) \rightarrow (r, \theta, z)$	$r = \rho\sin\phi, \quad \theta = \theta, \quad z = \rho\cos\phi$
$(r, \theta, z) \rightarrow (\rho, \theta, \phi)$	$\rho = \sqrt{r^2 + z^2}, \quad \theta = \theta, \quad \tan\phi = r/z$
$(\rho, \theta, \phi) \rightarrow (x, y, z)$	$x = \rho\sin\phi\cos\theta, \quad y = \rho\sin\phi\sin\theta, \quad z = \rho\cos\phi$
$(x, y, z) \rightarrow (\rho, \theta, \phi)$	$\rho = \sqrt{x^2 + y^2 + z^2}, \quad \tan\theta = y/x, \quad \cos\phi = z/\sqrt{x^2 + y^2 + z^2}$

The formulas in this table can be derived by considering the diagrams in Figure 14.8.3. Part (a) of the figure illustrates the relationship between the rectangular coordinates (x, y, z) and the cylindrical coordinates (r, θ, z) of a point P; it shows that the value of z is the same in

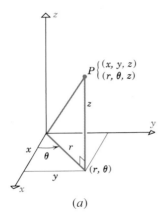

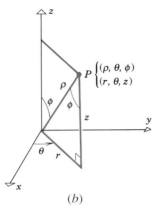

Figure 14.8.3

both coordinate systems and that (r, θ) is a pair of polar coordinates for the point (x, y) in the xy-plane. Thus, it follows from the relationship between polar and rectangular coordinates [Formulas (1a–b) of Section 13.1] that

$$x = r \cos \theta, \quad y = r \sin \theta, \quad z = z \tag{1}$$

Part (b) of Figure 14.8.3 illustrates the relationship between the spherical coordinates (ρ, θ, ϕ) and the cylindrical coordinates (r, θ, z) of a point P:

$$r = \rho \sin \phi, \quad \theta = \theta, \quad z = \rho \cos \phi \tag{2}$$

Substituting (2) into (1) yields the following relationships between the rectangular coordinates (x, y, z) and the spherical coordinates (ρ, θ, ϕ) of a point P:

$$x = \rho \sin \phi \cos \theta, \quad y = \rho \sin \phi \sin \theta, \quad z = \rho \cos \phi \tag{3}$$

We leave it as an exercise for the reader to deduce the remaining three conversion formulas in Table 14.8.1 from (1), (2), and (3).

Example 1   Find the rectangular coordinates of the point whose cylindrical coordinates are $(r, \theta, z) = (4, \pi/3, -3)$.

Solution.   From (1)

$$x = 4 \cos \frac{\pi}{3} = 2, \quad y = 4 \sin \frac{\pi}{3} = 2\sqrt{3}, \quad z = -3 \quad ◀$$

Example 2   Find an equation in cylindrical coordinates of the surface whose equation in rectangular coordinates is $z = x^2 + y^2 - 2x + y$.

Solution.   From (1)

$$z = r^2 - 2r \cos \theta + r \sin \theta \quad ◀$$

Example 3   Find an equation in rectangular coordinates of the surface whose equation in cylindrical coordinates is $r = 4 \cos \theta$.

Solution.   Multiplying both sides of the given equation by r yields the equation $r^2 = 4r \cos \theta$, then using the relationships $x^2 + y^2 = r^2$ and $x = r \cos \theta$, which follow from (1), yields

$$x^2 + y^2 = 4x \quad \text{or equivalently,} \quad (x - 2)^2 + y^2 = 4$$

This is a right-circular cylinder parallel to the z-axis.   ◀

Example 4   Find the rectangular coordinates of the point whose spherical coordinates (ρ, θ, ϕ) are $(4, \pi/3, \pi/4)$.

Solution.   From (3)

$$x = \rho \sin \phi \cos \theta = 4 \sin \frac{\pi}{4} \cos \frac{\pi}{3} = \sqrt{2}$$

$$y = \rho \sin \phi \sin \theta = 4 \sin \frac{\pi}{4} \sin \frac{\pi}{3} = \sqrt{6}$$

$$z = \rho \cos \phi = 4 \cos \frac{\pi}{4} = 2\sqrt{2} \quad ◀$$

Example 5   Find an equation of the paraboloid $z = x^2 + y^2$ in spherical coordinates.

Solution.   Substituting (3) in this equation yields

$$\rho \cos \phi = \rho^2 \sin^2 \phi \cos^2 \theta + \rho^2 \sin^2 \phi \sin^2 \theta$$

$$= \rho^2 \sin^2 \phi \, (\cos^2 \theta + \sin^2 \theta)$$

$$= \rho^2 \sin^2 \phi$$

which simplifies to $\rho \sin^2 \phi = \cos \phi$.   ◄

☐ SPHERICAL
COORDINATES IN
NAVIGATION

Spherical coordinates are related to longitude and latitude coordinates used in navigation. Let us construct a right-hand rectangular coordinate system with origin at the earth's center, positive z-axis passing through the north pole, and positive x-axis passing through the prime meridian (Figure 14.8.4). If we assume the earth to be a perfect sphere of radius $\rho = 4000$ miles, then each point on the earth has spherical coordinates of the form $(4000, \theta, \phi)$, where ϕ and θ determine the latitude and longitude of the point. It is useful to specify longitudes in degrees east or west of the prime meridian and latitudes in degrees north or south of the equator. However, it is a simple matter to determine ϕ and θ from such data.

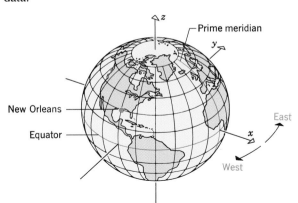

Figure 14.8.4

Example 6   The city of New Orleans is located at 90° West longitude and 30° North latitude. Find its spherical and rectangular coordinates relative to the coordinate axes of Figure 14.8.4. (Take miles as the unit of distance.)

Solution.   A longitude of 90° West corresponds to $\theta = 360° - 90° = 270°$ or $\theta = 3\pi/2$ radians; and a latitude of 30° North corresponds to $\phi = 90° - 30° = 60°$ or $\phi = \pi/3$ radians. Thus, the spherical coordinates (ρ, θ, ϕ) of New Orleans are $(4000, 3\pi/2, \pi/3)$.
  From (3), the rectangular coordinates of New Orleans are

$$x = 4000 \sin \frac{\pi}{3} \cos \frac{3\pi}{2} = 4000 \frac{\sqrt{3}}{2} (0) = 0 \text{ miles}$$

$$y = 4000 \sin \frac{\pi}{3} \sin \frac{3\pi}{2} = 4000 \frac{\sqrt{3}}{2} (-1) = -2000\sqrt{3} \text{ miles}$$

$$z = 4000 \cos \frac{\pi}{3} = 4000 \left(\frac{1}{2} \right) = 2000 \text{ miles}$$ ◄

▶ Exercise Set 14.8

1. Convert from rectangular to cylindrical coordinates.
    (a) $(4\sqrt{3}, 4, -4)$ (b) $(-5, 5, 6)$
    (c) $(0, 2, 0)$ (d) $(4, -4\sqrt{3}, 6)$
    (e) $(\sqrt{2}, -\sqrt{2}, 1)$ (f) $(0, 0, 1)$.

2. Convert from cylindrical to rectangular coordinates.
    (a) $(4, \pi/6, 3)$ (b) $(8, 3\pi/4, -2)$
    (c) $(5, 0, 4)$ (d) $(7, \pi, -9)$
    (e) $(6, 5\pi/3, 7)$ (f) $(1, \pi/2, 0)$.

3. Convert from rectangular to spherical coordinates.
    (a) $(1, \sqrt{3}, -2)$ (b) $(1, -1, \sqrt{2})$
    (c) $(0, 3\sqrt{3}, 3)$ (d) $(-5\sqrt{3}, 5, 0)$
    (e) $(4, 4, 4\sqrt{6})$ (f) $(1, -\sqrt{3}, -2)$.

4. Convert from spherical to rectangular coordinates.
    (a) $(5, \pi/6, \pi/4)$ (b) $(7, 0, \pi/2)$
    (c) $(1, \pi, 0)$ (d) $(2, 3\pi/2, \pi/2)$
    (e) $(1, 2\pi/3, 3\pi/4)$ (f) $(3, 7\pi/4, 5\pi/6)$.

5. Convert from cylindrical to spherical coordinates.
    (a) $(\sqrt{3}, \pi/6, 3)$ (b) $(1, \pi/4, -1)$
    (c) $(2, 3\pi/4, 0)$ (d) $(6, 1, -2\sqrt{3})$
    (e) $(4, 5\pi/6, 4)$ (f) $(2, 0, -2)$.

6. Convert from spherical to cylindrical coordinates.
    (a) $(5, \pi/4, 2\pi/3)$ (b) $(1, 7\pi/6, \pi)$
    (c) $(3, 0, 0)$ (d) $(4, \pi/6, \pi/2)$
    (e) $(5, \pi/2, 0)$ (f) $(6, 0, 3\pi/4)$.

In Exercises 7–14, an equation is given in cylindrical coordinates. Express the equation in rectangular coordinates and sketch the graph.

7. $r = 3$.   8. $\theta = \pi/4$.
9. $z = r^2$.   10. $z = r \cos \theta$.
11. $r = 4 \sin \theta$.   12. $r = 2 \sec \theta$.
13. $r^2 + z^2 = 1$.   14. $r^2 \cos 2\theta = z$.

In Exercises 15–22, an equation is given in spherical coordinates. Express the equation in rectangular coordinates and sketch the graph.

15. $\rho = 3$.   16. $\theta = \pi/3$.
17. $\phi = \pi/4$.   18. $\rho = 2 \sec \phi$.
19. $\rho = 4 \cos \phi$.   20. $\rho \sin \phi = 1$.
21. $\rho \sin \phi = 2 \cos \theta$.   22. $\rho - 2 \sin \phi \cos \theta = 0$.

In Exercises 23–34, an equation of a surface is given in rectangular coordinates. Find an equation of the surface in (a) cylindrical coordinates and (b) spherical coordinates.

23. $z = 3$.   24. $y = 2$.
25. $z = 3x^2 + 3y^2$.   26. $z = \sqrt{3x^2 + 3y^2}$.
27. $x^2 + y^2 = 4$.   28. $x^2 + y^2 - 6y = 0$.
29. $x^2 + y^2 + z^2 = 9$.   30. $z^2 = x^2 - y^2$.
31. $2x + 3y + 4z = 1$.   32. $x^2 + y^2 - z^2 = 1$.
33. $x^2 = 16 - z^2$.   34. $x^2 + y^2 + z^2 = 2z$.

In Exercises 35–38, describe the three-dimensional region that satisfies the given inequalities.

35. $r^2 \leq z \leq 4$.
36. $0 \leq r \leq 2 \sin \theta$, $0 \leq z \leq 3$.
37. $1 \leq \rho \leq 3$.
38. $0 \leq \phi \leq \pi/6$, $0 \leq \rho \leq 2$.
39. St. Petersburg (Leningrad), Russia, is located at 30° East longitude and 60° North latitude. Find its spherical and rectangular coordinates relative to the coordinate axes of Figure 14.8.4. Take miles as the unit of distance and assume the earth to be a sphere of radius 4000 miles.

40. (a) Show that the curve of intersection of the surfaces $z = \sin \theta$ and $r = a$ (cylindrical coordinates) is an ellipse.
     (b) Sketch a portion of the surface $z = \sin \theta$ for $0 \leq \theta \leq \pi/2$.

41. Sketch the surface whose equation in spherical coordinates is $\rho = a(1 - \cos \phi)$. [*Hint:* The surface is shaped like a familiar fruit.]

◆ TECHNOLOGY EXERCISES   Chapter 14

Most of these exercises require access to a graphing calculator or a computer algebra system (CAS) such as *Mathematica*, *Maple*, or *Derive*. When you are asked to *find* an answer or to *solve* an equation, you may choose to find an exact result or a numerical approximation, depending on the particular technology you are using and on your own imagination. The form of your answers may differ from those of other students or from those in the answer section of the text, depending on how you solve the problems and the accuracy you use in your numerical approximations. Those exercises that are more appropriate for a CAS than a graphing calculator are labeled with the icon ◆.

◆ **1. Minimum angle between two vectors:** For each x in $(-\infty, +\infty)$, let $\mathbf{u}(x)$ be the vector from the origin to the point $P(x, y)$ on the curve $y = x^2 + 1$, and $\mathbf{v}(x)$ the vector from the origin to the point $Q(x, y)$ on the line $y = -x - 1$.

  (a) Find, to the nearest degree, the minimum angle between $\mathbf{u}(x)$ and $\mathbf{v}(x)$ for x in $(-\infty, +\infty)$.

  (b) Find all values of x such that the vectors $\mathbf{u}(x)$ and $\mathbf{v}(x)$ are orthogonal.

◆ **2. Cones and vectors**

  (a) Find all unit vectors in 3-space that make an angle of 1 radian with both of the vectors $\mathbf{i} + 2\mathbf{j} + \mathbf{k}$ and $2\mathbf{i} - 2\mathbf{j} + \mathbf{k}$.

  (b) Let $\mathbf{v_0}$ be a vector with its initial point at the origin. All vectors with initial points at the origin that make an angle ϕ_0 with $\mathbf{v_0}$ lie on a cone whose axis is along $\mathbf{v_0}$ (see the accompanying figure). What does this imply about the vectors obtained in part (a)?

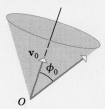

◆ **3. The angle between two vectors:** Let $\mathbf{u}$ and $\mathbf{v}$ be unit vectors. Assume that $\mathbf{u}$ is in the xy-plane and makes an angle θ_1 with the positive x-axis, and that $\mathbf{v}$ is in the yz-plane and makes an angle θ_2 with the positive z-axis, as shown in the accompanying figure.

  (a) Show that if θ is the angle between the vectors $\mathbf{u}$ and $\mathbf{v}$, then $\cos \theta = \sin \theta_1 \sin \theta_2$.

  (b) Use the formula in part (a) to find the angle θ if $\theta_1 = \theta_2 = 45°$.

  (c) Use the formula in part (a) to find, to the nearest degree, the angle θ_1 ($0° \leq \theta_1 < 360°$) if $\theta_2 = 60°$ and $\theta = 45°$.

  (d) Suppose that $\theta_1 = t$ and $\theta_2 = 2t$, where t is time ($t \geq 0$). Use the formula in part (a) to find, to the nearest degree, the maximum and minimum values of θ.

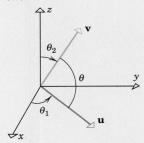

4. Distance between two moving points: Suppose that two particles moving in 3-space have equations of motion $x = 3 \cos 2t$, $y = 2 \sin 2t$, $z = 0$ ($t \geq 0$) and $x = 4 - 3t$, $y = -1 + t$, $z = -3 + t$ ($t \geq 0$). Given that x, y, and z are in meters and t is in seconds, how close do the particles get to one another?

◆ **5. Distance between a point and a line:** From Exercise 16, Section 14.4, if $\mathbf{u}$ is a vector from any point on a line to a point P not on the line, and $\mathbf{v}$ is a vector parallel to the line, then the distance between P and the line is given by $\|\mathbf{u} \times \mathbf{v}\| / \|\mathbf{v}\|$. Use this result to help find the minimum distance between a point on the curve $x = 3 \cos 2t$, $y = 2 \sin 2t$, $z = 0$ and the line $x = 4 - 3t$, $y = -1 + t$, $z = -3 + t$.

◆ **6. Extreme distances between a point and a curve:** Find the maximum and minimum distances between the point $(1, 1, 2)$ and a point on the curve of intersection of the plane $z = y + 2$ and the cylinder $x^2 + y^2 = 4$.

7. **Intersection of a line and a surface:** Find the coordinates of all points where the line $x = 2 + 3t$, $y = t$, $z = -1 + t$ intersects the cylindrical surface $2z = \sin x$.

◆ 8. **Area of a triangle:** Find the minimum area of a triangle if two of its vertices are $(2, -1, 0)$ and $(3, 2, 2)$ and its third vertex is on the curve $y = \ln x$ in the xy-plane.

9. **Maximum temperature on a line:** Suppose that the temperature T at a point (x, y, z) is given by $T = 25x^2yz$. Find the maximum value of T along the portion of the line $x = t$, $y = 1 + t$, $z = 3 - 2t$ that extends from the xz-plane to the xy-plane.

10. **Distance on the surface of the earth:** A ship at sea is at point A that is 60° West longitude and 40° North latitude. The ship travels to point B that is 40° West longitude and 20° North latitude. Assuming that the earth is a sphere with radius 6370 kilometers (km), find the shortest distance the ship can travel in going from A to B, given that the shortest distance between two points on a sphere is along the arc of the great circle joining the points. [*Hint:* See Figure 14.8.4 and Example 6, and consider the angle between the vectors from the center of the earth to the points A and B. (If you are not familiar with the term "great circle," consult a dictionary.)]

Karl Weierstrass (1815–1897)

15 VECTOR-VALUED FUNCTIONS

> *In this section we shall show how vectors can be used to express parametric equations in a more compact form. As part of our work we shall discuss functions that associate vectors with real numbers. This new category of functions has important applications in science and engineering.*

□ **PARAMETRIC CURVES IN 3-SPACE**

Recall from Section 13.4 that a parametric curve in 2-space is represented by a pair of parametric equations $x = x(t)$, $y = y(t)$, where the parameter t varies over some finite or infinite interval of real values. Similarly, we can represent a ***parametric curve*** in 3-space by three equations $x = x(t)$, $y = y(t)$, $z = z(t)$. Under appropriate conditions, the parametric curve will be traced in a specific direction as t increases; as in 2-space, we call this the ***direction of increasing parameter*** or the ***orientation*** of the parametric curve. If no restrictions on t are stated explicitly or implied by the equations, then it is understood that t varies over the interval $(-\infty, +\infty)$. In Section 14.5 we discussed parametric equations of lines in 3-space. Here we will be concerned with other kinds of parametric curves as well.

Example 1   Sketch the graph of the parametric equations

$$x = a \cos t, \quad y = a \sin t, \quad z = ct$$

where a and c are positive constants.

Solution.   As the parameter t increases, the value of $z = ct$ also increases, so the point (x, y, z) moves upward. However, as t increases, the point (x, y, z) also moves in a path directly over the circle

$$x = a \cos t, \quad y = a \sin t$$

in the xy-plane. The combination of these upward and circular motions produces a corkscrew-shaped curve that wraps around a right-circular cylinder of radius a centered on the z-axis (Figure 15.1.1). This curve is called a **circular helix**.   ◄

Example 2   Show that the graph of the parametric equations

$$x = t, \quad y = t^2, \quad z = t^3$$

is the intersection of the parabolic cylinder $y = x^2$ and the cubic cylinder $z = x^3$. Sketch the portion of the graph for $t \geq 0$. The curve is called a **twisted cubic**.

Solution.   Eliminating the parameter t in the equations for x and y yields $y = x^2$, so the curve lies on the parabolic cylinder with this equation. Similarly, eliminating t in the equations for x and z yields $z = x^3$, so the curve also lies on the cubic cylinder with this equation (Figure 15.1.2). The curve is traced by a point that starts at the origin when $t = 0$, and then moves upward as t increases, since x, y, and z increase with t.   ◄

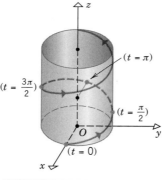

Figure 15.1.1 $x = a \cos t, \ y = a \sin t, \ z = ct$

Figure 15.1.2 $x = t, y = t^2, z = t^3$ \ $t \geq 0$

☐ **PARAMETRIC EQUATIONS IN VECTOR FORM**

Parametric equations in 2-space and 3-space can be expressed in a useful vector form. For example, in 2-space the two equations

$$x = x(t), \quad y = y(t)$$

can be written as the single vector equation

$$x\mathbf{i} + y\mathbf{j} = x(t)\mathbf{i} + y(t)\mathbf{j} \tag{1}$$

and in 3-space the three equations

$$x = x(t), \quad y = y(t), \quad z = z(t)$$

can be written as the single vector equation

$$x\mathbf{i} + y\mathbf{j} + z\mathbf{k} = x(t)\mathbf{i} + y(t)\mathbf{j} + z(t)\mathbf{k} \tag{2}$$

If we let $\mathbf{r} = x\mathbf{i} + y\mathbf{j}$ and $\mathbf{r}(t) = x(t)\mathbf{i} + y(t)\mathbf{j}$ in 2-space and let $\mathbf{r} = x\mathbf{i} + y\mathbf{j} + z\mathbf{k}$ and

$\mathbf{r}(t) = x(t)\mathbf{i} + y(t)\mathbf{j} + z(t)\mathbf{k}$ in 3-space, then both (1) and (2) can be expressed as

$$\mathbf{r} = \mathbf{r}(t) \tag{3}$$

For example, the vector form of the twisted cubic in Example 2 is

$$\mathbf{r} = t\mathbf{i} + t^2\mathbf{j} + t^3\mathbf{k}$$

and the parametric equations corresponding to the vector equation

$$\mathbf{r} = (t^3 + 1)\mathbf{i} + 3\mathbf{j} + e^t\mathbf{k}$$

are

$$x = t^3 + 1, \quad y = 3, \quad z = e^t$$

☐ **VECTOR-VALUED FUNCTIONS**

Recall that a function is a rule that assigns to each element in its domain one and only one element in its range. Thus far, we have considered primarily functions for which the domain and range are sets of real numbers; such functions are called ***real-valued functions of a real variable*** or sometimes simply ***real-valued functions***. In contrast, the function $\mathbf{r}(t)$ in (3) associates a vector in 2-space or 3-space with a real value of t; such functions are called ***vector-valued functions of a real variable*** or more simply ***vector-valued functions***. For example, if

$$\mathbf{r}(t) = t\mathbf{i} + t^2\mathbf{j} + t^3\mathbf{k}$$

then the vectors associated with $t = 1$, -2, and 0 are

$$\mathbf{r}(1) = \mathbf{i} + \mathbf{j} + \mathbf{k}, \quad \mathbf{r}(-2) = -2\mathbf{i} + 4\mathbf{j} - 8\mathbf{k}, \quad \mathbf{r}(0) = 0\mathbf{i} + 0\mathbf{j} + 0\mathbf{k} = \mathbf{0}$$

If $\mathbf{r}(t) = x(t)\mathbf{i} + y(t)\mathbf{j}$ in 2-space or if $\mathbf{r}(t) = x(t)\mathbf{i} + y(t)\mathbf{j} + z(t)\mathbf{k}$ in 3-space, then the real-valued functions $x(t)$, $y(t)$, and $z(t)$ are called the ***component functions*** or the ***components*** of $\mathbf{r}(t)$. The ***domain*** of $\mathbf{r}(t)$ is the set of allowable values for t. If the domain is not specified explicitly, then it is understood to consist of all values of t for which every component is defined and yields a real value; this is called the ***natural domain*** of $\mathbf{r}(t)$. Thus, the natural domain of $\mathbf{r}(t)$ is the intersection of the natural domains of its components. For example, the components of

$$\mathbf{r}(t) = \ln(t - 1)\mathbf{i} + e^t\mathbf{j} + \sqrt{t}\,\mathbf{k}$$

are

$$x(t) = \ln(t - 1), \quad y(t) = e^t, \quad z(t) = \sqrt{t}$$

and the natural domain of $\mathbf{r}(t)$ is the set of t values such that $t > 1$.

☐ **GRAPHS OF VECTOR-VALUED FUNCTIONS**

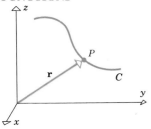

As t varies, the tip of the radius vector $\mathbf{r}$ traces out the curve C.

Figure 15.1.3

If $\mathbf{r}(t)$ is a vector-valued function in 2-space or 3-space, then we define the ***graph*** of $\mathbf{r}(t)$ [or of $\mathbf{r} = \mathbf{r}(t)$] to be the parametric curve represented by the parametric equations corresponding to $\mathbf{r}(t)$. For example, the graph of

$$\mathbf{r}(t) = t\mathbf{i} + t^2\mathbf{j} + t^3\mathbf{k}$$

is the parametric curve represented by the equations

$$x = t, \quad y = t^2, \quad z = t^3$$

which is the twisted cubic shown in Figure 15.1.2 (for $t \geq 0$).

Up to now, we have imagined a parametric curve C to be traced by moving point P. However, if the curve is viewed as the graph of a vector-valued function, then we can also imagine the curve to be traced by the tip of the vector $\mathbf{r}$ whose initial point is at the origin and whose terminal point is at P. We call $\mathbf{r}$ the ***radius vector*** or ***position vector*** for C (Figure 15.1.3).

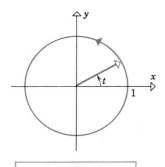

$$\mathbf{r} = (\cos t)\mathbf{i} + (\sin t)\mathbf{j}$$

Figure 15.1.4

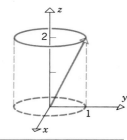

$$\mathbf{r} = (\cos t)\mathbf{i} + (\sin t)\mathbf{j} + 2\mathbf{k}$$

Figure 15.1.5

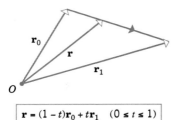

$$\mathbf{r} = (1 - t)\mathbf{r}_0 + t\mathbf{r}_1 \quad (0 \le t \le 1)$$

Figure 15.1.6

Example 3 Sketch the graph and a radius vector of

(a) $\mathbf{r}(t) = (\cos t)\mathbf{i} + (\sin t)\mathbf{j}, \quad 0 \le t \le 2\pi$

(b) $\mathbf{r}(t) = (\cos t)\mathbf{i} + (\sin t)\mathbf{j} + 2\mathbf{k}, \quad 0 \le t \le 2\pi$

Solution (a). The corresponding parametric equations are

$$x = \cos t, \quad y = \sin t \quad (0 \le t \le 2\pi)$$

so from Example 2 of Section 13.4 the graph is a circle of radius 1, centered at the origin, and oriented counterclockwise. The graph in 2-space and a radius vector are shown in Figure 15.1.4.

Solution (b). The corresponding parametric equations are

$$x = \cos t, \quad y = \sin t, \quad z = 2 \quad (0 \le t \le 2\pi)$$

From the last equation, the tip of the radius vector traces a curve in the plane $z = 2$, and from the first two equations and part (a), the curve is a circle of radius 1 centered on the z-axis and traced counterclockwise looking down the z-axis. The graph and a radius vector are shown in Figure 15.1.5. ◀

Example 4 Let $\mathbf{r}_0$ and $\mathbf{r}_1$ be distinct nonzero vectors in 2-space or 3-space with their initial points at the origin. Show that the graph of the equation

$$\mathbf{r} = (1 - t)\mathbf{r}_0 + t\mathbf{r}_1 \quad (0 \le t \le 1) \tag{4}$$

is the line segment joining the tips of $\mathbf{r}_0$ and $\mathbf{r}_1$ and oriented from $\mathbf{r}_0$ to $\mathbf{r}_1$ (Figure 15.1.6).

Solution. We shall give the solution in 2-space; the solution in 3-space is similar. Suppose that

$$\mathbf{r} = \langle x, y \rangle, \quad \mathbf{r}_0 = \langle x_0, y_0 \rangle, \quad \mathbf{r}_1 = \langle x_1, y_1 \rangle$$

Then (4) can be expressed as

$$\langle x, y \rangle = (1 - t)\langle x_0, y_0 \rangle + t\langle x_1, y_1 \rangle \quad (0 \le t \le 1)$$

or in parametric form as

$$x = x_0 + (x_1 - x_0)t, \quad y = y_0 + (y_1 - y_0)t \quad (0 \le t \le 1)$$

From part (a) of Theorem 14.5.1 these are parametric equations of some portion of the line that is parallel to $\mathbf{r}_1 - \mathbf{r}_0 = \langle x_1 - x_0, y_1 - y_0 \rangle$ and passes through the tip of $\mathbf{r}_0 = \langle x_0, y_0 \rangle$, which is the line through the tips of $\mathbf{r}_0$ and $\mathbf{r}_1$. As t varies from 0 to 1, the vector $\mathbf{r} = (1 - t)\mathbf{r}_0 + t\mathbf{r}_1$ varies from $\mathbf{r}_0$ to $\mathbf{r}_1$, and its tip traces out the line segment from the tip of $\mathbf{r}_0$ to the tip of $\mathbf{r}_1$. ◀

☐ **NORM OF A VECTOR-VALUED FUNCTION**

If $\mathbf{r}(t)$ is a vector-valued function, then $\|\mathbf{r}(t)\|$ is a real-valued function. Moreover, $\|\mathbf{r}(t)\|$ is continuous if $\mathbf{r}(t)$ is continuous (why?). For example, the vector-valued function

$$\mathbf{r}(t) = t\mathbf{i} + (t - 1)\mathbf{j}$$

is continuous, since its components are continuous, and the real-valued function

$$\|\mathbf{r}(t)\| = \sqrt{t^2 + (t-1)^2} = \sqrt{2t^2 - 2t + 1}$$

is continuous, since the function inside the radical is continuous and positive for all t.

▶ Exercise Set 15.1

In Exercises 1–4, find the domain of $\mathbf{r}(t)$ and the value of $\mathbf{r}(t_0)$.

1. $\mathbf{r}(t) = (\cos t)\mathbf{i} - 3t\mathbf{j}$; $t_0 = \pi$.

2. $\mathbf{r}(t) = \langle \sqrt{3t+1}, t^2 \rangle$; $t_0 = 1$.

3. $\mathbf{r}(t) = (\cos \pi t)\mathbf{i} - (\ln t)\mathbf{j} + \sqrt{t-2}\,\mathbf{k}$; $t_0 = 3$.

4. $\mathbf{r}(t) = \langle 2e^{-t}, \sin^{-1} t, \ln(1-t) \rangle$; $t_0 = 0$.

In Exercises 5–8, express the parametric equations as a single vector equation of the form $\mathbf{r} = x(t)\mathbf{i} + y(t)\mathbf{j}$ or $\mathbf{r} = x(t)\mathbf{i} + y(t)\mathbf{j} + z(t)\mathbf{k}$.

5. $x = 3\cos t$, $y = t + \sin t$.

6. $x = t^2 + 1$, $y = e^{-2t}$.

7. $x = 2t$, $y = 2\sin 3t$, $z = 5\cos 3t$.

8. $x = t\sin t$, $y = \ln t$, $z = \cos^2 t$.

In Exercises 9–12, find the parametric equations that correspond to the given vector equation.

9. $\mathbf{r} = 3t^2\mathbf{i} - 2\mathbf{j}$.

10. $\mathbf{r} = (\sin^2 t)\mathbf{i} + (1 - \cos 2t)\mathbf{j}$.

11. $\mathbf{r} = (2t - 1)\mathbf{i} - 3\sqrt{t}\mathbf{j} + (\sin 3t)\mathbf{k}$.

12. $\mathbf{r} = te^{-t}\mathbf{i} - 5t^2\mathbf{k}$.

In Exercises 13–18, describe the graph of the equation.

13. $\mathbf{r} = (2 - 3t)\mathbf{i} - 4t\mathbf{j}$.

14. $\mathbf{r} = (3\sin 2t)\mathbf{i} + (3\cos 2t)\mathbf{j}$.

15. $\mathbf{r} = 2t\mathbf{i} - 3\mathbf{j} + (1 + 3t)\mathbf{k}$.

16. $\mathbf{r} = 3\mathbf{i} + (2\cos t)\mathbf{j} + (2\sin t)\mathbf{k}$.

17. $\mathbf{r} = (3\cos t)\mathbf{i} + (2\sin t)\mathbf{j} - \mathbf{k}$.

18. $\mathbf{r} = -2\mathbf{i} + t\mathbf{j} + (t^2 - 1)\mathbf{k}$.

19. Find the slope of the line in 2-space that is represented by the vector equation $\mathbf{r} = (1 - 2t)\mathbf{i} - (2 - 3t)\mathbf{j}$.

20. Find the y-intercept of the line in 2-space that is represented by the vector equation $\mathbf{r} = (3 + 2t)\mathbf{i} + 5t\mathbf{j}$.

21. Find the coordinates of the point where the line $\mathbf{r} = (2 + t)\mathbf{i} + (1 - 2t)\mathbf{j} + 3t\mathbf{k}$ intersects the xz-plane.

22. Find the coordinates of the point where the line

$$\mathbf{r} = t\mathbf{i} + (1 + 2t)\mathbf{j} - 3t\mathbf{k}$$

intersects the plane $3x - y - z = 2$.

In Exercises 23–34, sketch the graph of $\mathbf{r}(t)$ and show the direction of increasing t.

23. $\mathbf{r}(t) = 2\mathbf{i} + t\mathbf{j}$.

24. $\mathbf{r}(t) = \langle 3t - 4, 6t + 2 \rangle$.

25. $\mathbf{r}(t) = (1 + \cos t)\mathbf{i} + (3 - \sin t)\mathbf{j}$, $0 \le t \le 2\pi$.

26. $\mathbf{r}(t) = \langle 2\cos t, 5\sin t \rangle$, $0 \le t \le 2\pi$.

27. $\mathbf{r}(t) = (\cosh t)\mathbf{i} + (\sinh t)\mathbf{j}$.

28. $\mathbf{r}(t) = \sqrt{t}\mathbf{i} + (2t + 4)\mathbf{j}$.

29. $\mathbf{r}(t) = t\mathbf{i} + t\mathbf{j} + t\mathbf{k}$.

30. $\mathbf{r}(t) = (1 + 3t)\mathbf{i} + (-1 + t)\mathbf{j} + 2t\mathbf{k}$.

31. $\mathbf{r}(t) = (2\cos t)\mathbf{i} + (2\sin t)\mathbf{j} + t\mathbf{k}$.

32. $\mathbf{r}(t) = (9\cos t)\mathbf{i} + (4\sin t)\mathbf{j} + t\mathbf{k}$.

33. $\mathbf{r}(t) = t\mathbf{i} + t^2\mathbf{j} + 2\mathbf{k}$.

34. $\mathbf{r}(t) = t\mathbf{i} + t\mathbf{j} + (\sin t)\mathbf{k}$, $0 \le t \le 2\pi$.

35. Show that the graph of

$$\mathbf{r} = (t\sin t)\mathbf{i} + (t\cos t)\mathbf{j} + t^2\mathbf{k}$$

lies on the paraboloid $z = x^2 + y^2$.

36. Show that the graph of

$$\mathbf{r} = t\mathbf{i} + \frac{1+t}{t}\mathbf{j} + \frac{1-t^2}{t}\mathbf{k}, \quad t > 0$$

lies in the plane $x - y + z + 1 = 0$.

37. Show that the graph of

$$\mathbf{r} = (\sin t)\mathbf{i} + (2\cos t)\mathbf{j} + (\sqrt{3}\sin t)\mathbf{k}$$

is a circle, and find its center and radius. [*Hint:* Show that the curve lies on both a sphere and a plane.]

38. Show that the graph of

$$\mathbf{r} = (3\cos t)\mathbf{i} + (3\sin t)\mathbf{j} + (3\sin t)\mathbf{k}$$

is an ellipse, and find the lengths of the major and minor axes. [*Hint:* Show that the graph lies on both a circular cylinder and a plane and use the result in Exercise 50 of Section 12.3.]

39. For the helix $\mathbf{r} = (a\cos t)\mathbf{i} + (a\sin t)\mathbf{j} + ct\mathbf{k}$, find c ($c > 0$) so that the helix will make one complete turn in a distance of 3 units measured along the z-axis.

40. How many revolutions will the circular helix

$$\mathbf{r} = (a\cos t)\mathbf{i} + (a\sin t)\mathbf{j} + 0.2t\mathbf{k}$$

make in a distance of 10 units measured along the z-axis?

41. Show that the curve $\mathbf{r} = (t \cos t)\mathbf{i} + (t \sin t)\mathbf{j} + t\mathbf{k}$, $t \geq 0$, lies on the cone $z = \sqrt{x^2 + y^2}$. Describe the curve.

42. Describe the curve $\mathbf{r} = (a \cos t)\mathbf{i} + (b \sin t)\mathbf{j} + ct\mathbf{k}$, where a, b, and c are positive constants such that $a \neq b$.

In Exercises 43–48, find parametric equations of the curve of intersection of the surfaces. [*Note:* The answer is not unique.]

43. The cone $z = \sqrt{x^2 + y^2}$ and the plane $z = y + 2$. Identify the curve.

44. The paraboloid $z = x^2 + y^2$ and the plane $x = -2$. Identify the curve.

45. The circular cylinder $x^2 + y^2 = 9$ and the parabolic cylinder $z = x^2$.

46. The paraboloid $z = 4 - x^2 - y^2$ and the circular cylinder $x^2 + y^2 = 1$. Identify the curve.

47. The elliptic paraboloid $z = x^2 + 4y^2$ and the plane $z = 2x$. [*Hint:* Find the orthogonal projection of the curve onto the xy-plane.]

48. The cone $z = \sqrt{x^2 + y^2}$ and the parabolic cylinder $z = 2\sqrt{y}$. [*Hint:* See the hint in Exercise 47.]

■ 15.2 CALCULUS OF VECTOR-VALUED FUNCTIONS

In this section we shall define limits, derivatives, and integrals of vector-valued functions and discuss their properties.

☐ **LIMITS, DERIVATIVES, AND INTEGRALS**

As shown in Table 15.2.1, limits, derivatives, and integrals of vector-valued functions can be defined by taking the limits, derivatives, and integrals of the components.

Table 15.2.1

2-SPACE $\mathbf{r}(t) = x(t)\mathbf{i} + y(t)\mathbf{j}$	3-SPACE $\mathbf{r}(t) = x(t)\mathbf{i} + y(t)\mathbf{j} + z(t)\mathbf{k}$
$\lim\limits_{t \to a} \mathbf{r}(t) = \left(\lim\limits_{t \to a} x(t)\right)\mathbf{i} + \left(\lim\limits_{t \to a} y(t)\right)\mathbf{j}$	$\lim\limits_{t \to a} \mathbf{r}(t) = \left(\lim\limits_{t \to a} x(t)\right)\mathbf{i} + \left(\lim\limits_{t \to a} y(t)\right)\mathbf{j} + \left(\lim\limits_{t \to a} z(t)\right)\mathbf{k}$
$\mathbf{r}'(t) = x'(t)\mathbf{i} + y'(t)\mathbf{j}$	$\mathbf{r}'(t) = x'(t)\mathbf{i} + y'(t)\mathbf{j} + z'(t)\mathbf{k}$
$\int \mathbf{r}(t)\,dt = \left(\int x(t)\,dt\right)\mathbf{i} + \left(\int y(t)\,dt\right)\mathbf{j}$	$\int \mathbf{r}(t)\,dt = \left(\int x(t)\,dt\right)\mathbf{i} + \left(\int y(t)\,dt\right)\mathbf{j} + \left(\int z(t)\,dt\right)\mathbf{k}$
$\int_a^b \mathbf{r}(t)\,dt = \left(\int_a^b x(t)\,dt\right)\mathbf{i} + \left(\int_a^b y(t)\,dt\right)\mathbf{j}$	$\int_a^b \mathbf{r}(t)\,dt = \left(\int_a^b x(t)\,dt\right)\mathbf{i} + \left(\int_a^b y(t)\,dt\right)\mathbf{j} + \left(\int_a^b z(t)\,dt\right)\mathbf{k}$

The definitions in Table 15.2.1 assume that the operations on the components can be performed. If, for example, the limit of any component of $\mathbf{r}(t)$ does not exist, then we shall agree that the limit of $\mathbf{r}(t)$ ***does not exist***. The limit definition is also applicable to one-sided and infinite limits. In addition, the standard terminology and notation relating to derivatives and integrals continues to apply. For example, a vector-valued function is said to be ***differentiable*** (***integrable***) if and only if each component is differentiable (integrable). Moreover, the derivative of $\mathbf{r}(t)$ can also be expressed in any of the following notations:

$$\frac{d}{dt}[\mathbf{r}(t)], \quad \frac{d\mathbf{r}}{dt}, \quad \mathbf{r}'(t), \quad \text{and} \quad \mathbf{r}'$$

Example 1 Let

$$\mathbf{r}(t) = t^2\mathbf{i} + e^t\mathbf{j} - 2 \cos \pi t\mathbf{k}$$

Then

$$\lim_{t \to 0} \mathbf{r}(t) = (\lim_{t \to 0} t^2)\mathbf{i} + (\lim_{t \to 0} e^t)\mathbf{j} - (\lim_{t \to 0} 2 \cos \pi t)\mathbf{k}$$

$$= \mathbf{j} - 2\mathbf{k}$$

$$\mathbf{r}'(t) = 2t\mathbf{i} + e^t\mathbf{j} + 2\pi \sin \pi t \mathbf{k}$$

$$\mathbf{r}'(1) = 2\mathbf{i} + e\mathbf{j}$$

$$\int_0^1 \mathbf{r}(t)\,dt = \frac{t^3}{3}\bigg]_0^1 \mathbf{i} + e^t\bigg]_0^1 \mathbf{j} - \frac{2}{\pi}\sin \pi t\bigg]_0^1 \mathbf{k}$$

$$= \tfrac{1}{3}\mathbf{i} + (e - 1)\mathbf{j} \qquad \blacktriangleleft$$

Recall that indefinite integration of a real-valued function produces a constant of integration C that is an arbitrary real number. Analogously, indefinite integration of a vector-valued function produces a constant of integration $\mathbf{C}$ that is an arbitrary vector. This is illustrated in the following example.

Example 2

$$\int (2t\mathbf{i} + 3t^2\mathbf{j})\,dt = \left(\int 2t\,dt\right)\mathbf{i} + \left(\int 3t^2\,dt\right)\mathbf{j}$$

$$= (t^2 + C_1)\mathbf{i} + (t^3 + C_2)\mathbf{j}$$

$$= (t^2\mathbf{i} + t^3\mathbf{j}) + C_1\mathbf{i} + C_2\mathbf{j} = t^2\mathbf{i} + t^3\mathbf{j} + \mathbf{C}$$

where $\mathbf{C} = C_1\mathbf{i} + C_2\mathbf{j}$ is an arbitrary vector constant of integration. $\qquad \blacktriangleleft$

☐ **PROPERTIES OF DERIVATIVES AND INTEGRALS**

Because limits, derivatives, and integrals of vector-valued functions are defined in terms of the same operations on components, most of the standard theorems on limits, derivatives, and integrals carry over to vector-valued functions. The following two theorems, whose proofs are left as exercises, list the standard properties of differentiation and integration of vector-valued functions.

15.2.1 THEOREM (*Rules of Differentiation*). *In either 2-space or 3-space let* $\mathbf{r}(t)$, $\mathbf{r}_1(t)$, *and* $\mathbf{r}_2(t)$ *be vector-valued functions,* $f(t)$ *a real-valued function,* k *a scalar, and* $\mathbf{c}$ *a fixed (constant) vector. Then the following rules of differentiation hold:*

(a) $\dfrac{d}{dt}[\mathbf{c}] = \mathbf{0}$

(b) $\dfrac{d}{dt}[k\mathbf{r}(t)] = k\dfrac{d}{dt}[\mathbf{r}(t)]$

(c) $\dfrac{d}{dt}[\mathbf{r}_1(t) + \mathbf{r}_2(t)] = \dfrac{d}{dt}[\mathbf{r}_1(t)] + \dfrac{d}{dt}[\mathbf{r}_2(t)]$

(d) $\dfrac{d}{dt}[\mathbf{r}_1(t) - \mathbf{r}_2(t)] = \dfrac{d}{dt}[\mathbf{r}_1(t)] - \dfrac{d}{dt}[\mathbf{r}_2(t)]$

(e) $\dfrac{d}{dt}[f(t)\mathbf{r}(t)] = f(t)\dfrac{d}{dt}[\mathbf{r}(t)] + \dfrac{d}{dt}[f(t)]\mathbf{r}(t)$

15.2.2 THEOREM (*Rules of Integration*).   *In either 2-space or 3-space let* $\mathbf{r}(t)$, $\mathbf{r}_1(t)$, *and* $\mathbf{r}_2(t)$ *be vector-valued functions, and let k be a scalar. Then the following rules of integration hold:*

(a) $\displaystyle\int k\mathbf{r}(t)\,dt = k\int \mathbf{r}(t)\,dt$

(b) $\displaystyle\int [\mathbf{r}_1(t) + \mathbf{r}_2(t)]\,dt = \int \mathbf{r}_1(t)\,dt + \int \mathbf{r}_2(t)\,dt$

(c) $\displaystyle\int [\mathbf{r}_1(t) - \mathbf{r}_2(t)]\,dt = \int \mathbf{r}_1(t)\,dt - \int \mathbf{r}_2(t)\,dt$

REMARK.   The results in the preceding theorem are also valid for definite integrals of vector-valued functions.

☐ **FURTHER PROPERTIES OF DERIVATIVES**

The derivative of a vector-valued function was defined in terms of the derivatives of its components. The following theorem provides a formula for the derivative of $\mathbf{r}(t)$ that does not require breaking up the function into components. [Compare the formula in this theorem to Formula (4) of Section 3.2.]

15.2.3 THEOREM.   *If* $\mathbf{r}(t)$ *is a vector-valued function in 2-space or 3-space, then the derivative of* $\mathbf{r}(t)$ *can be expressed as*

$$\mathbf{r}'(t) = \lim_{h \to 0} \frac{\mathbf{r}(t + h) - \mathbf{r}(t)}{h} \tag{1}$$

provided this limit exists.

Proof.   For simplicity, we give the proof in 2-space; the proof in 3-space is identical, except for the additional component. Assume that $\mathbf{r}(t) = x(t)\mathbf{i} + y(t)\mathbf{j}$, so

$$\mathbf{r}'(t) = x'(t)\mathbf{i} + y'(t)\mathbf{j}$$
$$= \lim_{h \to 0} \frac{[x(t + h) - x(t)]}{h}\mathbf{i} + \lim_{h \to 0} \frac{[y(t + h) - y(t)]}{h}\mathbf{j}$$
$$= \lim_{h \to 0} \frac{[x(t + h)\mathbf{i} + y(t + h)\mathbf{j}] - [x(t)\mathbf{i} + y(t)\mathbf{j}]}{h}$$
$$= \lim_{h \to 0} \frac{\mathbf{r}(t + h) - \mathbf{r}(t)}{h} \quad ■$$

☐ **FURTHER PROPERTIES OF INTEGRALS**

We leave it for the reader to show that the following analog of Formula (2) in Section 5.2 holds for vector-valued functions in 2-space or 3-space.

$$\frac{d}{dt}\left[\int \mathbf{r}(t)\,dt\right] = \mathbf{r}(t) \tag{2}$$

This shows that an indefinite integral of $\mathbf{r}(t)$ is, in fact, the set of antiderivatives of $\mathbf{r}(t)$, just as for real-valued functions. We also leave it as an exercise to show that if $\mathbf{R}(t)$ is any antiderivative of $\mathbf{r}(t)$ in the sense that $\mathbf{R}'(t) = \mathbf{r}(t)$, then

$$\int \mathbf{r}(t)\,dt = \mathbf{R}(t) + \mathbf{C} \tag{3}$$

where **C** is an arbitrary vector constant of integration. Moreover,

$$\int_a^b \mathbf{r}(t)\, dt = \mathbf{R}(t)\Big]_a^b = \mathbf{R}(b) - \mathbf{R}(a) \tag{4}$$

which is the extension of the First Fundamental Theorem of Calculus (Theorem 5.7.1) to vector-valued functions.

Example 3   In Example 2 we showed that $\mathbf{R}(t) = t^2\mathbf{i} + t^3\mathbf{j}$ is an antiderivative of $\mathbf{r}(t) = 2t\mathbf{i} + 3t^2\mathbf{j}$. (It is the antiderivative that results when $\mathbf{C} = \mathbf{0}$.) Thus, from (4)

$$\int_0^2 \mathbf{r}(t)\, dt = \int_0^2 (2t\mathbf{i} + 3t^2\mathbf{j})\, dt = \Big[\,\mathbf{R}(t)\,\Big]_0^2$$

$$= \Big[\, t^2\mathbf{i} + t^3\mathbf{j}\,\Big]_0^2 = (4\mathbf{i} + 8\mathbf{j}) - (0\mathbf{i} + 0\mathbf{j}) = 4\mathbf{i} + 8\mathbf{j}$$

The reader may want to check that the same result can be obtained by integrating the components of $\mathbf{r}(t)$ term by term.   ◄

Example 4   Find $\mathbf{r}(t)$ given that $\mathbf{r}'(t) = 3\mathbf{i} + 2t\mathbf{j}$ and $\mathbf{r}(1) = 2\mathbf{i} + 5\mathbf{j}$.

Solution.   Integrating $\mathbf{r}'(t)$ to obtain $\mathbf{r}(t)$ yields

$$\mathbf{r}(t) = \int \mathbf{r}'(t)\, dt = \int (3\mathbf{i} + 2t\mathbf{j})\, dt = 3t\mathbf{i} + t^2\mathbf{j} + \mathbf{C}$$

where **C** is a vector constant of integration. To find **C** we substitute $t = 1$ in this equation and use the given value of $\mathbf{r}(1)$ to obtain

$$\mathbf{r}(1) = 3\mathbf{i} + \mathbf{j} + \mathbf{C} = 2\mathbf{i} + 5\mathbf{j}$$

so that $\mathbf{C} = -\mathbf{i} + 4\mathbf{j}$. Thus,

$$\mathbf{r}(t) = 3t\mathbf{i} + t^2\mathbf{j} - \mathbf{i} + 4\mathbf{j} = (3t - 1)\mathbf{i} + (t^2 + 4)\mathbf{j}   ◄$$

☐ **GEOMETRIC INTERPRETATION OF LIMITS AND DERIVATIVES**

Although we have defined limits and derivatives of vector-valued functions in terms of the corresponding operations on components, it is desirable to have vector interpretations of limits and derivatives that do not require us to break the vectors into components. Limits of vector-valued functions can be interpreted geometrically as follows:

> **15.2.4**   GEOMETRIC INTERPRETATION OF LIMITS.   *If* $\mathbf{r}(t)$ *is a vector-valued function in 2-space or 3-space, then*
> $$\lim_{t \to a} \mathbf{r}(t) = \mathbf{L}$$
> *if and only if the radius vector* $\mathbf{r} = \mathbf{r}(t)$ *approaches* **L** *in both length and direction as* $t \to a$ *(Figure 15.2.1).*

This result should be evident from the fact that the components of $\mathbf{r}(t)$ approach the components of **L** as $t \to a$.

The definition of continuity for vector-valued functions in 2-space and 3-space is similar to that for real-valued functions: We shall say that $\mathbf{r}(t)$ is *continuous at t_0* if $\mathbf{r}(t_0)$ is defined and

$$\lim_{t \to t_0} \mathbf{r}(t) = \mathbf{r}(t_0)$$

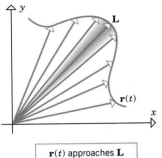

$\mathbf{r}(t)$ approaches **L** in length and direction if $\lim_{t \to a} \mathbf{r}(t) = \mathbf{L}$.

Figure 15.2.1

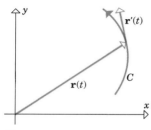

Figure 15.2.2

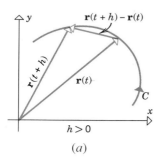

$h > 0$

(a)

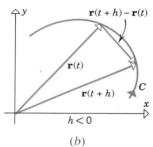

$h < 0$

(b)

Figure 15.2.3

It can be shown that $\mathbf{r}(t)$ is continuous at t_0 if and only if each component of $\mathbf{r}(t)$ is continuous at t_0 (Exercise 67). As with real-valued functions, we shall call $\mathbf{r}(t)$ **_continuous everywhere_** or simply **_continuous_** if $\mathbf{r}(t)$ is continuous at all real values of t. Geometrically, the graph of a continuous vector-valued function is an unbroken curve.

The derivative of a vector-valued function can be interpreted geometrically as follows:

> **15.2.5** GEOMETRIC INTERPRETATION OF THE DERIVATIVE. *Suppose that C is the graph of a vector-valued function $\mathbf{r}(t)$ in 2-space or 3-space and that $\mathbf{r}'(t)$ exists and is nonzero for a given value of t. If the vector $\mathbf{r}'(t)$ is positioned with its initial point at the terminal point of the radius vector $\mathbf{r}(t)$ (Figure 15.2.2), then $\mathbf{r}'(t)$ is tangent to C and points in the direction of increasing parameter.*

To make this result plausible, let C be the graph of $\mathbf{r}(t)$, and for a fixed value of the parameter t construct the position vectors $\mathbf{r}(t)$ and $\mathbf{r}(t + h)$. If $h > 0$, then the tip of the vector $\mathbf{r}(t + h)$ is in the direction of increasing parameter from the tip of $\mathbf{r}(t)$, and if $h < 0$, it is the other way around (Figure 15.2.3). In either case, the difference $\mathbf{r}(t + h) - \mathbf{r}(t)$ coincides with the secant line through the tips of the vectors $\mathbf{r}(t)$ and $\mathbf{r}(t + h)$. Moreover, since h is a scalar, the vector

$$\frac{1}{h}[\mathbf{r}(t + h) - \mathbf{r}(t)] = \frac{\mathbf{r}(t + h) - \mathbf{r}(t)}{h} \qquad (5)$$

also coincides with this secant line. If $h < 0$, then vector (5) is oppositely directed to $\mathbf{r}(t + h) - \mathbf{r}(t)$, and if $h > 0$, these vectors have the same direction. In both cases, vector (5) points in the direction of increasing parameter. If the limit of (5) exists as $h \rightarrow 0$, then the secant lines through the tips of $\mathbf{r}(t + h)$ and $\mathbf{r}(t)$ tend toward the tangent line to C at the tip of $\mathbf{r}(t)$. Thus, the vector

$$\mathbf{r}'(t) = \lim_{h \to 0} \frac{\mathbf{r}(t + h) - \mathbf{r}(t)}{h} \qquad (6)$$

is tangent to the curve C at the tip of $\mathbf{r}(t)$ and points in the direction of increasing parameter.

Motivated by the preceding discussion, we make the following definition.

> **15.2.6** DEFINITION. Let P be a point on the graph of a vector-valued function $\mathbf{r}(t)$, and let $\mathbf{r}(t_0)$ be the radius vector from the origin to P (Figure 15.2.4). If $\mathbf{r}'(t_0)$ exists and $\mathbf{r}'(t_0) \neq \mathbf{0}$, then we call $\mathbf{r}'(t_0)$ the **_tangent vector_** to the graph of $\mathbf{r}(t)$ at $\mathbf{r}(t_0)$.

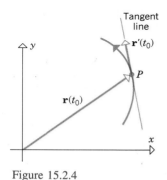

Figure 15.2.4

REMARK. Observe that the tangent vector can fail to exist at a point either because the derivative in (6) does not exist or because the derivative is the zero vector at the point.

If a vector-valued function $\mathbf{r}(t)$ has a tangent vector $\mathbf{r}'(t_0)$ at a point on its graph, then the line that is parallel to $\mathbf{r}'(t_0)$ and passes through the tip of the radius vector $\mathbf{r}(t_0)$ is called the **_tangent line_** to the graph of $\mathbf{r}(t)$ at $\mathbf{r}(t_0)$ (Figure 15.2.4). It follows from Formula (5) of Section 14.5 that a vector equation of the tangent line is

$$\mathbf{r} = \mathbf{r}(t_0) + t\mathbf{r}'(t_0) \qquad (7)$$

Example 5 Find parametric equations of the tangent line to the circular helix

$$x = \cos t, \quad y = \sin t, \quad z = t$$

at the point where $t = \pi/6$.

Solution. We shall first use Formula (7) to find a vector equation of the tangent line, then we shall equate components to obtain the parametric equations. A vector equation $\mathbf{r} = \mathbf{r}(t)$ of the helix is

$$x\mathbf{i} + y\mathbf{j} + z\mathbf{k} = (\cos t)\mathbf{i} + (\sin t)\mathbf{j} + t\mathbf{k}$$

Thus,

$$\mathbf{r}(t) = (\cos t)\mathbf{i} + (\sin t)\mathbf{j} + t\mathbf{k}$$

$$\mathbf{r}'(t) = (-\sin t)\mathbf{i} + (\cos t)\mathbf{j} + \mathbf{k}$$

At the point where $t = \pi/6$, these vectors are

$$\mathbf{r}\left(\frac{\pi}{6}\right) = \frac{\sqrt{3}}{2}\mathbf{i} + \frac{1}{2}\mathbf{j} + \frac{\pi}{6}\mathbf{k} \quad \text{and} \quad \mathbf{r}'\left(\frac{\pi}{6}\right) = -\frac{1}{2}\mathbf{i} + \frac{\sqrt{3}}{2}\mathbf{j} + \mathbf{k}$$

so from (7) with $t_0 = \pi/6$ a vector equation of the tangent line is

$$\mathbf{r} = \mathbf{r}\left(\frac{\pi}{6}\right) + t\mathbf{r}'\left(\frac{\pi}{6}\right) = \left(\frac{\sqrt{3}}{2}\mathbf{i} + \frac{1}{2}\mathbf{j} + \frac{\pi}{6}\mathbf{k}\right) + t\left(-\frac{1}{2}\mathbf{i} + \frac{\sqrt{3}}{2}\mathbf{j} + \mathbf{k}\right)$$

Simplifying, then equating the resulting components with the corresponding components of $\mathbf{r} = x\mathbf{i} + y\mathbf{j} + z\mathbf{k}$ yields the parametric equations

$$x = \frac{\sqrt{3}}{2} - \frac{1}{2}t, \quad y = \frac{1}{2} + \frac{\sqrt{3}}{2}t, \quad z = \frac{\pi}{6} + t \quad \blacktriangleleft$$

☐ **DERIVATIVES OF DOT AND CROSS PRODUCTS**

The following rules, which are derived in the exercises, provide a method for differentiating dot products in 2-space and 3-space and cross products in 3-space.

$$\frac{d}{dt}[\mathbf{r}_1(t) \cdot \mathbf{r}_2(t)] = \mathbf{r}_1(t) \cdot \frac{d\mathbf{r}_2}{dt} + \frac{d\mathbf{r}_1}{dt} \cdot \mathbf{r}_2(t) \tag{8}$$

$$\frac{d}{dt}[\mathbf{r}_1(t) \times \mathbf{r}_2(t)] = \mathbf{r}_1(t) \times \frac{d\mathbf{r}_2}{dt} + \frac{d\mathbf{r}_1}{dt} \times \mathbf{r}_2(t) \tag{9}$$

REMARK. In (8) the order of the factors in each term on the right does not matter, but in (9) it does.

In plane geometry one learns that a tangent line to a circle is perpendicular to the radius at the point of tangency. Consequently, if a point moves along a circular arc in 2-space, one would expect the radius vector and the tangent vector at any point on the arc to be perpendicular. This is the motivation for the following useful theorem, which is applicable in both 2-space and 3-space.

15.2.7 THEOREM. *If $\mathbf{r}(t)$ is a vector-valued function in 2-space or 3-space and $\|\mathbf{r}(t)\|$ is constant for all t, then*

$$\mathbf{r}(t) \cdot \mathbf{r}'(t) = 0 \tag{10}$$

that is, $\mathbf{r}(t)$ and $\mathbf{r}'(t)$ are orthogonal vectors for all t.

Proof. It follows from (8) with $\mathbf{r}_1(t) = \mathbf{r}_2(t) = \mathbf{r}(t)$ that

$$\frac{d}{dt}[\mathbf{r}(t) \cdot \mathbf{r}(t)] = \mathbf{r}(t) \cdot \frac{d\mathbf{r}}{dt} + \frac{d\mathbf{r}}{dt} \cdot \mathbf{r}(t)$$

or, equivalently,

$$\frac{d}{dt}[\|\mathbf{r}(t)\|^2] = 2\mathbf{r}(t) \cdot \frac{d\mathbf{r}}{dt} \tag{11}$$

But $\|\mathbf{r}(t)\|^2$ is constant, so its derivative is zero. Thus

$$2\mathbf{r}(t) \cdot \frac{d\mathbf{r}}{dt} = 0$$

from which (10) follows. ∎

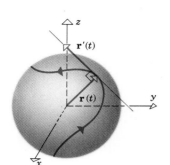

Figure 15.2.5

Example 6 Just as a tangent line to a circular arc in 2-space is perpendicular to the radius at the point of tangency, so a tangent vector to a curve on the surface of a sphere in 3-space is perpendicular to the radius vector at the point of tangency (Figure 15.2.5). To see that this is so, suppose that the graph of $\mathbf{r}(t)$ lies on the surface of the sphere of radius $k > 0$ centered at the origin. For each value of t we have $\|\mathbf{r}(t)\| = k$, so by Theorem 15.2.7

$$\mathbf{r}(t) \cdot \mathbf{r}'(t) = 0$$

and hence the radius vector $\mathbf{r}(t)$ and the tangent vector $\mathbf{r}'(t)$ are perpendicular. ◀

▶ Exercise Set 15.2   ⓒ 33, 35, 36

In Exercises 1–8, find the limit.

1. $\lim\limits_{t \to 3} (t^2\mathbf{i} + 2t\mathbf{j})$.

2. $\lim\limits_{t \to \pi/4} \langle \cos t, \sin t \rangle$.

3. $\lim\limits_{t \to 0^+} \left(\sqrt{t}\mathbf{i} + \dfrac{\sin t}{t}\mathbf{j} \right)$.

4. $\lim\limits_{t \to +\infty} \left\langle \dfrac{t^2 + 1}{3t^2 + 2}, \dfrac{1}{t} \right\rangle$.

5. $\lim\limits_{t \to 2} (t\mathbf{i} - 3\mathbf{j} + t^2\mathbf{k})$.

6. $\lim\limits_{t \to \pi} \langle \cos 3t, e^{-t}, \sqrt{t} \rangle$.

7. $\lim\limits_{t \to +\infty} \left(\tan^{-1} t\mathbf{i} + \dfrac{t}{t^2 + 3}\mathbf{j} + \cos\dfrac{2}{t}\mathbf{k} \right)$.

8. $\lim\limits_{t \to 1} \left\langle \dfrac{3}{t^2}, \dfrac{\ln t}{t^2 - 1}, \sin 2t \right\rangle$.

In Exercises 9 and 10, prove that $\mathbf{r}$ is continuous at t_0.

9. $\mathbf{r}(t) = (3 \sin t)\mathbf{i} - 2t\mathbf{j}$; $t_0 = \pi/2$.

10. $\mathbf{r}(t) = 5\mathbf{i} - \sqrt{3t + 1}\mathbf{j} + e^{2t}\mathbf{k}$; $t_0 = 1$.

In Exercises 11–14, find $\mathbf{r}'(t)$.

11. $\mathbf{r}(t) = (4 + 5t)\mathbf{i} + (t - t^2)\mathbf{j}$.

12. $\mathbf{r}(t) = 4\mathbf{i} - (\cos t)\mathbf{j}$.

13. $\mathbf{r}(t) = \dfrac{1}{t}\mathbf{i} + (\tan t)\mathbf{j} + e^{2t}\mathbf{k}$.

14. $\mathbf{r}(t) = (\tan^{-1} t)\mathbf{i} + (t \cos t)\mathbf{j} - \sqrt{t}\mathbf{k}$.

In Exercises 15–22, find $\mathbf{r}'(t_0)$; then sketch the graph of $\mathbf{r}(t)$ and the tangent vector $\mathbf{r}'(t_0)$.

15. $\mathbf{r}(t) = \langle t, t^2 \rangle$; $t_0 = 2$.

16. $\mathbf{r}(t) = (\cos t)\mathbf{i} + (\sin t)\mathbf{j}$; $t_0 = 3\pi/4$.

17. $\mathbf{r}(t) = \langle e^{-t}, e^{2t} \rangle$; $t_0 = \ln 2$.

18. $\mathbf{r}(t) = (\cos 2t)\mathbf{i} - (4 \sin t)\mathbf{j}$; $t_0 = \pi$.

19. $\mathbf{r}(t) = (2 \sin t)\mathbf{i} + \mathbf{j} + (2 \cos t)\mathbf{k}$; $t_0 = \pi/2$.

20. $\mathbf{r}(t) = (\cos t)\mathbf{i} + (\sin t)\mathbf{j} + t\mathbf{k}$; $t_0 = \pi/4$.

21. $\mathbf{r}(t) = 3\mathbf{i} + t\mathbf{j} + (2 - t^2)\mathbf{k}$; $t_0 = 1$.

22. $\mathbf{r}(t) = t\mathbf{i} + 2t\mathbf{j} + t^2\mathbf{k}$; $t_0 = 2$.

In Exercises 23–26, find parametric equations of the line tangent to the graph of $\mathbf{r}(t)$ at the point where $t = t_0$.

23. $\mathbf{r}(t) = t^2\mathbf{i} + (2 - \ln t)\mathbf{j}$; $t_0 = 1$.

24. $\mathbf{r}(t) = e^{2t}\mathbf{i} - (2 \cos 3t)\mathbf{j}$; $t_0 = 0$.

25. $\mathbf{r}(t) = (2 \cos \pi t)\mathbf{i} + (2 \sin \pi t)\mathbf{j} + 3t\mathbf{k}$; $t_0 = \frac{1}{3}$.

26. $\mathbf{r}(t) = (\ln t)\mathbf{i} + e^{-t}\mathbf{j} + t^3\mathbf{k}$; $t_0 = 2$.

In Exercises 27–30, find a vector equation of the line tangent to the graph of $\mathbf{r}(t)$ at the point P_0 on the curve.

27. $\mathbf{r}(t) = (2t - 1)\mathbf{i} + \sqrt{3t + 4}\mathbf{j}$; $P_0(-1, 2)$.

28. $\mathbf{r}(t) = (4 \cos t)\mathbf{i} - 3t\mathbf{j}$; $P_0(2, -\pi)$.

29. $\mathbf{r}(t) = t^2\mathbf{i} - \dfrac{1}{t + 1}\mathbf{j} + (4 - t^2)\mathbf{k}$; $P_0(4, 1, 0)$.

30. $\mathbf{r}(t) = (\sin t)\mathbf{i} + (\cosh t)\mathbf{j} + (\tan^{-1} t)\mathbf{k}$; $P_0(0, 1, 0)$.

31. Find an equation of the plane that is perpendicular to the curve $\mathbf{r} = (3 \sin t)\mathbf{i} - (2 \cos t)\mathbf{j} + t\mathbf{k}$ at the point where $t = \pi/2$. [*Note:* A plane is considered to be perpendicular to a curve at a point if it is perpendicular to the tangent line at that point.]

32. Find an equation of the plane that is perpendicular to the curve $\mathbf{r} = 3t^2\mathbf{i} + \sqrt{t + 5}\mathbf{j} - 2t\mathbf{k}$ at the point $P(3, 2, 2)$ on the curve. [See note in Exercise 31.]

33. (a) Find the points where the curve

$$\mathbf{r} = t\mathbf{i} + t^2\mathbf{j} - 3t\mathbf{k}$$

intersects the plane $2x - y + z = -2$.

(b) For the curve and plane in part (a), find, to the nearest degree, the acute angle that the tangent line to the curve makes with a line normal to the plane at each point of intersection.

34. Find where the tangent line to the curve

$$\mathbf{r} = e^{-2t}\mathbf{i} + (\cos t)\mathbf{j} + (3\sin t)\mathbf{k}$$

at the point $(1, 1, 0)$ intersects the yz-plane.

In Exercises 35 and 36, show that the graphs of $\mathbf{r}_1(t)$ and $\mathbf{r}_2(t)$ intersect at the point P. Find, to the nearest degree, the acute angle between the tangent lines to the graphs of $\mathbf{r}_1(t)$ and $\mathbf{r}_2(t)$ at the point P.

35. $\mathbf{r}_1(t) = t^2\mathbf{i} + t\mathbf{j} + 3t^3\mathbf{k}$,
$\mathbf{r}_2(t) = (t-1)\mathbf{i} + \frac{1}{4}t^2\mathbf{j} + (5-t)\mathbf{k};\ P(1, 1, 3)$.

36. $\mathbf{r}_1(t) = 2e^{-t}\mathbf{i} + (\cos t)\mathbf{j} + (t^2 + 3)\mathbf{k}$,
$\mathbf{r}_2(t) = (1-t)\mathbf{i} + t^2\mathbf{j} + (t^3 + 4)\mathbf{k};\ P(2, 1, 3)$.

In Exercises 37–48, evaluate the integral.

37. $\displaystyle\int (3\mathbf{i} + 4t\mathbf{j})\, dt.$ **38.** $\displaystyle\int [(\cos t)\mathbf{i} + (\sin t)\mathbf{j}]\, dt.$

39. $\displaystyle\int_0^{\pi/3} \langle \cos 3t, -\sin 3t \rangle\, dt.$

40. $\displaystyle\int_0^1 (t^2\mathbf{i} + t^3\mathbf{j})\, dt.$ **41.** $\displaystyle\int_1^9 (t^{1/2}\mathbf{i} + t^{-1/2}\mathbf{j})\, dt.$

42. $\displaystyle\int [(t\sin t)\mathbf{i} + \mathbf{j}]\, dt.$ **43.** $\displaystyle\int \langle te^t, \ln t \rangle\, dt.$

44. $\displaystyle\int_0^2 \|t\mathbf{i} + t^2\mathbf{j}\|\, dt.$ **45.** $\displaystyle\int \left[t^2\mathbf{i} - 2t\mathbf{j} + \frac{1}{t}\mathbf{k}\right] dt.$

46. $\displaystyle\int \langle e^{-t}, e^t, 3t^2 \rangle\, dt.$ **47.** $\displaystyle\int_0^1 (e^{2t}\mathbf{i} + e^{-t}\mathbf{j} + t\mathbf{k})\, dt.$

48. $\displaystyle\int_{-3}^3 \langle (3-t)^{3/2}, (3+t)^{3/2}, 1 \rangle\, dt.$

49. Suppose that a particle moves through 3-space along the curve $\mathbf{r} = t\mathbf{i} - 3t^2\mathbf{j} + \mathbf{k}$ and that it is subjected to a force of $\mathbf{F} = 3x\mathbf{i} - 2\mathbf{j} + yz\mathbf{k}$ when it is at the point (x, y, z).

(a) Find $\mathbf{F}$ in terms of t for points on the path.

(b) Find $\displaystyle\int_0^2 \mathbf{F} \cdot \frac{d\mathbf{r}}{dt}\, dt.$ [*Note:* Later, we shall see that this is the work done by the force as the particle moves along the curve from the point where $t = 0$ to the point where $t = 2$.]

50. Find $\mathbf{r}(t)$ given that $\mathbf{r}'(t) = t^2\mathbf{i} + 2t\mathbf{j}$ and $\mathbf{r}(0) = \mathbf{i} + \mathbf{j}$.

51. Find $\mathbf{r}(t)$ given that $\mathbf{r}'(t) = (\cos t)\mathbf{i} + (\sin t)\mathbf{j}$ and $\mathbf{r}(0) = \mathbf{i} - \mathbf{j}$.

52. Find $\mathbf{r}(t)$ given that $\mathbf{r}''(t) = \mathbf{i} + e^t\mathbf{j}$, $\mathbf{r}(0) = 2\mathbf{i}$, and $\mathbf{r}'(0) = \mathbf{j}$.

53. Find $\mathbf{r}(t)$ given that $\mathbf{r}''(t) = 12t^2\mathbf{i} - 2\mathbf{j}$, $\mathbf{r}'(0) = \mathbf{0}$, and $\mathbf{r}(0) = 2\mathbf{i} - 4\mathbf{j}$.

54. Find $\mathbf{r}(t)$ given that $\mathbf{r}'(t) = e^{-2t}\mathbf{i} + (\cos t)\mathbf{j} - \mathbf{k}$ and $\mathbf{r}(0) = 3\mathbf{j} + 2\mathbf{k}$.

55. Find $\mathbf{r}(t)$ given that $\mathbf{r}'(t) = 2\mathbf{i} + \dfrac{t}{t^2 + 1}\mathbf{j} + t\mathbf{k}$ and $\mathbf{r}(1) = \mathbf{0}$.

56. Find $\mathbf{r}(t)$ given that $\mathbf{r}''(t) = (4\sin 2t)\mathbf{i} + 6t\mathbf{j} + e^{-t}\mathbf{k}$, $\mathbf{r}(0) = 2\mathbf{i}$, and $\mathbf{r}'(0) = \mathbf{k}$.

57. Calculate $(d/dt)[\mathbf{r}_1(t) \cdot \mathbf{r}_2(t)]$ two ways: first using Formula (8), and then by differentiating $\mathbf{r}_1(t) \cdot \mathbf{r}_2(t)$ directly.

(a) $\mathbf{r}_1(t) = 2t\mathbf{i} + 3t^2\mathbf{j} + t^3\mathbf{k}$, $\mathbf{r}_2(t) = t^4\mathbf{k}$

(b) $\mathbf{r}_1(t) = 3\sec t\mathbf{i} - t\mathbf{j} + \ln t\mathbf{k}$, $\mathbf{r}_2(t) = 4t\mathbf{i} - \sin t\mathbf{k}$.

58. Calculate $(d/dt)[\mathbf{r}_1(t) \times \mathbf{r}_2(t)]$ two ways: first using Formula (9), and then by differentiating $\mathbf{r}_1(t) \times \mathbf{r}_2(t)$ directly.

(a) $\mathbf{r}_1(t) = 2t\mathbf{i} + 3t^2\mathbf{j} + t^3\mathbf{k}$, $\mathbf{r}_2(t) = t^4\mathbf{k}$

(b) $\mathbf{r}_1(t) = 3\sec t\mathbf{i} - t\mathbf{j} + \ln t\mathbf{k}$, $\mathbf{r}_2(t) = 4t\mathbf{i} - \sin t\mathbf{k}$.

59. Prove: $(d/dt)[\mathbf{r}(t) \times \mathbf{r}'(t)] = \mathbf{r}(t) \times \mathbf{r}''(t)$. [*Hint:* Use Formula (9).]

60. Let $\mathbf{r} = \mathbf{r}(t)$. Show that

$$\frac{d}{dt}[\|\mathbf{r}\|] = \frac{1}{\|\mathbf{r}\|}\mathbf{r} \cdot \mathbf{r}'$$

[*Hint:* Consider $\mathbf{r} \cdot \mathbf{r}$.]

61. Use part (*e*) of Theorem 15.2.1 and Exercise 60 to derive the formula

$$\frac{d}{dt}\left[\frac{\mathbf{r}}{\|\mathbf{r}\|}\right] = \frac{1}{\|\mathbf{r}\|}\mathbf{r}' - \frac{\mathbf{r} \cdot \mathbf{r}'}{\|\mathbf{r}\|^3}\mathbf{r}$$

62. Let $\mathbf{u} = \mathbf{u}(t)$, $\mathbf{v} = \mathbf{v}(t)$, and $\mathbf{w} = \mathbf{w}(t)$ be differentiable vector-valued functions. Use Formulas (8) and (9) to show that

$$\frac{d}{dt}[\mathbf{u} \cdot (\mathbf{v} \times \mathbf{w})]$$

$$= \frac{d\mathbf{u}}{dt} \cdot [\mathbf{v} \times \mathbf{w}] + \mathbf{u} \cdot \left[\frac{d\mathbf{v}}{dt} \times \mathbf{w}\right] + \mathbf{u} \cdot \left[\mathbf{v} \times \frac{d\mathbf{w}}{dt}\right]$$

63. Let $u_1, u_2, u_3, v_1, v_2, v_3, w_1, w_2,$ and w_3 be differentiable functions of t. Use Exercise 62 to show that

$$\frac{d}{dt}\begin{vmatrix} u_1 & u_2 & u_3 \\ v_1 & v_2 & v_3 \\ w_1 & w_2 & w_3 \end{vmatrix}$$

$$= \begin{vmatrix} u_1' & u_2' & u_3' \\ v_1 & v_2 & v_3 \\ w_1 & w_2 & w_3 \end{vmatrix} + \begin{vmatrix} u_1 & u_2 & u_3 \\ v_1' & v_2' & v_3' \\ w_1 & w_2 & w_3 \end{vmatrix} + \begin{vmatrix} u_1 & u_2 & u_3 \\ v_1 & v_2 & v_3 \\ w_1' & w_2' & w_3' \end{vmatrix}$$

64. Prove Theorem 15.2.1 for 2-space.

65. Derive Formulas (8) and (9) for 3-space.

66. Let lim stand for any one of the limit symbols $\lim\limits_{t \to a}$, $\lim\limits_{t \to a^+}$, $\lim\limits_{t \to a^-}$, $\lim\limits_{t \to +\infty}$, or $\lim\limits_{t \to -\infty}$. Prove for 2-space:

(a) If k is a scalar and $\lim \mathbf{r}(t)$ exists, then $\lim k\mathbf{r}(t) = k \lim \mathbf{r}(t)$.

(b) If $\lim \mathbf{r}_1(t)$ and $\lim \mathbf{r}_2(t)$ exist, then
$\lim [\mathbf{r}_1(t) + \mathbf{r}_2(t)] = \lim \mathbf{r}_1(t) + \lim \mathbf{r}_2(t)$,
$\lim [\mathbf{r}_1(t) - \mathbf{r}_2(t)] = \lim \mathbf{r}_1(t) - \lim \mathbf{r}_2(t)$.

67. Prove for 2-space: $\mathbf{r}$ is continuous at t_0 if and only if each component of $\mathbf{r}$ is continuous at t_0.

68. Prove for 2-space:

(a) $\int k\mathbf{r}(t)\,dt = k\int \mathbf{r}(t)\,dt$, where k is a scalar constant

(b) $\int [\mathbf{r}_1(t) + \mathbf{r}_2(t)]\,dt = \int \mathbf{r}_1(t)\,dt + \int \mathbf{r}_2(t)\,dt.$

69. Prove for 2-space: If $\mathbf{R}'(t) = \mathbf{r}(t)$ on an interval $[a, b]$, then

(a) $\int \mathbf{r}(t)\,dt = \mathbf{R}(t) + \mathbf{C}$, where $\mathbf{C}$ is an arbitrary vector constant

(b) $\int_a^b \mathbf{r}(t)\,dt = \mathbf{R}(b) - \mathbf{R}(a).$

■ 15.3 CHANGE OF PARAMETER; ARC LENGTH

We observed in earlier sections that a given curve in 2-space or 3-space can be represented parametrically in more than one way. For example, in Section 13.4 we gave two parametric representations of a circle—one in which the circle was traced clockwise and the other in which it was traced counterclockwise. Sometimes it is desirable to change a given parametric representation of a curve to an alternative representation that is better suited for the problem at hand. In this section we shall investigate issues associated with changes of parameter, and we shall show that arc length plays a special role in parametric representations of curves.

☐ **SMOOTH PARAMETRIZATIONS**

In classical applications, there is relatively little interest in parametric representations of curves in 2-space or 3-space in which the curve is traced in a discontinuous or erratic fashion. Thus, we impose restrictions on the kind of parametric representations that will be considered. We shall say that $\mathbf{r}(t)$ is *smoothly parametrized* or that $\mathbf{r}$ is a *smooth function of t* if $\mathbf{r}'(t)$ is continuous and $\mathbf{r}'(t) \neq \mathbf{0}$ for any value of t. Stated another way, the components of $\mathbf{r}(t)$ have continuous derivatives with respect to t and are not all zero for any value of t. Thus, in 3-space

$$\mathbf{r}(t) = x(t)\mathbf{i} + y(t)\mathbf{j} + z(t)\mathbf{k}$$

is a smooth function of t if $x'(t)$, $y'(t)$, and $z'(t)$ are continuous and there is no value of t at which all three derivatives are zero.

It can be shown that if $\mathbf{r}(t)$ is a smoothly parametrized function, then the angles between the tangent vector $\mathbf{r}'(t)$ and the unit vectors $\mathbf{i}$, $\mathbf{j}$, and $\mathbf{k}$ are continuous functions of t (Exercise 38). Thus, a smoothly parametrized function $\mathbf{r}(t)$ is said to have a *continuously turning tangent vector*.

Example 1   Determine whether the following vector-valued functions have continuously turning tangent vectors.

(a) $\mathbf{r}(t) = a\cos t\mathbf{i} + a\sin t\mathbf{j} + ct\mathbf{k}$ $(a > 0, c > 0)$

(b) $\mathbf{r}(t) = t^2\mathbf{i} + t^3\mathbf{j}$

Solution (a).   We have

$$\mathbf{r}'(t) = -a\sin t\mathbf{i} + a\cos t\mathbf{j} + c\mathbf{k}$$

All three components are continuous functions, and there is no value of t for which all three components are zero (verify), so $\mathbf{r}(t)$ has a continuously turning tangent vector. The graph of $\mathbf{r}(t)$ is the circular helix in Figure 15.1.1.

Solution (b). We have

$$\mathbf{r}'(t) = 2t\mathbf{i} + 3t^2\mathbf{j}$$

Although both components are continuous functions, they are both equal to zero if $t = 0$, so $\mathbf{r}(t)$ does not have a continuously turning tangent vector. The graph of $\mathbf{r}(t)$, which is shown in Figure 15.3.1, is a semicubical parabola traced in the upward direction (see Example 9 of Section 13.4). Observe that for values of t slightly less than zero the angle between $\mathbf{r}'(t)$ and $\mathbf{i}$ is near π, and for values of t slightly larger than zero the angle is near 0; hence there is a sudden reversal in the direction of the tangent vector as t increases through $t = 0$. ◀

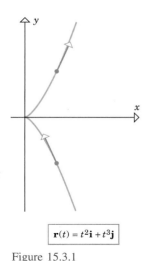

$\mathbf{r}(t) = t^2\mathbf{i} + t^3\mathbf{j}$

Figure 15.3.1

☐ **CHANGE OF PARAMETER**

A ***change of parameter*** in a vector-valued function $\mathbf{r}(t)$ is a substitution $t = g(\tau)$ which produces a new vector-valued function $\mathbf{r}(g(\tau))$ having the same graph as $\mathbf{r}(t)$, but possibly traced differently as the parameter τ increases.

Example 2 Find a change of parameter $t = g(\tau)$ for the circle

$$\mathbf{r}(t) = \cos t\,\mathbf{i} + \sin t\,\mathbf{j} \quad (0 \le t \le 2\pi)$$

such that

(a) the circle is traced counterclockwise as τ increases over the interval $[0, 1]$;

(b) the circle is traced clockwise as τ increases over the interval $[0, 1]$.

Solution (a). The circle is traced counterclockwise if t increases (Example 2 of Section 13.4). Thus, if we choose g to be an increasing function, then it will follow from the relationship $t = g(\tau)$ that t increases when τ increases, thereby ensuring that the circle is traced counterclockwise as τ increases. We also want to choose g so that t increases from 0 to 2π as τ increases from 0 to 1. A simple choice of g that satisfies all of the required criteria is the linear function graphed in Figure 15.3.2a. The equation of this line is

$$t = g(\tau) = 2\pi\tau \tag{1}$$

which is the desired change of parameter. The resulting representation of the circle in terms of the parameter τ is

$$\mathbf{r}(g(\tau)) = \cos 2\pi\tau\,\mathbf{i} + \sin 2\pi\tau\,\mathbf{j} \quad (0 \le \tau \le 1)$$

Solution (b). To ensure that the circle is traced clockwise, we shall choose g to be a decreasing function such that τ decreases from 2π to 0 as τ increases from 0 to 1. A simple choice of g that achieves this is the linear function

$$t = g(\tau) = 2\pi(1 - \tau) \tag{2}$$

graphed in Figure 15.3.2b. The resulting representation of the circle in terms of the parameter τ is

$$\mathbf{r}(g(\tau)) = \cos(2\pi(1 - \tau))\mathbf{i} + \sin(2\pi(1 - \tau))\mathbf{j} \quad (0 \le \tau \le 1)$$

which simplifies to (verify)

$$\mathbf{r}(g(\tau)) = \cos 2\pi\tau\,\mathbf{i} - \sin 2\pi\tau\,\mathbf{j} \quad (0 \le \tau \le 1) \quad ◀$$

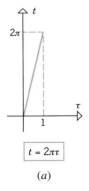

$t = 2\pi\tau$

(a)

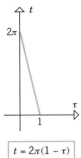

$t = 2\pi(1 - \tau)$

(b)

Figure 15.3.2

The following version of the chain rule for vector-valued functions relates the derivatives of $\mathbf{r}(t)$ and $\mathbf{r}(g(\tau))$ under a change of parameter $t = g(\tau)$. The proof is left as an exercise.

15.3.1   THEOREM (*Chain Rule*).   *Let $\mathbf{r}(t)$ be a vector-valued function in 2-space or 3-space that is differentiable with respect to t. If $t = g(\tau)$ is a change of parameter in which g is differentiable with respect to τ, then $\mathbf{r}(g(\tau))$ is differentiable with respect to τ and*

$$\frac{d\mathbf{r}}{d\tau} = \frac{d\mathbf{r}}{dt}\frac{dt}{d\tau} \tag{3}$$

When making a change of parameter $t = g(\tau)$ in a vector-valued function $\mathbf{r}(t)$, it will be important to ensure that $\mathbf{r}(g(\tau))$ is smooth if $\mathbf{r}(t)$ is smooth. A change of parameter for which this is true is called a ***smooth change of parameter***. It follows from (3) that $t = g(\tau)$ is a smooth change of parameter if $dt/d\tau$ is continuous and $dt/d\tau \neq 0$ for any value of τ, since these conditions imply that $d\mathbf{r}/d\tau$ is continuous and nonzero if $d\mathbf{r}/dt$ is continuous and nonzero. Smooth changes of parameter fall into two categories—those for which $dt/d\tau > 0$ for all τ and those for which $dt/d\tau < 0$ for all τ.

A smooth change of parameter for which $dt/d\tau > 0$ for all τ will be called a ***positive change of parameter***, and a smooth change of parameter for which $dt/d\tau < 0$ for all τ will be called a ***negative change of parameter***. A positive change of parameter preserves the orientation of a parametric curve and a negative change of parameter reverses the orientation.

Example 3   In Example 2 the change of parameter given by (1) is positive since $dt/d\tau = 2\pi > 0$, and the change of parameter given by (2) is negative since $dt/d\tau = -2\pi < 0$. The positive change of parameter preserved the orientation of the circle, and the negative change of parameter reversed it.   ◄

□ ARC LENGTH IN 3-SPACE

In Theorem 13.4.1 we showed that in 2-space the arc length L of a parametric curve

$$x = x(t), \quad y = y(t) \quad (a \leq t \leq b)$$

is given by

$$L = \int_a^b \sqrt{\left(\frac{dx}{dt}\right)^2 + \left(\frac{dy}{dt}\right)^2}\, dt \tag{4}$$

This result generalizes to curves in 3-space exactly as one would expect: If no segment of the curve

$$x = x(t), \quad y = y(t), \quad z = z(t) \quad (a \leq t \leq b)$$

is traced more than once as t increases from a to b, and if dx/dt, dy/dt, and dz/dt are continuous for $a \leq t \leq b$, then the arc length L of the curve is given by

$$L = \int_a^b \sqrt{\left(\frac{dx}{dt}\right)^2 + \left(\frac{dy}{dt}\right)^2 + \left(\frac{dz}{dt}\right)^2}\, dt \tag{5}$$

Example 4   Find the arc length of that portion of the circular helix

$$x = \cos t, \quad y = \sin t, \quad z = t$$

from $t = 0$ to $t = \pi$.

Solution. From (5) the arc length is

$$L = \int_0^\pi \sqrt{\left(\frac{dx}{dt}\right)^2 + \left(\frac{dy}{dt}\right)^2 + \left(\frac{dz}{dt}\right)^2}\, dt$$

$$= \int_0^\pi \sqrt{(-\sin t)^2 + (\cos t)^2 + 1}\, dt = \int_0^\pi \sqrt{2}\, dt = \sqrt{2}\,\pi \quad \blacktriangleleft$$

If the parametric equations corresponding to Formulas (4) and (5) are expressed in vector form, say

$$\mathbf{r}(t) = x(t)\mathbf{i} + y(t)\mathbf{j} \quad \text{or} \quad \mathbf{r}(t) = x(t)\mathbf{i} + y(t)\mathbf{j} + z(t)\mathbf{k}$$

then

$$\frac{d\mathbf{r}}{dt} = \frac{dx}{dt}\mathbf{i} + \frac{dy}{dt}\mathbf{j} \quad \text{or} \quad \frac{d\mathbf{r}}{dt} = \frac{dx}{dt}\mathbf{i} + \frac{dy}{dt}\mathbf{j} + \frac{dz}{dt}\mathbf{k}$$

Thus, Formulas (4) and (5) can both be expressed in vector form as

$$L = \int_a^b \left\| \frac{d\mathbf{r}}{dt} \right\| dt \tag{6}$$

☐ **ARC LENGTH AS A PARAMETER**

For many purposes the best parameter to use for representing a curve in 2-space or 3-space parametrically is the length of arc measured along the curve from some fixed reference point. This can be done as follows:

Step 1. Select an arbitrary point on the curve C to serve as a ***reference point***.

Step 2. Starting from the reference point, choose one direction along the curve to be the ***positive direction*** and the other to be the ***negative direction***.

Step 3. If P is a point on the curve, let s be the "signed" arc length along C from the reference point to P, where s is positive if P is in the positive direction from the reference point, and s is negative if P is in the negative direction. Figure 15.3.3 illustrates this idea.

By this procedure, a unique point P on the curve is determined when a value for s is given. For example, $s = 2$ determines the point that is 2 units along the curve in the positive direction from the reference point, and $s = -\frac{3}{2}$ determines the point that is $\frac{3}{2}$ units along the curve in the negative direction from the reference point.

Let us now treat s as a variable. As the value of s changes, the corresponding point P moves along C and the coordinates of P become functions of s. Thus, in 2-space the coordinates of P are $(x(s), y(s))$, and in 3-space they are $(x(s), y(s), z(s))$. Therefore, in 2-space or 3-space the curve C is given by the parametric equations

$$x = x(s), \quad y = y(s) \qquad \text{or} \qquad x = x(s), \quad y = y(s), \quad z = z(s)$$

A parametric representation of a curve with arc length as the parameter is called an ***arc-length parametrization*** of the curve. Note that a given curve will generally have infinitely many different arc-length parametrizations, since the reference point and orientation can be chosen arbitrarily.

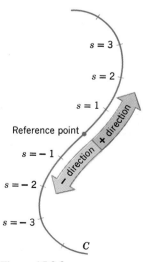

$s = 3$

$s = 2$

$s = 1$

Reference point

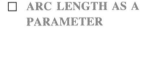

$s = -1$

$s = -2$

$s = -3$

C

Figure 15.3.3

Example 5 Find the arc-length parametrization of the circle $x^2 + y^2 = a^2$ with counterclockwise orientation and $(a, 0)$ as the reference point.

Solution. The circle with counterclockwise orientation can be represented by the parametric equations

$$x = a \cos t, \quad y = a \sin t \quad (0 \le t \le 2\pi) \tag{7}$$

in which t can be interpreted as the angle in radian measure from the positive x-axis to the radius from the origin to the point $P(x, y)$ (Figure 15.3.4). If we take the positive direction for measuring the arc length to be counterclockwise, and we take $(a, 0)$ to be the reference point, then s and t are related by

$$s = at \quad \text{or} \quad t = s/a$$

Making this change of variable in (7) and noting that s increases from 0 to $2\pi a$ as t increases from 0 to 2π yields the following arc-length parametrization of the circle

$$x = a \cos (s/a), \quad y = a \sin (s/a) \quad (0 \le s \le 2\pi a) \quad \blacktriangleleft$$

In the preceding example we used basic trigonometry to find a formula for changing the parameter from t to arc length s. However, it is only in the simplest cases that geometric methods can be used to find arc-length parametrizations. The following theorem will provide a more general method for doing this.

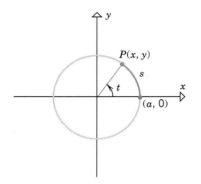

Figure 15.3.4

15.3.2 THEOREM. *Let C be the graph of a smooth vector-valued function* $\mathbf{r}(t)$ *in 2-space or 3-space, and let* $\mathbf{r}(t_0)$ *be any point on C. Then the following formula defines a positive change of parameter from t to s, where s is an arc-length parameter having* $\mathbf{r}(t_0)$ *as its reference point:*

$$s = \int_{t_0}^{t} \left\| \frac{d\mathbf{r}}{du} \right\| du \tag{8}$$

(Figure 15.3.5).

Proof. From (6) with u as the variable of integration instead of t, the integral represents the arc length of that portion of C between $\mathbf{r}(t_0)$ and $\mathbf{r}(t)$ if $t > t_0$ and the negative of that arc length if $t < t_0$. Thus, s is the arc-length parameter with $\mathbf{r}(t_0)$ as its reference point and its positive direction in the direction of increasing t. ■

For reference, we note that in 2-space and 3-space, respectively, Formula (8) can be expressed in component form as

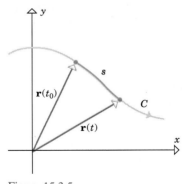

Figure 15.3.5

$$s = \int_{t_0}^{t} \sqrt{\left(\frac{dx}{du}\right)^2 + \left(\frac{dy}{du}\right)^2} \, du \tag{9}$$

$$s = \int_{t_0}^{t} \sqrt{\left(\frac{dx}{du}\right)^2 + \left(\frac{dy}{du}\right)^2 + \left(\frac{dz}{du}\right)^2} \, du \tag{10}$$

Example 6 Find the arc-length parametrization of the line

$$x = 2t + 1, \quad y = 3t - 2 \tag{11}$$

that has reference point $(1, -2)$ and the same orientation as the original line.

Solution.   In Formula (9) we used u as the variable of integration because t was needed as a limit of integration. To apply (9), we first rewrite the given parametric equations with u in place of t; this gives

$$x = 2u + 1, \quad y = 3u - 2$$

from which we obtain

$$\frac{dx}{du} = 2, \quad \frac{dy}{du} = 3$$

From (11) we see that the reference point $(1, -2)$ corresponds to $t = t_0 = 0$, so (9) yields

$$s = \int_{t_0}^{t} \sqrt{\left(\frac{dx}{du}\right)^2 + \left(\frac{dy}{du}\right)^2}\, du = \int_{0}^{t} \sqrt{13}\, du = \sqrt{13}\, u \Big]_{u=0}^{u=t} = \sqrt{13}\, t$$

Therefore, $t = s/\sqrt{13}$. Substituting this expression in (11) and simplifying yields the parametric equations

$$x = \frac{2}{\sqrt{13}} s + 1, \quad y = \frac{3}{\sqrt{13}} s - 2 \qquad \blacktriangleleft$$

Example 7   Let $\mathbf{u} = \langle u_1, u_2, u_3 \rangle$ be a vector of length 1, and let l be the line that is parallel to $\mathbf{u}$ and passes through the point $P_0(x_0, y_0, z_0)$. Find the arc-length parametrization of l with the reference point at P_0 and the positive direction in the direction of $\mathbf{u}$.

Solution.   If $P(x, y, z)$ is any point on the line, then the vector $\overrightarrow{P_0 P}$ is parallel to $\mathbf{u}$ and hence is some scalar multiple of $\mathbf{u}$, say

$$\overrightarrow{P_0 P} = s\mathbf{u} \tag{12}$$

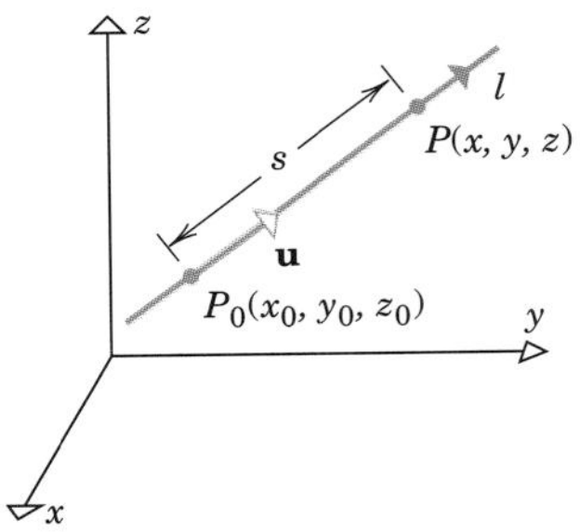

(Figure 15.3.6). Because $\mathbf{u}$ has length 1, it follows that s can be interpreted as the signed arc length from P_0 to P, where $s = 0$ if P_0 and P coincide, $s > 0$ if P is distinct from P_0 and $\overrightarrow{P_0 P}$ has the same direction as $\mathbf{u}$, and $s < 0$ if P is distinct from P_0 and $\overrightarrow{P_0 P}$ is oppositely directed to $\mathbf{u}$. It follows from (12) that

$$\langle x - x_0, y - y_0, z - z_0 \rangle = s\langle u_1, u_2, u_3 \rangle$$

from which we obtain the parametric equations

$$x = x_0 + su_1, \quad y = y_0 + su_2, \quad z = z_0 + su_3$$

(We leave it for the reader to show that the same result holds in 2-space with two components rather than three.) $\blacktriangleleft$

Figure 15.3.6

REMARK.   Formula (8) tends to produce integrals that cannot be evaluated easily, so it is only in relatively simple cases, such as those in the preceding examples, that explicit formulas for arc-length parametrizations can be obtained.

Because arc-length parameters for a curve C are intimately related to the geometric characteristics of C, arc-length parametrizations have properties that are not enjoyed by other parametrizations. For example, the following theorem shows that if a smooth curve is represented parametrically using an arc-length parameter, then the tangent vectors all have length 1.

15.3.3   THEOREM.   *If $\mathbf{r}(t)$ is a smooth vector-valued function in 2-space or 3-space, and if s is an arc-length parameter, then $\|d\mathbf{r}/ds\| = 1$ for all s.*

Proof.   It follows from the chain rule (3) with s in place of τ that

$$\frac{d\mathbf{r}}{ds} = \frac{d\mathbf{r}}{dt}\frac{dt}{ds} = \frac{1}{ds/dt}\frac{d\mathbf{r}}{dt} \tag{13}$$

But from (8) and the Second Fundamental Theorem of Calculus (5.9.1) we have

$$\frac{ds}{dt} = \left\|\frac{d\mathbf{r}}{dt}\right\| \tag{14}$$

so (13) can be written as

$$\frac{d\mathbf{r}}{ds} = \frac{1}{\|d\mathbf{r}/dt\|}\frac{d\mathbf{r}}{dt} \tag{15}$$

Recalling that multiplying a nonzero vector by the reciprocal of its length (normalizing) produces a vector of length 1, we conclude from (15) that $\|d\mathbf{r}/ds\| = 1$.   ∎

Example 8   In Example 5 we showed that the circle $x^2 + y^2 = a^2$ can be represented parametrically in terms of arc length as

$$x = a\cos(s/a), \quad y = a\sin(s/a) \qquad (0 \le s \le 2\pi a)$$

Writing these equations in vector form, and then differentiating with respect to s to obtain the tangent vector, yields

$$\mathbf{r}(s) = a\cos(s/a)\mathbf{i} + a\sin(s/a)\mathbf{j}$$

$$\frac{d\mathbf{r}}{ds} = -\sin(s/a)\mathbf{i} + \cos(s/a)\mathbf{j}$$

As guaranteed by Theorem 15.3.3, every tangent vector has length 1, since

$$\|d\mathbf{r}/ds\| = \sqrt{[-\sin(s/a)]^2 + [\cos(s/a)]^2} = \sqrt{1} = 1 \qquad \blacktriangleleft$$

The component forms of Formula (14) in 2-space and 3-space will be of sufficient interest in later sections that we provide them here for reference:

$$\frac{ds}{dt} = \left\|\frac{d\mathbf{r}}{dt}\right\| = \sqrt{\left(\frac{dx}{dt}\right)^2 + \left(\frac{dy}{dt}\right)^2} \tag{16}$$

$$\frac{ds}{dt} = \left\|\frac{d\mathbf{r}}{dt}\right\| = \sqrt{\left(\frac{dx}{dt}\right)^2 + \left(\frac{dy}{dt}\right)^2 + \left(\frac{dz}{dt}\right)^2} \tag{17}$$

REMARK.   It is of interest to note that (16) and (17) do not involve t_0, and hence do not depend on where the reference point for s is chosen. This is to be expected, since changing the position of the reference point shifts s by a constant (the arc length between the reference points), and this constant drops out on differentiating.

▶ Exercise Set 15.3   © *18*

1. Show that $\mathbf{r}_1(t) = t\mathbf{i} + t^2\mathbf{j}$ and $\mathbf{r}_2(t) = t^3\mathbf{i} + t^6\mathbf{j}$ have the same graphs, but that $\mathbf{r}_1'(t)$ is never zero, whereas $\mathbf{r}_2'(t) = \mathbf{0}$ for some value of t.

2. Show that the graphs of

$$\mathbf{r}_1(t) = (\cos t)\mathbf{i} + (\sin t)\mathbf{j} + t\mathbf{k}$$

and

$$\mathbf{r}_2(t) = \cos(t^3)\mathbf{i} + \sin(t^3)\mathbf{j} + t^3\mathbf{k}$$

are the same, but that $\mathbf{r}_1'(t)$ is never zero, whereas $\mathbf{r}_2'(t) = \mathbf{0}$ for some value of t.

In Exercises 3–6, determine whether $\mathbf{r}$ is a smooth function of the parameter t.

3. $\mathbf{r} = t^3\mathbf{i} + (3t^2 - 2t)\mathbf{j} + t^2\mathbf{k}$.

4. $\mathbf{r} = \cos(t^2)\mathbf{i} + \sin(t^2)\mathbf{j} + e^{-t}\mathbf{k}$.

5. $\mathbf{r} = te^{-t}\mathbf{i} + (t^2 - 2t)\mathbf{j} + \cos(\pi t)\mathbf{k}$.

6. $\mathbf{r} = \sin(\pi t)\mathbf{i} + (2t - \ln t)\mathbf{j} + (t^2 - t)\mathbf{k}$.

In Exercises 7–9, calculate $d\mathbf{r}/d\tau$ by the chain rule, and then check your result by expressing $\mathbf{r}$ in terms of τ and differentiating.

7. $\mathbf{r} = t\mathbf{i} + t^2\mathbf{j};\ t = 4\tau + 1$.

8. $\mathbf{r} = \langle 3\cos t, 3\sin t \rangle;\ t = \pi\tau$.

9. $\mathbf{r} = e^t\mathbf{i} + 4e^{-t}\mathbf{j};\ t = \tau^2$.

10. Prove Theorem 15.3.1 for 2-space.

In Exercises 11–16, find the arc length of the curve.

11. $\mathbf{r}(t) = (4 + 3t)\mathbf{i} + (2 - 2t)\mathbf{j} + (5 + t)\mathbf{k};\ 3 \le t \le 4$.

12. $\mathbf{r}(t) = 3\cos t\mathbf{i} + 3\sin t\mathbf{j} + t\mathbf{k};\ 0 \le t \le 2\pi$.

13. $\mathbf{r}(t) = t^3\mathbf{i} + t\mathbf{j} + \frac{1}{2}\sqrt{6}t^2\mathbf{k};\ 1 \le t \le 3$.

14. $x = \cos^3 t, y = \sin^3 t, z = 2;\ 0 \le t \le \pi/2$.

15. $\mathbf{r}(t) = \langle e^t, e^{-t}, \sqrt{2}t \rangle;\ 0 \le t \le 1$.

16. $x = \frac{1}{2}t, y = \frac{1}{3}(1 - t)^{3/2}, z = \frac{1}{3}(1 + t)^{3/2};\ -1 \le t \le 1$.

17. Find the arc length of the circular helix $x = a\cos t$, $y = a\sin t$, $z = ct$ for $0 \le t \le t_0$.

18. Copper tubing with an outside diameter of $\frac{1}{2}$ in. is to be wrapped in a circular helix around a cylindrical core that has a 12-in. diameter. What length of tubing will make one complete turn around the cylinder in a distance of 20 in. measured along the axis of the cylinder?

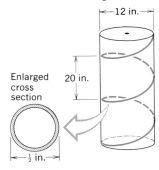

In Exercises 19–27, find parametric equations for the curve using arc length s as a parameter. Use the point on the curve where $t = 0$ as the reference point.

19. $\mathbf{r}(t) = (3t - 2)\mathbf{i} + (4t + 3)\mathbf{j}$.

20. $\mathbf{r}(t) = (3\cos 2t)\mathbf{i} + (3\sin 2t)\mathbf{j};\ 0 \le t \le \pi$.

21. $\mathbf{r}(t) = (3 + \cos t)\mathbf{i} + (2 + \sin t)\mathbf{j};\ 0 \le t \le 2\pi$.

22. $\mathbf{r}(t) = (\cos^3 t)\mathbf{i} + (\sin^3 t)\mathbf{j};\ 0 \le t \le \pi/2$.

23. $\mathbf{r}(t) = \frac{1}{3}t^3\mathbf{i} + \frac{1}{2}t^2\mathbf{j};\ t \ge 0$.

24. $\mathbf{r}(t) = (1 + t)^2\mathbf{i} + (1 + t)^3\mathbf{j};\ 0 \le t \le 1$.

25. $\mathbf{r}(t) = (e^t\cos t)\mathbf{i} + (e^t\sin t)\mathbf{j};\ 0 \le t \le \pi/2$.

26. $\mathbf{r}(t) = \sin(e^t)\mathbf{i} + \cos(e^t)\mathbf{j} + \sqrt{3}e^t\mathbf{k};\ t \ge 0$.

27. $\mathbf{r}(t) = (t\cos t)\mathbf{i} + (t\sin t)\mathbf{j} + \frac{2}{3}\sqrt{2}t^{3/2}\mathbf{k};\ t \ge 0$.

28. Find parametric equations for the cycloid
$$x = at - a\sin t$$
$$y = a - a\cos t$$
$(0 \le t \le 2\pi)$
using arc length as the parameter. Take $(0, 0)$ as the reference point.

29. Use the result in Exercise 17 to show that the circular helix
$$\mathbf{r} = (a\cos t)\mathbf{i} + (a\sin t)\mathbf{j} + ct\mathbf{k}$$
can be expressed as
$$\mathbf{r} = \left(a\cos\frac{s}{w}\right)\mathbf{i} + \left(a\sin\frac{s}{w}\right)\mathbf{j} + \frac{cs}{w}\mathbf{k}$$
where $w = \sqrt{a^2 + c^2}$ and s is an arc-length parameter with reference point at $(a, 0, 0)$.

30. Recall from Formula (5) of Section 14.5 that $\mathbf{r} = \mathbf{r}_0 + t\mathbf{v}$ is the vector equation of a line in 2-space or 3-space if $\mathbf{v} \ne \mathbf{0}$. Show that $\mathbf{r} = \mathbf{r}_0 + s\mathbf{v}/\|\mathbf{v}\|$ is a vector equation of the line in terms of an arc-length parameter s whose reference point is at the tip of $\mathbf{r}_0$.

31. A thread with negligible thickness is unwound from a spool of radius a, with the unwound piece always held straight.

(a) Show that for $\theta \ge 0$ the curve traced out by the free end of the thread has parametric equations
$$x = a(\cos\theta + \theta\sin\theta)$$
$$y = a(\sin\theta - \theta\cos\theta)$$
(see accompanying figure).

(b) Find parametric equations for the curve of part (a) using arc length as the parameter, where $\theta = 0$ is the reference point.

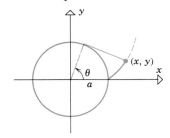

32. Show that in cylindrical coordinates a curve given by the parametric equations $r = r(t)$, $\theta = \theta(t)$, $z = z(t)$, for $a \le t \le b$, has arc length
$$L = \int_a^b \sqrt{\left(\frac{dr}{dt}\right)^2 + r^2\left(\frac{d\theta}{dt}\right)^2 + \left(\frac{dz}{dt}\right)^2}\, dt$$
[Hint: Use the relationships $x = r\cos\theta$, $y = r\sin\theta$.]

33. Use the formula in Exercise 32 to find the arc length of the following curves.

(a) $r = e^{2t}, \theta = t, z = e^{2t};\ 0 \le t \le \ln 2$

(b) $r = t^2, \theta = \ln t, z = \frac{1}{3}t^3;\ 1 \le t \le 2$.

34. Show that in spherical coordinates a curve given by the parametric equations $\rho = \rho(t)$, $\theta = \theta(t)$, $\phi = \phi(t)$, for $a \le t \le b$, has arc length

$$L = \int_a^b \sqrt{\left(\frac{d\rho}{dt}\right)^2 + \rho^2 \sin^2 \phi \left(\frac{d\theta}{dt}\right)^2 + \rho^2 \left(\frac{d\phi}{dt}\right)^2}\, dt$$

[*Hint:* $x = \rho \sin \phi \cos \theta$, $y = \rho \sin \phi \sin \theta$, $z = \rho \cos \phi$.]

35. Use the formula in Exercise 34 to find the arc length of the following curves.

(a) $\rho = e^{-t}, \theta = 2t, \phi = \pi/4;\ 0 \le t \le 2$

(b) $\rho = 2t, \theta = \ln t, \phi = \pi/6;\ 1 \le t \le 5$.

36. Let $x = \cos t$, $y = \sin t$, $z = t^{3/2}$. Find

(a) $\|\mathbf{r}'(t)\|$ (b) $\dfrac{ds}{dt}$ (c) $\displaystyle\int_0^2 \|\mathbf{r}'(t)\|\, dt$.

37. Let $\mathbf{r}(t) = (\ln t)\mathbf{i} + 2t\mathbf{j} + t^2\mathbf{k}$. Find

(a) $\|\mathbf{r}'(t)\|$ (b) $\dfrac{ds}{dt}$ (c) $\displaystyle\int_1^3 \|\mathbf{r}'(t)\|\, dt$.

38. Prove: If $\mathbf{r}(t)$ is a smoothly parametrized function, then the angles between $\mathbf{r}'(t)$ and the vectors $\mathbf{i}$, $\mathbf{j}$, and $\mathbf{k}$ are continuous functions of t.

■ 15.4 UNIT TANGENT AND NORMAL VECTORS

In this section we shall discuss some geometric properties of vector-valued functions. Our work here will have important applications to the study of motion along a curved path in 2-space or 3-space.

□ **UNIT TANGENT VECTORS**

Recall from Section 15.2 that if $\mathbf{r}(t)$ is a smooth vector-valued function in 2-space or 3-space with graph C, then the vector $\mathbf{r}'(t)$ is tangent to the graph of C if this vector is positioned so that its initial point is at the terminal point of the radius vector $\mathbf{r}(t)$. Moreover, $\mathbf{r}'(t)$ points in the direction of increasing t. Thus, if $\mathbf{r}'(t) \ne \mathbf{0}$, then the vector $\mathbf{T}(t)$ defined by

$$\mathbf{T}(t) = \frac{\mathbf{r}'(t)}{\|\mathbf{r}'(t)\|} \tag{1}$$

is tangent to C, points in the direction of increasing t, and has length 1 (Figure 15.4.1). We call $\mathbf{T}(t)$ the **unit tangent vector** to C at t. In the special case where C is parametrized by an arc-length parameter s, the tangent vectors have length 1 (Theorem 15.3.3), so (1) simplifies to

$$\mathbf{T}(s) = \mathbf{r}'(s) \tag{2}$$

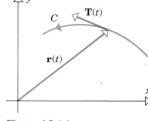

Figure 15.4.1

REMARK.   Unless stated otherwise, we shall always assume that the unit tangent vector $\mathbf{T}(t)$ is positioned with its initial point at the terminal point of $\mathbf{r}(t)$ as in Figure 15.4.1. This will ensure that $\mathbf{T}(t)$ is actually tangent to the graph of $\mathbf{r}(t)$ and not simply parallel to a tangent vector.

Example 1   Find the unit tangent vector to the graph of $\mathbf{r}(t) = t^2\mathbf{i} + t^3\mathbf{j}$ at the point where $t = 2$.

Solution.   Since

$$\mathbf{r}'(t) = 2t\mathbf{i} + 3t^2\mathbf{j}$$

we obtain

$$\mathbf{T}(2) = \frac{\mathbf{r}'(2)}{\|\mathbf{r}'(2)\|} = \frac{4\mathbf{i} + 12\mathbf{j}}{\sqrt{160}} = \frac{4\mathbf{i} + 12\mathbf{j}}{4\sqrt{10}} = \frac{1}{\sqrt{10}}\mathbf{i} + \frac{3}{\sqrt{10}}\mathbf{j}$$

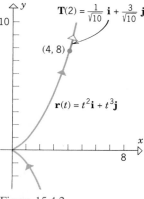

Figure 15.4.2

The graph of $\mathbf{r}(t)$ and the vector $\mathbf{T}(2)$ are shown in Figure 15.4.2.   ◄

☐ **UNIT NORMAL VECTORS**

Let C be the graph of a smooth vector-valued function $\mathbf{r}(t)$ in 2-space or 3-space. In 2-space there are two unit vectors that are perpendicular to the unit tangent vector $\mathbf{T}(t)$, and in 3-space there are infinitely many such vectors (Figure 15.4.3). Since $\|\mathbf{T}(t)\| = 1$ is constant, it follows from Theorem 15.2.7 with $\mathbf{T}$ in place of $\mathbf{r}$ that $\mathbf{T}'(t)$ is perpendicular to $\mathbf{T}(t)$. If $\mathbf{T}'(t) \neq \mathbf{0}$, then we define the **_principal unit normal vector_** to C at t by

$$\mathbf{N}(t) = \frac{\mathbf{T}'(t)}{\|\mathbf{T}'(t)\|} \tag{3}$$

The vector $\mathbf{N}(t)$ has length 1, is perpendicular to $\mathbf{T}(t)$, and has the same direction as $\mathbf{T}'(t)$. For simplicity we shall refer to $\mathbf{N}(t)$ as the **_unit normal vector_**.

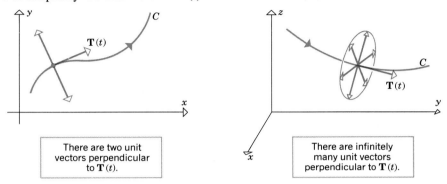

Figure 15.4.3

REMARK.   Unless stated otherwise, we shall always assume that $\mathbf{N}(t)$ is positioned with its initial point at the terminal point of $\mathbf{r}(t)$ (see Figure 15.4.4).

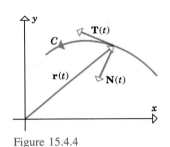

Figure 15.4.4

Example 2   Find $\mathbf{T}(t)$ and $\mathbf{N}(t)$ for the circular helix

$$x = a \cos t, \quad y = a \sin t, \quad z = ct$$

where $a > 0$.

Solution.   The radius vector for the helix is

$$\mathbf{r}(t) = (a \cos t)\mathbf{i} + (a \sin t)\mathbf{j} + (ct)\mathbf{k}$$

Thus,

$$\mathbf{r}'(t) = (-a \sin t)\mathbf{i} + (a \cos t)\mathbf{j} + c\mathbf{k}$$

$$\|\mathbf{r}'(t)\| = \sqrt{(-a \sin t)^2 + (a \cos t)^2 + c^2} = \sqrt{a^2 + c^2}$$

$$\mathbf{T}(t) = \frac{\mathbf{r}'(t)}{\|\mathbf{r}'(t)\|} = -\frac{a \sin t}{\sqrt{a^2 + c^2}}\mathbf{i} + \frac{a \cos t}{\sqrt{a^2 + c^2}}\mathbf{j} + \frac{c}{\sqrt{a^2 + c^2}}\mathbf{k}$$

$$\mathbf{T}'(t) = -\frac{a \cos t}{\sqrt{a^2 + c^2}}\mathbf{i} - \frac{a \sin t}{\sqrt{a^2 + c^2}}\mathbf{j}$$

$$\|\mathbf{T}'(t)\| = \sqrt{\left(-\frac{a \cos t}{\sqrt{a^2 + c^2}}\right)^2 + \left(-\frac{a \sin t}{\sqrt{a^2 + c^2}}\right)^2} = \sqrt{\frac{a^2}{a^2 + c^2}} = \frac{a}{\sqrt{a^2 + c^2}}$$

$$\mathbf{N}(t) = \frac{\mathbf{T}'(t)}{\|\mathbf{T}'(t)\|} = (-\cos t)\mathbf{i} - (\sin t)\mathbf{j}$$

Because the $\mathbf{k}$ component of $\mathbf{N}(t)$ is zero, this vector lies in a horizontal plane for every value of t. Moreover, in the exercises we ask the reader to show that $\mathbf{N}(t)$ points directly toward the z-axis for all t (Figure 15.4.5).   ◀

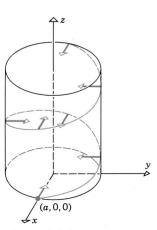

Figure 15.4.5

In the special case where the curve C is the graph of $\mathbf{r}(s)$, and s is an arc-length parameter, it follows from (2) that (3) simplifies to

$$\mathbf{N}(s) = \frac{\mathbf{r}''(s)}{\|\mathbf{r}''(s)\|} \tag{4}$$

Example 3 In Example 5 of the preceding section we showed that the circle $x^2 + y^2 = a^2$ can be parametrized in terms of arc length as

$$\mathbf{r}(s) = a \cos(s/a)\mathbf{i} + a \sin(s/a)\mathbf{j} \quad (0 \leq s \leq 2\pi a)$$

Find $\mathbf{N}(s)$.

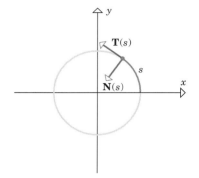

Figure 15.4.6

Solution. The first and second derivatives of $\mathbf{r}$ with respect to s are

$$\mathbf{r}'(s) = -\sin(s/a)\mathbf{i} + \cos(s/a)\mathbf{j}$$

$$\mathbf{r}''(s) = -(1/a)\cos(s/a)\mathbf{i} - (1/a)\sin(s/a)\mathbf{j}$$

so

$$\|\mathbf{r}''(s)\| = \sqrt{(-1/a)^2 \cos^2(s/a) + (-1/a)^2 \sin^2(s/a)} = 1/a$$

Thus, from (4)

$$\mathbf{N}(s) = \mathbf{r}''(s)/\|\mathbf{r}''(s)\| = -\cos(s/a)\mathbf{i} - \sin(s/a)\mathbf{j}$$

Observe that $\mathbf{N}(s)$ is a negative scalar multiple of the radius vector $\mathbf{r}(s)$, so that $\mathbf{N}(s)$ points toward the center of the circle for all s (Figure 15.4.6).   ◄

☐ **INWARD UNIT NORMAL VECTORS IN 2-SPACE**

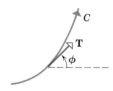

Figure 15.4.7

If $\mathbf{T}$ is a unit tangent vector to a curve C in 2-space, then as shown in Figure 15.4.3, there are two unit vectors perpendicular to $\mathbf{T}$. To determine which of these is the principal unit normal $\mathbf{N}$, we shall let $\phi = \phi(t)$ be the counterclockwise angle from the direction of the positive x-axis to $\mathbf{T}$ (Figure 15.4.7). Since $\mathbf{T}$ has length 1, it follows from Formula (8) of Section 14.2 that $\mathbf{T}$ can be expressed in terms of ϕ as

$$\mathbf{T} = (\cos\phi)\mathbf{i} + (\sin\phi)\mathbf{j} \tag{5}$$

From this formula and the chain rule we obtain

$$\mathbf{T}'(t) = \frac{d\mathbf{T}}{dt} = \frac{d\mathbf{T}}{d\phi}\frac{d\phi}{dt} = [(-\sin\phi)\mathbf{i} + (\cos\phi)\mathbf{j}]\frac{d\phi}{dt} = \mathbf{n}(t)\phi'(t) \tag{6}$$

where

$$\mathbf{n} = \mathbf{n}(t) = (-\sin\phi)\mathbf{i} + (\cos\phi)\mathbf{j} \tag{7}$$

Formula (6) implies that $\mathbf{n}(t)$ and $\mathbf{T}'(t)$ have the same direction if $\phi'(t) > 0$ (i.e., ϕ increasing) and opposite directions if $\phi'(t) < 0$ (i.e., ϕ decreasing). But $\mathbf{N}(t)$ has the same direction as $\mathbf{T}'(t)$ [see (3)], so $\mathbf{n}(t)$ and $\mathbf{N}(t)$ have the same direction if $\phi'(t) > 0$ and opposite directions if $\phi'(t) < 0$. However, $\mathbf{n}(t)$ lies 90° counterclockwise from $\mathbf{T}(t)$, since (7) can be written as (verify)

$$\mathbf{n} = \mathbf{n}(t) = \cos(\phi + \pi/2)\mathbf{i} + \sin(\phi + \pi/2)\mathbf{j}$$

and hence $\mathbf{N}(t)$ always points "inward" toward the concave side of the curve (Figure 15.4.8). For this reason $\mathbf{N}$ is sometimes referred to as the **inward unit normal**. Observe that this is consistent with Example 3, where we found that the principal unit normal for the circle always pointed inward toward the center.

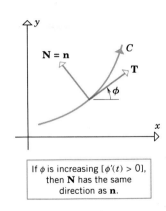

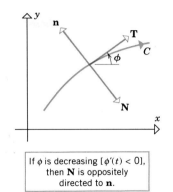

Figure 15.4.8

If φ is increasing [φ′(t) > 0], then **N** has the same direction as **n**.

If φ is decreasing [φ′(t) < 0], then **N** is oppositely directed to **n**.

☐ **BINORMAL VECTORS IN 3-SPACE**

If **r**(t) is a vector-valued function in 3-space with graph C, then for each value of t at which **T**(t) and **N**(t) exist, we define

$$\mathbf{B}(t) = \mathbf{T}(t) \times \mathbf{N}(t) \tag{8}$$

to be the **binormal vector** to C at t. For each value of t the binormal vector **B** has length 1 and is perpendicular to both **T** and **N**. At each point P on the curve C the vectors **T**, **N**, and **B** determine three mutually perpendicular planes through P whose names are shown in Figure 15.4.9. Moreover, as illustrated in Figure 14.4.4, the vectors are oriented by a "right-hand rule" in the sense that if the fingers of the right hand are cupped to point in the direction of rotation from **T** to **N**, then the thumb indicates the direction of **B**.

Three mutually perpendicular unit vectors in 3-space are sometimes called a *triad*. Whereas the **ijk**-triad is constant relative to the *xyz*-coordinate axes, the **TNB**-triad varies from point to point along the graph of **r**(t), since **T** = **T**(t), **N** = **N**(t), and **B** = **B**(t) are functions of t (Figure 15.4.10). The **TNB**-triad is sometimes described as a *moving* triad.

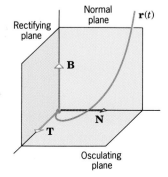

Figure 15.4.9

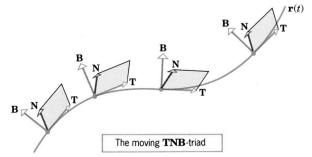

Figure 15.4.10

The moving **TNB**-triad

▶ Exercise Set 15.4

In Exercises 1–8, find the unit tangent vector **T** and the unit normal vector **N** for the given value of t. Then sketch the vectors and a portion of the curve containing the point of tangency.

1. $\mathbf{r}(t) = (5\cos t)\mathbf{i} + (5\sin t)\mathbf{j}$; $t = \pi/3$.

2. $\mathbf{r}(t) = 2t\mathbf{i} + 4t^2\mathbf{j}$; $t = 1$.

3. $\mathbf{r}(t) = (t^2 - 1)\mathbf{i} + t\mathbf{j}$; $t = 1$.

4. $\mathbf{r}(t) = e^t\mathbf{i} + e^{-t}\mathbf{j}$; $t = 0$.

5. $\mathbf{r}(t) = \frac{1}{2}t^2\mathbf{i} + \frac{1}{3}t^3\mathbf{j}$; $t = 1$.

6. $\mathbf{r}(t) = (\ln t)\mathbf{i} + t\mathbf{j}$; $t = e$.

7. $\mathbf{r}(t) = (4\cos t)\mathbf{i} + (9\sin t)\mathbf{j}$; $t = \pi/4$.

8. $\mathbf{r}(t) = (\ln \sin t)\mathbf{i} + (\ln \cos t)\mathbf{j}$; $t = \pi/6$.

In Exercises 9–16, find the unit tangent vector **T** and the unit normal vector **N** for the given value of t.

9. $\mathbf{r}(t) = 4 \cos t \mathbf{i} + 4 \sin t \mathbf{j} + t \mathbf{k}$; $t = \pi/2$.

10. $x = e^t, y = e^{-t}, z = t$; $t = 0$.

11. $\mathbf{r}(t) = t\mathbf{i} + \frac{1}{2}t^2\mathbf{j} + \frac{1}{3}t^3\mathbf{k}$; $t = 0$.

12. $x = \sin t, y = \cos t, z = \frac{1}{2}t^2$; $t = 0$.

13. $\mathbf{r}(t) = 3 \cos t\mathbf{i} + 4 \sin t\mathbf{j} + t\mathbf{k}$; $t = \pi/2$.

14. $x = e^t \cos t, y = e^t \sin t, z = e^t$; $t = 0$.

15. $\mathbf{r}(t) = \mathbf{i} + t\mathbf{j} + t^2\mathbf{k}$; $t = 1$.

16. $x = \cosh t, y = \sinh t, z = t$; $t = \ln 2$.

17. Let C be the curve given parametrically in terms of arc length by
$$x = \tfrac{3}{5}s + 1, \quad y = \tfrac{4}{5}s - 2 \quad (0 \le s \le 10)$$
   (a) Find $\mathbf{T} = \mathbf{T}(s)$.
   (b) Sketch the curve and the vector $\mathbf{T}(5)$.

18. Let C be the curve given parametrically in terms of arc length by
$$x = 2 \sin(s/2), \quad y = 2 \cos(s/2) \quad (0 \le s \le 4\pi)$$
   (a) Find $\mathbf{T} = \mathbf{T}(s)$ and $\mathbf{N} = \mathbf{N}(s)$.
   (b) Sketch the curve and the vectors $\mathbf{T}(4\pi/3)$ and $\mathbf{N}(4\pi/3)$.

19. Show that the vector $\mathbf{N}(t) = (-\cos t)\mathbf{i} - (\sin t)\mathbf{j}$ found in Example 2 points directly toward the z-axis.

20. From Formula (3), the vector **N** points in the direction of
$$\mathbf{T}' = \frac{d}{dt}[\mathbf{r}'/\|\mathbf{r}'\|]$$

Replace **r** by **r**′ in the formula of Exercise 61, Section

15.2, to show that **N** points in the direction of the vector $\mathbf{u} = \|\mathbf{r}'\|^2\mathbf{r}'' - (\mathbf{r}' \cdot \mathbf{r}'')\mathbf{r}'$ and hence that **N** can also be obtained from the formula $\mathbf{N} = \mathbf{u}/\|\mathbf{u}\|$.

In Exercises 21–28, use the formula in Exercise 20 to find **N** for the given value of t.

21. Exercise 1.   **22.** Exercise 2.

23. Exercise 3.   **24.** Exercise 4.

25. Exercise 9.   **26.** Exercise 10.

27. Exercise 11.   **28.** Exercise 12.

In Exercises 29 and 30, find $\mathbf{B} = \mathbf{T} \times \mathbf{N}$.

29. $\mathbf{r}(t) = (3 \sin t)\mathbf{i} + (3 \cos t)\mathbf{j} + 4t\mathbf{k}$.

30. $\mathbf{r}(t) = (a \cos t)\mathbf{i} + (a \sin t)\mathbf{j} + bt\mathbf{k}$ $(a \ne 0, b \ne 0)$.

31. Let S be the plane with the equation
$$a(x - x_0) + b(y - y_0) + c(z - z_0) = 0$$
and let $\mathbf{n} = a\mathbf{i} + b\mathbf{j} + c\mathbf{k}$ and $\mathbf{r}_0 = x_0\mathbf{i} + y_0\mathbf{j} + z_0\mathbf{k}$.
   (a) Show that a curve with the vector equation $\mathbf{r} = \mathbf{r}(t)$ lies in the plane S if and only if $\mathbf{n} \cdot (\mathbf{r}(t) - \mathbf{r}_0) = 0$ for all t.
   (b) Show that if the curve with the vector equation $\mathbf{r} = \mathbf{r}(t)$ lies in the plane S, then at each point where **T** and **N** are defined, these vectors also lie in S. [*Hint:* Use the result in part (a) to show that **r**′ and **r**″ are perpendicular to **n**, then use Formula (1) to show that $\mathbf{T} \cdot \mathbf{n} = 0$ and Exercise 20 to show that $\mathbf{N} \cdot \mathbf{n} = 0$.]

■ 15.5 CURVATURE

In this section we shall consider the problem of obtaining a numerical measure of how sharply a curve in 2-space or 3-space bends. Our results will have applications in geometry and in the study of motion along a curved path.

□ CURVATURE

Suppose that s is an arc-length parameter for a smooth curve C in 2-space or 3-space. Figure 15.5.1 suggests that for a curve in 2-space the "sharpness" of the bend in C is closely related to $d\mathbf{T}/ds$, which is the rate of change of the unit tangent vector **T** with respect to s. (Keep in mind that **T** has constant length, so only its direction changes.) If C is a straight line (no bend), then the direction of **T** remains constant (Figure 15.5.1a); if C bends slightly, then **T** undergoes a gradual change of direction (Figure 15.5.1b); and if C bends sharply, then **T** undergoes a rapid change of direction (Figure 15.5.1c).

In 3-space, bends in a curve are not limited to a single plane—they can occur in all possible directions. In 3-space, **T** is perpendicular to the normal plane (Figure 15.4.9), so $d\mathbf{T}/ds$ relates to the rate at which the normal plane turns as s increases. Similarly, $d\mathbf{N}/ds$

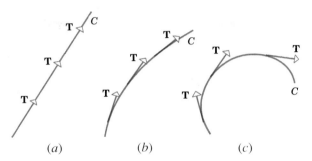

Figure 15.5.1   (a)   (b)   (c)

and $d\mathbf{B}/ds$ relate to the rates at which the rectifying and osculating planes turn as s increases. In this text we shall focus primarily on $d\mathbf{T}/ds$, which is the most important of these derivatives in applications. We make the following definition.

15.5.1 DEFINITION.   If C is a smooth curve in 2-space or 3-space, and if s is an arc-length parameter for C, then the **curvature** of C, denoted by $\kappa = \kappa(s)$ (κ = Greek "kappa"), is defined by

$$\kappa = \left\| \frac{d\mathbf{T}}{ds} \right\| \tag{1}$$

Observe that in (1) it is the *length* of the vector $d\mathbf{T}/ds$ that is the measure of curvature; thus, κ is a nonnegative real number. In general, the curvature varies from point to point along a curve; however, the following example shows that for circles in 2-space curvature is constant, as one might expect.

Example 1   In Example 5 of Section 15.3 we showed that the circle of radius a, centered at the origin, can be parametrized in terms of arc length as

$$\mathbf{r}(s) = a \cos (s/a)\mathbf{i} + a \sin (s/a)\mathbf{j} \quad (0 \leq s \leq 2\pi a)$$

Thus, from (1)

$$\frac{d\mathbf{T}}{ds} = \mathbf{r}''(s) = -\frac{1}{a}\cos (s/a)\mathbf{i} - \frac{1}{a}\sin (s/a)\mathbf{j}$$

$$\kappa = \left\| \frac{d\mathbf{T}}{ds} \right\| = \sqrt{\left[-\frac{1}{a}\cos (s/a) \right]^2 + \left[-\frac{1}{a}\sin (s/a) \right]^2} = \frac{1}{a}$$

so the circle has constant curvature $1/a$.   ◀

□ **FORMULAS FOR CURVATURE**

Formula (1) is not useful for computational purposes unless the curve happens to be expressed in terms of an arc-length parameter. The following theorem provides two formulas for curvature in terms of a general parameter t.

15.5.2 THEOREM.   *If $\mathbf{r}(t)$ is a smooth vector-valued function in 2-space or 3-space, then for each value of t at which $\mathbf{T}'(t)$ and $\mathbf{r}''(t)$ exist, the curvature κ can be expressed as*

(a) $\kappa = \kappa(t) = \dfrac{\|\mathbf{T}'(t)\|}{\|\mathbf{r}'(t)\|}$ (2)

(b) $\kappa = \kappa(t) = \dfrac{\|\mathbf{r}'(t) \times \mathbf{r}''(t)\|}{\|\mathbf{r}'(t)\|^3}$ (3)

Proof (a).   It follows from (1), Formula (13) of Section 15.3, and the chain rule that

$$\kappa = \left\| \frac{d\mathbf{T}}{ds} \right\| = \left\| \frac{d\mathbf{T}/dt}{ds/dt} \right\| = \left\| \frac{d\mathbf{T}/dt}{\|d\mathbf{r}/dt\|} \right\| = \frac{\|\mathbf{T}'(t)\|}{\|\mathbf{r}'(t)\|}$$

Proof (b).   It follows from Formula (1) of Section 15.4 that

$$\mathbf{r}'(t) = \|\mathbf{r}'(t)\| \, \mathbf{T}(t) \tag{4}$$

so

$$\mathbf{r}''(t) = \|\mathbf{r}'(t)\|' \, \mathbf{T}(t) + \|\mathbf{r}'(t)\| \, \mathbf{T}'(t) \tag{5}$$

But from Formula (3) of Section 15.4 and part (*a*) of this theorem we have

$$\mathbf{T}'(t) = \|\mathbf{T}'(t)\| \, \mathbf{N}(t) \quad \text{and} \quad \|\mathbf{T}'(t)\| = \kappa \, \|\mathbf{r}'(t)\|$$

so

$$\mathbf{T}'(t) = \kappa \, \|\mathbf{r}'(t)\| \, \mathbf{N}(t)$$

Substituting this into (5) yields

$$\mathbf{r}''(t) = \|\mathbf{r}'(t)\|' \, \mathbf{T}(t) + \kappa \, \|\mathbf{r}'(t)\|^2 \, \mathbf{N}(t) \tag{6}$$

Thus, from (4) and (6)

$$\mathbf{r}'(t) \times \mathbf{r}''(t) = \|\mathbf{r}'(t)\| \, \|\mathbf{r}'(t)\|' \, (\mathbf{T}(t) \times \mathbf{T}(t)) + \kappa \, \|\mathbf{r}'(t)\|^3 \, (\mathbf{T}(t) \times \mathbf{N}(t))$$

But the cross product of a vector with itself is zero [Theorem 14.4.4(*f*)], so this equation simplifies to

$$\mathbf{r}'(t) \times \mathbf{r}''(t) = \kappa \, \|\mathbf{r}'(t)\|^3 \, (\mathbf{T}(t) \times \mathbf{N}(t))$$

It follows from this equation and the fact that $\mathbf{T}(t) \times \mathbf{N}(t)$ is a unit vector (why?) that

$$\|\mathbf{r}'(t) \times \mathbf{r}''(t)\| = \kappa \, \|\mathbf{r}'(t)\|^3$$

Formula (3) now follows.   ∎

REMARKS.   Formula (2) is useful if $\mathbf{T}(t)$ happens to be known or is easy to obtain; however, Formula (3) will usually be easier to apply, since it involves only $\mathbf{r}(t)$ and its derivatives. We also note that cross products were defined only for vectors in 3-space, so to use Formula (4) in 2-space one must first write the 2-space function $\mathbf{r}(t) = x(t)\mathbf{i} + y(t)\mathbf{j}$ as the 3-space function $\mathbf{r}(t) = x(t)\mathbf{i} + y(t)\mathbf{j} + 0\mathbf{k}$ with a zero $\mathbf{k}$ component.

Example 2   Find $\kappa(t)$ for the circular helix

$$x = a \cos t, \quad y = a \sin t, \quad z = ct$$

where $a > 0$.

Solution.   The radius vector for the helix is

$$\mathbf{r}(t) = (a \cos t)\mathbf{i} + (a \sin t)\mathbf{j} + (ct)\mathbf{k}$$

Thus,

$$\mathbf{r}'(t) = (-a \sin t)\mathbf{i} + (a \cos t)\mathbf{j} + c\mathbf{k}$$

$$\mathbf{r}''(t) = (-a \cos t)\mathbf{i} + (-a \sin t)\mathbf{j}$$

so

$$\mathbf{r}'(t) \times \mathbf{r}''(t) = \begin{vmatrix} \mathbf{i} & \mathbf{j} & \mathbf{k} \\ -a \sin t & a \cos t & c \\ -a \cos t & -a \sin t & 0 \end{vmatrix}$$

$$= (ac \sin t)\mathbf{i} - (ac \cos t)\mathbf{j} + a^2\mathbf{k}$$

Therefore,

$$\|\mathbf{r}'(t)\| = \sqrt{(-a\sin t)^2 + (a\cos t)^2 + c^2} = \sqrt{a^2 + c^2}$$

and

$$\|\mathbf{r}'(t) \times \mathbf{r}''(t)\| = \sqrt{(ac\sin t)^2 + (-ac\cos t)^2 + a^4}$$
$$= \sqrt{a^2c^2 + a^4} = a\sqrt{a^2 + c^2}$$

so

$$\kappa(t) = \frac{\|\mathbf{r}'(t) \times \mathbf{r}''(t)\|}{\|\mathbf{r}'(t)\|^3} = \frac{a\sqrt{a^2 + c^2}}{(\sqrt{a^2 + c^2})^3} = \frac{a}{a^2 + c^2}$$

Note that κ does not depend on t, and hence the helix has constant curvature. ◀

Example 3 The graph of the vector equation

$$\mathbf{r} = (2\cos t)\mathbf{i} + (3\sin t)\mathbf{j} \quad (0 \le t \le 2\pi)$$

is the ellipse in Figure 15.5.2 (see Example 5 of Section 13.4). Find the curvature of the ellipse at the endpoints of the major and minor axes.

Solution. We have

$$\mathbf{r}'(t) = (-2\sin t)\mathbf{i} + (3\cos t)\mathbf{j} \quad \text{and} \quad \mathbf{r}''(t) = (-2\cos t)\mathbf{i} + (-3\sin t)\mathbf{j}$$

so

$$\mathbf{r}'(t) \times \mathbf{r}''(t) = \begin{vmatrix} \mathbf{i} & \mathbf{j} & \mathbf{k} \\ -2\sin t & 3\cos t & 0 \\ -2\cos t & -3\sin t & 0 \end{vmatrix} = [(6\sin^2 t) + (6\cos^2 t)]\mathbf{k} = 6\mathbf{k}$$

Therefore,

$$\|\mathbf{r}'(t)\| = \sqrt{(-2\sin t)^2 + (3\cos t)^2} = \sqrt{4\sin^2 t + 9\cos^2 t}$$

and

$$\|\mathbf{r}'(t) \times \mathbf{r}''(t)\| = 6$$

so

$$\kappa = \kappa(t) = \frac{\|\mathbf{r}'(t) \times \mathbf{r}''(t)\|}{\|\mathbf{r}'(t)\|^3} = \frac{6}{[4\sin^2 t + 9\cos^2 t]^{3/2}} \tag{7}$$

The endpoints of the minor axis are $(2, 0)$ and $(-2, 0)$, which correspond to $t = 0$ and $t = \pi$, respectively. Substituting these values in (7) yields the same curvature at both points, namely

$$\kappa = \kappa(0) = \kappa(\pi) = \frac{6}{9^{3/2}} = \frac{6}{27} = \frac{2}{9}$$

The endpoints of the major axis are $(0, 3)$ and $(0, -3)$, which correspond to $t = \pi/2$ and $t = 3\pi/2$, respectively; from (7) the curvature at these points is

$$\kappa = \kappa\left(\frac{\pi}{2}\right) = \kappa\left(\frac{3\pi}{2}\right) = \frac{6}{4^{3/2}} = \frac{3}{4}$$

Thus, the curvature is greater at the ends of the major axis than at the ends of the minor axis, as one would expect. ◀

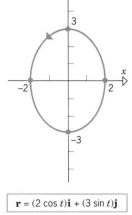

$\mathbf{r} = (2\cos t)\mathbf{i} + (3\sin t)\mathbf{j}$

Figure 15.5.2

☐ RADIUS OF CURVATURE

In the preceding example, we found the curvature at the ends of the minor axis to be $2/9$ and the curvature at the ends of the major axis to be $3/4$. To obtain a better understanding of the meaning of these numbers, recall from Example 1 that a circle of radius a has a constant curvature of $1/a$; thus, the curvature of the ellipse at the ends of the minor axis is the same as that of a circle of radius 9/2, and the curvature at the ends of the major axis is the same as that of a circle of radius 4/3 (Figure 15.5.3).

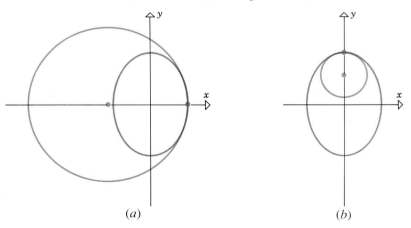

(a) (b) Figure 15.5.3

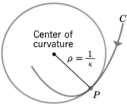

Osculating circle

Center of curvature $\rho = \frac{1}{\kappa}$

C

P

Figure 15.5.4

In general, if a curve C in 2-space has nonzero curvature κ at a point P, then the circle of radius $\rho = 1/\kappa$ sharing a common tangent with C at P, and centered on the concave side of the curve at P, is called the **circle of curvature** or **osculating circle** at P (Figure 15.5.4). The osculating circle and the curve C not only touch at P but they have equal curvatures at that point. In this sense, the osculating circle is the circle that best approximates the curve C near P. The radius ρ of the osculating circle at P is called the **radius of curvature** at P, and the center of the circle is called the **center of curvature** at P (Figure 15.5.4).

☐ **A BASIC RELATIONSHIP BETWEEN T, N, AND κ**

The following formula, which will be used in the next section, provides a fundamental relationship between **T**, **N**, and κ.

$$\frac{d\mathbf{T}}{ds} = \kappa \mathbf{N} \tag{8}$$

To derive this relationship, we use Formula (3) of Section 15.4 with s in place of t to obtain

$$\mathbf{N}(s) = \frac{d\mathbf{T}/ds}{\|d\mathbf{T}/ds\|} = \frac{1}{\kappa}\frac{d\mathbf{T}}{ds}$$

from which (8) follows.

☐ **INTERPRETATION OF CURVATURE IN 2-SPACE**

A simple geometric interpretation of curvature in 2-space can be obtained by considering the angle ϕ measured counterclockwise from the direction of the positive x-axis to **T** (Figure 15.4.7). By Formula (5) of Section 15.4, we can write $\mathbf{T} = (\cos \phi)\mathbf{i} + (\sin \phi)\mathbf{j}$ from which we obtain

$$\frac{d\mathbf{T}}{d\phi} = (-\sin \phi)\mathbf{i} + (\cos \phi)\mathbf{j}$$

By the chain rule

$$\frac{d\mathbf{T}}{ds} = \frac{d\mathbf{T}}{d\phi}\frac{d\phi}{ds}$$

so

$$\left\|\frac{d\mathbf{T}}{ds}\right\| = \left|\frac{d\phi}{ds}\right| \left\|\frac{d\mathbf{T}}{d\phi}\right\| = \left|\frac{d\phi}{ds}\right| \sqrt{(-\sin \phi)^2 + \cos^2 \phi} = \left|\frac{d\phi}{ds}\right|$$

It now follows from (1) that

$$\kappa = \left| \frac{d\phi}{ds} \right| \tag{9}$$

Thus, curvature in 2-space can be interpreted as the magnitude of the rate of change of ϕ with respect to s—the greater the curvature, the more rapidly ϕ changes with s (Figure 15.5.5). In the case of a straight line, the angle ϕ is constant (Figure 5.5.6) and consequently $\kappa = |d\phi/ds| = 0$, which means that a straight line has zero curvature at every point.

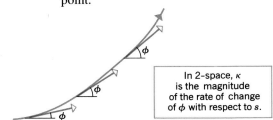

In 2-space, κ is the magnitude of the rate of change of ϕ with respect to s.

Figure 15.5.5

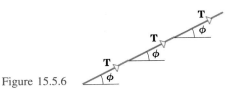

Figure 15.5.6

▶ Exercise Set 15.5   C̄ 57, 58

In Exercises 1–14, use Formula (3) to find the curvature at the indicated point.

1. $\mathbf{r}(t) = t^2\mathbf{i} + t^3\mathbf{j}$; $t = \frac{1}{2}$.
2. $\mathbf{r}(t) = (4 \cos t)\mathbf{i} + (\sin t)\mathbf{j}$; $t = \pi/2$.
3. $\mathbf{r}(t) = e^{3t}\mathbf{i} + e^{-t}\mathbf{j}$; $t = 0$.
4. $x = 1 - t^3, y = t - t^2$; $t = 1$.
5. $x = t \cos t, y = t \sin t$; $t = 1$.
6. $x = 2abt, y = b^2t^2$; $t = 1$ $(a > 0, b > 0)$.
7. $\mathbf{r}(t) = 4 \cos t\mathbf{i} + 4 \sin t\mathbf{j} + t\mathbf{k}$; $t = \pi/2$.
8. $x = e^t, y = e^{-t}, z = t$; $t = 0$.
9. $\mathbf{r}(t) = t\mathbf{i} + \frac{1}{2}t^2\mathbf{j} + \frac{1}{3}t^3\mathbf{k}$; $t = 0$.
10. $x = \sin t, y = \cos t, z = \frac{1}{2}t^2$; $t = 0$.
11. $\mathbf{r}(t) = 3 \cos t\mathbf{i} + 4 \sin t\mathbf{j} + t\mathbf{k}$; $t = \pi/2$.
12. $x = e^t \cos t, y = e^t \sin t, z = e^t$; $t = 0$.
13. $\mathbf{r}(t) = \mathbf{i} + t\mathbf{j} + t^2\mathbf{k}$; $t = 1$.
14. $x = \cosh t, y = \sinh t, z = t$; $t = \ln 2$.
15. Use Formula (3) to show that for a plane curve described by $y = f(x)$ the curvature $\kappa(x)$ is

$$\kappa(x) = \frac{|d^2y/dx^2|}{[1 + (dy/dx)^2]^{3/2}}$$

[Hint: Let x be the parameter so that $\mathbf{r}(x) = x\mathbf{i} + y\mathbf{j} = x\mathbf{i} + f(x)\mathbf{j}$.]

16. Show that the formula for κ in Exercise 15 can also be written as $\kappa = |y'' \cos^3 \phi|$, where ϕ is the angle of inclination of the tangent line to the graph of $y = f(x)$.

In Exercises 17–22, use the formula of Exercise 15 to find the curvature at the indicated point.

17. $y = \sin x$; $x = \pi/2$.
18. $y = x^3/3$; $x = 0$.
19. $y = 1/x$; $x = 1$.
20. $y = e^{-x}$; $x = 1$.
21. $y = \tan x$; $x = \pi/4$.
22. $y^2 - 4x^2 = 9$; $(2, 5)$.
23. Use the formula of Exercise 15 to find the curvature of $y = \ln \cos x$ for $-\pi/2 < x < \pi/2$. For what value of x is the curvature maximum?
24. Use Formula (3) to show that for a plane curve described by $\mathbf{r} = x(t)\mathbf{i} + y(t)\mathbf{j}$ the curvature is

$$\kappa = \frac{|x'y'' - y'x''|}{(x'^2 + y'^2)^{3/2}}$$

where a prime denotes differentiation with respect to t.

In Exercises 25–30, use the formula of Exercise 24 to find the curvature in the indicated problem.

25. Exercise 1.   26. Exercise 2.
27. Exercise 3.   28. Exercise 4.
29. Exercise 5.   30. Exercise 6.

31. Use the formula of Exercise 24 to find the curvature at $t = 0$ and $t = \pi/2$ of the curve $x = a \cos t$, $y = b \sin t$ $(a > 0, b > 0)$.
32. Use Formula (3) to show that for a curve in polar coordinates described by $r = f(\theta)$ the curvature is

$$\kappa = \frac{\left| r^2 + 2\left(\frac{dr}{d\theta}\right)^2 - r\frac{d^2r}{d\theta^2} \right|}{\left[r^2 + \left(\frac{dr}{d\theta}\right)^2 \right]^{3/2}}$$

[Hint: Let θ be the parameter and use the relationships $x = r \cos \theta$, $y = r \sin \theta$.]

In Exercises 33–36, use the formula of Exercise 32 to find the curvature at the indicated point.

33. $r = 2 \sin \theta;\ \theta = \pi/6.$ **34.** $r = \theta;\ \theta = 1.$

35. $r = a(1 + \cos \theta);\ \theta = \pi/2.$

36. $r = e^{2\theta};\ \theta = 1.$

37. Find $\kappa(t)$ for the cycloid
$$x = a(t - \sin t),\ y = a(1 - \cos t)$$
for $0 < t < \pi$ and sketch the graph of $\kappa(t)$.

38. Find $\kappa(t)$ for the curve $x = e^{-t}\cos t,\ y = e^{-t}\sin t$ and sketch the graph of $\kappa(t)$.

In Exercises 39–45, sketch the curve, calculate the radius of curvature at the indicated point, and sketch the osculating circle.

39. $y = \sin x;\ x = \pi/2.$ **40.** $x = t, y = t^2;\ t = 1.$

41. $y = \frac{1}{2}x^2;\ x = -1.$ **42.** $x = t, y = \sqrt{t};\ t = 2.$

43. $y = \ln x;\ x = 1.$ **44.** $x = 2t, y = 4/t;\ t = 1.$

45. $x = t - \sin t, y = 1 - \cos t;\ t = \pi.$

46. Find the radius of curvature of $y = \cos x$ at $x = 0$ and $x = \pi$. Sketch the osculating circles at those points.

47. Find the radius of curvature of the ellipse $x = 2\cos t$, $y = \sin t,\ 0 \le t \le 2\pi$ at $t = 0$ and $t = \pi/2$. Sketch the osculating circles at those points.

48. Consider the curve $y = x^4 - 2x^2$.

(a) Find the radius of curvature at each relative extremum.

(b) Sketch the curve and show the osculating circles at the relative extrema.

49. Find the radius of curvature of the parabola $y^2 = 4px$ at $(0, 0)$.

50. At what point(s) does $y = e^x$ have maximum curvature?

51. At what point(s) does $4x^2 + 9y^2 = 36$ have minimum radius of curvature?

52. Find the value of x, $x > 0$, where $y = x^3$ has maximum curvature.

53. Find the maximum and minimum values of the radius of curvature for the curve $x = \cos t,\ y = \sin t,\ z = \cos t$.

54. Find the minimum value of the radius of curvature for the curve $x = e^t,\ y = e^{-t},\ z = \sqrt{2}t$.

55. Show that the curvature of the polar curve $r = e^{a\theta}$ is inversely proportional to r.

56. Show that the curvature of the polar curve $r^2 = a^2 \cos 2\theta$ is directly proportional to r $(r > 0)$.

57. Given that $d\mathbf{T}/ds = 0.03\mathbf{i} - 0.04\mathbf{j}$ at a certain point P on a curve, find the value of $|d\phi/ds|$, where ϕ is the angle of inclination of the tangent line to the curve at P. Assuming that s is measured in centimeters, express your answer in units of °/cm.

58. Assuming that arc length is measured in inches along the curve $y = x^3$, find the magnitude of the rate of change with respect to arc length of the angle of inclination of the tangent line to the curve in units of °/in. at the point where $x = 1$.

In Exercises 59–62, we shall say that a **smooth transition** occurs at a point P on a curve if the curvature κ is continuous at P. Use the formula for κ given in Exercise 15 to help solve the problem.

59. Show that the transition at $x = 0$ from the horizontal line $y = 0$ for $x \le 0$ to the parabola $y = x^2$ for $x > 0$ is not smooth, whereas the transition to $y = x^3$ for $x > 0$ is smooth.

60. (a) Sketch the graph of the curve defined piecewise by $y = x^2$ for $x < 0$, $y = x^4$ for $x \ge 0$.

(b) Show that for the curve in part (a) the transition at $x = 0$ is not smooth.

61. The figure below shows the arc of a circle of radius r with center at $(0, r)$. Find the value of a so that there is a smooth transition from the circle to the parabola $y = ax^2$ at the point where $x = 0$.

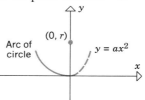

62. Find a, b, and c so that there is a smooth transition at $x = 0$ from the curve $y = e^x$ for $x \le 0$ to the parabola $y = ax^2 + bx + c$ for $x > 0$. [*Hint:* The curvature is continuous at those points where y'' is continuous.]

63. Let C be a smooth curve in the xy-plane that is tangent to the x-axis at the origin, and let s be an arc-length parameter for C with its reference point at the origin ($x = 0$ and $y = 0$ if $s = 0$). Suppose that the curvature κ is proportional to s and that s is chosen so that the constant of proportionality is positive ($\kappa = as$ where $a > 0$).

(a) Assuming that $d\phi/ds \ge 0$, use Formula (9) to show that $\phi = \frac{1}{2}as^2$.

(b) Use the result in part (a) and the fact that $\mathbf{T} = (\cos \phi)\mathbf{i} + (\sin \phi)\mathbf{j}$ to show that
$$x = \int_0^s \cos\left(\tfrac{1}{2}au^2\right) du, \quad y = \int_0^s \sin\left(\tfrac{1}{2}au^2\right) du$$
are parametric equations of the curve.

In Exercises 64–66, we assume that s is an arc-length parameter for a smooth curve C in 3-space and that $d\mathbf{T}/ds$ and $d\mathbf{N}/ds$ exist at each point on the curve. (This implies that $d\mathbf{B}/ds$ exists as well, since $\mathbf{B} = \mathbf{T} \times \mathbf{N}$.)

64. (a) Show that $d\mathbf{B}/ds$ is perpendicular to $\mathbf{B}$. [*Hint:* See Theorem 15.2.7.]

(b) Show that $d\mathbf{B}/ds$ is perpendicular to $\mathbf{T}$. [*Hint:* Use the fact that $\mathbf{B}$ is perpendicular to both $\mathbf{T}$ and $\mathbf{N}$, and differentiate $\mathbf{B} \cdot \mathbf{T}$ with respect to s.]

(c) Use the results in parts (a) and (b) to show that $d\mathbf{B}/ds$ is a scalar multiple of $\mathbf{N}$.

(d) It follows from part (c) that there is a real-valued function $\tau = \tau(s)$ ($\tau =$ Greek "tau"), called the *torsion*, such that $d\mathbf{B}/ds = -\tau\mathbf{N}$. Use the result in Exercise 31(b), Section 15.4, to show that if C lies entirely in a plane, then $\tau(s) = 0$ for all values of s. [*Note:* $d\mathbf{B}/ds$ is related to the rate at which the osculating plane turns (Figure 15.4.9), so the torsion is sometimes viewed as a measure of the tendency for C to twist out of the osculating plane.]

65. Let κ be the curvature of C and τ the torsion (defined in Exercise 64). By differentiating $\mathbf{N} = \mathbf{B} \times \mathbf{T}$ with respect to s, show that $d\mathbf{N}/ds = -\kappa\mathbf{T} + \tau\mathbf{B}$.

66. The following derivatives, known as the *Frenet–Serret formulas*, are fundamental in the theory of curves in 3-space:

$d\mathbf{T}/ds = \kappa\mathbf{N}$ [Formula (8)]
$d\mathbf{N}/ds = -\kappa\mathbf{T} + \tau\mathbf{B}$ [Exercise 65]
$d\mathbf{B}/ds = -\tau\mathbf{N}$ [Exercise 64(d)]

Use the first two Frenet–Serret formulas and the fact that $\mathbf{r}'(s) = \mathbf{T}$ if $\mathbf{r} = \mathbf{r}(s)$ to show that

$$\tau = \frac{[\mathbf{r}'(s) \times \mathbf{r}''(s)] \cdot \mathbf{r}'''(s)}{\|\mathbf{r}''(s)\|^2} \quad \text{and} \quad \mathbf{B} = \frac{\mathbf{r}'(s) \times \mathbf{r}''(s)}{\|\mathbf{r}''(s)\|}$$

67. Use the results in Exercise 66 and the results in Exercise 29 of Section 15.3 to show that for the circular helix

$$\mathbf{r} = (a \cos t)\mathbf{i} + (a \sin t)\mathbf{j} + ct\mathbf{k}$$

with $a > 0$ the torsion and the binormal vector are

$$\tau = \frac{c}{w^2}$$

and

$$\mathbf{B} = \left(\frac{c}{w} \sin \frac{s}{w}\right)\mathbf{i} - \left(\frac{c}{w} \cos \frac{s}{w}\right)\mathbf{j} + \left(\frac{a}{w}\right)\mathbf{k}$$

where $w = \sqrt{a^2 + c^2}$ and s has its reference point at $(a, 0, 0)$.

68. Suppose that the arc-length parameter in the Frenet–Serret formulas of Exercise 66 is a function $s = s(t)$ of a general parameter t and that primes are used to denote derivatives with respect to t.

(a) Use the first two Frenet–Serret formulas to show that $\mathbf{T}' = \kappa s'\mathbf{N}$ and $\mathbf{N}' = -\kappa s'\mathbf{T} + \tau s'\mathbf{B}$. [*Hint:* Use the chain rule.]

(b) Show that Formulas (4) and (6) can be written in the form

$$\mathbf{r}'(t) = s'\mathbf{T} \quad \text{and} \quad \mathbf{r}''(t) = s''\mathbf{T} + \kappa(s')^2\mathbf{N}$$

(c) Use the results in parts (a) and (b) to show that

$$\mathbf{r}'''(t) = [s''' - \kappa^2(s')^3]\mathbf{T} + [3\kappa s's'' + \kappa'(s')^2]\mathbf{N} + \kappa\tau(s')^3\mathbf{B}$$

(d) Use the results in parts (b) and (c) to show that

$$\tau = \frac{[\mathbf{r}'(t) \times \mathbf{r}''(t)] \cdot \mathbf{r}'''(t)}{\|\mathbf{r}'(t) \times \mathbf{r}''(t)\|^2}$$

In Exercises 69–72, use the formula in Exercise 68(d) to find the torsion $\tau = \tau(t)$.

69. The twisted cubic $\mathbf{r}(t) = 2t\mathbf{i} + t^2\mathbf{j} + \frac{1}{3}t^3\mathbf{k}$.

70. The circular helix $\mathbf{r}(t) = (a \cos t)\mathbf{i} + (a \sin t)\mathbf{j} + bt\mathbf{k}$.

71. $\mathbf{r}(t) = e^t\mathbf{i} + e^{-t}\mathbf{j} + \sqrt{2}t\mathbf{k}$.

72. $\mathbf{r}(t) = (t - \sin t)\mathbf{i} + (1 - \cos t)\mathbf{j} + t\mathbf{k}$.

■ **15.6** MOTION ALONG A CURVE

In Section 4.10 we considered the motion of a particle moving along a coordinate line. For such particles, there are only two possible directions of motion—the positive direction or the negative direction. For particles moving in 2-space or 3-space the situation is much more complicated, since there are infinitely many directions of motion possible. In this section we shall show how vectors can be used to study the motion of a particle moving along a curve in 2-space or 3-space.

□ VELOCITY, ACCELERATION, AND SPEED

Suppose that a particle is moving through 2-space or 3-space along a smooth curve C. At each instant of time we will regard the direction of motion to be the direction of the unit tangent vector, and we will regard the speed of the particle to be the instantaneous rate of change with respect to time of the arc length traveled by the particle, as measured from some arbitrary reference point (Figure 15.6.1*a*). The speed and the direction of motion together determine what is called the "velocity vector" of the particle. Mathematically, this vector can be written in the form

$$\mathbf{v}(t) = \frac{ds}{dt}\,\mathbf{T}(t) \tag{1}$$

which is a vector of length ds/dt with the same direction as $\mathbf{T}(t)$ (Figure 15.6.1*b*).

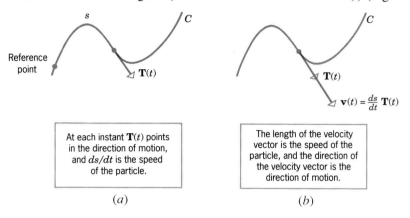

At each instant $\mathbf{T}(t)$ points in the direction of motion, and ds/dt is the speed of the particle.

The length of the velocity vector is the speed of the particle, and the direction of the velocity vector is the direction of motion.

Figure 15.6.1   (*a*)   (*b*)

Recall that for a particle moving on a coordinate line, the velocity function is the derivative of the position function. The same is true for a particle moving along a smooth curve in 2-space or 3-space. To see that this is so, suppose that the position function of the particle is $\mathbf{r}(t)$. From the definition of the unit tangent vector [Formula (1) of Section 15.4] and Formula (17) of Section 15.3 we have

$$\mathbf{T}(t) = \frac{\mathbf{r}'(t)}{\|\mathbf{r}'(t)\|} \quad \text{and} \quad \frac{ds}{dt} = \|\mathbf{r}'(t)\|$$

Substituting these expressions in (1) and simplifying yields

$$\mathbf{v}(t) = \mathbf{r}'(t)$$

which shows that the velocity function is the derivative of the position function. As was the case for motion along a coordinate line, the *acceleration function* of a particle moving along a curve in 2-space or 3-space is defined to be the derivative of the velocity function. Thus, we have the following definitions.

> **15.6.1**   DEFINITION.   If a particle moves along a curve *C* in 2-space or 3-space so that its position vector at time *t* is $\mathbf{r}(t)$, then the *instantaneous velocity*, *instantaneous acceleration*, and *instantaneous speed* of the particle at time *t* are defined by
>
> $$\text{velocity} = \mathbf{v}(t) = \frac{d\mathbf{r}}{dt} \tag{2}$$
>
> $$\text{acceleration} = \mathbf{a}(t) = \frac{d\mathbf{v}}{dt} = \frac{d^2\mathbf{r}}{dt^2} \tag{3}$$
>
> $$\text{speed} = \|\mathbf{v}(t)\| = \frac{ds}{dt} \tag{4}$$

where *s* is the total arc length traveled by the particle at time *t*, measured from some arbitrary reference point on *C*.

Example 1   A particle moves along a circular path in such a way that its x- and y-coordinates at time t are

$$x = 2 \cos t, \quad y = 2 \sin t$$

(a)   Find the instantaneous velocity and speed of the particle at time t.

(b)   Sketch the path of the particle, and show the position and velocity vectors at time $t = \pi/4$ with the velocity vector drawn so that its initial point is at the tip of the position vector.

(c)   Show that at each instant the acceleration vector is perpendicular to the velocity vector.

Solution (a).   At time t, the position vector is

$$\mathbf{r}(t) = 2 \cos t\,\mathbf{i} + 2 \sin t\,\mathbf{j}$$

so the instantaneous velocity and speed are

$$\mathbf{v}(t) = \frac{d\mathbf{r}}{dt} = -2 \sin t\,\mathbf{i} + 2 \cos t\,\mathbf{j}$$

$$\|\mathbf{v}(t)\| = \sqrt{(-2 \sin t)^2 + (2 \cos t)^2} = 2$$

Solution (b).   The graph of the parametric equations is a circle of radius 2 centered at the origin. At time $t = \pi/4$ the position and velocity vectors of the particles are

$$\mathbf{r}(\pi/4) = 2 \cos (\pi/4)\mathbf{i} + 2 \sin (\pi/4)\mathbf{j} = \sqrt{2}\,\mathbf{i} + \sqrt{2}\,\mathbf{j}$$

$$\mathbf{v}(\pi/4) = -2 \sin (\pi/4)\mathbf{i} + 2 \cos (\pi/4)\mathbf{j} = -\sqrt{2}\,\mathbf{i} + \sqrt{2}\,\mathbf{j}$$

These vectors and the circle are shown in Figure 15.6.2.

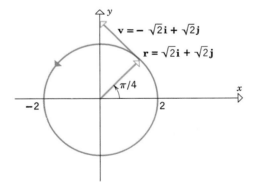

Figure 15.6.2

Solution (c).   At time t, the acceleration vector is

$$\mathbf{a}(t) = \frac{d\mathbf{v}}{dt} = -2 \cos t\,\mathbf{i} - 2 \sin t\,\mathbf{j}$$

One way of showing that $\mathbf{v}(t)$ and $\mathbf{a}(t)$ are perpendicular is to show that their dot product is zero (try it). However, it is easier to observe that $\mathbf{a}(t)$ is the negative of $\mathbf{r}(t)$, which implies that $\mathbf{v}(t)$ and $\mathbf{a}(t)$ are perpendicular, since at each point on a circle the radius and tangent line are perpendicular.   ◀

REMARKS.   In Example 1 we omitted all units for simplicity. We shall follow this practice throughout this section, except in some applied problems that appear later. Moreover, as in Example 1, we shall follow the standard convention of drawing velocity and acceleration vectors with their initial points at the tip of the position vector, rather than at the origin.

Since the velocity function $\mathbf{v}(t)$ of a particle moving in 2-space or 3-space can be obtained by differentiating the position function $\mathbf{r}(t)$, it follows that $\mathbf{r}(t)$ can be obtained by integrating $\mathbf{v}(t)$. Similarly, $\mathbf{v}(t)$ can be obtained by integrating $\mathbf{a}(t)$. However, integrating $\mathbf{v}(t)$ does not produce a unique position function, and integrating $\mathbf{a}(t)$ does not produce a unique velocity function, since constants of integration occur. To determine these constants, some additional information is needed. For example, if the position vector of the particle is known at some specific point in time, then the constant resulting from integrating $\mathbf{v}(t)$ can be determined from this information, and if the velocity vector of the particle is known at some point in time, then the constant resulting from integrating $\mathbf{a}(t)$ can be determined.

Example 2   A particle moves through 3-space in such a way that its velocity at time t is

$$\mathbf{v}(t) = \mathbf{i} + t\mathbf{j} + t^2\mathbf{k}$$

Find the coordinates of the particle at time $t = 1$ given that the particle is at the point $(-1, 2, 4)$ at time $t = 0$.

Solution.   Integrating the velocity function to obtain the position function yields

$$\mathbf{r}(t) = \int \mathbf{v}(t)\, dt = \int (\mathbf{i} + t\mathbf{j} + t^2\mathbf{k})\, dt = t\mathbf{i} + \frac{t^2}{2}\mathbf{j} + \frac{t^3}{3}\mathbf{k} + \mathbf{C} \tag{5}$$

where $\mathbf{C}$ is a vector constant of integration. Since the coordinates of the particle at time $t = 0$ are $(-1, 2, 4)$, the position vector at time $t = 0$ is

$$\mathbf{r}(0) = -\mathbf{i} + 2\mathbf{j} + 4\mathbf{k} \tag{6}$$

It follows on substituting $t = 0$ in (5) and equating the result with (6) that $\mathbf{C} = -\mathbf{i} + 2\mathbf{j} + 4\mathbf{k}$. Substituting this value of $\mathbf{C}$ in (5) and simplifying yields

$$\mathbf{r}(t) = (t - 1)\mathbf{i} + \left(\frac{t^2}{2} + 2\right)\mathbf{j} + \left(\frac{t^3}{3} + 4\right)\mathbf{k}$$

Thus, at time $t = 1$ the position vector of the particle is

$$\mathbf{r}(1) = 0\mathbf{i} + \frac{5}{2}\mathbf{j} + \frac{13}{3}\mathbf{k}$$

so its coordinates at that instant are $(0, \frac{5}{2}, \frac{13}{3})$.   ◄

☐ **DISPLACEMENT AND DISTANCE TRAVELED**

If a particle travels along a curve C in 2-space or 3-space, the **displacement** of the particle over the time interval $t_1 \le t \le t_2$ is commonly denoted by $\Delta\mathbf{r}$, and is defined as

$$\Delta\mathbf{r} = \mathbf{r}(t_2) - \mathbf{r}(t_1) \tag{7}$$

(Figure 15.6.3). The displacement vector, which describes the change in position of the particle during the time interval, can be obtained by integrating the velocity function from t_1 to t_2:

$$\Delta\mathbf{r} = \int_{t_1}^{t_2} \mathbf{v}(t)\, dt = \int_{t_1}^{t_2} \frac{d\mathbf{r}}{dt}\, dt = \mathbf{r}(t)\Big]_{t_1}^{t_2} = \mathbf{r}(t_2) - \mathbf{r}(t_1) \tag{8}$$

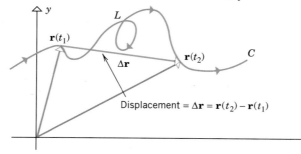

Displacement $= \Delta\mathbf{r} = \mathbf{r}(t_2) - \mathbf{r}(t_1)$

Figure 15.6.3

The distance L traveled by the particle along C during the time interval $t_1 \leq t \leq t_2$ can be obtained by integrating the magnitude of the velocity function from t_1 to t_2 [Formula (6) of Section 15.3]:

$$L = \int_{t_1}^{t_2} \|\mathbf{v}(t)\|\, dt = \int_{t_1}^{t_2} \left\| \frac{d\mathbf{r}}{dt} \right\|\, dt \tag{9}$$

Example 3 Suppose that a particle travels along a circular helix in 3-space so that its position vector at time t is

$$\mathbf{r}(t) = (4 \cos \pi t)\mathbf{i} + (4 \sin \pi t)\mathbf{j} + t\mathbf{k}$$

Find the displacement and distance traveled by the particle during the time interval $1 \leq t \leq 5$.

Solution. We have

$$\mathbf{v}(t) = \frac{d\mathbf{r}}{dt} = (-4\pi \sin \pi t)\mathbf{i} + (4\pi \cos \pi t)\mathbf{j} + \mathbf{k}$$

$$\|\mathbf{v}(t)\| = \sqrt{(-4\pi \sin \pi t)^2 + (4\pi \cos \pi t)^2 + 1} = \sqrt{16\pi^2 + 1}$$

Since $\mathbf{r}(t)$ is known, the displacement can be found directly from (7); it is

$$\begin{aligned}
\Delta \mathbf{r} &= \mathbf{r}(5) - \mathbf{r}(1) \\
&= [(4 \cos 5\pi)\mathbf{i} + (4 \sin 5\pi)\mathbf{j} + 5\mathbf{k}] - [(4 \cos \pi)\mathbf{i} + (4 \sin \pi)\mathbf{j} + \mathbf{k}] \\
&= (-4\mathbf{i} + 5\mathbf{k}) - (-4\mathbf{i} + \mathbf{k}) = 4\mathbf{k}
\end{aligned}$$

which tells us that the change in the position of the particle over the time interval was 4 units straight up.

From (9), the distance L traveled by the particle during the time interval is

$$L = \int_1^5 \sqrt{16\pi^2 + 1}\, dt = 4\sqrt{16\pi^2 + 1} \text{ units} \qquad \blacktriangleleft$$

☐ **NORMAL AND TANGENTIAL COMPONENTS OF ACCELERATION**

In applications, it is often desirable to resolve the velocity and acceleration vectors into vector components that are parallel to the unit tangent and unit normal vectors. The following theorem explains how this can be done.

15.6.2 THEOREM. *For a particle moving along a curve C in 2-space or 3-space, the velocity and acceleration vectors can be written as*

$$\mathbf{v} = \frac{ds}{dt}\mathbf{T} \tag{10}$$

$$\mathbf{a} = \frac{d^2 s}{dt^2}\mathbf{T} + \kappa \left(\frac{ds}{dt} \right)^2 \mathbf{N} \tag{11}$$

where s is an arc-length parameter for the curve, and $\mathbf{T} = \mathbf{T}(t)$, $\mathbf{N} = \mathbf{N}(t)$, and $\kappa = \kappa(t)$ denote the unit tangent vector, unit normal vector, and curvature (Figure 15.6.4).

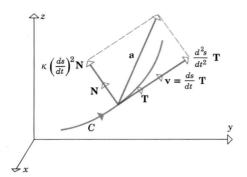

Figure 15.6.4

Proof. Formula (10) is just a restatement of (1). To obtain (11), we differentiate both sides of (10) with respect to t; this yields

$$\mathbf{a} = \frac{d}{dt}\left(\frac{ds}{dt}\mathbf{T}\right) = \frac{d^2s}{dt^2}\mathbf{T} + \frac{ds}{dt}\frac{d\mathbf{T}}{dt}$$

$$= \frac{d^2s}{dt^2}\mathbf{T} + \frac{ds}{dt}\frac{d\mathbf{T}}{ds}\frac{ds}{dt}$$

$$= \frac{d^2s}{dt^2}\mathbf{T} + \left(\frac{ds}{dt}\right)^2\frac{d\mathbf{T}}{ds}$$

$$= \frac{d^2s}{dt^2}\mathbf{T} + \left(\frac{ds}{dt}\right)^2\kappa\mathbf{N} \qquad \boxed{\begin{array}{l}\text{Formula (8)}\\ \text{of Section 15.5}\end{array}}$$

from which (11) follows. ∎

Formula (11) should make sense from your experience as an automobile passenger. If a car increases its speed rapidly, the effect is to press the passenger back against the seat. This occurs because the rapid increase in speed results in a large value for

$$\frac{d}{dt}\left(\frac{ds}{dt}\right) = \frac{d^2s}{dt^2}$$

which produces a large tangential component of acceleration in Formula (11). On the other hand, if the car rounds a turn in the road, the passenger is thrown to the side: the greater the curvature in the road or the greater the speed of the car, the greater the force with which the passenger is thrown. This occurs because ds/dt (speed) and κ (curvature) both enter as factors in the normal component of acceleration in (11).

The coefficients of $\mathbf{T}$ and $\mathbf{N}$ in (11) are commonly denoted by

$$a_T = \frac{d^2s}{dt^2} \qquad a_N = \kappa\left(\frac{ds}{dt}\right)^2 \tag{12a–b}$$

These numbers are called the *scalar tangential component of acceleration* and the *scalar normal component of acceleration*, respectively. Using this notation, (11) can be expressed as

$$\mathbf{a} = a_T\mathbf{T} + a_N\mathbf{N} \tag{13}$$

The vectors $a_T\mathbf{T}$ and $a_N\mathbf{N}$ are called the *vector tangential component of acceleration* and *vector normal component of acceleration*, respectively.

It should be noted that Formula (13) applies to motion in both 2-space and 3-space. What is interesting is that the 3-space formula does not involve the binormal vector $\mathbf{B}$, so the acceleration vector always lies in the plane of $\mathbf{T}$ and $\mathbf{N}$ (the osculating plane), even for the most twisting paths of motion (Figure 15.6.5).

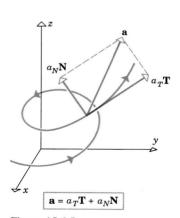

$$\boxed{\mathbf{a} = a_T\mathbf{T} + a_N\mathbf{N}}$$

Figure 15.6.5

Although Formulas (12a) and (12b) provide useful insight into the physical properties of the tangential and normal components of acceleration, they are not always the best formulas to use for computations. We shall now derive some more useful formulas that express the scalar normal and tangential components of acceleration directly in terms of the velocity and acceleration of the particle. If, as shown in Figure 15.6.6, we let θ be the angle between the vector $\mathbf{a}$ and the vector $a_T\mathbf{T}$, then we obtain

$$a_T = \|\mathbf{a}\| \cos\theta \quad \text{and} \quad a_N = \|\mathbf{a}\| \sin\theta$$

With the help of Definition 14.3.1 and Formula (7) of Section 14.4, these equations can be rewritten as

$$a_T = \frac{\|\mathbf{v}\|\,\|\mathbf{a}\| \cos\theta}{\|\mathbf{v}\|} = \frac{\mathbf{v}\cdot\mathbf{a}}{\|\mathbf{v}\|} \quad \text{and} \quad a_N = \frac{\|\mathbf{v}\|\,\|\mathbf{a}\| \sin\theta}{\|\mathbf{v}\|} = \frac{\|\mathbf{v}\times\mathbf{a}\|}{\|\mathbf{v}\|}$$

Summarizing, we have

$$a_T = \frac{\mathbf{v}\cdot\mathbf{a}}{\|\mathbf{v}\|} \quad \text{and} \quad a_N = \frac{\|\mathbf{v}\times\mathbf{a}\|}{\|\mathbf{v}\|} \tag{14a–b}$$

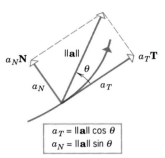

$$a_T = \|\mathbf{a}\| \cos\theta$$
$$a_N = \|\mathbf{a}\| \sin\theta$$

Figure 15.6.6

These formulas are applicable in both 2-space and 3-space, but to compute the cross product in 2-space, the vectors $\mathbf{v}$ and $\mathbf{a}$ must be treated as vectors in 3-space with a zero $\mathbf{k}$ component. Note also that a_N is nonnegative. Thus, for motion in 2-space the vector $a_N\mathbf{N}$ points "inward" toward the concave side of the curve because $\mathbf{N}$ does. (See the subsection of 15.4 on Inward Unit Normal Vectors in 2-Space.) For this reason, the vector $a_N\mathbf{N}$ is also called the ***inward vector component of acceleration*** in 2-space.

Before proceeding to a numerical example, it is worth noting that by rewriting Formula (12b) as

$$\kappa = \frac{a_N}{(ds/dt)^2} = \frac{a_N}{\|\mathbf{v}\|^2} \tag{15}$$

and combining it with (14b) we can write

$$\kappa = \frac{\|\mathbf{v}\times\mathbf{a}\|}{\|\mathbf{v}\|^3} \tag{16}$$

which expresses curvature in terms of velocity and acceleration. [Compare this result with Formula (3) of Section 15.5.]

Example 4   Suppose that a particle moves through 3-space so that its position vector at time t is

$$\mathbf{r}(t) = t\mathbf{i} + t^2\mathbf{j} + t^3\mathbf{k}$$

(The path is the twisted cubic shown in Figure 15.1.2.)

(a) Find the scalar tangential and normal components of acceleration at time t.

(b) Find the scalar tangential and normal components of acceleration at time $t = 1$.

(c) Find the vector tangential and normal components of acceleration at time $t = 1$.

(d) Find the curvature of the path at the point where the particle is located at time $t = 1$.

Solution (*a*).   We have

$$\mathbf{v}(t) = \mathbf{r}'(t) = \mathbf{i} + 2t\mathbf{j} + 3t^2\mathbf{k}$$

$$\mathbf{a}(t) = \mathbf{v}'(t) = 2\mathbf{j} + 6t\mathbf{k}$$

$$\|\mathbf{v}(t)\| = \sqrt{1 + 4t^2 + 9t^4}$$

$$\mathbf{v}(t)\cdot\mathbf{a}(t) = 4t + 18t^3$$

$$\mathbf{v}(t) \times \mathbf{a}(t) = \begin{vmatrix} \mathbf{i} & \mathbf{j} & \mathbf{k} \\ 1 & 2t & 3t^2 \\ 0 & 2 & 6t \end{vmatrix} = 6t^2\mathbf{i} - 6t\mathbf{j} + 2\mathbf{k}$$

Thus, from (14a) and (14b)

$$a_T = \frac{\mathbf{v} \cdot \mathbf{a}}{\|\mathbf{v}\|} = \frac{4t + 18t^3}{\sqrt{1 + 4t^2 + 9t^4}}$$

$$a_N = \frac{\|\mathbf{v} \times \mathbf{a}\|}{\|\mathbf{v}\|} = \frac{\sqrt{36t^4 + 36t^2 + 4}}{\sqrt{1 + 4t^2 + 9t^4}} = 2\sqrt{\frac{9t^4 + 9t^2 + 1}{9t^4 + 4t^2 + 1}}$$

Solution (b).   At time $t = 1$, the components a_T and a_N in part (a) are

$$a_T = \frac{22}{\sqrt{14}} \approx 5.88 \quad \text{and} \quad a_N = 2\sqrt{\frac{19}{14}} \approx 2.33$$

Solution (c).   Since $\mathbf{T}$ and $\mathbf{v}$ have the same direction, $\mathbf{T}$ can be obtained by normalizing $\mathbf{v}$, that is,

$$\mathbf{T}(t) = \frac{\mathbf{v}(t)}{\|\mathbf{v}(t)\|}$$

At time $t = 1$ we have

$$\mathbf{T}(1) = \frac{\mathbf{v}(1)}{\|\mathbf{v}(1)\|} = \frac{\mathbf{i} + 2\mathbf{j} + 3\mathbf{k}}{\|\mathbf{i} + 2\mathbf{j} + 3\mathbf{k}\|} = \frac{1}{\sqrt{14}}(\mathbf{i} + 2\mathbf{j} + 3\mathbf{k})$$

From this and part (b) we obtain the vector tangential component of acceleration:

$$a_T\mathbf{T}(1) = \frac{22}{\sqrt{14}}\mathbf{T}(1) = \tfrac{11}{7}(\mathbf{i} + 2\mathbf{j} + 3\mathbf{k}) = \tfrac{11}{7}\mathbf{i} + \tfrac{22}{7}\mathbf{j} + \tfrac{33}{7}\mathbf{k}$$

To find the normal vector component of acceleration, we rewrite $\mathbf{a} = a_T\mathbf{T} + a_N\mathbf{N}$ as

$$a_N\mathbf{N} = \mathbf{a} - a_T\mathbf{T}$$

Thus, at time $t = 1$ the normal vector component of acceleration is

$$a_N\mathbf{N}(1) = \mathbf{a}(1) - a_T\mathbf{T}(1)$$
$$= (2\mathbf{j} + 6\mathbf{k}) - (\tfrac{11}{7}\mathbf{i} + \tfrac{22}{7}\mathbf{j} + \tfrac{33}{7}\mathbf{k})$$
$$= -\tfrac{11}{7}\mathbf{i} - \tfrac{8}{7}\mathbf{j} + \tfrac{9}{7}\mathbf{k}$$

Solution (d).   We shall apply Formula (16) with $t = 1$. From part (a)

$$\|\mathbf{v}(1)\| = \sqrt{14} \quad \text{and} \quad \mathbf{v}(1) \times \mathbf{a}(1) = 6\mathbf{i} - 6\mathbf{j} + 2\mathbf{k}$$

Thus, at time $t = 1$

$$\kappa = \frac{\|\mathbf{v} \times \mathbf{a}\|}{\|\mathbf{v}\|^3} = \frac{\sqrt{76}}{(\sqrt{14})^3} = \frac{1}{14}\sqrt{\frac{38}{7}} \approx 0.17 \qquad \blacktriangleleft$$

☐ **MOTION OF A PROJECTILE**

Our work in this section provides many of the mathematical tools required to study such fundamental physical phenomena as planetary motion and the motion of objects in the earth's gravitational field. As an illustration, we shall investigate the following problem.

15.6.3   PROBLEM.   *An object of mass m is fired, thrown, or released at some known point near the surface of the earth with a known initial velocity vector. Find the trajectory of the object.*

To solve this problem we need a principle from physics known as ***Newton's Second Law of Motion***; it states that when an object of mass m is subjected to a force $\mathbf{F}$, it undergoes an acceleration $\mathbf{a}$ satisfying

$$\mathbf{F} = m\mathbf{a} \tag{17}$$

We shall make three simplifying assumptions:

1. The mass of the object is constant.
2. The only force acting on the object is the force of the earth's gravity. (Thus, air resistance and the gravitational effect of other planets and celestial objects are ignored.)
3. The object remains sufficiently close to the earth that the earth's gravity can be assumed constant.

We shall assume that the mass of the object is m and that t (measured in seconds) is the time elapsed from the initial firing or release, so that this firing or release occurs at time $t = 0$. We shall denote the known initial velocity vector by $\mathbf{v}_0$ and the known initial position vector by $\mathbf{r}_0$.

As shown in Figure 15.6.7, we shall introduce an xy-coordinate system whose origin is on the surface of the earth and whose positive y-axis points up and passes through the initial position of the object. Thus, at time $t = 0$ the object has coordinates $(0, s_0)$, which are assumed known, and the initial position vector is $\mathbf{r}_0 = s_0\mathbf{j}$.

It is shown in physics that the downward force of the earth's gravity on an object of mass m is

$$\mathbf{F} = -mg\mathbf{j} \tag{18}$$

where g is a constant approximately equal to 32 ft/sec^2 if distance is measured in feet and 9.8 m/sec^2 if distance is measured in meters. Substituting (18) into (17) yields

$$m\mathbf{a} = -mg\mathbf{j}$$

or on canceling m from both sides

$$\mathbf{a} = -g\mathbf{j} \tag{19}$$

Observe that the acceleration vector $\mathbf{a}$ is constant because (19) does not involve t. Moreover, since the value of g is known, the acceleration vector $\mathbf{a}$ is also known; thus, we can find the position function of the object by integrating the known acceleration twice, once to obtain the velocity $\mathbf{v}(t)$, and again to obtain $\mathbf{r}(t)$. Thus, Problem 15.6.3 has been reduced to solving the vector differential equation

$$\frac{d^2\mathbf{r}}{dt^2} = -g\mathbf{j} \tag{20}$$

subject to the initial conditions

$$\mathbf{r}(0) = \mathbf{r}_0 = s_0\mathbf{j} \tag{21}$$

$$\mathbf{v}(0) = \mathbf{v}_0 \tag{22}$$

Integrating (20) with respect to t (keeping in mind that $-g\mathbf{j}$ is constant) yields

$$\mathbf{v}(t) = -gt\mathbf{j} + \mathbf{c}_1 \tag{23}$$

where $\mathbf{c}_1$ is a vector constant of integration. Substituting $t = 0$ in (23) and using initial condition (22) yields

$$\mathbf{v}_0 = \mathbf{c}_1$$

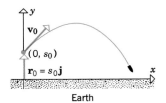

Earth

Figure 15.6.7

so that (23) may be written as

$$\mathbf{v}(t) = -gt\mathbf{j} + \mathbf{v_0} \tag{24}$$

This formula specifies the velocity of the object at any time t. Integrating $\mathbf{v}(t)$ with respect to t (keeping in mind that $\mathbf{v_0}$ is constant) yields

$$\mathbf{r}(t) = -\tfrac{1}{2}gt^2\mathbf{j} + \mathbf{v_0}t + \mathbf{c_2} \tag{25}$$

where $\mathbf{c_2}$ is another vector constant of integration. Substituting $t = 0$ in (25) and using initial condition (21) yields

$$s_0\mathbf{j} = \mathbf{c_2}$$

so that (25) can be written as

$$\mathbf{r}(t) = (-\tfrac{1}{2}gt^2 + s_0)\mathbf{j} + \mathbf{v_0}t \tag{26}$$

This is the result we were seeking—the position function of the object, expressed in terms of the initial velocity and position.

REMARK.   Observe that the mass of the object does not enter into the final formulas for velocity and position. Physically, this means that the mass has no influence on the trajectory or the velocity of the object—these are completely determined by the initial position and velocity. This explains the famous observation of Galileo that two objects of different mass, released from the same height, will reach the ground at the same time if air resistance is neglected.

There is a useful alternative form of (26) that expresses the position function of the object in terms of its initial speed and the angle that the initial velocity vector makes with the x-axis. As shown in Figure 15.6.8, suppose that the initial speed $\|\mathbf{v_0}\|$ is denoted by v_0 and the angle that $\mathbf{v_0}$ makes with the x-axis is denoted by α. Thus, the vector $\mathbf{v_0}$ can be expressed as

$$\mathbf{v_0} = (v_0 \cos \alpha)\mathbf{i} + (v_0 \sin \alpha)\mathbf{j}$$

Substituting this expression in (26) and combining like components yields

$$\mathbf{r}(t) = (v_0 \cos \alpha)t\mathbf{i} + (s_0 + (v_0 \sin \alpha)t - \tfrac{1}{2}gt^2)\mathbf{j} \tag{27}$$

which is equivalent to the parametric equations

$$x = (v_0 \cos \alpha)t, \quad y = s_0 + (v_0 \sin \alpha)t - \tfrac{1}{2}gt^2 \tag{28}$$

The parametric equations reveal that the trajectory of the object is a parabolic arc. To see that this is so, we can solve the first equation for t, then substitute in the second to eliminate the parameter. This yields (verify)

$$y = s_0 + (\tan \alpha)x - \left(\frac{g}{2v_0^2 \cos^2 \alpha}\right)x^2$$

which is an equation of a parabola since the right side is a quadratic polynomial in x.

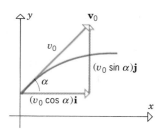

Figure 15.6.8

Example 5   A shell, fired from a cannon, has a muzzle speed (the speed as it leaves the barrel) of 800 ft/sec. The barrel makes an angle of 45° with the horizontal and, for simplicity, the barrel opening is assumed to be at ground level.

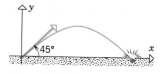

Figure 15.6.9

(a)   Find parametric equations for the shell's trajectory relative to the coordinate system in Figure 15.6.9.

(b)   How high does the shell rise?

(c) How far does the shell travel horizontally?

(d) What is the speed of the shell at its point of impact with the ground?

Solution (a). From (28) with $v_0 = 800$, $\alpha = 45°$, $s_0 = 0$ (since the shell starts at ground level), and $g = 32$ (since distance is measured in feet), we obtain the parametric equations

$$x = (800 \cos 45°)t, \quad y = (800 \sin 45°)t - 16t^2 \qquad (t \geq 0)$$

which simplify to

$$x = 400\sqrt{2}t, \quad y = 400\sqrt{2}t - 16t^2 \qquad (t \geq 0) \qquad (29)$$

Solution (b). The maximum height of the shell is the maximum value of y in (29), which occurs when $dy/dt = 0$, that is, when

$$400\sqrt{2} - 32t = 0 \quad \text{or} \quad t = \frac{25\sqrt{2}}{2}$$

Substituting this value of t in (29) yields

$$y = 5000 \text{ ft}$$

as the maximum height of the shell.

Solution (c). The shell will hit the ground when $y = 0$. From (29), this occurs when

$$400\sqrt{2}t - 16t^2 = 0 \quad \text{or} \quad t(400\sqrt{2} - 16t) = 0$$

The solution $t = 0$ corresponds to the initial position of the shell and the solution $t = 25\sqrt{2}$ to the time of impact. Substituting the latter value in the equation for x in (29) yields

$$x = 20{,}000 \text{ ft}$$

as the horizontal distance traveled by the shell.

Solution (d). From (29), the position function of the shell is

$$\mathbf{r}(t) = (400\sqrt{2}t)\mathbf{i} + (400\sqrt{2}t - 16t^2)\mathbf{j}$$

so that the velocity function is

$$\mathbf{v}(t) = \mathbf{r}'(t) = 400\sqrt{2}\,\mathbf{i} + (400\sqrt{2} - 32t)\mathbf{j}$$

From part (c), impact occurs when $t = 25\sqrt{2}$, so that the velocity vector at this point is

$$\mathbf{v}(25\sqrt{2}) = 400\sqrt{2}\,\mathbf{i} + [400\sqrt{2} - 32(25\sqrt{2})]\mathbf{j} = 400\sqrt{2}\,\mathbf{i} - 400\sqrt{2}\,\mathbf{j}$$

Thus, the speed at impact is

$$\|\mathbf{v}(25\sqrt{2})\| = \sqrt{(400\sqrt{2})^2 + (-400\sqrt{2})^2} = 800 \text{ ft/sec} \qquad \blacktriangleleft$$

▶ Exercise Set 15.6   🅒 *17, 61, 65–68, 72, 74, 75*

In Exercises 1–6, $\mathbf{r}(t)$ is the position vector of a particle moving in the plane. Find the velocity, acceleration, and speed at an arbitrary time t; then sketch the path of the particle together with the velocity and acceleration vectors at the indicated time t.

1. $\mathbf{r}(t) = 3 \cos t\mathbf{i} + 3 \sin t\mathbf{j}$; $t = \pi/3$.

2. $\mathbf{r}(t) = t\mathbf{i} + t^2\mathbf{j}$; $t = 2$.

3. $\mathbf{r}(t) = e^t\mathbf{i} + e^{-t}\mathbf{j}$; $t = 0$.

4. $\mathbf{r}(t) = (2 + 4t)\mathbf{i} + (1 - t)\mathbf{j}$; $t = 1$.

5. $\mathbf{r}(t) = \cosh t\mathbf{i} + \sinh t\mathbf{j}$; $t = \ln 2$.

6. $\mathbf{r}(t) = 4 \cos t\mathbf{i} + 9 \sin t\mathbf{j}$; $t = \pi/2$.

In Exercises 7–12, find the velocity, speed, and acceleration at the given time t of a particle moving along the given curve.

7. $\mathbf{r}(t) = t\mathbf{i} + \frac{1}{2}t^2\mathbf{j} + \frac{1}{3}t^3\mathbf{k}$; $t = 1$.

8. $x = 1 + 3t, y = 2 - 4t, z = 7 + t$; $t = 2$.

9. $\mathbf{r}(t) = 2 \cos t\mathbf{i} + 2 \sin t\mathbf{j} + t\mathbf{k}$; $t = \pi/4$.

10. $\mathbf{r}(t) = 3t\mathbf{i} + 2t^2\mathbf{j} + \ln t\mathbf{k}$; $t = 1$.

11. $\mathbf{r}(t) = e^t \sin t\mathbf{i} + e^t \cos t\mathbf{j} + t\mathbf{k}$; $t = \pi/2$.

12. $\mathbf{r}(t) = 2t\mathbf{i} + t^2\mathbf{j} + \ln t\mathbf{k}$; $t = 2$.

13. Suppose that the position vector of a particle moving in the plane is $\mathbf{r} = 12\sqrt{t}\mathbf{i} + t^{3/2}\mathbf{j}$, $t > 0$. Find the minimum speed of the particle and its location when it has this speed.

14. Suppose that the motion of a particle is described by $\mathbf{r} = (t - t^2)\mathbf{i} - t^2\mathbf{j}$. Find the minimum speed of the particle and its location when it has this speed.

15. Find the maximum and minimum speeds of a particle whose motion is described by the position vector $\mathbf{r} = \sin 3t\mathbf{i} - 2 \cos 3t\mathbf{j}$.

16. Find the maximum and minimum speeds of a particle whose motion is described by the position vector $\mathbf{r} = 3 \cos 2t\mathbf{i} + \sin 2t\mathbf{j} + 4t\mathbf{k}$.

17. Find, to the nearest degree, the angle between $\mathbf{v}$ and $\mathbf{a}$ for $\mathbf{r} = t^3\mathbf{i} + t^2\mathbf{j}$ when $t = 1$.

18. Show that the angle between $\mathbf{v}$ and $\mathbf{a}$ is constant for $\mathbf{r} = e^t \cos t\mathbf{i} + e^t \sin t\mathbf{j}$. Find the angle.

19. Where on the path

$$\mathbf{r} = (t^2 - 5t)\mathbf{i} + (2t + 1)\mathbf{j} + 3t^2\mathbf{k}$$

are the velocity and acceleration vectors orthogonal?

20. Prove: If the speed of a particle is constant, then the acceleration and velocity vectors are orthogonal. [*Hint:* Consider $\mathbf{v} \cdot \mathbf{v}$.]

In Exercises 21 and 22, the position vectors of two particles are given. Show that the particles move along the same path but the speed of the first is constant and the speed of the second is not.

21. $\mathbf{r}_1 = 2 \cos 3t\mathbf{i} + 2 \sin 3t\mathbf{j}$,
    $\mathbf{r}_2 = 2 \cos (t^2)\mathbf{i} + 2 \sin (t^2)\mathbf{j}$ $(t \geq 0)$.

22. $\mathbf{r}_1 = (3 + 2t)\mathbf{i} + t\mathbf{j} + (1 - t)\mathbf{k}$,
    $\mathbf{r}_2 = (5 - 2t^3)\mathbf{i} + (1 - t^3)\mathbf{j} + t^3\mathbf{k}$.

In Exercises 23–30, use the given information to find the position and velocity vectors of the particle.

23. $\mathbf{a}(t) = -32\mathbf{j}$; $\mathbf{v}(0) = \mathbf{0}$; $\mathbf{r}(0) = \mathbf{0}$.

24. $\mathbf{a}(t) = t\mathbf{j}$; $\mathbf{v}(0) = \mathbf{i} + \mathbf{j}$; $\mathbf{r}(0) = \mathbf{0}$.

25. $\mathbf{a}(t) = -\cos t\mathbf{i} - \sin t\mathbf{j}$; $\mathbf{v}(0) = \mathbf{i}$; $\mathbf{r}(0) = \mathbf{j}$.

26. $\mathbf{a}(t) = \mathbf{i} + e^{-t}\mathbf{j}$; $\mathbf{v}(0) = 2\mathbf{i} + \mathbf{j}$; $\mathbf{r}(0) = \mathbf{i} - \mathbf{j}$.

27. $\mathbf{a}(t) = \mathbf{i} + t\mathbf{k}$; $\mathbf{v}(0) = \mathbf{0}$; $\mathbf{r}(0) = \mathbf{j}$.

28. $\mathbf{a}(t) = (\sin 2t)\mathbf{j}$; $\mathbf{v}(0) = \mathbf{i} - \mathbf{k}$; $\mathbf{r}(0) = \mathbf{0}$.

29. $\mathbf{a}(t) = \sin t\mathbf{i} + \cos t\mathbf{j} + e^t\mathbf{k}$; $\mathbf{v}(0) = \mathbf{k}$; $\mathbf{r}(0) = -\mathbf{i} + \mathbf{k}$.

30. $\mathbf{a}(t) = (t + 1)^{-2}\mathbf{j} - e^{-2t}\mathbf{k}$, $t > -1$; $\mathbf{v}(0) = 3\mathbf{i} - \mathbf{j}$; $\mathbf{r}(0) = 2\mathbf{k}$.

31. Prove: If the acceleration of a moving particle is zero for all t, then the particle moves on a straight line.

32. Show that for a particle in motion over the time interval $[t_1, t_2]$ the change in velocity $\mathbf{v}(t_2) - \mathbf{v}(t_1)$ is given by

$$\mathbf{v}(t_2) - \mathbf{v}(t_1) = \int_{t_1}^{t_2} \mathbf{a}(t)\, dt$$

In Exercises 33–38, find the displacement and the distance traveled over the indicated time interval.

33. $\mathbf{r}(t) = t^2\mathbf{i} + \frac{1}{3}t^3\mathbf{j}$; $1 \leq t \leq 3$.

34. $\mathbf{r}(t) = (1 - 3 \sin t)\mathbf{i} + 3 \cos t\mathbf{j}$; $0 \leq t \leq 3\pi/2$.

35. $\mathbf{r}(t) = 2 \sin 3t\mathbf{i} + 2 \cos 3t\mathbf{j}$; $0 \leq t \leq 2\pi$.

36. $\mathbf{r}(t) = 3 \sin 2t\mathbf{i} - 3 \cos 2t\mathbf{j} + 8t\mathbf{k}$; $0 \leq t \leq 3\pi/4$.

37. $\mathbf{r}(t) = e^t\mathbf{i} + e^{-t}\mathbf{j} + \sqrt{2}t\mathbf{k}$; $0 \leq t \leq \ln 3$.

38. $\mathbf{r}(t) = \cos 2t\mathbf{i} + (1 - \cos 2t)\mathbf{j} + (3 + \frac{1}{2} \cos 2t)\mathbf{k}$; $0 \leq t \leq \pi$.

In Exercises 39–48, find the scalar tangential and normal components of acceleration at the indicated time t.

39. $\mathbf{r}(t) = 2 \cos t\mathbf{i} + 2 \sin t\mathbf{j}$; $t = \pi/3$.

40. $\mathbf{r}(t) = t\mathbf{i} + t^2\mathbf{j}$; $t = 1$.

41. $\mathbf{r}(t) = e^{-t}\mathbf{i} + e^t\mathbf{j}$; $t = 0$.

42. $\mathbf{r}(t) = \cos (t^2)\mathbf{i} + \sin (t^2)\mathbf{j}$; $t = \sqrt{\pi}/2$.

43. $\mathbf{r}(t) = (t^3 - 2t)\mathbf{i} + (t^2 - 4)\mathbf{j}$; $t = 1$.

44. $\mathbf{r}(t) = e^t \cos t\mathbf{i} + e^t \sin t\mathbf{j}$; $t = \pi/4$.

45. $\mathbf{r}(t) = t\mathbf{i} + t^2\mathbf{j} + t^3\mathbf{k}$; $t = 1$.

46. $\mathbf{r}(t) = e^t\mathbf{i} + e^{-2t}\mathbf{j} + t\mathbf{k}$; $t = 0$.

47. $\mathbf{r}(t) = 3 \sin t\mathbf{i} + 2 \cos t\mathbf{j} - \sin 2t\mathbf{k}$; $t = \pi/2$.

48. $\mathbf{r}(t) = 2\mathbf{i} + t^3\mathbf{j} - 16 \ln t\mathbf{k}$; $t = 1$.

In Exercises 49–52, $\mathbf{v}$ and $\mathbf{a}$ are given at a certain instant of time. Find a_T, a_N, $\mathbf{T}$, and $\mathbf{N}$ at this instant.

49. $\mathbf{v} = -4\mathbf{j}$, $\mathbf{a} = 2\mathbf{i} + 3\mathbf{j}$.

50. $\mathbf{v} = \mathbf{i} + 2\mathbf{j}$, $\mathbf{a} = 3\mathbf{i}$.

51. $\mathbf{v} = 2\mathbf{i} + 2\mathbf{j} + \mathbf{k}$, $\mathbf{a} = \mathbf{i} + 2\mathbf{k}$.

52. $\mathbf{v} = 3\mathbf{i} - 4\mathbf{k}$, $\mathbf{a} = \mathbf{i} - \mathbf{j} + 2\mathbf{k}$.

In Exercises 53–56, the speed $\|\mathbf{v}\|$ of a particle at an arbitrary time t is given. Find the scalar tangential component of acceleration at the indicated time.

53. $\|\mathbf{v}\| = \sqrt{3t^2 + 4}$; $t = 2$.

54. $\|\mathbf{v}\| = \sqrt{t^2 + e^{-3t}}$; $t = 0$.

55. $\|\mathbf{v}\| = \sqrt{(4t - 1)^2 + \cos^2 \pi t}$; $t = \frac{1}{4}$.

56. $\|\mathbf{v}\| = \sqrt{t^4 + 5t^2 + 3}$; $t = 1$.

In Exercises 57–60, find the curvature of the path of motion of the particle at the instant at which the velocity and acceleration vectors are given.

57. Exercise 49.   58. Exercise 50.

59. Exercise 51.   60. Exercise 52.

61. The nuclear accelerator at the Enrico Fermi Laboratory is circular with a radius of 1 km. Find the scalar normal component of acceleration of a proton moving around the accelerator with a constant speed of 2.9×10^5 km/sec.

In Exercises 62–64, use the formula for $\kappa(x)$ in Exercise 15 of Section 15.5 to help solve the problem.

---

62. Suppose that a particle moves with nonzero acceleration along the curve $y = f(x)$. Show that the acceleration vector is tangent to the curve at each point where $f''(x) = 0$.

63. A particle moves along the parabola $y = x^2$ with a constant speed of 3 units/sec. Find the normal component of acceleration as a function of x.

64. A particle moves along the curve $y = e^x$ with a constant speed of 2 units/sec. Find the normal component of acceleration as a function of x.

65. A shell is fired from ground level with a muzzle speed of 320 ft/sec and elevation angle of 60°. Find

    (a) parametric equations for the shell's trajectory

    (b) the maximum height reached by the shell

    (c) the horizontal distance traveled by the shell

    (d) the speed of the shell at impact.

66. Solve Exercise 65 assuming that the muzzle speed is 980 m/sec and the elevation angle is 45°.

67. A rock is thrown downward from the top of a building, 168 ft high, at an angle of 60° with the horizontal. How far from the base of the building will the rock land if its initial speed is 80 ft/sec?

68. Solve Exercise 67 assuming that the rock is thrown horizontally at a speed of 80 ft/sec.

69. A shell is to be fired from ground level at an elevation angle of 30°. What should the muzzle speed be in order for the maximum height of the shell to be 2500 ft?

70. A shell, fired from ground level at an elevation angle of 45°, hits the ground 24,500 m away. Calculate the muzzle speed of the shell.

71. Find two elevation angles that will enable a shell, fired from ground level with a muzzle speed of 800 ft/sec, to hit a ground-level target 10,000 ft away.

72. A ball rolls off a table 4 ft high while moving at a constant speed of 5 ft/sec.

    (a) How long does it take for the ball to hit the floor after it leaves the table?

    (b) At what speed does the ball hit the floor?

    (c) If a ball were dropped from table height (initial speed 0) at the same time the rolling ball leaves the table, which ball would hit the ground first?

73. A shell is fired from ground level at an elevation angle of α and a muzzle speed of v_0.

    (a) Show that the maximum height reached by the shell is

$$\text{maximum height} = \frac{(v_0 \sin \alpha)^2}{2g}$$

    (b) The **horizontal range** R of the shell is the horizontal distance traveled when the shell returns to ground level. Show that $R = (v_0^2 \sin 2\alpha)/g$. For what elevation angle will the range be maximum? What is the maximum range?

74. A shell is fired from ground level with an elevation angle α and a muzzle speed of v_0. Find the angle that should be used to hit a target at ground level that is at a distance of 3/4 the maximum range of the shell. Express your answer to the nearest tenth of a degree. [*Hint:* See Exercise 73(b).]

75. At time $t = 0$ a baseball that is 5 ft above the ground is hit with a bat. The ball leaves the bat with a speed of 80 ft/sec at an angle of 30° above the horizontal.

    (a) How long will it take for the baseball to hit the ground? Express your answer to the nearest hundredth of a second.

    (b) Use the result in part (a) to find the horizontal distance traveled by the ball. Express your answer to the nearest tenth of a foot.

76. At time $t = 0$ a projectile is fired from a height h above level ground at an elevation angle of α with a speed v. Let R be the horizontal distance to the point where the projectile hits the ground.

    (a) Show that α and R must satisfy the equation

$$g(\sec^2 \alpha)R^2 - 2v^2(\tan \alpha)R - 2v^2h = 0$$

    (b) If g, h, and v are constant, then the equation in part (a) defines R implicitly as a function of α. Let R_0 be the maximum value of R and α_0 the value of α when $R = R_0$. Use implicit differentiation to find $dR/d\alpha$ and show that

$$\tan \alpha_0 = \frac{v^2}{gR_0}$$

    [*Hint:* Assume that $dR/d\alpha = 0$ when R is maximum.]

    (c) Use the results in parts (a) and (b) to show that

$$R_0 = \frac{v}{g}\sqrt{v^2 + 2gh}$$

    and

$$\alpha_0 = \tan^{-1}\frac{v}{\sqrt{v^2 + 2gh}}$$

■ 15.7 KEPLER'S LAWS OF PLANETARY MOTION

> *One of the great advances in the history of astronomy occurred in the early 1600s when Johannes Kepler* deduced from empirical data that all planets in our solar system move in elliptical orbits with the sun at a focus. Subsequently, Isaac Newton showed mathematically that such planetary motion is the consequence of an inverse-square law of gravitational attraction. In this section we shall use the concepts developed in the preceding sections of this chapter to derive three basic laws of planetary motion, known as **Kepler's laws**.*

□ **KEPLER'S LAWS**

In 1609 Johannes Kepler published a book known as *Astronomia Nova* (or sometimes as *Commentaries on the Motions of Mars*) in which he succeeded in distilling thousands of years of observational astronomy into three beautiful laws of planetary motion.

15.7.1 KEPLER'S LAWS.

- First law (**Law of Orbits**).   Each planet moves in an elliptical orbit with the sun at a focus.
- Second law (**Law of Areas**).   Equal areas are swept out in equal times by the line from the sun to a planet.
- Third law (**Law of Periods**).   The square of a planet's period (the time it takes the planet to complete one orbit about the sun) is proportional to the cube of the length of the semimajor axis of its elliptical orbit.

□ **CENTRAL FORCES**

To derive Kepler's laws, we shall assume that the force exerted by the sun on a planet is always directed toward the sun's center. In general, a force that is always directed toward a fixed point is called a **central force**.

Suppose that a particle P of mass m moves with acceleration $\mathbf{a}$ under the influence of a central force $\mathbf{F}$ that is directed toward a fixed point O, and let $\mathbf{r} = \mathbf{r}(t)$ be the position vector for the particle from O to P (Figure 15.7.1). We will show that the particle moves in a *plane* containing O.

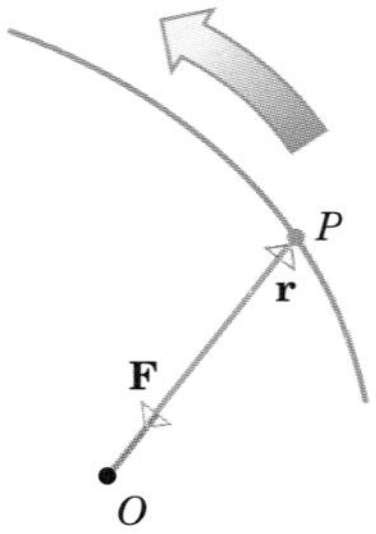

It follows from Newton's Second Law of Motion ($\mathbf{F} = m\mathbf{a}$) that $\mathbf{a}$ is in the same direction as $\mathbf{F}$, and consequently $\mathbf{a}$ and $\mathbf{r}$ are oppositely directed vectors. Thus, it follows from Theorem 14.4.3 that

$$\mathbf{r} \times \mathbf{a} = \mathbf{0}$$

Since the velocity and acceleration of the particle are given by $\mathbf{v} = d\mathbf{r}/dt$ and $\mathbf{a} = d\mathbf{v}/dt$, respectively, we have

$$\frac{d}{dt}(\mathbf{r} \times \mathbf{v}) = \mathbf{r} \times \frac{d\mathbf{v}}{dt} + \frac{d\mathbf{r}}{dt} \times \mathbf{v} = (\mathbf{r} \times \mathbf{a}) + (\mathbf{v} \times \mathbf{v}) = \mathbf{0} + \mathbf{0} = \mathbf{0}$$

Integrating the left and right sides of this equation with respect to t yields

$$\mathbf{r} \times \mathbf{v} = \mathbf{b} \tag{1}$$

where $\mathbf{b}$ is a constant vector. This implies that $\mathbf{r}$ is perpendicular to $\mathbf{b}$, and hence the particle moves along a path that lies in the plane that is perpendicular to $\mathbf{b}$ and contains the fixed point O. This shows that each planet moves in a plane through the center of the sun. Astronomers call this plane the **ecliptic** of the planet.

Figure 15.7.1

---

*See biography on p. 633.

We shall now derive Kepler's laws: first the Law of Areas, then the Law of Orbits, and finally the Law of Periods. In all of these derivations we shall assume that the sun and planets are homogeneous spheres, so that they can be treated as if their masses were concentrated at their centers. When we speak of the distance between the sun and a planet, we will mean the distance between their centers. Moreover, we will ignore the gravitational effects of all other celestial bodies on the sun and the planet.

☐ **KEPLER'S SECOND LAW**

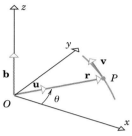

Figure 15.7.2

To establish Kepler's second law we introduce a Cartesian xyz-coordinate system with its origin O at the sun's center and its positive z-axis in the direction of $\mathbf{b} = \mathbf{r} \times \mathbf{v}$. Because $\mathbf{b}$, $\mathbf{r}$, and $\mathbf{v}$ are related by the right-hand rule, the polar angle θ from the positive x-axis to P increases with time t (Figure 15.7.2). It will be convenient to express $\mathbf{r}$ as $\mathbf{r} = r\mathbf{u}$, where $r = \|\mathbf{r}\|$ and

$$\mathbf{u} = \cos\theta\mathbf{i} + \sin\theta\mathbf{j} \tag{2}$$

is a unit vector in the direction of $\mathbf{r}$. It then follows that

$$\mathbf{v} = \frac{d\mathbf{r}}{dt} = \frac{d}{dt}(r\mathbf{u}) = r\frac{d\mathbf{u}}{dt} + \frac{dr}{dt}\mathbf{u}$$

and hence from (1)

$$\mathbf{b} = \mathbf{r} \times \mathbf{v} = (r\mathbf{u}) \times \left(r\frac{d\mathbf{u}}{dt} + \frac{dr}{dt}\mathbf{u}\right) = r^2\mathbf{u} \times \frac{d\mathbf{u}}{dt} + r\frac{dr}{dt}\mathbf{u} \times \mathbf{u} = r^2\mathbf{u} \times \frac{d\mathbf{u}}{dt} \tag{3}$$

But (2) implies that

$$\frac{d\mathbf{u}}{dt} = \frac{d\mathbf{u}}{d\theta}\frac{d\theta}{dt} = (-\sin\theta\mathbf{i} + \cos\theta\mathbf{j})\frac{d\theta}{dt}$$

so

$$\mathbf{u} \times \frac{d\mathbf{u}}{dt} = \frac{d\theta}{dt}\mathbf{k} \tag{4}$$

Substituting (4) in (3) yields

$$\mathbf{b} = r^2\frac{d\theta}{dt}\mathbf{k} \tag{5}$$

which shows that $r^2(d\theta/dt)$ must be constant because $\mathbf{b}$ is constant.

If the path of P is described in polar coordinates by an equation $r = f(\theta)$, then from Formula (2) of Section 13.3 the area A swept out by the radius vector as it varies from some initial angle θ_0 to an arbitrary angle θ can be expressed as a function of θ by

$$A = \int_{\theta_0}^{\theta} \frac{1}{2}[f(\phi)]^2 \, d\phi$$

where we have used ϕ as the variable of integration rather than θ, since θ is already being used as an independent variable in the upper limit of integration. Applying the Second Fundamental Theorem of Calculus (Theorem 5.9.1) and the chain rule we obtain

$$\frac{dA}{dt} = \frac{dA}{d\theta}\frac{d\theta}{dt} = \frac{1}{2}[f(\theta)]^2\frac{d\theta}{dt} = \frac{1}{2}r^2\frac{d\theta}{dt} \tag{6}$$

Thus dA/dt must be constant because $r^2(d\theta/dt)$ is constant. This shows that area is swept out at a constant rate and hence that equal areas are swept out in equal times by the line from the sun to the planet (Figure 15.7.3). It also follows from this that a planet moves more rapidly when it is closer to the sun.

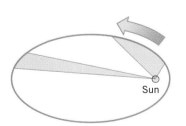

Sun

Equal areas are swept out in equal lengths of time.

Figure 15.7.3

☐ **KEPLER'S FIRST LAW**

Kepler's second law made no assumptions about the magnitude of the force $\mathbf{F}$. However, to obtain Kepler's first law we need *Newton's inverse-square law of gravitational attraction*, namely, that *every particle of matter in the universe attracts every other particle of matter with a force that is proportional to the product of their masses and inversely proportional to the square of the distance between them.* Thus, the central force $\mathbf{F}$ exerted by the sun on a planet can be expressed as

$$\mathbf{F} = -\frac{GmM}{r^2}\mathbf{u} \tag{7}$$

where m is the mass of the planet, M is the mass of the sun, G is a constant (called the *universal gravitational constant*), and $\mathbf{r} = r\mathbf{u}$ is the position vector from the center of the sun to the center of the planet. The negative sign in (7) indicates that the force $\mathbf{F}$ acts opposite to $\mathbf{r}$ (toward the sun's center).

It follows from (7) and Newton's Second Law of Motion ($\mathbf{F} = m\mathbf{a}$) that the acceleration of the planet is given by

$$\mathbf{a} = -\frac{GM}{r^2}\mathbf{u} \tag{8}$$

Observe that the acceleration does not depend on the mass of the planet.

It will be convenient to orient the coordinate axes so that at time $t = 0$ the planet is at its closest to the sun and on the positive x-axis (i.e., $\theta = 0$ if $t = 0$). Under this assumption, the velocity vector at time $t = 0$ is perpendicular to the radius vector and consequently the radius vector and the velocity vector at this time are of the form

$$\mathbf{r} = r_0\mathbf{i} \quad \text{and} \quad \mathbf{v} = v_0\mathbf{j} \tag{9}$$

(Figure 15.7.4). It follows from this assumption that at time $t = 0$ we have

$$\mathbf{b} = \mathbf{r} \times \mathbf{v} = (r_0\mathbf{i}) \times (v_0\mathbf{j}) = r_0v_0\mathbf{k} \tag{10}$$

and hence from (5) that

$$r^2\frac{d\theta}{dt} = r_0v_0 \tag{11}$$

It follows from (2), (5), and (8) that

$$\mathbf{a} \times \mathbf{b} = -\frac{GM}{r^2}(\cos\theta\mathbf{i} + \sin\theta\mathbf{j}) \times \left(r^2\frac{d\theta}{dt}\mathbf{k}\right)$$

$$= GM(-\sin\theta\mathbf{i} + \cos\theta\mathbf{j})\frac{d\theta}{dt}$$

$$= GM\frac{d\mathbf{u}}{dt}$$

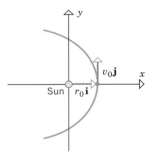

Figure 15.7.4

From this formula and the fact that $d\mathbf{b}/dt = \mathbf{0}$ (since $\mathbf{b}$ is constant) and $d\mathbf{v}/dt = \mathbf{a}$, we obtain

$$\frac{d}{dt}(\mathbf{v} \times \mathbf{b}) = \mathbf{v} \times \frac{d\mathbf{b}}{dt} + \frac{d\mathbf{v}}{dt} \times \mathbf{b} = \mathbf{a} \times \mathbf{b} = GM\frac{d\mathbf{u}}{dt}$$

and integrating the left and right sides of this equation with respect to t yields

$$\mathbf{v} \times \mathbf{b} = GM\mathbf{u} + \mathbf{C} \tag{12}$$

where $\mathbf{C}$ is a vector constant of integration. We leave it for the reader to evaluate $\mathbf{v} \times \mathbf{b}$ at time $t = 0$ to show that

$$\mathbf{C} = (r_0v_0^2 - GM)\mathbf{i} \tag{13}$$

Next, we take the dot product of both sides of (12) with $\mathbf{r}$ to obtain

$$\mathbf{r} \cdot (\mathbf{v} \times \mathbf{b}) = GM\mathbf{r} \cdot \mathbf{u} + \mathbf{r} \cdot \mathbf{C} \tag{14}$$

But

$$\mathbf{r} \cdot (\mathbf{v} \times \mathbf{b}) = (\mathbf{r} \times \mathbf{v}) \cdot \mathbf{b} = \mathbf{b} \cdot \mathbf{b} = r_0^2v_0^2$$

$$\mathbf{r} \cdot \mathbf{u} = (r\mathbf{u}) \cdot \mathbf{u} = r(\mathbf{u} \cdot \mathbf{u}) = r$$

$$\mathbf{r} \cdot \mathbf{C} = r(\cos\theta\mathbf{i} + \sin\theta\mathbf{j}) \cdot (r_0v_0^2 - GM)\mathbf{i} = r(r_0v_0^2 - GM)\cos\theta$$

so (14) can be written as

$$r_0^2 v_0^2 = GMr + r(r_0 v_0^2 - GM) \cos \theta$$

which when solved for r gives

$$r = \frac{r_0^2 v_0^2}{GM + (r_0 v_0^2 - GM) \cos \theta} = \frac{\dfrac{r_0^2 v_0^2}{GM}}{1 + \left(\dfrac{r_0 v_0^2}{GM} - 1\right) \cos \theta} \tag{15}$$

or simply

$$r = \frac{d}{1 + e \cos \theta} \tag{16}$$

where

$$d = \frac{r_0^2 v_0^2}{GM} \quad \text{and} \quad e = \frac{r_0 v_0^2}{GM} - 1 \tag{17–18}$$

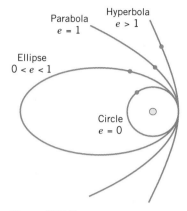

Parabola
$e = 1$

Hyperbola
$e > 1$

Ellipse
$0 < e < 1$

Circle
$e = 0$

Figure 15.7.5

From (16) and the fact that r is smallest when $\theta = 0$, it follows that $e \ge 0$ (why?), so Equation (16) is the polar form of a conic with eccentricity e and a focus at the origin (Section 13.2). This implies that the path of an object subjected only to the sun's gravitational force is a circle if $e = 0$, an ellipse if $0 < e < 1$, a parabola if $e = 1$, or a hyperbola if $e > 1$ (Figure 15.7.5). As indicated by (18), the value of e depends on r_0 and v_0, so at the time the planets in our solar system were created those position and velocity vectors were such that elliptical orbits resulted.

☐ **KEPLER'S THIRD LAW**

To derive Kepler's third law, we let a and b be the semimajor and semiminor axes of the planet's elliptical orbit, and we recall that the area of such an ellipse is $\pi a b$. It follows from (6) and (11) that the radius vector from the sun to a planet sweeps out area at the constant rate of $\frac{1}{2} r_0 v_0$ (units of area per unit of time). Thus, if one revolution around the sun is completed in T units of time (the period), then the area swept out during that revolution is

$$\pi a b = \tfrac{1}{2} r_0 v_0 T \tag{19}$$

The vertices of a planet's elliptical orbit are reached when $\theta = 0$ and when $\theta = \pi$. When $\theta = 0$ the planet is at its closest point to the sun, and when $\theta = \pi$ it is at its farthest point. If we denote the distances from the planet to the center of the sun at those points by r_{min} and r_{max}, respectively, then substituting $\theta = 0$ and $\theta = \pi$ in (16) yields

$$r_{min} = \frac{d}{1 + e} \quad \text{and} \quad r_{max} = \frac{d}{1 - e}$$

so the length of the major axis is

$$2a = r_{min} + r_{max} = \frac{d}{1 + e} + \frac{d}{1 - e} = \frac{2d}{1 - e^2}$$

and hence

$$a = \frac{d}{1 - e^2} \tag{20}$$

Also, because $b^2 = a^2 - c^2$ and $e = c/a$ for an ellipse (Section 12.3) and $d = a(1 - e^2)$ from (20), it follows that $b^2 = a^2(1 - e^2) = ad$; thus, from (17) and (19)

$$T^2 = \frac{4\pi^2 a^2 (ad)}{r_0^2 v_0^2} = \frac{4\pi^2 a^3}{r_0^2 v_0^2} d = \frac{4\pi^2 a^3}{r_0^2 v_0^2} \frac{r_0^2 v_0^2}{GM} = \frac{4\pi^2}{GM} a^3 \tag{21}$$

which establishes Kepler's third law in which $4\pi^2/GM$ is the constant of proportionality.

□ **ARTIFICIAL SATELLITES**

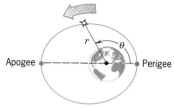

Apogee ● - - - - - - - ● Perigee

Figure 15.7.6

Kepler's laws apply to all celestial bodies subjected to a central gravitational force; we need only interpret the value of M in (7) to be the mass of the body exerting the central force and m to be the mass of the body subjected to the force. In particular, Kepler's laws apply to artificial satellites under the central force of the earth's gravity. For elliptical orbits about the sun, the point where the distance from the planet to the sun's center is minimum is called the **perihelion**, and the point where the distance is maximum is called the **aphelion**. We will refer to the actual distances as the **perihelion and aphelion distances**. For elliptical orbits about the earth, the minimum and maximum distances to the earth's center occur at points called the **perigee** and **apogee**, respectively (Figure 15.7.6), and we will refer to the actual distances as the **perigee and apogee distances**. We will refer to the distances to the **surface** of the earth as the **perigee and apogee altitudes**.

For orbits about the sun it is often convenient to measure distances in **astronomical units**, where one astronomical unit (AU) is the length of the semimajor axis a of the earth's orbit (approximately 150×10^6 km or 92.9×10^6 miles). From Kepler's third law the ratio T^2/a^3 is constant for all elliptical orbits about a central point with a given mass M. Since the period T of rotation of the earth about the sun is 1 year and $a = 1$ AU for the earth's orbit, we find that $T^2/a^3 = 1$, so that

$$T^2 = a^3 \tag{22}$$

for all elliptical orbits about the sun, provided that T is measured in years and a in astronomical units.

Example 1   The value of a for the elliptical orbit of Halley's comet (last seen in 1986) about the sun is approximately 18.1 AU. Find the period of its orbit in years.

Solution.   We are given that $a = 18.1$, so from (22)

$$T = a^{3/2} = (18.1)^{3/2} \approx 77 \text{ years} \quad ◄$$

Example 2   The eccentricity of the orbit of Halley's comet is approximately 0.97. Find the perihelion and the aphelion distances from the sun in miles.

Solution.   The perihelion and aphelion distances, r_{min} and r_{max}, can be obtained from the formulas (see Exercise 1)

$$r_{min} = a(1 - e) \quad \text{and} \quad r_{max} = a(1 + e)$$

From Example 1 $a = 18.1$, so the perihelion distance is

$$r_{min} = a(1 - e) = 18.1(1 - 0.97) \approx 0.54 \text{ AU} \approx 5 \times 10^7 \text{ miles}$$

and the aphelion distance is

$$r_{max} = a(1 + e) = 18.1(1 + 0.97) \approx 35.7 \text{ AU} \approx 3.3 \times 10^9 \text{ miles} \quad ◄$$

► **Exercise Set 15.7**   Ⓒ 3, 6–15

In exercises that require numerical calculations, use the following values:

   radius of the earth = 6370 km = 3960 mi
   1 AU = 150×10^6 km = 92.9×10^6 mi
   GM (for the earth) = 3.99×10^5 km³/sec²
                = 1.238×10^{12} mi³/hr²

1. Using the notation in Example 2, show that $r_{min} = a(1 - e)$ and $r_{max} = a(1 + e)$.

2. (a) Use the results in Exercise 1 to show that

$$e = \frac{r_{max} - r_{min}}{r_{max} + r_{min}}$$

   (b) Show that

$$r_{max} = r_{min} \frac{1 + e}{1 - e}$$

3. (a) Use Formula (18) to show that

$$v_0 = \sqrt{\frac{2GM}{r_0}}$$

is the minimum speed needed when $\theta = 0$ so that an object will escape from the pull of a central force due to mass M.

(b) Use the result in part (a) to find the minimum speed needed when $\theta = 0$ for a space probe at an altitude of 300 km above the surface of the earth to escape from the gravitational pull of the earth.

4. If $\mathbf{v}$ is the velocity of an object at any point in its orbit, then $v = \|\mathbf{v}\|$ is its speed.

(a) Show that

$$v = \frac{v_0}{1 + e} \sqrt{e^2 + 2e \cos \theta + 1}$$

[*Hint:* Use (12) along with (2), (10), (13), and (18). Note that $\|\mathbf{v} \times \mathbf{b}\| = \|\mathbf{v}\| \|\mathbf{b}\|$ because $\mathbf{v}$ and $\mathbf{b}$ are perpendicular.]

(b) Use the result in part (a) to find the speed of an object when it reaches an end of the minor axis of an elliptical orbit.

5. Use the result in part (a) of Exercise 4 to show that for an object moving in an elliptical orbit

$$v_{max} = v_{min} \frac{1 + e}{1 - e}$$

where v_{min} and v_{max} are the minimum and maximum speeds of the object respectively.

6. (a) Use the result in part (a) of Exercise 4, and (18), to show that for an object moving in a circular orbit of radius r_0 the speed v is constant and

$$v = v_0 = \sqrt{\frac{GM}{r_0}}$$

(b) Use the result in part (a) to find the speed of a spacecraft in a circular orbit about the earth at an altitude of 200 km above the surface of the earth.

7. (a) A *geosynchronous orbit* is a circular orbit about the equator of the earth in which an object appears to remain stationary over some point on the equator. Find the altitude in miles of a communications satellite that is in geosynchronous orbit about the earth. [*Hint:* The earth makes one revolution about its axis of rotation in 24 hours.]

(b) Use the result in part (a) of Exercise 6 to find the speed in miles per hour of a satellite that is in geosynchronous orbit about the earth.

8. (a) Assume that T is measured in days and a in kilometers (km). Find the value of T^2/a^3 for planets that orbit the sun. [*Hint:* The period of the earth's orbit is about 365 days.]

(b) Use the result in part (a) to find the period in days for the orbit of the planet Mercury, given that the length of its semimajor axis is 57.95×10^6 km.

9. Find the period in years for the orbit of the planet Pluto, given that $a = 39.5$ AU.

10. (a) The eccentricity of the moon's orbit about the earth is 0.055, and the length of the semimajor axis is about 238,900 miles. Find the perigee and apogee distances of the moon from the earth.

(b) Find the period in days of the moon's orbit.

11. (a) Vanguard 1 was launched in March 1958 with perigee and apogee altitudes above the earth of 649 km and 4340 km, respectively. Find the length of the semimajor axis of its orbit.

(b) Use the result in part (a) of Exercise 2 to find the eccentricity of its orbit.

(c) Find its period of rotation in minutes.

12. (a) Intelsat 5 was launched in December 1980 with perigee and apogee altitudes above the earth of 35,143 km and 35,707 km, respectively. Find the length of the semimajor axis of its orbit.

(b) Use the result in part (a) of Exercise 2 to find the eccentricity of its orbit.

(c) Find its period of rotation in hours.

13. (a) Suppose that a space probe is in a circular orbit at an altitude of 180 miles above the surface of the earth. Use the result in part (a) of Exercise 6 to find its speed.

(b) During a very short period of time, a thruster rocket on the space probe is fired to increase the speed of the probe by 600 mi/hr in its direction of motion. Find the eccentricity of the resulting elliptical orbit, and use the result in part (b) of Exercise 2 to find the apogee altitude.

14. Suppose that the perigee altitude of an earth satellite is 600 km and that the period is 100 min. Find the eccentricity of the orbit and the apogee altitude.

15. (a) Use Formula (20) to show that (16) can be expressed as

$$r = \frac{a(1 - e^2)}{1 + e \cos \theta}$$

(b) Use the result in part (a) to find an equation of the orbit of Mercury given that $a = 57.9 \times 10^6$ km, $e = 0.206$.

Most of these exercises require access to a graphing calculator or a computer algebra system (CAS) such as *Mathematica*, *Maple*, or *Derive*. When you are asked to *find* an answer or to *solve* an equation, you may choose to find an exact result or a numerical approximation, depending on the particular technology you are using and on your own imagination. The form of your answers may differ from those of other students or from those in the answer section of the text, depending on how you solve the problems and the accuracy you use in your numerical approximations. Those exercises that are more appropriate for a CAS than a graphing calculator are labeled with the icon ◆.

1. **Distance from a point to a curve:** Find the shortest distance between the point $(1, 2, 1)$ and a point on the curve

$$\mathbf{r}(t) = \frac{1}{t}\mathbf{i} + \ln t\mathbf{j} + \sqrt{t}\mathbf{k}$$

2. **Distance from a point to a curve:** Find the maximum and minimum distances from the point $(1, 2, -1)$ to a point on the curve of intersection of the plane $z = y/2$ and the ellipsoid $x^2/4 + y^2/9 + z^2/4 = 1$. [*Suggestion:* Find parametric equations for the curve of intersection.]

3. **Distance between moving particles:** Given that two particles moving in 3-space have equations of motion $x = 2\cos t$, $y = 3\sin t$, $z = t$ $(t \geq 0)$ and $x = t$, $y = t^2$, $z = t^3$ $(t \geq 0)$, how close do the particles get to one another and when are they closest? (Assume that x, y, and z are in feet and t is in minutes.)

4. **Speed of a particle:** Suppose that the equation of motion of a moving particle is given by

$$\mathbf{r}(t) = (\sin t + \cos t)\mathbf{i} + (\sin t - t)\mathbf{j}$$

where $\mathbf{r}$ is in meters and t is in seconds.

   (a) Graph $\mathbf{r}(t)$ for $0 \leq t \leq 2\pi$.

   (b) Find the maximum speed of the particle over the interval $0 \leq t \leq 2\pi$.

5. **Maximum temperature along a curve:** Suppose that the temperature T at a point (x, y, z) is given by $T = 30ze^{-x^2-2y^2}$. Find the maximum value of T along the curve $x = t\cos t$, $y = t$, $z = e^{-t}$.

◆ 6. **Trajectory of a ski jump:** At time $t = 0$ a skier leaves the end of a ski jump with a speed of v_0 ft/sec at an angle α with the horizontal (see the accompanying figure). The

skier lands 259 ft down the incline 2.9 sec later. Use (28) of Section 15.6 to find

   (a) v_0 and α

   (b) the distance traveled by the skier.

(Use $g = 32$ ft/sec^2 as the acceleration due to gravity.)

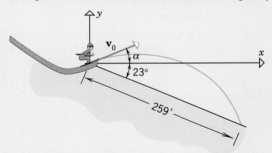

◆ 7. **Tangential and normal components of acceleration:** Suppose that a particle moves through 3-space so that its position vector at time t is

$$\mathbf{r}(t) = t\cos t\mathbf{i} + t\sin t\mathbf{j} + 3t\mathbf{k}, \quad t \geq 0$$

where $\mathbf{r}$ is in meters and t is in seconds.

   (a) Show that the particle moves on a circular cone.

   (b) Find, to the nearest degree, the angle between the velocity and acceleration vectors when $t = 1.5$.

   (c) Find the scalar tangential and normal components of acceleration when $t = 1.5$.

◆ 8. **Curvature:** Consider the curve

$$\mathbf{r}(t) = 3\cos t\mathbf{i} + 4\sin t\mathbf{j} + \sin 2t\mathbf{k}$$

   (a) Show that the curve lies on an elliptic cylinder.

   (b) Find the maximum value and the minimum value of the curvature.

◆ 9. **Path along a hill:** Suppose that a hill has the shape of the circular paraboloid $z = 1 - x^2 - y^2$, where x, y, and z are in miles. At time $t = 0$ a hiker at the point $(1, 0, 0)$ starts to walk up the hill with equations of motion given by

$$x = e^{-0.2t} \cos t, \quad y = e^{-0.2t} \sin t, \quad z = 1 - e^{-0.4t}$$

where t is in hours.

(a) Verify that the path of motion lies on the hill.

(b) Find, to the nearest tenth of a degree, the acute angle that the tangent line to the path makes with the xy-plane at time $t = 1$.

(c) How long does it take the hiker to travel 2 miles along the path?

◆ 10. **An amusement park ride:** An amusement park ride has cars that move along a track as shown in the accompanying figure. The cars are attached to arms that are 20 feet long and are connected to a common point on the axis of rotation (the z-axis). The arms swing up and down as the cars move along a track whose elevation is given by $z = 4 \sin^2 \theta$. Assuming that all distances are measured in feet, find the length of the track. [*Suggestion:* Convert to spherical coordinates and use the formula for arc length in Exercise 34, Section 15.3, with θ as the parameter.]

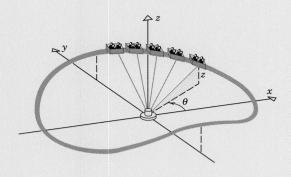

Joseph Louis Lagrange (1736–1813)

16 PARTIAL DERIVATIVES

◼ 16.1 FUNCTIONS OF TWO OR MORE VARIABLES

In previous sections we studied real-valued functions of a real variable and vector-valued functions of a real variable. In this section we shall consider real-valued functions of two or more real variables.

□ NOTATION AND
TERMINOLOGY

There are many familiar formulas in which a given variable depends on two or more other variables. For example, the area A of a triangle depends on the base length b and height h by the formula $A = \frac{1}{2}bh$; the volume V of a rectangular box depends on the length l, the width w, and the height h by the formula $V = lwh$; and the arithmetic average $\bar{x}$ of n real numbers, $x_1, x_2, \ldots, x_n$, depends on those numbers by the formula $\bar{x} = (x_1 + x_2 + \cdots + x_n)/n$. Thus, we say that

> A is a function of the two variables b and h;
>
> V is a function of the three variables l, w, and h;
>
> $\bar{x}$ is a function of the n variables $x_1, x_2, \ldots, x_n$.

The terminology and notation for functions of two or more variables is similar to that used for functions of one variable. For example, the expression

$$z = f(x, y)$$

means that z is a function of x and y in the sense that a unique value of the *dependent variable* z is determined by specifying values for the *independent variables* x and y. Similarly,

$$w = f(x, y, z)$$

expresses w as a function of x, y, and z, and

$$u = f(x_1, x_2, \ldots, x_n)$$

expresses u as a function of $x_1, x_2, \ldots, x_n$.

The functional relationship $z = f(x, y)$ has a useful geometric interpretation. When values of the independent variables x and y are specified, a point (x, y) in the xy-plane is determined. Thus, the dependent variable z may be viewed as a numerical value associated with the point (x, y). Similarly, the functional relationship $w = f(x, y, z)$ associates the numerical value w with the point (x, y, z) in 3-space.

The following definitions summarize this discussion.

16.1.1 DEFINITION. A *function f of two real variables*, x and y, is a rule that assigns a unique real number $f(x, y)$ to each point (x, y) in some set D of the xy-plane.

16.1.2 DEFINITION. A *function f of three variables*, x, y, and z, is a rule that assigns a unique real number $f(x, y, z)$ to each point (x, y, z) in some set D of three-dimensional space.

The set D in these definitions is the **domain** of the function; it is the set of points at which the function is defined. If a function f is specified by a formula and the domain of f is not stated explicitly, then it is understood that the domain consists of all points at which the formula has no divisions by zero and produces only real numbers; this is called the **natural domain** of the function.

REMARK. In more advanced courses the notion of ''n-dimensional space'' for $n > 3$ is defined, and a **function f of n real variables**, $x_1, x_2, \ldots, x_n$, is regarded as a rule that assigns a unique real number $f(x_1, x_2, \ldots, x_n)$ to each ''point'' $(x_1, x_2, \ldots, x_n)$ in some set of n-dimensional space. However, we shall not pursue that idea in this text.

Example 1 Let

$$f(x, y) = 3x^2\sqrt{y} - 1$$

Find $f(1, 4)$, $f(0, 9)$, $f(t^2, t)$, $f(ab, 9b)$, and the natural domain of f.

Solution. By substitution

$$f(1, 4) = 3(1)^2\sqrt{4} - 1 = 5$$

$$f(0, 9) = 3(0)^2\sqrt{9} - 1 = -1$$

$$f(t^2, t) = 3(t^2)^2\sqrt{t} - 1 = 3t^4\sqrt{t} - 1$$

$$f(ab, 9b) = 3(ab)^2\sqrt{9b} - 1 = 9a^2b^2\sqrt{b} - 1$$

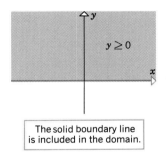

The solid boundary line is included in the domain.

Figure 16.1.1

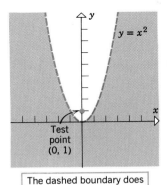

The dashed boundary does not belong to the domain.

Figure 16.1.2

Because of the $\sqrt{y}$, we must have $y \geq 0$ to avoid imaginary values for $f(x, y)$. Thus, the natural domain of f consists of all points in the xy-plane that are on or above the x-axis. (See Figure 16.1.1.) ◄

Example 2 Sketch the natural domain of the function $f(x, y) = \ln{(x^2 - y)}$.

Solution. $\ln{(x^2 - y)}$ is defined only when $0 < x^2 - y$ or $y < x^2$. To sketch this region, we use the fact that the curve $y = x^2$ separates the region where $y < x^2$ from the region where $y > x^2$. To determine the region where $y < x^2$ holds, we can select an arbitrary "test point" off the boundary $y = x^2$ and determine whether $y < x^2$ or $y > x^2$ at the test point. For example, if we choose the test point $(x, y) = (0, 1)$, then $x^2 = 0$, $y = 1$, so that this point lies in the region where $y > x^2$. Thus, the region where $y < x^2$ is the one that does *not* contain the test point (Figure 16.1.2). ◄

Example 3 Let
$$f(x, y, z) = \sqrt{1 - x^2 - y^2 - z^2}$$
Find $f(0, \frac{1}{2}, -\frac{1}{2})$ and the natural domain of f.

Solution. By substitution,
$$f(0, \tfrac{1}{2}, -\tfrac{1}{2}) = \sqrt{1 - (0)^2 - (\tfrac{1}{2})^2 - (-\tfrac{1}{2})^2} = \sqrt{\tfrac{1}{2}}$$
Because of the square root sign, we must have $0 \leq 1 - x^2 - y^2 - z^2$ in order to have a real value for $f(x, y, z)$. Rewriting this inequality in the form
$$x^2 + y^2 + z^2 \leq 1$$
we see that the natural domain of f consists of all points on or within the sphere $x^2 + y^2 + z^2 = 1$. ◄

☐ **GRAPHS OF FUNCTIONS OF TWO VARIABLES**

Recall that for a function f of one variable, the graph of $f(x)$ in the xy-plane was defined to be the graph of the equation $y = f(x)$. Similarly, if f is a function of two variables, we define the **graph** of $f(x, y)$ in xyz-space to be the graph of the equation $z = f(x, y)$. In general, such a graph will be a surface in 3-space.

Example 4 Describe the graph of the function $f(x, y) = 1 - x - \frac{1}{2}y$ in xyz-space.

Solution. By definition, the graph of the given function is the graph of the equation
$$z = 1 - x - \tfrac{1}{2}y \quad \text{or equivalently,} \quad x + \tfrac{1}{2}y + z = 1$$
which is a plane. A triangular portion of the plane can be sketched by plotting the intersections with the coordinate axes and joining them with line segments (Figure 16.1.3). ◄

Example 5 Sketch the graphs of the following functions in xyz-space.
(a) $f(x, y) = \sqrt{1 - x^2 - y^2}$ (b) $f(x, y) = -\sqrt{x^2 + y^2}$

Solution (a). By definition, the graph of the given function is the graph of the equation
$$z = \sqrt{1 - x^2 - y^2} \tag{1}$$
After squaring both sides, this can be rewritten as
$$x^2 + y^2 + z^2 = 1$$

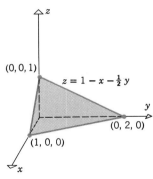

Figure 16.1.3

which represents a sphere of radius 1, centered at the origin. Since (1) imposes the added condition that $z \geq 0$, the graph is just the upper hemisphere (Figure 16.1.4).

Solution (b).   The graph of the given function is the graph of the equation

$$z = -\sqrt{x^2 + y^2} \tag{2}$$

After squaring, we obtain

$$z^2 = x^2 + y^2$$

which is the equation of a right-circular cone [(9) of Section 14.7]. Since (2) imposes the condition that $z \leq 0$, the graph is just the lower nappe of the cone (Figure 16.1.5).   ◀

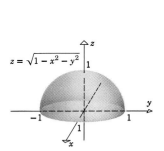

Figure 16.1.4

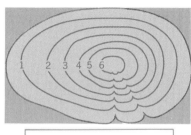

Figure 16.1.5

☐ **LEVEL CURVES**

When it is necessary to introduce a dependent variable for a function $f(x, y, z)$ of three variables, we shall usually use the letter w and write $w = f(x, y, z)$. Because this equation involves four variables, it cannot be graphed in three dimensions—"four dimensions" are needed. Thus, there is no *direct* way to represent a function of three or more variables geometrically. However, we shall now discuss some methods for representing functions geometrically that can be applied to functions of three variables.

We are all familiar with topographic (or contour) maps in which a three-dimensional landscape, such as a mountain range, is represented by two-dimensional contour lines or curves of constant elevation. Consider, for example, the model hill and its contour map shown in Figure 16.1.6. The contour map is constructed by passing planes of constant elevation through the hill, projecting the resulting contours onto a flat surface, and labeling the contours with their elevations. In Figure 16.1.6, note how the two gullies appear as indentations in the contour lines and how the curves are close together on the contour map where the hill has a steep slope and become more widely spaced where the slope is gradual.

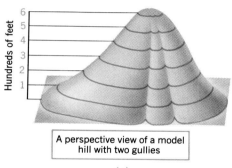

A perspective view of a model
hill with two gullies

A contour map of the model hill

Figure 16.1.6   (a)   (b)

Contour maps are useful for studying functions of two variables. If the surface $z = f(x, y)$ is cut by the horizontal plane $z = k$, then for points on the intersection we have $f(x, y) = k$. The projection of this intersection onto the xy-plane is called the ***level curve of height k*** or the ***level curve with constant k*** (Figure 16.1.7).

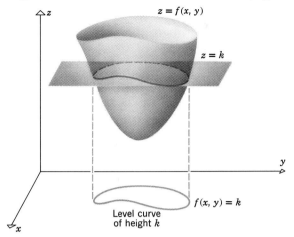

Figure 16.1.7

Example 6   The graph of the function $f(x, y) = x^2 + \frac{1}{4}y^2$ in xyz-space is the elliptic paraboloid shown in Figure 16.1.8a. The level curves have equations of the form

$$x^2 + \tfrac{1}{4}y^2 = k \tag{3}$$

For $k > 0$ these are ellipses; for $k = 0$ it is the single point $(0, 0)$; and for $k < 0$ there are no level curves, since (3) is not satisfied by any real values of x and y. Some sample level curves are shown in Figure 16.1.8b. ◀

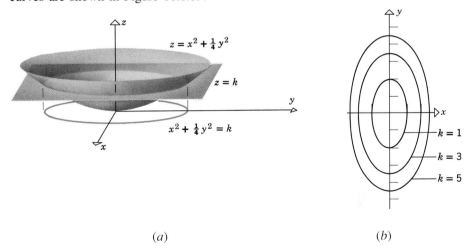

Figure 16.1.8   (a)   (b)

Example 7   The graph of the function $f(x, y) = 2 - x - y$ in xyz-space is the plane shown in Figure 16.1.9a. The level curves have equations of the form $2 - x - y = k$, or $y = -x + (2 - k)$. These form a family of parallel lines each of which has a slope of -1 (Figure 16.1.9b). ◀

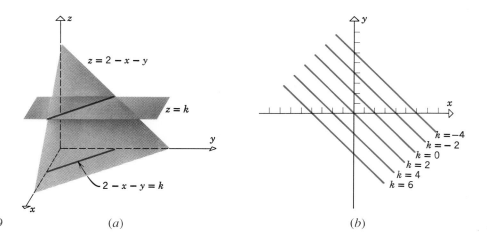

Figure 16.1.9

(a)

(b)

Example 8 The graph of the function $f(x, y) = y^2 - x^2$ in xyz-space is the hyperbolic paraboloid (saddle curve) shown in Figure 16.1.10a. The level curves have equations of the form $y^2 - x^2 = k$. For $k > 0$ these curves are hyperbolas opening along lines parallel to the y-axis; for $k < 0$ they are hyperbolas opening along lines parallel to the x-axis; and for $k = 0$ the level curve consists of the intersecting lines $y + x = 0$ and $y - x = 0$ (Figure 16.1.10b). ◄

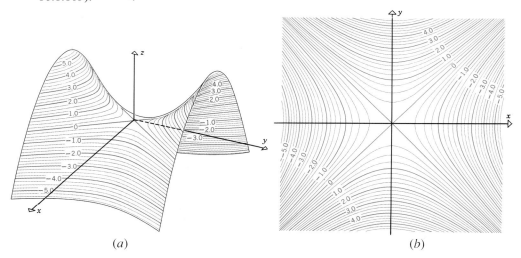

Figure 16.1.10

(a)

(b)

☐ **LEVEL SURFACES**

The concept of a level curve for a function of two variables can be extended to functions of three variables. If k is a constant, then an equation of the form $f(x, y, z) = k$ will, in general, represent a surface in three-dimensional space (e.g., $x^2 + y^2 + z^2 = 1$ represents a sphere). The graph of this surface is called the **level surface with constant k** for the function f.

REMARK. The term ''level surface'' can be confusing. A level surface need *not* be level in the sense of being horizontal. It is simply a surface on which all values of f are the same.

Example 9 Describe the level surfaces of

    (a) $f(x, y, z) = x^2 + y^2 + z^2$ (b) $f(x, y, z) = z^2 - x^2 - y^2$

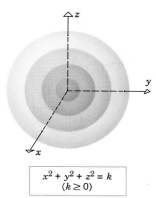

$$x^2 + y^2 + z^2 = k$$
$$(k \geq 0)$$

Figure 16.1.11

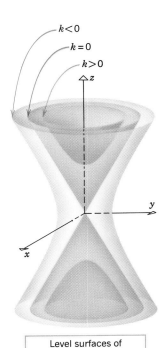

Level surfaces of
$f(x, y, z) = z^2 - x^2 - y^2$

Figure 16.1.12

Solution (a). The level surfaces have equations of the form

$$x^2 + y^2 + z^2 = k$$

For $k > 0$ the graph of this equation is a sphere of radius $\sqrt{k}$, centered at the origin; for $k = 0$ the graph is the single point $(0, 0, 0)$; and for $k < 0$ there is no level surface (Figure 16.1.11).

Solution (b). The level surfaces have equations of the form

$$z^2 - x^2 - y^2 = k$$

As discussed in Section 14.7, this equation represents a cone if $k = 0$, a hyperboloid of two sheets if $k > 0$, and a hyperboloid of one sheet if $k < 0$ (Figure 16.1.12). ◀

□ **OPEN AND CLOSED SETS**

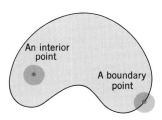

Figure 16.1.13

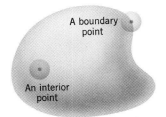

Figure 16.1.14

In our study of functions of one variable, the domains of the functions we encountered were generally intervals. For functions of two or three variables the situation is more complicated, so we shall need to discuss some terminology about sets in 2-space and 3-space that will be helpful when we want to accurately describe the domain of a function of two or three variables.

If D is a set of points in 2-space, then a point (x_0, y_0) is called an ***interior point*** of D if there is *some* circular disk with positive radius, centered at (x_0, y_0), and containing only points in D (Figure 16.1.13). A point (x_0, y_0) is called a ***boundary point*** of D if *every* circular disk with positive radius and centered at (x_0, y_0) contains both points in D and points not in D (Figure 16.1.13). Similarly, if D is a set of points in 3-space, then a point (x_0, y_0, z_0) is called an ***interior point*** of D if there is some spherical ball with positive radius, centered at (x_0, y_0, z_0), and containing only points in D (Figure 16.1.14). A point (x_0, y_0, z_0) is called a ***boundary point*** of D if *every* spherical ball with positive radius and centered at (x_0, y_0, z_0) contains both points in D and points not in D (Figure 16.1.14).

For a set D in either 2-space or 3-space, the set of all boundary points of D is called the ***boundary*** of D and the set of all interior points of D is called the ***interior*** of D.

Recall that an open interval (a, b) on a coordinate line contains *neither* of its endpoints and a closed interval $[a, b]$ contains *both* of its endpoints. Analogously, a set D in 2-space or 3-space is called ***open*** if it contains *none* of its boundary points and ***closed*** if it contains *all* of its boundary points. The set D of all points in 2-space has no boundary; it is regarded as both open and closed. Similarly, the set D of all points in 3-space is both open and closed.

Example 10   Let D be the set of points in the xy-plane that are inside or on the circle of radius 1 centered at the origin. The set D, its interior I, and its boundary B can be expressed in set notation as

$$D = \{(x, y): x^2 + y^2 \leq 1\}, \quad I = \{(x, y): x^2 + y^2 < 1\}, \quad B = \{(x, y): x^2 + y^2 = 1\}$$

respectively (Figure 16.1.15). The set D is closed and the set I is open.   ◀

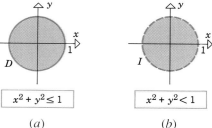

Figure 16.1.15

$$x^2 + y^2 \leq 1$$ $$x^2 + y^2 < 1$$ $$x^2 + y^2 = 1$$

(a)   (b)   (c)

☐ **BOUNDED SETS**

Just as we distinguished between finite intervals and infinite intervals on the real line, so we shall want to distinguish between regions of "finite extent" and regions of "infinite extent" in 2-space and 3-space. A set of points in 2-space is called **bounded** if the entire set can be contained within some rectangle, and is called **unbounded** if there is no rectangle that contains all the points of the set. Similarly, a set of points in 3-space is **bounded** if the entire set can be contained within some box, and is unbounded otherwise (Figure 16.1.16).

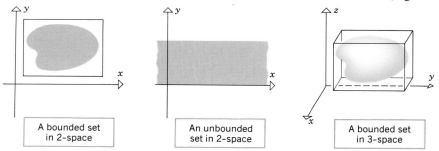

Figure 16.1.16

A bounded set in 2-space

An unbounded set in 2-space

A bounded set in 3-space

☐ **PARAMETRIC REPRESENTATION OF SURFACES**

We have seen that curves in 3-space can be represented parametrically by three equations involving one parameter. Similarly, surfaces in 3-space can be represented by three equations involving two parameters, say u and v, as

$$x = f(u, v), \quad y = g(u, v), \quad z = h(u, v)$$

or by a single vector-valued function

$$\mathbf{r}(u, v) = x\mathbf{i} + y\mathbf{j} + z\mathbf{k} = f(u, v)\mathbf{i} + g(u, v)\mathbf{j} + h(u, v)\mathbf{k}$$

We can view $\mathbf{r}(u, v) = x\mathbf{i} + y\mathbf{j} + z\mathbf{k}$ as a **radius vector** from the origin to a point (x, y, z) that moves over the surface as u and v vary (Figure 16.1.17).

Example 11   Consider the portion of the paraboloid $z = 4 - x^2 - y^2$ that lies in the first octant (Figure 16.1.18). We can obtain a parametric representation of this surface by letting $x = u$ and $y = v$, from which it follows that $z = 4 - u^2 - v^2$. Thus, the paraboloid can be represented parametrically as

$$x = u, \quad y = v, \quad z = 4 - u^2 - v^2$$

or in vector form as

$$\mathbf{r}(u, v) = u\mathbf{i} + v\mathbf{j} + (4 - u^2 - v^2)\mathbf{k}$$

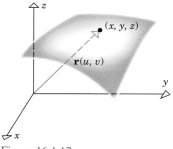

Figure 16.1.17

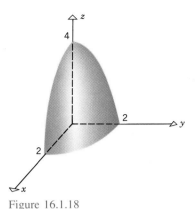

Figure 16.1.18

where u and v satisfy

$$u^2 + v^2 \leq 4, \quad u \geq 0, \quad \text{and} \quad v \geq 0 \qquad (4)$$

(These conditions ensure that $x^2 + y^2 \leq 4$, $x \geq 0$, and $y \geq 0$, which gives the portion of the paraboloid in the first octant, rather than the entire paraboloid.) ◀

If we think of (u, v) as a point that varies over some region in a uv-plane, then a vector function $\mathbf{r}(u, v)$ associates points on a surface with points in that region, just as $\mathbf{r}(t)$ associates points on a curve with points in an interval on the t-axis. If u is allowed to change, while v is kept constant, then the radius vector traces out a curve on the surface called a ***u-curve***; and if v is allowed to change, while u is kept constant, the radius vector traces out a ***v-curve***. The u-curves (v constant) are associated with horizontal lines in the uv-plane, and the v-curves (u constant) are associated with vertical lines (Figure 16.1.19). The u-curves and v-curves usually cover the surface like a net and are useful for visualizing the shape of the surface.

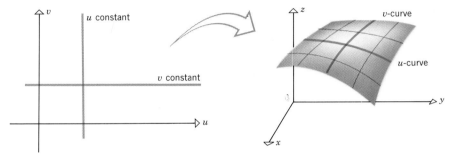

Figure 16.1.19

Example 12   Describe the u-curves and v-curves for the surface in Example 11.

Solution.   From (4), the allowable values of u and v correspond to points in the quarter-circle of the uv-plane shown in Figure 16.1.20. If v is constant, say $v = v_0$, where $0 \leq v_0 \leq 2$, then u varies from 0 to $\sqrt{4 - v_0^2}$, resulting in the u-curve

$$x = u, \quad y = v_0, \quad z = 4 - u^2 - v_0^2$$

or equivalently,

$$z - (4 - v_0^2) = -x^2 \quad (y = v_0, 0 \leq x \leq \sqrt{4 - v_0^2})$$

These u-curves are portions of parabolas that are parallel to the xz-plane, open down, and have their vertices over the y-axis at the points $(0, v_0, 4 - v_0^2)$. They are the green curves in Figure 16.1.20. Similarly, if u is constant, say $u = u_0$, where $0 \leq u_0 \leq 2$, then v varies from 0 to $\sqrt{4 - u_0^2}$, resulting in the v-curve

$$z - (4 - u_0^2) = -y^2 \quad (x = u_0, 0 \leq y \leq \sqrt{4 - u_0^2})$$

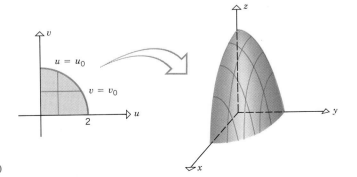

Figure 16.1.20

These v-curves are portions of parabolas that are parallel to the yz-plane, open down, and have their vertices over the x-axis at the points $(u_0, 0, 4 - u_0^2)$. They are the blue curves in Figure 16.1.20.   ◄

Example 13   The surface in Example 11 can be parametrized in another way by expressing its equation in cylindrical coordinates (r, θ, z). To convert $z = 4 - x^2 - y^2$ to cylindrical coordinates, we substitute $x = r \cos \theta$ and $y = r \sin \theta$, which yields $z = 4 - (r \cos \theta)^2 - (r \sin \theta)^2 = 4 - r^2$. Thus, the surface can be represented parametrically in terms of the parameters r and θ as

$$x = r \cos \theta, \quad y = r \sin \theta, \quad z = 4 - r^2$$

or in vector form as

$$\mathbf{r}(r, \theta) = r \cos \theta \mathbf{i} + r \sin \theta \mathbf{j} + (4 - r^2)\mathbf{k}$$

where r and θ satisfy $0 \leq r \leq 2$ and $0 \leq \theta \leq \pi/2$, since we have only the portion of the paraboloid in the first octant. These parametric equations associate points on the surface with points (r, θ) in an $r\theta$-plane (Figure 16.1.21). The r-curves (θ constant, say $\theta = \theta_0$), which are shown in green, are portions of parabolas that extend from the top of the paraboloid down to the xy-plane and lie in the plane making an angle θ_0 with the x-axis. The θ-curves (r constant, say $r = r_0$), which are shown in blue in Figure 16.1.21, are quarter-circles of radius r_0 centered on the z-axis and lying in planes parallel to the xy-plane.   ◄

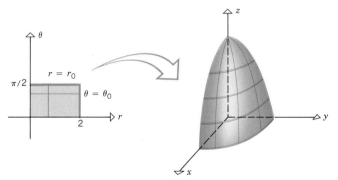

Figure 16.1.21

Example 14   Find a parametric representation of the portion of the cylindrical surface $x^2 + z^2 = 9$ for which $0 \leq y \leq 5$ in terms of the parameters u and v shown in Figure 16.1.22. The parameter u is the distance from a point $P(x, y, z)$ on the surface to the xz-plane, and v is the angle shown in the figure.

Solution.   The radius of the cylinder is 3, so it is evident from the figure that $y = u$, $x = 3 \cos v$, and $z = 3 \sin v$. Thus, the surface can be represented parametrically as

$$x = 3 \cos v, \quad y = u, \quad z = 3 \sin v$$

or in vector form as

$$\mathbf{r}(u, v) = 3 \cos v \mathbf{i} + u \mathbf{j} + 3 \sin v \mathbf{k}$$

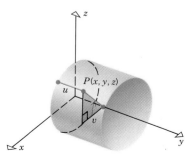

Figure 16.1.22

where $0 \leq u \leq 5$ and $0 \leq v < 2\pi$ (Figure 16.1.23). The u-curves (v constant, say $v = v_0$) are line segments parallel to the y-axis that extend from the xz-plane to the plane $y = 5$, and the v-curves (u constant, say $u = u_0$) are circles of radius 3 centered on the y-axis and lying in planes parallel to the xz-plane (Figure 16.1.23).   ◄

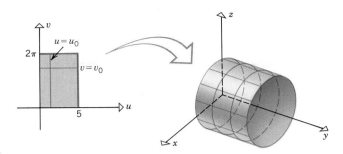

Figure 16.1.23

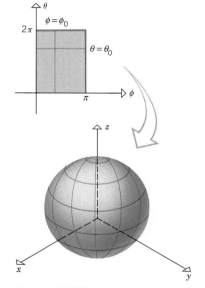

Figure 16.1.24

☐ **COMPUTER-GENERATED SURFACES**

Example 15 By eliminating the parameters, show that if $a > 0$, then

$$\mathbf{r}(\phi, \theta) = a \sin \phi \cos \theta \mathbf{i} + a \sin \phi \sin \theta \mathbf{j} + a \cos \phi \mathbf{k}$$

for $0 \le \phi \le \pi$ and $0 \le \theta \le 2\pi$ represents a sphere of radius a centered at the origin, and describe the ϕ-curves and θ-curves.

Solution. It follows from the formula for $\mathbf{r}(\phi, \theta)$ that

$$x = a \sin \phi \cos \theta, \quad y = a \sin \phi \sin \theta, \quad z = a \cos \phi \tag{5}$$

so

$$x^2 + y^2 + z^2 = a^2 \sin^2 \phi \cos^2 \theta + a^2 \sin^2 \phi \sin^2 \theta + a^2 \cos^2 \phi$$
$$= a^2 \sin^2 \phi + a^2 \cos^2 \phi = a^2$$

which, in Cartesian coordinates, is the equation of a sphere of radius a centered at the origin. By comparing the formulas in (5) to the third set of formulas in Table 14.8.1, we observe that ϕ and θ can be interpreted geometrically as spherical coordinates of a point on the surface. It follows that a ϕ-curve (θ constant, say $\theta = \theta_0$) is a semicircle of radius a, centered at the origin and lying in a plane that makes an angle of θ_0 with the xz-plane; and a θ-curve (ϕ constant) is a circle centered on the z-axis and lying in a plane parallel to the xy-plane (Figure 16.1.24). In the language of cartographers, the ϕ-curves are *lines of longitude* and the θ-curves are *lines of latitude*. ◄

In recent years computers have extended our ability to visualize complicated three-dimensional surfaces. Many computer programs draw surfaces within a box whose edges are parallel to the coordinate axes. As illustrated with the elliptic paraboloid in Figure 16.1.25, this sometimes produces artificial-looking cuts in the surface; however, this can be

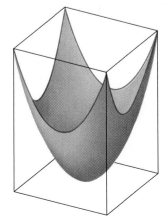

The box cuts the elliptic paraboloid, producing parabolic traces in the sides of the box.

Figure 16.1.25

useful for visualizing the surface since the cuts are traces of the surface parallel to the coordinate planes. Some examples of other computer-generated surfaces in three-dimensional space are shown in Figure 16.1.26.

SOME COMPUTER GENERATED SURFACES IN 3-SPACE

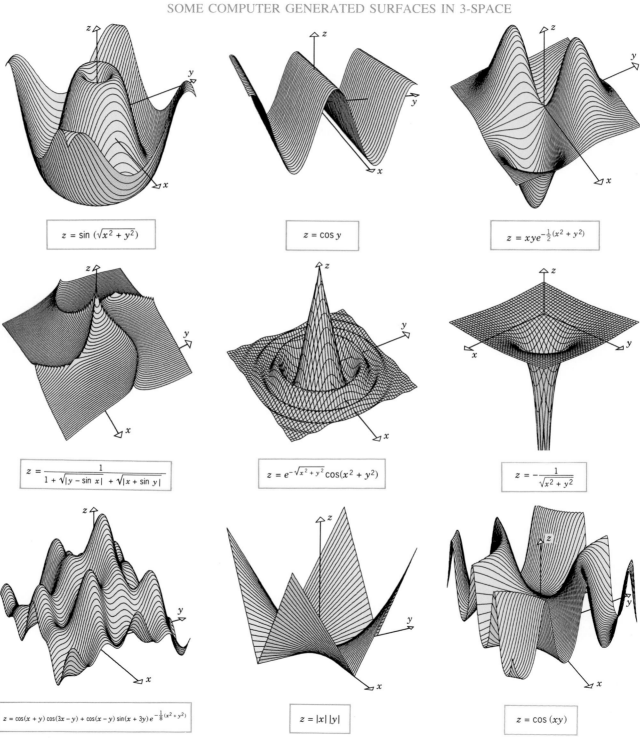

$$z = \sin\left(\sqrt{x^2 + y^2}\right)$$

$$z = \cos y$$

$$z = xy e^{-\frac{1}{2}(x^2 + y^2)}$$

$$z = \frac{1}{1 + \sqrt{|y - \sin x|} + \sqrt{|x + \sin y|}}$$

$$z = e^{-\sqrt{x^2 + y^2}} \cos(x^2 + y^2)$$

$$z = -\frac{1}{\sqrt{x^2 + y^2}}$$

$$z = \cos(x + y)\cos(3x - y) + \cos(x - y)\sin(x + 3y)\,e^{-\frac{1}{8}(x^2 + y^2)}$$

$$z = |x|\,|y|$$

$$z = \cos(xy)$$

Figure 16.1.26

▶ Exercise Set 16.1

1. Let $f(x, y) = x^2y + 1$. Find
   (a) $f(2, 1)$ (b) $f(1, 2)$ (c) $f(0, 0)$
   (d) $f(1, -3)$ (e) $f(3a, a)$ (f) $f(ab, a - b)$.

2. Let $f(x, y) = x + \sqrt[3]{xy}$. Find
   (a) $f(t, t^2)$ (b) $f(x, x^2)$ (c) $f(2y^2, 4y)$.

3. Let $f(x, y) = xy + 3$. Find
   (a) $f(x + y, x - y)$ (b) $f(xy, 3x^2y^3)$.

4. Let $g(x) = x \sin x$. Find
   (a) $g(x/y)$ (b) $g(xy)$ (c) $g(x - y)$.

5. Find $F(g(x), h(y))$ if $F(x, y) = xe^{xy}$, $g(x) = x^3$, and $h(y) = 3y + 1$.

6. Find $g(u(x, y), v(x, y))$ if $g(x, y) = y \sin(x^2y)$, $u(x, y) = x^2y^3$, and $v(x, y) = \pi xy$.

7. Let $f(x, y) = x + 3x^2y^2$, $x(t) = t^2$, and $y(t) = t^3$. Find
   (a) $f(x(t), y(t))$ (b) $f(x(0), y(0))$
   (c) $f(x(2), y(2))$.

8. Let $g(x, y) = ye^{-3x}$, $x(t) = \ln(t^2 + 1)$, and $y(t) = \sqrt{t}$. Find $g(x(t), y(t))$.

9. Let $f(x, y, z) = xy^2z^3 + 3$. Find
   (a) $f(2, 1, 2)$ (b) $f(-3, 2, 1)$
   (c) $f(0, 0, 0)$ (d) $f(a, a, a)$
   (e) $f(t, t^2, -t)$ (f) $f(a + b, a - b, b)$.

10. Let $f(x, y, z) = zxy + x$. Find
    (a) $f(x + y, x - y, x^2)$ (b) $f(xy, y/x, xz)$.

11. Find $F(f(x), g(y), h(z))$ if $F(x, y, z) = ye^{xyz}$, $f(x) = x^2$, $g(y) = y + 1$, and $h(z) = z^2$.

12. Find $g(u(x, y, z), v(x, y, z), w(x, y, z))$ if $g(x, y, z) = z \sin xy$, $u(x, y, z) = x^2z^3$, $v(x, y, z) = \pi xyz$, and $w(x, y, z) = xy/z$.

13. Let $f(x, y, z) = x^2y^2z^4$, $x(t) = t^3$, $y(t) = t^2$, and $z(t) = t$. Find
    (a) $f(x(t), y(t), z(t))$ (b) $f(x(0), y(0), z(0))$
    (c) $f(x(2), y(2), z(2))$.

In Exercises 14–19, sketch the domain of f. Use solid lines for portions of the boundary included in the domain and dashed lines for portions not included.

14. $f(x, y) = xy\sqrt{y - 1}$.

15. $f(x, y) = \ln(1 - x^2 - y^2)$.

16. $f(x, y) = \sqrt{x^2 + y^2 - 4}$.

17. $f(x, y) = \dfrac{1}{x - y^2}$.

18. $f(x, y) = \ln xy$.

19. $f(x, y) = \sqrt{\dfrac{x^2 + y^2}{x^2 - y^2}}$.

In Exercises 20–27, describe the domain of f.

20. $f(x, y) = \sin^{-1}(x + y)$.   21. $f(x, y) = xe^{-\sqrt{y+2}}$.

22. $f(x, y) = \dfrac{\sqrt{4 - x^2}}{y^2 + 3}$.   23. $f(x, y) = \ln(y - 2x)$.

24. $f(x, y, z) = \sqrt{25 - x^2 - y^2 - z^2}$.

25. $f(x, y, z) = \dfrac{xyz}{x + y + z}$.   26. $f(x, y, z) = e^{xyz}$.

27. $f(x, y, z) = z + \ln(1 - x^2 - y^2)$.

In Exercises 28–39, sketch the graph of f.

28. $f(x, y) = 3$.   29. $f(x, y) = 4 - 2x - 4y$.

30. $f(x, y) = \sqrt{9 - x^2 - y^2}$.

31. $f(x, y) = \sqrt{x^2 + y^2}$.   32. $f(x, y) = x^2 + y^2$.

33. $f(x, y) = x^2 - y^2$.   34. $f(x, y) = 4 - x^2 - y^2$.

35. $f(x, y) = -\sqrt{1 - x^2/4 - y^2/9}$.

36. $f(x, y) = \sqrt{x^2 + y^2 - 1}$.

37. $f(x, y) = \sqrt{x^2 + y^2 + 1}$.

38. $f(x, y) = x^2$.   39. $f(x, y) = y + 1$.

In Exercises 40–47, sketch the level curve $z = k$ for the specified values of k.

40. $z = 3x + y$; $k = -2, -1, 0, 1, 2$.

41. $z = x^2 + y^2$; $k = 0, 1, 2, 3, 4$.

42. $z = y/x$; $k = -2, -1, 0, 1, 2$.

43. $z = x^2 + y$; $k = -2, -1, 0, 1, 2$.

44. $z = x^2 + 9y^2$; $k = 0, 1, 2, 3, 4$.

45. $z = x^2 - y^2$; $k = -2, -1, 0, 1, 2$.

46. $z = y \csc x$; $k = -2, -1, 0, 1, 2$.

47. $z = \sqrt{\dfrac{x + y}{x - y}}$; $k = 0, 1, 2, 3, 4$.

In Exercises 48–51, sketch the level surface $f(x, y, z) = k$.

48. $f(x, y, z) = 4x - 2y + z$; $k = 1$.

49. $f(x, y, z) = 4x^2 + y^2 + 4z^2$; $k = 16$.

50. $f(x, y, z) = x^2 + y^2 - z^2$; $k = 0$.

51. $f(x, y, z) = z - x^2 - y^2 + 4$; $k = 7$.

In Exercises 52–55, describe the level surfaces.

52. $f(x, y, z) = 3x - y + 2z$.

53. $f(x, y, z) = (x - 2)^2 + y^2 + z^2$.

54. $f(x, y, z) = z - x^2 - y^2$.

55. $f(x, y, z) = x^2 + z^2$.

56. Let $f(x, y) = yx^2 + 1$. Find an equation of the level curve that passes through the point
    (a) $(1, 2)$ (b) $(-2, 4)$ (c) $(0, 0)$.

57. Let $f(x, y) = x^2 - 2x^3 + 3xy$. Find an equation of the level curve that passes through the point

    (a) $(-1, 1)$ (b) $(0, 0)$ (c) $(2, -1)$.

58. Let $f(x, y) = ye^x$. Find an equation of the level curve that passes through the point

    (a) $(\ln 2, 1)$ (b) $(0, 3)$ (c) $(1, -2)$.

59. Let $f(x, y, z) = x^2 + y^2 - z$. Find an equation of the level surface that passes through the point

    (a) $(1, -2, 0)$ (b) $(1, 0, 3)$ (c) $(0, 0, 0)$.

60. Let $f(x, y, z) = xyz + 3$. Find an equation of the level surface that passes through the point

    (a) $(1, 0, 2)$ (b) $(-2, 4, 1)$ (c) $(0, 0, 0)$.

61. If $V(x, y)$ is the voltage or potential at a point (x, y) in the xy-plane, then the level curves of V are called *equipotential curves*. Along such a curve, the voltage remains constant. Given that

$$V(x, y) = \frac{8}{\sqrt{16 + x^2 + y^2}}$$

sketch the equipotential curves at which $V = 2.0$, $V = 1.0$, and $V = 0.5$.

62. If $T(x, y)$ is the temperature at a point (x, y) on a thin metal plate in the xy-plane, then the level curves of T are called *isothermal curves*. All points on such a curve are at the same temperature. Suppose that a plate occupies the first quadrant and $T(x, y) = xy$.

    (a) Sketch the isothermal curves on which $T = 1$, $T = 2$, and $T = 3$.

    (b) An ant, initially at $(1, 4)$, wants to walk on the plate so that the temperature along its path remains constant. What path should the ant take?

In Exercises 63–65, use the contour map shown in Figure 16.1.27 (all elevations in hundreds of feet).

63. For points A and B,

    (a) which one is higher

    (b) which one is on the steeper slope?

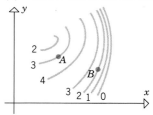

Figure 16.1.27

64. Starting at point A, will the elevation begin to increase or decrease if we travel so that

    (a) y remains constant and x increases

    (b) y remains constant and x decreases

    (c) x remains constant and y increases

    (d) x remains constant and y decreases?

65. Starting at point B, will the elevation begin to increase or decrease if we travel so that

    (a) y remains constant and x increases

    (b) y remains constant and x decreases

    (c) x remains constant and y increases

    (d) x remains constant and y decreases?

In Exercises 66 and 67, classify the given set of points as open, closed, or neither.

66. The set of points (x, y) in the xy-plane that satisfy the inequality

    (a) $1 \leq x^2 + y^2 \leq 3$

    (b) $0 \leq y < 2x + 1$ and $0 \leq x \leq 1$

    (c) $\sqrt{y} < x < 1$

    (d) $0 < y < e^x$ and $0 < x < 2$.

67. The set of points (x, y, z) in 3-space that satisfy the inequality

    (a) $x^2 + y^2 + z^2 < 3$

    (b) $0 \leq z \leq x^2 + y^2$ for $x^2 + y^2 < 1$

    (c) $0 \leq z \leq 5 - 3x - 2y$ for $x \geq 0$ and $y \geq 0$

    (d) $y^2 + z^2 \leq 5$ for $1 \leq x \leq 2$.

In Exercises 68 and 69, classify the given set as bounded or unbounded.

68. The set of points (x, y) in the xy-plane that satisfy the inequality

    (a) $x^2 + y^2 < 100$ (b) $x^2 + y^2 \geq 100$

    (c) $-1 < y < 2$

    (d) $y \leq 3 - x$ and $-1 < x < 1$.

69. The set of points in 3-space that satisfy the inequality

    (a) $x^2 + y^2 + z^2 \leq 3$

    (b) $0 < z < 2 - y$ and $0 < x < 1$

    (c) $x^2 + y^2 < 4$ (d) $z \geq 2 - x - 3y$.

In Exercises 70–74, find a parametric representation of the surface in terms of the parameters u and v, where $u = x$ and $v = y$.

70. $2z - 3x + 4y = 5$.   **71.** $z = x^2$.   **72.** $y^2 - 3z = 5$.

73. The portion of the cylinder $x^2 + z^2 = 4$ on or above the xy-plane and on or between the planes $y = 1$ and $y = 3$.

74. The portion of the sphere $x^2 + y^2 + z^2 = 25$ on or above the plane $z = 3$.

In Exercises 75–82, find a parametric representation of the surface in terms of the parameters r and θ, where (r, θ, z) are cylindrical coordinates of a point on the surface.

75. $z = \dfrac{1}{1 + x^2 + y^2}$.   **76.** $z = e^{-(x^2 + y^2)}$.

77. $z = 2xy$.   **78.** $z = x^2 - y^2$.

79. The portion of the sphere $x^2 + y^2 + z^2 = 9$ on or above the plane $z = 2$.

80. The portion of the cone $z = \sqrt{x^2 + y^2}$ on or below the plane $z = 3$.

81. The portion of the plane $z + 2y = 3$ on or inside the cylinder $x^2 + y^2 = 4$.

82. The surface of revolution that is generated by revolving about the z-axis the curve $z = 1/x^2$ (in the plane $y = 0$) for $\frac{1}{2} \le x \le 2$.

83. Find a parametric representation of the cone

$$z = \sqrt{3x^2 + 3y^2}$$

in terms of parameters ρ and θ, where (ρ, θ, ϕ) are spherical coordinates of a point on the surface.

84. Find a parametric representation of the cylinder $x^2 + y^2 = 9$ in terms of parameters θ and ϕ, where (ρ, θ, ϕ) are spherical coordinates.

85. Find a parametric representation of the portion of the elliptic cylinder $x^2 + 4y^2 = 9$ for $0 \le z \le 5$.

86. Find a parametric representation of the *torus* that is generated by revolving the circle $(x - a)^2 + z^2 = b^2$ (in the plane $y = 0$), where $0 < b < a$, about the z-axis. Use the angles u and v shown in Figure 16.1.28 as parameters.

In Exercises 87–92, describe the surface by eliminating the parameters to obtain an equation for the surface in rectangular coordinates.

87. $x = 2u + v$, $y = u - v$, $z = 3v$ for $-\infty < u < +\infty$ and $-\infty < v < +\infty$.

88. $x = u \cos v, y = u^2, z = u \sin v$ for $0 \le u \le 2$ and $0 \le v < 2\pi$.

89. $x = 3 \sin u, y = 2 \cos u, z = 2v$ for $0 \le u < 2\pi$ and $1 \le v \le 2$.

90. $x = \sqrt{u} \cos v, y = \sqrt{u} \sin v, z = u$ for $0 \le u \le 4$ and $0 \le v < 2\pi$.

91. $\mathbf{r}(u, v) = 3u \cos v \mathbf{i} + 4u \sin v \mathbf{j} + u \mathbf{k}$ for $0 \le u \le 1$ and $0 \le v < 2\pi$.

92. $\mathbf{r}(u, v) = \sin u \cos v \mathbf{i} + 2 \sin u \sin v \mathbf{j} + 3 \cos u \mathbf{k}$ for $0 \le u \le \pi$ and $0 \le v < 2\pi$.

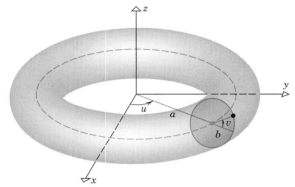

Figure 16.1.28

16.2 LIMITS AND CONTINUITY

In this section we shall introduce the notions of limit and continuity for functions of two or more variables. We shall not go into great detail; our objective is to develop the basic ideas accurately, and to obtain results needed in later sections. A more extensive study of these topics is usually given in advanced calculus.

☐ **LIMITS ALONG CURVES**

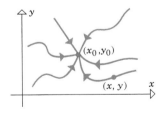

Figure 16.2.1

For a function of one variable there are two one-sided limits at a point x_0, namely

$$\lim_{x \to x_0^+} f(x) \quad \text{and} \quad \lim_{x \to x_0^-} f(x)$$

reflecting the fact that there are only two directions from which x can approach x_0, the right or the left. For functions of two or three variables the situation is more complicated because there are infinitely many different curves along which one point can approach another (Figure 16.2.1). Our first objective in this section is to define the limit of $f(x, y)$ as (x, y) approaches a point (x_0, y_0) along a curve C (and similarly for functions of three variables). Limits along curves are the two-variable and three-variable analogs of one-sided limits for functions of one variable. Later in this section we shall define analogs of two-sided limits.

If C is a smooth parametric curve in 2-space or 3-space that is represented by the equations

$$x = x(t), \quad y = y(t) \qquad \text{or} \qquad x = x(t), \quad y = y(t), \quad z = z(t)$$

and if $x_0 = x(t_0)$, $y_0 = y(t_0)$, and $z_0 = z(t_0)$, then the limits

$$\lim_{\substack{(x, y) \to (x_0, y_0) \\ \text{(along } C\text{)}}} f(x, y) \quad \text{and} \quad \lim_{\substack{(x, y, z) \to (x_0, y_0, z_0) \\ \text{(along } C\text{)}}} f(x, y, z)$$

are defined by

$$\lim_{\substack{(x, y) \to (x_0, y_0) \\ \text{(along } C\text{)}}} f(x, y) = \lim_{t \to t_0} f(x(t), y(t)) \tag{1}$$

$$\lim_{\substack{(x, y, z) \to (x_0, y_0, z_0) \\ \text{(along } C\text{)}}} f(x, y, z) = \lim_{t \to t_0} f(x(t), y(t), z(t)) \tag{2}$$

Simply stated, limits along parametric curves are obtained by substituting the parametric equations into the formula for the function f and computing the appropriate limit of the resulting function of one variable. A geometric interpretation of the limit along a curve for a function of two variables is shown in Figure 16.2.2: As the point $(x(t), y(t))$ moves along the curve C in the xy-plane toward (x_0, y_0), the point $(x(t), y(t), f(x(t), y(t)))$ moves directly above it along the graph of $z = f(x, y)$ with $f(x(t), y(t))$ approaching the limiting value L. In that figure we followed a common practice of omitting the zero z-coordinate for points in the xy-plane.

REMARK. In both (1) and (2), the limit of the function of t has to be treated as a one-sided limit if (x_0, y_0) or (x_0, y_0, z_0) is an endpoint of C.

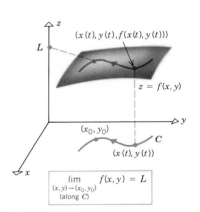

Figure 16.2.2

Example 1 Let

$$f(x, y) = \frac{xy}{x^2 + y^2}$$

Find the limit of $f(x, y)$ as $(x, y) \to (0, 0)$ along

(a) the x-axis (b) the y-axis
(c) the parabola $y = x^2$ (d) the line $y = x$

Solution (a). The x-axis has parametric equations $x = t$, $y = 0$, with $(0, 0)$ corresponding to $t = 0$, so

$$\lim_{\substack{(x, y) \to (0,0) \\ \text{(along } y=0\text{)}}} f(x, y) = \lim_{t \to 0} f(t, 0) = \lim_{t \to 0} \frac{0}{t^2} = \lim_{t \to 0} 0 = 0$$

Solution (b). The y-axis has parametric equations $x = 0$, $y = t$, with $(0, 0)$ corresponding to $t = 0$, so

$$\lim_{\substack{(x, y) \to (0,0) \\ \text{(along } x=0\text{)}}} f(x, y) = \lim_{t \to 0} f(0, t) = \lim_{t \to 0} \frac{0}{t^2} = \lim_{t \to 0} 0 = 0$$

Solution (c). The parabola $y = x^2$ has parametric equations $x = t$, $y = t^2$, with $(0, 0)$ corresponding to $t = 0$, so

$$\lim_{\substack{(x, y) \to (0,0) \\ \text{(along } y=x^2\text{)}}} f(x, y) = \lim_{t \to 0} f(t, t^2) = \lim_{t \to 0} \frac{t^3}{t^2 + t^4} = \lim_{t \to 0} \frac{t}{1 + t^2} = 0$$

Solution (d).   The line $y = x$ has parametric equations $x = t$, $y = t$, with $(0, 0)$ corresponding to $t = 0$, so

$$\lim_{\substack{(x,y)\to(0,0) \\ \text{(along } y=x)}} f(x, y) = \lim_{t\to 0} f(t, t) = \lim_{t\to 0} \frac{t^2}{2t^2} = \lim_{t\to 0} \frac{1}{2} = \frac{1}{2} \quad \blacktriangleleft$$

REMARK.   The graph of the function f in the preceding example is shown in Figure 16.2.3 with the axes drawn in red for clarity. The limits in parts (a), (b), and (d) can be visualized from this graph if you keep in mind that, except at the origin, where f is undefined, the x- and y-axes lie on the surface (why?). The limit in part (c) can also be visualized from the graph, but it may take some effort.

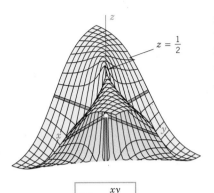

$$z = \frac{xy}{x^2 + y^2}$$

Figure 16.2.3

Example 2   Find

$$\lim_{\substack{(x,y,z)\to(-1,0,\pi) \\ \text{(along } C)}} \frac{x^2 + y^2 + x}{z - \pi}$$

where C is the circular helix with parametric equations $x = \cos t$, $y = \sin t$, $z = t$.

Solution.   The point $(-1, 0, \pi)$ corresponds to $t = \pi$, so

$$\lim_{\substack{(x,y,z)\to(-1,0,\pi) \\ \text{(along } C)}} \frac{x^2 + y^2 + x}{z - \pi} = \lim_{t\to\pi} \frac{\cos^2 t + \sin^2 t + \cos t}{t - \pi}$$

$$= \lim_{t\to\pi} \frac{1 + \cos t}{t - \pi} = 0$$

where L'Hôpital's rule was applied to evaluate the last limit. $\blacktriangleleft$

☐ **GENERAL LIMITS OF FUNCTIONS OF TWO AND THREE VARIABLES**

Although limits along specific curves are useful for many purposes, they do not always tell the complete story about the limiting behavior of a function; what is required is a limit concept that accounts for the behavior of the function in an *entire vicinity* of a point, not just along smooth curves passing through the point. As illustrated in Figure 16.2.4, we will want the statement

$$\lim_{(x, y)\to(x_0, y_0)} f(x, y) = L$$

to mean that the value of $f(x, y)$ can be made as close as we like to L (say within ϵ units of L) by restricting (x, y) to lie within (but not at the center of) some sufficiently small circle

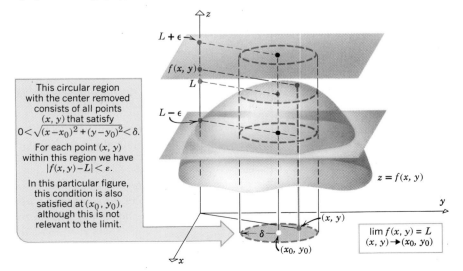

This circular region with the center removed consists of all points (x, y) that satisfy $0 < \sqrt{(x - x_0)^2 + (y - y_0)^2} < \delta$.

For each point (x, y) within this region we have $|f(x, y) - L| < \varepsilon$.

In this particular figure, this condition is also satisfied at (x_0, y_0), although this is not relevant to the limit.

$$\lim f(x, y) = L \\ (x, y) \to (x_0, y_0)$$

Figure 16.2.4

centered at (x_0, y_0) (say a circle of radius δ). This idea is conveyed by Definition 16.2.1 and for functions of three variables by Definition 16.2.2.

16.2.1   DEFINITION.   Let f be a function of two variables. We shall write

$$\lim_{(x,y)\to(x_0,y_0)} f(x, y) = L \tag{3}$$

if given any number $\epsilon > 0$, we can find a number $\delta > 0$ such that $f(x, y)$ satisfies

$$|f(x, y) - L| < \epsilon$$

whenever (x, y) lies in the domain of f and the distance between (x, y) and (x_0, y_0) satisfies

$$0 < \sqrt{(x - x_0)^2 + (y - y_0)^2} < \delta$$

16.2.2   DEFINITION.   Let f be a function of three variables. We shall write

$$\lim_{(x,y,z)\to(x_0,y_0,z_0)} f(x, y, z) = L \tag{4}$$

if given any number $\epsilon > 0$, we can find a number $\delta > 0$ such that $f(x, y, z)$ satisfies

$$|f(x, y, z) - L| < \epsilon$$

whenever (x, y, z) lies in the domain of f and the distance between (x, y, z) and (x_0, y_0, z_0) satisfies

$$0 < \sqrt{(x - x_0)^2 + (y - y_0)^2 + (z - z_0)^2} < \delta$$

When convenient, (3) and (4) can also be written in the alternative notations

$$f(x, y) \to L \quad \text{as} \quad (x, y) \to (x_0, y_0)$$

and

$$f(x, y, z) \to L \quad \text{as} \quad (x, y, z) \to (x_0, y_0, z_0)$$

□ **PROPERTIES OF LIMITS**

We note without proof that the basic properties of limits given in Theorem 2.5.1 hold for limits along curves and for the limits in the preceding definitions, so that computations involving such limits can be performed in the usual way.

Example 3

$$\lim_{(x,y)\to(1,4)} [5x^3y^2 - 9] = \lim_{(x,y)\to(1,4)} [5x^3y^2] - \lim_{(x,y)\to(1,4)} 9$$

$$= 5\left[\lim_{(x,y)\to(1,4)} x\right]^3 \left[\lim_{(x,y)\to(1,4)} y\right]^2 - 9$$

$$= 5(1)^3(4)^2 - 9 = 71 \quad \blacktriangleleft$$

□ **RELATIONSHIPS BETWEEN GENERAL LIMITS AND LIMITS ALONG SMOOTH CURVES**

The following theorem, which we state without proof, establishes an important relationship between general limits and limits along smooth curves.

16.2.3   THEOREM.   *If a function $f(x, y)$ has a limit L as (x, y) approaches a point (x_0, y_0), then $f(x, y)$ approaches the same limit L as (x, y) approaches (x_0, y_0) along any smooth curve that lies in the domain of f. Similarly, for functions of three variables.*

It follows from this theorem that if one can find two different smooth curves containing (x_0, y_0) along which $f(x, y)$ has different limits as (x, y) approaches (x_0, y_0), or if one can find any single smooth curve containing (x_0, y_0) such that the limit of $f(x, y)$ does not exist as (x, y) approaches (x_0, y_0), then

$$\lim_{(x,y) \to (x_0, y_0)} f(x, y)$$

does not exist. Similarly for functions of three variables.

Example 4   The limit

$$\lim_{(x,y) \to (0,0)} \frac{xy}{x^2 + y^2}$$

does not exist because in Example 1 we found two different smooth curves along which this limit had different values. For example, we saw that

$$\lim_{\substack{(x,y) \to (0,0) \\ (\text{along } x=0)}} \frac{xy}{x^2 + y^2} = 0 \quad \text{and} \quad \lim_{\substack{(x,y) \to (0,0) \\ (\text{along } y=x)}} \frac{xy}{x^2 + y^2} = \frac{1}{2} \quad \blacktriangleleft$$

☐ **CONTINUITY**

Informally stated, a function of one variable is continuous if its graph is an unbroken curve without jumps or holes. To extend this idea to functions of two variables, imagine that the graph of $z = f(x, y)$ is molded from a thin sheet of clay that has been hollowed or pinched into peaks and valleys. We shall consider f to be continuous if the clay surface has no tears or holes. The functions graphed in Figure 16.2.5 fail to be continuous because of their behavior at $(0, 0)$.

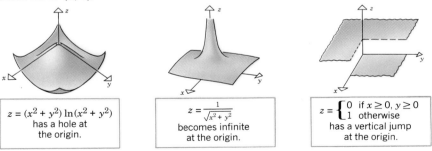

$z = (x^2 + y^2) \ln(x^2 + y^2)$
has a hole at
the origin.

$z = \dfrac{1}{\sqrt{x^2 + y^2}}$
becomes infinite
at the origin.

$z = \begin{cases} 0 & \text{if } x \geq 0, \, y \geq 0 \\ 1 & \text{otherwise} \end{cases}$
has a vertical jump
at the origin.

Figure 16.2.5

The precise definitions of continuity for functions of two and three variables are similar to the definition for a function of one variable. Recall that $f(x)$ is continuous at a point x_0 if

$$\lim_{x \to x_0} f(x) = f(x_0)$$

Analogously, we define $f(x, y)$ to be **continuous at (x_0, y_0)** and $f(x, y, z)$ to be **continuous at (x_0, y_0, z_0)** if

$$\lim_{(x, y) \to (x_0, y_0)} f(x, y) = f(x_0, y_0) \quad \text{and} \quad \lim_{(x, y, z) \to (x_0, y_0, z_0)} f(x, y, z) = f(x_0, y_0, z_0)$$

respectively. Moreover, a function f of two or three variables that is continuous at each point of a region R in 2-space or 3-space is said to be **continuous on R**, and if f is continuous at every point in 2-space or 3-space, then f is said to be **continuous everywhere** or simply **continuous**.

☐ **COMPOSITIONS OF CONTINUOUS FUNCTIONS**

The following theorem, which we state without proof, will help us to identify continuous functions of two variables.

16.2.4   THEOREM.

(a)   If g and h are continuous functions of one variable, then $f(x, y) = g(x)h(y)$ is a continuous function of x and y.

(b)   If g is a continuous function of one variable and h is a continuous function of two variables, then their composition $f(x, y) = g(h(x, y))$ is a continuous function of x and y.

Example 5   The function $f(x, y) = 3x^2y^5$ is continuous because it is the product of the continuous functions $g(x) = 3x^2$ and $h(y) = y^5$.

In general, any function of the form $f(x, y) = Ax^my^n$ (m and n nonnegative integers) is continuous because it is the product of the continuous functions Ax^m and y^n.   ◀

Example 6   Since $g(x) = \sin x$ is a continuous function of one variable and $h(x, y) = xy^2$ is a continuous function of two variables, it follows that $g(h(x, y)) = g(xy^2) = \sin(xy^2)$ is a continuous function of x and y. By a similar argument, each of the following is continuous:

$$(x^4y^5)^{1/3}, \quad e^{xy}, \quad \cosh(x^3y) \quad ◀$$

Example 7   By Example 6, e^{xy} is continuous. Thus, $\cos(e^{xy})$ is continuous by part (b) of Theorem 16.2.4.   ◀

Theorem 16.2.4 is one of a whole class of theorems about continuity of functions in any number of variables. The content of these theorems can be summarized informally with three basic principles:

- A composition of continuous functions is continuous.
- A sum, difference, or product of continuous functions is continuous.
- A quotient of continuous functions is continuous, except where the denominator is zero.

Example 8   The following functions are continuous since they are sums, differences, products, and compositions of continuous functions:

$$3 - 2x^2yz + 9x^4y^8z^3, \quad e^{xy}\cos(xy^2 + 1), \quad (3z + ye^x)^{17} \quad ◀$$

Example 9   Since the function

$$f(x, y) = \frac{x^3y^2}{1 - xy}$$

is a quotient of continuous functions, it is continuous except where $1 - xy = 0$. Thus, $f(x, y)$ is continuous everywhere except on the hyperbola $xy = 1$.   ◀

Example 10   Evaluate

$$\lim_{(x,y)\to(-1,2)} \frac{xy}{x^2 + y^2}$$

Solution.   Since $f(x, y) = xy/(x^2 + y^2)$ is continuous at $(-1, 2)$ (why?), it follows from the definition of continuity for functions of two variables that

$$\lim_{(x,y)\to(-1,2)} \frac{xy}{x^2 + y^2} = \frac{(-1)(2)}{(-1)^2 + (2)^2} = -\frac{2}{5} \quad ◀$$

☐ **LIMITS AT POINTS OF DISCONTINUITY**

At a point of discontinuity the method of Example 10 cannot be used. However, sometimes such limits can be obtained by converting the given function to polar coordinates.

Example 11   Find
$$\lim_{(x,y)\to(0,0)} (x^2 + y^2) \ln (x^2 + y^2)$$

Solution.   Let (r, θ) be the coordinates of the point (x, y) with $r \ge 0$. Then we have
$$x = r \cos \theta, \quad y = r \sin \theta, \quad r^2 = x^2 + y^2$$
Moreover, since $r \ge 0$ we have $r = \sqrt{x^2 + y^2}$, so that $r \to 0^+$ if and only if $(x, y) \to (0, 0)$. Thus, we can rewrite the given limit as
$$\lim_{(x,y)\to(0,0)} (x^2 + y^2) \ln (x^2 + y^2) = \lim_{r\to0^+} r^2 \ln r^2$$

$$= \lim_{r\to0^+} \frac{2 \ln r}{1/r^2} \qquad \boxed{\text{This converts the limit to an indeterminate form of type } \infty/\infty.}$$

$$= \lim_{r\to0^+} \frac{2/r}{-2/r^3} \qquad \boxed{\text{L'Hôpital's rule}}$$

$$= \lim_{r\to0^+} (-r^2) = 0$$

This result is consistent with the graph of f shown in Figure 16.2.5.   ◀

▶ Exercise Set 16.2

In Exercises 1–8, sketch the region where the function f is continuous.

1. $f(x, y) = y \ln (1 + x)$.   **2.** $f(x, y) = \sqrt{x - y}$.

3. $f(x, y) = \dfrac{x^2 y}{\sqrt{25 - x^2 - y^2}}$.

4. $f(x, y) = \ln (2x - y + 1)$.

5. $f(x, y) = \cos \left(\dfrac{xy}{1 + x^2 + y^2} \right)$.

6. $f(x, y) = e^{(1-xy)}$.   **7.** $f(x, y) = \sin^{-1} (xy)$.

8. $f(x, y) = \tan^{-1} (y - x)$.

In Exercises 9–12, describe the region on which the function f is continuous.

9. $f(x, y, z) = 3x^2 e^{yz} \cos (xyz)$.

10. $f(x, y, z) = \ln (4 - x^2 - y^2 - z^2)$.

11. $f(x, y, z) = \dfrac{y + 1}{x^2 + z^2 - 1}$.

12. $f(x, y, z) = \sin \sqrt{x^2 + y^2 + 3z^2}$.

In Exercises 13–35, find the limit, if it exists.

13. $\lim\limits_{(x,y)\to(1,3)} (4xy^2 - x)$.   **14.** $\lim\limits_{(x,y)\to(1/2,\pi)} (xy^2 \sin xy)$.

15. $\lim\limits_{(x,y)\to(-1,2)} \dfrac{xy^3}{x + y}$.   **16.** $\lim\limits_{(x,y)\to(1,-3)} e^{2x-y^2}$.

17. $\lim\limits_{(x,y)\to(0,0)} \ln (1 + x^2 y^3)$.   **18.** $\lim\limits_{(x,y)\to(4,-2)} x \sqrt[3]{y^3 + 2x}$.

19. $\lim\limits_{(x,y)\to(0,0)} \dfrac{x - y}{x^2 + y^2}$. [*Hint:* Let $(x, y) \to (0, 0)$ along the line $y = 0$.]

20. $\lim\limits_{(x,y)\to(0,0)} \dfrac{3}{x^2 + 2y^2}$.   **21.** $\lim\limits_{(x,y)\to(0,0)} \dfrac{\sin (x^2 + y^2)}{x^2 + y^2}$.

22. $\lim\limits_{(x,y)\to(0,0)} \dfrac{1 - \cos (x^2 + y^2)}{x^2 + y^2}$.

23. $\lim\limits_{(x,y)\to(0,0)} \dfrac{x^4 - y^4}{x^2 + y^2}$.   **24.** $\lim\limits_{(x,y)\to(0,0)} \dfrac{x^4 - 16y^4}{x^2 + 4y^2}$.

25. $\lim\limits_{(x,y)\to(0,0)} \dfrac{xy}{3x^2 + 2y^2}$.   **26.** $\lim\limits_{(x,y)\to(0,0)} \dfrac{1 - x^2 - y^2}{x^2 + y^2}$.

27. $\lim\limits_{(x,y)\to(0,0)} e^{-1/(x^2+y^2)}$.   **28.** $\lim\limits_{(x,y)\to(0,0)} \dfrac{e^{-1/\sqrt{x^2+y^2}}}{\sqrt{x^2 + y^2}}$.

29. $\lim\limits_{(x,y)\to(0,0)} y \ln (x^2 + y^2)$.

30. $\lim\limits_{(x,y)\to(0,0)} x \ln (|x| + |y|)$.

31. $\lim\limits_{(x,y,z)\to(2,-1,2)} \dfrac{xz^2}{\sqrt{x^2 + y^2 + z^2}}$.

32. $\lim\limits_{(x,y,z)\to(2,0,-1)} \ln(2x + y - z)$.

33. $\lim\limits_{(x,y,z)\to(0,0,0)} \dfrac{\sin(x^2 + y^2 + z^2)}{\sqrt{x^2 + y^2 + z^2}}$.

34. $\lim\limits_{(x,y,z)\to(0,0,0)} \dfrac{\sin\sqrt{x^2 + y^2 + z^2}}{x^2 + y^2 + z^2}$.

35. $\lim\limits_{(x,y,z)\to(0,0,0)} \dfrac{yz}{x^2 + y^2 + z^2}$.

[*Hint:* First let $(x, y, z)\to(0, 0, 0)$ along the z-axis and then along the line $x = t$, $y = t$, $z = t$.]

36. Show that $\dfrac{xy}{x^2 + y^2}$ can be made to approach any value in the interval $[-\frac{1}{2}, \frac{1}{2}]$ by letting $(x, y)\to(0, 0)$ along some line $y = mx$.

37. (a) Show that the value of $\dfrac{x^2 y}{x^4 + y^2}$ approaches zero as $(x, y)\to(0, 0)$ along any straight line $y = mx$.

(b) Show that $\lim\limits_{(x,y)\to(0,0)} \dfrac{x^2 y}{x^4 + y^2}$ does not exist by letting $(x, y)\to(0, 0)$ along the parabola $y = x^2$.

38. (a) Show that the value of $\dfrac{x^3 y}{2x^6 + y^2}$ approaches 0 as $(x, y)\to(0, 0)$ along any straight line $y = mx$, or along any parabola $y = kx^2$.

(b) Show that $\lim\limits_{(x,y)\to(0,0)} \dfrac{x^3 y}{2x^6 + y^2}$ does not exist by letting $(x, y)\to(0, 0)$ along the curve $y = x^3$.

39. (a) Show that the value of $\dfrac{xyz}{x^2 + y^4 + z^4}$ approaches 0 as $(x, y, z)\to(0, 0, 0)$ along any line $x = at$, $y = bt$, $z = ct$.

(b) Show that the limit $\lim\limits_{(x,y,z)\to(0,0,0)} \dfrac{xyz}{x^2 + y^4 + z^4}$ does

not exist by letting $(x, y, z)\to(0, 0, 0)$ along the curve $x = t^2$, $y = t$, $z = t$.

40. Find
$$\lim\limits_{(x,y)\to(0,1)} \tan^{-1}\left[\frac{x^2 + 1}{x^2 + (y - 1)^2}\right]$$

41. Find
$$\lim\limits_{(x,y)\to(0,1)} \tan^{-1}\left[\frac{x^2 - 1}{x^2 + (y - 1)^2}\right]$$

42. Let $f(x, y) = \begin{cases} \dfrac{\sin(x^2 + y^2)}{x^2 + y^2}, & (x, y) \neq (0, 0) \\ 1, & (x, y) = (0, 0). \end{cases}$

Show that f is continuous at $(0, 0)$.

43. Let $f(x, y) = \dfrac{x^2}{x^2 + y^2}$. Is it possible to define $f(0, 0)$ so that f will be continuous at $(0, 0)$?

44. Let $f(x, y) = xy \ln(x^2 + y^2)$. Is it possible to define $f(0, 0)$ so that f will be continuous at $(0, 0)$?

In Exercises 45 and 46, use Definition 16.2.1 to prove the given statement. [*Hint:* Let $r = \sqrt{x^2 + y^2}$.]

45. $\lim\limits_{(x,y)\to(0,0)} (x^2 + y^2) = 0$.

46. $\lim\limits_{(x,y)\to(0,0)} \dfrac{x^2 y^2}{\sqrt{x^2 + y^2}} = 0$.

In Exercises 47 and 48, use Definition 16.2.2 to prove the given statement. [*Hint:* Let $\rho = \sqrt{x^2 + y^2 + z^2}$.]

47. $\lim\limits_{(x,y,z)\to(0,0,0)} (x^2 + y^2 + z^2) = 0$.

48. $\lim\limits_{(x,y,z)\to(0,0,0)} e^{\sqrt{x^2 + y^2 + z^2}} = 1$.

■ **16.3** PARTIAL DERIVATIVES

*If f is a function of two or more independent variables and all but one of those variables are held fixed, then the derivative of f with respect to that one remaining independent variable is called a **partial derivative** of f. In this section we shall show how to compute partial derivatives and discuss their geometric significance.*

☐ **PARTIAL DERIVATIVES OF FUNCTIONS OF TWO VARIABLES**

Let f be a function of x and y. If we hold y constant, say $y = y_0$, and view x as a variable, then $f(x, y_0)$ is a function of x alone. If this function is differentiable at $x = x_0$, then the value of this derivative is denoted by

$$f_x(x_0, y_0) \tag{1}$$

and is called the ***partial derivative of f with respect to x*** at the point (x_0, y_0). Similarly, if we hold x constant, say $x = x_0$, then $f(x_0, y)$ is a function of y alone. If this function is differentiable at $y = y_0$, then the value of this derivative is denoted by

$$f_y(x_0, y_0) \tag{2}$$

and is called the ***partial derivative of f with respect to y*** at (x_0, y_0).

The partial derivatives of $f(x, y)$ have a simple geometric interpretation. Let P be a point on the intersection of the surface $z = f(x, y)$ and the plane $y = y_0$. If y is held constant at $y = y_0$ and x is allowed to vary, then the point P moves along the curve C_1 that is the intersection of the surface with the vertical plane $y = y_0$ (Figure 16.3.1a). Thus, the partial derivative $f_x(x_0, y_0)$ can be interpreted as the slope (change in z per unit increase in x) of the tangent line to the curve C_1 at the point (x_0, y_0). Similarly, if x is held constant, say $x = x_0$, and y is allowed to vary, then the point P moves along the curve C_2 that is the intersection of the surface with the vertical plane $x = x_0$. Thus, the partial derivative $f_y(x_0, y_0)$ can be interpreted as the slope of the tangent line (change in z per unit increase in y) to the curve C_2 at the point (x_0, y_0) (Figure 16.3.1b).

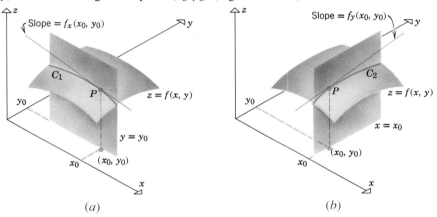

Figure 16.3.1   (a)   (b)

REMARK.   In this text we shall only consider partial derivatives at *interior points* of the domain of f. Partial derivatives at boundary points lead to complications that are best left for more advanced courses.

The values of $f_x(x_0, y_0)$ and $f_y(x_0, y_0)$ are usually obtained by finding expressions for $f_x(x, y)$ and $f_y(x, y)$ at a general point (x, y) and then substituting $x = x_0$ and $y = y_0$ in these expressions. To obtain $f_x(x, y)$ we differentiate $f(x, y)$ with respect to x, *treating y as a constant*; and to obtain $f_y(x, y)$ we differentiate $f(x, y)$ with respect to y, *treating x as a constant*.

Example 1   Find $f_x(1, 2)$ and $f_y(1, 2)$ if $f(x, y) = 2x^3y^2 + 2y + 4x$.

Solution.   Treating y as a constant and differentiating with respect to x, we obtain

$$f_x(x, y) = 6x^2y^2 + 4$$

Treating x as a constant and differentiating with respect to y, we obtain

$$f_y(x, y) = 4x^3y + 2$$

Substituting $x = 1$ and $y = 2$ in these partial-derivative formulas yields

$$f_x(1, 2) = 6(1)^2(2)^2 + 4 = 28$$
$$f_y(1, 2) = 4(1)^3(2) + 2 = 10 \quad \blacktriangleleft$$

PARTIAL-DERIVATIVE NOTATION

If $z = f(x, y)$, then the partial derivatives f_x and f_y can also be denoted by the symbols*

$$\frac{\partial f}{\partial x}, \quad \frac{\partial z}{\partial x} \quad \text{and} \quad \frac{\partial f}{\partial y}, \quad \frac{\partial z}{\partial y}$$

Some typical notations for the partial derivatives at a point (x_0, y_0) are

$$\left.\frac{\partial f}{\partial x}\right|_{x=x_0, \, y=y_0}, \quad \left.\frac{\partial z}{\partial y}\right|_{(x_0, \, y_0)}, \quad \left.\frac{\partial f}{\partial x}\right|_{(x_0, \, y_0)}, \quad \frac{\partial f}{\partial x}(x_0, y_0)$$

Example 2   Find $\partial z/\partial x$ and $\partial z/\partial y$ if $z = x^4 \sin(xy^3)$.

Solution.

$$\frac{\partial z}{\partial x} = \frac{\partial}{\partial x}[x^4 \sin(xy^3)] = x^4 \frac{\partial}{\partial x}[\sin(xy^3)] + \sin(xy^3) \cdot \frac{\partial}{\partial x}(x^4)$$

$$= x^4 \cos(xy^3) \cdot y^3 + \sin(xy^3) \cdot 4x^3 = x^4 y^3 \cos(xy^3) + 4x^3 \sin(xy^3)$$

$$\frac{\partial z}{\partial y} = \frac{\partial}{\partial y}[x^4 \sin(xy^3)] = x^4 \frac{\partial}{\partial y}[\sin(xy^3)] + \sin(xy^3) \cdot \frac{\partial}{\partial y}(x^4)$$

$$= x^4 \cos(xy^3) \cdot 3xy^2 + \sin(xy^3) \cdot 0 = 3x^5 y^2 \cos(xy^3) \quad \blacktriangleleft$$

Example 3   Suppose that a point Q moves along the intersection of the sphere $x^2 + y^2 + z^2 = 1$ with the plane $x = \frac{2}{3}$. At what rate is z changing with respect to y when the point is at $P(\frac{2}{3}, \frac{1}{3}, \frac{2}{3})$?

Solution.   Since the z-coordinate of the point $P(\frac{2}{3}, \frac{1}{3}, \frac{2}{3})$ is positive, this point lies on the upper hemisphere

$$z = \sqrt{1 - x^2 - y^2} \tag{3}$$

and hence for each fixed value of x, the rate of change of z with respect to y on the upper hemisphere is

$$\frac{\partial z}{\partial y} = \frac{\partial}{\partial y}[(1 - x^2 - y^2)^{1/2}] = \frac{1}{2}(1 - x^2 - y^2)^{-1/2}(-2y) = -\frac{y}{\sqrt{1 - x^2 - y^2}}$$

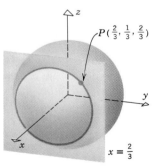

Figure 16.3.2

In particular, if $x = \frac{2}{3}$ (Figure 16.3.2), then it follows from this equation that the rate of change of z with respect to y at the point P is

$$\left.\frac{\partial z}{\partial y}\right|_{x=\frac{2}{3}, \, y=\frac{1}{3}} = -\frac{\frac{1}{3}}{\sqrt{1 - (\frac{2}{3})^2 - (\frac{1}{3})^2}} = -\frac{1}{2}$$

Alternative Solution.   Instead of solving $x^2 + y^2 + z^2 = 1$ explicitly for z as a function of x and y, we can obtain $\partial z/\partial y$ by implicit differentiation. Differentiating both sides of $x^2 + y^2 + z^2 = 1$ with respect to y and treating z as a function of x and y yields

$$\frac{\partial}{\partial y}[x^2 + y^2 + z^2] = \frac{\partial}{\partial y}[1]$$

$$2y + 2z\frac{\partial z}{\partial y} = 0$$

$$\frac{\partial z}{\partial y} = -\frac{y}{z}$$

Substituting the y- and z-coordinates of the point $(\frac{2}{3}, \frac{1}{3}, \frac{2}{3})$ yields $-\frac{1}{2}$ as before. $\blacktriangleleft$

---

*The symbol ∂ is called a partial derivative sign. It is derived from the Cyrillic alphabet.

Example 4 Suppose that $D = \sqrt{x^2 + y^2}$ is the length of the diagonal of a rectangle whose sides have lengths x and y that are allowed to vary. Find a formula for the rate of change of D with respect to x if x varies with y held constant, and use this formula to find the rate of change of D with respect to x at the instant when $x = 3$ and $y = 4$.

Solution. The instantaneous rate of change of D with respect to x with y held constant is

$$\frac{\partial D}{\partial x} = \frac{1}{2}(x^2 + y^2)^{-1/2}(2x) = \frac{x}{\sqrt{x^2 + y^2}}$$

from which it follows that

$$\frac{\partial D}{\partial x}\bigg|_{x=3, \, y=4} = \frac{3}{\sqrt{3^2 + 4^2}} = \frac{3}{5}$$

Thus, D is increasing at a rate of $\frac{3}{5}$ unit per unit increase in x at the point $(3, 4)$. ◀

☐ **HIGHER-ORDER PARTIAL DERIVATIVES**

Since the partial derivatives $\partial f/\partial x$ and $\partial f/\partial y$ are functions of x and y, each can have partial derivatives. This gives rise to four possible *second-order* partial derivatives of f, which are defined by

$$\frac{\partial^2 f}{\partial x^2} = \frac{\partial}{\partial x}\left(\frac{\partial f}{\partial x}\right), \qquad \frac{\partial^2 f}{\partial y^2} = \frac{\partial}{\partial y}\left(\frac{\partial f}{\partial y}\right)$$

$$\frac{\partial^2 f}{\partial x \, \partial y} = \frac{\partial}{\partial x}\left(\frac{\partial f}{\partial y}\right), \qquad \frac{\partial^2 f}{\partial y \, \partial x} = \frac{\partial}{\partial y}\left(\frac{\partial f}{\partial x}\right)$$

Henceforth, we shall call $\partial f/\partial x$ and $\partial f/\partial y$ the *first-order* partial derivatives of f.

Example 5 Find the second-order partial derivatives of $f(x, y) = x^2 y^3 + x^4 y$.

Solution. We have

$$\frac{\partial f}{\partial x} = 2xy^3 + 4x^3 y \quad \text{and} \quad \frac{\partial f}{\partial y} = 3x^2 y^2 + x^4$$

so that

$$\frac{\partial^2 f}{\partial x^2} = \frac{\partial}{\partial x}\left(\frac{\partial f}{\partial x}\right) = \frac{\partial}{\partial x}(2xy^3 + 4x^3 y) = 2y^3 + 12x^2 y$$

$$\frac{\partial^2 f}{\partial y^2} = \frac{\partial}{\partial y}\left(\frac{\partial f}{\partial y}\right) = \frac{\partial}{\partial y}(3x^2 y^2 + x^4) = 6x^2 y$$

$$\frac{\partial^2 f}{\partial x \, \partial y} = \frac{\partial}{\partial x}\left(\frac{\partial f}{\partial y}\right) = \frac{\partial}{\partial x}(3x^2 y^2 + x^4) = 6xy^2 + 4x^3$$

$$\frac{\partial^2 f}{\partial y \, \partial x} = \frac{\partial}{\partial y}\left(\frac{\partial f}{\partial x}\right) = \frac{\partial}{\partial y}(2xy^3 + 4x^3 y) = 6xy^2 + 4x^3 \quad ◀$$

REMARK. The derivatives

$$\frac{\partial^2 f}{\partial y \, \partial x} \quad \text{and} \quad \frac{\partial^2 f}{\partial x \, \partial y}$$

are called the *mixed second-order partial derivatives* or *mixed second-partials*. For most functions that arise in applications, these mixed partial derivatives are equal (as in the last example). In the next section we shall state precise conditions under which equality holds.

By successively differentiating, we can obtain third-order partial derivatives and higher. Some possibilities are

$$\frac{\partial^3 f}{\partial x^3} = \frac{\partial}{\partial x}\left(\frac{\partial^2 f}{\partial x^2}\right), \qquad \frac{\partial^3 f}{\partial y^2\, \partial x} = \frac{\partial}{\partial y}\left(\frac{\partial^2 f}{\partial y\, \partial x}\right)$$

$$\frac{\partial^3 f}{\partial y\, \partial x^2} = \frac{\partial}{\partial y}\left(\frac{\partial^2 f}{\partial x^2}\right), \qquad \frac{\partial^4 f}{\partial y^2\, \partial x^2} = \frac{\partial}{\partial y}\left(\frac{\partial^3 f}{\partial y\, \partial x^2}\right)$$

Higher-order partial derivatives can be denoted more compactly with subscript notation. For example,

$$\frac{\partial^2 f}{\partial y\, \partial x} = \frac{\partial}{\partial y}\left(\frac{\partial f}{\partial x}\right) = \frac{\partial}{\partial y}(f_x) = (f_x)_y$$

It is usual to drop the parentheses and write simply

$$\frac{\partial^2 f}{\partial y\, \partial x} = f_{xy}$$

Note that in "∂" notation the sequence of differentiations is obtained by reading from right to left, but in the subscript notation it is left to right. Some other examples are

$$f_{xx} = \frac{\partial^2 f}{\partial x^2}, \quad f_{yyx} = \frac{\partial^3 f}{\partial x\, \partial y^2}, \quad f_{xxyy} = \frac{\partial^4 f}{\partial y^2\, \partial x^2}$$

Example 6   Let $f(x, y) = y^2 e^x + y$. Find f_{xyy}.

Solution.

$$f_{xyy} = \frac{\partial^3 f}{\partial y^2\, \partial x} = \frac{\partial^2}{\partial y^2}\left(\frac{\partial f}{\partial x}\right) = \frac{\partial^2}{\partial y^2}(y^2 e^x) = \frac{\partial}{\partial y}(2y e^x) = 2e^x \quad \blacktriangleleft$$

☐ **PARTIAL DERIVATIVES OF FUNCTIONS WITH MORE THAN TWO VARIABLES**

For a function $f(x, y, z)$ of three variables, there are three **partial derivatives**:

$$f_x(x, y, z), \quad f_y(x, y, z), \quad f_z(x, y, z)$$

The partial derivative f_x is calculated by holding y and z constant and differentiating with respect to x. For f_y the variables x and z are held constant, and for f_z the variables x and y are held constant. If a dependent variable

$$w = f(x, y, z)$$

is used, then the three partial derivatives of f may be denoted by

$$\frac{\partial w}{\partial x}, \quad \frac{\partial w}{\partial y}, \quad \text{and} \quad \frac{\partial w}{\partial z}$$

Example 7   If $f(x, y, z) = x^3 y^2 z^4 + 2xy + z$, then

$$f_x(x, y, z) = 3x^2 y^2 z^4 + 2y$$
$$f_y(x, y, z) = 2x^3 yz^4 + 2x$$
$$f_z(x, y, z) = 4x^3 y^2 z^3 + 1$$
$$f_z(-1, 1, 2) = 4(-1)^3(1)^2(2)^3 + 1 = -31 \quad \blacktriangleleft$$

Example 8   If $f(\rho, \theta, \phi) = \rho^2 \cos \phi \sin \theta$, then

$$f_\rho(\rho, \theta, \phi) = 2\rho \cos \phi \sin \theta$$
$$f_{\rho\phi}(\rho, \theta, \phi) = -2\rho \sin \phi \sin \theta$$
$$f_{\rho\phi\theta}(\rho, \theta, \phi) = -2\rho \sin \phi \cos \theta \quad \blacktriangleleft$$

In general, if $f(v_1, v_2, \ldots, v_n)$ is a function of n variables, there are n partial derivatives of f, each of which is obtained by holding $n - 1$ of the variables fixed and differentiating the function f with respect to the remaining variable. If $w = f(v_1, v_2, \ldots, v_n)$, then these partial derivatives are denoted by

$$\frac{\partial w}{\partial v_1}, \frac{\partial w}{\partial v_2}, \ldots, \frac{\partial w}{\partial v_n}$$

where $\partial w / \partial v_i$ is obtained by holding all variables except v_i fixed and differentiating with respect to v_i.

Example 9   Find

$$\frac{\partial}{\partial x_i} \left[\sqrt{x_1^2 + x_2^2 + \cdots + x_n^2} \right]$$

for $i = 1, 2, \ldots, n$.

Solution.   For each $i = 1, 2, \ldots, n$ we obtain

$$\frac{\partial}{\partial x_i} \left[\sqrt{x_1^2 + x_2^2 + \cdots + x_n^2} \right] = \frac{1}{2\sqrt{x_1^2 + x_2^2 + \cdots + x_n^2}} \cdot \frac{\partial}{\partial x_i} \left[x_1^2 + x_2^2 + \cdots + x_n^2 \right]$$

$$= \frac{1}{2\sqrt{x_1^2 + x_2^2 + \cdots + x_n^2}} [2x_i] \qquad \boxed{\text{All terms except } x_i^2 \text{ are constant.}}$$

$$= \frac{x_i}{\sqrt{x_1^2 + x_2^2 + \cdots + x_n^2}} \qquad \blacktriangleleft$$

☐ **PARTIAL DERIVATIVES OF VECTOR-VALUED FUNCTIONS**

Partial derivatives of vector-valued functions of two or more variables are defined by taking partial derivatives of the components. For example, if

$$\mathbf{r}(u, v) = x(u, v)\mathbf{i} + y(u, v)\mathbf{j} + z(u, v)\mathbf{k}$$

then

$$\frac{\partial \mathbf{r}}{\partial u} = \frac{\partial x}{\partial u}\mathbf{i} + \frac{\partial y}{\partial u}\mathbf{j} + \frac{\partial z}{\partial u}\mathbf{k}$$

$$\frac{\partial \mathbf{r}}{\partial v} = \frac{\partial x}{\partial v}\mathbf{i} + \frac{\partial y}{\partial v}\mathbf{j} + \frac{\partial z}{\partial v}\mathbf{k}$$

These derivatives can also be expressed as limits:

$$\frac{\partial \mathbf{r}}{\partial u} = \lim_{h \to 0} \frac{\mathbf{r}(u + h, v) - \mathbf{r}(u, v)}{h} \tag{4}$$

$$\frac{\partial \mathbf{r}}{\partial v} = \lim_{k \to 0} \frac{\mathbf{r}(u, v + k) - \mathbf{r}(u, v)}{k} \tag{5}$$

Suppose now that σ denotes the surface represented parametrically by $\mathbf{r}(u, v)$ for (u, v) in a region R of the uv-plane and that $\partial \mathbf{r}/\partial u$ and $\partial \mathbf{r}/\partial v$ are continuous on R. It follows from 15.2.5 that if $\partial \mathbf{r}/\partial u \neq \mathbf{0}$ at a point (u_0, v_0), then $\partial \mathbf{r}/\partial u$ is tangent to the u-curve at the point $\mathbf{r}(u_0, v_0)$ and points in the direction of increasing u; and similarly, if $\partial \mathbf{r}/\partial v \neq \mathbf{0}$ at (u_0, v_0), then $\partial \mathbf{r}/\partial v$ is tangent to the v-curve at $\mathbf{r}(u_0, v_0)$ and points in the direction of increasing v. Thus, if $\partial \mathbf{r}/\partial u \times \partial \mathbf{r}/\partial v \neq \mathbf{0}$, then the vector

$$\mathbf{n} = \frac{\dfrac{\partial \mathbf{r}}{\partial u} \times \dfrac{\partial \mathbf{r}}{\partial v}}{\left\| \dfrac{\partial \mathbf{r}}{\partial u} \times \dfrac{\partial \mathbf{r}}{\partial v} \right\|} \tag{6}$$

is a unit vector that is perpendicular to both $\partial \mathbf{r}/\partial u$ and $\partial \mathbf{r}/\partial v$ at the point $\mathbf{r}(u_0, v_0)$. We call $\mathbf{n}$ the **unit normal** to σ at the point $\mathbf{r}(u_0, v_0)$, and we define the **tangent plane** to σ at $\mathbf{r}(u_0, v_0)$ to be the plane through $\mathbf{r}(u_0, v_0)$ that has $\mathbf{n}$ as a normal (Figure 16.3.3).

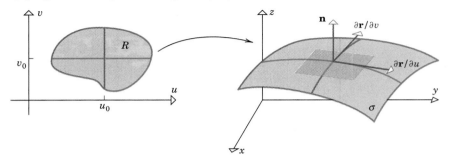

Figure 16.3.3

Example 10   Show that the tangent plane is perpendicular to the radius vector at each point of the sphere of radius a

$$\mathbf{r}(\phi, \theta) = a \sin \phi \cos \theta \mathbf{i} + a \sin \phi \sin \theta \mathbf{j} + a \cos \phi \mathbf{k}$$

where $0 \le \phi \le \pi$ and $0 \le \theta \le 2\pi$.

Solution.   It is sufficient to show that at each point of the sphere the unit normal vector $\mathbf{n}$ is a scalar multiple of $\mathbf{r}$ (and hence parallel to $\mathbf{r}$). But

$$\frac{\partial \mathbf{r}}{\partial \phi} \times \frac{\partial \mathbf{r}}{\partial \theta} = \begin{vmatrix} \mathbf{i} & \mathbf{j} & \mathbf{k} \\ \frac{\partial x}{\partial \phi} & \frac{\partial y}{\partial \phi} & \frac{\partial z}{\partial \phi} \\ \frac{\partial x}{\partial \theta} & \frac{\partial y}{\partial \theta} & \frac{\partial z}{\partial \theta} \end{vmatrix} = \begin{vmatrix} \mathbf{i} & \mathbf{j} & \mathbf{k} \\ a \cos \phi \cos \theta & a \cos \phi \sin \theta & -a \sin \phi \\ -a \sin \phi \sin \theta & a \sin \phi \cos \theta & 0 \end{vmatrix}$$

$$= a^2 \sin^2 \phi \cos \theta \mathbf{i} + a^2 \sin^2 \phi \sin \theta \mathbf{j} + a^2 \sin \phi \cos \phi \mathbf{k}$$

and hence

$$\left\| \frac{\partial \mathbf{r}}{\partial \phi} \times \frac{\partial \mathbf{r}}{\partial \theta} \right\| = \sqrt{a^4 \sin^4 \phi \cos^2 \theta + a^4 \sin^4 \phi \sin^2 \theta + a^4 \sin^2 \phi \cos^2 \phi}$$

$$= \sqrt{a^4 \sin^4 \phi + a^4 \sin^2 \phi \cos^2 \phi}$$

$$= a^2 \sqrt{\sin^2 \phi} = a^2 |\sin \phi| = a^2 \sin \phi$$

Thus, it follows from (6) that

$$\mathbf{n} = \sin \phi \cos \theta \mathbf{i} + \sin \phi \sin \theta \mathbf{j} + \cos \phi \mathbf{k} = \frac{1}{a} \mathbf{r}$$

which is also evident geometrically.   ◄

▶ Exercise Set 16.3

In Exercises 1–6, find $\partial z/\partial x$ and $\partial z/\partial y$.

1. $z = 3x^3y^2$.

2. $z = 4x^2 - 2y + 7x^4y^5$.

3. $z = 4e^{x^2y^3}$.

4. $z = \cos (x^5y^4)$.

5. $z = x^3 \ln (1 + xy^{-3/5})$.

6. $z = e^{xy} \sin 4y^2$.

In Exercises 7–12, find $f_x(x, y)$ and $f_y(x, y)$.

7. $f(x, y) = \sqrt{3x^5y - 7x^3y}$.

8. $f(x, y) = \dfrac{x + y}{x - y}$.

9. $f(x, y) = y^{-3/2} \tan^{-1} (x/y)$.

10. $f(x, y) = x^3 e^{-y} + y^3 \sec \sqrt{x}$.

11. $f(x, y) = (y^2 \tan x)^{-4/3}$.

12. $f(x, y) = \cosh(\sqrt{x}) \sinh^2(xy^2)$.

13. Given $f(x, y) = 9 - x^2 - 7y^3$, find

   (a) $f_x(3, 1)$ (b) $f_y(3, 1)$.

14. Given $f(x, y) = x^2 y e^{xy}$, find

   (a) $\left. \dfrac{\partial f}{\partial x} \right|_{(1, 1)}$ (b) $\left. \dfrac{\partial f}{\partial y} \right|_{(1, 1)}$.

15. Given $z = \sqrt{x^2 + 4y^2}$, find

   (a) $\left. \dfrac{\partial z}{\partial x} \right|_{(1, 2)}$ (b) $\left. \dfrac{\partial z}{\partial y} \right|_{(1, 2)}$.

16. Given $w = x^2 \cos xy$, find

   (a) $\dfrac{\partial w}{\partial x} \left(\tfrac{1}{2}, \pi \right)$ (b) $\dfrac{\partial w}{\partial y} \left(\tfrac{1}{2}, \pi \right)$.

In Exercises 17–20, calculate $\partial z/\partial x$ and $\partial z/\partial y$ using implicit differentiation. Leave your answers in terms of x, y, and z.

---

17. $(x^2 + y^2 + z^2)^{3/2} = 1$.   **18.** $\ln(2x^2 + y - z^3) = x$.

19. $x^2 + z \sin xyz = 0$.   **20.** $e^{xy} \sinh z - z^2 x + 1 = 0$.

In Exercises 21–26, find f_{xx}, f_{yy}, f_{xy}, and f_{yx}.

---

21. $f(x, y) = 4x^2 - 8xy^4 + 7y^5 - 3$.

22. $f(x, y) = \sqrt{x^2 + y^2}$.   **23.** $f(x, y) = e^x \cos y$.

24. $f(x, y) = e^{x - y^2}$.   **25.** $f(x, y) = \ln(4x - 5y)$.

26. $f(x, y) = (x^2 - y^2)/(x^2 + y^2)$.

27. Given $f(x, y) = x^3 y^5 - 2x^2 y + x$, find

   (a) f_{xxy} (b) f_{yxy} (c) f_{yyy}.

28. Given $z = (2x - y)^5$, find

   (a) $\dfrac{\partial^3 z}{\partial y \, \partial x \, \partial y}$ (b) $\dfrac{\partial^3 z}{\partial x^2 \, \partial y}$ (c) $\dfrac{\partial^4 z}{\partial x^2 \, \partial y^2}$.

29. Given $f(x, y) = y^3 e^{-5x}$, find

   (a) $f_{xyy}(0, 1)$ (b) $f_{xxx}(0, 1)$ (c) $f_{yyxx}(0, 1)$.

30. Given $w = e^y \cos x$, find

   (a) $\left. \dfrac{\partial^3 w}{\partial y^2 \, \partial x} \right|_{(\pi/4, 0)}$ (b) $\left. \dfrac{\partial^3 w}{\partial x^2 \, \partial y} \right|_{(\pi/4, 0)}$.

31. Express the following derivatives in "∂" notation.

   (a) f_{xxx} (b) f_{xyy} (c) f_{yyxx} (d) f_{xyyy}.

32. Express the derivatives in "subscript" notation.

   (a) $\dfrac{\partial^3 f}{\partial y^2 \, \partial x}$ (b) $\dfrac{\partial^4 f}{\partial x^4}$

   (c) $\dfrac{\partial^4 f}{\partial y^2 \, \partial x^2}$ (d) $\dfrac{\partial^5 f}{\partial x^2 \, \partial y^3}$.

In Exercises 33–37, find $\partial w/\partial x$, $\partial w/\partial y$, and $\partial w/\partial z$.

---

33. $w = x^2 y^4 z^3 + xy + z^2 + 1$.

34. $w = ye^z \sin x$.   **35.** $w = \dfrac{x^2 - y^2}{y^2 + z^2}$.

36. $w = y^3 e^{2x + 3z}$.   **37.** $w = \sqrt{x^2 + y^2 + z^2}$.

In Exercises 38–42, find f_x, f_y, and f_z.

---

38. $f(x, y, z) = z \ln(x^2 y \cos z)$.

39. $f(x, y, z) = \tan^{-1} \left(\dfrac{1}{xy^2 z^3} \right)$.

40. $f(x, y, z) = y^{-3/2} \sec \left(\dfrac{xz}{y} \right)$.

41. $f(x, y, z) = \cosh(\sqrt{z}) \sinh^2(x^2 yz)$.

42. $f(x, y, z) = \left(\dfrac{xz}{1 - z^2 - y^2} \right)^{-3/4}$.

43. Let $f(x, y, z) = 5x^2 yz^3$. Find

   (a) $f_x(1, -1, 2)$ (b) $f_y(1, -1, 2)$

   (c) $f_z(1, -1, 2)$.

44. Let $f(x, y, z) = y^2 e^{xz}$. Find

   (a) $\partial f/\partial x|_{(1, 1, 1)}$ (b) $\partial f/\partial y|_{(1, 1, 1)}$

   (c) $\partial f/\partial z|_{(1, 1, 1)}$.

45. Let $w = \sqrt{x^2 + 4y^2 - z^2}$. Find

   (a) $\partial w/\partial x|_{(2, 1, -1)}$ (b) $\partial w/\partial y|_{(2, 1, -1)}$

   (c) $\partial w/\partial z|_{(2, 1, -1)}$.

46. Let $w = x \sin xyz$. Find

   (a) $\dfrac{\partial w}{\partial x} \left(1, \tfrac{1}{2}, \pi \right)$ (b) $\dfrac{\partial w}{\partial y} \left(1, \tfrac{1}{2}, \pi \right)$

   (c) $\dfrac{\partial w}{\partial z} \left(1, \tfrac{1}{2}, \pi \right)$.

In Exercises 47–50, find $\partial w/\partial x$, $\partial w/\partial y$, and $\partial w/\partial z$ using implicit differentiation. Leave your answers in terms of x, y, z, and w.

---

47. $(x^2 + y^2 + z^2 + w^2)^{3/2} = 4$.

48. $\ln(2x^2 + y - z^3 + 3w) = z$.

49. $w^2 + w \sin xyz = 1$.

50. $e^{xy} \sinh w - z^2 w + 1 = 0$.

51. Let $f(x, y, z) = x^3 y^5 z^7 + xy^2 + y^3 z$. Find

   (a) f_{xy} (b) f_{yz} (c) f_{xz} (d) f_{zz}

   (e) f_{zyy} (f) f_{xxy} (g) f_{zyx} (h) f_{xxyz}.

52. Let $w = (4x - 3y + 2z)^5$. Find

   (a) $\dfrac{\partial^2 w}{\partial x \, \partial z}$ (b) $\dfrac{\partial^3 w}{\partial x \, \partial y \, \partial z}$ (c) $\dfrac{\partial^4 w}{\partial z^2 \, \partial y \, \partial x}$.

53. Show that the functions in parts (a)–(c) satisfy *Laplace's equation*

$$\frac{\partial^2 f}{\partial x^2} + \frac{\partial^2 f}{\partial y^2} = 0$$

(a) $f(x, y) = e^x \sin y + e^y \cos x$

(b) $f(x, y) = \ln (x^2 + y^2)$

(c) $f(x, y) = \tan^{-1} \dfrac{2xy}{x^2 - y^2}$.

54. Show that $u(x, y)$ and $v(x, y)$ satisfy

$$\frac{\partial u}{\partial x} = \frac{\partial v}{\partial y} \quad \text{and} \quad \frac{\partial u}{\partial y} = -\frac{\partial v}{\partial x}$$

(These are called the **Cauchy–Riemann equations**.)

(a) $u = x^2 - y^2$, $v = 2xy$

(b) $u = e^x \cos y$, $v = e^x \sin y$

(c) $u = \ln (x^2 + y^2)$, $v = 2 \tan^{-1} (y/x)$.

55. A point moves along the intersection of the elliptic paraboloid $z = x^2 + 3y^2$ and the plane $x = 2$. At what rate is z changing with y when the point is at $(2, 1, 7)$?

56. A point moves along the intersection of the surface $z = \sqrt{29 - x^2 - y^2}$ and the plane $y = 3$. At what rate is z changing with x when the point is at $(4, 3, 2)$?

57. Find the slope of the tangent line at $(-1, 1, 5)$ to the curve of intersection of the surface $z = x^2 + 4y^2$ and

(a) the plane $x = -1$ (b) the plane $y = 1$.

58. Find the slope of the tangent line at $(2, 1, 2)$ to the curve of intersection of the surface $x^2 + y^2 + z^2 = 9$ and

(a) the plane $x = 2$ (b) the plane $y = 1$.

59. The volume V of a right-circular cylinder is given by $V = \pi r^2 h$, where r is the radius and h is the height.

(a) Find a formula for the instantaneous rate of change of V with respect to r if h remains constant.

(b) Find a formula for the instantaneous rate of change of V with respect to h if r remains constant.

(c) Suppose that h has a constant value of 4 in., but r varies. Find the rate of change of V with respect to r at the instant when $r = 6$ in.

(d) Suppose that r has a constant value of 8 in., but h varies. Find the instantaneous rate of change of V with respect to h at the instant when $h = 10$ in.

60. The volume V of a right-circular cone is given by

$$V = \frac{\pi}{24} d^2 \sqrt{4s^2 - d^2}$$

where s is the slant height and d is the diameter of the base.

(a) Find a formula for the instantaneous rate of change of V with respect to s if d remains constant.

(b) Find a formula for the instantaneous rate of change of V with respect to d if s remains constant.

(c) Suppose that d has a constant value of 16 cm, but s varies. Find the rate of change of V with respect to s at the instant when $s = 10$ cm.

(d) Suppose that s has a constant value of 10 cm, but d varies. Find the rate of change of V with respect to d at the instant when $d = 16$ cm.

61. According to the ideal gas law, the pressure, temperature, and volume of a gas are related by $P = kT/V$. Suppose that for a certain gas, $k = 10$.

(a) Find the instantaneous rate of change of pressure (lb/in^2) with respect to temperature if the temperature is 80 K and the volume remains fixed at 50 in^3.

(b) Find the instantaneous rate of change of volume with respect to pressure if the volume is 50 in^3 and the temperature remains fixed at 80 K.

62. Find parametric equations for the tangent line at $(1, 3, 3)$ to the curve of intersection of the surface $z = x^2 y$ and

(a) the plane $x = 1$ (b) the plane $y = 3$.

63. Suppose that $\sin (x + z) + \sin (x - y) = 1$. Use implicit differentiation to find $\partial z/\partial x$, $\partial z/\partial y$, and $\partial^2 z/\partial x \, \partial y$ in terms of x, y, and z.

64. The volume of a right-circular cone of radius r and height h is $V = \frac{1}{3}\pi r^2 h$. Show that if the height remains constant while the radius changes, then the volume satisfies

$$\frac{\partial V}{\partial r} = \frac{2V}{r}$$

65. The temperature at a point (x, y) on a metal plate in the xy-plane is $T(x, y) = x^3 + 2y^2 + x$ degrees. Assume that distance is measured in centimeters and find the rate at which temperature changes with distance if we start at the point $(1, 2)$ and move

(a) to the right and parallel to the x-axis

(b) upward and parallel to the y-axis.

66. When two resistors having resistances R_1 ohms and R_2 ohms are connected in parallel, their combined resistance R in ohms is $R = R_1 R_2/(R_1 + R_2)$. Show:

$$\frac{\partial^2 R}{\partial R_1^2} \frac{\partial^2 R}{\partial R_2^2} = \frac{4R^2}{(R_1 + R_2)^4}$$

67. Prove: If $u(x, y)$ and $v(x, y)$ each have equal mixed second partials, and if u and v satisfy the Cauchy–Riemann equations (Exercise 54), then u and v both satisfy Laplace's equation (Exercise 53).

68. Recall that for a function of one variable the derivative $f'(x)$ can be expressed as the limit

$$f'(x) = \lim_{h \to 0} \frac{f(x + h) - f(x)}{h}$$

Express $f_x(x, y)$ and $f_y(x, y)$ as limits.

In Exercises 69 and 70, the figures show some of the isotherms (curves of constant temperature) of a temperature function $T(x, y)$ for a thin metal plate in the xy-plane. Based on the figures, determine at the point P

(a) the signs of $\partial T/\partial x$ and $\partial T/\partial y$, and which is largest in absolute value

(b) the signs of $\partial^2 T/\partial x^2$, $\partial^2 T/\partial y^2$, $\partial^2 T/\partial y \, \partial x$, and $\partial^2 T/\partial x \, \partial y$.

69.

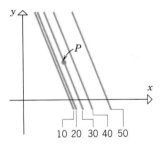

70.

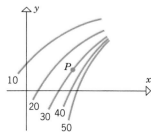

71. Let $f(x, y) = (x^2 + y^2)^{2/3}$. Show that

$$f_x(x, y) = \begin{cases} \dfrac{4x}{3(x^2 + y^2)^{1/3}}, & (x, y) \neq (0, 0) \\ 0, & (x, y) = (0, 0) \end{cases}$$

[This problem, due to Don Cohen, appeared in *Mathematics and Computer Education*, Vol. 25, No. 2, 1991, p. 179.]

72. Let $f(x, y) = (x^3 + y^3)^{1/3}$.

(a)   Show that $f_y(0, 0) = 1$.

(b)   At what points, if any, does $f_y(x, y)$ fail to exist?

■ **16.4   DIFFERENTIABILITY AND CHAIN RULES FOR FUNCTIONS OF TWO VARIABLES**

> *In this section we shall extend the notion of differentiability to functions of two variables and derive versions of the chain rule for these functions. We have restricted the discussion in this section to functions of two variables because some of the results we shall discuss have geometric interpretations that only apply to such functions. In a later section we shall extend the concepts developed here to functions of three or more variables.*

□ **DIFFERENTIABILITY OF FUNCTIONS OF TWO VARIABLES**

Recall that a function f of one variable is called differentiable at x_0 if it has a derivative at x_0, or, in other words, if the limit

$$f'(x_0) = \lim_{\Delta x \to 0} \frac{f(x_0 + \Delta x) - f(x_0)}{\Delta x} \tag{1}$$

exists. A function f that is differentiable at a point x_0 enjoys two important properties:

- $f(x)$ is continuous at x_0.
- The curve $y = f(x)$ has a nonvertical tangent line at x_0.

Our primary objective in this section is to extend the notion of differentiability to functions of two variables in such a way that the natural analogs of these two properties hold. More precisely, when $f(x, y)$ is differentiable at (x_0, y_0), we shall want it to be the case that

- $f(x, y)$ is continuous at (x_0, y_0).
- The surface $z = f(x, y)$ has a nonvertical tangent plane at (x_0, y_0) (Figure 16.4.1). (A precise definition of a tangent plane will be given later.)

It would not be unreasonable to guess that a function f of two variables should be called differentiable at (x_0, y_0) if the two partial derivatives $f_x(x_0, y_0)$ and $f_y(x_0, y_0)$ exist at

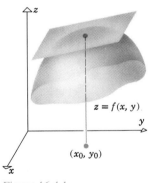

Figure 16.4.1

(x_0, y_0). Unfortunately, this condition is not strong enough to meet our objectives, since there are functions that have partial derivatives at a point, but are not continuous at that point. For example, the function

$$f(x, y) = \begin{cases} -1 & \text{if } x > 0 \text{ and } y > 0 \\ 0 & \text{otherwise} \end{cases}$$

is discontinuous at $(0, 0)$, but has partial derivatives at $(0, 0)$; these derivatives are $f_x(0, 0) = 0$ and $f_y(0, 0) = 0$. The discontinuity and the values of the partial derivatives should be evident from Figure 16.4.2.

To motivate an appropriate definition of differentiability for functions of two variables, it will be helpful to reexamine the concept of differentiability for functions of one variable. Assuming, for the moment, that f is a function of one variable that is differentiable at $x = x_0$, it follows from Formula (8) of Section 3.7 that (1) can be rewritten as

$$f'(x_0) = \lim_{\Delta x \to 0} \frac{\Delta f}{\Delta x} \tag{2}$$

or, equivalently, as

$$\lim_{\Delta x \to 0} \left[\frac{\Delta f}{\Delta x} - f'(x_0) \right] = 0 \tag{3}$$

where

$$\Delta f = f(x_0 + \Delta x) - f(x_0)$$

If we now define ϵ by

$$\epsilon = \frac{\Delta f}{\Delta x} - f'(x_0) \tag{4}$$

then it follows from this formula that

$$\Delta f = f'(x_0) \Delta x + \epsilon \Delta x \tag{5}$$

where ϵ is a function of Δx. Using (4), the limit in (3) can be rewritten as

$$\lim_{\Delta x \to 0} \epsilon = 0 \tag{6}$$

Formulas (5) and (6) suggest the following alternative definition of differentiability for functions of one variable.

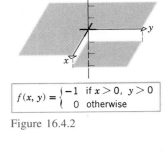

$$f(x, y) = \begin{cases} -1 & \text{if } x > 0,\ y > 0 \\ 0 & \text{otherwise} \end{cases}$$

Figure 16.4.2

16.4.1   DEFINITION.   A function f of one variable is said to be **differentiable** at x_0 if there exists a number $f'(x_0)$ such that Δf can be written in the form

$$\Delta f = f'(x_0) \Delta x + \epsilon \Delta x \tag{7}$$

where ϵ is a function of Δx such that $\epsilon \to 0$ as $\Delta x \to 0$.

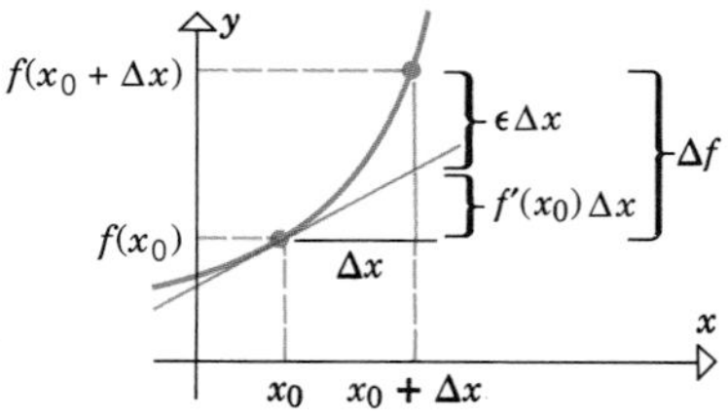

Although this definition of differentiability is more complicated than that given earlier in the text, it provides the basis for extending the notion of differentiability to functions of two or more variables. A geometric interpretation of the terms appearing in (7) is shown in Figure 16.4.3. The term Δf represents the change in height that results when a point moves along the graph of f as the x-coordinate changes from x_0 to $x_0 + \Delta x$; the term $f'(x_0) \Delta x$ represents the change in height that results when a point moves along the tangent line at $(x_0, f(x_0))$ as the x-coordinate changes from x_0 to $x_0 + \Delta x$; finally, the term $\epsilon \Delta x$ represents the difference between Δf and $f'(x_0) \Delta x$. It is evident from Figure 16.4.3 that $\epsilon \Delta x \to 0$ as $\Delta x \to 0$. However, (7) actually makes the stronger statement that $\epsilon \to 0$ as $\Delta x \to 0$. This is not at all evident from Figure 16.4.3, but it follows from (6).

Figure 16.4.3

If f is a function of x and y, then the symbol Δf, called the **increment** of f, denotes the change in the value of $f(x, y)$ that results when (x, y) varies from some initial position (x_0, y_0) to some new position $(x_0 + \Delta x, y_0 + \Delta y)$; thus,

$$\Delta f = f(x_0 + \Delta x, y_0 + \Delta y) - f(x_0, y_0) \tag{8}$$

(See Figure 16.4.4.) If a dependent variable $z = f(x, y)$ is used, then we shall sometimes write Δz rather than Δf.

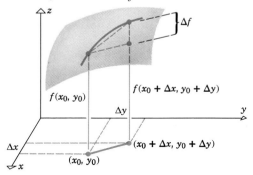

Figure 16.4.4

Motivated by Definition 16.4.1, we now make the following definition of differentiability for functions of two variables.

16.4.2 DEFINITION. A function f of two variables is said to be **differentiable** at (x_0, y_0) if $f_x(x_0, y_0)$ and $f_y(x_0, y_0)$ exist and Δf can be written in the form

$$\Delta f = f_x(x_0, y_0)\, \Delta x + f_y(x_0, y_0)\, \Delta y + \epsilon_1\, \Delta x + \epsilon_2\, \Delta y \tag{9}$$

where ϵ_1 and ϵ_2 are functions of Δx and Δy such that $\epsilon_1 \to 0$ and $\epsilon_2 \to 0$ as $(\Delta x, \Delta y) \to (0, 0)$.

A function is called **differentiable on a region** R of the xy-plane if it is differentiable at each point of R. A function that is differentiable on the entire xy-plane is called **everywhere differentiable** or simply **differentiable**.

REMARK. Before proceeding further, we note that for functions of one variable, the terms "differentiable" and "has a derivative" are synonymous. However, for functions of two variables, differentiability requires more than the existence of partial derivatives.

□ **RELATIONSHIP BETWEEN DIFFERENTIABILITY AND CONTINUITY**

Earlier, we set two goals for our definition of differentiability: We wanted a function that is differentiable at (x_0, y_0) to be continuous at (x_0, y_0), and we wanted its graph to have a nonvertical tangent plane at (x_0, y_0). The next theorem shows that the continuity criterion is met; the existence of a nonvertical tangent plane will be demonstrated in the next section.

16.4.3 THEOREM. *If f is differentiable at (x_0, y_0), then f is continuous at (x_0, y_0).*

Proof. We must prove that

$$\lim_{(x, y) \to (x_0, y_0)} f(x, y) = f(x_0, y_0)$$

which, on letting $x = x_0 + \Delta x$ and $y = y_0 + \Delta y$, is equivalent to

$$\lim_{(\Delta x, \Delta y) \to (0, 0)} f(x_0 + \Delta x, y_0 + \Delta y) = f(x_0, y_0)$$

which from (8) is equivalent to

$$\lim_{(\Delta x, \Delta y) \to (0,0)} \Delta f = 0$$

But f is assumed to be differentiable at (x_0, y_0), so it follows from (9) that

$$\Delta f = f_x(x_0, y_0) \Delta x + f_y(x_0, y_0) \Delta y + \epsilon_1 \Delta x + \epsilon_2 \Delta y$$

where $\epsilon_1 \to 0$, $\epsilon_2 \to 0$ as $(\Delta x, \Delta y) \to (0, 0)$. Thus,

$$\lim_{(\Delta x, \Delta y) \to (0,0)} \Delta f = \lim_{(\Delta x, \Delta y) \to (0,0)} [f_x(x_0, y_0) \Delta x + f_y(x_0, y_0) \Delta y + \epsilon_1 \Delta x + \epsilon_2 \Delta y] = 0$$

which completes the proof. ▮

The next theorem, whose proof is usually studied in advanced calculus courses, provides simple conditions for a function of two variables to be differentiable at a point.

> **16.4.4** THEOREM. *If f has first-order partial derivatives at each point in some circular region centered at (x_0, y_0), and if these partial derivatives are continuous at (x_0, y_0), then f is differentiable at (x_0, y_0).*

Example 1 Show that $f(x, y) = x^3 y^4$ is a differentiable function.

Solution. The partial derivatives $f_x = 3x^2 y^4$ and $f_y = 4x^3 y^3$ are defined and continuous everywhere in the xy-plane. Thus, the hypotheses of Theorem 16.4.4 are satisfied at each point (x_0, y_0) in the xy-plane, so $f(x, y) = x^3 y^4$ is everywhere differentiable. ◀

The following important corollary follows from Theorems 16.4.3 and 16.4.4.

> **16.4.5** COROLLARY. *If f has first-order partial derivatives at each point of some circular region centered at (x_0, y_0), and if these partial derivatives are continuous at (x_0, y_0), then f is continuous at (x_0, y_0).*

☐ **EQUALITY OF MIXED PARTIALS**

The following theorem, which we state without proof, shows that with appropriate continuity restrictions the mixed second-order partial derivatives of a function of two variables are equal.

> **16.4.6** THEOREM. *Let f be a function of two variables. If f_x, f_y, f_{xy}, and f_{yx} are continuous on an open set, then $f_{xy} = f_{yx}$ at each point of the set.*

Example 2 Let $f(x, y) = 2e^{xy} \sin y$. It should be evident from the form of this function and from Theorem 16.2.4 that f and all its partial derivatives are continuous everywhere. Thus, Theorem 16.4.6 guarantees that $f_{xy} = f_{yx}$ everywhere. This is, in fact, the case since

$$f_x(x, y) = 2ye^{xy} \sin y = (2y \sin y)e^{xy}$$

$$f_{xy}(x, y) = (2y \sin y)(xe^{xy}) + e^{xy}(2y \cos y + 2 \sin y)$$

$$= 2e^{xy}(xy \sin y + y \cos y + \sin y)$$

$$f_y(x, y) = 2e^{xy} \cos y + 2xe^{xy} \sin y$$

$$f_{yx}(x, y) = 2ye^{xy} \cos y + 2xye^{xy} \sin y + 2e^{xy} \sin y$$

$$= 2e^{xy}(xy \sin y + y \cos y + \sin y)$$

so $f_{xy}(x, y) = f_{yx}(x, y)$ for all (x, y). ◀

In general, the order of differentiation in an nth-order partial derivative can be changed without affecting the final result whenever the function and all its partial derivatives of order n or less are continuous. For example, if f and its partial derivatives of the first, second, and third orders are continuous on an open set, then at each point of that set,

$$f_{xyy} = f_{yxy} = f_{yyx}$$

or in another notation,

$$\frac{\partial^3 f}{\partial y^2 \, \partial x} = \frac{\partial^3 f}{\partial y \, \partial x \, \partial y} = \frac{\partial^3 f}{\partial x \, \partial y^2}$$

☐ **CHAIN RULES**

If y is a differentiable function of x and x is a differentiable function of t, then the chain rule for functions of one variable states that

$$\frac{dy}{dt} = \frac{dy}{dx}\frac{dx}{dt}$$

We shall now derive versions of the chain rule for functions of two variables.

Assume that z is a function of x and y, say

$$z = f(x, y) \tag{10}$$

and suppose that x and y, in turn, are functions of a single variable t, say

$$x = x(t), \quad y = y(t)$$

On substituting these functions of t in (10), we obtain the relationship

$$z = f(x(t), y(t))$$

which expresses z as a function of the single variable t. Thus, we may ask for the derivative dz/dt, and we may inquire about its relationship to the derivatives $\partial z/\partial x$, $\partial z/\partial y$, dx/dt, and dy/dt.

16.4.7 THEOREM (*Chain Rule*). *If $x = x(t)$ and $y = y(t)$ are differentiable at t, and if $z = f(x, y)$ is differentiable at the point $(x(t), y(t))$, then $z = f(x(t), y(t))$ is differentiable at t, and*

$$\frac{dz}{dt} = \frac{\partial z}{\partial x}\frac{dx}{dt} + \frac{\partial z}{\partial y}\frac{dy}{dt} \tag{11}$$

Proof. From the derivative definition for functions of one variable

$$\frac{dz}{dt} = \lim_{\Delta t \to 0} \frac{\Delta z}{\Delta t} \tag{12}$$

Since $z = f(x, y)$ is differentiable at the point $(x, y) = (x(t), y(t))$, we can express Δz in the form

$$\Delta z = \frac{\partial z}{\partial x}\Delta x + \frac{\partial z}{\partial y}\Delta y + \epsilon_1 \Delta x + \epsilon_2 \Delta y \tag{13}$$

where the partial derivatives are evaluated at the point $(x(t), y(t))$ and $\epsilon_1 \to 0$, $\epsilon_2 \to 0$ as $(\Delta x, \Delta y) \to (0, 0)$. Thus, from (12) and (13),

$$\frac{dz}{dt} = \lim_{\Delta t \to 0} \frac{\Delta z}{\Delta t} = \lim_{\Delta t \to 0} \left[\frac{\partial z}{\partial x}\frac{\Delta x}{\Delta t} + \frac{\partial z}{\partial y}\frac{\Delta y}{\Delta t} + \epsilon_1 \frac{\Delta x}{\Delta t} + \epsilon_2 \frac{\Delta y}{\Delta t} \right] \tag{14}$$

But

$$\lim_{\Delta t \to 0} \frac{\Delta x}{\Delta t} = \frac{dx}{dt} \quad \text{and} \quad \lim_{\Delta t \to 0} \frac{\Delta y}{\Delta t} = \frac{dy}{dt}$$

Therefore, if we can show that $\epsilon_1 \to 0$, $\epsilon_2 \to 0$ as $\Delta t \to 0$, then the proof will be complete, since (14) will reduce to (11). But $\Delta x \to 0$ and $\Delta y \to 0$ as $\Delta t \to 0$, since

$$\lim_{\Delta t \to 0} \Delta x = \lim_{\Delta t \to 0} \frac{\Delta x}{\Delta t} \Delta t = \frac{dx}{dt} \cdot 0 = 0$$

and similarly for Δy. Thus, as Δt tends to zero, $(\Delta x, \Delta y) \to (0, 0)$, which implies that $\epsilon_1 \to 0$, $\epsilon_2 \to 0$. ∎

Example 3   Suppose that

$$z = x^2 y, \quad x = t^2, \quad y = t^3$$

Use the chain rule to find dz/dt, and check the result by expressing z as a function of t and differentiating directly.

Solution.   By the chain rule

$$\frac{dz}{dt} = \frac{\partial z}{\partial x}\frac{dx}{dt} + \frac{\partial z}{\partial y}\frac{dy}{dt} = (2xy)(2t) + (x^2)(3t^2)$$

$$= (2t^5)(2t) + (t^4)(3t^2) = 7t^6$$

Alternatively, we may express z directly as a function of t,

$$z = x^2 y = (t^2)^2(t^3) = t^7$$

and then differentiate to obtain $dz/dt = 7t^6$. However, this procedure is not always convenient. ◀

Example 4   Suppose that

$$z = \sqrt{xy + y}, \quad x = \cos \theta, \quad y = \sin \theta$$

Use the chain rule to find $dz/d\theta$ when $\theta = \pi/2$.

Solution.   From the chain rule with θ in place of t,

$$\frac{dz}{d\theta} = \frac{\partial z}{\partial x}\frac{dx}{d\theta} + \frac{\partial z}{\partial y}\frac{dy}{d\theta}$$

we obtain

$$\frac{dz}{d\theta} = \frac{1}{2}(xy + y)^{-1/2}(y)(-\sin \theta) + \frac{1}{2}(xy + y)^{-1/2}(x + 1)(\cos \theta)$$

When $\theta = \pi/2$, we have

$$x = \cos \frac{\pi}{2} = 0, \quad y = \sin \frac{\pi}{2} = 1$$

Substituting $x = 0$, $y = 1$, $\theta = \pi/2$ in the formula for $dz/d\theta$ yields

$$\frac{dz}{d\theta}\bigg|_{\theta = \pi/2} = \frac{1}{2}(1)(1)(-1) + \frac{1}{2}(1)(1)(0) = -\frac{1}{2} \quad ◀$$

REMARK.   There are many variations in derivative notations, each of which gives the chain rule a different look. If $z = f(x, y)$, where x and y are functions of t, then some possibilities are

$$\frac{dz}{dt} = f_x\frac{dx}{dt} + f_y\frac{dy}{dt}$$

$$\frac{df}{dt} = \frac{\partial f}{\partial x}\frac{dx}{dt} + \frac{\partial f}{\partial y}\frac{dy}{dt}$$

$$\frac{df}{dt} = f_x x'(t) + f_y y'(t)$$

The reader may be able to construct other variations as well.

In the special case where $z = F(x, y)$ and y is a differentiable function of x, chain-rule formula (11) yields

$$\frac{dz}{dx} = \frac{\partial F}{\partial x}\frac{dx}{dx} + \frac{\partial F}{\partial y}\frac{dy}{dx} = \frac{\partial F}{\partial x} + \frac{\partial F}{\partial y}\frac{dy}{dx} \tag{15}$$

This result can be used to find derivatives of functions that are defined implicitly. Suppose that the equation

$$F(x, y) = 0 \tag{16}$$

defines y implicitly as a differentiable function of x, and we are interested in finding dy/dx. Differentiating both sides of (16) with respect to x and applying (15) yields

$$\frac{\partial F}{\partial x} + \frac{\partial F}{\partial y}\frac{dy}{dx} = 0$$

or

$$\frac{dy}{dx} = -\frac{\partial F/\partial x}{\partial F/\partial y} \tag{17}$$

provided $\partial F/\partial y \neq 0$.

Example 5   Given that

$$x^3 + y^2 x - 3 = 0$$

find dy/dx using (17) and check the result using implicit differentiation.

Solution.   By (17) with $F(x, y) = x^3 + y^2 x - 3$

$$\frac{dy}{dx} = -\frac{\partial F/\partial x}{\partial F/\partial y} = -\frac{3x^2 + y^2}{2yx}$$

On the other hand, differentiating the given equation implicitly yields

$$3x^2 + y^2 + x\left(2y\frac{dy}{dx}\right) - 0 = 0 \quad \text{or} \quad \frac{dy}{dx} = -\frac{3x^2 + y^2}{2yx}$$

which agrees with the result obtained by (17).   ◄

In Theorem 16.4.7 the variables x and y are each functions of a single variable t. We now consider the case where x and y are each functions of two variables. Let

$$z = f(x, y) \tag{18}$$

and suppose that x and y are functions of u and v, say

$$x = x(u, v), \quad y = y(u, v)$$

On substituting these functions of u and v into (18), we obtain the relationship

$$z = f(x(u, v), y(u, v))$$

which expresses z as a function of the two variables u and v. Thus, we may ask for the partial derivatives $\partial z/\partial u$ and $\partial z/\partial v$; and we may inquire about the relationship between these derivatives and the derivatives $\partial z/\partial x$, $\partial z/\partial y$, $\partial x/\partial u$, $\partial x/\partial v$, $\partial y/\partial u$, and $\partial y/\partial v$.

16.4.8   THEOREM (*Chain Rule*).   *If $x = x(u, v)$ and $y = y(u, v)$ have first-order partial derivatives at the point (u, v), and if $z = f(x, y)$ is differentiable at the point $(x(u, v), y(u, v))$, then $z = f(x(u, v), y(u, v))$ has first-order partial derivatives at (u, v) given by*

$$\frac{\partial z}{\partial u} = \frac{\partial z}{\partial x}\frac{\partial x}{\partial u} + \frac{\partial z}{\partial y}\frac{\partial y}{\partial u} \qquad \text{and} \qquad \frac{\partial z}{\partial v} = \frac{\partial z}{\partial x}\frac{\partial x}{\partial v} + \frac{\partial z}{\partial y}\frac{\partial y}{\partial v}$$

Proof.   If v is held fixed, then $x = x(u, v)$ and $y = y(u, v)$ become functions of u alone. Thus, we are back to the case of Theorem 16.4.7. If we apply that theorem with u in place of t, and if we use ∂ rather than d to indicate that the variable v is fixed, we obtain

$$\frac{\partial z}{\partial u} = \frac{\partial z}{\partial x}\frac{\partial x}{\partial u} + \frac{\partial z}{\partial y}\frac{\partial y}{\partial u}$$

The formula for $\partial z / \partial v$ is derived similarly.   ∎

Example 6   Given that

$$z = e^{xy}, \quad x = 2u + v, \quad y = u/v$$

find $\partial z / \partial u$ and $\partial z / \partial v$ using the chain rule.

Solution.

$$\frac{\partial z}{\partial u} = \frac{\partial z}{\partial x}\frac{\partial x}{\partial u} + \frac{\partial z}{\partial y}\frac{\partial y}{\partial u} = (ye^{xy})(2) + (xe^{xy})(1/v) = \left[2y + \frac{x}{v}\right]e^{xy}$$

$$= \left[\frac{2u}{v} + \frac{2u + v}{v}\right]e^{(2u+v)(u/v)} = \left[\frac{4u}{v} + 1\right]e^{(2u+v)(u/v)}$$

$$\frac{\partial z}{\partial v} = \frac{\partial z}{\partial x}\frac{\partial x}{\partial v} + \frac{\partial z}{\partial y}\frac{\partial y}{\partial v} = (ye^{xy})(1) + (xe^{xy})\left(-\frac{u}{v^2}\right)$$

$$= \left[y - x\left(\frac{u}{v^2}\right)\right]e^{xy} = \left[\frac{u}{v} - (2u + v)\left(\frac{u}{v^2}\right)\right]e^{(2u+v)(u/v)}$$

$$= -\frac{2u^2}{v^2}e^{(2u+v)(u/v)} \qquad ◀$$

The chain rules are useful in related rates problems.

Example 7   At what rate is the area of a rectangle changing if its length is 15 ft and increasing at 3 ft/sec while its width is 6 ft and increasing at 2 ft/sec?

Solution.   Let

$x = $ length of the rectangle in feet

$y = $ width of the rectangle in feet

$A = $ area of the rectangle in square feet

$t = $ time in seconds

We are given that

$$\frac{dx}{dt} = 3 \quad \text{and} \quad \frac{dy}{dt} = 2 \tag{19}$$

at the instant when

$$x = 15, \quad y = 6 \tag{20}$$

We want to find dA/dt at that instant.

From the area formula $A = xy$, we obtain

$$\frac{dA}{dt} = \frac{\partial A}{\partial x}\frac{dx}{dt} + \frac{\partial A}{\partial y}\frac{dy}{dt} = y\frac{dx}{dt} + x\frac{dy}{dt}$$

Substituting (19) and (20) in this equation yields

$$\frac{dA}{dt} = 6(3) + 15(2) = 48$$

Thus, the area is increasing at a rate of 48 ft^2/sec at the given instant.   ◀

▶ Exercise Set 16.4

1. Find Δf given that $f(x, y) = x^2 y$, $(x_0, y_0) = (1, 3)$, $\Delta x = 0.1$, and $\Delta y = 0.2$.

2. Find Δz given that $z = 3x^2 - 2y$, $(x_0, y_0) = (-2, 4)$, $\Delta x = 0.02$, and $\Delta y = -0.03$.

3. Find the increment of $f(x, y) = x/y$ as (x, y) varies from $(-1, 2)$ to $(3, 1)$.

4. Find the increment of $g(u, v) = 2uv - v^3$ as (u, v) varies from $(0, 1)$ to $(4, -2)$.

In Exercises 5–8, use Definition 16.4.2 to establish the differentiability of the given function. [*Remark:* ϵ_1 and ϵ_2 are not unique.]

5. $f(x, y) = xy$.   6. $f(x, y) = x^2 + y^2$.

7. $f(x, y) = x^2 y$.   8. $f(x, y) = 3x + y^2$.

9. Let $f(x, y) = \sqrt{x^2 + y^2}$.

   (a) Show that f is continuous at $(0, 0)$.

   (b) Show that $f_x(0, 0)$ does not exist, and hence that f is not differentiable at $(0, 0)$. [*Hint:* Express $f_x(0, 0)$ as a limit (see Exercise 68, Section 16.3).]

10. Let

$$f(x, y) = \begin{cases} 5 - 3x - 2y, & x \ge 0 \text{ or } y \ge 0 \\ 0, & x < 0 \text{ and } y < 0 \end{cases}$$

Show that $f_x(0, 0)$ and $f_y(0, 0)$ exist, but f is not continuous at $(0, 0)$.

11. Let

$$f(x, y) = \begin{cases} \dfrac{xy}{x^2 + y^2}, & (x, y) \ne (0, 0) \\ 0, & (x, y) = (0, 0) \end{cases}$$

Prove: $f_x(0, 0)$ and $f_y(0, 0)$ exist, but f is not continuous at $(0, 0)$. [*Hint:* Show that $f_x(0, 0) = 0$ and $f_y(0, 0) = 0$ by expressing these derivatives as limits (see Exercise 68, Section 16.3). To prove that f is not continuous at $(0, 0)$, show that

$$\lim_{(x,y)\to(0,0)} f(x, y)$$

does not exist by letting $(x, y) \to (0, 0)$ along $y = 0$ and along $y = x$.]

In Exercises 12–16, find f_{xy} and f_{yx} and verify their equality.

12. $f(x, y) = 2xy - 3y^2$.   13. $f(x, y) = 4x^3 y + 3x^2 y$.

14. $f(x, y) = x^3/y$.   15. $f(x, y) = \sin(x^2 + y^3)$.

16. $f(x, y) = \sqrt{x^2 + y^2 - 1}$.

17. Let f be a function of two variables with continuous third- and fourth-order partial derivatives.

    (a) How many of the third-order partial derivatives can be distinct?

    (b) How many of the fourth order?

18. Let $f(x, y) = e^{xy^2}$. Find f_{xyx}, f_{xxy}, and f_{yxx} and verify their equality.

In Exercises 19–24, find dz/dt using the chain rule.

19. $z = 3x^2 y^3$; $x = t^4$, $y = t^2$.

20. $z = \ln(2x^2 + y)$; $x = \sqrt{t}$, $y = t^{2/3}$.

21. $z = 3\cos x - \sin xy$; $x = 1/t$, $y = 3t$.

22. $z = \sqrt{1 + x - 2xy^4}$; $x = \ln t$, $y = t$.

23. $z = e^{1-xy}$; $x = t^{1/3}$, $y = t^3$.

24. $z = \cosh^2 xy$; $x = t/2$, $y = e^t$.

In Exercises 25–31, find $\partial z/\partial u$ and $\partial z/\partial v$ by the chain rule.

25. $z = 8x^2 y - 2x + 3y$; $x = uv$, $y = u - v$.

26. $z = x^2 - y\tan x$; $x = u/v$, $y = u^2 v^2$.

27. $z = x/y$; $x = 2\cos u$, $y = 3\sin v$.

28. $z = 3x - 2y$; $x = u + v\ln u$, $y = u^2 - v\ln v$.

29. $z = e^{x^2 y}$; $x = \sqrt{uv}$, $y = 1/v$.

30. $z = \cos x \sin y$; $x = u - v$, $y = u^2 + v^2$.

31. $z = \tan^{-1}(x^2 + y^2)$; $x = e^u \sin v$, $y = e^u \cos v$.

32. Let $w = rs/(r^2 + s^2)$; $r = uv$, $s = u - 2v$. Use the chain rule to find $\partial w/\partial u$ and $\partial w/\partial v$.

33. Let $T = x^2 y - xy^3 + 2$; $x = r\cos\theta$, $y = r\sin\theta$. Use the chain rule to find $\partial T/\partial r$ and $\partial T/\partial\theta$.

34. Let $R = e^{2s-t^2}$; $s = 3\phi$, $t = \phi^{1/2}$. Use the chain rule to find $dR/d\phi$.

35. Let $t = u/v$; $u = x^2 - y^2$, $v = 4xy^3$. Use the chain rule to find $\partial t/\partial x$ and $\partial t/\partial y$.

36. Use the chain rule to find $\left.\dfrac{dz}{dt}\right|_{t=3}$ if $z = x^2y$; $x = t^2$, $y = t + 7$.

37. Use the chain rule to find the value of $\left.\dfrac{dw}{ds}\right|_{s=1/4}$ if $w = r^2 - r \tan \theta$; $r = \sqrt{s}$, $\theta = \pi s$.

38. Use the chain rule to find the value of

$$\left.\frac{\partial f}{\partial u}\right|_{u=1,\, v=-2} \quad \text{and} \quad \left.\frac{\partial f}{\partial v}\right|_{u=1,\, v=-2}$$

if $f(x, y) = x^2y^2 - x + 2y$; $x = \sqrt{u}$, $y = uv^3$.

39. Use the chain rule to find the value of

$$\left.\frac{\partial z}{\partial r}\right|_{r=2,\, \theta=\pi/6} \quad \text{and} \quad \left.\frac{\partial z}{\partial \theta}\right|_{r=2,\, \theta=\pi/6}$$

if $z = xye^{x/y}$; $x = r \cos \theta$, $y = r \sin \theta$.

In Exercises 40–43, use (17) to find dy/dx and check your result using implicit differentiation.

---

40. $x^2y^3 + \cos y = 0$.   **41.** $x^3 - 3xy^2 + y^3 = 5$.

42. $e^{xy} + ye^y = 1$.   **43.** $x - \sqrt{xy} + 3y = 4$.

44. The portion of a tree that is usable for lumber may be viewed as a right-circular cylinder. If the height of a tree increases 2 ft per year and the diameter increases 3 in. per year, how fast is the volume of usable lumber increasing when the tree is 20 ft high and the diameter is 30 in.?

45. Two straight roads intersect at right angles. Car A, moving on one of the roads, approaches the intersection at 25 mi/hr and car B, moving on the other road, approaches the intersection at 30 mi/hr. At what rate is the distance between the cars changing when A is 0.3 mile from the intersection and B is 0.4 mile from the intersection?

46. Use the ideal gas law, $P = kT/V$, with $k = 10$ to find the rate at which the temperature of a gas is changing when the volume is 200 in^3 and increasing at the rate of 4 in^3/sec, while the pressure is 5 lb/in^2 and decreasing at the rate of 1 lb/in^2/sec.

47. Two sides of a triangle have lengths $a = 4$ cm and $b = 3$ cm, but are increasing at the rate of 1 cm/sec. If the area of the triangle remains constant, at what rate is the angle θ between a and b changing when $\theta = \pi/6$?

48. Two sides of a triangle have lengths $a = 5$ cm and $b = 10$ cm, and the included angle is $\theta = \pi/3$. If a is increasing at a rate of 2 cm/sec, b is increasing at a rate of 1 cm/sec, and θ remains constant, at what rate is the third side changing? Is it increasing or decreasing? [*Hint:* Use the law of cosines.]

49. Suppose that a particle moving along a metal plate in the xy-plane has velocity $\mathbf{v} = \mathbf{i} - 4\mathbf{j}$ (cm/sec) at the point

$(3, 2)$. If the temperature of the plate at points in the xy-plane is $T(x, y) = y^2 \ln x$, $x \geq 1$, in degrees Celsius, find dT/dt at $(3, 2)$.

In Exercises 50–67, assume that the derivatives satisfy all continuity requirements needed for the problem.

---

50. Let $z = f(u)$ where $u = g(x, y)$. Express $\partial z/\partial x$ and $\partial z/\partial y$ in terms of dz/du, $\partial u/\partial x$, and $\partial u/\partial y$.

51. Let $z = f(x^2 - y^2)$. Show that $y\partial z/\partial x + x\partial z/\partial y = 0$.

52. Let $z = f(xy)$. Show that $x\,\partial z/\partial x - y\,\partial z/\partial y = 0$.

53. Let $z = f(u)$ where $u = g(x, y)$. Show that

(a) $\dfrac{\partial^2 z}{\partial x^2} = \dfrac{dz}{du}\dfrac{\partial^2 u}{\partial x^2} + \dfrac{d^2z}{du^2}\left(\dfrac{\partial u}{\partial x}\right)^2$

and $\dfrac{\partial^2 z}{\partial y^2} = \dfrac{dz}{du}\dfrac{\partial^2 u}{\partial y^2} + \dfrac{d^2z}{du^2}\left(\dfrac{\partial u}{\partial y}\right)^2$

(b) $\dfrac{\partial^2 z}{\partial y\, \partial x} = \dfrac{dz}{du}\dfrac{\partial^2 u}{\partial y\, \partial x} + \dfrac{d^2z}{du^2}\dfrac{\partial u}{\partial x}\dfrac{\partial u}{\partial y}$.

54. Let $r = \sqrt{x^2 + y^2}$. Show that

(a) $\dfrac{\partial r}{\partial x} = \dfrac{x}{r}$ (b) $\dfrac{\partial r}{\partial y} = \dfrac{y}{r}$

(c) $\dfrac{\partial^2 r}{\partial x^2} = \dfrac{y^2}{r^3}$ (d) $\dfrac{\partial^2 r}{\partial y^2} = \dfrac{x^2}{r^3}$.

55. Let $z = f(r)$ where $r = \sqrt{x^2 + y^2}$, $r \neq 0$.

(a) Use the formulas in Exercises 53 and 54 to show that if z satisfies Laplace's equation, $\dfrac{\partial^2 z}{\partial x^2} + \dfrac{\partial^2 z}{\partial y^2} = 0$, then

$$r\frac{d^2z}{dr^2} + \frac{dz}{dr} = 0.$$

(b) Use the result of part (a) to solve Laplace's equation for $z = f(r)$. [*Hint:* Write

$$r\frac{d^2z}{dr^2} + \frac{dz}{dr}$$

as the derivative of a product.]

56. Let $z = f(x - y, y - x)$. Show that $\partial z/\partial x + \partial z/\partial y = 0$.

57. Let $z = f(y + cx) + g(y - cx)$, where $c \neq 0$. Show that

$$\frac{\partial^2 z}{\partial x^2} = c^2 \frac{\partial^2 z}{\partial y^2}$$

58. Let $z = f(x, y)$ where $x = g(t)$ and $y = h(t)$.

(a) Show that

$$\frac{d}{dt}\left(\frac{\partial z}{\partial x}\right) = \frac{\partial^2 z}{\partial x^2}\frac{dx}{dt} + \frac{\partial^2 z}{\partial y\, \partial x}\frac{dy}{dt}$$

and

$$\frac{d}{dt}\left(\frac{\partial z}{\partial y}\right) = \frac{\partial^2 z}{\partial x\, \partial y}\frac{dx}{dt} + \frac{\partial^2 z}{\partial y^2}\frac{dy}{dt}$$

(b) Use the formulas in part (a) to help find a formula for d^2z/dt^2.

59. Let $z = f(x, y)$ where $x = g(u, v)$ and $y = h(u, v)$. Show that

$$\frac{\partial}{\partial u}\left(\frac{\partial z}{\partial x}\right) = \frac{\partial^2 z}{\partial x^2}\frac{\partial x}{\partial u} + \frac{\partial^2 z}{\partial y \, \partial x}\frac{\partial y}{\partial u}$$

60. Let $z = f(x, y)$ where $x = u + v$ and $y = u - v$. Show that

$$\frac{\partial^2 z}{\partial v \, \partial u} = \frac{\partial^2 z}{\partial x^2} - \frac{\partial^2 z}{\partial y^2}$$

61. The equations $x = r\cos\theta$ and $y = r\sin\theta$, which relate Cartesian and polar coordinates, define r and θ implicitly as functions of x and y.

(a) Use implicit differentiation with respect to the variable x on both equations to show that

$$\frac{\partial r}{\partial x} = \cos\theta \quad \text{and} \quad \frac{\partial\theta}{\partial x} = -\frac{\sin\theta}{r}$$

(b) Use implicit differentiation with respect to y on both equations to show that

$$\frac{\partial r}{\partial y} = \sin\theta \quad \text{and} \quad \frac{\partial\theta}{\partial y} = \frac{\cos\theta}{r}$$

62. Let $z = f(r, \theta)$, where r and θ are defined implicitly as functions of x and y by the equations $x = r\cos\theta$ and $y = r\sin\theta$. Use the results in Exercise 61 to show that

(a) $\dfrac{\partial z}{\partial x} = \dfrac{\partial z}{\partial r}\cos\theta - \dfrac{1}{r}\dfrac{\partial z}{\partial\theta}\sin\theta$

(b) $\dfrac{\partial z}{\partial y} = \dfrac{\partial z}{\partial r}\sin\theta + \dfrac{1}{r}\dfrac{\partial z}{\partial\theta}\cos\theta$.

63. Use the formulas in Exercise 62 to show that

$$\left(\frac{\partial z}{\partial x}\right)^2 + \left(\frac{\partial z}{\partial y}\right)^2 = \left(\frac{\partial z}{\partial r}\right)^2 + \frac{1}{r^2}\left(\frac{\partial z}{\partial\theta}\right)^2$$

64. Use the formulas in Exercises 61 and 62 to show that Laplace's equation

$$\frac{\partial^2 z}{\partial x^2} + \frac{\partial^2 z}{\partial y^2} = 0$$

becomes

$$\frac{\partial^2 z}{\partial r^2} + \frac{1}{r^2}\frac{\partial^2 z}{\partial\theta^2} + \frac{1}{r}\frac{\partial z}{\partial r} = 0$$

in polar coordinates.

65. A function $f(x, y)$ is said to be *homogeneous of degree n* if $f(tx, ty) = t^n f(x, y)$ for $t > 0$. Show that the following functions are homogeneous and find the degree of each.

(a) $f(x, y) = 3x^2 + y^2$

(b) $f(x, y) = \sqrt{x^2 + y^2}$

(c) $f(x, y) = x^2 y - 2y^3$

(d) $f(x, y) = \dfrac{5}{(x^2 + 2y^2)^2}$, $(x, y) \neq (0, 0)$.

66. Show that for a homogeneous function $f(x, y)$ of degree n (see Exercise 65)

$$x\frac{\partial f}{\partial x} + y\frac{\partial f}{\partial y} = nf$$

[*Hint:* Let $u = tx$ and $v = ty$ in $f(tx, ty)$, and differentiate both sides of $f(u, v) = t^n f(x, y)$ with respect to t.]

67. Let $w = f(x, y)$ where $y = g(x, z)$. Taking x and z as the independent variables, express each of the following in terms of $\partial f/\partial x$, $\partial f/\partial y$, $\partial y/\partial x$, and $\partial y/\partial z$.

(a) $\dfrac{\partial w}{\partial x}$ \qquad\qquad (b) $\dfrac{\partial w}{\partial z}$.

68. Show that if $u(x, y)$ and $v(x, y)$ satisfy the Cauchy–Riemann equations (Exercise 54, Section 16.3), and if $x = r\cos\theta$ and $y = r\sin\theta$, then

$$\frac{\partial u}{\partial r} = \frac{1}{r}\frac{\partial v}{\partial\theta} \quad \text{and} \quad \frac{\partial v}{\partial r} = -\frac{1}{r}\frac{\partial u}{\partial\theta}$$

69. Prove: If f, f_x, and f_y are continuous on a circular region containing $A(x_0, y_0)$ and $B(x_1, y_1)$, then there is a point (x^*, y^*) on the line segment joining A and B such that

$$f(x_1, y_1) - f(x_0, y_0) \\ = f_x(x^*, y^*)(x_1 - x_0) + f_y(x^*, y^*)(y_1 - y_0)$$

This result is the two-dimensional version of the Mean-Value Theorem. [*Hint:* Express the line segment joining A and B in parametric form and use the Mean-Value Theorem for functions of one variable.]

70. Prove: If $f_x(x, y) = 0$ and $f_y(x, y) = 0$ throughout a circular region, then $f(x, y)$ is constant on that region. [*Hint:* Use the result of Exercise 69.]

■ **16.5 TANGENT PLANES; TOTAL DIFFERENTIALS FOR FUNCTIONS OF TWO VARIABLES**

In this section we shall discuss tangent planes to surfaces in three-dimensional space. We are concerned with three main questions: What is a tangent plane? When do tangent planes exist? How do we find equations of tangent planes? We shall use our results on tangent planes to extend the concept of a differential to functions of two variables.

☐ **TANGENT PLANES**

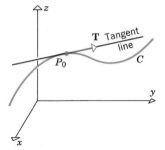

Figure 16.5.1

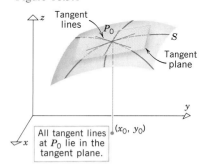

Figure 16.5.2

Recall that if C is a smooth parametric curve in 3-space, then the tangent line to C at a point P_0 is the line through P_0 along the unit tangent vector to C at P_0 (Figure 16.5.1). The concept of a *tangent plane* builds on this definition. If $P_0(x_0, y_0, z_0)$ is a point on a surface S, and if the tangent lines at P_0 to all smooth curves that pass through P_0 and lie on the surface S all lie in a common plane, then we shall regard that plane to be the **tangent plane** to the surface at P_0 (Figure 16.5.2).

The following theorem states conditions that ensure the existence of a tangent plane and tells us how to find its equation.

16.5.1   THEOREM.   *Let $P_0(x_0, y_0, z_0)$ be any point on the surface $z = f(x, y)$. If $f(x, y)$ is differentiable at (x_0, y_0), then the surface has a tangent plane at P_0, and this plane has the equation*

$$f_x(x_0, y_0)(x - x_0) + f_y(x_0, y_0)(y - y_0) - (z - z_0) = 0 \qquad (1)$$

Proof.   To prove the existence of a tangent plane at P_0, we must show that all smooth curves on the surface $z = f(x, y)$ that pass through P_0 have tangent lines that lie in a common plane. We shall do this by showing that these curves all have unit tangent vectors at P_0 that are perpendicular to the vector

$$\mathbf{n} = \langle f_x(x_0, y_0), f_y(x_0, y_0), -1 \rangle \qquad (2)$$

This will force all the tangent lines at P_0 to lie in the plane through P_0 with $\mathbf{n}$ as a normal. Moreover, it will follow from (2) that the point-normal equation of this plane is (1), thereby completing the proof.

Assume that C is any smooth curve that lies on the surface $z = f(x, y)$ and passes through $P_0(x_0, y_0, z_0)$. Moreover, assume that C has parametric equations

$$x = x(s), \quad y = y(s), \quad z = z(s)$$

where s is an arc-length parameter, and assume that $P_0(x_0, y_0, z_0)$ is the point on C that corresponds to the parameter value $s = s_0$. Thus, $x_0 = x(s_0)$, $y_0 = y(s_0)$, and $z_0 = z(s_0)$. Because C lies on the surface $z = f(x, y)$, every point $(x(s), y(s), z(s))$ on C must satisfy this equation for all s, so

$$z(s) = f(x(s), y(s))$$

for all s. If we differentiate both sides of this equation and apply the version of the chain rule in Theorem 16.4.7 with s replacing t, we obtain

$$\frac{dz}{ds} = \frac{\partial f}{\partial x}\frac{dx}{ds} + \frac{\partial f}{\partial y}\frac{dy}{ds}$$

or equivalently,

$$\frac{\partial f}{\partial x}\frac{dx}{ds} + \frac{\partial f}{\partial y}\frac{dy}{ds} - \frac{dz}{ds} = 0$$

The left side of this equation can be rewritten as a dot product of vectors:

$$\left\langle \frac{\partial f}{\partial x}, \frac{\partial f}{\partial y}, -1 \right\rangle \cdot \left\langle \frac{dx}{ds}, \frac{dy}{ds}, \frac{dz}{ds} \right\rangle = 0$$

or in an alternative notation as

$$\langle f_x(x, y), f_y(x, y), -1 \rangle \cdot \langle x'(s), y'(s), z'(s) \rangle = 0$$

In particular, if $s = s_0$, we have

$$\langle f_x(x_0, y_0), f_y(x_0, y_0), -1 \rangle \cdot \langle x'(s_0), y'(s_0), z'(s_0) \rangle = 0 \qquad (3)$$

But the second vector in the dot product is the unit tangent vector to C at the point $P_0(x_0, y_0, z_0)$ [see Formula (2) of Section 15.4], so that from (3) the unit tangent vector to C at P_0 is perpendicular to the vector

$$\mathbf{n} = \langle f_x(x_0, y_0), f_y(x_0, y_0), -1 \rangle$$

which completes the proof. ∎

If $f(x, y)$ is differentiable at (x_0, y_0), then the vector

$$\mathbf{n} = \langle f_x(x_0, y_0), f_y(x_0, y_0), -1 \rangle \qquad (4)$$

is called a ***normal vector*** to the surface $z = f(x, y)$ at $P_0(x_0, y_0, z_0)$, and the line through P_0 parallel to $\mathbf{n}$ is called the ***normal line*** to the surface at P_0 (Figure 16.5.3). Parametric equations of the normal line are

$$x = x_0 + f_x(x_0, y_0)t, \quad y = y_0 + f_y(x_0, y_0)t, \quad z = z_0 - t \qquad (5)$$

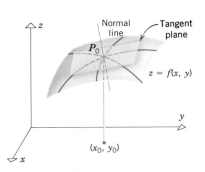

Figure 16.5.3

Example 1 Find equations of the tangent plane and normal line to the surface $z = x^2 y$ at the point $(2, 1, 4)$.

Solution. Since $f(x, y) = x^2 y$, it follows that

$$f_x(x, y) = 2xy \quad \text{and} \quad f_y(x, y) = x^2$$

so that with $x = 2$, $y = 1$,

$$f_x(2, 1) = 4 \quad \text{and} \quad f_y(2, 1) = 4$$

Thus, a vector normal to the surface at $(2, 1, 4)$ is

$$\mathbf{n} = f_x(2, 1)\mathbf{i} + f_y(2, 1)\mathbf{j} - \mathbf{k} = 4\mathbf{i} + 4\mathbf{j} - \mathbf{k}$$

Therefore, the tangent plane has the equation

$$4(x - 2) + 4(y - 1) - (z - 4) = 0 \quad \text{or} \quad 4x + 4y - z = 8$$

and the normal line has equations

$$x = 2 + 4t, \quad y = 1 + 4t, \quad z = 4 - t \qquad ◀$$

REMARK. In the preceding section we set two goals for the definition of differentiability of a function $f(x, y)$ of two variables at a point (x_0, y_0)—we wanted f to be continuous at (x_0, y_0) and we wanted the surface $z = f(x, y)$ to have a nonvertical tangent plane at (x_0, y_0). Both of these goals have now been achieved, for we showed in Theorem 16.4.3 that differentiability implies continuity and now Theorem 16.5.1 shows that differentiability implies the existence of a nonvertical tangent plane. [The tangent plane given by (1) is nonvertical because the third component of the normal vector $\mathbf{n}$ in (2) is not zero.]

☐ **DIFFERENTIALS**

Recall that if $y = f(x)$ is a function of one variable, then the differential

$$dy = f'(x_0)\, dx$$

represents the change in y along the *tangent line* at (x_0, y_0) produced by a change dx in x and

$$\Delta y = f(x_0 + \Delta x) - f(x_0)$$

represents the change in y along the *curve* $y = f(x)$ produced by a change Δx in x. Analogously, if $z = f(x, y)$ is a function of two variables, we shall define dz to be the change in z along the *tangent plane* at (x_0, y_0, z_0) to the surface $z = f(x, y)$ produced by changes dx and dy in x and y, respectively. This is in contrast to

$$\Delta z = f(x_0 + \Delta x, y_0 + \Delta y) - f(x_0, y_0)$$

which represents the change in z *along the surface* produced by changes Δx and Δy in x and y. A comparison of dz and Δz is shown in Figure 16.5.4 in the case where $dx = \Delta x$ and $dy = \Delta y$.

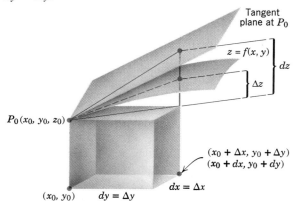

Figure 16.5.4

To derive a formula for dz, let $P_0(x_0, y_0, z_0)$ be a fixed point on the surface $z = f(x, y)$. If we assume f to be differentiable at (x_0, y_0), then the surface has a tangent plane at P_0, given by the equation

$$f_x(x_0, y_0)(x - x_0) + f_y(x_0, y_0)(y - y_0) - (z - z_0) = 0$$

or

$$z = z_0 + f_x(x_0, y_0)(x - x_0) + f_y(x_0, y_0)(y - y_0) \tag{6}$$

From (6) it follows that the tangent plane has height z_0 when $x = x_0$, $y = y_0$, and it has height

$$z_0 + f_x(x_0, y_0)\, dx + f_y(x_0, y_0)\, dy \tag{7}$$

when $x = x_0 + dx$, $y = y_0 + dy$. Thus, the change dz in the height of the tangent plane as (x, y) varies from (x_0, y_0) to $(x_0 + dx, y_0 + dy)$ is obtained by subtracting z_0 from expression (7). This yields

$$dz = f_x(x_0, y_0)\, dx + f_y(x_0, y_0)\, dy$$

This quantity is called the ***total differential*** of z at (x_0, y_0). Often, we omit the subscripts on x_0 and y_0 and write

$$dz = f_x(x, y)\, dx + f_y(x, y)\, dy \tag{8}$$

In this formula, dx and dy are usually viewed as variables and x and y as fixed numbers. Formula (8) can also be written using df in place of dz.

If $z = f(x, y)$ is differentiable at the point (x, y), then the increment Δz can be written as

$$\Delta z = f_x(x, y)\, \Delta x + f_y(x, y)\, \Delta y + \epsilon_1 \Delta x + \epsilon_2 \Delta y \tag{9}$$

where $\epsilon_1 \to 0$, $\epsilon_2 \to 0$ as $(\Delta x, \Delta y) \to (0, 0)$. In the case where $dx = \Delta x$ and $dy = \Delta y$, it follows from (8) and (9) that

$$\Delta z = dz + \epsilon_1 \Delta x + \epsilon_2 \Delta y$$

Thus, when $\Delta x = dx$ and $\Delta y = dy$ are small, we can approximate Δz by

$$\Delta z \approx dz \tag{10}$$

Geometrically, this approximation tells us that the change in z along the surface and the change in z along the tangent plane are approximately equal when $\Delta x = dx$ and $\Delta y = dy$ are small (see Figure 16.5.4).

Example 2 Let $z = 4x^3y^2$. Find dz.

Solution. Since $f(x, y) = 4x^3y^2$,

$$f_x(x, y) = 12x^2y^2 \quad \text{and} \quad f_y(x, y) = 8x^3y$$

so

$$dz = 12x^2y^2\, dx + 8x^3y\, dy \quad \blacktriangleleft$$

Example 3 Let $f(x, y) = \sqrt{x^2 + y^2}$. Use a total differential to approximate the change in $f(x, y)$ as (x, y) varies from the point $(3, 4)$ to the point $(3.04, 3.98)$.

Solution. We shall approximate Δf (the change in f) by

$$df = f_x(x, y)\, dx + f_y(x, y)\, dy = \frac{x}{\sqrt{x^2 + y^2}}\, dx + \frac{y}{\sqrt{x^2 + y^2}}\, dy$$

Since $x = 3$, $y = 4$, $dx = 0.04$, and $dy = -0.02$, we obtain

$$\Delta f \approx df = \frac{3}{\sqrt{3^2 + 4^2}}\,(0.04) + \frac{4}{\sqrt{3^2 + 4^2}}\,(-0.02)$$

$$= \tfrac{3}{5}\,(0.04) - \tfrac{4}{5}\,(0.02) = 0.008$$

The reader may want to use a calculator to show that the true value of Δf to five decimal places is

$$\Delta f = \sqrt{(3.04)^2 + (3.98)^2} - \sqrt{3^2 + 4^2} \approx 0.00819 \quad \blacktriangleleft$$

Example 4 The radius of a right-circular cylinder is measured with an error of at most 2%, and the height is measured with an error of at most 4%. Approximate the maximum possible percentage error in the volume V calculated from these measurements.

Solution. Let r, h, and V be the true radius, height, and volume of the cylinder, and let Δr, Δh, and ΔV be the errors in these quantities. We are given that

$$\left|\frac{\Delta r}{r}\right| \le 0.02 \quad \text{and} \quad \left|\frac{\Delta h}{h}\right| \le 0.04$$

We want to find the maximum possible value of $|\Delta V/V|$. Since the volume of the cylinder is $V = \pi r^2 h$, it follows from (8) that

$$dV = \frac{\partial V}{\partial r}\, dr + \frac{\partial V}{\partial h}\, dh = 2\pi rh\, dr + \pi r^2\, dh$$

If we choose $dr = \Delta r$ and $dh = \Delta h$, then we can use the approximations

$$\Delta V \approx dV \quad \text{and} \quad \frac{\Delta V}{V} \approx \frac{dV}{V}$$

But

$$\frac{dV}{V} = \frac{2\pi rh\, dr + \pi r^2\, dh}{\pi r^2 h} = 2\frac{dr}{r} + \frac{dh}{h}$$

so by the triangle inequality (Theorem 1.2.5)

$$\left|\frac{dV}{V}\right| = \left|2\frac{dr}{r} + \frac{dh}{h}\right| \le 2\left|\frac{dr}{r}\right| + \left|\frac{dh}{h}\right| \le 2(0.02) + (0.04) = 0.08$$

Thus, the maximum percentage error in V is approximately 8%. $\quad \blacktriangleleft$

▶ Exercise Set 16.5   © 24, 32, 34

In Exercises 1–8, find equations for the tangent plane and normal line to the given surface at the point P.

1. $z = 4x^3y^2 + 2y$; $P(1, -2, 12)$.

2. $z = \frac{1}{2}x^7y^{-2}$; $P(2, 4, 4)$.

3. $z = xe^{-y}$; $P(1, 0, 1)$.

4. $z = \ln\sqrt{x^2 + y^2}$; $P(-1, 0, 0)$.

5. $z = e^{3y}\sin 3x$; $P(\pi/6, 0, 1)$.

6. $z = x^{1/2} + y^{1/2}$; $P(4, 9, 5)$.

7. $x^2 + y^2 + z^2 = 25$; $P(-3, 0, 4)$.

8. $x^2y - 4z^2 = -7$; $P(-3, 1, -2)$.

In Exercises 9–12, find dz.

9. $z = 7x - 2y$.

10. $z = 5x^2y^5 - 2x + 4y + 7$.

11. $z = \tan^{-1} xy$.   12. $z = \sec^2(x - 3y)$.

In Exercises 13–16, use a total differential to approximate the change in $f(x, y)$ as (x, y) varies from P to Q.

13. $f(x, y) = x^2 + 2xy - 4x$; $P(1, 2)$, $Q(1.01, 2.04)$.

14. $f(x, y) = x^{1/3}y^{1/2}$; $P(8, 9)$, $Q(7.78, 9.03)$.

15. $f(x, y) = \dfrac{x + y}{xy}$; $P(-1, -2)$, $Q(-1.02, -2.04)$.

16. $f(x, y) = \ln\sqrt{1 + xy}$; $P(0, 2)$, $Q(-0.09, 1.98)$.

17. Find all points on the surface at which the tangent plane is horizontal.

    (a) $z = x^3y^2$

    (b) $z = x^2 - xy + y^2 - 2x + 4y$.

18. Find a point on the surface $z = 3x^2 - y^2$ at which the tangent plane is parallel to the plane $6x + 4y - z = 5$.

19. Find a point on the surface $z = 8 - 3x^2 - 2y^2$ at which the tangent plane is perpendicular to the line $x = 2 - 3t$, $y = 7 + 8t$, $z = 5 - t$.

20. Show that the surfaces

$$z = \sqrt{x^2 + y^2} \quad \text{and} \quad z = \tfrac{1}{10}(x^2 + y^2) + \tfrac{5}{2}$$

intersect at $(3, 4, 5)$ and have a common tangent plane at that point.

21. Show that the surfaces $z = \sqrt{16 - x^2 - y^2}$ and $z = \sqrt{x^2 + y^2}$ intersect at $(2, 2, 2\sqrt{2})$ and have tangent planes that are perpendicular at that point.

22. (a) Show that every line normal to the cone $z = \sqrt{x^2 + y^2}$ passes through the z-axis.

    (b) Show that every tangent plane intersects the cone in a line passing through the origin.

23. One leg of a right triangle increases from 3 cm to 3.2 cm, while the other leg decreases from 4 cm to 3.96 cm. Use a total differential to approximate the change in the length of the hypotenuse.

24. The volume V of a right-circular cone of radius r and height h is given by $V = \frac{1}{3}\pi r^2 h$. Suppose that the height decreases from 20 in. to 19.95 in., while the radius increases from 4 in. to 4.05 in. Use a total differential to approximate the change in volume.

25. The length and width of a rectangle are measured with errors of at most 3% and 5%, respectively. Use differentials to approximate the maximum percentage error in the calculated area.

26. The radius and height of a right-circular cone are measured with errors of at most 1% and 4%, respectively. Use differentials to approximate the maximum percentage error in the calculated volume.

27. The length and width of a rectangle are measured with errors of at most $r\%$. Use differentials to approximate the maximum percentage error in the calculated length of the diagonal.

28. The legs of a right triangle are measured to be 3 cm and 4 cm, with a maximum error of 0.05 cm in each measurement. Use differentials to approximate the maximum possible error in the calculated value of (a) the hypotenuse and (b) the area of the triangle.

29. The total resistance R of two resistances R_1 and R_2, connected in parallel, is

$$R = \frac{R_1 R_2}{R_1 + R_2}$$

Suppose that R_1 and R_2 are measured to be 200 ohms and 400 ohms, respectively, with a maximum error of 2% in each. Use differentials to approximate the maximum percentage error in the calculated value of R.

30. According to the ideal gas law, the pressure, temperature, and volume of a confined gas are related by $P = kT/V$, where k is a constant. Use differentials to approximate the percentage change in pressure if the temperature of a gas is increased 3% and the volume is increased 5%.

31. An angle θ of a right triangle is calculated by the formula

$$\theta = \sin^{-1}\frac{a}{c}$$

where a is the length of the side opposite to θ and c is the length of the hypotenuse. Suppose that the measurements $a = 3$ in. and $c = 5$ in. each have a maximum possible error of 0.01 in. Use differentials to approximate the maximum possible error in the the calculated value of θ.

32. An open cylindrical can has an inside radius of 2 cm and an inside height of 5 cm. Use differentials to approximate the volume of metal in the can if it is 0.01 cm thick. [*Hint:* The volume of metal is the difference, ΔV, in the volumes of two cylinders.]

33. The period T of a simple pendulum with small oscillations is calculated from the formula $T = 2\pi\sqrt{L/g}$, where L is the length of the pendulum and g is the acceleration due to gravity. Suppose that values of L and g have errors of at most 0.5% and 0.1%, respectively. Use differentials to approximate the maximum percentage error in the calculated value of T.

34. The angle of elevation from a point on the ground to the top of a building is measured as 60° with a maximum possible error of 0.2°. Suppose that the distance from the point to the building is measured as 100 ft with a maximum possible error of 2 in. Use differentials to approximate the maximum possible error in the calculated height of the building.

35. Suppose that x and y have errors of at most $r\%$ and $s\%$, respectively. For each of the following formulas in x and y, use differentials to approximate the maximum possible error in the calculated result.

(a) xy (b) x/y (c) $x^2 y^3$ (d) $x^3\sqrt{y}$.

36. Show that the volume of the solid bounded by the coordinate planes and the plane tangent to the portion of the surface $xyz = k$, $k > 0$, in the first octant does not depend on the point of tangency.

37. (a) Find all points of intersection of the line $x = -1 + t$, $y = 2 + t$, $z = 2t + 7$ and the surface $z = x^2 + y^2$.

(b) At each point of intersection, find the cosine of the acute angle between the given line and the line normal to the surface.

38. Show that if f is differentiable and $z = xf(x/y)$, then all tangent planes to the graph of this equation pass through a common point.

39. Show that the equation of the plane that is tangent to the ellipsoid

$$\frac{x^2}{a^2} + \frac{y^2}{b^2} + \frac{z^2}{c^2} = 1$$

at (x_0, y_0, z_0) can be written in the form

$$\frac{x_0 x}{a^2} + \frac{y_0 y}{b^2} + \frac{z_0 z}{c^2} = 1$$

40. Show that the equation of the plane that is tangent to the paraboloid

$$z = \frac{x^2}{a^2} + \frac{y^2}{b^2}$$

at (x_0, y_0, z_0) can be written in the form

$$z + z_0 = \frac{2x_0 x}{a^2} + \frac{2y_0 y}{b^2}$$

41. Prove: If the surfaces $z = f(x, y)$ and $z = g(x, y)$ intersect at $P(x_0, y_0, z_0)$, and if f and g are differentiable at (x_0, y_0), then the normal lines at P are perpendicular if and only if

$$f_x(x_0, y_0)g_x(x_0, y_0) + f_y(x_0, y_0)g_y(x_0, y_0) = -1$$

16.6 DIRECTIONAL DERIVATIVES AND GRADIENTS FOR FUNCTIONS OF TWO VARIABLES

The partial derivatives $f_x(x, y)$ and $f_y(x, y)$ represent the rates of change of $f(x, y)$ in directions parallel to the x- and y-axes. In this section we shall investigate rates of change of $f(x, y)$ in other directions.

□ **DIRECTIONAL DERIVATIVES**

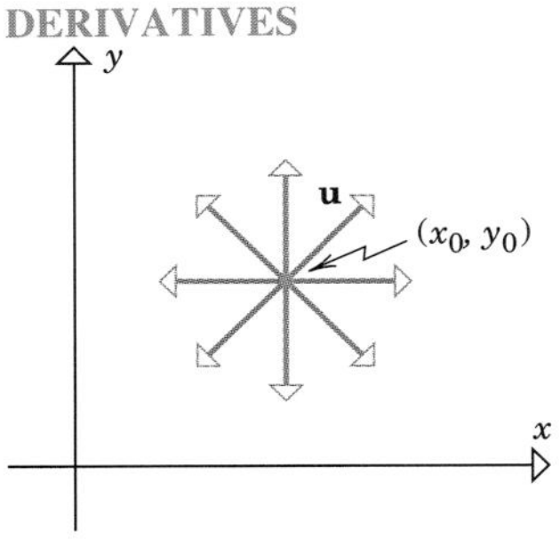

Figure 16.6.1

Just as a person standing on the slope of a hill with mounds and gullies may encounter different grades by walking in different directions, so a surface $z = f(x, y)$ can have different slopes in different directions from a point (x_0, y_0, z_0) on the surface.

To make this idea more precise, we shall use a *unit vector* $\mathbf{u} = \langle u_1, u_2 \rangle$ in the xy-plane to designate the direction in which the slope is to be examined (Figure 16.6.1), and we shall let l be the line in the xy-plane that is parallel to $\mathbf{u}$ and passes through the point $P_0(x_0, y_0)$. As shown in Example 7 of Section 15.3, this line can be represented by the parametric equations

$$x = x_0 + su_1, \quad y = y_0 + su_2 \tag{1}$$

where s is an arc-length parameter with its reference point at $P_0(x_0, y_0)$, and the positive direction is in the direction of $\mathbf{u}$ (Figure 16.6.2). As s increases, the point $P(x, y)$ moves in the direction of **u** along l, and a companion point Q with z-coordinate

$$z = f(x, y) = f(x_0 + su_1, y_0 + su_2)$$

moves directly above (or below) along the surface, tracing out a curve C (Figure 16.6.3).

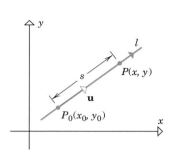

Figure 16.6.2

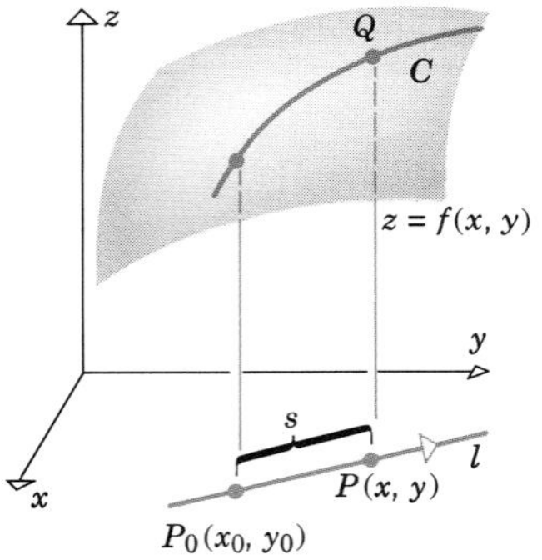

Figure 16.6.3

The rate of change of z with respect to s can be calculated using the chain rule. We obtain

$$\frac{dz}{ds} = f_x(x, y)\frac{dx}{ds} + f_y(x, y)\frac{dy}{ds}$$

or from (1)

$$\frac{dz}{ds} = f_x(x, y)u_1 + f_y(x, y)u_2 \tag{2}$$

where x and y in this formula are expressed in terms of s by the equations in (1). But $P_0(x_0, y_0)$ is the point on l that corresponds to $s = 0$ [see (1)], so setting $s = 0$ in (2) yields the instantaneous rate of change of z with respect to s at the point $P_0(x_0, y_0)$:

$$\left.\frac{dz}{ds}\right|_{s=0} = f_x(x_0, y_0)u_1 + f_y(x_0, y_0)u_2 \tag{3}$$

This quantity is called a *directional derivative of f* and is commonly denoted by $D_{\mathbf{u}}z(x_0, y_0)$ or $D_{\mathbf{u}}f(x_0, y_0)$. More precisely, we make the following definition.

16.6.1   DEFINITION.   If $f(x, y)$ is differentiable at (x_0, y_0), and if $\mathbf{u} = \langle u_1, u_2 \rangle$ is a unit vector, then the ***directional derivative*** of f at (x_0, y_0) in the direction of **u** is defined by

$$D_{\mathbf{u}}f(x_0, y_0) = f_x(x_0, y_0)u_1 + f_y(x_0, y_0)u_2 \tag{4}$$

REMARK.   It should be kept in mind that there are infinitely many directional derivatives of $z = f(x, y)$ at a point (x_0, y_0), one for each possible choice of the direction vector **u** (Figure 16.6.4). In the special case where $\mathbf{u} = \mathbf{i} = \langle 1, 0 \rangle$, it follows from (4) that $D_{\mathbf{i}}f(x_0, y_0) = f_x(x_0, y_0)$, and in the case where $\mathbf{u} = \mathbf{j} = \langle 0, 1 \rangle$, it follows from (4) that $D_{\mathbf{j}}f(x_0, y_0) = f_y(x_0, y_0)$, so that the partial derivatives with respect to x and with respect to y can be viewed as directional derivatives in the positive x- and positive y-directions, respectively.

Example 1   Find the directional derivative of $f(x, y) = 3x^2y$ at the point $(1, 2)$ in the direction of the vector $\mathbf{a} = 3\mathbf{i} + 4\mathbf{j}$.

Solution.   The vector **a** is not a unit vector, so we must normalize it to apply Formula (4). This yields

$$\mathbf{u} = \frac{\mathbf{a}}{\|\mathbf{a}\|} = \frac{1}{\sqrt{25}}(3\mathbf{i} + 4\mathbf{j}) = \tfrac{3}{5}\mathbf{i} + \tfrac{4}{5}\mathbf{j}$$

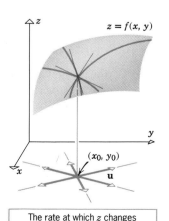

The rate at which z changes with s at (x_0, y_0) depends on the direction of **u**.

Figure 16.6.4

from which we obtain $u_1 = \frac{3}{5}$ and $u_2 = \frac{4}{5}$. Since the partial derivatives of f are

$$f_x(x, y) = 6xy, \quad f_y(x, y) = 3x^2$$

we have $f_x(1, 2) = 12$ and $f_y(1, 2) = 3$, so from (4)

$$D_{\mathbf{u}}f(1, 2) = f_x(1, 2)u_1 + f_y(1, 2)u_2 = 12u_1 + 3u_2$$

$$= 12\left(\tfrac{3}{5}\right) + 3\left(\tfrac{4}{5}\right) = \tfrac{48}{5} \quad \blacktriangleleft$$

The following alternative version of Formula (4) can be obtained by expressing the unit direction vector as $\mathbf{u} = \cos\theta\,\mathbf{i} + \sin\theta\,\mathbf{j}$, where θ is the angle that $\mathbf{u}$ makes with the positive x-axis [Formula (8) of Section 14.2]:

$$D_{\mathbf{u}}f(x_0, y_0) = f_x(x_0, y_0)\cos\theta + f_y(x_0, y_0)\sin\theta \tag{5}$$

Example 2 Find the directional derivative of $f(x, y) = e^{xy}$ at $(-2, 0)$ in the direction of the unit vector $\mathbf{u}$ that makes an angle of $\pi/3$ with the positive x-axis.

Solution. Let $f(x, y) = e^{xy}$ so that

$$f_x(x, y) = ye^{xy}, \quad f_y(x, y) = xe^{xy}$$

$$f_x(-2, 0) = 0, \quad f_y(-2, 0) = -2$$

From (5)

$$D_{\mathbf{u}}f(-2, 0) = f_x(-2, 0)\cos\frac{\pi}{3} + f_y(-2, 0)\sin\frac{\pi}{3}$$

$$= 0\left(\frac{1}{2}\right) + (-2)\left(\frac{\sqrt{3}}{2}\right) = -\sqrt{3} \quad \blacktriangleleft$$

REMARK. It is worth noting that reversing the direction of $\mathbf{u}$ reverses the sign of the directional derivative $D_{\mathbf{u}}$, since

$$D_{-\mathbf{u}}f(x_0, y_0) = f_x(x_0, y_0)(-u_1) + f_y(x_0, y_0)(-u_2) = -D_{\mathbf{u}}f(x_0, y_0)$$

Thus, we can conclude from the result in Example 2 that the directional derivative of $f(x, y) = e^{xy}$ at $(-2, 0)$ in the direction of the vector making an angle of $4\pi/3$ with the positive x-axis is $\sqrt{3}$.

☐ **THE GRADIENT**

The directional derivative formula

$$D_{\mathbf{u}}f(x, y) = f_x(x, y)u_1 + f_y(x, y)u_2$$

can be expressed in the form of a dot product by writing

$$D_{\mathbf{u}}f(x, y) = (f_x(x, y)\mathbf{i} + f_y(x, y)\mathbf{j}) \cdot (u_1\mathbf{i} + u_2\mathbf{j}) \tag{6}$$

The second vector in the dot product is $\mathbf{u}$. However, the first vector is new; it is called the *gradient of f* and is denoted by the symbol ∇f or $\nabla f(x, y)$.*

16.6.2 DEFINITION. If f is a function of x and y, then the **gradient of f** is defined by

$$\nabla f(x, y) = f_x(x, y)\mathbf{i} + f_y(x, y)\mathbf{j} \tag{7}$$

---

*The symbol ∇ (read, "del") is an inverted delta. In older books this symbol is sometimes called a "nabla" because of its similarity in form to an ancient Hebrew ten-stringed harp of that name.

With the gradient notation, Formula (6) for the directional derivative can be written in the following compact form:

$$D_{\mathbf{u}}f(x, y) = \nabla f(x, y) \cdot \mathbf{u} \tag{8}$$

In words, the dot product of the gradient of f with a unit vector $\mathbf{u}$ produces the directional derivative of f in the direction of $\mathbf{u}$.

Example 3 Find the gradient of $f(x, y) = 3x^2y$ at the point $(1, 2)$ and use it to calculate the directional derivative of f at $(1, 2)$ in the direction of the vector $\mathbf{a} = 3\mathbf{i} + 4\mathbf{j}$.

Solution. From (7)

$$\nabla f(x, y) = f_x(x, y)\mathbf{i} + f_y(x, y)\mathbf{j} = 6xy\mathbf{i} + 3x^2\mathbf{j}$$

so that the gradient of f at $(1, 2)$ is

$$\nabla f(1, 2) = 12\mathbf{i} + 3\mathbf{j}$$

The unit vector in the direction of $\mathbf{a}$ is

$$\mathbf{u} = \frac{\mathbf{a}}{\|\mathbf{a}\|} = \tfrac{1}{5}(3\mathbf{i} + 4\mathbf{j}) = \tfrac{3}{5}\mathbf{i} + \tfrac{4}{5}\mathbf{j}$$

Thus, from (9)

$$D_{\mathbf{u}}f(1, 2) = \nabla f(1, 2) \cdot \mathbf{u} = (12\mathbf{i} + 3\mathbf{j}) \cdot (\tfrac{3}{5}\mathbf{i} + \tfrac{4}{5}\mathbf{j}) = \tfrac{48}{5}$$

which agrees with the result obtained in Example 1. ◄

□ **PROPERTIES OF THE GRADIENT**

The gradient is not merely a notational device to simplify the formula for the directional derivative; the length and direction of the gradient ∇f provide important information about the function f.

16.6.3 THEOREM. *Let f be a function of two variables that is differentiable at (x_0, y_0).*

(a) *If $\nabla f(x_0, y_0) = \mathbf{0}$, then all directional derivatives of f at (x_0, y_0) are zero.*

(b) *If $\nabla f(x_0, y_0) \neq \mathbf{0}$, then among all possible directional derivatives of f at (x_0, y_0), the derivative in the direction of $\nabla f(x_0, y_0)$ has the largest value. The value of that directional derivative is $\|\nabla f(x_0, y_0)\|$.*

(c) *If $\nabla f(x_0, y_0) \neq \mathbf{0}$, then among all possible directional derivatives of f at (x_0, y_0), the derivative in the direction opposite to that of $\nabla f(x_0, y_0)$ has the smallest value. The value of that directional derivative is $-\|\nabla f(x_0, y_0)\|$.*

Proof (a). If $\nabla f(x_0, y_0) = \mathbf{0}$, then for all choices of $\mathbf{u}$ we have

$$D_{\mathbf{u}}f(x_0, y_0) = \nabla f(x_0, y_0) \cdot \mathbf{u} = \mathbf{0} \cdot \mathbf{u} = 0$$

Proofs (b) and (c). Assume that $\nabla f(x_0, y_0) \neq \mathbf{0}$ and let θ be the angle between $\nabla f(x_0, y_0)$ and an arbitrary unit vector $\mathbf{u}$. By the definition of dot product,

$$D_{\mathbf{u}}f(x_0, y_0) = \nabla f(x_0, y_0) \cdot \mathbf{u} = \|\nabla f(x_0, y_0)\| \, \|\mathbf{u}\| \cos \theta$$

or, since $\|\mathbf{u}\| = 1$,

$$D_{\mathbf{u}}f(x_0, y_0) = \|\nabla f(x_0, y_0)\| \cos \theta$$

Thus, the maximum value of $D_{\mathbf{u}}f(x_0, y_0)$ is $\|\nabla f(x_0, y_0)\|$, and this occurs when $\cos \theta = 1$, or when $\mathbf{u}$ has the same direction as $\nabla f(x_0, y_0)$ (since $\theta = 0$). The minimum value of

$D_{\mathbf{u}}f(x_0, y_0)$ is $-\|\nabla f(x_0, y_0)\|$, and this occurs when $\cos\theta = -1$ or when $\mathbf{u}$ and $\nabla f(x_0, y_0)$ are oppositely directed (since $\theta = \pi$).   ∎

Example 4   For the function $f(x, y) = x^2 e^y$, find the maximum value of a directional derivative at $(-2, 0)$, and give a unit vector in the direction in which the maximum value occurs.

Solution.   Since

$$\nabla f(x, y) = f_x(x, y)\mathbf{i} + f_y(x, y)\mathbf{j} = 2xe^y\mathbf{i} + x^2 e^y\mathbf{j}$$

the gradient of f at $(-2, 0)$ is

$$\nabla f(-2, 0) = -4\mathbf{i} + 4\mathbf{j}$$

By Theorem 16.6.3, the maximum value of the directional derivative is

$$\|\nabla f(-2, 0)\| = \sqrt{(-4)^2 + 4^2} = \sqrt{32} = 4\sqrt{2}$$

This maximum occurs in the direction of $\nabla f(-2, 0)$. A unit vector in this direction is

$$\frac{\nabla f(-2, 0)}{\|\nabla f(-2, 0)\|} = \frac{1}{4\sqrt{2}}(-4\mathbf{i} + 4\mathbf{j}) = -\frac{1}{\sqrt{2}}\mathbf{i} + \frac{1}{\sqrt{2}}\mathbf{j} \qquad \blacktriangleleft$$

Our next objective is to establish a geometric relationship between the level curves and the gradient of a function f of two variables. For this purpose we note that if (x_0, y_0) is any point in the domain of f, and if $f(x_0, y_0) = c$, then under appropriate conditions the equation

$$f(x, y) = c$$

defines a unique level curve of f that can be smoothly parametrized in terms of arc length and that passes through (x_0, y_0).* It follows that under these conditions the level curve has a unit tangent vector at (x_0, y_0) (Figure 16.6.5).

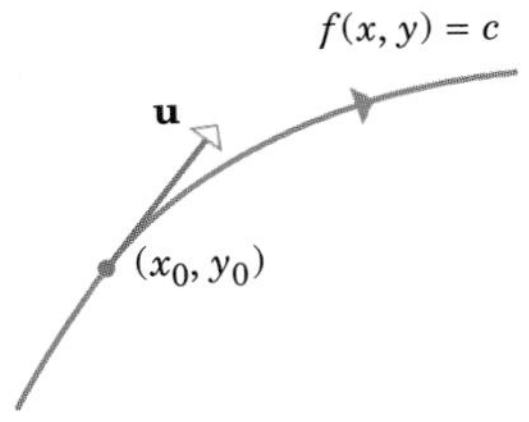

If $f(x, y) = c$ is the level curve for f that passes through the point (x_0, y_0), then the value of $f(x, y)$ remains constant as (x, y) moves away from (x_0, y_0) along the level curve. It seems plausible, therefore, that the directional derivative of f at (x_0, y_0) is zero in the direction of the unit tangent vector $\mathbf{u}$ at (x_0, y_0) to the level curve $f(x, y) = c$. To see that this is so, suppose that $f(x, y) = c$ is smoothly parametrized in terms of an arc-length parameter s by the equations $x = x(s)$, $y = y(s)$, so that

$$f(x(s), y(s)) = c$$

Differentiating this equation with respect to s by the chain rule yields

$$\frac{\partial f}{\partial x}\frac{dx}{ds} + \frac{\partial f}{\partial y}\frac{dy}{ds} = 0$$

which can be rewritten as

$$\left(\frac{\partial f}{\partial x}\mathbf{i} + \frac{\partial f}{\partial y}\mathbf{j}\right) \cdot \left(\frac{dx}{ds}\mathbf{i} + \frac{dy}{ds}\mathbf{j}\right) = 0 \qquad (9)$$

At (x_0, y_0) the first vector is $\nabla f(x_0, y_0)$, and from Formula (2) of Section 15.4 the second vector is $\mathbf{u}$. Thus, from (8) and (9)

$$D_{\mathbf{u}}f(x_0, y_0) = \nabla f(x_0, y_0) \cdot \mathbf{u} = 0 \qquad (10)$$

which is what we wanted to show.

$f(x, y) = c$

$\mathbf{u}$

(x_0, y_0)

Figure 16.6.5

---

*The conditions under which this occurs are specified in a theorem from advanced calculus called the "implicit function theorem." We shall assume in the remainder of this section that f is differentiable at (x_0, y_0) and that a smooth arc-length parametrization of the level curve through (x_0, y_0) is possible.

The last equality in (10) implies that the gradient of f is perpendicular to the tangent vector $\mathbf{u}$ at the point (x_0, y_0), and hence is normal to the level curve at that point. In summary, we have the following theorem.

> **16.6.4** THEOREM. *If f is differentiable at (x_0, y_0), then $\nabla f(x_0, y_0)$ is normal to the level curve of f through (x_0, y_0).*

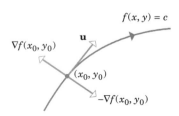

$f(x, y) = c$

$\nabla f(x_0, y_0)$

$\mathbf{u}$

(x_0, y_0)

$-\nabla f(x_0, y_0)$

Figure 16.6.6

REMARK.   Figure 16.6.6 shows the relationship between the vectors $\mathbf{u}$, $\nabla f(x_0, y_0)$, and $-\nabla f(x_0, y_0)$ for the level curve of f that passes through the point (x_0, y_0). If the gradient of f is nonzero at that point, then the function f has its maximum rate of increase in the direction of $\nabla f(x_0, y_0)$, its maximum rate of decrease in the direction of $-\nabla f(x_0, y_0)$, and is neither increasing nor decreasing in the direction of $\mathbf{u}$, since $D_{\mathbf{u}} f(x_0, y_0) = 0$. Good skiers use these facts intuitively to control their speed by zigzagging down ski slopes: They ski across the slope with their skis tangential to a level curve to stop their downhill motion and turn their skis down the slope and perpendicular to the level curves to obtain the most rapid downhill motion.

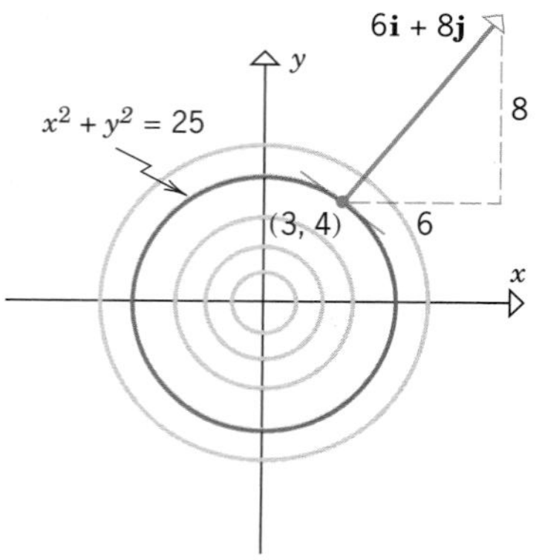

Example 5   For the function $f(x, y) = x^2 + y^2$, sketch the level curve through the point $(3, 4)$, and draw the gradient vector at this point.

Solution.   Since $f(3, 4) = 25$, the level curve through the point $(3, 4)$ has the equation $f(x, y) = 25$, which is the circle

$$x^2 + y^2 = 25$$

Since

$$\nabla f(x, y) = f_x(x, y)\mathbf{i} + f_y(x, y)\mathbf{j} = 2x\mathbf{i} + 2y\mathbf{j}$$

the gradient vector at $(3, 4)$ is

$$\nabla f(3, 4) = 6\mathbf{i} + 8\mathbf{j}$$

(Figure 16.6.7). Note that the gradient vector is perpendicular to the circle at $(3, 4)$, as guaranteed by Theorem 16.6.4.   ◀

$6\mathbf{i} + 8\mathbf{j}$

y

$x^2 + y^2 = 25$

8

$(3, 4)$ 6

x

Figure 16.6.7

▶ Exercise Set 16.6

In Exercises 1–4, find ∇z.

1. $z = 4x - 8y$.

2. $z = e^{-3y} \cos 4x$.

3. $z = \ln \sqrt{x^2 + y^2}$.

4. $z = e^{-5x} \sec x^2 y$.

In Exercises 5–8, find the gradient of f at the indicated point.

5. $f(x, y) = (x^2 + xy)^3$; $(-1, -1)$.

6. $f(x, y) = (x^2 + y^2)^{-1/2}$; $(3, 4)$.

7. $f(x, y) = y \ln (x + y)$; $(-3, 4)$.

8. $f(x, y) = y^2 \tan^3 x$; $(\pi/4, -3)$.

In Exercises 9–12, find $D_{\mathbf{u}} f$ at P.

9. $f(x, y) = (1 + xy)^{3/2}$; $P(3, 1)$; $\mathbf{u} = \dfrac{1}{\sqrt{2}}\mathbf{i} + \dfrac{1}{\sqrt{2}}\mathbf{j}$.

10. $f(x, y) = e^{2xy}$; $P(4, 0)$; $\mathbf{u} = -\frac{3}{5}\mathbf{i} + \frac{4}{5}\mathbf{j}$.

11. $f(x, y) = \ln (1 + x^2 + y)$; $P(0, 0)$;

$$\mathbf{u} = -\frac{1}{\sqrt{10}}\mathbf{i} - \frac{3}{\sqrt{10}}\mathbf{j}.$$

12. $f(x, y) = \dfrac{cx + dy}{x - y}$; $P(3, 4)$; $\mathbf{u} = \frac{4}{5}\mathbf{i} + \frac{3}{5}\mathbf{j}$.

In Exercises 13–18, find the directional derivative of f at P in the direction of $\mathbf{a}$.

13. $f(x, y) = 4x^3 y^2$; $P(2, 1)$; $\mathbf{a} = 4\mathbf{i} - 3\mathbf{j}$.

14. $f(x, y) = x^2 - 3xy + 4y^3$; $P(-2, 0)$; $\mathbf{a} = \mathbf{i} + 2\mathbf{j}$.

15. $f(x, y) = y^2 \ln x$; $P(1, 4)$; $\mathbf{a} = -3\mathbf{i} + 3\mathbf{j}$.

16. $f(x, y) = e^x \cos y$; $P(0, \pi/4)$; $\mathbf{a} = 5\mathbf{i} - 2\mathbf{j}$.

17. $f(x, y) = \tan^{-1} (y/x)$; $P(-2, 2)$; $\mathbf{a} = -\mathbf{i} - \mathbf{j}$.

18. $f(x, y) = xe^y - ye^x$; $P(0, 0)$; $\mathbf{a} = 5\mathbf{i} - 2\mathbf{j}$.

In Exercises 19–22, find the directional derivative of f at P in the direction of a vector making the angle θ with the positive x-axis.

19. $f(x, y) = \sqrt{xy}$; $P(1, 4)$; $\theta = \pi/3$.

20. $f(x, y) = \dfrac{x - y}{x + y}$; $P(-1, -2)$; $\theta = \pi/2$.

21. $f(x, y) = \tan(2x + y)$; $P(\pi/6, \pi/3)$; $\theta = 7\pi/4$.

22. $f(x, y) = \sinh x \cosh y$; $P(0, 0)$; $\theta = \pi$.

In Exercises 23–26, sketch the level curve of $f(x, y)$ that passes through P and draw the gradient vector at P.

23. $f(x, y) = 4x - 2y + 3$; $P(1, 2)$.

24. $f(x, y) = y/x^2$; $P(-2, 2)$.

25. $f(x, y) = x^2 + 4y^2$; $P(-2, 0)$.

26. $f(x, y) = x^2 - y^2$; $P(2, -1)$.

In Exercises 27–30, find a unit vector in the direction in which f increases most rapidly at P; and find the rate of change of f at P in that direction.

27. $f(x, y) = 4x^3y^2$; $P(-1, 1)$.

28. $f(x, y) = 3x - \ln y$; $P(2, 4)$.

29. $f(x, y) = \sqrt{x^2 + y^2}$; $P(4, -3)$.

30. $f(x, y) = \dfrac{x}{x + y}$; $P(0, 2)$.

In Exercises 31–34, find a unit vector in the direction in which f decreases most rapidly at P; and find the rate of change of f at P in that direction.

31. $f(x, y) = 20 - x^2 - y^2$; $P(-1, -3)$.

32. $f(x, y) = e^{xy}$; $P(2, 3)$.

33. $f(x, y) = \cos(3x - y)$; $P(\pi/6, \pi/4)$.

34. $f(x, y) = \sqrt{\dfrac{x - y}{x + y}}$; $P(3, 1)$.

35. Find the directional derivative of $f(x, y) = \dfrac{x}{x + y}$ at $P(1, 0)$ in the direction to $Q(-1, -1)$.

36. Find the directional derivative of $f(x, y) = e^{-x} \sec y$ at $P(0, \pi/4)$ in the direction of the origin.

37. Find the directional derivative of $f(x, y) = \sqrt{xy}\, e^y$ at $P(1, 1)$ in the direction of the negative y-axis.

38. Let $f(x, y) = \dfrac{y}{x + y}$. Find a unit vector $\mathbf{u}$ for which $D_{\mathbf{u}}f(2, 3) = 0$.

39. Find a unit vector $\mathbf{u}$ that is perpendicular at $P(1, -2)$ to the level curve of $f(x, y) = 4x^2y$ through P.

40. Find a unit vector $\mathbf{u}$ that is perpendicular at $P(2, -3)$ to the level curve of $f(x, y) = 3x^2y - xy$ through P.

41. Given that $D_{\mathbf{u}}f(1, 2) = -5$ if $\mathbf{u} = \frac{3}{5}\mathbf{i} - \frac{4}{5}\mathbf{j}$ and $D_{\mathbf{v}}f(1, 2) = 10$ if $\mathbf{v} = \frac{4}{5}\mathbf{i} + \frac{3}{5}\mathbf{j}$, find

(a) $f_x(1, 2)$ (b) $f_y(1, 2)$

(c) the directional derivative of f at $(1, 2)$ in the direction of the origin.

42. Given that $f_x(-5, 1) = -3$ and $f_y(-5, 1) = 2$, find the directional derivative of f at $P(-5, 1)$ in the direction of the vector from P to $Q(-4, 3)$.

43. Given that $\nabla f(4, -5) = 2\mathbf{i} - \mathbf{j}$, find the directional derivative of the function f at $(4, -5)$ in the direction of $\mathbf{a} = 5\mathbf{i} + 2\mathbf{j}$.

44. Given that $\nabla f(x_0, y_0) = \mathbf{i} - 2\mathbf{j}$ and $D_{\mathbf{u}}f(x_0, y_0) = -2$, find $\mathbf{u}$.

45. Let $z = 3x^2 - y^2$. Find all points where $\|\nabla z\| = 6$.

46. Given that $z = 3x + y^2$, find the maximum value of $d\|\nabla z\|/ds$ at the point $(5, 2)$ and a unit vector in the direction in which the maximum is attained.

47. A particle moves along a path C given by $x = t$ and $y = -t^2$. If $z = x^2 + y^2$, find dz/ds along C at the instant when the particle is at the point $(2, -4)$.

48. The temperature at a point (x, y) on a metal plate in the xy-plane is $T(x, y) = \dfrac{xy}{1 + x^2 + y^2}$ degrees Celsius.

(a) Find the rate of change of temperature at $(1, 1)$ in the direction of $\mathbf{a} = 2\mathbf{i} - \mathbf{j}$.

(b) An ant at $(1, 1)$ wants to walk in the direction in which the temperature drops most rapidly. Find a unit vector in that direction.

49. If the electric potential at a point (x, y) in the xy-plane is $V(x, y)$, then the **electric intensity vector** at (x, y) is $\mathbf{E} = -\nabla V(x, y)$. Suppose that $V(x, y) = e^{-2x} \cos 2y$.

(a) Find the electric intensity vector at $(\pi/4, 0)$.

(b) Show that at each point in the plane, the electric potential decreases most rapidly in the direction of the vector $\mathbf{E}$.

50. On a certain mountain, the elevation z above a point (x, y) in a horizontal xy-plane that lies at sea level is $z = 2000 - 2x^2 - 4y^2$ ft. The positive x-axis points east, and the positive y-axis north. A climber is at the point $(-20, 5, 1100)$.

(a) If the climber uses a compass reading to walk due west, will he begin to ascend or descend?

(b) If the climber uses a compass reading to walk northeast, will he ascend or descend? At what rate?

(c) In what compass direction should the climber walk to travel a level path?

51. Let $r = \sqrt{x^2 + y^2}$.

(a) Show that $\nabla r = \dfrac{\mathbf{r}}{r}$, where $\mathbf{r} = x\mathbf{i} + y\mathbf{j}$.

(b) Show that $\nabla f(r) = f'(r)\nabla r = \dfrac{f'(r)}{r}\mathbf{r}$.

52. Use the formula of part (b) in Exercise 51 to find

(a) $\nabla f(r)$ if $f(r) = re^{-3r}$

(b) $f(r)$ if $\nabla f(r) = 3r^2 \mathbf{r}$ and $f(2) = 1$.

53. Let $\mathbf{u}_r$ be a unit vector making an angle θ with the positive x-axis, and let $\mathbf{u}_\theta$ be a unit vector 90° counterclockwise from $\mathbf{u}_r$. Show that if $z = f(x, y)$, $x = r\cos\theta$, and $y = r\sin\theta$, then

$$\nabla z = \frac{\partial z}{\partial r}\mathbf{u}_r + \frac{1}{r}\frac{\partial z}{\partial \theta}\mathbf{u}_\theta$$

[*Hint:* Use parts (a) and (b) of Exercise 62, Section 16.4.]

54. Prove: If f and g are differentiable, then

(a) $\nabla(f + g) = \nabla f + \nabla g$

(b) $\nabla(cf) = c\nabla f$ (c constant)

(c) $\nabla(fg) = f\nabla g + g\nabla f$

(d) $\nabla\left(\dfrac{f}{g}\right) = \dfrac{g\nabla f - f\nabla g}{g^2}$

(e) $\nabla(f^p) = pf^{p-1}\nabla f$.

55. Prove: If $x = x(t)$ and $y = y(t)$ are differentiable at t, and if $z = f(x, y)$ is differentiable at the point $(x(t), y(t))$, then

$$\frac{dz}{dt} = \nabla z \cdot \mathbf{r}'(t)$$

where $\mathbf{r}(t) = x(t)\mathbf{i} + y(t)\mathbf{j}$.

56. Prove: If f, f_x, and f_y are continuous on a circular region, and if $\nabla f(x, y) = \mathbf{0}$ throughout the region, then $f(x, y)$ is constant on the region. [*Hint:* See Exercise 70, Section 16.4.]

57. Prove: If $D_{\mathbf{u}} f(x, y) = 0$ in two nonparallel directions, then $D_{\mathbf{u}} f(x, y) = 0$ in all directions.

■ 16.7 DIFFERENTIABILITY, DIRECTIONAL DERIVATIVES, AND GRADIENTS FOR FUNCTIONS OF THREE VARIABLES

> *In this section we shall extend most of the results obtained in the last two sections to functions of three variables. The main difference between functions of two and three variables is geometric: The graph of $z = f(x, y)$ represents a surface in 3-space, whereas $w = f(x, y, z)$ has no analogous interpretation.*

□ DIFFERENTIABILITY

The definition of differentiability for functions of three variables and the basic theorems about differentiability are direct generalizations of the corresponding results for functions of two variables. (Compare the following with Definition 16.4.2, Theorem 16.4.4, and Theorem 16.4.7).

16.7.1 DEFINITION. A function f of three variables is defined to be **differentiable** at the point (x_0, y_0, z_0) if the partial derivatives $f_x(x_0, y_0, z_0)$, $f_y(x_0, y_0, z_0)$, and $f_z(x_0, y_0, z_0)$ exist and

$$\Delta f = f(x_0 + \Delta x, y_0 + \Delta y, z_0 + \Delta z) - f(x_0, y_0, z_0)$$

can be written in the form

$$\Delta f = f_x(x_0, y_0, z_0)\,\Delta x + f_y(x_0, y_0, z_0)\,\Delta y + f_z(x_0, y_0, z_0)\,\Delta z + \epsilon_1\,\Delta x + \epsilon_2\,\Delta y + \epsilon_3\,\Delta z$$

where ϵ_1, ϵ_2, and ϵ_3 are functions of Δx, Δy, and Δz such that $\epsilon_1 \to 0$, $\epsilon_2 \to 0$, and $\epsilon_3 \to 0$ as $(\Delta x, \Delta y, \Delta z) \to (0, 0, 0)$.

16.7.2 THEOREM. *If f has first-order partial derivatives at each point of some spherical region centered at (x_0, y_0, z_0), and if these partial derivatives are continuous at (x_0, y_0, z_0), then f is differentiable at (x_0, y_0, z_0).*

> **16.7.3** THEOREM (*Chain Rule*). *If* $x = x(t)$, $y = y(t)$, *and* $z = z(t)$, *are differentiable at* t *and* $w = f(x, y, z)$ *is differentiable at the point* $(x(t), y(t), z(t))$, *then* $w = f(x(t), y(t), z(t))$ *is differentiable at* t, *and*
>
> $$\frac{dw}{dt} = \frac{\partial w}{\partial x}\frac{dx}{dt} + \frac{\partial w}{\partial y}\frac{dy}{dt} + \frac{\partial w}{\partial z}\frac{dz}{dt}$$

The meaning of such terms as **differentiable on a region R** and **differentiable** (i.e., **differentiable everywhere**) should be clear, keeping in mind the regions involved are in 3-space. Moreover, as for functions of one and two variables, a function of three variables is continuous at a point if it is differentiable at that point.

Example 1 Suppose that

$$w = x^3 y^2 z, \quad x = t^2, \quad y = t^3, \quad z = t^4$$

Use the chain rule to find dw/dt.

Solution. By the chain rule,

$$\frac{dw}{dt} = \frac{\partial w}{\partial x}\frac{dx}{dt} + \frac{\partial w}{\partial y}\frac{dy}{dt} + \frac{\partial w}{\partial z}\frac{dz}{dt}$$
$$= (3x^2 y^2 z)(2t) + (2x^3 yz)(3t^2) + (x^3 y^2)(4t^3)$$
$$= (3t^{14})(2t) + (2t^{13})(3t^2) + (t^{12})(4t^3) = 16t^{15} \quad ◀$$

Other variations of the chain rule for functions of three variables will be considered in the next section.

REMARK. The most significant difference between working with functions of two variables and functions of three variables is geometric. For a function of two variables the equation $z = f(x, y)$ can be graphed as a surface in three-dimensional space. However, for a function of three variables, there is no direct way to graph $w = f(x, y, z)$, since "four dimensions" would be required (one dimension for each variable). This is not devastating, however; it simply means that we must rely more heavily on the analytic formulas than on the geometry.

☐ **DIRECTIONAL DERIVATIVES**

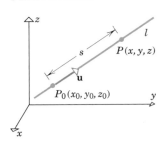

Figure 16.7.1

To define a directional derivative at a point (x_0, y_0, z_0) for a function f of three variables, we shall use a unit vector $\mathbf{u} = \langle u_1, u_2, u_3 \rangle$ to designate the direction, and we shall let l be the line through (x_0, y_0, z_0) that is parallel to $\mathbf{u}$ (Figure 16.7.1). This line can be represented parametrically as

$$x = x_0 + su_1, \quad y = y_0 + su_2, \quad z = z_0 + su_3$$

where s is an arc-length parameter with its reference point at $P_0(x_0, y_0, z_0)$ and the positive direction in the direction of $\mathbf{u}$ [compare with Formula (1) of Section 16.6]. As s increases, the point $P(x, y, z)$ moves in the direction of $\mathbf{u}$ along l, and the value of $w = f(x, y, z)$ changes with s. As for functions of two variables, we define $D_{\mathbf{u}} f(x_0, y_0, z_0)$ to be the instantaneous rate of change of w with respect to s at (x_0, y_0, z_0). Proceeding as in the derivation of Formula (4) in Section 16.6 leads to the following definition.

16.7.4   DEFINITION.   If f is differentiable at (x_0, y_0, z_0), and if $\mathbf{u} = \langle u_1, u_2, u_3 \rangle$ is a unit vector, then the **directional derivative** of f at (x_0, y_0, z_0) in the direction of $\mathbf{u}$ is defined by

$$D_{\mathbf{u}}f(x_0, y_0, z_0) = f_x(x_0, y_0, z_0)u_1 + f_y(x_0, y_0, z_0)u_2 + f_z(x_0, y_0, z_0)u_3 \tag{1}$$

☐ **THE GRADIENT**

The definition of the gradient of a function of three variables is

$$\nabla f(x, y, z) = f_x(x, y, z)\mathbf{i} + f_y(x, y, z)\mathbf{j} + f_z(x, y, z)\mathbf{k} \tag{2}$$

which is identical to Definition 16.6.2, except for the additional third component. It follows from (1) and (2) that

$$D_{\mathbf{u}}f(x, y, z) = \nabla f(x, y, z) \cdot \mathbf{u} \tag{3}$$

which parallels Formula (8) of Section 16.6. Using this result, the reader should have no trouble proving the following extension of Theorem 16.6.3.

16.7.5   THEOREM.   *Let f be a function of three variables that is differentiable at (x_0, y_0, z_0).*

(a)   *If $\nabla f(x_0, y_0, z_0) = \mathbf{0}$, then all directional derivatives of f at (x_0, y_0, z_0) are zero.*

(b)   *If $\nabla f(x_0, y_0, z_0) \neq \mathbf{0}$, then among all possible directional derivatives of f at (x_0, y_0, z_0), the derivative in the direction of $\nabla f(x_0, y_0, z_0)$ has the largest value. The value of that directional derivative is $\|\nabla f(x_0, y_0, z_0)\|$.*

(c)   *If $\nabla f(x_0, y_0, z_0) \neq \mathbf{0}$, then among all possible directional derivatives of f at (x_0, y_0, z_0), the derivative in the direction opposite to that of $\nabla f(x_0, y_0, z_0)$ has the smallest value. The value of that directional derivative is $-\|\nabla f(x_0, y_0, z_0)\|$.*

Example 2   Find the directional derivative of $f(x, y, z) = x^2 y - yz^3 + z$ at the point $P(1, -2, 0)$ in the direction of the vector $\mathbf{a} = 2\mathbf{i} + \mathbf{j} - 2\mathbf{k}$, and find the maximum rate of increase of f at P.

Solution.   Since

$$f_x(x, y, z) = 2xy, \quad f_y(x, y, z) = x^2 - z^3, \quad f_z(x, y, z) = -3yz^2 + 1$$

it follows that

$$\nabla f(x, y, z) = 2xy\mathbf{i} + (x^2 - z^3)\mathbf{j} + (-3yz^2 + 1)\mathbf{k}$$

$$\nabla f(1, -2, 0) = -4\mathbf{i} + \mathbf{j} + \mathbf{k}$$

A unit vector in the direction of $\mathbf{a}$ is

$$\mathbf{u} = \frac{\mathbf{a}}{\|\mathbf{a}\|} = \frac{1}{\sqrt{9}}(2\mathbf{i} + \mathbf{j} - 2\mathbf{k}) = \tfrac{2}{3}\mathbf{i} + \tfrac{1}{3}\mathbf{j} - \tfrac{2}{3}\mathbf{k}$$

Therefore

$$D_{\mathbf{u}}f(1, -2, 0) = \nabla f(1, -2, 0) \cdot \mathbf{u} = (-4)(\tfrac{2}{3}) + (1)(\tfrac{1}{3}) + (1)(-\tfrac{2}{3}) = -3$$

The maximum rate of increase of f at P is

$$\|\nabla f(1, -2, 0)\| = \sqrt{(-4)^2 + (1)^2 + (1)^2} = 3\sqrt{2} \quad \blacktriangleleft$$

Our next objective is to establish a geometric relationship between the level surfaces and the gradient of a function f of three variables. For this purpose we note that if (x_0, y_0, z_0)

is any point in the domain of f, and if $f(x_0, y_0, z_0) = c$, then under appropriate conditions the equation

$$f(x, y, z) = c$$

defines a unique level surface of f through (x_0, y_0, z_0) that has a tangent plane at (x_0, y_0, z_0).*

If $f(x, y, z) = c$ is the level surface of f that passes through the point (x_0, y_0, z_0), then the value of $f(x, y, z)$ remains constant as (x, y, z) moves away from (x_0, y_0, z_0) along the level surface. It seems plausible, therefore, that the directional derivative of f at (x_0, y_0, z_0) is zero in the direction of any unit vector $\mathbf{u}$ at (x_0, y_0, z_0) that lies in the tangent plane to the level surface at (x_0, y_0, z_0). That is, for any such $\mathbf{u}$

$$D_{\mathbf{u}}f(x_0, y_0, z_0) = \nabla f(x_0, y_0, z_0) \cdot \mathbf{u} = 0 \qquad (4)$$

which is analogous to Formula (10) in Section 16.6. We shall omit the formal proof.

The last equality in (4) implies that the gradient of f is perpendicular to every unit vector $\mathbf{u}$ at (x_0, y_0, z_0) that lies in the tangent plane and hence is perpendicular to the tangent plane itself. This means that the gradient at (x_0, y_0, z_0) is normal to the level surface at (x_0, y_0, z_0) (Figure 16.7.2). In summary, we have the following theorem.

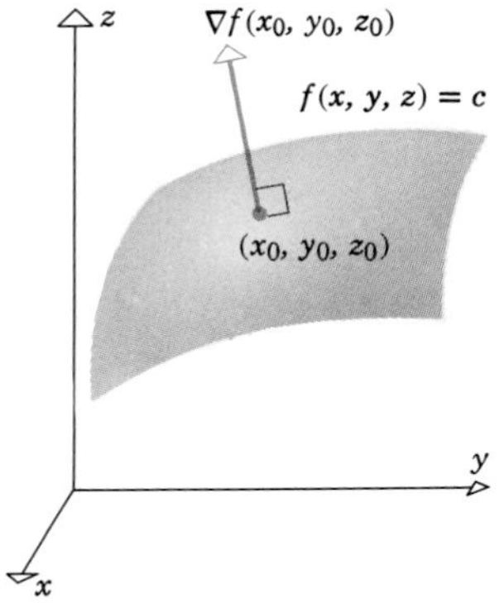

16.7.6 THEOREM. *If f is differentiable at (x_0, y_0, z_0), then $\nabla f(x_0, y_0, z_0)$ is normal to the level surface of $f(x, y, z)$ through (x_0, y_0, z_0).*

Example 3 Find an equation of the plane that is tangent to the ellipsoid $x^2 + 4y^2 + z^2 = 18$ at the point $(1, 2, -1)$.

Solution. If we let $F(x, y, z) = x^2 + 4y^2 + z^2$, then the given equation has the form $F(x, y, z) = 18$, which may be viewed as the equation of a level surface for F. Thus, the vector $\nabla F(1, 2, -1)$ is normal to the ellipsoid at the point $(1, 2, -1)$. To find this vector we write

$$\nabla F(x, y, z) = \frac{\partial F}{\partial x}\mathbf{i} + \frac{\partial F}{\partial y}\mathbf{j} + \frac{\partial F}{\partial z}\mathbf{k} = 2x\mathbf{i} + 8y\mathbf{j} + 2z\mathbf{k}$$

$$\nabla F(1, 2, -1) = 2\mathbf{i} + 16\mathbf{j} - 2\mathbf{k}$$

Using this normal and the point $(1, 2, -1)$, we obtain as the equation of the tangent plane

$$2(x - 1) + 16(y - 2) - 2(z + 1) = 0$$

or

$$x + 8y - z = 18 \qquad \blacktriangleleft$$

□ **TOTAL DIFFERENTIALS**

If $w = f(x, y, z)$, then we define the ***increment*** Δw (also written Δf) to be

$$\Delta w = f(x + \Delta x, y + \Delta y, z + \Delta z) - f(x, y, z)$$

and we define the ***total differential*** dw (also written df) to be

$$dw = f_x(x, y, z)\,dx + f_y(x, y, z)\,dy + f_z(x, y, z)\,dz$$

where Δx, Δy, Δz, dx, dy, and dz are all variables representing changes in the values of x, y, and z.

Figure 16.7.2

(figure labels: z, $\nabla f(x_0, y_0, z_0)$, $f(x, y, z) = c$, (x_0, y_0, z_0), y, x)

---

*The conditions under which this occurs are specified in a theorem from advanced calculus called the "implicit function theorem." We shall assume in the remainder of this section that f is differentiable at (x_0, y_0, z_0) and that the level surface through the point (x_0, y_0, z_0) has a tangent plane at that point.

The increment Δw represents the change in the value of $w = f(x, y, z)$ when x, y, and z are changed by amounts Δx, Δy, and Δz, respectively. However, for functions of three variables, the total differential dw has no natural geometric interpretation.

If we let

$$dx = \Delta x, \quad dy = \Delta y, \quad dz = \Delta z$$

and if $w = f(x, y, z)$ is differentiable at (x, y, z), then it follows from Definition 16.7.1 that

$$\Delta w = dw + \epsilon_1 \Delta x + \epsilon_2 \Delta y + \epsilon_3 \Delta z \tag{5}$$

where $\epsilon_1 \to 0$, $\epsilon_2 \to 0$, $\epsilon_3 \to 0$ as $(\Delta x, \Delta y, \Delta z) \to (0, 0, 0)$. Thus, when Δx, Δy, and Δz are small, it follows from (5) that

$$\Delta w \approx dw$$

Example 4   The length, width, and height of a rectangular box are each measured with an error of at most 5%. Find an upper bound on the maximum possible percentage error that results if these quantities are used to calculate the diagonal of the box.

Solution.   Let x, y, z, and D be the true length, width, height, and diagonal of the box, respectively; and let Δx, Δy, Δz, and ΔD be the errors in these quantities. We are given that

$$|\Delta x/x| \le 0.05, \quad |\Delta y/y| \le 0.05, \quad |\Delta z/z| \le 0.05$$

We want to estimate $|\Delta D/D|$. Since the diagonal D is related to the length, width, and height by

$$D = \sqrt{x^2 + y^2 + z^2}$$

it follows that

$$dD = \frac{\partial D}{\partial x} dx + \frac{\partial D}{\partial y} dy + \frac{\partial D}{\partial z} dz$$

$$= \frac{x}{\sqrt{x^2 + y^2 + z^2}} dx + \frac{y}{\sqrt{x^2 + y^2 + z^2}} dy + \frac{z}{\sqrt{x^2 + y^2 + z^2}} dz$$

If we choose $\Delta x = dx$, $\Delta y = dy$, $\Delta z = dz$, then we can use the approximation $\Delta D/D \approx dD/D$. But,

$$\frac{dD}{D} = \frac{x}{x^2 + y^2 + z^2} dx + \frac{y}{x^2 + y^2 + z^2} dy + \frac{z}{x^2 + y^2 + z^2} dz$$

or

$$\frac{dD}{D} = \frac{x^2}{x^2 + y^2 + z^2} \frac{dx}{x} + \frac{y^2}{x^2 + y^2 + z^2} \frac{dy}{y} + \frac{z^2}{x^2 + y^2 + z^2} \frac{dz}{z}$$

Thus,

$$\left| \frac{dD}{D} \right| = \left| \frac{x^2}{x^2 + y^2 + z^2} \frac{dx}{x} + \frac{y^2}{x^2 + y^2 + z^2} \frac{dy}{y} + \frac{z^2}{x^2 + y^2 + z^2} \frac{dz}{z} \right|$$

$$\le \left| \frac{x^2}{x^2 + y^2 + z^2} \frac{dx}{x} \right| + \left| \frac{y^2}{x^2 + y^2 + z^2} \frac{dy}{y} \right| + \left| \frac{z^2}{x^2 + y^2 + z^2} \frac{dz}{z} \right|$$

$$\le \frac{x^2}{x^2 + y^2 + z^2} (0.05) + \frac{y^2}{x^2 + y^2 + z^2} (0.05) + \frac{z^2}{x^2 + y^2 + z^2} (0.05)$$

$$= 0.05$$

Therefore, the maximum percentage error in D is at most 5%.   ◀

▶ Exercise Set 16.7 $\boxed{C}$ 41

In Exercises 1–4, find dw/dt using the chain rule.

1. $w = 5x^2y^3z^4$; $x = t^2, y = t^3, z = t^5$.

2. $w = \ln(3x^2 - 2y + 4z^3)$; $x = t^{1/2}, y = t^{2/3}, z = t^{-2}$.

3. $w = 5\cos xy - \sin xz$; $x = 1/t, y = t, z = t^3$.

4. $w = \sqrt{1 + x - 2yz^4x}$; $x = \ln t, y = t, z = 4t$.

5. Use the chain rule to find $\left.\dfrac{dw}{dt}\right|_{t=1}$ if $w = x^3y^2z^4$; $x = t^2$, $y = t + 2, z = 2t^4$.

6. Use the chain rule to find $\left.\dfrac{dw}{dt}\right|_{t=0}$ if $w = x\sin yz^2$; $x = \cos t, y = t^2, z = e^t$.

In Exercises 7–10, find the gradient of f at P, and then use the gradient to calculate $D_{\mathbf{u}}f$ at P.

7. $f(x, y, z) = 4x^5y^2z^3$; $P(2, -1, 1)$; $\mathbf{u} = \frac{1}{3}\mathbf{i} + \frac{2}{3}\mathbf{j} - \frac{2}{3}\mathbf{k}$.

8. $f(x, y, z) = ye^{xz} + z^2$; $P(0, 2, 3)$; $\mathbf{u} = \frac{2}{7}\mathbf{i} - \frac{3}{7}\mathbf{j} + \frac{6}{7}\mathbf{k}$.

9. $f(x, y, z) = \ln(x^2 + 2y^2 + 3z^2)$; $P(-1, 2, 4)$; $\mathbf{u} = -\frac{3}{13}\mathbf{i} - \frac{4}{13}\mathbf{j} - \frac{12}{13}\mathbf{k}$.

10. $f(x, y, z) = \sin xyz$; $P(\frac{1}{2}, \frac{1}{3}, \pi)$; $\mathbf{u} = \dfrac{1}{\sqrt{3}}\mathbf{i} - \dfrac{1}{\sqrt{3}}\mathbf{j} + \dfrac{1}{\sqrt{3}}\mathbf{k}$.

In Exercises 11–14, find the directional derivative of f at P in the direction of $\mathbf{a}$.

11. $f(x, y, z) = x^3z - yx^2 + z^2$; $P(2, -1, 1)$; $\mathbf{a} = 3\mathbf{i} - \mathbf{j} + 2\mathbf{k}$.

12. $f(x, y, z) = y - \sqrt{x^2 + z^2}$; $P(-3, 1, 4)$; $\mathbf{a} = 2\mathbf{i} - 2\mathbf{j} - \mathbf{k}$.

13. $f(x, y, z) = \dfrac{z - x}{z + y}$; $P(1, 0, -3)$; $\mathbf{a} = -6\mathbf{i} + 3\mathbf{j} - 2\mathbf{k}$.

14. $f(x, y, z) = e^{x+y+3z}$; $P(-2, 2, -1)$; $\mathbf{a} = 20\mathbf{i} - 4\mathbf{j} + 5\mathbf{k}$.

In Exercises 15–18, find a unit vector in the direction in which f increases most rapidly at P, and find the rate of increase of f in that direction.

15. $f(x, y, z) = x^3z^2 + y^3z + z - 1$; $P(1, 1, -1)$.

16. $f(x, y, z) = \sqrt{x - 3y + 4z}$; $P(0, -3, 0)$.

17. $f(x, y, z) = \dfrac{x}{z} + \dfrac{z}{y^2}$; $P(1, 2, -2)$.

18. $f(x, y, z) = \tan^{-1}\left(\dfrac{x}{y + z}\right)$; $P(4, 2, 2)$.

In Exercises 19 and 20, find a unit vector in the direction in which f decreases most rapidly at P, and find the rate of change of f in that direction.

19. $f(x, y, z) = \dfrac{x + z}{z - y}$; $P(5, 7, 6)$.

20. $f(x, y, z) = 4e^{xy}\cos z$; $P(0, 1, \pi/4)$.

21. Find the directional derivative of
$$f(x, y, z) = \frac{y}{x + z}$$
at $P(2, 1, -1)$ in the direction from P to $Q(-1, 2, 0)$.

22. Find the directional derivative of the function
$$f(x, y, z) = x^3y^2z^5 - 2xz + yz + 3x$$
at $P(-1, -2, 1)$ in the direction of the negative z-axis.

23. Given that the directional derivative of $f(x, y, z)$ at the point $(3, -2, 1)$ in the direction of $\mathbf{a} = 2\mathbf{i} - \mathbf{j} - 2\mathbf{k}$ is -5 and that $\|\nabla f(3, -2, 1)\| = 5$, find $\nabla f(3, -2, 1)$.

24. The temperature (in degrees Celsius) at a point (x, y, z) in a metal solid is
$$T(x, y, z) = \frac{xyz}{1 + x^2 + y^2 + z^2}$$
   (a) Find the rate of change of temperature at $(1, 1, 1)$ in the direction of the origin.

   (b) Find the direction in which the temperature rises most rapidly at $(1, 1, 1)$. (Express your answer as a unit vector.)

   (c) Find the rate at which the temperature rises moving from $(1, 1, 1)$ in the direction obtained in part (b).

In Exercises 25–28, find equations for the tangent plane and the line that is normal to the given surface at the point P.

25. $x^2 + y^2 + z^2 = 49$; $P(-3, 2, -6)$.

26. $xz - yz^3 + yz^2 = 2$; $P(2, -1, 1)$.

27. $\sqrt{\dfrac{z + x}{y - 1}} = z^2$; $P(3, 5, 1)$.

28. $\sin xz - 4\cos yz = 4$; $P(\pi, \pi, 1)$.

29. Show that every line that is normal to the sphere
$$x^2 + y^2 + z^2 = 1$$
passes through the origin.

30. Find all points on the ellipsoid $2x^2 + 3y^2 + 4z^2 = 9$ at which the tangent plane is parallel to the plane $x - 2y + 3z = 5$.

31. Find all points on the surface $x^2 + y^2 - z^2 = 1$ at which the normal line is parallel to the line through $P(1, -2, 1)$ and $Q(4, 0, -1)$.

32. Show that the ellipsoid $2x^2 + 3y^2 + z^2 = 9$ and the sphere
$$x^2 + y^2 + z^2 - 6x - 8y - 8z + 24 = 0$$
have a common tangent plane at the point $(1, 1, 2)$.

In Exercises 33–36, find dw.

33. $w = 8x - 3y + 4z$.

34. $w = 4x^2y^3z^7 - 3xy + z + 5$.

35. $w = \tan^{-1}(xyz)$.   **36.** $w = \sqrt{x} + \sqrt{y} + \sqrt{z}$.

37. Use a total differential to approximate the change in $f(x, y, z) = 2xy^2z^3$ as (x, y, z) varies from $P(1, -1, 2)$ to $Q(0.99, -1.02, 2.02)$.

38. Use a total differential to approximate the change in $f(x, y, z) = xyz/(x + y + z)$ as (x, y, z) varies from $P(-1, -2, 4)$ to $Q(-1.04, -1.98, 3.97)$.

39. The length, width, and height of a rectangular box are measured to be 3 cm, 4 cm, and 5 cm, respectively, with a maximum error of 0.05 cm in each measurement. Use differentials to approximate the maximum error in the calculated volume.

40. The total resistance R of three resistances R_1, R_2, and R_3, connected in parallel, is given by

$$\frac{1}{R} = \frac{1}{R_1} + \frac{1}{R_2} + \frac{1}{R_3}$$

Suppose that R_1, R_2, and R_3 are measured to be 100 ohms, 200 ohms, and 500 ohms, respectively, with a maximum error of 10% in each. Use differentials to approximate the maximum percentage error in the calculated value of R.

41. The area of a triangle is to be computed from the formula $A = \frac{1}{2}ab \sin \theta$, where a and b are the lengths of two sides and θ is the included angle. Suppose that a, b, and θ are measured to be 40 ft, 50 ft, and 30°, respectively. Use differentials to approximate the maximum error in the calculated value of A if the maximum errors in a, b, and θ are $\frac{1}{2}$ ft, $\frac{1}{4}$ ft, and 2°, respectively.

42. The length, width, and height of a rectangular box are measured with errors of at most $r\%$. Use differentials to approximate the maximum percentage error in the computed value of the volume.

43. Use differentials to approximate the maximum percentage error in $w = xy^2z^3$ if x, y, and z have errors of at most 1%, 2%, and 3%, respectively.

44. Two surfaces are said to be **orthogonal** at a point of intersection if their normal lines are perpendicular at that point. Prove that the surfaces $f(x, y, z) = 0$ and $g(x, y, z) = 0$ are orthogonal at a point of intersection, (x_0, y_0, z_0), if and only if

$$f_x g_x + f_y g_y + f_z g_z = 0$$

at (x_0, y_0, z_0). [Assume that $\nabla f(x_0, y_0, z_0) \neq \mathbf{0}$ and $\nabla g(x_0, y_0, z_0) \neq \mathbf{0}$.]

45. Use the result of Exercise 44 to show that the sphere $x^2 + y^2 + z^2 = a^2$ and the cone $z^2 = x^2 + y^2$ are orthogonal at every point of intersection.

■ **16.8   FUNCTIONS OF n VARIABLES; MORE ON THE CHAIN RULE**

In this section we shall discuss functions involving more than three variables. Our main objective is to develop forms of the chain rule for such functions.

Most of the definitions and theorems we have stated for functions of two and three variables can be extended to functions of four or more variables. Recall that if

$$w = f(v_1, v_2, \ldots, v_n)$$

is a function of n variables, then there are n partial derivatives

$$\frac{\partial w}{\partial v_1}, \frac{\partial w}{\partial v_2}, \ldots, \frac{\partial w}{\partial v_n}$$

each of which is calculated by holding $n - 1$ of the variables fixed and differentiating with respect to the remaining variable. In order to define directional derivatives for a function of n variables, it is first necessary to define the notion of a vector in "n-dimensional space." This topic is studied in a branch of mathematics called **linear algebra**, and will not be considered in this text.

☐ **TOTAL DIFFERENTIALS**

If $w = f(v_1, v_2, \ldots, v_n)$, we define the ***increment*** Δw and the ***total differential*** dw to be

$$\Delta w = f(v_1 + \Delta v_1, v_2 + \Delta v_2, \ldots, v_n + \Delta v_n) - f(v_1, v_2, \ldots, v_n) \tag{1}$$

and

$$dw = \frac{\partial w}{\partial v_1} dv_1 + \frac{\partial w}{\partial v_2} dv_2 + \cdots + \frac{\partial w}{\partial v_n} dv_n \tag{2}$$

where $\Delta v_1, \Delta v_2, \ldots, \Delta v_n$ and $dv_1, dv_2, \ldots, dv_n$ are variables representing changes in the values of $v_1, v_2, \ldots, v_n$.

If $v_1, v_2, \ldots, v_n$ are functions of a single variable t, then $w = f(v_1, v_2, \ldots, v_n)$ is a function of t, and a chain-rule formula for dw/dt is

$$\frac{dw}{dt} = \frac{\partial w}{\partial v_1} \frac{dv_1}{dt} + \frac{\partial w}{\partial v_2} \frac{dv_2}{dt} + \cdots + \frac{\partial w}{\partial v_n} \frac{dv_n}{dt} \tag{3}$$

which is a natural extension of the chain-rule formulas in Theorems 16.4.7 and 16.7.3. Observe that (3) results if we formally divide both sides of (2) by dt.

☐ **CHAIN RULES**

Other forms of the chain rule arise, depending on the number of variables involved. For example, in Theorem 16.4.8 we obtained the chain-rule formulas

$$\frac{\partial z}{\partial u} = \frac{\partial z}{\partial x} \frac{\partial x}{\partial u} + \frac{\partial z}{\partial y} \frac{\partial y}{\partial u} \tag{4}$$

$$\frac{\partial z}{\partial v} = \frac{\partial z}{\partial x} \frac{\partial x}{\partial v} + \frac{\partial z}{\partial y} \frac{\partial y}{\partial v} \tag{5}$$

for the case where z is a function of two variables $z = f(x, y)$, and x and y in turn are functions of two other variables, $x = x(u, v)$, $y = y(u, v)$.

Formulas (4) and (5) can be represented schematically by a "tree diagram" constructed as follows (Figure 16.8.1). Starting with z at the top of the diagram and moving downward, join each variable by lines (or branches) to those variables on which it depends *directly*. Thus, z is joined to x and y and these in turn are each joined to u and v. Next, label each branch with a derivative whose "numerator" contains the variable at the top end of that branch, and whose "denominator" contains the variable at the bottom end of that branch. This completes the "tree." To find the formula for $\partial z/\partial u$ trace all paths through the tree that start at z and end at u. Each such path produces one of the terms in the formula for $\partial z/\partial u$ (Figure 16.8.1a). Similarly, each term in the formula for $\partial z/\partial v$ corresponds to a path starting at z and ending at v (Figure 16.8.1b).

The following examples illustrate how tree diagrams can be used to construct other forms of the chain rule.

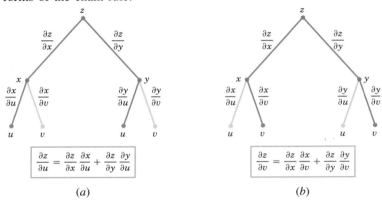

Figure 16.8.1   (a)   (b)

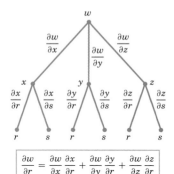

$$\frac{\partial w}{\partial r} = \frac{\partial w}{\partial x}\frac{\partial x}{\partial r} + \frac{\partial w}{\partial y}\frac{\partial y}{\partial r} + \frac{\partial w}{\partial z}\frac{\partial z}{\partial r}$$

$$\frac{\partial w}{\partial s} = \frac{\partial w}{\partial x}\frac{\partial x}{\partial s} + \frac{\partial w}{\partial y}\frac{\partial y}{\partial s} + \frac{\partial w}{\partial z}\frac{\partial z}{\partial s}$$

Figure 16.8.2

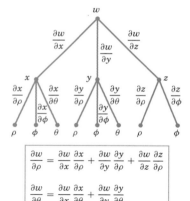

$$\frac{\partial w}{\partial \rho} = \frac{\partial w}{\partial x}\frac{\partial x}{\partial \rho} + \frac{\partial w}{\partial y}\frac{\partial y}{\partial \rho} + \frac{\partial w}{\partial z}\frac{\partial z}{\partial \rho}$$

$$\frac{\partial w}{\partial \theta} = \frac{\partial w}{\partial x}\frac{\partial x}{\partial \theta} + \frac{\partial w}{\partial y}\frac{\partial y}{\partial \theta}$$

Figure 16.8.3

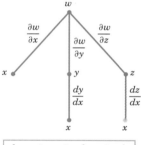

$$\frac{dw}{dx} = \frac{\partial w}{\partial x} + \frac{\partial w}{\partial y}\frac{dy}{dx} + \frac{\partial w}{\partial z}\frac{dz}{dx}$$

Figure 16.8.4

Example 1   Suppose that

$$w = e^{xyz}, \quad x = 3r + s, \quad y = 3r - s, \quad z = r^2 s$$

Use appropriate forms of the chain rule to find $\partial w/\partial r$ and $\partial w/\partial s$.

Solution.   From the tree diagram and corresponding formulas in Figure 16.8.2 we obtain

$$\frac{\partial w}{\partial r} = yze^{xyz}(3) + xze^{xyz}(3) + xye^{xyz}(2rs) = e^{xyz}(3yz + 3xz + 2xyrs)$$

and

$$\frac{\partial w}{\partial s} = yze^{xyz}(1) + xze^{xyz}(-1) + xye^{xyz}(r^2) = e^{xyz}(yz - xz + xyr^2)$$

If desired, we can express $\partial w/\partial r$ and $\partial w/\partial s$ in terms of r and s alone by replacing x, y, and z by their expressions in terms of r and s.   ◄

Example 2   Suppose that $w = x^2 + y^2 - z^2$ and

$$x = \rho \sin \phi \cos \theta, \quad y = \rho \sin \phi \sin \theta, \quad z = \rho \cos \phi$$

Use appropriate forms of the chain rule to find $\partial w/\partial \rho$ and $\partial w/\partial \theta$.

Solution.   From the tree diagram and corresponding formulas in Figure 16.8.3 we obtain

$$\frac{\partial w}{\partial \rho} = (2x) \sin \phi \cos \theta + 2y \sin \phi \sin \theta - 2z \cos \phi$$

$$= 2\rho \sin^2 \phi \cos^2 \theta + 2\rho \sin^2 \phi \sin^2 \theta - 2\rho \cos^2 \phi$$

$$= 2\rho \sin^2 \phi (\cos^2 \theta + \sin^2 \theta) - 2\rho \cos^2 \phi$$

$$= 2\rho (\sin^2 \phi - \cos^2 \phi)$$

$$= -2\rho \cos 2\phi$$

$$\frac{\partial w}{\partial \theta} = (2x)(-\rho \sin \phi \sin \theta) + (2y) \rho \sin \phi \cos \theta$$

$$= -2\rho^2 \sin^2 \phi \sin \theta \cos \theta + 2\rho^2 \sin^2 \phi \sin \theta \cos \theta$$

$$= 0$$

This result is explained by the fact that w does not vary with θ. We may see this directly by expressing w in terms of ρ, ϕ, and θ. If this is done, the expressions involving θ will cancel, leaving w as a function of ρ and ϕ alone. (Verify that $w = -\rho^2 \cos 2\phi$.)   ◄

In many applications of the chain rule, some of the variables in the function $w = f(v_1, v_2, \ldots, v_n)$ are functions of the remaining variables. Tree diagrams are especially helpful in such situations.

Example 3   Suppose that

$$w = xy + yz, \quad y = \sin x, \quad z = e^x$$

Use an appropriate form of the chain rule to find dw/dx.

Solution.   From the tree diagram and corresponding formulas in Figure 16.8.4 we obtain

$$\frac{dw}{dx} = y + (x + z) \cos x + ye^x$$

$$= \sin x + (x + e^x) \cos x + e^x \sin x$$

This same result can be obtained by first expressing w explicitly in terms of x,

$$w = x \sin x + e^x \sin x$$

and then differentiating with respect to x; however, such direct substitution is not always convenient.   ◄

REMARK.   Unlike the differential dz, a partial symbol ∂z has no meaning of its own. For example, if we were to "cancel" partial symbols in the chain-rule formula

$$\frac{\partial z}{\partial u} = \frac{\partial z}{\partial x}\frac{\partial x}{\partial u} + \frac{\partial z}{\partial y}\frac{\partial y}{\partial u}$$

we would obtain

$$\frac{\partial z}{\partial u} = \frac{\partial z}{\partial u} + \frac{\partial z}{\partial u}$$

which is false in cases where $\partial z/\partial u \neq 0$.

In each of the expressions

$$z = \sin xy, \quad z = \frac{xy}{1 + xy}, \quad z = e^{xy}$$

the independent variables occur only in the combination xy, so the substitution $t = xy$ reduces the expression to a function of one variable:

$$z = \sin t, \quad z = \frac{t}{1 + t}, \quad z = e^t$$

Conversely, if we begin with a function of one variable $z = f(t)$ and substitute $t = xy$, we obtain a function $z = f(xy)$ in which the variables appear only in the combination xy. Functions whose variables occur in fixed combinations arise frequently in applications.

Example 4   Show that a function of the form $z = f(xy)$ satisfies the equation

$$x\frac{\partial z}{\partial x} - y\frac{\partial z}{\partial y} = 0$$

(assuming the derivatives exist).

Solution.   Let $t = xy$, so that $z = f(t)$. From the tree diagram in Figure 16.8.5 we obtain the formulas

$$\frac{\partial z}{\partial x} = \frac{dz}{dt}\frac{\partial t}{\partial x} = y\frac{dz}{dt} \quad \text{and} \quad \frac{\partial z}{\partial y} = \frac{dz}{dt}\frac{\partial t}{\partial y} = x\frac{dz}{dt}$$

from which it follows that

$$x\frac{\partial z}{\partial x} - y\frac{\partial z}{\partial y} = xy\frac{dz}{dt} - yx\frac{dz}{dt} = 0 \quad ◄$$

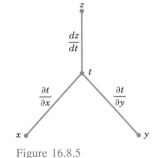

Figure 16.8.5

▶ Exercise Set 16.8

1. Let $f(v, w, x, y) = 4v^2w^3x^4y^5$. Find $\partial f/\partial v$, $\partial f/\partial w$, $\partial f/\partial x$, and $\partial f/\partial y$.

2. Let $w = r\cos st + e^u \sin ur$. Find $\partial w/\partial r$, $\partial w/\partial s$, $\partial w/\partial t$, and $\partial w/\partial u$.

3. Let $f(v_1, v_2, v_3, v_4) = \dfrac{v_1^2 - v_2^2}{v_3^2 + v_4^2}$. Find $\partial f/\partial v_1$, $\partial f/\partial v_2$, $\partial f/\partial v_3$, and $\partial f/\partial v_4$.

4. Let $V = xe^{2x-y} + we^{zw} + yw$. Find V_x, V_y, V_z, and V_w.

5. Let $f(v, w, x, y) = 2v^{1/2}w^4x^{1/2}y^{2/3}$. Find $f_v(1, -2, 4, 8)$, $f_w(1, -2, 4, 8)$, $f_x(1, -2, 4, 8)$, and $f_y(1, -2, 4, 8)$.

6. Let $u(w, x, y, z) = xe^{yw}\sin^2 z$. Find

   (a) $\dfrac{\partial u}{\partial x}(0, 0, 1, \pi)$ (b) $\dfrac{\partial u}{\partial y}(0, 0, 1, \pi)$

   (c) $\dfrac{\partial u}{\partial w}(0, 0, 1, \pi)$ (d) $\dfrac{\partial u}{\partial z}(0, 0, 1, \pi)$

   (e) $\dfrac{\partial^4 u}{\partial x\, \partial y\, \partial w\, \partial z}$ (f) $\dfrac{\partial^4 u}{\partial w\, \partial z\, \partial y^2}$.

In Exercises 7–13, use appropriate forms of the chain rule to find the derivatives.

7. Let $v = 7w^2x^3y^4z^5$, where $w = t^4$, $x = t^3$, $y = t^2$, $z = t$. Find dv/dt.

8. Let $z = u^7$, where $u = 5x^2 - 2y^3$. Find $\partial z/\partial x$ and $\partial z/\partial y$.

9. Let $z = \ln(x^2 + 1)$, where $x = r\cos\theta$. Find $\partial z/\partial r$ and $\partial z/\partial\theta$.

10. Let $u = rs^2\ln t$, $r = x^2$, $s = 4y + 1$, $t = xy^3$. Find $\partial u/\partial x$ and $\partial u/\partial y$.

11. Let $w = 4x^2 + 4y^2 + z^2$, $x = \rho\sin\phi\cos\theta$, $y = \rho\sin\phi\sin\theta$, $z = \rho\cos\phi$. Find $\partial w/\partial\rho$, $\partial w/\partial\phi$, and $\partial w/\partial\theta$.

12. Let $w = 3xy^2z^3$, $y = 3x^2 + 2$, $z = \sqrt{x - 1}$. Find dw/dx.

13. Let $w = \sqrt{x^2 + y^2 + z^2}$, $x = \cos 2y$, and $z = \sqrt{y}$. Find dw/dy.

14. The length, width, and height of a rectangular box are increasing at rates of 1 in/sec, 2 in/sec, and 3 in/sec, respectively.

    (a) At what rate is the volume increasing when the length is 2 in., the width is 3 in., and the height is 6 in.?

    (b) At what rate is the length of the diagonal increasing at that instant?

15. Angle A of triangle ABC is increasing at a rate of $\pi/60$ rad/sec, side AB is increasing at a rate of 2 cm/sec, and side AC is increasing at a rate of 4 cm/sec. At what rate is the length of BC changing when angle A is $\pi/3$ rad, $AB = 20$ cm, and $AC = 10$ cm? Is the length of BC increasing or decreasing? [*Hint:* Use the law of cosines.]

16. The area A of a triangle is given by $A = \frac{1}{2}ab\sin\theta$, where a and b are the lengths of two sides and θ is the angle between these sides. Suppose that $a = 5$, $b = 10$, and $\theta = \pi/3$. Find

    (a) the rate at which A changes with a if b and θ are held constant

    (b) the rate at which A changes with θ if a and b are held constant

    (c) the rate at which b changes with a if A and θ are held constant.

17. Suppose that $x^2 + 4xz + z^2 - 3yz + 5 = 0$. Find $\partial z/\partial x$ and $\partial z/\partial y$ by implicit differentiation.

18. Suppose that $e^{xy}\cos yz - e^{yz}\sin xz + 2 = 0$. Find $\partial z/\partial x$ and $\partial z/\partial y$ by implicit differentiation.

19. Let $f(w, x, y, z) = wz\tan^{-1}\dfrac{x}{y} + 5w$. Show that

$$f_{ww} + f_{xx} + f_{yy} + f_{zz} = 0$$

In the remaining exercises, you may assume that all derivatives mentioned exist.

20. Let f be a function of one variable, and let $z = f(x + 2y)$. Show that

$$2\frac{\partial z}{\partial x} - \frac{\partial z}{\partial y} = 0$$

21. Let f be a function of one variable and let $z = f(x^2 + y^2)$. Show that

$$y\frac{\partial z}{\partial x} - x\frac{\partial z}{\partial y} = 0$$

22. Let f be a function of one variable, and let $w = f(\rho)$, where $\rho = (x^2 + y^2 + z^2)^{1/2}$. Show that

$$\left(\frac{\partial w}{\partial x}\right)^2 + \left(\frac{\partial w}{\partial y}\right)^2 + \left(\frac{\partial w}{\partial z}\right)^2 = \left(\frac{dw}{d\rho}\right)^2$$

23. Let f be a function of three variables and suppose that $w = f(x - y, y - z, z - x)$. Show that

$$\frac{\partial w}{\partial x} + \frac{\partial w}{\partial y} + \frac{\partial w}{\partial z} = 0$$

24. Let $w = f(x, y, z)$, $x = \rho\sin\phi\cos\theta$, $y = \rho\sin\phi\sin\theta$, and $z = \rho\cos\phi$. Express $\partial w/\partial\rho$, $\partial w/\partial\phi$, and $\partial w/\partial\theta$ in terms of $\partial w/\partial x$, $\partial w/\partial y$, and $\partial w/\partial z$.

25. Assume that $F(x, y, z) = 0$ defines z implicitly as a function of x and y. Show that

$$\frac{\partial z}{\partial x} = -\frac{\partial F/\partial x}{\partial F/\partial z} \quad\text{and}\quad \frac{\partial z}{\partial y} = -\frac{\partial F/\partial y}{\partial F/\partial z}$$

In Exercises 26–28, use the formulas in Exercise 25 to find $\partial z/\partial x$ and $\partial z/\partial y$.

26. $x^2 - 3yz^2 + xyz - 2 = 0$.

27. $ye^x - 5\sin 3z = 3z$.

28. $\ln(1 + z) + xy^2 + z = 1$.

29. Given that $u = u(x, y, z)$, $v = v(x, y, z)$, and $w = w(x, y, z)$, show that

$$\nabla f(u, v, w) = \frac{\partial f}{\partial u}\nabla u + \frac{\partial f}{\partial v}\nabla v + \frac{\partial f}{\partial w}\nabla w$$

30. Let $w = f(x, y, z)$ where $z = g(x, y)$. Taking x and y as the independent variables, express each of the following in terms of $\partial f/\partial x$, $\partial f/\partial y$, $\partial f/\partial z$, $\partial z/\partial x$, and $\partial z/\partial y$.

    (a) $\partial w/\partial x$ (b) $\partial w/\partial y$.

31. Let $w = \ln(e^r + e^s + e^t + e^u)$. Show that

$$w_{rstu} = -6e^{r+s+t+u-4w}$$

[*Hint:* Take advantage of the relationship $e^w = e^r + e^s + e^t + e^u$.]

32. Suppose that w is a function of x_1, x_2, and x_3, and

$$x_1 = a_1y_1 + b_1y_2$$
$$x_2 = a_2y_1 + b_2y_2$$
$$x_3 = a_3y_1 + b_3y_2$$

where the a's and b's are constants. Express $\partial w/\partial y_1$ and $\partial w/\partial y_2$ in terms of $\partial w/\partial x_1$, $\partial w/\partial x_2$, and $\partial w/\partial x_3$.

33. (a) Let w be a function of x_1, x_2, x_3, and x_4, and let each x_i be a function of t. Find a chain-rule formula for dw/dt.

   (b) Let w be a function of x_1, x_2, x_3, and x_4, and let each x_i be a function of v_1, v_2, and v_3. Find chain-rule formulas for $\partial w/\partial v_1$, $\partial w/\partial v_2$, and $\partial w/\partial v_3$.

34. Let $w = (x_1^2 + x_2^2 + \cdots + x_n^2)^k$, where $n > 2$. For what values of k does

$$\frac{\partial^2 w}{\partial x_1^2} + \frac{\partial^2 w}{\partial x_2^2} + \cdots + \frac{\partial^2 w}{\partial x_n^2} = 0$$

hold?

35. Show that

$$\frac{d}{dx}\left[\int_{a(x)}^{b(x)} f(t)\, dt\right] = f(b(x))b'(x) - f(a(x))a'(x)$$

This result is called **Leibniz' rule**. [*Hint:* Let $u = a(x)$, $v = b(x)$, and

$$F(u, v) = \int_u^v f(t)\, dt$$

Then use a chain rule and Theorem 5.9.1.]

36. Use the result of Exercise 35 to compute the following derivatives without performing the integrations.

   (a) $\dfrac{d}{dx} \displaystyle\int_x^{x^2} e^{t^2}\, dt$

   (b) $\dfrac{d}{dx} \displaystyle\int_{\sin x}^{\cos x} (t^3 + 2)^{2/3}\, dt$

   (c) $\dfrac{d}{dx} \displaystyle\int_{3x}^{x^3} \sin^5 t\, dt$

   (d) $\dfrac{d}{dx} \displaystyle\int_{e^x}^{e^{2x}} (\ln t)^4\, dt$.

■ 16.9 MAXIMA AND MINIMA OF FUNCTIONS OF TWO VARIABLES

Earlier in this text we learned how to find maximum and minimum values of a function of one variable. In this section we shall develop similar techniques for functions of two variables.

□ **EXTREMA**

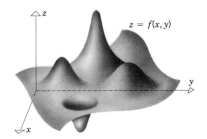

Figure 16.9.1

If we imagine the graph of a function f of two variables to be a portion of the earth's terrain (Figure 16.9.1), then the mountaintops, which are the high points in their immediate vicinity, are called *relative maxima* of f, and the valley bottoms, which are the low points in their immediate vicinity, are called *relative minima* of f.

Just as a geologist might be interested in finding the highest mountain and deepest valley in an entire mountain range, so a mathematician might be interested in finding the largest and smallest values of $f(x, y)$ over the *entire* domain of f. These are called the *absolute maximum* and *absolute minimum values* of f. The following definitions make these informal ideas precise.

16.9.1 DEFINITION. A function f of two variables is said to have a **relative maximum** at a point (x_0, y_0) if there is a circle centered at (x_0, y_0) such that $f(x_0, y_0) \geq f(x, y)$ for all points (x, y) in the domain of f that lie inside the circle, and f is said to have an **absolute maximum** at (x_0, y_0) if $f(x_0, y_0) \geq f(x, y)$ for all points (x, y) in the domain of f.

16.9.2 DEFINITION. A function f of two variables is said to have a **relative minimum** at a point (x_0, y_0) if there is a circle centered at (x_0, y_0) such that $f(x_0, y_0) \leq f(x, y)$ for all points (x, y) in the domain of f that lie inside the circle, and f is said to have an **absolute minimum** at (x_0, y_0) if $f(x_0, y_0) \leq f(x, y)$ for all points (x, y) in the domain of f.

If f has a relative maximum or a relative minimum at (x_0, y_0), then we say that f has a **relative extremum** at (x_0, y_0), and if f has an absolute maximum or absolute minimum at (x_0, y_0), then we say that f has an **absolute extremum** at (x_0, y_0).

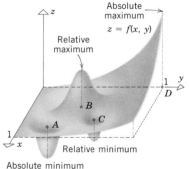

Figure 16.9.2

In Figure 16.9.2 we have sketched the graph of a function f whose domain is the closed square region in the xy-plane whose points satisfy the inequalities $0 \le x \le 1$, $0 \le y \le 1$. The function f has relative minima at the points A and C and a relative maximum at B. There is an absolute minimum at A and an absolute maximum at D.

For functions of two variables we shall be concerned with two important questions:

• Are there any relative or absolute extrema?

• If so, where are they located?

For functions of one variable that are continuous on a closed interval, the Extreme-Value Theorem (4.6.4) answered the existence question for absolute extrema. The following theorem, which we state without proof, is the corresponding result for functions of two variables.

> **16.9.3** **THEOREM** (*Extreme-Value Theorem*). *If $f(x, y)$ is continuous on a closed and bounded set R, then f has both an absolute maximum and an absolute minimum on R.*

Example 1 The square region R whose points satisfy the inequalities

$$0 \le x \le 1 \quad \text{and} \quad 0 \le y \le 1$$

is a closed and bounded set in the xy-plane. The function f whose graph is sketched in Figure 16.9.2 is continuous on R; thus, it is guaranteed to have an absolute maximum and minimum on R by the preceding theorem. These occur at points D and A that are shown in the figure. ◄

REMARK. If any of the conditions in the Extreme-Value Theorem fail to hold, then there is no guarantee that an absolute maximum or absolute minimum exists on the region R. Thus, a discontinuous function on a closed and bounded set need not have any absolute extrema, and a continuous function on a set that is not closed and bounded also need not have any absolute extrema. (See Exercise 49.)

☐ **FINDING RELATIVE EXTREMA**

Recall that if a function g of one variable has a relative extremum at a point x_0 where g is differentiable, then $g'(x_0) = 0$. To obtain the analog of this result for functions of two variables, suppose that $f(x, y)$ has a relative maximum at a point (x_0, y_0) and that the partial derivatives of f exist at (x_0, y_0). It seems plausible geometrically that the traces of the surface $z = f(x, y)$ on the planes $x = x_0$ and $y = y_0$ have horizontal tangent lines at (x_0, y_0) (Figure 16.9.3), so

$$f_x(x_0, y_0) = 0 \quad \text{and} \quad f_y(x_0, y_0) = 0$$

The same conclusion holds if f has a relative minimum at (x_0, y_0), all of which suggests the following result.

> **16.9.4** **THEOREM.** *If f has a relative extremum at a point (x_0, y_0), and if the first-order partial derivatives of f exist at this point, then*
>
> $$f_x(x_0, y_0) = 0 \quad \text{and} \quad f_y(x_0, y_0) = 0$$

Proof. Assume that f has a relative maximum at (x_0, y_0) and that both partial derivatives of f exist at (x_0, y_0). We leave it as an exercise to show that $G(x) = f(x, y_0)$ has a relative maximum at $x = x_0$ and $H(y) = f(x_0, y)$ has a relative maximum at $y = y_0$ (Figure 16.9.3). It follows from these results that

$$G'(x_0) = f_x(x_0, y_0) = 0$$

and

$$H'(y_0) = f_y(x_0, y_0) = 0$$

(Exercise 48). The proof for a relative minimum is similar. ∎

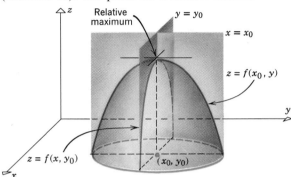

Figure 16.9.3

Recall that for a function f of one variable we defined a critical point to be a point x_0 at which $f'(x_0) = 0$ or at which $f'(x_0)$ does not exist. Analogously, for a function f of two variables we define a ***critical point*** of f to be a point (x_0, y_0) in the interior of the domain of f at which

$$f_x(x_0, y_0) = 0 \quad \text{and} \quad f_y(x_0, y_0) = 0$$

or at which one or both of the partial derivatives does not exist.

For a function of one variable, the condition $f'(x_0) = 0$ is *not* sufficient to guarantee that f has a relative extremum at x_0. (The graph of f may have an inflection point with a horizontal tangent line at x_0.) Similarly, the conditions $f_x(x_0, y_0) = 0$ and $f_y(x_0, y_0) = 0$ are not sufficient to guarantee that a function f of two variables has a relative extremum at (x_0, y_0).

Example 2 The graphs of the functions

$$z = f(x, y) = x^2 + y^2 \qquad \boxed{\text{Paraboloid}}$$
$$z = g(x, y) = 1 - x^2 - y^2 \qquad \boxed{\text{Paraboloid}}$$
$$z = h(x, y) = y^2 - x^2 \qquad \boxed{\text{Hyperbolic paraboloid}}$$

are the quadric surfaces graphed in Figure 16.9.4. We have

$$f_x(x, y) = 2x, \qquad f_y(x, y) = 2y$$
$$g_x(x, y) = -2x, \qquad g_y(x, y) = -2y$$
$$h_x(x, y) = -2x, \qquad h_y(x, y) = 2y$$

so in all three cases the partial derivatives are zero at $(0, 0)$, which means that $(0, 0)$ is a critical point for all three functions. The function f has a relative minimum at $(0, 0)$ and the function g a relative maximum. However, the function h has neither. To see this, observe that inside any circle in the xy-plane centered at $(0, 0)$, there exist points where $h(x, y)$ is positive (points on the y-axis) and there exist points where $h(x, y)$ is negative (points on the x-axis). Thus, $h(0, 0) = 0$ is neither the largest nor the smallest value of $h(x, y)$ in the circle. ◄

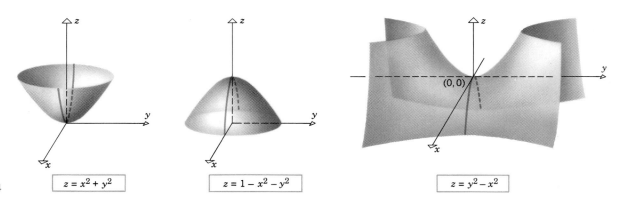

Figure 16.9.4

$$z = x^2 + y^2$$ $$z = 1 - x^2 - y^2$$ $$z = y^2 - x^2$$

A critical point at which a function does not have a relative extremum is called a *saddle point* of the function. Thus, the point $(0, 0)$ is a saddle point of the function $h(x, y) = y^2 - x^2$. (More advanced books distinguish between various types of saddle points. We shall not do this, however.)

Example 3 The graph of the function

$$f(x, y) = \sqrt{x^2 + y^2}$$

is the circular cone shown in Figure 16.9.5. The point $(0, 0)$ is a critical point of f because the partial derivatives do not both exist there. It is evident geometrically that $f_x(0, 0)$ does not exist because the trace of the cone in the plane $y = 0$ has a corner at the origin. The fact that $f_x(0, 0)$ does not exist can also be seen algebraically by noting that $f_x(0, 0)$ can be interpreted as the derivative with respect to x of the function

$$f(x, 0) = \sqrt{x^2 + 0} = |x|$$

at $x = 0$. But $|x|$ is not differentiable at $x = 0$, so $f_x(0, 0)$ does not exist. Similarly, $f_y(0, 0)$ does not exist. The function f has a relative minimum at the critical point $(0, 0)$. ◀

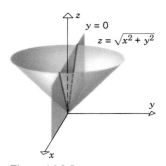

Figure 16.9.5

For functions of one variable the second derivative test (Theorem 4.3.7) was used to determine the behavior of a function at a critical point. The following theorem, which is usually proved in advanced calculus, is the analog of that theorem for functions of two variables.

16.9.5   THEOREM (*The Second Partials Test*).   *Let f be a function of two variables with continuous second-order partial derivatives in some circle centered at a critical point (x_0, y_0), and let*

$$D = f_{xx}(x_0, y_0)f_{yy}(x_0, y_0) - f_{xy}^2(x_0, y_0)$$

(a)   If $D > 0$ and $f_{xx}(x_0, y_0) > 0$, then f has a relative minimum at (x_0, y_0).
(b)   If $D > 0$ and $f_{xx}(x_0, y_0) < 0$, then f has a relative maximum at (x_0, y_0).
(c)   If $D < 0$, then f has a saddle point at (x_0, y_0).
(d)   If $D = 0$, then no conclusion can be drawn.

Example 4 Locate all relative extrema and saddle points of

$$f(x, y) = 3x^2 - 2xy + y^2 - 8y$$

Solution.   Since $f_x(x, y) = 6x - 2y$ and $f_y(x, y) = -2x + 2y - 8$, the critical points of f satisfy the equations

$$6x - 2y = 0$$
$$-2x + 2y - 8 = 0$$

Solving these for x and y yields $x = 2$, $y = 6$ (verify), so $(2, 6)$ is the only critical point. To apply Theorem 16.9.5 we need the second-order partial derivatives

$$f_{xx}(x, y) = 6, \quad f_{yy}(x, y) = 2, \quad f_{xy}(x, y) = -2$$

At the point $(2, 6)$ we have

$$D = f_{xx}(2, 6)f_{yy}(2, 6) - f^2_{xy}(2, 6) = (6)(2) - (-2)^2 = 8 > 0$$

and

$$f_{xx}(2, 6) = 6 > 0$$

so that f has a relative minimum at $(2, 6)$ by part (a) of the second partials test.   ◀

Example 5   Locate all relative extrema and saddle points of

$$f(x, y) = 4xy - x^4 - y^4$$

Solution.   Since

$$f_x(x, y) = 4y - 4x^3 \tag{1}$$
$$f_y(x, y) = 4x - 4y^3$$

the critical points of f have coordinates satisfying the equations

$$\begin{array}{ccc} 4y - 4x^3 = 0 & & y = x^3 \\ & \text{or} & \\ 4x - 4y^3 = 0 & & x = y^3 \end{array} \tag{2}$$

Substituting the top equation in the bottom yields $x = (x^3)^3$ or $x^9 - x = 0$ or $x(x^8 - 1) = 0$, which has solutions $x = 0$, $x = 1$, $x = -1$. Substituting these values in the top equation of (2) we obtain the corresponding y values $y = 0$, $y = 1$, $y = -1$. Thus, the critical points of f are $(0, 0)$, $(1, 1)$, and $(-1, -1)$.

From (1),

$$f_{xx}(x, y) = -12x^2, \quad f_{yy}(x, y) = -12y^2, \quad f_{xy}(x, y) = 4$$

which yields the following table:

CRITICAL POINT (x_0, y_0)	$f_{xx}(x_0, y_0)$	$f_{yy}(x_0, y_0)$	$f_{xy}(x_0, y_0)$	$D = f_{xx}f_{yy} - f^2_{xy}$
$(0, 0)$	0	0	4	-16
$(1, 1)$	-12	-12	4	128
$(-1, -1)$	-12	-12	4	128

At the points $(1, 1)$ and $(-1, -1)$, we have $D > 0$ and $f_{xx} < 0$, so relative maxima occur at these critical points. At $(0, 0)$ there is a saddle point since $D < 0$. The surface is shown in Figure 16.9.6.   ◀

The following analog of Theorem 4.6.5 is a key result for locating the absolute extreme values of a function of two variables.

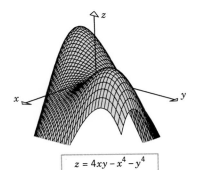

$$z = 4xy - x^4 - y^4$$

Figure 16.9.6

> **16.9.6** THEOREM. *If a function f of two variables has an absolute extremum (either an absolute maximum or an absolute minimum) at an interior point of its domain, then this extremum occurs at a critical point.*

Proof. We shall prove the result for an absolute maximum. The proof for an absolute minimum is similar.

If f has an absolute maximum at the point (x_0, y_0) in the interior of the domain of f, then f has a relative maximum at (x_0, y_0) (why?). If both partial derivatives exist at (x_0, y_0), then

$$f_x(x_0, y_0) = 0 \quad \text{and} \quad f_y(x_0, y_0) = 0$$

by Theorem 16.9.4, so (x_0, y_0) is a critical point of f. If either partial derivative does not exist, then again (x_0, y_0) is a critical point, so (x_0, y_0) is a critical point in all cases. ∎

☐ **FINDING ABSOLUTE EXTREMA ON CLOSED AND BOUNDED SETS**

If $f(x, y)$ is continuous on a closed and bounded set R, then the Extreme-Value Theorem (16.9.3) guarantees the existence of an absolute maximum and an absolute minimum of f on R. These absolute extrema can occur either on the boundary of R or in the interior of R, but if an absolute extremum occurs in the interior, then it occurs at a critical point by Theorem 16.9.6. Thus, we are led to the following procedure for finding absolute extrema:

> ### How to Find the Absolute Extrema of a Continuous Function f of Two Variables on a Closed and Bounded Set R
>
> **Step 1.** Find the critical points of f that lie in the interior of R.
>
> **Step 2.** Find all boundary points at which the absolute extrema can occur.
>
> **Step 3.** Evaluate $f(x, y)$ at the points obtained in the preceding steps. The largest of these values is the absolute maximum and the smallest the absolute minimum.

Example 6 Find the absolute maximum and minimum values of

$$f(x, y) = 3xy - 6x - 3y + 7 \tag{3}$$

on the closed triangular region R with vertices $(0, 0)$, $(3, 0)$, and $(0, 5)$.

Solution. The region R is shown in Figure 16.9.7. We have

$$\frac{\partial f}{\partial x} = 3y - 6 \quad \text{and} \quad \frac{\partial f}{\partial y} = 3x - 3$$

so all critical points occur where

$$3y - 6 = 0 \quad \text{and} \quad 3x - 3 = 0$$

Solving these equations yields $x = 1$ and $y = 2$, so $(1, 2)$ is the only critical point. As shown in Figure 16.9.7, this critical point is in the interior of R.

Next, we want to determine the location of the points on the boundary of R at which the absolute extrema might occur. The boundary of R consists of three line segments, each of which we shall treat separately:

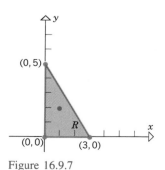

Figure 16.9.7

The line segment between $(0, 0)$ *and* $(3, 0)$:   On this line segment we have $y = 0$, so (3) simplifies to a function of the single variable x,

$$u(x) = f(x, 0) = -6x + 7, \quad 0 \leq x \leq 3$$

This function has no critical points because $u'(x) = -6$ is nonzero for all x. Thus the extreme values of $u(x)$ occur at the endpoints $x = 0$ and $x = 3$, which correspond to the points $(0, 0)$ and $(3, 0)$ of R.

The line segment between $(0, 0)$ *and* $(0, 5)$:   On this line segment we have $x = 0$, so (3) simplifies to a function of the single variable y,

$$v(y) = f(0, y) = -3y + 7, \quad 0 \leq y \leq 5$$

This function has no critical points because $v'(y) = -3$ is nonzero for all y. Thus, the extreme values of $v(y)$ occur at the endpoints $y = 0$ and $y = 5$, which correspond to the points $(0, 0)$ and $(0, 5)$ of R.

The line segment between $(3, 0)$ *and* $(0, 5)$:   In the xy-plane, an equation for this line segment is

$$y = -\tfrac{5}{3}x + 5, \quad 0 \leq x \leq 3 \tag{4}$$

so (3) simplifies to a function of the single variable x,

$$w(x) = f(x, -\tfrac{5}{3}x + 5) = 3x(-\tfrac{5}{3}x + 5) - 6x - 3(-\tfrac{5}{3}x + 5) + 7$$

$$= -5x^2 + 14x - 8, \quad 0 \leq x \leq 3$$

Since $w'(x) = -10x + 14$, the equation $w'(x) = 0$ yields $x = \tfrac{7}{5}$ as the only critical point of w. Thus, the extreme values of w occur either at the critical point $x = \tfrac{7}{5}$ or at the endpoints $x = 0$ and $x = 3$. The endpoints correspond to the points $(0, 5)$ and $(3, 0)$ of R, and from (4) the critical point corresponds to $(\tfrac{7}{5}, \tfrac{8}{3})$.

Finally, Table 16.9.1 lists the values of $f(x, y)$ at the interior critical point and at the points on the boundary where an absolute extremum can occur. From the table we conclude that the absolute maximum value of f is $f(0, 0) = 7$ and the absolute minimum value is $f(3, 0) = -11$.   ◀

Table 16.9.1

(x, y)	$(0, 0)$	$(3, 0)$	$(0, 5)$	$(\tfrac{7}{5}, \tfrac{8}{3})$	$(1, 2)$
$f(x, y)$	7	-11	-8	$\tfrac{9}{5}$	1

Example 7   Determine the dimensions of a rectangular box, open at the top, having a volume of 32 ft^3, and requiring the least amount of material for its construction.

Solution.   Let

$x = $ length of the box (in feet)

$y = $ width of the box (in feet)

$z = $ height of the box (in feet)

$S = $ surface area of the box (in square feet)

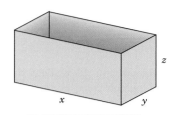

Two sides have area xz.
Two sides have area yz.
The base has area xy.

Figure 16.9.8

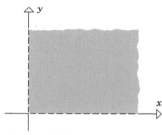

Figure 16.9.9

We may reasonably assume that the box with least surface area requires the least amount of material, so our objective is to minimize the surface area

$$S = xy + 2xz + 2yz \tag{5}$$

(see Figure 16.9.8) subject to the volume requirement

$$xyz = 32 \tag{6}$$

From (6) we obtain $z = 32/xy$, so (5) can be rewritten as

$$S = xy + \frac{64}{y} + \frac{64}{x} \tag{7}$$

which expresses S as a function of two variables. The dimensions x and y in this formula must be positive, but otherwise have no limitation, so our problem reduces to finding the absolute minimum value of S over the region for which $x > 0$ and $y > 0$ (Figure 16.9.9). Because this region is not bounded, we have no mathematical guarantee at this stage that an absolute minimum exists. However, if it does, then it occurs at a critical point of S, so we shall begin by finding the critical points. Differentiating (7) we obtain

$$\frac{\partial S}{\partial x} = y - \frac{64}{x^2}, \quad \frac{\partial S}{\partial y} = x - \frac{64}{y^2} \tag{8}$$

so the coordinates of the critical points of S satisfy

$$y - \frac{64}{x^2} = 0, \quad x - \frac{64}{y^2} = 0$$

Solving the first equation for y yields

$$y = \frac{64}{x^2} \tag{9}$$

and substituting this expression in the second equation yields

$$x - \frac{64}{(64/x^2)^2} = 0$$

which can be rewritten as

$$x\left(1 - \frac{x^3}{64}\right) = 0$$

The solutions of this equation are $x = 0$ and $x = 4$. Since we require $x > 0$, the only solution of significance is $x = 4$. Substituting this value in (9) yields $y = 4$. To see that we have located a relative minimum, we use the second partials test. From (8),

$$\frac{\partial^2 S}{\partial x^2} = \frac{128}{x^3}, \quad \frac{\partial^2 S}{\partial y^2} = \frac{128}{y^3}, \quad \frac{\partial^2 S}{\partial y \, \partial x} = 1$$

Thus, when $x = 4$ and $y = 4$, we have

$$\frac{\partial^2 S}{\partial x^2} = 2, \quad \frac{\partial^2 S}{\partial y^2} = 2, \quad \frac{\partial^2 S}{\partial y \, \partial x} = 1$$

and

$$D = \frac{\partial^2 S}{\partial x^2} \frac{\partial^2 S}{\partial y^2} - \left(\frac{\partial^2 S}{\partial y \, \partial x}\right)^2 = (2)(2) - (1)^2 = 3$$

Since $\partial^2 S / \partial x^2 > 0$ and $D > 0$, it follows from the second partials test that a relative minimum occurs when $x = y = 4$. Substituting these values in (6) yields $z = 2$, so the box using least material has a height of 2 ft and a square base whose edges are 4 ft long. ◄

REMARK.   Strictly speaking, the solution in the last example is incomplete since we have not shown that an *absolute minimum* for S occurs when $x = y = 4$ and $z = 2$, only a relative minimum. The problem of showing that a relative extremum is also an absolute extremum can be difficult for functions of two or more variables and will not be considered in this text. However, in applied problems such as this it is often obvious from physical or geometric considerations that an absolute extremum has been found.

☐ **RELATIVE EXTREMA FOR FUNCTIONS OF THREE OR MORE VARIABLES**

Definitions of relative extrema can be given for functions of three or more variables. For example, if f is a function of three variables, then f has a ***relative maximum*** at (x_0, y_0, z_0) if $f(x_0, y_0, z_0) \geq f(x, y, z)$ for all points (x, y, z) in the domain of f that lie within some sphere centered at (x_0, y_0, z_0), and for a ***relative minimum*** the inequality is reversed. If $f(x, y, z)$ has first partial derivatives, then the relative extrema occur at ***critical points***, that is, points where

$$f_x(x_0, y_0, z_0) = f_y(x_0, y_0, z_0) = f_z(x_0, y_0, z_0) = 0$$

The extension of the second partials test (Theorem 16.9.5) to functions of three or more variables is given in advanced calculus texts.

▶ **Exercise Set 16.9**

In Exercises 1–20, locate all relative maxima, relative minima, and saddle points.

1. $f(x, y) = 3x^2 + 2xy + y^2$.

2. $f(x, y) = x^3 - 3xy - y^3$.

3. $f(x, y) = y^2 + xy + 3y + 2x + 3$.

4. $f(x, y) = x^2 + xy - 2y - 2x + 1$.

5. $f(x, y) = x^2 + xy + y^2 - 3x$.

6. $f(x, y) = xy - x^3 - y^2$.

7. $f(x, y) = x^2 + 2y^2 - x^2y$.

8. $f(x, y) = 2x^2 - 4xy + y^4 + 2$.

9. $f(x, y) = x^2 + y^2 + 2/(xy)$.

10. $f(x, y) = x^3 + y^3 - 3x - 3y$.

11. $f(x, y) = x^2 + y - e^y$.

12. $f(x, y) = xe^y$.

13. $f(x, y) = e^x \sin y$.

14. $f(x, y) = xy + 2/x + 4/y$.

15. $f(x, y) = 2y^2x - yx^2 + 4xy$.

16. $f(x, y) = y \sin x$.

17. $f(x, y) = e^{-(x^2+y^2+2x)}$.

18. $f(x, y) = xy + \dfrac{a^3}{x} + \dfrac{b^3}{y}$ $(a \neq 0, b \neq 0)$.

19. $f(x, y) = \sin x + \sin y, 0 < x < \pi, 0 < y < \pi$.

20. $f(x, y) = \sin x + \sin y + \sin(x + y), 0 < x < \pi/2, 0 < y < \pi/2$.

21. (a) Show that the second partials test provides no information about the critical points of $f(x, y) = x^4 + y^4$.

(b) Classify all critical points of f as relative maxima, relative minima, or saddle points.

22. (a) Show that the second partials test provides no information about the critical points of $f(x, y) = x^4 - y^4$.

(b) Classify all critical points of f as relative maxima, relative minima, or saddle points.

23. If f is a function of one variable, and f is continuous on an interval I and has exactly one relative extremum on I, say at x_0, then f has an absolute extremum at x_0 (Theorem 4.6.6). This exercise shows that a similar result does not hold for functions of two variables.

(a) Show that $f(x, y) = 3xe^y - x^3 - e^{3y}$ has only one critical point and that a relative maximum occurs there. (See Figure 16.9.10.)

(b) Show that f does not have an absolute maximum.

[This exercise is based on the article ''The Only Critical Point in Town Test'' by Ira Rosenholtz and Lowell Smylie, *Mathematics Magazine*, Vol. 58, No. 3, May 1985, pp. 149–150.]

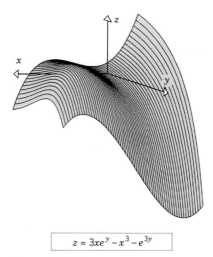

$$z = 3xe^y - x^3 - e^{3y}$$

Figure 16.9.10

24. Let f be a function of two variables that is continuous everywhere. One might think that if f has relative maxima at two points, then f must have another critical point because it is impossible to have two mountains without some sort of valley in between. This exercise shows that this is not true. Let $f(x, y) = 4x^2e^y - 2x^4 - e^{4y}$. Show that f has exactly two critical points and that a relative maximum occurs at each one (Figure 16.9.11).

[This exercise is based on the problem Two Mountains Without a Valley, proposed and solved by Ira Rosenholtz, *Mathematics Magazine*, Vol. 60, No. 1, February 1987, p. 48.]

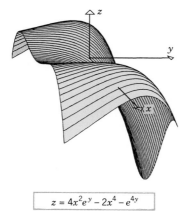

$$z = 4x^2e^y - 2x^4 - e^{4y}$$

Figure 16.9.11

In Exercises 25–30, find the absolute extrema of the given function on the indicated closed and bounded set R.

25. $f(x, y) = xy - x - 3y$; R is the triangular region with vertices $(0, 0)$, $(0, 4)$, and $(5, 0)$.

26. $f(x, y) = xy - 2x$; R is the triangular region with vertices $(0, 0)$, $(0, 4)$, and $(4, 0)$.

27. $f(x, y) = x^2 - 3y^2 - 2x + 6y$; R is the square region with vertices $(0, 0)$, $(0, 2)$, $(2, 2)$, and $(2, 0)$.

28. $f(x, y) = xe^y - x^2 - e^y$; R is the rectangular region with vertices $(0, 0)$, $(0, 1)$, $(2, 1)$, and $(2, 0)$.

29. $f(x, y) = x^2 + 2y^2 - x$; R is the circular region $x^2 + y^2 \leq 4$.

30. $f(x, y) = xy^2$; R is the region that satisfies the inequalities $x \geq 0$, $y \geq 0$, and $x^2 + y^2 \leq 1$.

31. Show that $f(x, y) = y^2 - 2xy + x^2$ has an absolute minimum at each point on the line $y = x$.

32. Find three positive numbers whose sum is 48 and such that their product is as large as possible.

33. Find three positive numbers whose sum is 27 and such that the sum of their squares is as small as possible.

34. Find all points on the plane $x + y + z = 5$ in the first octant at which $f(x, y, z) = xy^2z^2$ has a maximum value.

35. Find the points on the surface $x^2 - yz = 5$ that are closest to the origin.

36. Find the dimensions of the rectangular box of maximum volume that can be inscribed in a sphere of radius a.

37. Find the maximum volume of a rectangular box with three faces in the coordinate planes and a vertex in the first octant on the plane $x + y + z = 1$.

38. A manufacturer makes two models of an item, standard and deluxe. It costs \$40 to manufacture the standard model and \$60 for the deluxe. A market research firm estimates that if the standard model is priced at x dollars and the deluxe at y dollars, then the manufacturer will sell $500(y - x)$ of the standard items and $45,000 + 500(x - 2y)$ of the deluxe each year. How should the items be priced to maximize the profit?

39. A closed rectangular box with a volume of 16 ft^3 is made from two kinds of materials. The top and bottom are made of material costing 10¢ per square foot and the sides from material costing 5¢ per square foot. Find the dimensions of the box so that the cost of materials is minimized.

40. Use the methods of this section to find the distance between the lines

$$\begin{array}{ll} x = 3t & x = 2t \\ y = 2t \quad \text{and} & y = 2t + 3 \\ z = t & z = 2t \end{array}$$

41. Use the methods of this section to find the distance from the point $(-1, 3, 2)$ to the plane $x - 2y + z = 4$.

42. Show that among all parallelograms with perimeter l, a square with sides of length $l/4$ has maximum area. [*Hint:* The area of a parallelogram is given by the formula $A = ab \sin \alpha$, where a and b are the lengths of two adjacent sides and α is the angle between them.]

43. Determine the dimensions of a rectangular box, open at the top, having volume V, and requiring the least amount of material for its construction.

44. A length of sheet metal 27 in. wide is to be made into a water trough by bending up two sides as shown in Figure 16.9.12. Find x and ϕ so that the trapezoid-shaped cross section has a maximum area.

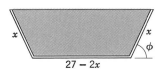

27 − 2x   Figure 16.9.12

45. A common problem in experimental work is to obtain a mathematical relationship $y = f(x)$ between two variables x and y by "fitting" a curve to points in the plane corresponding to various experimentally determined values of x and y, say

$$(x_1, y_1), (x_2, y_2), \ldots, (x_n, y_n)$$

Based on theoretical considerations, or simply on the pattern of the points, one decides on the general form of the curve $y = f(x)$ to be fitted. Often, the "curve" to be fitted is a straight line, $y = ax + b$. However, because of experimental errors in the data it is often impossible to find a line that passes through all of the points (Figure 16.9.13), so one looks for a line that "best fits" the data.

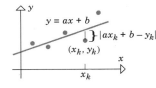

Figure 16.9.13

One criterion for selecting a line of "best fit" is to choose a and b to minimize the function

$$f(a, b) = \sum_{k=1}^{n} (ax_k + b - y_k)^2$$

Geometrically, $|ax_k + b - y_k|$ is the vertical distance between the data point (x_k, y_k) and the line $y = ax + b$, so in effect, minimizing $f(a, b)$ minimizes the sum of the squares of these vertical distances (Figure 16.9.13). This procedure is called the method of **least squares**.

(a) Show that the conditions $\dfrac{\partial f}{\partial a} = 0$ and $\dfrac{\partial f}{\partial b} = 0$ result in the equations

$$(\Sigma x_k^2)a + (\Sigma x_k)b = \Sigma x_k y_k$$
$$(\Sigma x_k)a + nb = \Sigma y_k$$

where $\Sigma = \sum_{k=1}^{n}$.

(b) Solve the equations in part (a) for a and b to show that

$$a = \frac{n(\Sigma x_k y_k) - (\Sigma x_k)(\Sigma y_k)}{n(\Sigma x_k^2) - (\Sigma x_k)^2}$$

and

$$b = \frac{(\Sigma x_k^2)(\Sigma y_k) - (\Sigma x_k)(\Sigma x_k y_k)}{n(\Sigma x_k^2) - (\Sigma x_k)^2}$$

46. This exercise shows that the values of a and b obtained in Exercise 45 produce the absolute minimum value of $f(a, b)$.

(a) Given that the arithmetic average

$$\bar{x} = \frac{1}{n} \sum_{k=1}^{n} x_k$$

minimizes $\Sigma(x_k - \bar{x})^2$ (see Exercise 60, Section 4.7), show that

$$n(\Sigma x_k^2) - (\Sigma x_k)^2 > 0$$

[*Note:* $\Sigma(x_k - \bar{x})^2 > 0$ if the x_k's are not all the same.]

(b) Find $f_{aa}(a, b)$, $f_{bb}(a, b)$, and $f_{ab}(a, b)$.

(c) Use Theorem 16.9.5 and the results of parts (a) and (b) to show that f has a relative minimum at the critical point found in Exercise 45.

(d) Based on the result of part (c) and the nature of the function $f(a, b)$, show that f takes on its absolute minimum value at the critical point.

47. Use the formulas in part (b) of Exercise 45 to find the equation of the least squares line $y = ax + b$ for the following data:

x	1	2	3	4
y	1.5	1.6	2.1	3.0

48. Suppose that f has a relative maximum at (x_0, y_0), and both $f_x(x_0, y_0)$ and $f_y(x_0, y_0)$ exist. Prove that $G(x) = f(x, y_0)$ has a relative maximum at $x = x_0$ and $H(y) = f(x_0, y)$ has a relative maximum at $y = y_0$.

49. Find an example to show that a function that has a discontinuity on a closed and bounded set need not have any absolute extrema and give an example to show that a continuous function on a set that is not closed and bounded also need not have any absolute extrema.

■ **16.10**   LAGRANGE MULTIPLIERS

In this section we shall study a powerful method for solving certain types of optimization problems that is sometimes simpler to apply than the methods studied in the last section.

☐ **EXTREMUM PROBLEMS WITH CONSTRAINTS**

In the preceding section, we considered the problem of minimizing

$$S = xy + 2xz + 2yz$$

subject to the constraint

$$xyz - 32 = 0$$

This is a special case of the following general problem, which we shall study in this section:

Three-Variable Extremum Problem with One Constraint
Maximize or minimize the function $f(x, y, z)$ subject to the constraint $g(x, y, z) = 0$.

We shall also be interested in the two-variable version of this problem.

Two-Variable Extremum Problem with One Constraint
Maximize or minimize the function $f(x, y)$ subject to the constraint $g(x, y) = 0$.

☐ **LAGRANGE MULTIPLIERS**

One way to attack these problems is to solve the constraint equation for one of the variables in terms of the rest and substitute the result into f. The resulting function of one or two variables can then be maximized or minimized by finding its critical points. (See the solution of Example 7 in the preceding section.) However, if the constraint equation is too complicated to solve for one of the variables in terms of the rest, then other techniques must be used. We shall discuss one such technique, called **the method of Lagrange multipliers**. (See p. 569 for biography.) Since a rigorous discussion of this topic requires results from advanced calculus, we shall not be too concerned about all the technical details; we shall emphasize computational techniques.

Let us begin with the two-variable problem of maximizing or minimizing $f(x, y)$ subject to the constraint $g(x, y) = 0$. The graph of $g(x, y) = 0$ is usually some curve C in the xy-plane. Geometrically, we are concerned with finding the maximum or minimum value of $f(x, y)$ as (x, y) varies over the constraint curve C. If (x_0, y_0) is a point on the constraint curve C, then we shall say that $f(x, y)$ has a **constrained relative maximum** at (x_0, y_0) if there is a circle centered at (x_0, y_0) such that

$$f(x_0, y_0) \geq f(x, y) \tag{1}$$

for all points (x, y) on C within the circle (Figure 16.10.1). For a **constrained relative minimum** at (x_0, y_0), the inequality in (1) is reversed. We shall say that f has a **constrained relative extremum** at (x_0, y_0) if f has either a constrained relative maximum or a constrained relative minimum at (x_0, y_0). The following result is the key to finding constrained relative extrema.

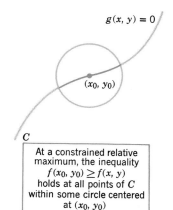

$g(x, y) = 0$

(x_0, y_0)

C

At a constrained relative maximum, the inequality $f(x_0, y_0) \geq f(x, y)$ holds at all points of C within some circle centered at (x_0, y_0)

Figure 16.10.1

> **16.10.1** THEOREM (*The Constrained-Extremum Principle for Two Variables*). *Let f and g be functions of two variables with continuous first partial derivatives on some open set containing the constraint curve g(x, y) = 0, and assume that $\nabla g \neq \mathbf{0}$ at any point on this curve. If f has a constrained relative extremum, then this extremum occurs at a point (x_0, y_0) on the constraint curve where the gradient vectors $\nabla f(x_0, y_0)$ and $\nabla g(x_0, y_0)$ are parallel; that is, there is some number λ, called a **Lagrange multiplier**, such that*
>
> $$\nabla f(x_0, y_0) = \lambda \nabla g(x_0, y_0) \qquad (2)$$

To make this result plausible, let $z = f(x, y)$, and assume that the constraint curve $g(x, y) = 0$ can be smoothly parametrized in terms of an arc-length parameter s as

$$x = x(s), \quad y = y(s)$$

where the point (x_0, y_0) corresponds to $s = s_0$. Thus, for a point (x, y) on the constraint curve we have

$$z = f(x(s), y(s))$$

Since a relative extremum occurs at (x_0, y_0), we must have $dz/ds = 0$ if $s = s_0$. Thus, from the chain rule we have

$$\frac{dz}{ds} = \frac{\partial f}{\partial x}\frac{dx}{ds} + \frac{\partial f}{\partial y}\frac{dy}{ds} = \left(\frac{\partial f}{\partial x}\mathbf{i} + \frac{\partial f}{\partial y}\mathbf{j}\right) \cdot \left(\frac{dx}{ds}\mathbf{i} + \frac{dy}{ds}\mathbf{j}\right) = 0$$

At $s = s_0$, the first vector is the gradient of f at (x_0, y_0) and the second vector is the unit tangent vector $\mathbf{u}$ to $g(x, y) = 0$ at (x_0, y_0). Thus, at $s = s_0$ we have

$$\nabla f(x_0, y_0) \cdot \mathbf{u} = 0$$

which implies that the gradient of f at (x_0, y_0) is perpendicular to the unit tangent vector to the constraint curve at (x_0, y_0). But the unit tangent vector to the constraint curve $g(x, y) = 0$ is perpendicular to the gradient of g, and hence $\nabla f(x_0, y_0)$ is parallel to $\nabla g(x_0, y_0)$, which is what we wanted to show.

Example 1 At what point on the circle $x^2 + y^2 = 1$ does the product xy have a maximum?

Solution. We want to maximize $f(x, y) = xy$ subject to the constraint

$$g(x, y) = x^2 + y^2 - 1 = 0 \qquad (3)$$

We have

$$\nabla f = y\mathbf{i} + x\mathbf{j} \quad \text{and} \quad \nabla g = 2x\mathbf{i} + 2y\mathbf{j}$$

From the formula for ∇g we see that $\nabla g = \mathbf{0}$ if and only if $x = 0$ and $y = 0$, so $\nabla g \neq \mathbf{0}$ at any point on the circle $x^2 + y^2 = 1$. Thus, it follows from Theorem 16.10.1 that at a constrained relative extremum we must have

$$\nabla f = \lambda \nabla g \quad \text{or} \quad y\mathbf{i} + x\mathbf{j} = \lambda(2x\mathbf{i} + 2y\mathbf{j})$$

which is equivalent to the pair of equations

$$y = 2x\lambda \quad \text{and} \quad x = 2y\lambda \qquad (4)$$

Since the maximum value of xy on the circle $x^2 + y^2 = 1$ is obviously greater than zero, we must have $x \neq 0$ and $y \neq 0$ at a constrained maximum. Thus, the equations in (4) can be rewritten as

$$\lambda = \frac{y}{2x} \quad \text{and} \quad \lambda = \frac{x}{2y}$$

from which we obtain

$$\frac{y}{2x} = \frac{x}{2y}$$

or

$$y^2 = x^2 \tag{5}$$

Substituting this in (3) yields

$$2x^2 - 1 = 0$$

or

$$x = 1/\sqrt{2} \quad \text{and} \quad x = -1/\sqrt{2}$$

Substituting $x = 1/\sqrt{2}$ in (5) yields $y = \pm 1/\sqrt{2}$ and substituting $x = -1/\sqrt{2}$ in (5) yields $y = \pm 1/\sqrt{2}$, so there are four candidates for the location of a maximum:

$$(1/\sqrt{2}, 1/\sqrt{2}), \quad (1/\sqrt{2}, -1/\sqrt{2}), \quad (-1/\sqrt{2}, 1/\sqrt{2}), \quad (-1/\sqrt{2}, -1/\sqrt{2})$$

At the first and fourth points the function $f(x, y) = xy$ has value $\frac{1}{2}$, while at the second and third points the value is $-\frac{1}{2}$. Thus, the constrained maximum value of $\frac{1}{2}$ occurs at $(1/\sqrt{2}, 1/\sqrt{2})$ and $(-1/\sqrt{2}, -1/\sqrt{2})$. ◀

REMARK. If c is a constant, then the functions $g(x, y)$ and $g(x, y) - c$ have the same gradient since the constant c drops out when we differentiate. Consequently, it is *not* essential to rewrite a constraint of the form $g(x, y) = c$ as $g(x, y) - c = 0$ in order to apply the constrained-extremum principle. Thus, in the last example, we could have kept the constraint in the form $x^2 + y^2 = 1$ and then taken $g(x, y) = x^2 + y^2$ rather than $g(x, y) = x^2 + y^2 - 1$.

In Exercise 12 of Section 4.7, it was stated that among all rectangles of perimeter p, a square has maximum area. This result can be obtained using Lagrange multipliers.

Example 2 Find the dimensions of a rectangle having perimeter p and maximum area.

Solution. Let

$$x = \text{length of the rectangle}$$
$$y = \text{width of the rectangle}$$
$$A = \text{area of the rectangle}$$

We want to maximize $A = xy$ subject to the perimeter constraint

$$2x + 2y = p \tag{6}$$

If we let $f(x, y) = xy$ and $g(x, y) = 2x + 2y$, then we have

$$\nabla f = y\mathbf{i} + x\mathbf{j} \quad \text{and} \quad \nabla g = 2\mathbf{i} + 2\mathbf{j}$$

Noting that $\nabla g \neq \mathbf{0}$, it follows from the constrained-extremum principle (Theorem 16.10.1) that

$$\nabla f = \lambda \nabla g \quad \text{or} \quad y\mathbf{i} + x\mathbf{j} = \lambda(2\mathbf{i} + 2\mathbf{j})$$

at a constrained relative maximum. This is equivalent to the two equations

$$y = 2\lambda \quad \text{and} \quad x = 2\lambda$$

Eliminating λ from these equations we obtain $x = y$, which shows that the rectangle is actually a square. Using this condition and constraint (6), we obtain $x = p/4$, $y = p/4$. ◀

Lagrange multipliers can also be used in the three-variable problem of maximizing or minimizing $f(x, y, z)$ subject to the constraint $g(x, y, z) = 0$. The graph of $g(x, y, z) = 0$ is generally some surface σ in 3-space. Geometrically, we are concerned with finding the maximum or minimum of the function $f(x, y, z)$ as (x, y, z) varies over the surface σ. We shall say that $f(x, y, z)$ has a **constrained relative maximum** at (x_0, y_0, z_0) if there is a sphere centered at (x_0, y_0, z_0) such that

$$f(x_0, y_0, z_0) \geq f(x, y, z)$$

for all points (x, y, z) on σ within the sphere (Figure 16.10.2). The meaning of the terms **constrained relative minimum** and **constrained relative extremum** should be clear. It can be shown that if f and g have continuous first partial derivatives and $\nabla g \neq \mathbf{0}$ on the surface $g(x, y, z) = 0$, then a constrained relative extremum can only occur at a point (x_0, y_0, z_0) where $\nabla f(x_0, y_0, z_0)$ and $\nabla g(x_0, y_0, z_0)$ are parallel, that is,

$$\nabla f(x_0, y_0, z_0) = \lambda \nabla g(x_0, y_0, z_0)$$

for some number λ.

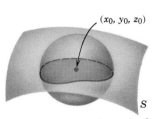

(x_0, y_0, z_0)

S

$g(x, y, z) = 0$

At a constrained relative maximum, the inequality $f(x_0, y_0, z_0) \geq f(x, y, z)$ holds at all points of S inside some sphere centered at (x_0, y_0, z_0).

Figure 16.10.2

Example 3 Find the points on the sphere $x^2 + y^2 + z^2 = 36$ that are closest to and farthest from $(1, 2, 2)$.

Solution. To avoid radicals, we shall find points on the sphere that minimize and maximize the *square* of the distance to $(1, 2, 2)$. Thus, we want to find the extrema of

$$f(x, y, z) = (x - 1)^2 + (y - 2)^2 + (z - 2)^2$$

subject to the constraint

$$x^2 + y^2 + z^2 = 36 \tag{7}$$

Therefore, with $g(x, y, z) = x^2 + y^2 + z^2$, we must have

$$\nabla f(x, y, z) = \lambda \nabla g(x, y, z)$$

at a constrained relative extremum; that is,

$$2(x - 1)\mathbf{i} + 2(y - 2)\mathbf{j} + 2(z - 2)\mathbf{k} = \lambda(2x\mathbf{i} + 2y\mathbf{j} + 2z\mathbf{k})$$

which leads to the equations

$$2(x - 1) = 2x\lambda, \quad 2(y - 2) = 2y\lambda, \quad 2(z - 2) = 2z\lambda \tag{8}$$

We may assume that x, y, and z are nonzero since $x = 0$ does not satisfy the first equation, $y = 0$ does not satisfy the second, and $z = 0$ does not satisfy the third. Thus, we can rewrite (8) as

$$\frac{x - 1}{x} = \lambda, \quad \frac{y - 2}{y} = \lambda, \quad \frac{z - 2}{z} = \lambda$$

The first two equations imply that

$$\frac{x - 1}{x} = \frac{y - 2}{y}$$

from which it follows that

$$y = 2x \tag{9}$$

Similarly, the first and third equations imply that

$$z = 2x \tag{10}$$

Substituting (9) and (10) in the constraint equation (7), we obtain

$$9x^2 = 36, \quad \text{or} \quad x = \pm 2$$

Substituting these values in (9) and (10) yields two points

$$(2, 4, 4) \quad \text{and} \quad (-2, -4, -4)$$

Since $f(2, 4, 4) = 9$ and $f(-2, -4, -4) = 81$, it follows that $(2, 4, 4)$ is the point on the sphere closest to $(1, 2, 2)$, and $(-2, -4, -4)$ is the point that is farthest (Figure 16.10.3).   ◄

Next we shall use Lagrange multipliers to solve the problem of Example 7 in the preceding section.

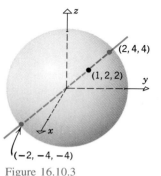

Figure 16.10.3

Example 4   Use Lagrange multipliers to determine the dimensions of a rectangular box, open at the top, having a volume of 32 ft^3, and requiring the least amount of material for its construction.

Solution.   With the notation of Example 7 in Section 16.9, the problem is to minimize the surface area

$$S = xy + 2xz + 2yz$$

subject to the volume constraint

$$xyz = 32 \tag{11}$$

If we let $f(x, y, z) = xy + 2xz + 2yz$ and $g(x, y, z) = xyz$, then

$$\nabla f = (y + 2z)\mathbf{i} + (x + 2z)\mathbf{j} + (2x + 2y)\mathbf{k} \quad \text{and} \quad \nabla g = yz\mathbf{i} + xz\mathbf{j} + xy\mathbf{k}$$

It follows that $\nabla g \neq \mathbf{0}$ at any point on the surface $xyz = 32$, since x, y, and z are all nonzero on this surface. Thus, at a constrained relative extremum we must have $\nabla f = \lambda \nabla g$, that is,

$$(y + 2z)\mathbf{i} + (x + 2z)\mathbf{j} + (2x + 2y)\mathbf{k} = \lambda(yz\mathbf{i} + xz\mathbf{j} + xy\mathbf{k})$$

This condition yields the three equations

$$y + 2z = \lambda yz, \quad x + 2z = \lambda xz, \quad 2x + 2y = \lambda xy$$

Because x, y, and z are nonzero these equations can be rewritten as

$$\frac{1}{z} + \frac{2}{y} = \lambda, \quad \frac{1}{z} + \frac{2}{x} = \lambda, \quad \frac{2}{y} + \frac{2}{x} = \lambda$$

From the first two equations,

$$y = x \tag{12}$$

and from the second and third equations, $z = \frac{1}{2}y$. This and (12) imply that

$$z = \frac{1}{2}x \tag{13}$$

Substituting (12) and (13) in the volume constraint (11) yields

$$\tfrac{1}{2}x^3 = 32$$

This equation, together with (12) and (13), yields

$$x = 4, \quad y = 4, \quad z = 2$$

which agrees with the result that was obtained in Example 7 of the preceding section.   ◄

Lagrange multipliers can also be used in problems involving two or more constraints. However, we shall not pursue this topic here.

▶ Exercise Set 16.10

In Exercises 1–7, use Lagrange multipliers to find the maximum and minimum values of f subject to the given constraint. Also, find the points at which these extreme values occur.

1. $f(x, y) = xy$; $4x^2 + 8y^2 = 16$.

2. $f(x, y) = x^2 - y$; $x^2 + y^2 = 25$.

3. $f(x, y) = 4x^3 + y^2$; $2x^2 + y^2 = 1$.

4. $f(x, y) = x - 3y - 1$; $x^2 + 3y^2 = 16$.

5. $f(x, y, z) = 2x + y - 2z$; $x^2 + y^2 + z^2 = 4$.

6. $f(x, y, z) = 3x + 6y + 2z$; $2x^2 + 4y^2 + z^2 = 70$.

7. $f(x, y, z) = xyz$; $x^2 + y^2 + z^2 = 1$.

In Exercises 8–16, solve using Lagrange multipliers.

8. Find the point on the line $2x - 4y = 3$ that is closest to the origin.

9. Find the point on the line $y = 2x + 3$ that is closest to $(4, 2)$.

10. Find the point on the plane $x + 2y + z = 1$ that is closest to the origin.

11. Find the point on the plane $4x + 3y + z = 2$ that is closest to $(1, -1, 1)$.

12. Find the points on the surface $xy - z^2 = 1$ that are closest to the origin.

13. Find the maximum value of $\sin x \sin y$, where x and y denote the acute angles of a right triangle.

14. Find a vector in 3-space whose length is 5 and whose components have the largest possible sum.

15. Find the points on the circle $x^2 + y^2 = 45$ that are closest to and farthest from $(1, 2)$.

16. The temperature at a point (x, y) on a metal plate is $T(x, y) = 4x^2 - 4xy + y^2$. An ant, walking on the plate, traverses a circle of radius 5 centered at the origin. What are the highest and lowest temperatures encountered by the ant?

In Exercises 17–24, use Lagrange multipliers to solve the indicated problems from Section 16.9.

17. Exercise 33.
18. Exercise 34.
19. Exercise 35.
20. Exercise 36.
21. Exercise 39.
22. Exercise 41.
23. Exercise 42.
24. Exercise 43.

◆ **TECHNOLOGY EXERCISES**   Chapter 16

Most of these exercises require access to a graphing calculator or a computer algebra system (CAS) such as *Mathematica*, *Maple*, or *Derive*. When you are asked to *find* an answer or to *solve* an equation, you may choose to find an exact result or a numerical approximation, depending on the particular technology you are using and on your own imagination. The form of your answers may differ from those of other students or from those in the answer section of the text, depending on how you solve the problems and the accuracy you use in your numerical approximations. Those exercises that are more appropriate for a CAS than a graphing calculator are labeled with the icon ◆.

◆ 1. **Surfaces defined parametrically—a helicoid:** Graph the portion of the helicoid (spiral ramp)

$$x = u \cos v$$
$$y = u \sin v$$
$$z = v$$

for $0 \le u \le 2$ and $0 \le v \le \pi$.

◆ 2. **Surfaces defined parametrically—a Möbius strip:** Graph the Möbius strip

$$x = \cos v + u \cos (v/2) \cos v$$
$$y = \sin v + u \cos (v/2) \sin v$$
$$z = u \sin (v/2)$$

for $-0.2 \le u \le 0.2$ and $0 \le v \le 2\pi$.

3. **Collision of a particle with a surface:** Suppose that the equations of motion of a particle are $x = t - 1$, $y = 4e^{-t}$, $z = 2 - \sqrt{t}$, where $t > 0$. Find, to the nearest tenth of a degree, the acute angle between the velocity vector and the normal line to the surface $x^2/4 + y^2 + z^2 = 1$ at the point where the particle collides with the surface.

4. **Directional derivative:** Let $f(x, y) = x + \sin(xy)$. Find all unit vectors **u** such that the directional derivative of f at the point $(2, 3)$ in the direction of **u** is 1.

5. **Maximum rate of change of temperature at a point:** Assume that the temperature T at a point (x, y, z) in 3-space is given by $T = f(x, y, z)$. If a particle starts at the point $P(-1, 3, 2)$, then the instantaneous rate of change of T with respect to distance will depend on the direction in which the particle moves. Suppose that the temperature increases at the rate of 2°/m if the particle moves toward the origin, increases at the rate of 1°/m if it moves toward the point $(-2, 4, 1)$, and decreases at the rate of 3°/m if it moves toward the point $(1, 2, 1)$. Find the unit vector that points in the direction in which T increases most rapidly starting at P. What is the maximum rate of increase of T at the point P?

6. **Motion on a surface:** Suppose that a particle is moving on the surface $z = 20/(3 + 2x^2 + y^2)$, where x, y, and z are in meters, and that the speed of the particle is 2 m/sec at the instant when the particle is at the point $P(2, 3, 1)$.

(a) Find the two possible velocity vectors at point P if the instantaneous rate of change of the elevation z at the point P is 0.3 m/sec.

(b) Find the velocity vector at P for which the elevation z increases most rapidly. For this velocity vector, what is the instantaneous rate of increase of elevation?

7. **Differentiation under the integral sign:** Let

$$F(x) = \int_c^d f(x, y)\, dy, \quad a \le x \le b$$

It can be shown that if $f(x, y)$ and $\partial f/\partial x$ are continuous for $a \le x \le b$ and $c \le y \le d$, then

$$F'(x) = \int_c^d \frac{\partial f}{\partial x}\, dy$$

(a) Use this result to find $F'(x)$ if

$$F(x) = \int_0^1 \sin(xe^y)\, dy, \quad 0 \le x \le 2$$

[*Note:* Express $F'(x)$ as an integral and perform the integration.]

(b) Use the result in part (a) to find the maximum value of $F(x)$ for $0 \le x \le 2$.

8. **Minimum distance between two curves:** Find the minimum distance between a point on the graph of $y = x^2$ and a point on the graph of $y = \ln x$. [*Suggestion:* Let (a, a^2) be a point on the graph of $y = x^2$ and $(b, \ln b)$ a point on the graph of $y = \ln x$.]

9. **Manufacturing cost:** A metal box with no lid is to be assembled from rectangular pieces of sheet metal and is to have a volume of 1 ft^3. Suppose that material costs $2/ft^2 for the base, $3/ft^2 for the sides, and that it costs $1/ft to weld each of the eight seams. Find the dimensions and cost of the most economical box.

10. **Minimum distance from a point to a surface:** Find the shortest distance from the point $(1, 0, 0)$ to a point on the surface $z = 1/(xy)$, where $x > 0$ and $y > 0$.

Pierre-Simon de Laplace (1749–1827)

17 MULTIPLE INTEGRALS

◼ 17.1 DOUBLE INTEGRALS

The notion of a definite integral can be extended to functions of two or more variables. In this section we shall discuss the double integral, which is the extension to functions of two variables.

☐ **DEFINITION OF A DOUBLE INTEGRAL**

Recall that the definite integral of a function of one variable

$$\int_a^b f(x)\, dx = \lim_{n \to +\infty} \sum_{k=1}^{n} f(x_k^*)\, \Delta x_k \qquad (1)$$

arose from the problem of finding areas under curves. Integrals of functions of two variables arise from a volume problem, which can be stated as follows:

17.1.1   PROBLEM.   *Find the volume of the solid consisting of all points that lie between a region R in the xy-plane and a surface z = f(x, y), where f is continuous on R and f(x, y) ≥ 0 for all (x, y) in R (Figure 17.1.1).*

Figure 17.1.1

Later, we shall place more restrictions on the region R, but for now let us just assume that the entire region R can be enclosed within some suitably large rectangle with sides parallel to the coordinate axes. This ensures that R does not extend indefinitely in any direction.

The procedure for finding the volume V of the solid in Figure 17.1.1 will be similar to the limiting process used for finding areas, except that now the approximating elements will be rectangular parallelepipeds rather than rectangles. We proceed as follows:

- Using lines parallel to the coordinate axes, divide the rectangle enclosing the region R into subrectangles, and exclude from consideration all those subrectangles that contain any points outside of R. This leaves only rectangles that are subsets of R (Figure 17.1.2). Assume that there are n such rectangles, and denote the area of the kth such rectangle by ΔA_k.

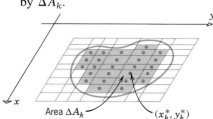

Figure 17.1.2

- Choose any arbitrary point in each subrectangle, and denote the point in the kth subrectangle by (x_k^*, y_k^*). As shown in Figure 17.1.3, the product $f(x_k^*, y_k^*)\,\Delta A_k$ is the volume of a rectangular parallelepiped with base area ΔA_k and height $f(x_k^*, y_k^*)$, so the sum

$$\sum_{k=1}^{n} f(x_k^*, y_k^*)\,\Delta A_k$$

can be viewed as an approximation to the volume V of the entire solid.

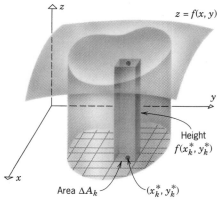

Figure 17.1.3

- There are two sources of error in the approximation: First, the parallelepipeds have flat tops, whereas the surface $z = f(x, y)$ may be curved; second, the rectangles that form the bases of the parallelepipeds do not completely cover the region R. However, if we repeat the above process with more and more subdivisions in such a way that the lengths and widths of the base rectangles approach zero, then it is plausible that the errors of both types approach zero, and the exact volume of the solid is

$$V = \lim_{n \to +\infty} \sum_{k=1}^{n} f(x_k^*, y_k^*)\,\Delta A_k \tag{2}$$

The sums in (2) are called **Riemann sums**, and the limit of the Riemann sums is denoted by

$$\iint\limits_{R} f(x, y)\, dA = \lim_{n \to +\infty} \sum_{k=1}^{n} f(x_k^*, y_k^*)\, \Delta A_k \tag{3}$$

which is called the **double integral** of $f(x, y)$ over R. With this notation, the volume V of the solid can be expressed as

$$V = \iint\limits_{R} f(x, y)\, dA \tag{4}$$

This result was derived under the assumption that f is continuous and nonnegative on the region R. In the case where $f(x, y)$ can have both positive and negative values on R, the double integral over R represents a difference of two volumes: the volume of the solid that is above R but below $z = f(x, y)$ minus the volume of the solid that is below R but above $z = f(x, y)$. We call this difference the **net signed volume** between $z = f(x, y)$ and R; it is analogous to the concept of net signed area defined in Section 5.6. Thus, a positive value for a double integral means that there is more volume above R than below, a negative value means that there is more volume below than above, and a value of zero means that the two volumes are the same.

A precise definition of expression (3) (in terms of ϵ's and δ's) and conditions under which the double integral exists are studied in advanced calculus. However, for our purposes it suffices to say that existence is ensured when f is continuous on R and the region R is not too "complicated."

Observe the similarity between (1) and (3). Because of this it should not be surprising that double integrals have many of the same properties as definite integrals (which we now also call **single integrals**):

$$\iint\limits_{R} cf(x, y)\, dA = c \iint\limits_{R} f(x, y)\, dA \quad (c \text{ a constant}) \tag{5}$$

$$\iint\limits_{R} [f(x, y) + g(x, y)]\, dA = \iint\limits_{R} f(x, y)\, dA + \iint\limits_{R} g(x, y)\, dA \tag{6}$$

$$\iint\limits_{R} [f(x, y) - g(x, y)]\, dA = \iint\limits_{R} f(x, y)\, dA - \iint\limits_{R} g(x, y)\, dA \tag{7}$$

It is evident intuitively that if $f(x, y)$ is nonnegative on a region R, then subdividing R into two regions R_1 and R_2 has the effect of subdividing the solid between R and $z = f(x, y)$ into two solids, the sum of whose volumes is the volume of the entire solid (Figure 17.1.4). This suggests the following result, which holds even if f has negative values:

$$\iint\limits_{R} f(x, y)\, dA = \iint\limits_{R_1} f(x, y)\, dA + \iint\limits_{R_2} f(x, y)\, dA \tag{8}$$

The proofs of these results will be omitted.

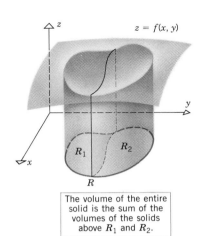

$z = f(x, y)$

The volume of the entire solid is the sum of the volumes of the solids above R_1 and R_2.

Figure 17.1.4

☐ **EVALUATING DOUBLE INTEGRALS**

Except in the simplest cases, it is impractical to obtain the value of a double integral from the limit in (3). However, we shall now show how to evaluate double integrals by calculating two successive single integrals. For the remainder of this section, we shall limit our discussion to the case where R is a rectangle. In the next section we shall consider double integrals over more complicated regions.

The partial derivatives of a function $f(x, y)$ are calculated by holding one of the variables fixed and differentiating with respect to the other variable. Let us consider the reverse of this process, *partial integration*. The symbols

$$\int_a^b f(x, y)\, dx \quad \text{and} \quad \int_c^d f(x, y)\, dy$$

denote *partial definite integrals*; the first integral, called the *partial definite integral with respect to x*, is evaluated by holding y fixed and integrating with respect to x, and the second integral, called the *partial definite integral with respect to y*, is evaluated by holding x fixed and integrating with respect to y. As the following example shows, the partial definite integral with respect to x is a function of y, and the partial definite integral with respect to y is a function of x.

Example 1

$$\int_0^1 xy^2\, dx = y^2 \int_0^1 x\, dx = \left. \frac{y^2 x^2}{2} \right]_{x=0}^1 = \frac{y^2}{2}$$

$$\int_0^1 xy^2\, dy = x \int_0^1 y^2\, dy = \left. \frac{xy^3}{3} \right]_{y=0}^1 = \frac{x}{3} \qquad \blacktriangleleft$$

A partial definite integral with respect to x is a function of y and hence can be integrated with respect to y; similarly, a partial definite integral with respect to y can be integrated with respect to x. This two-stage integration process is called *iterated* (or *repeated*) *integration*. We introduce the following notation:

$$\int_c^d \int_a^b f(x, y)\, dx\, dy = \int_c^d \left[\int_a^b f(x, y)\, dx \right] dy \qquad (9a)$$

$$\int_a^b \int_c^d f(x, y)\, dy\, dx = \int_a^b \left[\int_c^d f(x, y)\, dy \right] dx \qquad (9b)$$

These integrals are called *iterated integrals*.

Example 2 Evaluate

(a) $\displaystyle \int_0^3 \int_1^2 (1 + 8xy)\, dy\, dx$ (b) $\displaystyle \int_1^2 \int_0^3 (1 + 8xy)\, dx\, dy$

Solution (a).

$$\int_0^3 \int_1^2 (1 + 8xy)\, dy\, dx = \int_0^3 \left[\int_1^2 (1 + 8xy)\, dy \right] dx$$

$$= \int_0^3 \left[y + 4xy^2 \right]_{y=1}^2 dx$$

$$= \int_0^3 [(2 + 16x) - (1 + 4x)]\, dx$$

$$= \int_0^3 (1 + 12x)\, dx$$

$$= x + 6x^2 \Big]_0^3 = 57$$

Solution (b).

$$\int_1^2 \int_0^3 (1 + 8xy)\, dx\, dy = \int_1^2 \left[\int_0^3 (1 + 8xy)\, dx \right] dy$$

$$= \int_1^2 \left[x + 4x^2 y \right]_{x=0}^3 dy$$

$$= \int_1^2 (3 + 36y)\, dy$$

$$= 3y + 18y^2 \Big]_1^2 = 57 \qquad \blacktriangleleft$$

It is no accident that the two iterated integrals in the last example have the same value; it is a consequence of the following theorem.

17.1.2 THEOREM. *Let R be the rectangle defined by the inequalities*

$$a \le x \le b, \quad c \le y \le d$$

If f(x, y) is continuous on this rectangle, then

$$\iint_R f(x, y)\, dA = \int_c^d \int_a^b f(x, y)\, dx\, dy = \int_a^b \int_c^d f(x, y)\, dy\, dx$$

This major theorem enables us to evaluate a double integral over a rectangle by calculating an iterated integral. Moreover, the theorem tells us that the order of integration in the iterated integral does not matter. We shall not formally prove this result; however, we will give a geometric argument for the case where $f(x, y)$ is nonnegative on R. In this case the double integral can be interpreted as the volume of the solid S bounded above by the surface $z = f(x, y)$ and below by the region R, so it suffices to show that the two iterated integrals also represent this volume.

For a fixed value of y, the function $f(x, y)$ is a function of x, and hence the integral

$$A(y) = \int_a^b f(x, y)\, dx$$

represents the area under the graph of this function of x. This area, shown in yellow in Figure 17.1.5, is the cross-sectional area at y of the solid S bounded above by $z = f(x, y)$ and below by the region R. Thus, by the method of slicing discussed in Section 6.2, the volume of the solid S is

$$\text{Vol}(S) = \int_c^d A(y)\, dy = \int_c^d \left[\int_a^b f(x, y)\, dx \right] dy = \int_c^d \int_a^b f(x, y)\, dx\, dy \qquad (10)$$

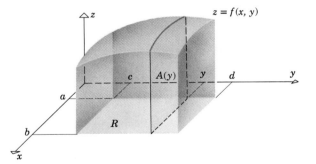

Figure 17.1.5

Similarly, the integral

$$A(x) = \int_c^d f(x, y)\,dy$$

represents the area of the cross section of S at x (Figure 17.1.6), and the method of slicing again yields

$$\text{Vol}(S) = \int_a^b A(x)\,dx = \int_a^b \left[\int_c^d f(x, y)\,dy \right] dx = \int_a^b \int_c^d f(x, y)\,dy\,dx \qquad (11)$$

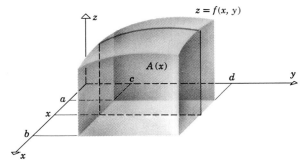

$z = f(x, y)$

$A(x)$

Figure 17.1.6

This establishes the result in Theorem 17.1.2 for the case where $f(x, y)$ is nonnegative on R.

Example 3   Evaluate the double integral

$$\iint_R y^2 x\,dA$$

over the rectangle $R = \{ (x, y): -3 \le x \le 2,\ 0 \le y \le 1 \}$.

Solution.   In view of Theorem 17.1.2, the value of the double integral may be obtained from either of the iterated integrals

$$\int_{-3}^2 \int_0^1 y^2 x\,dy\,dx \quad \text{or} \quad \int_0^1 \int_{-3}^2 y^2 x\,dx\,dy \qquad (12)$$

Using the first of these, we obtain

$$\iint_R y^2 x\,dA = \int_{-3}^2 \int_0^1 y^2 x\,dy\,dx = \int_{-3}^2 \left[\frac{1}{3} y^3 x \right]_{y=0}^1 dx$$

$$= \int_{-3}^2 \frac{1}{3} x\,dx = \frac{x^2}{6} \Bigg]_{-3}^2 = -\frac{5}{6}$$

The reader can check this result by evaluating the second integral in (12).   ◄

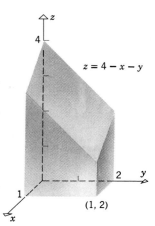

$z = 4 - x - y$

$(1, 2)$

Figure 17.1.7

Example 4   Use a double integral to find the volume of the solid that is bounded above by the plane $z = 4 - x - y$ and below by the rectangle $R = \{ (x, y): 0 \le x \le 1,\ 0 \le y \le 2 \}$ (Figure 17.1.7).

Solution.

$$V = \iint\limits_{R} (4 - x - y)\, dA = \int_0^2 \int_0^1 (4 - x - y)\, dx\, dy$$

$$= \int_0^2 \left[4x - \frac{x^2}{2} - xy \right]_{x=0}^{1} dy = \int_0^2 \left(\frac{7}{2} - y \right) dy$$

$$= \left[\frac{7}{2}y - \frac{y^2}{2} \right]_0^2 = 5$$

The volume can also be obtained by first integrating with respect to y and then with respect to x. ◄

► Exercise Set 17.1

In Exercises 1–14, evaluate the iterated integrals.

1. $\int_0^1 \int_0^2 (x + 3)\, dy\, dx.$ **2.** $\int_1^3 \int_{-1}^1 (2x - 4y)\, dy\, dx.$

3. $\int_2^4 \int_0^1 x^2 y\, dx\, dy.$ **4.** $\int_{-2}^0 \int_{-1}^2 (x^2 + y^2)\, dx\, dy.$

5. $\int_0^{\ln 3} \int_0^{\ln 2} e^{x+y}\, dy\, dx.$ **6.** $\int_0^2 \int_0^1 y \sin x\, dy\, dx.$

7. $\int_0^3 \int_0^1 x(x^2 + y)^{1/2}\, dx\, dy.$

8. $\int_{-1}^2 \int_2^4 (2x^2 y + 3xy^2)\, dx\, dy.$

9. $\int_{-1}^0 \int_2^5 dx\, dy.$ **10.** $\int_4^6 \int_{-3}^7 dy\, dx.$

11. $\int_0^1 \int_0^1 \frac{x}{(xy + 1)^2}\, dy\, dx.$ **12.** $\int_{\pi/2}^\pi \int_1^2 x \cos xy\, dy\, dx.$

13. $\int_0^{\ln 2} \int_0^1 xy\, e^{y^2 x}\, dy\, dx.$ **14.** $\int_3^4 \int_1^2 \frac{1}{(x + y)^2}\, dy\, dx.$

In Exercises 15–19, evaluate the double integral over the rectangular region R.

15. $\iint\limits_{R} 4xy^3\, dA;\ R = \{(x, y): -1 \le x \le 1, -2 \le y \le 2\}.$

16. $\iint\limits_{R} \frac{xy}{\sqrt{x^2 + y^2 + 1}}\, dA;$

$R = \{(x, y): 0 \le x \le 1, 0 \le y \le 1\}.$

17. $\iint\limits_{R} x\sqrt{1 - x^2}\, dA;\ R = \{(x, y): 0 \le x \le 1, 2 \le y \le 3\}.$

18. $\iint\limits_{R} (x \sin y - y \sin x)\, dA;$

$R = \{(x, y): 0 \le x \le \pi/2, 0 \le y \le \pi/3\}.$

19. $\iint\limits_{R} \cos(x + y)\, dA;$

$R = \{(x, y): -\pi/4 \le x \le \pi/4, 0 \le y \le \pi/4\}.$

In Exercises 20–25, the iterated integral represents the volume of a solid. Make an accurate sketch of the solid. (You do *not* have to find the volume.)

20. $\int_0^5 \int_1^2 4\, dx\, dy.$ **21.** $\int_0^1 \int_0^1 (2 - x - y)\, dy\, dx.$

22. $\int_2^3 \int_3^4 y\, dx\, dy.$

23. $\int_0^3 \int_0^4 \sqrt{25 - x^2 - y^2}\, dy\, dx.$

24. $\int_{-2}^2 \int_{-2}^2 (x^2 + y^2)\, dx\, dy.$ **25.** $\int_0^1 \int_{-1}^1 \sqrt{4 - x^2}\, dy\, dx.$

In Exercises 26–30, use a double integral to find the volume.

26. The volume under the plane $z = 2x + y$ and over the rectangle $R = \{(x, y): 3 \le x \le 5, 1 \le y \le 2\}.$

27. The volume under the surface $z = 3x^3 + 3x^2 y$ and over the rectangle $R = \{(x, y): 1 \le x \le 3, 0 \le y \le 2\}.$

28. The volume in the first octant bounded by the coordinate planes, the plane $y = 4$, and the plane $x/3 + z/5 = 1.$

29. The volume of the solid in the first octant enclosed by the surface $z = x^2$ and the planes $x = 2$, $y = 3$, $y = 0$, and $z = 0.$

30. The volume of the solid in the first octant that is enclosed by the planes $x = 0$, $z = 0$, $x = 5$, $z - y = 0$, and $z = -2y + 6$. [*Hint:* Break the solid into two parts.]

31. Suppose that $f(x, y) = g(x)h(y)$ and
$R = \{(x, y) : a \le x \le b,\ c \le y \le d\}$. Show that

$$\iint_R f(x, y)\,dA = \left[\int_a^b g(x)\,dx\right]\left[\int_c^d h(y)\,dy\right]$$

32. Evaluate

$$\iint_R x\cos(xy)\cos^2 \pi x\,dA$$

where $R = \{(x, y) : 0 \le x \le \frac{1}{2},\ 0 \le y \le \pi\}$. [*Hint:* One order of integration leads to a simpler solution than the other.]

▪ 17.2 DOUBLE INTEGRALS OVER NONRECTANGULAR REGIONS

In this section we shall show how to evaluate double integrals over regions other than rectangles.

☐ **ITERATED INTEGRALS WITH NONCONSTANT LIMITS OF INTEGRATION**

Later in this section we will see that double integrals over nonrectangular regions can often be evaluated as iterated integrals of the following types:

$$\int_a^b \int_{g_1(x)}^{g_2(x)} f(x, y)\,dy\,dx = \int_a^b \left[\int_{g_1(x)}^{g_2(x)} f(x, y)\,dy\right] dx \tag{1a}$$

$$\int_c^d \int_{h_1(y)}^{h_2(y)} f(x, y)\,dx\,dy = \int_c^d \left[\int_{h_1(y)}^{h_2(y)} f(x, y)\,dx\right] dy \tag{1b}$$

We begin with an example that illustrates how to evaluate such integrals.

Example 1 Evaluate

(a) $\displaystyle\int_0^2 \int_{x^2}^x y^2 x\,dy\,dx$ (b) $\displaystyle\int_0^\pi \int_0^{\cos y} x \sin y\,dx\,dy$

Solution (a).

$$\int_0^2 \int_{x^2}^x y^2 x\,dy\,dx = \int_0^2 \left[\int_{x^2}^x y^2 x\,dy\right] dx = \int_0^2 \left[\frac{y^3 x}{3}\right]_{y=x^2}^x dx$$

$$= \int_0^2 \left(\frac{x^4}{3} - \frac{x^7}{3}\right) dx = \left[\frac{x^5}{15} - \frac{x^8}{24}\right]_0^2$$

$$= \frac{32}{15} - \frac{256}{24} = -\frac{128}{15}$$

Solution (b).

$$\int_0^\pi \int_0^{\cos y} x \sin y\,dx\,dy = \int_0^\pi \left[\int_0^{\cos y} x \sin y\,dx\right] dy$$

$$= \int_0^\pi \left[\frac{x^2}{2}\sin y\right]_{x=0}^{\cos y} dy$$

$$= \int_0^\pi \frac{1}{2}\cos^2 y \sin y\,dy$$

$$= \left[-\frac{1}{6}\cos^3 y\right]_0^\pi = \frac{1}{3} \qquad ◀$$

☐ **DOUBLE INTEGRALS OVER NONRECTANGULAR REGIONS**

Plane regions can be extremely complex, and the theory of double integrals over very general regions is a topic for advanced courses in mathematics. We shall limit our study of double integrals to two basic types of regions, which we shall call *type I* and *type II*; they are defined as follows:

17.2.1   DEFINITION.

(a)   A *type I region* is bounded on the left and right by vertical lines $x = a$ and $x = b$ and is bounded below and above by continuous curves $y = g_1(x)$ and $y = g_2(x)$, where $g_1(x) \leq g_2(x)$ for $a \leq x \leq b$ (Figure 17.2.1a).

(b)   A *type II region* is bounded below and above by horizontal lines $y = c$ and $y = d$ and is bounded on the left and right by continuous curves $x = h_1(y)$ and $x = h_2(y)$ satisfying $h_1(y) \leq h_2(y)$ for $c \leq y \leq d$ (Figure 17.2.1b).

The following theorem will enable us to evaluate double integrals over type I and type II regions using iterated integrals.

17.2.2   THEOREM.

(a)   *If R is a type I region on which $f(x, y)$ is continuous, then*

$$\iint_R f(x, y)\, dA = \int_a^b \int_{g_1(x)}^{g_2(x)} f(x, y)\, dy\, dx \tag{2a}$$

(b)   *If R is a type II region on which $f(x, y)$ is continuous, then*

$$\iint_R f(x, y)\, dA = \int_c^d \int_{h_1(y)}^{h_2(y)} f(x, y)\, dx\, dy \tag{2b}$$

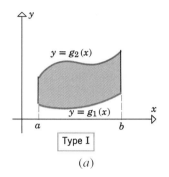

Type I

(a)

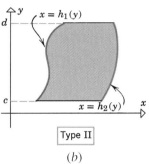

Type II

(b)

Figure 17.2.1

We will not prove this theorem, but for the case where $f(x, y)$ is nonnegative on the region R, it can be made plausible by a geometric argument that is similar to that given for Theorem 17.1.2. Since $f(x, y)$ is nonnegative, the double integral can be interpreted as the volume of the solid S that is bounded above by the surface $z = f(x, y)$ and below by the region R, so it suffices to show that the iterated integrals also represent this volume. Consider the iterated integral in (2a), for example. For a fixed value of x, the function $f(x, y)$ is a function of y, and hence the integral

$$A(x) = \int_{g_1(x)}^{g_2(x)} f(x, y)\, dy$$

represents the area under the graph of this function of y between the points $y = g_1(x)$ and $y = g_2(x)$. This area, shown in yellow in Figure 17.2.2, is the cross-sectional area at x of the solid S, and hence by the method of slicing, the volume of the solid S is

$$\text{Vol}\,(S) = \int_a^b \int_{g_1(x)}^{g_2(x)} f(x, y)\, dy\, dx$$

which shows that in (2a) the iterated integral is equal to the double integral. Similarly for (2b).

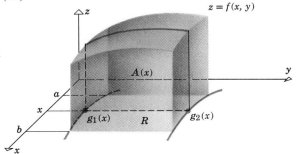

Figure 17.2.2

□ **SETTING UP LIMITS OF INTEGRATION FOR EVALUATING DOUBLE INTEGRALS**

To apply Theorem 17.2.2, it is usual to start with a two-dimensional sketch of the region R. [It is not necessary to graph $f(x, y)$.] For a type I region, the limits of integration in Formula (2a) can be obtained as follows:

Step 1. Since x is held fixed for the first integration, we draw a vertical line through the region R at an arbitrary fixed point x (Figure 17.2.3). This line crosses the boundary of R twice. The lower point of intersection is on the curve $y = g_1(x)$ and the higher point is on the curve $y = g_2(x)$. These two intersections determine the lower and upper y-limits of integration in Formula (2a).

Step 2. Imagine moving the line drawn in Step 1 first to the left and then to the right (Figure 17.2.3). The leftmost position where the line intersects the region R is $x = a$ and the rightmost position where the line intersects the region R is $x = b$. This yields the limits for the x-integration in Formula (2a).

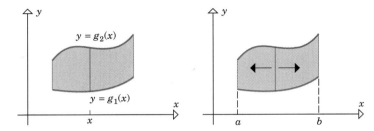

Figure 17.2.3

Example 2 Evaluate

$$\iint\limits_{R} xy \, dA$$

over the region R enclosed between $y = \frac{1}{2}x$, $y = \sqrt{x}$, $x = 2$, and $x = 4$.

Solution. We view R as a type I region. The region R and a vertical line corresponding to a fixed x are shown in Figure 17.2.4. This line meets the region R at the lower boundary $y = \frac{1}{2}x$ and the upper boundary $y = \sqrt{x}$. These are the y-limits of integration. Moving this line first left and then right yields the x-limits of integration, $x = 2$ and $x = 4$. Thus,

$$\iint\limits_{R} xy \, dA = \int_{2}^{4} \int_{x/2}^{\sqrt{x}} xy \, dy \, dx = \int_{2}^{4} \left[\frac{xy^2}{2} \right]_{y=x/2}^{\sqrt{x}} dx = \int_{2}^{4} \left(\frac{x^2}{2} - \frac{x^3}{8} \right) dx$$

$$= \left[\frac{x^3}{6} - \frac{x^4}{32} \right]_{2}^{4} = \left(\frac{64}{6} - \frac{256}{32} \right) - \left(\frac{8}{6} - \frac{16}{32} \right) = \frac{11}{6} \quad \blacktriangleleft$$

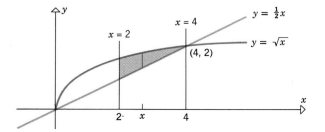

Figure 17.2.4

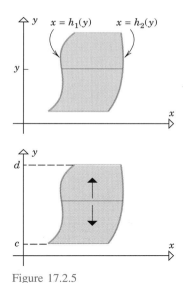

Figure 17.2.5

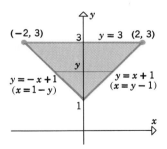

Figure 17.2.6

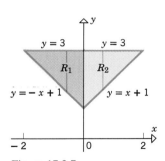

Figure 17.2.7

If R is a type II region, then the limits of integration in Formula (2b) can be obtained as follows:

Step 1. Since y is held fixed for the first integration, we draw a horizontal line through the region R at a fixed point y (Figure 17.2.5). This line crosses the boundary of R twice. The leftmost point of intersection is on the curve $x = h_1(y)$ and the rightmost point is on the curve $x = h_2(y)$. These intersections determine the x-limits of integration in (2b).

Step 2. Imagine moving the line drawn in Step 1 first down and then up (Figure 17.2.5). The lowest position where the line intersects the region R is $y = c$, and the highest position where the line intersects the region R is $y = d$. This yields the y-limits of integration in (2b).

Example 3 Evaluate

$$\iint_R (2x - y^2)\, dA$$

over the triangular region R enclosed between the lines $y = -x + 1$, $y = x + 1$, and $y = 3$.

Solution. We view R as a type II region. The region R and a horizontal line corresponding to a fixed y are shown in Figure 17.2.6. This line meets the region R at its left-hand boundary $x = 1 - y$ and its right-hand boundary $x = y - 1$. These are the x-limits of integration. Moving this line first down and then up yields the y-limits, $y = 1$ and $y = 3$. Thus,

$$\iint_R (2x - y^2)\, dA = \int_1^3 \int_{1-y}^{y-1} (2x - y^2)\, dx\, dy = \int_1^3 \left[x^2 - y^2 x \right]_{x=1-y}^{y-1} dy$$

$$= \int_1^3 [(1 - 2y + 2y^2 - y^3) - (1 - 2y + y^3)]\, dy$$

$$= \int_1^3 (2y^2 - 2y^3)\, dy = \left[\frac{2y^3}{3} - \frac{y^4}{2} \right]_1^3 = -\frac{68}{3} \quad \blacktriangleleft$$

REMARK.   To integrate over a type II region, the left- and right-hand boundaries must be expressed in the form $x = h_1(y)$ and $x = h_2(y)$. This is why we rewrote the boundary equations $y = -x + 1$ and $y = x + 1$ as $x = 1 - y$ and $x = y - 1$ in the last example.

In Example 3 we could have treated R as a type I region, but with an added complication. Viewed as a type I region, the upper boundary of R is the line $y = 3$ (Figure 17.2.7) and the lower boundary consists of two parts, the line $y = -x + 1$ to the left of the origin and the line $y = x + 1$ to the right of the origin. To carry out the integration it is necessary to decompose the region R into two parts, R_1 and R_2, as shown in Figure 17.2.7, and write

$$\iint_R (2x - y^2)\, dA = \iint_{R_1} (2x - y^2)\, dA + \iint_{R_2} (2x - y^2)\, dA$$

$$= \int_{-2}^0 \int_{-x+1}^3 (2x - y^2)\, dy\, dx + \int_0^2 \int_{x+1}^3 (2x - y^2)\, dy\, dx$$

This will yield the same result that was obtained in Example 3.

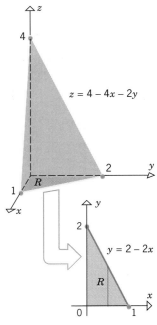

Example 4   Use a double integral to find the volume of the tetrahedron bounded by the coordinate planes and the plane $z = 4 - 4x - 2y$.

Solution.   The tetrahedron in question is bounded above by the plane

$$z = 4 - 4x - 2y \tag{3}$$

and below by the triangular region R shown in Figure 17.2.8. Thus, the volume is given by

$$V = \iint_R (4 - 4x - 2y)\, dA$$

The region R is bounded by the x-axis, the y-axis, and the line $y = 2 - 2x$ [set $z = 0$ in (3)], so that treating R as a type I region yields

$$V = \iint_R (4 - 4x - 2y)\, dA = \int_0^1 \int_0^{2-2x} (4 - 4x - 2y)\, dy\, dx$$

$$= \int_0^1 \left[4y - 4xy - y^2 \right]_{y=0}^{2-2x} dx = \int_0^1 (4 - 8x + 4x^2)\, dx = \frac{4}{3} \quad \blacktriangleleft$$

Figure 17.2.8

Example 5   Find the volume of the solid bounded by the cylinder $x^2 + y^2 = 4$ and the planes $y + z = 4$ and $z = 0$.

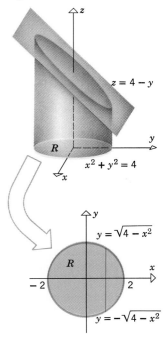

Figure 17.2.9

Solution.   The solid shown in Figure 17.2.9 is bounded above by the plane $z = 4 - y$ and below by the region R within the circle $x^2 + y^2 = 4$. The volume is given by

$$V = \iint_R (4 - y)\, dA$$

Treating R as a type I region we obtain

$$V = \int_{-2}^{2} \int_{-\sqrt{4-x^2}}^{\sqrt{4-x^2}} (4 - y)\, dy\, dx = \int_{-2}^{2} \left[4y - \frac{1}{2}y^2 \right]_{y=-\sqrt{4-x^2}}^{\sqrt{4-x^2}} dx$$

$$= \int_{-2}^{2} 8\sqrt{4 - x^2}\, dx = 8(2\pi) = 16\pi \qquad \boxed{\text{See Example 4 of Section 9.5.}} \quad \blacktriangleleft$$

☐ **REVERSING THE ORDER OF INTEGRATION**

Sometimes the evaluation of an iterated integral can be simplified by reversing the order of integration. The next example illustrates how this is done.

Example 6   Since there is no elementary antiderivative of e^{x^2}, the integral

$$\int_0^2 \int_{y/2}^1 e^{x^2}\, dx\, dy$$

cannot be evaluated by performing the x-integration first. Evaluate this integral by expressing it as an equivalent iterated integral with the order of integration reversed.

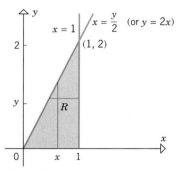

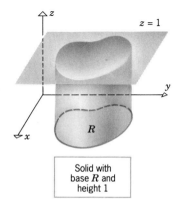

Figure 17.2.10

☐ **AREA CALCULATED AS A
DOUBLE INTEGRAL**

Figure 17.2.11

Solution.   For the inside integration, y is fixed and x varies from the line $x = y/2$ to the line $x = 1$ (Figure 17.2.10). For the outside integration, y varies from 0 to 2, so the given iterated integral is equal to a double integral over the triangular region R in Figure 17.2.10.

To reverse the order of integration, we treat R as a type I region, which enables us to write the given integral as

$$\int_0^2 \int_{y/2}^1 e^{x^2}\, dx\, dy = \iint_R e^{x^2}\, dA = \int_0^1 \int_0^{2x} e^{x^2}\, dy\, dx = \int_0^1 \left[e^{x^2} y \right]_{y=0}^{2x} dx$$

$$= \int_0^1 2xe^{x^2}\, dx = e^{x^2} \Big]_0^1 = e - 1 \quad \blacktriangleleft$$

Although double integrals arose in the context of calculating volumes, they can also be used to calculate areas. For this purpose, we consider the solid consisting of the points between the plane $z = 1$ and a region R in the xy-plane (Figure 17.2.11). The volume V of this solid is

$$V = \iint_R 1\, dA = \iint_R dA \tag{4}$$

However, the solid has congruent cross sections taken parallel to the xy-plane, so that

$$V = \text{area of base} \cdot \text{height} = \text{area of } R \cdot 1 = \text{area of } R$$

Combining this with (4) yields the area formula

$$\text{area of } R = \iint_R dA \tag{5}$$

REMARK.   Formula (5) is sometimes confusing because it equates an area and a volume; the formula is intended to equate only the *numerical values* of the area and volume and not the units, which must, of course, be different.

Example 7   Use a double integral to find the area of the region R enclosed between the parabola $y = \frac{1}{2}x^2$ and the line $y = 2x$.

Solution.   The region R may be treated equally well as type I (Figure 17.2.12a) or type II (Figure 17.2.12b). Treating R as type I yields

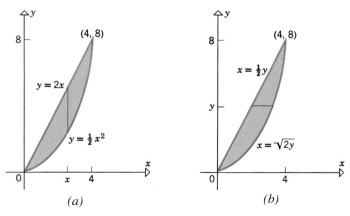

Figure 17.2.12 (a) (b)

$$\text{area of } R = \iint\limits_{R} dA = \int_{0}^{4}\int_{x^2/2}^{2x} dy\, dx = \int_{0}^{4}\left[y\right]_{y=x^2/2}^{2x} dx$$

$$= \int_{0}^{4}\left(2x - \frac{1}{2}x^2\right) dx = \left[x^2 - \frac{x^3}{6}\right]_{0}^{4} = \frac{16}{3}$$

Treating R as type II yields

$$\text{area of } R = \iint\limits_{R} dA = \int_{0}^{8}\int_{y/2}^{\sqrt{2y}} dx\, dy = \int_{0}^{8}\left[x\right]_{x=y/2}^{\sqrt{2y}} dy$$

$$= \int_{0}^{8}\left(\sqrt{2y} - \frac{1}{2}y\right) dy = \left[\frac{2\sqrt{2}}{3}y^{3/2} - \frac{y^2}{4}\right]_{0}^{8} = \frac{16}{3} \quad \blacktriangleleft$$

▶ Exercise Set 17.2

In Exercises 1–12, evaluate the iterated integral.

1. $\displaystyle\int_{0}^{1}\int_{x^2}^{x} xy^2\, dy\, dx.$ **2.** $\displaystyle\int_{1}^{2}\int_{y}^{3-y} y\, dx\, dy.$

3. $\displaystyle\int_{0}^{3}\int_{0}^{\sqrt{9-y^2}} y\, dx\, dy.$ **4.** $\displaystyle\int_{1/4}^{1}\int_{x^2}^{x}\sqrt{\frac{x}{y}}\, dy\, dx.$

5. $\displaystyle\int_{\sqrt{\pi}}^{\sqrt{2\pi}}\int_{0}^{x^3}\sin\frac{y}{x}\, dy\, dx.$ **6.** $\displaystyle\int_{-1}^{1}\int_{-x^2}^{x^2}(x^2 - y)\, dy\, dx.$

7. $\displaystyle\int_{\pi/2}^{\pi}\int_{0}^{x^2}\frac{1}{x}\cos\frac{y}{x}\, dy\, dx.$

8. $\displaystyle\int_{0}^{\pi/2}\int_{0}^{\sin y} e^x \cos y\, dx\, dy.$

9. $\displaystyle\int_{0}^{a}\int_{0}^{\sqrt{a^2-x^2}}(x + y)\, dy\, dx \quad (a > 0).$

10. $\displaystyle\int_{1}^{2}\int_{0}^{y^2} e^{x/y^2}\, dx\, dy.$ **11.** $\displaystyle\int_{0}^{1}\int_{0}^{x} y\sqrt{x^2 - y^2}\, dy\, dx.$

12. $\displaystyle\int_{0}^{1}\int_{0}^{x} e^{x^2}\, dy\, dx.$

In Exercises 13–28, evaluate the double integral.

13. $\displaystyle\iint\limits_{R} 6xy\, dA$; R is the region bounded by $y = 0$, $x = 2$, and $y = x^2$.

14. $\displaystyle\iint\limits_{R} xy\, dA$; R is the region bounded by the trapezoid with vertices $(1, 3)$, $(5, 3)$, $(2, 1)$, and $(4, 1)$.

15. $\displaystyle\iint\limits_{R} x\cos xy\, dA$; R is the region enclosed by $x = 1$, $x = 2$, $y = \pi/2$, and $y = 2\pi/x$.

16. $\displaystyle\iint\limits_{R}(x + y)\, dA$; R is the region enclosed between the curves $y = x^2$ and $y = \sqrt{x}$.

17. $\displaystyle\iint\limits_{R} x^2\, dA$; R is the region bounded by $y = 16/x$, $y = x$, and $x = 8$.

18. $\displaystyle\iint\limits_{R} xy^2\, dA$; R is the region enclosed by $y = 1$, $y = 2$, $x = 0$, and $y = x$.

19. $\displaystyle\iint\limits_{R} x(1 + y^2)^{-1/2}\, dA$; R is the region in the first quadrant enclosed by $y = x^2$, $y = 4$, and $x = 0$. [*Hint:* Choose your order of integration carefully.]

20. $\displaystyle\iint\limits_{R} x\cos y\, dA$; R is the triangular region bounded by $y = x$, $y = 0$, and $x = \pi$.

21. $\displaystyle\iint\limits_{R}(3x - 2y)\, dA$; R is the region enclosed by the circle $x^2 + y^2 = 1$.

22. $\displaystyle\iint\limits_{R} y\, dA$; R is the region in the first quadrant enclosed between the circle $x^2 + y^2 = 25$ and the line $x + y = 5$.

23. $\displaystyle\iint\limits_{R}\frac{1}{1 + x^2}\, dA$; R is the triangular region with vertices $(0, 0)$, $(1, 1)$, and $(0, 1)$.

24. $\displaystyle\iint\limits_{R}(x^2 - xy)\, dA$; R is the region enclosed by $y = x$ and $y = 3x - x^2$.

25. $\iint\limits_R xy \, dA$; R is the region enclosed by $y = \sqrt{x}$, $y = 6 - x$, and $y = 0$.

26. $\iint\limits_R x \, dA$; R is the region enclosed by $y = \sin^{-1} x$, $x = 1/\sqrt{2}$, and $y = 0$.

27. $\iint\limits_R (x - 1) \, dA$; R is the region enclosed between $y = x$ and $y = x^3$.

28. $\iint\limits_R x^2 \, dA$; R is the region in the first quadrant enclosed by $xy = 1$, $y = x$, and $y = 2x$.

In Exercises 29–34, use double integration to find the area of the plane region enclosed by the given curves.

29. $x + y = 5$, $x = 0$, and $y = 0$.
30. $y = x^2$ and $y = 4x$.
31. $y = \sin x$ and $y = \cos x$, for $0 \le x \le \pi/4$.
32. $y^2 = -x$ and $3y - x = 4$.
33. $y^2 = 9 - x$ and $y^2 = 9 - 9x$.
34. $y = \cosh x$, $y = \sinh x$, $x = 0$, and $x = 1$.

In Exercises 35–48, use double integration to find the volume of each solid.

35. The tetrahedron that lies in the first octant that is bounded by the three coordinate planes and the plane $z = 5 - 2x - y$.

36. The solid bounded by the cylinder $x^2 + y^2 = 9$ and the planes $z = 0$ and $z = 3 - x$.

37. The solid that is bounded above by the plane $z = x + 2y + 2$, below by the xy-plane, and laterally by $y = 0$ and $y = 1 - x^2$.

38. The solid in the first octant bounded above by the paraboloid $z = x^2 + 3y^2$, below by the plane $z = 0$, and laterally by $y = x^2$ and $y = x$.

39. The solid that is bounded above by the paraboloid $z = 9x^2 + y^2$, below by the plane $z = 0$, and laterally by the planes $x = 0$, $y = 0$, $x = 3$, and $y = 2$.

40. The solid enclosed by $y^2 = x$, $z = 0$, and $x + z = 1$.

41. The wedge cut from the cylinder $4x^2 + y^2 = 9$ by the planes $z = 0$ and $z = y + 3$.

42. The solid in the first octant bounded above by $z = 9 - x^2$, below by $z = 0$, and laterally by $y^2 = 3x$.

43. The solid in the first octant bounded by the three coordinate planes and the planes

$$x + 2y = 4 \quad \text{and} \quad x + 8y - 4z = 0$$

44. The solid in the first octant bounded by the surface $z = e^{y-x}$, the plane $x + y = 1$, and the coordinate planes.

45. The solid bounded above by the paraboloid $z = 1 - x^2 - y^2$ and below by the xy-plane. [*Hint:* Use a trigonometric substitution to evaluate the integral.]

46. The solid in the first octant bounded by the paraboloid $z = x^2 + y^2$ and the cylinder $x^2 + y^2 = 4$. [*Hint:* Use a trigonometric substitution to evaluate the integral.]

47. The solid common to the cylinders $x^2 + y^2 = 25$ and $x^2 + z^2 = 25$.

48. The solid that is bounded above by the paraboloid $z = x^2 + y^2$, bounded laterally by the circular cylinder $x^2 + (y - 1)^2 = 1$, and bounded below by the xy-plane.

In Exercises 49–56, express the integral as an equivalent integral with the order of integration reversed.

49. $\displaystyle\int_0^2 \int_0^{\sqrt{x}} f(x, y) \, dy \, dx$.   50. $\displaystyle\int_0^4 \int_{2y}^8 f(x, y) \, dx \, dy$.

51. $\displaystyle\int_0^2 \int_1^{e^y} f(x, y) \, dx \, dy$.   52. $\displaystyle\int_1^e \int_0^{\ln x} f(x, y) \, dy \, dx$.

53. $\displaystyle\int_{-2}^2 \int_{-\sqrt{1-(x^2/4)}}^{\sqrt{1-(x^2/4)}} f(x, y) \, dy \, dx$.

54. $\displaystyle\int_0^1 \int_{y^2}^{\sqrt{y}} f(x, y) \, dx \, dy$.   55. $\displaystyle\int_0^1 \int_{\sin^{-1}y}^{\pi/2} f(x, y) \, dx \, dy$.

56. $\displaystyle\int_{-3}^1 \int_{x^2+6x}^{4x+3} f(x, y) \, dy \, dx$.

In Exercises 57–62, evaluate the integral by first reversing the order of integration.

57. $\displaystyle\int_0^1 \int_{4x}^4 e^{-y^2} \, dy \, dx$.   58. $\displaystyle\int_0^2 \int_{y/2}^1 \cos(x^2) \, dx \, dy$.

59. $\displaystyle\int_0^4 \int_{\sqrt{y}}^2 e^{x^3} \, dx \, dy$.   60. $\displaystyle\int_1^3 \int_0^{\ln x} x \, dy \, dx$.

61. $\displaystyle\int_0^1 \int_0^{\cos^{-1}x} x \, dy \, dx$.

62. $\displaystyle\int_0^1 \int_{\sin^{-1}y}^{\pi/2} \sec^2(\cos x) \, dx \, dy$.

63. Evaluate $\displaystyle\iint\limits_R \sin(y^3) \, dA$, where R is the region bounded by $y = \sqrt{x}$, $y = 2$, and $x = 0$. [*Hint:* Choose the order of integration carefully.]

64. Evaluate $\displaystyle\iint\limits_R x \, dA$, where R is the region bounded by $x = \ln y$, $x = 0$, and $y = e$. [*Hint:* Choose the order of integration carefully.]

65. In each part evaluate $\displaystyle\iint_R xy^2\,dA$.

(a)

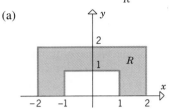

(b)

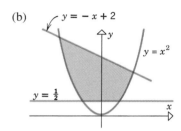

66. Assume that the region R shown in the following figure is symmetric about the y-axis and that the y-axis divides R into the subregions R_1 and R_2 shown. Suppose also that

$$\iint_{R_1} f(x, y)\,dA = 3.$$

(a) Find $\displaystyle\iint_R f(x, y)\,dA$ if $f(-x, y) = f(x, y)$.

(b) Find $\displaystyle\iint_R f(x, y)\,dA$ if $f(-x, y) = -f(x, y)$.

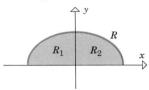

In Exercises 67–69, use symmetry to evaluate the integral without performing an integration. [*Hint:* See Exercise 66.]

67. $\displaystyle\int_{-1}^{1}\int_{0}^{\sqrt{1-x^2}} (2 + x\sqrt{9 - y^2})\,dy\,dx$.

68. $\displaystyle\int_{0}^{2}\int_{x-2}^{2-x} \sin(xy^3)\,dy\,dx$.

69. $\displaystyle\iint_R x^3 y\,dA$, where R is the semicircular region bounded by $y = \sqrt{4 - x^2}$ and $x = 0$.

17.3 DOUBLE INTEGRALS IN POLAR COORDINATES

> *In this section we shall study double integrals in which the integrand and the region of integration are expressed in polar coordinates. Such integrals are important for two reasons: First, they arise naturally in many applications, and second, many double integrals in rectangular coordinates are more easily evaluated if they are converted to polar coordinates.*

☐ **DEFINITION OF DOUBLE INTEGRALS IN POLAR COORDINATES**

Some problems that lead to double integrals are most easily solved if the integrand and the region of integration are expressed in polar coordinates rather than rectangular coordinates. This is often the case when the region of integration has the form described in the following definition.

17.3.1 DEFINITION. A *simple polar region* is a region (in polar coordinates) enclosed between two rays $\theta = \alpha$ and $\theta = \beta$ and two continuous polar curves $r = r_1(\theta)$ and $r = r_2(\theta)$, where

(i) $\alpha \le \beta$ and $\beta - \alpha \le 2\pi$
(ii) $0 \le r_1(\theta) \le r_2(\theta)$

Figure 17.3.1*a* shows a typical simple polar region R. Figure 17.3.1*b* shows the special case where $r_1(\theta)$ is identically zero, so the inner boundary reduces to a single point (the

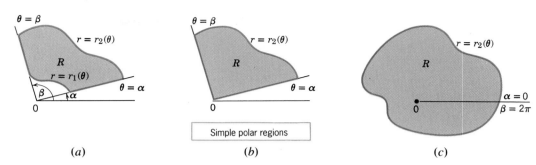

Simple polar regions

Figure 17.3.1

(a)

(b)

(c)

origin); and Figure 17.3.1c shows the special case where $r_1(\theta)$ is identically zero and the rays are $\alpha = 0$ and $\beta = 2\pi$, so the side boundaries coincide.

The concept of a double integral in polar coordinates can be motivated by the following volume problem:

17.3.2 PROBLEM. *Find the volume of the solid consisting of all points that lie between a simple polar region R in the xy-plane and a surface whose equation in cylindrical coordinates is $z = f(r, \theta)$, where f is continuous on R and $f(r, \theta) \geq 0$ for all (r, θ) in R (Figure 17.3.2).*

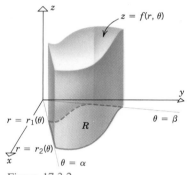

Figure 17.3.2

To find the volume V of the solid in Figure 17.3.2, we shall use a limiting process similar to that used to derive Formula (4) of Section 17.1, except that here we shall use circular arcs and rays to subdivide the region R into blocks, called *polar rectangles*. As shown in Figure 17.3.3, we exclude from consideration all polar rectangles that contain any points outside of R, leaving only polar rectangles that are subsets of R. Assume that there are n such polar rectangles, and denote the area of the kth polar rectangle by ΔA_k. Let (r_k^*, θ_k^*) be any point in this polar rectangle. As shown in Figure 17.3.4, the product $f(r_k^*, \theta_k^*)\,\Delta A_k$ is the volume of a solid with base area ΔA_k and height $f(r_k^*, \theta_k^*)$, so the sum

$$\sum_{k=1}^{n} f(r_k^*, \theta_k^*)\,\Delta A_k$$

can be viewed as an approximation to the volume V of the entire solid.

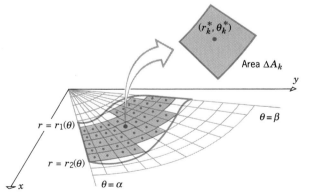

Figure 17.3.3

If we now increase the number of subdivisions in such a way that the dimensions of the polar rectangles approach zero, then it seems plausible that the errors in the approximations approach zero, and the exact volume of the solid is

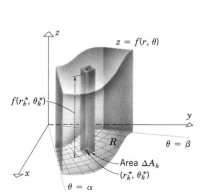

Figure 17.3.4

$$V = \lim_{n \to +\infty} \sum_{k=1}^{n} f(r_k^*, \theta_k^*)\,\Delta A_k \tag{1}$$

The sums in this equation are called **polar Riemann sums**, and the limit is called a **polar double integral**, which we denote as

$$\iint\limits_{R} f(r, \theta)\, dA = \lim_{n \to +\infty} \sum_{k=1}^{n} f(r_k^*, \theta_k^*)\, \Delta A_k \tag{2}$$

so the volume V of the solid can be expressed as

$$V = \iint\limits_{R} f(r, \theta)\, dA \tag{3}$$

For computational purposes, it will be helpful to express the double polar integral in (3) as an iterated integral. For this purpose, let us choose the arbitrary point (r_k^*, θ_k^*) in (2) to be at the "center" of the kth polar rectangle as shown in Figure 17.3.5. Suppose also that this polar rectangle has a central angle $\Delta \theta_k$ and a "radial thickness" Δr_k. Thus, the inner radius of this polar rectangle is $r_k^* - \frac{1}{2}\Delta r_k$ and the outer radius is $r_k^* + \frac{1}{2}\Delta r_k$. Treating the area ΔA_k of this polar rectangle as the difference in area of two sectors, we obtain

$$\Delta A_k = \tfrac{1}{2}(r_k^* + \tfrac{1}{2}\Delta r_k)^2\, \Delta \theta_k - \tfrac{1}{2}(r_k^* - \tfrac{1}{2}\Delta r_k)^2\, \Delta \theta_k$$

which simplifies to

$$\Delta A_k = r_k^*\, \Delta r_k\, \Delta \theta_k \tag{4}$$

Thus, from (2) and (3)

$$V = \iint\limits_{R} f(r, \theta)\, dA = \lim_{n \to +\infty} \sum_{k=1}^{n} f(r_k^*, \theta_k^*) r_k^*\, \Delta r_k\, \Delta \theta_k$$

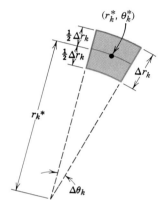

Figure 17.3.5

which suggests that the volume V can be expressed as the iterated integral

$$V = \iint\limits_{R} f(r, \theta)\, dA = \int_{\alpha}^{\beta} \int_{r_1(\theta)}^{r_2(\theta)} f(r, \theta) r\, dr\, d\theta \tag{5}$$

in which the limits of integration are chosen to cover the region R; that is, with θ fixed between α and β, the value of r varies from $r_1(\theta)$ to $r_2(\theta)$ (Figure 17.3.6).

If $f(r, \theta)$ can have both positive and negative values on R, then, in keeping with past experience, the polar double integral over R represents the **net signed volume** between $z = f(r, \theta)$ and R, that is, the difference between the volume of the solid above R but below $z = f(r, \theta)$ and the volume of the solid that is below R but above $z = f(r, \theta)$. Moreover, the relationship between the double polar integral and the iterated integral in (5) holds even if $f(r, \theta)$ is not positive on R.

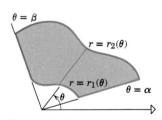

Figure 17.3.6

17.3.3 THEOREM. *If R is a region of the type shown in* Figure 17.3.1, *and if $f(r, \theta)$ is continuous on R, then*

$$\iint\limits_{R} f(r, \theta)\, dA = \int_{\alpha}^{\beta} \int_{r_1(\theta)}^{r_2(\theta)} f(r, \theta) r\, dr\, d\theta \tag{6}$$

REMARK. We mention without proof that polar double integrals satisfy the usual properties of integrals such as (5), (6), (7), and (8) in Section 17.1.

☐ **EVALUATION OF POLAR DOUBLE INTEGRALS**

To apply Theorem 17.3.3, we start with a sketch of the region R. From this sketch the limits of integration in (6) can be obtained as follows:

Step 1. Since θ is held fixed for the first integration, draw a radial line from the origin through the region R at a fixed angle θ (Figure 17.3.6). This line crosses the boundary of R at most twice. The innermost point of intersection is on the curve $r = r_1(\theta)$ and the outermost point is on the curve $r = r_2(\theta)$. These intersections determine the r-limits of integration in (6).

Step 2. Imagine rotating a ray along the polar x-axis one revolution counter-clockwise about the origin. The smallest angle at which this ray intersects the region R is $\theta = \alpha$ and the largest angle is $\theta = \beta$. This yields the θ-limits of integration.

Example 1   Evaluate

$$\iint\limits_{R} \sin\theta \, dA$$

where R is the region in the first quadrant that is outside the circle $r = 2$ and inside the cardioid $r = 2(1 + \cos\theta)$.

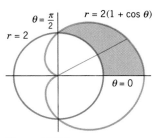

Figure 17.3.7

Solution.   The region R is sketched in Figure 17.3.7. Following the two steps outlined above we obtain

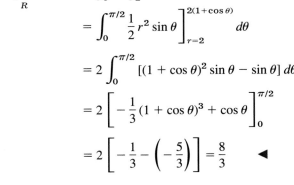

$$\iint\limits_{R} \sin\theta \, dA = \int_0^{\pi/2} \int_2^{2(1+\cos\theta)} (\sin\theta)r \, dr \, d\theta$$

$$= \int_0^{\pi/2} \frac{1}{2} r^2 \sin\theta \Big]_{r=2}^{2(1+\cos\theta)} d\theta$$

$$= 2\int_0^{\pi/2} [(1+\cos\theta)^2 \sin\theta - \sin\theta] \, d\theta$$

$$= 2\left[-\frac{1}{3}(1+\cos\theta)^3 + \cos\theta\right]_0^{\pi/2}$$

$$= 2\left[-\frac{1}{3} - \left(-\frac{5}{3}\right)\right] = \frac{8}{3} \quad \blacktriangleleft$$

Example 2   In cylindrical coordinates, the equation $r^2 + z^2 = a^2$ represents a sphere of radius a centered at the origin. (It is the sphere whose equation in rectangular coordinates is $x^2 + y^2 + z^2 = a^2$.) Use a double polar integral to calculate the volume of this sphere.

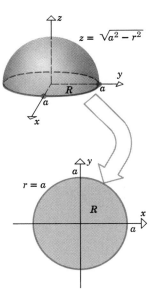

Figure 17.3.8

Solution.   In cylindrical coordinates the upper hemisphere is given by the equation

$$z = \sqrt{a^2 - r^2}$$

so the volume enclosed by the entire sphere is

$$V = 2\iint\limits_{R} \sqrt{a^2 - r^2} \, dA$$

where R is the circular region shown in Figure 17.3.8. Thus,

$$V = 2 \iint_R \sqrt{a^2 - r^2}\, dA = \int_0^{2\pi} \int_0^a \sqrt{a^2 - r^2}\,(2r)\, dr\, d\theta$$

$$= \int_0^{2\pi} \left[-\frac{2}{3}(a^2 - r^2)^{3/2} \right]_{r=0}^a d\theta = \int_0^{2\pi} \frac{2}{3} a^3\, d\theta$$

$$= \left[\frac{2}{3} a^3 \theta \right]_0^{2\pi} = \frac{4}{3}\pi a^3 \quad \blacktriangleleft$$

The argument used to derive Formula (5) of Section 17.2 also applies to polar double integrals, so the area of a simple polar region R can be expressed as

$$\text{area of } R = \iint_R dA \tag{7}$$

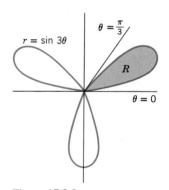

Figure 17.3.9

Example 3   Use a double polar integral to find the area enclosed by the three-petaled rose $r = \sin 3\theta$.

Solution.   The rose is sketched in Figure 17.3.9. We shall use Formula (7) to calculate the area of the petal R in the first quadrant and multiply by three.

$$A = 3 \iint_R dA = 3 \int_0^{\pi/3} \int_0^{\sin 3\theta} r\, dr\, d\theta$$

$$= \frac{3}{2} \int_0^{\pi/3} \sin^2 3\theta\, d\theta = \frac{3}{4} \int_0^{\pi/3} (1 - \cos 6\theta)\, d\theta$$

$$= \frac{3}{4} \left[\theta - \frac{\sin 6\theta}{6} \right]_0^{\pi/3} = \frac{1}{4}\pi \quad \blacktriangleleft$$

☐ **CONVERTING DOUBLE INTEGRALS FROM RECTANGULAR TO POLAR COORDINATES**

Polar double integrals are sometimes called *double integrals in polar coordinates* in contrast to the double integrals studied in the preceding two sections, which are called *double integrals in rectangular coordinates*. Sometimes a double integral that is difficult to evaluate in rectangular coordinates can be evaluated more easily by making the substitution $x = r\cos\theta$, $y = r\sin\theta$ to convert it to an integral in polar coordinates. Under such a substitution, the rectangular and polar integrals are related by the equation

$$\iint_R f(x, y)\, dA = \iint_{\substack{\text{appropriate} \\ \text{limits}}} f(r, \theta)\, r\, dr\, d\theta \tag{8}$$

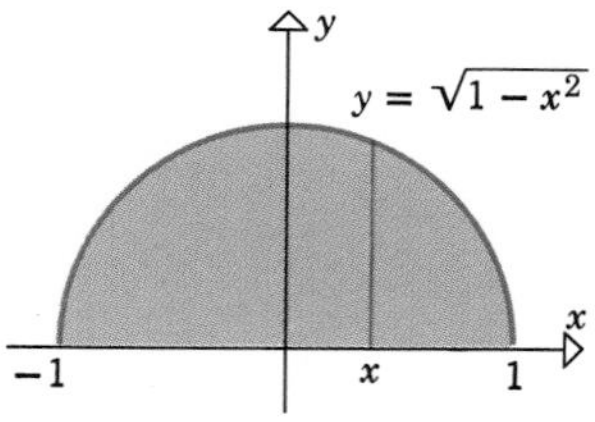

Example 4   Use polar coordinates to evaluate $\displaystyle\int_{-1}^1 \int_0^{\sqrt{1-x^2}} (x^2 + y^2)^{3/2}\, dy\, dx$.

Solution.   The first step is to express this iterated integral as a double integral in rectangular coordinates. For fixed x, the y-integration runs from the lower boundary $y = 0$ up to the semicircle $y = \sqrt{1 - x^2}$. The fixed x can extend from -1 on the left to $+1$ on the right, so the region of integration R for the double integral is as shown in Figure 17.3.10. In

Figure 17.3.10

polar coordinates, this is the region swept out as r varies between 0 and 1 and θ varies between 0 and π. Thus,

$$\int_{-1}^{1} \int_{0}^{\sqrt{1-x^2}} (x^2 + y^2)^{3/2} \, dy \, dx = \iint_{R} (x^2 + y^2)^{3/2} \, dA$$

$$= \int_{0}^{\pi} \int_{0}^{1} (r^3)r \, dr \, d\theta = \int_{0}^{\pi} \frac{1}{5} \, d\theta = \frac{\pi}{5} \quad \blacktriangleleft$$

▶ Exercise Set 17.3

In Exercises 1–6, evaluate the iterated integral.

1. $\displaystyle\int_{0}^{\pi/2} \int_{0}^{\sin\theta} r\cos\theta \, dr \, d\theta.$ 2. $\displaystyle\int_{0}^{\pi} \int_{0}^{1+\cos\theta} r \, dr \, d\theta.$

3. $\displaystyle\int_{-\pi/2}^{\pi/2} \int_{0}^{a\sin\theta} r^2 \, dr \, d\theta.$ 4. $\displaystyle\int_{0}^{\pi/3} \int_{0}^{\cos 3\theta} r \, dr \, d\theta.$

5. $\displaystyle\int_{0}^{\pi} \int_{0}^{1-\sin\theta} r^2 \cos\theta \, dr \, d\theta.$

6. $\displaystyle\int_{0}^{\pi} \int_{0}^{\cos\theta} r^3 \, dr \, d\theta.$

In Exercises 7–12, use a double integral in polar coordinates to find the area of the region described.

7. The region enclosed by the cardioid $r = 1 - \cos\theta$.

8. The region enclosed by the rose $r = \sin 2\theta$.

9. The region in the first quadrant bounded by $r = 1$ and $r = \sin 2\theta$, with $\pi/4 \leq \theta \leq \pi/2$.

10. The region inside the circle $x^2 + y^2 = 4$ and to the right of the line $x = 1$.

11. The region inside the circle $r = 4\sin\theta$ and outside the circle $r = 2$.

12. The region inside the circle $r = 1$ and outside the cardioid $r = 1 + \cos\theta$.

In Exercises 13–18, use a double integral in polar coordinates to find the volume of the solid.

13. The solid enclosed by the sphere $x^2 + y^2 + z^2 = 9$ and the cylinder $x^2 + y^2 = 1$.

14. The solid enclosed by the sphere $r^2 + z^2 = 4$ and the cylinder $r = 2\cos\theta$.

15. The solid that is bounded above by the cone $z = \sqrt{x^2 + y^2}$, below by the xy-plane, and laterally by the cylinder $x^2 + y^2 = 2y$.

16. The solid that is bounded above by the surface $z = (x^2 + y^2)^{-1/2}$, below by the xy-plane, and enclosed between the cylinders $x^2 + y^2 = 1$ and $x^2 + y^2 = 9$.

17. The solid that is bounded above by the paraboloid $z = 1 - x^2 - y^2$, below by the xy-plane, and laterally by the cylinder $x^2 + y^2 - x = 0$.

18. The solid in the first octant bounded above by the plane $z = r\sin\theta$, below by the xy-plane, and laterally by the plane $x = 0$ and the cylinder $r = 3\sin\theta$.

In Exercises 19–22, use polar coordinates to evaluate the double integral.

19. $\displaystyle\iint_{R} e^{-(x^2+y^2)} \, dA$, where R is the region enclosed by the circle $x^2 + y^2 = 1$.

20. $\displaystyle\iint_{R} \sqrt{9 - x^2 - y^2} \, dA$, where R is the region in the first quadrant within the circle $x^2 + y^2 = 9$.

21. $\displaystyle\iint_{R} \frac{1}{1 + x^2 + y^2} \, dA$, where R is the sector in the first quadrant bounded by $y = 0$, $y = x$, and $x^2 + y^2 = 4$.

22. $\displaystyle\iint_{R} 2y \, dA$, where R is the region in the first quadrant bounded above by the circle $(x - 1)^2 + y^2 = 1$ and below by the line $y = x$.

In Exercises 23–30, evaluate the iterated integral by converting to polar coordinates.

23. $\displaystyle\int_{0}^{1} \int_{0}^{\sqrt{1-x^2}} (x^2 + y^2) \, dy \, dx.$

24. $\displaystyle\int_{-2}^{2} \int_{-\sqrt{4-y^2}}^{\sqrt{4-y^2}} e^{-(x^2+y^2)} \, dx \, dy.$

25. $\displaystyle\int_{0}^{2} \int_{0}^{\sqrt{2x-x^2}} \sqrt{x^2 + y^2} \, dy \, dx.$

26. $\displaystyle\int_{0}^{1} \int_{0}^{\sqrt{1-y^2}} \cos(x^2 + y^2) \, dx \, dy.$

27. $\displaystyle\int_{0}^{a} \int_{0}^{\sqrt{a^2-x^2}} \frac{dy \, dx}{(1 + x^2 + y^2)^{3/2}} \quad (a > 0).$

28. $\displaystyle\int_{0}^{1} \int_{y}^{\sqrt{y}} \sqrt{x^2 + y^2} \, dx \, dy.$

29. $\int_0^{\sqrt{2}} \int_y^{\sqrt{4-y^2}} \dfrac{1}{\sqrt{1+x^2+y^2}} \, dx \, dy.$

30. $\int_0^4 \int_3^{\sqrt{25-x^2}} dy \, dx.$

31. Use polar coordinates to find the volume of the solid that is bounded above by the ellipsoid

$$x^2/a^2 + y^2/a^2 + z^2/c^2 = 1$$

below by the xy-plane, and laterally by the cylinder $x^2 + y^2 - ay = 0$.

32. Find the area of the region enclosed by the lemniscate $r^2 = 2a^2 \cos 2\theta$.

33. Find the area in the first quadrant inside the circle $r = 4 \sin \theta$ and outside the lemniscate $r^2 = 8 \cos 2\theta$.

34. Show that the shaded area in the following figure is $a^2\phi - \frac{1}{2}a^2 \sin 2\phi$.

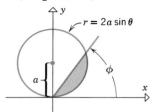

35. The integral $\int_0^{+\infty} e^{-x^2} \, dx$, which arises in probability theory, can be evaluated using the following method. Let the value of the integral be I. Thus,

$$I = \int_0^{+\infty} e^{-x^2} \, dx = \int_0^{+\infty} e^{-y^2} \, dy$$

since the letter used for the variable of integration in a definite integral does not matter.

(a) Show that

$$I^2 = \int_0^{+\infty} \int_0^{+\infty} e^{-(x^2+y^2)} \, dx \, dy$$

(b) Evaluate the iterated integral in part (a) by converting to polar coordinates.

(c) Use the result in part (b) to find I.

36. In each part evaluate $\displaystyle\iint_R x^2 \, dA$.

(a)

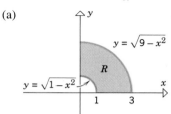

(b)

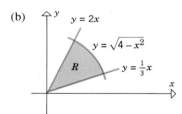

■ **17.4   SURFACE AREA**

In Section 6.5 we showed how to find the surface area of a surface of revolution. In this section we shall consider more general surface area problems.

☐ **THE SURFACE AREA FORMULA**

The first surface area problem of concern in this section can be posed as follows:

> **17.4.1**   PROBLEM.   *Let f be a function defined on a closed region R of the xy-plane. Find the surface area of that portion of the surface $z = f(x, y)$ whose projection on the xy-plane is the region R* (Figure 17.4.1).

We will attack this problem in stages, first solving it in the special case where the surface is a plane and the region R is a rectangle, then using the special case to solve the general problem.

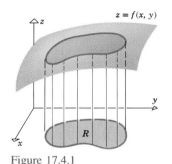

Figure 17.4.1

17.4.2 THEOREM. *Let R be a closed rectangular region in the xy-plane with sides parallel to the coordinate axes. If the lengths of the sides are l and w, then the surface area S of that portion of the plane z = ax + by + c that projects onto the region R is given by*

$$S = \sqrt{a^2 + b^2 + 1} \, lw \tag{1}$$

Proof. Suppose that the four corners of R are

$$A(x_0, y_0), \quad B(x_0 + l, y_0), \quad C(x_0, y_0 + w), \quad D(x_0 + l, y_0 + w)$$

from which it follows that the points on the plane $z = ax + by + c$ that project onto A, B, and C are

$$A'(x_0, y_0, ax_0 + by_0 + c)$$
$$B'(x_0 + l, y_0, a(x_0 + l) + by_0 + c)$$
$$C'(x_0, y_0 + w, ax_0 + b(y_0 + w) + c)$$

(Figure 17.4.2).

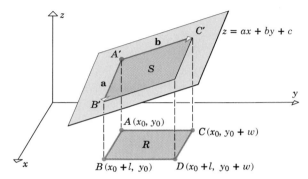

Figure 17.4.2

These points determine a parallelogram with sides

$$\mathbf{a} = \overrightarrow{A'B'} = l\mathbf{i} + 0\mathbf{j} + al\mathbf{k}$$
$$\mathbf{b} = \overrightarrow{A'C'} = 0\mathbf{i} + w\mathbf{j} + bw\mathbf{k}$$

which constitutes the portion of the plane that projects onto the region R. The cross-product of these vectors is

$$\mathbf{a} \times \mathbf{b} = \begin{vmatrix} \mathbf{i} & \mathbf{j} & \mathbf{k} \\ l & 0 & al \\ 0 & w & bw \end{vmatrix} = -alw\mathbf{i} - lbw\mathbf{j} + lw\mathbf{k}$$

and hence from Formula (7) of Section 14.4 the area S of the parallelogram determined by $\mathbf{a}$ and $\mathbf{b}$ is

$$S = \|\mathbf{a} \times \mathbf{b}\| = \sqrt{(-alw)^2 + (-lbw)^2 + (lw)^2} = \sqrt{a^2 + b^2 + 1} \, lw \quad ■$$

We shall now show how Theorem 17.4.2 can be used to derive the following more general result.

17.4.3 SURFACE AREA FORMULA. *If f has continuous first partial derivatives on a closed region R of the xy-plane, then the area S of that portion of the surface z = f(x, y) that projects onto R is*

$$S = \iint_R \sqrt{\left(\frac{\partial z}{\partial x}\right)^2 + \left(\frac{\partial z}{\partial y}\right)^2 + 1} \, dA \tag{2}$$

To motivate Formula (2) we shall use a now familiar limiting process:

- As shown in Figure 17.4.3, enclose R within a rectangle whose sides are parallel to the x- and y-axes, and subdivide that rectangle into smaller rectangles by lines parallel to these axes. Exclude from consideration all rectangles that contain points outside of R, leaving only rectangles that are subsets of R. Assume that there are n such rectangles, and denote the kth rectangle by R_k. Suppose that the sides of R_k have dimensions Δx_k and Δy_k, and let ΔS_k be the area of the patch on the surface $z = f(x, y)$ determined by projecting R_k onto this surface.

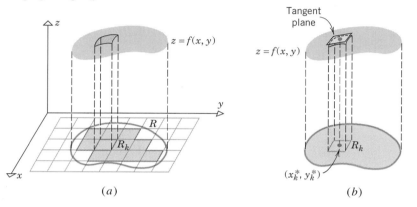

Figure 17.4.3   (a)   (b)

- If we let (x_k^*, y_k^*) be any point in R_k, then the area ΔS_k can be approximated by the area of that portion of the tangent plane to the surface $z = f(x, y)$ at (x_k^*, y_k^*) that projects onto R_k. It follows from Theorem 16.5.1 that this tangent plane has an equation of the form

$$z = f_x(x_k^*, y_k^*)x + f_y(x_k^*, y_k^*)y + c$$

for an appropriate constant c, and hence from Theorem 17.4.2

$$\Delta S_k \approx \sqrt{f_x(x_k^*, y_k^*)^2 + f_y(x_k^*, y_k^*)^2 + 1} \, \Delta x_k \, \Delta y_k \tag{3}$$

Thus, if we let $\Delta A_k = \Delta x_k \, \Delta y_k$ and add the areas of the patches, we obtain the following approximation to the entire area S:

$$S \approx \sum_{k=1}^{n} \sqrt{f_x(x_k^*, y_k^*)^2 + f_y(x_k^*, y_k^*)^2 + 1} \, \Delta A_k$$

- There are two sources of error in this approximation, one resulting from the fact that rectangles $R_1, R_2, \ldots, R_n$ do not fill up the region R completely, and the other resulting from the fact that the patches of area on the surface have been approximated by areas on tangent planes. However, it seems plausible that if we repeat the subdivision process using more and more rectangles with dimensions approaching zero, then the errors will diminish and the exact surface area S will be given by

$$S = \lim_{n \to +\infty} \sum_{k=1}^{n} \sqrt{f_x(x_k^*, y_k^*)^2 + f_y(x_k^*, y_k^*)^2 + 1} \, \Delta A_k$$

$$= \iint\limits_{R} \sqrt{f_x(x, y)^2 + f_y(x, y)^2 + 1} \, dA$$

which is just Formula (2) with a different notation for the partial derivatives.

Example 1   Find the surface area of the portion of the cylinder $x^2 + z^2 = 4$ above the rectangle $R = \{(x, y) : 0 \le x \le 1, 0 \le y \le 4\}$ in the xy-plane.

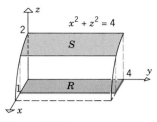

Figure 17.4.4

Solution. The surface is shown in Figure 17.4.4. The portion of the cylinder $x^2 + z^2 = 4$ that lies above the xy-plane has the equation $z = \sqrt{4 - x^2}$. Thus, from (2)

$$
S = \iint_R \sqrt{\left(\frac{\partial z}{\partial x}\right)^2 + \left(\frac{\partial z}{\partial y}\right)^2 + 1}\; dA
$$

$$
= \iint_R \sqrt{\left(-\frac{x}{\sqrt{4 - x^2}}\right)^2 + 0 + 1}\; dA = \int_0^4 \int_0^1 \frac{2}{\sqrt{4 - x^2}}\, dx\, dy
$$

$$
= 2 \int_0^4 \left[\sin^{-1}\left(\frac{1}{2}x\right) \right]_{x=0}^1 dy = 2 \int_0^4 \frac{\pi}{6}\, dy = \frac{4}{3}\pi \qquad \blacktriangleleft
$$

Formula (21)
of Section 8.2

Example 2 Find the surface area of the portion of the paraboloid $z = x^2 + y^2$ below the plane $z = 1$.

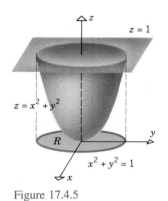

Figure 17.4.5

Solution. The surface is shown in Figure 17.4.5. From the equations $z = 1$ and $z = x^2 + y^2$, we see that the plane and the paraboloid intersect in a circle whose projection on the xy-plane has the equation $x^2 + y^2 = 1$. Consequently, the surface whose area we seek projects onto the region R enclosed by this circle. Since the surface has the equation

$$
z = x^2 + y^2
$$

it follows from (2) that

$$
S = \iint_R \sqrt{4x^2 + 4y^2 + 1}\; dA
$$

This integral is best evaluated in polar coordinates. Replacing $x^2 + y^2$ by r^2 and substituting $r\, dr\, d\theta$ for dA we obtain

$$
S = \int_0^{2\pi} \int_0^1 \sqrt{4r^2 + 1}\; r\, dr\, d\theta = \int_0^{2\pi} \left[\frac{1}{12}(4r^2 + 1)^{3/2} \right]_{r=0}^1 d\theta
$$

$$
= \int_0^{2\pi} \frac{1}{12}(5\sqrt{5} - 1)\, d\theta = \frac{1}{6}\pi(5\sqrt{5} - 1) \qquad \blacktriangleleft
$$

☐ **SURFACE AREA OF PARAMETRIC SURFACES**

We now consider the problem of finding the surface area S of a surface σ that is represented parametrically as

$$
\mathbf{r}(u, v) = x(u, v)\mathbf{i} + y(u, v)\mathbf{j} + z(u, v)\mathbf{k}
$$

where (u, v) varies over some region R of the uv-plane. We shall say that $\mathbf{r}$ is a *smooth function* of u and v or that σ is a *smooth parametric surface* if $\partial\mathbf{r}/\partial u$ and $\partial\mathbf{r}/\partial v$ are continuous and $\partial\mathbf{r}/\partial u \times \partial\mathbf{r}/\partial v \neq \mathbf{0}$ on R. Geometrically, this means that σ has a unit normal (and hence a tangent plane) at each point of σ and that the unit normal $\mathbf{n} = \mathbf{n}(u, v)$ is continuous on R [see (6) of Section 16.3]. Thus, on a smooth parametric surface the unit normal varies continuously with no abrupt changes in direction. We shall derive the formula for the surface area of a smooth parametric surface that has no self-intersections.

We begin by subdividing R into rectangular regions by lines parallel to the u and v axes, discarding any nonrectangular portions that contain points of the boundary. Assume that there are n rectangles, and let R_k denote the kth rectangle. Let (u_k, v_k) be the lower left corner of R_k, and assume that R_k has area $\Delta A_k = \Delta u_k \Delta v_k$, where Δu_k and Δv_k are the dimensions of R_k (Figure 17.4.6a). The image of R_k will be some *curvilinear patch* σ_k on the surface σ that has a corner at $\mathbf{r}(u_k, v_k)$; denote the area of this patch by ΔS_k (Figure 17.4.6b).

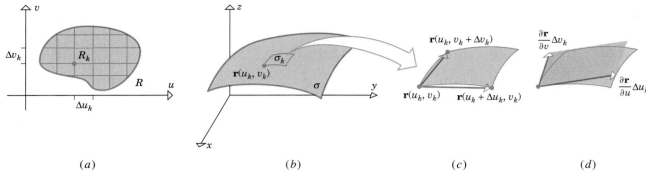

Figure 17.4.6

As suggested by Figure 17.4.6c, the two edges of the patch that meet at $\mathbf{r}(u_k, v_k)$ can be approximated by the "secant" vectors

$$\mathbf{r}(u_k + \Delta u_k, v_k) - \mathbf{r}(u_k, v_k)$$

$$\mathbf{r}(u_k, v_k + \Delta v_k) - \mathbf{r}(u_k, v_k)$$

and hence the area of σ_k can be approximated by the area of the parallelogram determined by these vectors. However, it follows from Formulas (4) and (5) of Section 16.3 that if Δu_k and Δv_k are small, then these secant vectors can in turn be approximated by the tangent vectors

$$\frac{\partial \mathbf{r}}{\partial u} \Delta u_k \quad \text{and} \quad \frac{\partial \mathbf{r}}{\partial v} \Delta v_k$$

where the partial derivatives are evaluated at (u_k, v_k). Thus, the area of the patch σ_k can be approximated by the area of the parallelogram determined by these vectors (Figure 17.4.6d); that is,

$$\Delta S_k \approx \left\| \frac{\partial \mathbf{r}}{\partial u} \Delta u_k \times \frac{\partial \mathbf{r}}{\partial v} \Delta v_k \right\| = \left\| \frac{\partial \mathbf{r}}{\partial u} \times \frac{\partial \mathbf{r}}{\partial v} \right\| \Delta u_k \Delta v_k = \left\| \frac{\partial \mathbf{r}}{\partial u} \times \frac{\partial \mathbf{r}}{\partial v} \right\| \Delta A_k \tag{4}$$

It follows that the surface area S of the entire surface σ can be approximated as

$$S \approx \sum_{k=1}^{n} \left\| \frac{\partial \mathbf{r}}{\partial u} \times \frac{\partial \mathbf{r}}{\partial v} \right\| \Delta A_k$$

Thus, if we assume that the errors in the approximations approach zero as n increases in such a way that the dimensions of the rectangles approach zero, then it is plausible that the exact value of S is

$$S = \lim_{n \to +\infty} \sum_{k=1}^{n} \left\| \frac{\partial \mathbf{r}}{\partial u} \times \frac{\partial \mathbf{r}}{\partial v} \right\| \Delta A_k$$

or equivalently,

$$S = \iint\limits_{R} \left\| \frac{\partial \mathbf{r}}{\partial u} \times \frac{\partial \mathbf{r}}{\partial v} \right\| dA \tag{5}$$

REMARK.   If the surface σ is given by $z = f(x, y)$, then with x and y as parameters, the surface can be represented parametrically by

$$\mathbf{r}(x, y) = x\mathbf{i} + y\mathbf{j} + f(x, y)\mathbf{k}$$

We leave it for the reader to use this parametrization to obtain Formula (2) as a special case of Formula (5).

Example 3   The surface represented by

$$\mathbf{r}(u, v) = u \cos v\mathbf{i} + u \sin v\mathbf{j} + 3u\mathbf{k}$$

where $0 \le u \le 2$ and $0 \le v \le 2\pi$ is a portion of a cone (verify). Find its surface area.

Solution.

$$\frac{\partial \mathbf{r}}{\partial u} = \cos v\mathbf{i} + \sin v\mathbf{j} + 3\mathbf{k} \quad \text{and} \quad \frac{\partial \mathbf{r}}{\partial v} = -u \sin v\mathbf{i} + u \cos v\mathbf{j}$$

$$\frac{\partial \mathbf{r}}{\partial u} \times \frac{\partial \mathbf{r}}{\partial v} = \begin{vmatrix} \mathbf{i} & \mathbf{j} & \mathbf{k} \\ \cos v & \sin v & 3 \\ -u \sin v & u \cos v & 0 \end{vmatrix} = -3u \cos v\mathbf{i} - 3u \sin v\mathbf{j} + u\mathbf{k}$$

Thus, from (5)

$$S = \iint_R \left\| \frac{\partial \mathbf{r}}{\partial u} \times \frac{\partial \mathbf{r}}{\partial v} \right\| dA = \int_0^{2\pi} \int_0^2 \sqrt{10}u \, du \, dv = 2\sqrt{10} \int_0^{2\pi} dv = 4\sqrt{10}\pi \quad \blacktriangleleft$$

▶ Exercise Set 17.4

1. Find the surface area of the portion of the cylinder $y^2 + z^2 = 9$ that is above the rectangle $R = \{(x, y): 0 \le x \le 2, -3 \le y \le 3\}$.

2. By integration, find the surface area of the portion of the plane $2x + 2y + z = 8$ in the first octant that is cut off by the three coordinate planes.

3. Find the surface area of the portion of the cone $z^2 = 4x^2 + 4y^2$ that is above the region in the first quadrant bounded by the line $y = x$ and the parabola $y = x^2$.

4. Find the surface area of the portion of the cone $z = \sqrt{x^2 + y^2}$ that lies inside the cylinder $x^2 + y^2 = 2x$.

5. Find the surface area of the portion of the paraboloid $z = 1 - x^2 - y^2$ that is above the xy-plane.

6. Find the area of the portion of the surface $z = 2x + y^2$ that is above the triangular region with vertices $(0, 0)$, $(0, 1)$, and $(1, 1)$.

7. Find the area of the portion of the surface $z = xy$ that is above the sector in the first quadrant bounded by the lines $y = x/\sqrt{3}$, $y = 0$, and the circle $x^2 + y^2 = 9$.

8. Find the surface area of the portion of the paraboloid $2z = x^2 + y^2$ that is inside the cylinder $x^2 + y^2 = 8$.

9. Find the surface area of the portion of the sphere $x^2 + y^2 + z^2 = 16$ between the planes $z = 1$ and $z = 2$.

10. Find the surface area of the portion of the sphere $x^2 + y^2 + z^2 = 8$ that is cut out by the cone $z = \sqrt{x^2 + y^2}$.

11. Find the total surface area of the portion of the sphere $x^2 + y^2 + z^2 = a^2$ inside the cylinder $x^2 + y^2 = ay$.

12. Use a double integral in Cartesian coordinates to derive the formula for the surface area of a sphere of radius a.

13. Find the surface area of the portion of the cylinder $x^2 + z^2 = 16$ that lies inside the circular cylinder $x^2 + y^2 = 16$. [*Hint:* Find the area in the first octant and use symmetry.]

14. Find the surface area of the portion of the cylinder $x^2 + z^2 = 5x$ inside the sphere $x^2 + y^2 + z^2 = 25$.

15. The portion of the surface

$$z = \frac{h}{a}\sqrt{x^2 + y^2} \qquad (a, h > 0)$$

between the xy-plane and the plane $z = h$ is a right-circular cone of height h and radius a. Use a double integral to show that the lateral surface area of this cone is $S = \pi a\sqrt{a^2 + h^2}$.

16. Find the surface area of the portion of the cone $z = \sqrt{x^2 + y^2}$ that lies between the planes $z = 0$ and $z = y/2 + 3$.

17. Find the surface area of the portion of the paraboloid

$$\mathbf{r}(u, v) = u \cos v\mathbf{i} + u \sin v\mathbf{j} + u^2\mathbf{k}$$

for which $1 \le u \le 2$, $0 \le v \le 2\pi$.

18. Find the surface area of the portion of the cone

$$\mathbf{r}(u, v) = u \cos v\mathbf{i} + u \sin v\mathbf{j} + u\mathbf{k}$$

for which $0 \le u \le 2v$, $0 \le v \le \pi/2$.

19. Find the surface area of the sphere of radius a

$$\mathbf{r}(u, v) = a \sin u \cos v\mathbf{i} + a \sin u \sin v\mathbf{j} + a \cos u\mathbf{k}$$

for which $0 \le u \le \pi$, $0 \le v \le 2\pi$.

20. Find the surface area of the portion of the hyperbolic paraboloid

$$\mathbf{r}(u, v) = (u + v)\mathbf{i} + (u - v)\mathbf{j} + uv\mathbf{k}$$

for which $u^2 + v^2 \le 4$.

21. Find the surface area of the portion of the spiral ramp

$$\mathbf{r}(u, v) = u \cos v\mathbf{i} + u \sin v\mathbf{j} + v\mathbf{k}$$

for which $0 \le u \le 2$, $0 \le v \le 3u$.

22. The torus in Figure 16.1.28 can be represented by

$$\mathbf{r}(u, v) = (a + b \cos v) \cos u\mathbf{i} + (a + b \cos v) \sin u\mathbf{j} + b \sin v\mathbf{k}$$

where

$$0 < b \le a \quad \text{and} \quad 0 \le u \le 2\pi, 0 \le v \le 2\pi$$

Find the surface area.

■ **17.5** TRIPLE INTEGRALS

In the preceding sections we defined and discussed properties of double integrals for functions of two variables. In this section we shall define triple integrals for functions of three variables.

☐ **DEFINITION OF A TRIPLE INTEGRAL**

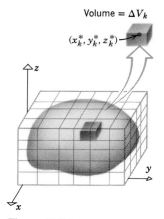

Volume = ΔV_k

(x_k^*, y_k^*, z_k^*)

Figure 17.5.1

Whereas a double integral of a function $f(x, y)$ is defined over a closed region R in the xy-plane, a triple integral of a function $f(x, y, z)$ is defined over a closed three-dimensional solid region G. To ensure that G does not extend indefinitely in any direction, we shall assume that G can be enclosed within some suitably large box (rectangular parallelepiped) with sides parallel to the coordinate planes (Figure 17.5.1).

To define the triple integral of $f(x, y, z)$ over G, we first divide the box into n "subboxes" by planes parallel to the coordinate planes. We then discard those subboxes that contain any points outside of G and choose an arbitrary point in each of the remaining subboxes. As shown in Figure 17.5.1, we denote the volume of the kth remaining subbox by ΔV_k and the point selected in the kth subbox by (x_k^*, y_k^*, z_k^*). Next, we form the product

$$f(x_k^*, y_k^*, z_k^*) \Delta V_k$$

for each subbox, then add the products for all of the subboxes to obtain the **Riemann sum**

$$\sum_{k=1}^{n} f(x_k^*, y_k^*, z_k^*) \Delta V_k$$

Finally, we repeat this process with more and more subdivisions in such a way that the length, width, and height of each subbox approaches zero, and n approaches $+\infty$. The limit

$$\iiint\limits_{G} f(x, y, z) \, dV = \lim_{n \to +\infty} \sum_{k=1}^{n} f(x_k^*, y_k^*, z_k^*) \Delta V_k \tag{1}$$

is called the **triple integral** of $f(x, y, z)$ over the region G. Conditions under which the triple integral exists are studied in advanced calculus. However, for our purposes it suffices to say that existence is ensured when f is continuous on G and the region G is not too "complicated."

□ **PROPERTIES OF TRIPLE INTEGRALS**

Triple integrals enjoy many properties of single and double integrals:

$$\iiint\limits_{G} cf(x, y, z) \, dV = c \iiint\limits_{G} f(x, y, z) \, dV \quad (c \text{ a constant})$$

$$\iiint\limits_{G} [f(x, y, z) + g(x, y, z)] \, dV = \iiint\limits_{G} f(x, y, z) \, dV + \iiint\limits_{G} g(x, y, z) \, dV$$

$$\iiint\limits_{G} [f(x, y, z) - g(x, y, z)] \, dV = \iiint\limits_{G} f(x, y, z) \, dV - \iiint\limits_{G} g(x, y, z) \, dV$$

Moreover, if the region G is subdivided into two subregions G_1 and G_2 (Figure 17.5.2), then

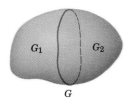

Figure 17.5.2

$$\iiint\limits_{G} f(x, y, z) \, dV = \iiint\limits_{G_1} f(x, y, z) \, dV + \iiint\limits_{G_2} f(x, y, z) \, dV$$

We omit the proofs.

□ **EVALUATING TRIPLE INTEGRALS OVER RECTANGULAR BOXES**

Just as a double integral can be evaluated by two successive single integrations, so a triple integral can be evaluated by three successive integrations. The following theorem, which we state without proof, is the analog of Theorem 17.1.2.

17.5.1   THEOREM.   *Let G be the rectangular box defined by the inequalities*

$$a \le x \le b, \quad c \le y \le d, \quad k \le z \le l$$

If f is continuous on the region G, then

$$\iiint\limits_{G} f(x, y, z) \, dV = \int_a^b \int_c^d \int_k^l f(x, y, z) \, dz \, dy \, dx \qquad (2)$$

Moreover, the iterated integral on the right can be replaced with any of the five other iterated integrals that result by altering the order of integration.

Example 1   Evaluate the triple integral

$$\iiint\limits_{G} 12xy^2 z^3 \, dV$$

over the rectangular box G defined by the inequalities $-1 \le x \le 2$, $0 \le y \le 3$, $0 \le z \le 2$.

Solution.   Of the six possible iterated integrals we might use, we shall choose the one in (2). Thus, we shall first integrate with respect to z, holding x and y fixed, then with respect to y, holding x fixed, and finally with respect to x.

$$\iiint\limits_{G} 12xy^2 z^3 \, dV = \int_{-1}^{2} \int_{0}^{3} \int_{0}^{2} 12xy^2 z^3 \, dz \, dy \, dx$$

$$= \int_{-1}^{2} \int_{0}^{3} \left[3xy^2 z^4 \right]_{z=0}^{2} dy \, dx = \int_{-1}^{2} \int_{0}^{3} 48xy^2 \, dy \, dx$$

$$= \int_{-1}^{2} \left[16xy^3 \right]_{y=0}^{3} dx = \int_{-1}^{2} 432x \, dx$$

$$= 216x^2 \Big]_{-1}^{2} = 648 \qquad \blacktriangleleft$$

☐ **EVALUATING TRIPLE INTEGRALS OVER MORE GENERAL REGIONS**

We shall also be concerned with evaluating triple integrals over solid regions other than rectangular boxes. For simplicity, we shall restrict our discussion to solid regions constructed as follows. Let R be a closed region in the xy-plane and let $g_1(x, y)$ and $g_2(x, y)$ be continuous functions such that $g_1(x, y) \leq g_2(x, y)$ for all (x, y) in R. Geometrically, this condition states that the surface $z = g_2(x, y)$ does not dip below the surface $z = g_1(x, y)$ over R (Figure 17.5.3a). We shall call $z = g_1(x, y)$ the **lower surface** and $z = g_2(x, y)$ the **upper surface**. Let G be the solid consisting of all points above or below the region R that lie between the upper surface and the lower surface (Figure 17.5.3b). A solid G constructed in this way will be called a **simple solid**, and the region R will be called the **projection** of G on the xy-plane.

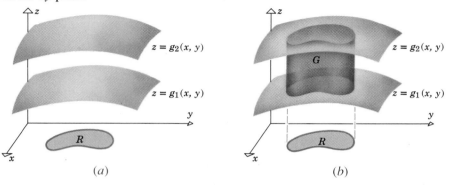

Figure 17.5.3   (a)   (b)

The following theorem, which we state without proof, will enable us to evaluate triple integrals over simple solids.

17.5.2   THEOREM.   *Let G be a simple solid with upper surface $z = g_2(x, y)$ and lower surface $z = g_1(x, y)$, and let R be the projection of G on the xy-plane. If $f(x, y, z)$ is continuous on G, then*

$$\iiint_G f(x, y, z)\, dV = \iint_R \left[\int_{g_1(x, y)}^{g_2(x, y)} f(x, y, z)\, dz \right] dA \tag{3}$$

In (3), the first integration is with respect to z, after which a function of x and y remains. This function of x and y is then integrated over the region R in the xy-plane. To apply (3), it is usual to begin with a three-dimensional sketch of the solid G, from which the limits of integration can be obtained as follows:

Step 1.   Find an equation $z = g_2(x, y)$ for the upper surface and an equation $z = g_1(x, y)$ for the lower surface of G. The functions $g_1(x, y)$ and $g_2(x, y)$ determine the lower and upper z-limits of integration.

Step 2.   Make a two-dimensional sketch of the projection R of the solid on the xy-plane. From this sketch determine the limits of integration for the double integral over R in (3).

Example 2   Let G be the wedge in the first octant cut from the cylindrical solid $y^2 + z^2 \leq 1$ by the planes $y = x$ and $x = 0$. Evaluate

$$\iiint_G z\, dV$$

Solution.   The solid G and its projection R on the xy-plane are shown in Figure 17.5.4. The upper surface of the solid is formed by the cylinder and the lower surface by the xy-plane.

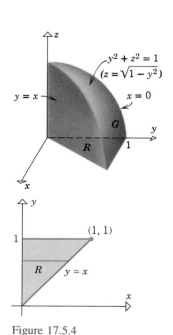

Figure 17.5.4

Since the portion of the cylinder $y^2 + z^2 = 1$ that lies above the xy-plane has the equation $z = \sqrt{1 - y^2}$, and the xy-plane has the equation $z = 0$, it follows from (3) that

$$\iiint\limits_{G} z\, dV = \iint\limits_{R} \left[\int_0^{\sqrt{1-y^2}} z\, dz \right] dA \tag{4}$$

For the double integral over R, the x and y integrations can be performed in either order, since R is both a type I and type II region. We shall integrate with respect to x first. With this choice, (4) yields

$$\iiint\limits_{G} z\, dV = \int_0^1 \int_0^y \int_0^{\sqrt{1-y^2}} z\, dz\, dx\, dy = \int_0^1 \int_0^y \frac{1}{2} z^2 \Big]_{z=0}^{\sqrt{1-y^2}} dx\, dy$$

$$= \int_0^1 \int_0^y \frac{1}{2}(1 - y^2)\, dx\, dy = \frac{1}{2} \int_0^1 (1 - y^2)x \Big]_{x=0}^{y} dy$$

$$= \frac{1}{2} \int_0^1 (y - y^3)\, dy = \frac{1}{2} \left[\frac{1}{2}y^2 - \frac{1}{4}y^4 \right]_0^1 = \frac{1}{8} \quad \blacktriangleleft$$

☐ **VOLUME CALCULATED AS A TRIPLE INTEGRAL**

Triple integrals have a number of physical interpretations, some of which we shall consider in the next section. However, in the special case where $f(x, y, z) = 1$, Formula (1) yields

$$\iiint\limits_{G} dV = \lim_{n \to +\infty} \sum_{k=1}^{n} \Delta V_k$$

which Figure 17.5.1 suggests is the volume of G; that is,

$$\text{volume of } G = \iiint\limits_{G} dV \tag{5}$$

Example 3 Use a triple integral to find the volume of the solid enclosed between the cylinder $x^2 + y^2 = 9$ and the planes $z = 1$ and $x + z = 5$.

Solution. The solid G and its projection R on the xy-plane are shown in Figure 17.5.5. The lower surface of the solid is the plane $z = 1$, and the upper surface is the plane $x + z = 5$, or equivalently, $z = 5 - x$. Thus, from (3) and (5)

$$\text{volume of } G = \iiint\limits_{G} dV = \iint\limits_{R} \left[\int_1^{5-x} dz \right] dA \tag{6}$$

For the double integral over R, we shall integrate with respect to y first. Thus, (6) yields

$$\text{volume of } G = \int_{-3}^{3} \int_{-\sqrt{9-x^2}}^{\sqrt{9-x^2}} \int_1^{5-x} dz\, dy\, dx = \int_{-3}^{3} \int_{-\sqrt{9-x^2}}^{\sqrt{9-x^2}} z \Big]_{z=1}^{5-x} dy\, dx$$

$$= \int_{-3}^{3} \int_{-\sqrt{9-x^2}}^{\sqrt{9-x^2}} (4 - x)\, dy\, dx = \int_{-3}^{3} (8 - 2x)\sqrt{9 - x^2}\, dx$$

$$= 8 \int_{-3}^{3} \sqrt{9 - x^2}\, dx - \int_{-3}^{3} 2x\sqrt{9 - x^2}\, dx \qquad \boxed{\text{For the first integral, see Example 4 of Section 9.5.}}$$

$$= 8\left(\frac{9}{2}\pi \right) - \int_{-3}^{3} 2x\sqrt{9 - x^2}\, dx \qquad \boxed{\text{Let } u = 9 - x^2, \text{ or better, apply the result in Exercise 48(a) of Section 5.8.}}$$

$$= 8\left(\frac{9}{2}\pi \right) - 0 = 36\pi \quad \blacktriangleleft$$

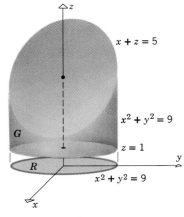

z
$x + z = 5$
$x^2 + y^2 = 9$
G
$z = 1$
R
y
$x^2 + y^2 = 9$
x

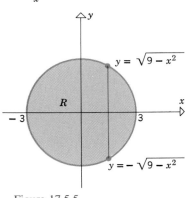

y
$y = \sqrt{9 - x^2}$
R
-3 3 x
$y = -\sqrt{9 - x^2}$

Figure 17.5.5

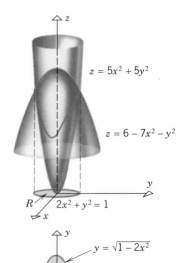

$z = 5x^2 + 5y^2$

$z = 6 - 7x^2 - y^2$

R $2x^2 + y^2 = 1$

$y = \sqrt{1 - 2x^2}$

R

$-1/\sqrt{2}$ $1/\sqrt{2}$

$y = -\sqrt{1 - 2x^2}$

Figure 17.5.7

Example 4   Find the volume of the solid enclosed by the paraboloids

$$z = 5x^2 + 5y^2 \quad \text{and} \quad z = 6 - 7x^2 - y^2$$

Solution.   The solid G and its projection R on the xy-plane are shown in Figure 17.5.6. The projection R is obtained by solving the given equations simultaneously to determine where the paraboloids intersect. We obtain

$$5x^2 + 5y^2 = 6 - 7x^2 - y^2$$

or

$$2x^2 + y^2 = 1 \tag{7}$$

which tells us that the paraboloids intersect in a curve on the elliptic cylinder given by (7).

The projection of this intersection on the xy-plane is an ellipse with this same equation. Therefore,

$$\text{volume of } G = \iiint_G dV = \iint_R \left[\int_{5x^2+5y^2}^{6-7x^2-y^2} dz \right] dA$$

$$\text{volume of } G = \int_{-1/\sqrt{2}}^{1/\sqrt{2}} \int_{-\sqrt{1-2x^2}}^{\sqrt{1-2x^2}} \int_{5x^2+5y^2}^{6-7x^2-y^2} dz\, dy\, dx$$

$$= \int_{-1/\sqrt{2}}^{1/\sqrt{2}} \int_{-\sqrt{1-2x^2}}^{\sqrt{1-2x^2}} (6 - 12x^2 - 6y^2)\, dy\, dx$$

$$= \int_{-1/\sqrt{2}}^{1/\sqrt{2}} \left[6(1 - 2x^2)y - 2y^3 \right]_{y=-\sqrt{1-2x^2}}^{\sqrt{1-2x^2}} dx$$

$$= 8 \int_{-1/\sqrt{2}}^{1/\sqrt{2}} (1 - 2x^2)^{3/2}\, dx = \frac{8}{\sqrt{2}} \int_{-\pi/2}^{\pi/2} \cos^4 \theta\, d\theta = \frac{3\pi}{\sqrt{2}} \quad \blacktriangleleft$$

Let $x = \dfrac{1}{\sqrt{2}} \sin\theta.$

Exercise 36, Section 9.3

☐ **INTEGRATION IN OTHER ORDERS**

For certain regions, triple integrals are best evaluated by integrating first with respect to x or y rather than z. For example, if the solid G is bounded on the left and right by the surfaces $y = g_1(x, z)$ and $y = g_2(x, z)$ and bounded laterally by a cylinder extending in the y-direction (Figure 17.5.7a), then

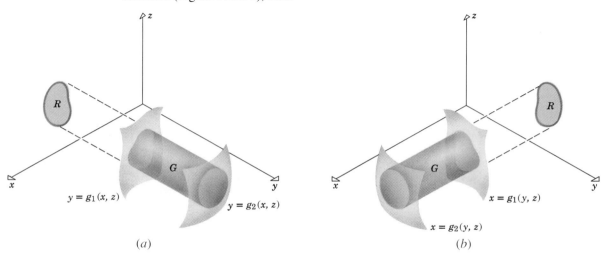

Figure 17.5.7

(a)   (b)

$$\iiint\limits_G f(x, y, z)\, dV = \iint\limits_R \left[\int_{g_1(x,z)}^{g_2(x,z)} f(x, y, z)\, dy \right] dA$$

where R is the projection of the solid G on the xz-plane.

Similarly, if the solid G is bounded in the back and front by the surfaces $x = g_1(y, z)$ and $x = g_2(y, z)$ and bounded laterally by a cylinder extending in the x-direction, then

$$\iiint\limits_G f(x, y, z)\, dV = \iint\limits_R \left[\int_{g_1(y,z)}^{g_2(y,z)} f(x, y, z)\, dx \right] dA$$

where R is the projection of the solid on the yz-plane (Figure 17.5.7b).

Example 5 In Example 2, we evaluated

$$\iiint\limits_G z\, dV$$

over the wedge in Figure 17.5.4 by integrating first with respect to z. Evaluate this integral by integrating first with respect to x.

Solution. The solid is bounded in the back by the plane $x = 0$ and in the front by the plane $x = y$, so

$$\iiint\limits_G z\, dV = \iint\limits_R \left[\int_0^y z\, dx \right] dA$$

where R is the projection of G on the yz-plane (Figure 17.5.8). The integration over R can be performed first with respect to z and then y or vice versa. Performing the z-integration first yields

$$\iiint\limits_G z\, dV = \int_0^1 \int_0^{\sqrt{1-y^2}} \int_0^y z\, dx\, dz\, dy = \int_0^1 \int_0^{\sqrt{1-y^2}} zx \bigg]_{x=0}^y dz\, dy$$

$$= \int_0^1 \int_0^{\sqrt{1-y^2}} zy\, dz\, dy = \int_0^1 \frac{1}{2} z^2 y \bigg]_{z=0}^{\sqrt{1-y^2}} dy = \int_0^1 \frac{1}{2}(1 - y^2) y\, dy = \frac{1}{8}$$

which agrees with the result in Example 2. ◄

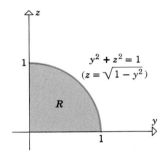

$y^2 + z^2 = 1$
$(z = \sqrt{1 - y^2})$

R

Figure 17.5.8

► **Exercise Set 17.5**

In Exercises 1–8, evaluate the iterated integral.

1. $\displaystyle\int_{-1}^1 \int_0^2 \int_0^1 (x^2 + y^2 + z^2)\, dx\, dy\, dz.$

2. $\displaystyle\int_{1/3}^{1/2} \int_0^\pi \int_0^1 zx \sin xy\, dz\, dy\, dx.$

3. $\displaystyle\int_0^2 \int_{-1}^{y^2} \int_1^z yz\, dx\, dz\, dy.$

4. $\displaystyle\int_0^{\pi/4} \int_0^1 \int_0^{x^2} x \cos y\, dz\, dx\, dy.$

5. $\displaystyle\int_0^3 \int_0^{\sqrt{9-z^2}} \int_0^x xy\, dy\, dx\, dz.$

6. $\displaystyle\int_1^3 \int_x^{x^2} \int_0^{\ln z} xe^y\, dy\, dz\, dx.$

7. $\displaystyle\int_0^2 \int_0^{\sqrt{4-x^2}} \int_{-5+x^2+y^2}^{3-x^2-y^2} x\, dz\, dy\, dx.$

8. $\displaystyle\int_1^2 \int_z^2 \int_0^{\sqrt{3}y} \frac{y}{x^2 + y^2}\, dx\, dy\, dz.$

In Exercises 9–12, evaluate the triple integral.

9. $\iiint\limits_{G} xy \sin yz \, dV$, where G is the rectangular box defined by the inequalities $0 \le x \le \pi$, $0 \le y \le 1$, $0 \le z \le \pi/6$.

10. $\iiint\limits_{G} y \, dV$, where G is the solid enclosed by the plane $z = y$, the xy-plane, and the parabolic cylinder $y = 1 - x^2$.

11. $\iiint\limits_{G} xyz \, dV$, where G is the solid in the first octant that is bounded by the parabolic cylinder $z = 2 - x^2$ and the planes $z = 0$, $y = x$, and $y = 0$.

12. $\iiint\limits_{G} \cos(z/y) \, dV$, where G is the solid defined by the inequalities, $\pi/6 \le y \le \pi/2$, $y \le x \le \pi/2$, $0 \le z \le xy$.

In Exercises 13–21, use a triple integral to find the volume of the solid.

13. The solid in the first octant bounded by the coordinate planes and the plane $3x + 6y + 4z = 12$.

14. The solid bounded by the surface $z = \sqrt{y}$ and the planes $x + y = 1$, $x = 0$, and $z = 0$.

15. The solid bounded by the surface $y = x^2$ and the planes $y + z = 4$ and $z = 0$.

16. The wedge in the first octant cut from the cylinder $y^2 + z^2 \le 1$ by the planes $y = x$ and $x = 0$.

17. The solid enclosed between the elliptic cylinder $x^2 + 9y^2 = 9$ and the planes $z = 0$ and $z = x + 3$.

18. The solid enclosed by the cylinders $x^2 + y^2 = 1$ and $x^2 + z^2 = 1$.

19. The solid bounded by the paraboloid $z = 4x^2 + y^2$ and the parabolic cylinder $z = 4 - 3y^2$.

20. The solid that is enclosed between the paraboloids $z = 8 - x^2 - y^2$ and $z = 3x^2 + y^2$.

21. The solid that is enclosed between the surfaces $x^2 + y^2 + z^2 = 2a^2$ and $az = x^2 + y^2$ $(a > 0)$.

22. Sketch the solid whose volume is given by the integral.

(a) $\displaystyle\int_0^3 \int_{x^2}^9 \int_0^2 dz \, dy \, dx$

(b) $\displaystyle\int_0^2 \int_0^{2-y} \int_0^{2-x-y} dz \, dx \, dy$.

23. Sketch the solid whose volume is given by the integral.

(a) $\displaystyle\int_{-1}^1 \int_{-\sqrt{1-x^2}}^{\sqrt{1-x^2}} \int_0^{y+1} dz \, dy \, dx$

(b) $\displaystyle\int_0^9 \int_0^{y/3} \int_0^{\sqrt{y^2-9x^2}} dz \, dx \, dy$.

24. Sketch the solid whose volume is given by the integral.

(a) $\displaystyle\int_0^1 \int_0^{\sqrt{1-x^2}} \int_0^2 dy \, dz \, dx$

(b) $\displaystyle\int_{-2}^2 \int_0^{4-y^2} \int_0^2 dx \, dz \, dy$.

25. Let G be the tetrahedron in the first octant bounded by the coordinate planes and the plane

$$\frac{x}{a} + \frac{y}{b} + \frac{z}{c} = 1 \quad (a > 0, b > 0, c > 0)$$

(a) List six different iterated integrals that represent the volume of G.

(b) Evaluate any one of the six to show that the volume of G is $\frac{1}{6}abc$.

26. In parts (a)–(c), express the integral as an equivalent integral in which the z-integration is performed first, the y-integration second, and the x-integration last.

(a) $\displaystyle\int_0^3 \int_0^{\sqrt{9-z^2}} \int_0^{\sqrt{9-y^2-z^2}} f(x, y, z) \, dx \, dy \, dz$

(b) $\displaystyle\int_0^4 \int_0^2 \int_0^{x/2} f(x, y, z) \, dy \, dz \, dx$

(c) $\displaystyle\int_0^4 \int_0^{4-y} \int_0^{\sqrt{z}} f(x, y, z) \, dx \, dz \, dy$.

27. Use a triple integral to find the volume of the tetrahedron with vertices $(0, 0, 0)$, $(a, a, 0)$, $(a, 0, 0)$, and $(a, 0, a)$.

28. Let G be the rectangular box defined by the inequalities $a \le x \le b$, $c \le y \le d$, $k \le z \le l$. Show that

$$\iiint\limits_{G} f(x)g(y)h(z) \, dV$$
$$= \left[\int_a^b f(x) \, dx \right] \left[\int_c^d g(y) \, dy \right] \left[\int_k^l h(z) \, dz \right]$$

29. Use the result of Exercise 28 to evaluate

(a) $\iiint\limits_{G} xy^2 \sin z \, dV$, where G is the set of points satisfying $-1 \le x \le 1$, $0 \le y \le 1$, $0 \le z \le \pi/2$

(b) $\iiint\limits_{G} e^{2x+y-z} \, dV$, where G is the set of points satisfying $0 \le x \le 1$, $0 \le y \le \ln 3$, $0 \le z \le \ln 2$.

30. Use a triple integral to find the volume of the ellipsoid

$$\frac{x^2}{a^2} + \frac{y^2}{b^2} + \frac{z^2}{c^2} = 1$$

31. Let G be a region that is symmetric about the xy-plane, and let G_1 and G_2 be the portions of G above and below the xy-plane, respectively. Suppose that

$$\iiint\limits_{G_1} f(x, y, z) \, dV = 5$$

(a) Find $\iiint\limits_G f(x, y, z)\, dV$ given that

$f(x, y, -z) = f(x, y, z)$.

(b) Find $\iiint\limits_G f(x, y, z)\, dV$ given that

$f(x, y, -z) = -f(x, y, z)$.

In Exercises 32–34, use symmetry and Exercise 31 to evaluate the integral without performing an integration.

32. $\displaystyle\int_{-2}^{2}\int_{0}^{\sqrt{4-x^2}}\int_{-\sqrt{4-x^2-y^2}}^{\sqrt{4-x^2-y^2}} [z^3 \cos(xyz) - 3]\, dz\, dy\, dx.$

33. $\displaystyle\int_{-3}^{3}\int_{-\sqrt{9-y^2}}^{\sqrt{9-y^2}}\int_{x^2+y^2}^{9} x^2 y^3 e^z\, dz\, dx\, dy.$

34. $\displaystyle\iiint\limits_G e^{xy} \sin z\, dV$, where G is the spherical solid enclosed by $x^2 + y^2 + z^2 = R^2$.

17.6 CENTROID, CENTER OF GRAVITY, THEOREM OF PAPPUS

*Suppose that a physical body is acted on by a gravitational field. Because the body is composed of many particles, each of which is affected by gravity, the action of the gravitational field on the body consists of a large number of forces distributed over the entire body. However, these individual forces can be replaced by a single force acting at a point called the **center of gravity** of the body. In this section we shall show how double and triple integrals can be used to locate centers of gravity.*

☐ **DENSITY OF A LAMINA**

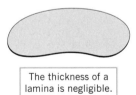

The thickness of a lamina is negligible.

Figure 17.6.1

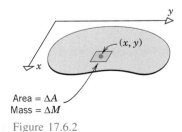

Area $= \Delta A$
Mass $= \Delta M$

Figure 17.6.2

To begin, let us consider an idealized flat object that is sufficiently thin that it can be viewed as two-dimensional (Figure 17.6.1). Such an object is called a **lamina**. A lamina is called **homogeneous** if its composition and structure are uniform throughout and **inhomogeneous** otherwise. The **density** of a *homogeneous* lamina is defined to be its mass per unit area. Thus, the density δ of a homogeneous lamina of mass M and area A is given by $\delta = M/A$.

For an inhomogeneous lamina the composition may vary from point to point, and hence an appropriate definition of "density" must reflect this. To motivate such a definition, suppose that the lamina is placed in an xy-plane. The density at a point (x, y) can be specified by a function $\delta(x, y)$, called the **density function**, which can be interpreted as follows. Construct a small rectangle centered at (x, y) and let ΔM and ΔA be the mass and area of the portion of the lamina enclosed by this rectangle (Figure 17.6.2). If the ratio $\Delta M/\Delta A$ approaches a limiting value as the dimensions (and hence the area) of the rectangle approach zero, then this limit is considered to be the density of the lamina at (x, y). Symbolically,

$$\delta(x, y) = \lim_{\Delta A \to 0} \frac{\Delta M}{\Delta A} \tag{1}$$

From this relationship we obtain the approximation

$$\Delta M \approx \delta(x, y)\, \Delta A \tag{2}$$

which relates the mass and area of a small rectangular portion of the lamina centered at (x, y). It is assumed that as the dimensions of the rectangle tend to zero, the error in this approximation also tends to zero.

REMARK. Note that a lamina with a constant density function is homogeneous.

☐ **MASS OF A LAMINA**

The mass of a lamina can be found from its density function by the following formula.

17.6.1   MASS OF A LAMINA.   If a lamina with a continuous density function $\delta(x, y)$ occupies a region R in the xy-plane, then its total mass M is given by

$$M = \iint\limits_{R} \delta(x, y)\, dA \tag{3}$$

This formula can be motivated by a familiar limiting process that can be outlined as follows: Imagine the lamina to be subdivided into rectangular pieces using lines parallel to the coordinate axes and excluding from consideration any nonrectangular parts at the boundary (Figure 17.6.3). Assume that there are n such rectangular pieces, and suppose that the kth piece has area ΔA_k. If we let (x_k^*, y_k^*) denote the center of the kth piece, then from Formula (2), the mass ΔM_k of this piece can be approximated by

$$\Delta M_k \approx \delta(x_k^*, y_k^*)\, \Delta A_k \tag{4}$$

and hence the mass M of the entire lamina can be approximated by

$$M \approx \sum_{k=1}^{n} \delta(x_k^*, y_k^*)\, \Delta A_k$$

If we now increase n in such a way that the dimensions of the rectangles tend to zero, then it is plausible that the errors in our approximations will approach zero, so

$$M = \lim_{n \to +\infty} \sum_{k=1}^{n} \delta(x_k^*, y_k^*)\, \Delta A_k = \iint\limits_{R} \delta(x, y)\, dA$$

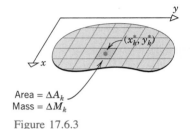

Area $= \Delta A_k$
Mass $= \Delta M_k$

Figure 17.6.3

Example 1   A triangular lamina with vertices $(0, 0)$, $(0, 1)$, and $(1, 0)$ has density function $\delta(x, y) = xy$. Find its total mass.

Solution.   Referring to (3) and Figure 17.6.4, the mass M of the lamina is

$$M = \iint\limits_{R} \delta(x, y)\, dA = \iint\limits_{R} xy\, dA = \int_{0}^{1} \int_{0}^{-x+1} xy\, dy\, dx$$

$$= \int_{0}^{1} \left[\frac{1}{2} xy^2 \right]_{y=0}^{-x+1} dx = \int_{0}^{1} \left[\frac{1}{2} x^3 - x^2 + \frac{1}{2} x \right] dx$$

$$= \frac{1}{24} \text{ (unit of mass)} \quad \blacktriangleleft$$

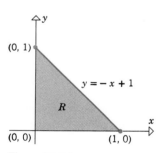

Figure 17.6.4

☐ **CENTER OF GRAVITY OF A LAMINA**

Assuming that the force of gravity is constant and acts downward, consider the following problem.

17.6.2   PROBLEM.   *Suppose that a lamina with a continuous density function $\delta(x, y)$ occupies a region R in a horizontal xy-plane. Find the coordinates $(\bar{x}, \bar{y})$ of the center of gravity.*

To motivate the solution, consider what happens if we try to balance the lamina on a knife-edge parallel to the x-axis. Suppose the lamina in Figure 17.6.5 is placed on a knife-

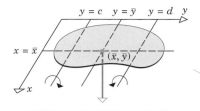

Force of gravity acting on the center of gravity of the lamina

Figure 17.6.5

edge along a line $y = c$ that does not pass through the center of gravity. Because the lamina behaves as if its entire mass is concentrated at the center of gravity $(\bar{x}, \bar{y})$, the lamina will be rotationally unstable and the force of gravity will cause a rotation about $y = c$. Similarly, the lamina will undergo a rotation if placed on a knife-edge along $y = d$. However, if the knife-edge runs along the line $y = \bar{y}$ through the center of gravity, the lamina will be in perfect balance. Similarly, the lamina will be in perfect balance on a knife-edge along the line $x = \bar{x}$ through the center of gravity. This suggests that the center of gravity of a lamina can be determined as the intersection of two lines of balance, one parallel to the x-axis and the other parallel to the y-axis. In order to find these lines of balance, we shall need some preliminary results about rotations.

Children on a seesaw learn by experience that a lighter child can balance a heavier one by sitting farther from the fulcrum or pivot point. This is because the tendency for an object to produce rotation is proportional not only to its mass but also to the distance between the object and the fulcrum. To be precise, if a point-mass m is located on a coordinate axis at a point x, then the tendency for that mass to produce a rotation about a point a on the axis is measured by the following quantity, called the ***moment of m about $x = a$***:

$$\begin{bmatrix} \text{moment of } m \\ \text{about } a \end{bmatrix} = m(x - a)$$

The number $x - a$ is called the ***lever arm***. Depending on whether the mass is to the right or left of a, the lever arm is either the distance between x and a or the negative of this distance (Figure 17.6.6). Positive lever arms result in positive moments and clockwise rotations, while negative lever arms result in negative moments and counterclockwise rotations.

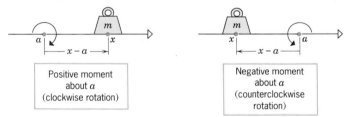

Figure 17.6.6

Suppose that masses $m_1, m_2, \ldots, m_n$ are located at points $x_1, x_2, \ldots, x_n$ on a coordinate axis and a fulcrum is positioned at the point a (Figure 17.6.7). Depending on whether the sum of the moments about a,

$$\sum_{k=1}^{n} m_k(x_k - a) = m_1(x_1 - a) + m_2(x_2 - a) + \cdots + m_n(x_n - a)$$

is positive, negative, or zero, the axis will rotate clockwise about a, rotate counterclockwise about a, or balance perfectly. In the last case, the system of masses is said to be in ***equilibrium***.

Figure 17.6.7

The preceding ideas can be extended to masses distributed in two-dimensional space: If we imagine the xy-plane to be a weightless sheet supporting a point-mass m located at a point (x, y), then the tendency for the mass to produce a rotation of the sheet about the line $x = a$ is $m(x - a)$, called the ***moment of m about $x = a$***, and the tendency for the mass to

produce a rotation about the line $y = c$ is $m(y - c)$, called the ***moment of m about y = c***
(Figure 17.6.8). In summary,

$$\begin{bmatrix} \text{moment of } m \\ \text{about the} \\ \text{line } x = a \end{bmatrix} = m(x - a) \quad \text{and} \quad \begin{bmatrix} \text{moment of } m \\ \text{about the} \\ \text{line } y = c \end{bmatrix} = m(y - c) \tag{5--6}$$

If a number of masses are distributed throughout the xy-plane, then the plane (viewed as a weightless sheet) will balance on a knife-edge along the line $x = a$ if the sum of the moments about the line is zero. Similarly for the line $y = c$.

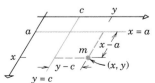

Figure 17.6.8

We are now ready to solve Problem 17.6.2. We imagine the lamina to be subdivided into rectangular pieces using lines parallel to the coordinate axes and excluding from consideration any nonrectangular pieces at the boundary (Figure 17.6.3). We assume that there are n such rectangular pieces and that the kth piece has area ΔA_k and mass ΔM_k. We shall let (x_k^*, y_k^*) be the center of the kth piece, and we shall assume that the entire mass of the kth piece is concentrated at its center. From (4), the mass of the kth piece can be approximated by

$$\Delta M_k \approx \delta(x_k^*, y_k^*)\, \Delta A_k$$

Since the lamina balances on the lines $x = \bar{x}$ and $y = \bar{y}$, the sum of the moments of the rectangular pieces about those lines should be close to zero; that is,

$$\sum_{k=1}^{n} (x_k^* - \bar{x})\, \Delta M_k = \sum_{k=1}^{n} (x_k^* - \bar{x})\, \delta(x_k^*, y_k^*)\, \Delta A_k \approx 0$$

$$\sum_{k=1}^{n} (y_k^* - \bar{y})\, \Delta M_k = \sum_{k=1}^{n} (y_k^* - \bar{y})\, \delta(x_k^*, y_k^*)\, \Delta A_k \approx 0$$

If we now increase n in such a way that the dimensions of the rectangles tend to zero, then it is plausible that the errors in our approximations will approach zero, so that

$$\lim_{n \to +\infty} \sum_{k=1}^{n} (x_k^* - \bar{x})\, \delta(x_k^*, y_k^*)\, \Delta A_k = 0$$

$$\lim_{n \to +\infty} \sum_{k=1}^{n} (y_k^* - \bar{y})\, \delta(x_k^*, y_k^*)\, \Delta A_k = 0$$

from which we obtain

$$\iint_R (x - \bar{x})\, \delta(x, y)\, dA = 0 \quad \text{and} \quad \iint_R (y - \bar{y})\, \delta(x, y)\, dA = 0$$

Since $\bar{x}$ and $\bar{y}$ are constant, these equations can be rewritten as

$$\iint_R x\delta(x, y)\, dA = \bar{x} \iint_R \delta(x, y)\, dA$$

$$\iint_R y\delta(x, y)\, dA = \bar{y} \iint_R \delta(x, y)\, dA$$

from which we obtain the following formulas for the center of gravity of the lamina:

Center of Gravity $(\bar{x}, \bar{y})$ of a Lamina

$$\bar{x} = \frac{\displaystyle\iint_R x\delta(x, y)\, dA}{\displaystyle\iint_R \delta(x, y)\, dA}, \qquad \bar{y} = \frac{\displaystyle\iint_R y\delta(x, y)\, dA}{\displaystyle\iint_R \delta(x, y)\, dA} \tag{7–8}$$

Observe that in both formulas the denominator is the mass M of the lamina [see (3)]. The numerator in the formula for $\bar{x}$ is denoted by M_y and is called the *first moment of the lamina about the y-axis*; the numerator in the formula for $\bar{y}$ is denoted by M_x and is called the *first moment of the lamina about the x-axis*. Thus, Formulas (7) and (8) can be expressed as

$$\bar{x} = \frac{M_y}{M} = \frac{1}{\text{mass of } R}\iint_R x\delta(x, y)\, dA \tag{9a}$$

$$\bar{y} = \frac{M_x}{M} = \frac{1}{\text{mass of } R}\iint_R y\delta(x, y)\, dA \tag{9b}$$

Example 2 Find the center of gravity of the triangular lamina with vertices $(0, 0)$, $(0, 1)$, and $(1, 0)$ and density function $\delta(x, y) = xy$.

Solution. The lamina is shown in Figure 17.6.4. In Example 1 we found the mass of the lamina to be

$$M = \iint_R \delta(x, y)\, dA = \iint_R xy\, dA = \frac{1}{24}$$

The moment of the lamina about the y-axis is

$$M_y = \iint_R x\delta(x, y)\, dA = \iint_R x^2 y\, dA = \int_0^1 \int_0^{-x+1} x^2 y\, dy\, dx$$

$$= \int_0^1 \left[\frac{1}{2}x^2 y^2\right]_{y=0}^{-x+1} dx = \int_0^1 \left(\frac{1}{2}x^4 - x^3 + \frac{1}{2}x^2\right) dx = \frac{1}{60}$$

and the moment about the x-axis is

$$M_x = \iint_R y\delta(x, y)\, dA = \iint_R xy^2\, dA = \int_0^1 \int_0^{-x+1} xy^2\, dy\, dx$$

$$= \int_0^1 \left[\frac{1}{3}xy^3\right]_{y=0}^{-x+1} dx = \int_0^1 \left(-\frac{1}{3}x^4 + x^3 - x^2 + \frac{1}{3}x\right) dx = \frac{1}{60}$$

From (9a) and (9b),

$$\bar{x} = \frac{M_y}{M} = \frac{1/60}{1/24} = \frac{2}{5}, \qquad \bar{y} = \frac{M_x}{M} = \frac{1/60}{1/24} = \frac{2}{5}$$

so the center of gravity is $\left(\frac{2}{5}, \frac{2}{5}\right)$. ◄

□ **CENTROIDS**

In the special case of a *homogeneous* lamina, the center of gravity is called the *centroid of the lamina* or sometimes the *centroid of the region R*. Because the density function δ is constant for a homogeneous lamina, the factor δ may be moved through the integral signs in (7) and (8) and canceled. Thus, the centroid $(\bar{x}, \bar{y})$ of a region R is given by the following formulas:

Centroid of a Region R

$$\bar{x} = \frac{\displaystyle\iint_R x\, dA}{\displaystyle\iint_R dA} = \frac{1}{\text{area of } R}\iint_R x\, dA \tag{10}$$

$$\bar{y} = \frac{\displaystyle\iint_R y\, dA}{\displaystyle\iint_R dA} = \frac{1}{\text{area of } R}\iint_R y\, dA \tag{11}$$

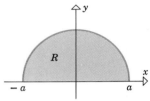

Figure 17.6.9

Example 3 Find the centroid of the semicircular region in Figure 17.6.9.

Solution. By symmetry, $\bar{x} = 0$ since the y-axis is obviously a line of balance. From (11),

$$\bar{y} = \frac{1}{\text{area of } R}\iint_R y\, dA = \frac{1}{\frac{1}{2}\pi a^2}\iint_R y\, dA$$

$$= \frac{1}{\frac{1}{2}\pi a^2}\int_0^\pi \int_0^a (r\sin\theta) r\, dr\, d\theta = \frac{1}{\frac{1}{2}\pi a^2}\int_0^\pi \left[\frac{1}{3} r^3 \sin\theta\right]_{r=0}^a d\theta$$

Evaluating in polar coordinates

$$= \frac{1}{\frac{1}{2}\pi a^2}\left(\frac{2}{3} a^3\right) = \frac{4a}{3\pi}$$

so the centroid is $\left(0, \dfrac{4a}{3\pi}\right)$. ◀

☐ **CENTER OF GRAVITY OF A SOLID**

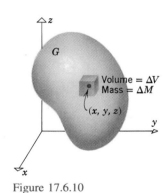

Figure 17.6.10

For a three-dimensional solid G, the formulas for moments, center of gravity, and centroid are similar to those for laminas. If G is *homogeneous,* then its **density** is defined to be its mass per unit volume. Thus, if G is a homogeneous solid of mass M and volume V, then its density δ is given by $\delta = M/V$. If G is inhomogeneous and is in an xyz-coordinate system, then its density at a general point (x, y, z) is specified by a **density function** $\delta(x, y, z)$ whose value at a point can be viewed as a limit:

$$\delta(x, y, z) = \lim_{\Delta V \to 0} \frac{\Delta M}{\Delta V}$$

where ΔM and ΔV represent the mass and volume of a rectangular parallelepiped, centered at (x, y, z), whose dimensions tend to zero (Figure 17.6.10).

Using the discussion of laminas as a model, the reader should be able to show that the mass M of a solid with a continuous density function $\delta(x, y, z)$ is

$$M = \text{mass of } G = \iiint_G \delta(x, y, z)\, dV \tag{12}$$

The formulas for center of gravity and centroid are

Center of Gravity $(\bar{x}, \bar{y}, \bar{z})$ **of a Solid G**	**Centroid** $(\bar{x}, \bar{y}, \bar{z})$ **of a Solid G**
$\bar{x} = \dfrac{1}{M} \iiint_G x \delta(x, y, z)\, dV$	$\bar{x} = \dfrac{1}{V} \iiint_G x\, dV$
$\bar{y} = \dfrac{1}{M} \iiint_G y \delta(x, y, z)\, dV$	$\bar{y} = \dfrac{1}{V} \iiint_G y\, dV$
$\bar{z} = \dfrac{1}{M} \iiint_G z \delta(x, y, z)\, dV$	$\bar{z} = \dfrac{1}{V} \iiint_G z\, dV$

$(13\text{--}14)$

Example 4 Find the mass and the center of gravity of a cylindrical solid of height h and radius a (Figure 17.6.11), assuming that the density at each point is proportional to the distance between the point and the base of the solid.

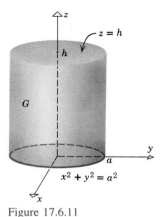

Figure 17.6.11

Solution. Since the density is proportional to the distance z from the base, the density function has the form $\delta(x, y, z) = kz$, where k is some (unknown) positive constant of proportionality. From (12) the mass of the solid is

$$M = \iiint_G \delta(x, y, z)\, dV = \int_{-a}^{a} \int_{-\sqrt{a^2-x^2}}^{\sqrt{a^2-x^2}} \int_0^h kz\, dz\, dy\, dx$$

$$= k \int_{-a}^{a} \int_{-\sqrt{a^2-x^2}}^{\sqrt{a^2-x^2}} \tfrac{1}{2}h^2\, dy\, dx$$

$$= kh^2 \int_{-a}^{a} \sqrt{a^2 - x^2}\, dx$$

$$= \tfrac{1}{2} kh^2 \pi a^2 \qquad \boxed{\text{See Example 4} \atop \text{of Section 9.5.}}$$

Without additional information, the constant k cannot be determined. However, as we shall now see, the value of k does not affect the center of gravity.

From (14),

$$\bar{z} = \frac{1}{\text{mass } G} \iiint_G z \delta(x, y, z)\, dV = \frac{1}{\tfrac{1}{2}kh^2\pi a^2} \iiint_G z \delta(x, y, z)\, dV$$

$$= \frac{1}{\tfrac{1}{2}kh^2\pi a^2} \int_{-a}^{a} \int_{-\sqrt{a^2-x^2}}^{\sqrt{a^2-x^2}} \int_0^h z(kz)\, dz\, dy\, dx$$

$$= \frac{k}{\tfrac{1}{2}kh^2\pi a^2} \int_{-a}^{a} \int_{-\sqrt{a^2-x^2}}^{\sqrt{a^2-x^2}} \tfrac{1}{3}h^3\, dy\, dx = \frac{\tfrac{1}{3}kh^3}{\tfrac{1}{2}kh^2\pi a^2} \int_{-a}^{a} 2\sqrt{a^2 - x^2}\, dx$$

$$= \frac{\tfrac{1}{3}kh^3\pi a^2}{\tfrac{1}{2}kh^2\pi a^2} = \tfrac{2}{3}h$$

Similar calculations using (13) will yield $\bar{x} = \bar{y} = 0$. However, this is evident by inspection, since it follows from the symmetry of the solid and the form of its density function that the center of gravity is on the z-axis. Thus, the center of gravity is $(0, 0, \tfrac{2}{3}h)$. ◄

☐ **THEOREM OF PAPPUS**

The next theorem, due to the Greek mathematician Pappus,* gives an interesting and useful relationship between the centroid of a plane region R and the volume of the solid generated when the region is revolved about a line.

> **17.6.3** THEOREM. *If R is a plane region and L is a line that lies in the plane of R, but does not intersect R, then the volume of the solid formed by revolving R about L is given by*
>
> $$\text{volume} = (\text{area of } R) \cdot \left(\begin{array}{c} \text{distance traveled} \\ \text{by the centroid} \end{array} \right)$$

Proof. Introduce an xy-coordinate system so that L is along the y-axis and the region R is in the first quadrant (Figure 17.6.12). Let R be partitioned into subregions in the usual way and let R_k be a typical rectangle interior to R. If (x_k^*, y_k^*) is the center of R_k, and if the area of R_k is $\Delta A_k = \Delta x_k \, \Delta y_k$, then the volume generated by R_k as it revolves about L is

$$2\pi x_k^* \, \Delta x_k \, \Delta y_k = 2\pi x_k^* \, \Delta A_k$$

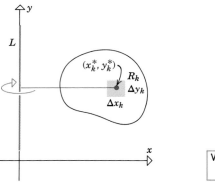
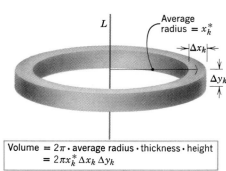

Volume $= 2\pi \cdot$ average radius $\cdot$ thickness $\cdot$ height
$\qquad = 2\pi x_k^* \, \Delta x_k \, \Delta y_k$

Figure 17.6.12

[See Figure 17.6.12 and Formula (1) of Section 6.3.] Therefore, the total volume of the solid is approximately

$$V \approx \sum_{k=1}^{n} 2\pi x_k^* \, \Delta A_k$$

from which it follows that the exact volume is

$$V = \iint_R 2\pi x \, dA = 2\pi \iint_R x \, dA$$

Thus, from (10)

$$V = 2\pi \cdot \bar{x} \cdot [\text{area of } R]$$

---

*PAPPUS OF ALEXANDRIA (4th century A.D.). Greek mathematician. Pappus lived during the early Christian era when mathematical activity was in a period of decline. His main contributions to mathematics appeared in a series of eight books called *The Collection* (written about 340 A.D.). This work, which survives only partially, contained some original results, but was devoted mostly to statements, refinements, and proofs of results by earlier mathematicians. Pappus' Theorem, stated without proof in Book VII of *The Collection*, was probably known and proved in earlier times. This result is sometimes called Guldin's Theorem in recognition of the Swiss mathematician, Paul Guldin (1577–1643), who rediscovered it independently.

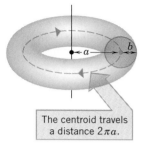

The centroid travels a distance $2\pi a$.

Figure 17.6.13

This completes the proof since $2\pi\bar{x}$ is the distance traveled by the centroid when R is revolved about the y-axis. ∎

Example 5   Use Pappus' Theorem to find the volume V of the torus generated by revolving a circular region of radius b about a line at a distance a (greater than b) from the center of the circle (Figure 17.6.13).

Solution.   By symmetry, the centroid of a circular region is its center. Thus, the distance traveled by the centroid is $2\pi a$. Since the area of a circle is πr^2, it follows from Pappus' Theorem that the volume of the torus is

$$V = (2\pi a)(\pi b^2) = 2\pi^2 ab^2 \quad \blacktriangleleft$$

► Exercise Set 17.6

1. Where should the fulcrum be placed so that the seesaw is in equilibrium?

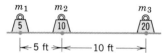

2. A rectangular lamina with density $\delta(x, y) = xy^2$ has vertices $(0, 0), (0, 2), (3, 0), (3, 2)$. Find its mass and center of gravity.

3. A lamina with density $\delta(x, y) = x + y$ is bounded by the x-axis, the line $x = 1$, and the curve $y = \sqrt{x}$. Find its mass and center of gravity.

4. A lamina with density $\delta(x, y) = y$ is bounded by $y = \sin x$, $y = 0$, $x = 0$, and $x = \pi$. Find its mass and center of gravity.

5. A lamina with density $\delta(x, y) = xy$ is in the first quadrant and is bounded by the circle $x^2 + y^2 = a^2$ and the coordinate axes. Find its mass and center of gravity.

6. A lamina with density $\delta(x, y) = x^2 + y^2$ is bounded by the x-axis and the upper half of the circle $x^2 + y^2 = 1$. Find its mass and center of gravity.

In Exercises 7–13, find the centroid of the region.

7. The triangular region bounded by $y = x$, $x = 1$, and the x-axis.

8. The region bounded by $y = x^2$, $x = 1$, and the x-axis.

9. The region that is enclosed between $y = x$ and $y = 2 - x^2$.

10. The region enclosed by the parabolas $x = 3y^2 - 6y$ and $x = 2y - y^2$.

11. The region above the x-axis and between the circles $x^2 + y^2 = a^2$ and $x^2 + y^2 = b^2$ $(a < b)$.

12. The region enclosed between the y-axis and the right half of the circle $x^2 + y^2 = a^2$.

13. The region enclosed between $y = |x|$ and the line $y = 4$.

14. A rectangular box that is defined by the three inequalities $0 \le x \le 1$, $0 \le y \le 1$, and $0 \le z \le 1$ has density function $\delta(x, y, z) = 3xyz$. Find its mass and center of gravity.

15. A cube that is defined by the three inequalities $0 \le x \le a$, $0 \le y \le a, 0 \le z \le a$ has density $\delta(x, y, z) = a - x$. Find its mass and center of gravity.

16. The cylindrical solid enclosed by $x^2 + y^2 = a^2$, $z = 0$, and $z = h$ has density $\delta(x, y, z) = h - z$. Find its mass and center of gravity.

17. The solid enclosed by the surface $z = 1 - y^2$ (where $y \ge 0$) and the planes $z = 0$, $x = -1$, $x = 1$ has density $\delta(x, y, z) = yz$. Find its mass and center of gravity.

18. The solid enclosed by the surface $y = 9 - x^2$ (where $x \ge 0$) and the planes $y = 0$, $z = 0$, and $z = 1$ has density $\delta(x, y, z) = xz$. Find its mass and center of gravity.

In Exercises 19–23, find the centroid of the solid.

19. The tetrahedron in the first octant enclosed by the coordinate planes and the plane $x + y + z = 1$.

20. The solid enclosed by the xy-plane and the hemisphere $z = \sqrt{a^2 - x^2 - y^2}$.

21. The solid bounded by the surface $z = y^2$, and the planes $x = 0$, $x = 1$, and $z = 1$.

22. The solid in the first octant bounded by the surface $z = xy$ and the planes $z = 0$, $x = 2$, and $y = 2$.

23. The solid in the first octant bounded by the sphere $x^2 + y^2 + z^2 = a^2$ and the coordinate planes.

24. Find the center of gravity of the square lamina with vertices $(0, 0), (1, 0), (0, 1),$ and $(1, 1)$ if

    (a) the density is proportional to the square of the distance from the origin

    (b) the density is proportional to the distance from the y-axis.

25. Find the mass of a circular lamina of radius a whose density at a point is proportional to the distance from the center. (Let k denote the constant of proportionality.)

26. Find the center of gravity of the cube that is determined by the inequalities $0 \le x \le 1$, $0 \le y \le 1, 0 \le z \le 1$ if

    (a) the density is proportional to the square of the distance to the origin

(b) the density is proportional to the sum of the distances to the faces that lie in the coordinate planes.

27. Show that in polar coordinates the formulas for the centroid $(\bar{x}, \bar{y})$ of a region R are

$$\bar{x} = \frac{1}{\text{area of } R} \iint_R r^2 \cos \theta \, dr \, d\theta$$

$$\bar{y} = \frac{1}{\text{area of } R} \iint_R r^2 \sin \theta \, dr \, d\theta$$

28. Use the result of Exercise 27 to find the centroid $(\bar{x}, \bar{y})$ of the region enclosed by the cardioid $r = a(1 + \sin \theta)$.

29. Use the result of Exercise 27 to find the centroid $(\bar{x}, \bar{y})$ of the petal of the rose $r = \sin 2\theta$ in the first quadrant.

30. Let R be the rectangle bounded by the lines $x = 0$, $x = 3$, $y = 0$, and $y = 2$. By inspection, find the centroid of R and use it to evaluate

$$\iint_R x \, dA \quad \text{and} \quad \iint_R y \, dA$$

31. Use the Theorem of Pappus and the fact that the volume of a sphere of radius a is $V = \frac{4}{3}\pi a^3$ to show that the centroid of the lamina that is bounded by the x-axis and the semicircle $y = \sqrt{a^2 - x^2}$ is $(0, 4a/(3\pi))$. (This problem was solved directly in Example 3.)

32. Use the Theorem of Pappus and the result of Exercise 31 to find the volume of the solid generated when the region bounded by the x-axis and the semicircle $y = \sqrt{a^2 - x^2}$ is revolved about

(a) the line $y = -a$ (b) the line $y = x - a$.

33. Use the Theorem of Pappus and the fact that the area of an ellipse with semiaxes a and b is πab to find the volume of the elliptical torus generated by revolving the ellipse

$$\frac{(x - k)^2}{a^2} + \frac{y^2}{b^2} = 1$$

about the y-axis. Assume that $k > a$.

34. Use the Theorem of Pappus to find the volume of the solid generated when the region enclosed by $y = x^2$ and $y = 8 - x^2$ is revolved about the x-axis.

35. Use the Theorem of Pappus to find the centroid of the triangular region with vertices $(0, 0)$, $(a, 0)$, and $(0, b)$, where $a > 0$ and $b > 0$. [*Hint:* Revolve the region about the x-axis to obtain $\bar{y}$ and about the y-axis to obtain $\bar{x}$.]

36. If a lamina with a continuous density function $\delta(x, y)$ occupies a region R in the xy-plane, then the **second moments** (or **moments of inertia**) of the lamina about the x-axis, y-axis, and z-axis, respectively, are defined by

$$I_x = \iint_R y^2 \delta \, dA, \quad I_y = \iint_R x^2 \delta \, dA,$$

$$I_z = \iint_R (x^2 + y^2) \delta \, dA$$

The corresponding definitions for a solid with density function $\delta(x, y, z)$ that occupies a region G are

$$I_x = \iiint_G (y^2 + z^2) \delta(x, y, z) \, dV$$

$$I_y = \iiint_G (x^2 + z^2) \delta(x, y, z) \, dV$$

$$I_z = \iiint_G (x^2 + y^2) \delta(x, y, z) \, dV$$

Find I_x, I_y, and I_z for the homogeneous (constant δ) rectangular lamina described by the inequalities $0 \leq x \leq a$ and $0 \leq y \leq b$. Express your answers in terms of the mass M of the lamina. [*Note:* Moments of inertia are important in problems involving rotational motion.]

In Exercises 37–44, a homogeneous lamina or solid is given. Use the formulas in Exercise 36 to find the indicated moment of inertia. Express your answers in terms of the mass M of the lamina or solid.

---

37. The moment I_z for the rectangular lamina $-a/2 \leq x \leq a/2$, $-b/2 \leq y \leq b/2$.

38. The moment I_x for the rectangular lamina $-a/2 \leq x \leq a/2$, $-b/2 \leq y \leq b/2$.

39. The circular lamina $x^2 + y^2 \leq R^2$; I_y.

40. The circular lamina $x^2 + y^2 \leq R^2$; I_z.

41. The circular lamina $x^2 + (y - R)^2 \leq R^2$; I_z.

42. The triangular lamina $0 \leq y \leq 2 - 2x$, $x \geq 0$; I_y.

43. The solid $0 \leq x \leq a$, $0 \leq y \leq a$, $0 \leq z \leq a$; I_z.

44. The rectangular solid $0 \leq x \leq a$, $-b/2 \leq y \leq b/2$, $-c/2 \leq z \leq c/2$; I_x.

■ **17.7 TRIPLE INTEGRALS IN CYLINDRICAL AND SPHERICAL COORDINATES**

In Section 17.3 we saw that some double integrals are easier to evaluate in polar coordinates. Similarly, some triple integrals are easier to evaluate in cylindrical or spherical coordinates. In this section we shall study triple integrals in these coordinate systems.

☐ **TRIPLE INTEGRALS IN CYLINDRICAL COORDINATES**

Recall that in rectangular coordinates the triple integral of a continuous function f over a solid region G is defined as

$$\iiint\limits_{G} f(x, y, z)\, dV = \lim_{n \to +\infty} \sum_{k=1}^{n} f(x_k^*, y_k^*, z_k^*)\, \Delta V_k$$

where ΔV_k denotes the volume of a rectangular parallelepiped interior to G and (x_k^*, y_k^*, z_k^*) is a point in this parallelepiped (Figure 17.5.1). Triple integrals in cylindrical and spherical coordinates are defined similarly, except that the region G is divided not into rectangular parallelepipeds but into regions more appropriate to these coordinate systems.

In cylindrical coordinates, the simplest equations are of the form

$$r = \text{constant}, \quad \theta = \text{constant}, \quad z = \text{constant}$$

As indicated in Figure 14.8.2b, the first equation represents a right-circular cylinder centered on the z-axis, the second a vertical half-plane hinged on the z-axis, and the third a horizontal plane. These surfaces can be paired up to determine solids called *cylindrical wedges* or *cylindrical elements of volume*. To be precise, a cylindrical wedge is a solid enclosed between six surfaces of the following type:

two cylinders $\quad r = r_1, \quad r = r_2 \quad (r_1 < r_2)$

two half-planes $\quad \theta = \theta_1, \quad \theta = \theta_2 \quad (\theta_1 < \theta_2)$

two planes $\quad\quad z = z_1, \quad z = z_2 \quad (z_1 < z_2)$

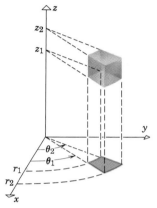

Figure 17.7.1

(Figure 17.7.1). The dimensions $\theta_2 - \theta_1$, $r_2 - r_1$, and $z_2 - z_1$ are called the *central angle*, *thickness*, and *height* of the wedge.

To define the triple integral over G of a function $f(r, \theta, z)$ in cylindrical coordinates we proceed as follows:

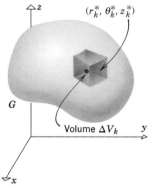

Figure 17.7.2

- Subdivide G into pieces by a three-dimensional grid consisting of concentric circular cylinders centered on the z-axis, half-planes hinged on the z-axis, and horizontal planes. Exclude from consideration all pieces that contain any points outside of G, thereby leaving only cylindrical wedges that are subsets of G.

- Assume that there are n such cylindrical wedges, and denote the volume of the kth cylindrical wedge by ΔV_k. As indicated in Figure 17.7.2, let $(r_k^*, \theta_k^*, z_k^*)$ be any point in the kth cylindrical wedge.

- Repeat this process with more and more subdivisions, so that as n increases the height, thickness, and central angle of the cylindrical wedges approach zero. Define

$$\iiint\limits_{G} f(r, \theta, z)\, dV = \lim_{n \to +\infty} \sum_{k=1}^{n} f(r_k^*, \theta_k^*, z_k^*)\, \Delta V_k \tag{1}$$

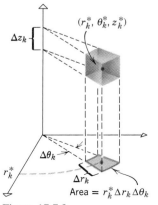

Figure 17.7.3

For computational purposes, it will be helpful to express (1) as an iterated integral. Toward this end we note that the volume ΔV_k of the kth cylindrical wedge can be expressed as

$$\Delta V_k = [\text{area of base}] \cdot [\text{height}] \tag{2}$$

If we denote the thickness, central angle, and height of this wedge by Δr_k, $\Delta \theta_k$, and Δz_k, and if we choose the arbitrary point $(r_k^*, \theta_k^*, z_k^*)$ to lie above the "center" of the base (Figures 17.3.5 and 17.7.3), then it follows from (4) of Section 17.3 that the base has area $r_k^* \, \Delta r_k \, \Delta \theta_k$. Thus, (2) can be written as

$$\Delta V_k = r_k^* \, \Delta r_k \, \Delta \theta_k \, \Delta z_k = r_k^* \, \Delta z_k \, \Delta r_k \, \Delta \theta_k$$

Substituting this expression in (1) yields

$$\iiint\limits_{G} f(r, \theta, z)\, dV = \lim_{n \to +\infty} \sum_{k=1}^{n} f(r_k^*, \theta_k^*, z_k^*) r_k^* \, \Delta z_k \, \Delta r_k \, \Delta \theta_k$$

which suggests that a triple integral in cylindrical coordinates can be evaluated as an iterated integral of the form

$$\iiint\limits_{G} f(r, \theta, z)\, dV = \iiint\limits_{\substack{\text{appropriate}\\ \text{limits}}} f(r, \theta, z)\, r\, dz\, dr\, d\theta \tag{3}$$

REMARK.   Note the extra factor of r that appears in the integrand on converting from the triple integral to the iterated integral. In this formula the integration with respect to z is done first, then with respect to r, and then with respect to θ, but any order of integration is allowable.

The following theorem, which we state without proof, makes the preceding ideas more precise.

17.7.1   **THEOREM.**   *Let G be a simple solid whose upper surface has the equation $z = g_2(r, \theta)$ and whose lower surface has the equation $z = g_1(r, \theta)$ in cylindrical coordinates. If R is the projection of the solid on the xy-plane, and if $f(r, \theta, z)$ is continuous on G, then*

$$\iiint\limits_{G} f(r, \theta, z)\, dV = \iint\limits_{R} \left[\int_{g_1(r,\theta)}^{g_2(r,\theta)} f(r, \theta, z)\, dz \right] dA \tag{4}$$

where the double integral over R is evaluated in polar coordinates. In particular, if the projection R is as shown in Figure 17.7.4, then (4) can be written as

$$\iiint\limits_{G} f(r, \theta, z)\, dV = \int_{\theta_1}^{\theta_2} \int_{r_1(\theta)}^{r_2(\theta)} \int_{g_1(r,\theta)}^{g_2(r,\theta)} f(r, \theta, z)\, r\, dz\, dr\, d\theta \tag{5}$$

The type of solid to which Formula (5) applies is illustrated in Figure 17.7.4.

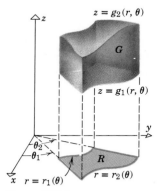

Figure 17.7.4

To apply (4) and (5) it is best to begin with a three-dimensional sketch of the solid G, from which the limits of integration can be obtained as follows:

Step 1.   Identify the upper surface $z = g_2(r, \theta)$ and the lower surface $z = g_1(r, \theta)$ of the solid. The functions $g_1(r, \theta)$ and $g_2(r, \theta)$ determine the z-limits of integration. (If the upper and lower surfaces are given in rectangular coordinates, convert them to cylindrical coordinates.)

Step 2.   Make a two-dimensional sketch of the projection R of the solid on the xy-plane. From this sketch the r- and θ-limits of integration may be obtained exactly as with double integrals in polar coordinates.

Example 1   Use triple integration in cylindrical coordinates to find the volume and the centroid of the solid G that is bounded above by the hemisphere $z = \sqrt{25 - x^2 - y^2}$, below by the xy-plane, and laterally by the cylinder $x^2 + y^2 = 9$.

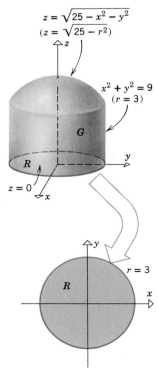

$z = \sqrt{25 - x^2 - y^2}$
$(z = \sqrt{25 - r^2})$

$x^2 + y^2 = 9$
$(r = 3)$

G

y

R

$z = 0$

x

Figure 17.7.5

$r = 3$

R

x

Solution.   The solid G and its projection R on the xy-plane are shown in Figure 17.7.5. In cylindrical coordinates, the upper surface of G is the hemisphere $z = \sqrt{25 - r^2}$ and the lower surface is the plane $z = 0$. Thus, from (4), the volume of G is

$$V = \iiint\limits_{G} dV = \iint\limits_{R} \left[\int_{0}^{\sqrt{25-r^2}} dz \right] dA$$

For the double integral over R, we use polar coordinates:

$$V = \int_{0}^{2\pi} \int_{0}^{3} \int_{0}^{\sqrt{25-r^2}} r \, dz \, dr \, d\theta = \int_{0}^{2\pi} \int_{0}^{3} \left[rz \right]_{z=0}^{\sqrt{25-r^2}} dr \, d\theta$$

$$= \int_{0}^{2\pi} \int_{0}^{3} r\sqrt{25 - r^2} \, dr \, d\theta = \int_{0}^{2\pi} \left[-\frac{1}{3}(25 - r^2)^{3/2} \right]_{r=0}^{3} d\theta$$

$$= \int_{0}^{2\pi} \frac{61}{3} \, d\theta = \frac{122}{3}\pi$$

$u = 25 - r^2$
$du = -2r \, dr$

From this result and (14) of Section 17.6,

$$\bar{z} = \frac{1}{V}\iiint\limits_{G} z \, dV = \frac{3}{122\pi} \iiint\limits_{G} z \, dV = \frac{3}{122\pi} \iint\limits_{R} \left[\int_{0}^{\sqrt{25-r^2}} z \, dz \right] dA$$

$$= \frac{3}{122\pi} \int_{0}^{2\pi} \int_{0}^{3} \int_{0}^{\sqrt{25-r^2}} zr \, dz \, dr \, d\theta = \frac{3}{122\pi} \int_{0}^{2\pi} \int_{0}^{3} \left[\frac{1}{2}rz^2 \right]_{z=0}^{\sqrt{25-r^2}} dr \, d\theta$$

$$= \frac{3}{244\pi} \int_{0}^{2\pi} \int_{0}^{3} (25r - r^3) \, dr \, d\theta = \frac{3}{244\pi} \int_{0}^{2\pi} \frac{369}{4} \, d\theta = \frac{1107}{488}$$

By symmetry, the centroid $(\bar{x}, \bar{y}, \bar{z})$ of G lies on the z-axis, so $\bar{x} = \bar{y} = 0$. Thus, the centroid is at the point $(0, 0, 1107/488)$.   ◄

☐ CONVERTING TRIPLE INTEGRALS FROM RECTANGULAR TO CYLINDRICAL COORDINATES

Sometimes a triple integral that is difficult to integrate in rectangular coordinates can be evaluated more easily by making the substitution $x = r\cos\theta$, $y = r\sin\theta$, $z = z$ to convert it to an integral in cylindrical coordinates. Under such a substitution, the rectangular and cylindrical triple integrals are related by the equation

$$\iiint\limits_{G} f(x, y, z) \, dV = \iiint\limits_{\substack{\text{appropriate} \\ \text{limits}}} f(r\cos\theta, r\sin\theta, z) \, r \, dz \, dr \, d\theta \tag{6}$$

REMARK.   In (6), the order of integration is first with respect to z, then r, and then θ. However, when appropriate, the order of integration can be changed, provided the limits of integration are adjusted accordingly.

Example 2   Use cylindrical coordinates to evaluate

$$\int_{-3}^{3} \int_{-\sqrt{9-x^2}}^{\sqrt{9-x^2}} \int_{0}^{9-x^2-y^2} x^2 \, dz \, dy \, dx$$

Solution.   In problems of this type, it is helpful to sketch the region of integration G and its projection R on the xy-plane. From the z-limits of integration, the upper surface of G is the paraboloid $z = 9 - x^2 - y^2$ and the lower surface is the xy-plane $z = 0$. From the x-

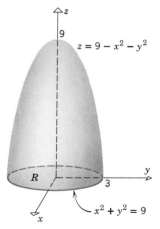

Figure 17.7.6

and y-limits of integration, the projection R is the region in the xy-plane enclosed by the circle $x^2 + y^2 = 9$ (Figure 17.7.6). Thus,

$$\int_{-3}^{3} \int_{-\sqrt{9-x^2}}^{\sqrt{9-x^2}} \int_{0}^{9-x^2-y^2} x^2 \, dz \, dy \, dx = \iiint_{G} x^2 \, dV$$

$$= \iint_{R} \left[\int_{0}^{9-r^2} r^2 \cos^2 \theta \, dz \right] dA = \int_{0}^{2\pi} \int_{0}^{3} \int_{0}^{9-r^2} (r^2 \cos^2 \theta) \, r \, dz \, dr \, d\theta$$

$$= \int_{0}^{2\pi} \int_{0}^{3} \int_{0}^{9-r^2} r^3 \cos^2 \theta \, dz \, dr \, d\theta = \int_{0}^{2\pi} \int_{0}^{3} \left[zr^3 \cos^2 \theta \right]_{z=0}^{9-r^2} dr \, d\theta$$

$$= \int_{0}^{2\pi} \int_{0}^{3} (9r^3 - r^5) \cos^2 \theta \, dr \, d\theta = \int_{0}^{2\pi} \left[\left(\frac{9r^4}{4} - \frac{r^6}{6} \right) \cos^2 \theta \right]_{r=0}^{3} d\theta$$

$$= \frac{243}{4} \int_{0}^{2\pi} \cos^2 \theta \, d\theta = \frac{243}{4} \int_{0}^{2\pi} \frac{1}{2}(1 + \cos 2\theta) \, d\theta = \frac{243\pi}{4} \quad \blacktriangleleft$$

☐ **TRIPLE INTEGRALS IN SPHERICAL COORDINATES**

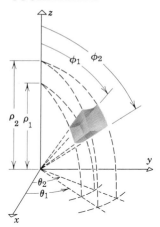

Figure 17.7.7

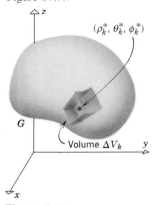

Figure 17.7.8

In spherical coordinates, the simplest equations are of the form

$$\rho = \text{constant}, \quad \theta = \text{constant}, \quad \phi = \text{constant}$$

As indicated in Figure 14.8.2c, the first equation represents a sphere centered at the origin, the second a half-plane hinged on the z-axis, and the third a right-circular cone with its vertex at the origin and its line of symmetry along the z-axis. By a *spherical wedge* or *spherical element of volume* we mean a solid enclosed between six surfaces of the following form:

two spheres	$\rho = \rho_1,$	$\rho = \rho_2$	$(\rho_1 < \rho_2)$
two half-planes	$\theta = \theta_1,$	$\theta = \theta_2$	$(\theta_1 < \theta_2)$
two right-circular cones	$\phi = \phi_1,$	$\phi = \phi_2$	$(\phi_1 < \phi_2)$

(Figure 17.7.7). We shall refer to the numbers $\rho_2 - \rho_1$, $\theta_2 - \theta_1$, and $\phi_2 - \phi_1$ as the *dimensions* of a spherical wedge.

If G is a solid region in three-dimensional space, then the triple integral over G of a continuous function $f(\rho, \theta, \phi)$ in spherical coordinates is similar in definition to the triple integral in cylindrical coordinates, except that the solid G is partitioned into *spherical wedges* by a three-dimensional grid consisting of spheres centered at the origin, half-planes hinged on the z-axis, and right-circular cones with vertices at the origin and lines of symmetry along the z-axis (Figure 17.7.8).

The defining equation of a triple integral in spherical coordinates is

$$\iiint_{G} f(\rho, \theta, \phi) \, dV = \lim_{n \to +\infty} \sum_{k=1}^{n} f(\rho_k^*, \theta_k^*, \phi_k^*) \, \Delta V_k \tag{7}$$

where ΔV_k is the volume of the kth spherical wedge that is interior to G, $(\rho_k^*, \theta_k^*, \phi_k^*)$ is an arbitrary point in this wedge, and n increases in such a way that the dimensions of each interior spherical wedge tend to zero.

For computational purposes, it will be desirable to express (7) as an iterated integral. In the exercises we will help the reader to show that if the point $(\rho_k^*, \theta_k^*, \phi_k^*)$ is suitably chosen, then the volume ΔV_k in (7) can be written as

$$\Delta V_k = \rho_k^{*2} \sin \phi_k^* \, \Delta \rho_k \, \Delta \phi_k \, \Delta \theta_k$$

where $\Delta\rho_k$, $\Delta\phi_k$, and $\Delta\theta_k$ are the dimensions of the wedge. Substituting this in (7) we obtain

$$\iiint\limits_{G} f(\rho, \theta, \phi)\, dV = \lim_{n \to +\infty} \sum_{k=1}^{n} f(\rho_k^*, \theta_k^*, \phi_k^*) \rho_k^{*2} \sin \phi_k^* \, \Delta\rho_k\, \Delta\phi_k\, \Delta\theta_k$$

which suggests that a triple integral in spherical coordinates can be evaluated as an iterated integral of the form

$$\iiint\limits_{G} f(\rho, \theta, \phi)\, dV = \iiint\limits_{\substack{\text{appropriate}\\ \text{limits}}} f(\rho, \theta, \phi) \rho^2 \sin \phi \, d\rho\, d\phi\, d\theta \tag{8}$$

REMARK.   Note the extra factor of $\rho^2 \sin \phi$ that appears in the integrand of the iterated integral. This is analogous to the extra factor of r that appeared when we integrated in cylindrical coordinates.

The analog of Theorem 17.7.1 for triple integrals in spherical coordinates is tedious to state, so instead we shall give some examples that illustrate techniques for obtaining the limits of integration. In all of our examples we shall use the same order of integration—first with respect to ρ, then ϕ, and then θ. Once the reader has mastered the basic ideas, there should be no trouble using other orders of integration.

Suppose that we want to integrate $f(\rho, \theta, \phi)$ over the spherical solid G enclosed by the sphere $\rho = \rho_0$. The basic idea is to choose the limits of integration so that every point of the solid is accounted for in the integration process. Figure 17.7.9 illustrates one way of doing this:

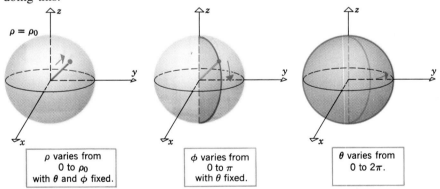

| ρ varies from 0 to ρ_0 with θ and ϕ fixed. | ϕ varies from 0 to π with θ fixed. | θ varies from 0 to 2π. |

Figure 17.7.9

Holding θ and ϕ fixed for the first integration, we let ρ vary from 0 to ρ_0. This covers a radial line from the origin to the surface of the sphere. Next, keeping θ fixed, we let ϕ vary from 0 to π so that the radial line sweeps out a fan-shaped region. Finally, we let θ vary from 0 to 2π so that the fan-shaped region makes a complete revolution, thereby sweeping out the entire sphere. Thus, the triple integral of $f(\rho, \theta, \phi)$ over the spherical solid G may be evaluated by writing

$$\iiint\limits_{G} f(\rho, \theta, \phi)\, dV = \int_0^{2\pi} \int_0^{\pi} \int_0^{\rho_0} f(\rho, \theta, \phi) \rho^2 \sin \phi \, d\rho\, d\phi\, d\theta$$

Table 17.7.1 (pages 900–901) suggests how the limits of integration in spherical coordinates can be obtained for some other common solids.

Table 17.7.1

DETERMINATION OF LIMITS	INTEGRAL

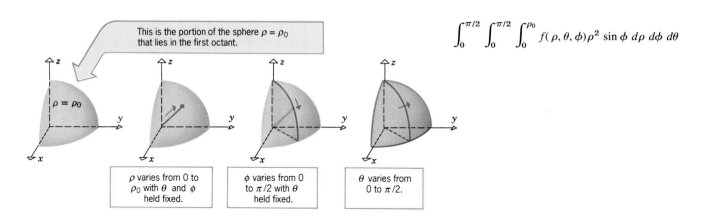

This is the portion of the sphere $\rho = \rho_0$ that lies in the first octant.

ρ varies from 0 to ρ_0 with θ and ϕ held fixed.

ϕ varies from 0 to $\pi/2$ with θ held fixed.

θ varies from 0 to $\pi/2$.

$$\int_0^{\pi/2} \int_0^{\pi/2} \int_0^{\rho_0} f(\rho, \theta, \phi)\rho^2 \sin\phi \, d\rho \, d\phi \, d\theta$$

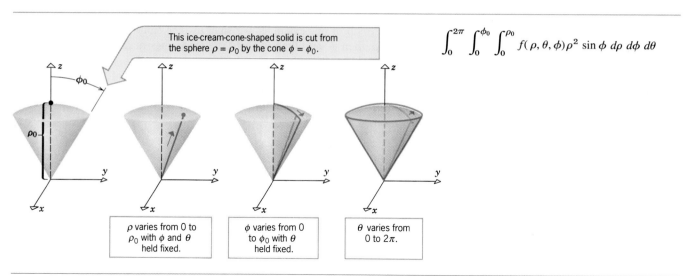

This ice-cream-cone-shaped solid is cut from the sphere $\rho = \rho_0$ by the cone $\phi = \phi_0$.

ρ varies from 0 to ρ_0 with ϕ and θ held fixed.

ϕ varies from 0 to ϕ_0 with θ held fixed.

θ varies from 0 to 2π.

$$\int_0^{2\pi} \int_0^{\phi_0} \int_0^{\rho_0} f(\rho, \theta, \phi)\rho^2 \sin\phi \, d\rho \, d\phi \, d\theta$$

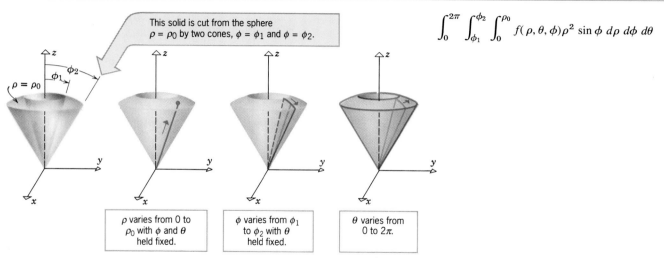

This solid is cut from the sphere $\rho = \rho_0$ by two cones, $\phi = \phi_1$ and $\phi = \phi_2$.

ρ varies from 0 to ρ_0 with ϕ and θ held fixed.

ϕ varies from ϕ_1 to ϕ_2 with θ held fixed.

θ varies from 0 to 2π.

$$\int_0^{2\pi} \int_{\phi_1}^{\phi_2} \int_0^{\rho_0} f(\rho, \theta, \phi)\rho^2 \sin\phi \, d\rho \, d\phi \, d\theta$$

Table 17.7.1

DETERMINATION OF LIMITS	INTEGRAL

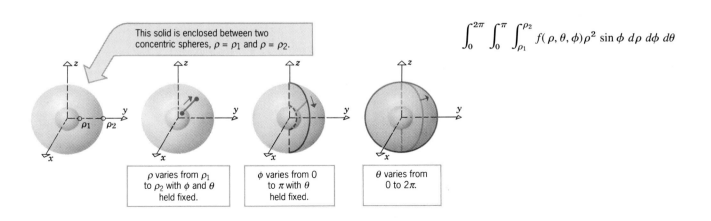

This solid is enclosed laterally by the cone $\phi = \phi_0$ and on top by the horizontal plane $z = a$.

$$\int_0^{2\pi} \int_0^{\phi_0} \int_0^{a\sec\phi} f(\rho, \theta, \phi)\rho^2 \sin\phi \, d\rho \, d\phi \, d\theta$$

ρ varies from 0 to $a \sec\phi$ with ϕ and θ held fixed.

ϕ varies from 0 to ϕ_0 with θ held fixed.

θ varies from 0 to 2π.

This solid is enclosed between two concentric spheres, $\rho = \rho_1$ and $\rho = \rho_2$.

$$\int_0^{2\pi} \int_0^{\pi} \int_{\rho_1}^{\rho_2} f(\rho, \theta, \phi)\rho^2 \sin\phi \, d\rho \, d\phi \, d\theta$$

ρ varies from ρ_1 to ρ_2 with ϕ and θ held fixed.

ϕ varies from 0 to π with θ held fixed.

θ varies from 0 to 2π.

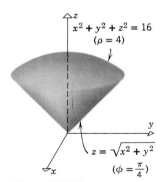

Figure 17.7.10

Example 3 Use spherical coordinates to find the volume and the centroid of the solid G bounded above by the sphere $x^2 + y^2 + z^2 = 16$ and below by the cone $z = \sqrt{x^2 + y^2}$.

Solution. The solid G is sketched in Figure 17.7.10.

In spherical coordinates, the equation of the sphere $x^2 + y^2 + z^2 = 16$ is $\rho = 4$ and the equation of the cone $z = \sqrt{x^2 + y^2}$ is

$$\rho\cos\phi = \sqrt{\rho^2\sin^2\phi\cos^2\theta + \rho^2\sin^2\phi\sin^2\theta}$$

which simplifies to

$$\rho\cos\phi = \rho\sin\phi$$

or, on dividing both sides by $\rho\cos\phi$,

$$\tan\phi = 1$$

Thus $\phi = \pi/4$, so the volume of G is

$$V = \iiint\limits_G dV = \int_0^{2\pi} \int_0^{\pi/4} \int_0^4 \rho^2 \sin \phi \, d\rho \, d\phi \, d\theta$$

$$= \int_0^{2\pi} \int_0^{\pi/4} \left[\frac{\rho^3}{3} \sin \phi \right]_{\rho=0}^4 d\phi \, d\theta$$

$$= \int_0^{2\pi} \int_0^{\pi/4} \frac{64}{3} \sin \phi \, d\phi \, d\theta$$

$$= \frac{64}{3} \int_0^{2\pi} \left[-\cos \phi \right]_{\phi=0}^{\pi/4} d\theta = \frac{64}{3} \int_0^{2\pi} \left(1 - \frac{\sqrt{2}}{2} \right) d\theta$$

$$= \frac{64\pi}{3} (2 - \sqrt{2})$$

By symmetry, the centroid $(\bar{x}, \bar{y}, \bar{z})$ is on the z-axis, so $\bar{x} = \bar{y} = 0$. From (14) of Section 17.6 and the volume calculated above,

$$\bar{z} = \frac{1}{V} \iiint\limits_G z \, dV = \frac{1}{V} \int_0^{2\pi} \int_0^{\pi/4} \int_0^4 (\rho \cos \phi) \rho^2 \sin \phi \, d\rho \, d\phi \, d\theta$$

$$= \frac{1}{V} \int_0^{2\pi} \int_0^{\pi/4} \left[\frac{\rho^4}{4} \cos \phi \sin \phi \right]_{\rho=0}^4 d\phi \, d\theta$$

$$= \frac{64}{V} \int_0^{2\pi} \int_0^{\pi/4} \sin \phi \cos \phi \, d\phi \, d\theta = \frac{64}{V} \int_0^{2\pi} \left[\frac{1}{2} \sin^2 \phi \right]_{\phi=0}^{\pi/4} d\theta$$

$$= \frac{16}{V} \int_0^{2\pi} d\theta = \frac{32\pi}{V} = \frac{3}{2(2 - \sqrt{2})}$$

With the help of a calculator, $\bar{z} \approx 2.56$ (to two decimal places), so the approximate location of the centroid in the xyz-coordinate system is $(0, 0, 2.56)$. ◀

□ CONVERTING TRIPLE
INTEGRALS FROM
RECTANGULAR TO
SPHERICAL
COORDINATES

Referring to Table 14.8.1, triple integrals can be converted from rectangular coordinates to spherical coordinates by making the substitution $x = \rho \sin \phi \cos \theta$, $y = \rho \sin \phi \sin \theta$, $z = \rho \cos \phi$. The two integrals are related by the equation

$$\iiint\limits_G f(x, y, z) \, dV = \iiint\limits_{\substack{\text{appropriate} \\ \text{limits}}} f(\rho \sin \phi \cos \theta, \rho \sin \phi \sin \theta, \rho \cos \phi) \, \rho^2 \sin \phi \, d\rho \, d\phi \, d\theta$$

(9)

Example 4   Use spherical coordinates to evaluate

$$\int_{-2}^2 \int_{-\sqrt{4-x^2}}^{\sqrt{4-x^2}} \int_0^{\sqrt{4-x^2-y^2}} z^2 \sqrt{x^2 + y^2 + z^2} \, dz \, dy \, dx$$

Solution.   In problems like this, it is helpful to begin (when possible) with a sketch of the region G of integration. From the z-limits of integration, the upper surface of G is the hemisphere $z = \sqrt{4 - x^2 - y^2}$ and the lower surface is the xy-plane $z = 0$. From the x- and y-limits of integration, the projection of the solid G on the xy-plane is the region

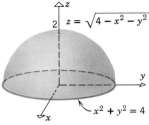

Figure 17.7.11

enclosed by the circle $x^2 + y^2 = 4$. From this information we obtain the sketch of G in Figure 17.7.11. Thus,

$$\int_{-2}^{2} \int_{-\sqrt{4-x^2}}^{\sqrt{4-x^2}} \int_{0}^{\sqrt{4-x^2-y^2}} z^2 \sqrt{x^2 + y^2 + z^2} \, dz \, dy \, dx$$

$$= \iiint_{G} z^2 \sqrt{x^2 + y^2 + z^2} \, dV$$

$$= \int_{0}^{2\pi} \int_{0}^{\pi/2} \int_{0}^{2} \rho^5 \cos^2 \phi \sin \phi \, d\rho \, d\phi \, d\theta$$

$$= \int_{0}^{2\pi} \int_{0}^{\pi/2} \frac{32}{3} \cos^2 \phi \sin \phi \, d\phi \, d\theta$$

$$= \frac{32}{3} \int_{0}^{2\pi} \left[-\frac{1}{3} \cos^3 \phi \right]_{\phi=0}^{\pi/2} d\theta = \frac{32}{9} \int_{0}^{2\pi} d\theta = \frac{64}{9} \pi \qquad \blacktriangleleft$$

▶ Exercise Set 17.7

In Exercises 1–4, evaluate the iterated integral.

1. $\displaystyle\int_{0}^{2\pi} \int_{0}^{1} \int_{0}^{\sqrt{1-r^2}} zr \, dz \, dr \, d\theta$.

2. $\displaystyle\int_{0}^{\pi/2} \int_{0}^{\cos\theta} \int_{0}^{r^2} r \sin \theta \, dz \, dr \, d\theta$.

3. $\displaystyle\int_{0}^{\pi/2} \int_{0}^{\pi/2} \int_{0}^{1} \rho^3 \sin \phi \cos \phi \, d\rho \, d\phi \, d\theta$.

4. $\displaystyle\int_{0}^{2\pi} \int_{0}^{\pi/4} \int_{0}^{a \sec \phi} \rho^2 \sin \phi \, d\rho \, d\phi \, d\theta$ $(a > 0)$.

In Exercises 5–9, use cylindrical coordinates to find the volume of the solid.

5. The solid bounded by the paraboloid $z = x^2 + y^2$ and the plane $z = 9$.

6. The solid that is bounded above and below by the sphere $x^2 + y^2 + z^2 = 9$ and inside the cylinder $x^2 + y^2 = 4$.

7. The solid that is enclosed between the surface $r^2 + z^2 = 20$ and the surface $z = r^2$.

8. The solid that is enclosed between the cone $z = hr/a$ and the plane $z = h$.

9. The solid in the first octant that is below by the sphere $x^2 + y^2 + z^2 = 16$ and lies inside the sphere $x^2 + y^2 = 4x$.

In Exercises 10–14, use spherical coordinates to find the volume of the solid.

10. The solid bounded above by the sphere $\rho = 4$ and below by the cone $\phi = \pi/3$.

11. The solid in the first octant bounded by the sphere $\rho = 2$, the coordinate planes, and the cones $\phi = \pi/6$ and $\phi = \pi/3$.

12. The solid within the cone $\phi = \pi/4$ and between the spheres $\rho = 1$ and $\rho = 2$.

13. The solid within the sphere $x^2 + y^2 + z^2 = 9$, outside the cone $z = \sqrt{x^2 + y^2}$, and above the xy-plane.

14. The solid bounded by the sphere $x^2 + y^2 + z^2 = 4a^2$ and the planes $z = 0$ and $z = a$.

15. Find the volume of $x^2 + y^2 + z^2 = a^2$ using

   (a) cylindrical coordinates

   (b) spherical coordinates.

In Exercises 16–18, use cylindrical coordinates.

16. Find the mass of the solid that is bounded by the cone $z = \sqrt{x^2 + y^2}$ and the plane $z = 3$ if the density of the solid is $\delta(x, y, z) = 3 - z$.

17. Find the mass of the solid in the first octant bounded above by the paraboloid $z = 4 - x^2 - y^2$, below by the plane $z = 0$, and laterally by the cylinder $x^2 + y^2 = 2x$ and the planes $x = 0$ and $y = 0$, assuming the density to be $\delta(x, y, z) = z$. [*Hint:* The Wallis formulas in Exercises 34 and 36 of Section 9.3 will help with the integration.]

18. Find the mass of a right-circular cylinder of radius a and height h if the density is proportional to the distance from the base. (Let k be the constant of proportionality.)

In Exercises 19–21, use spherical coordinates.

19. Find the mass of the solid that is enclosed by the sphere $x^2 + y^2 + z^2 = 1$ and lies within the cone $z = \sqrt{x^2 + y^2}$ if the density is $\delta(x, y, z) = \sqrt{x^2 + y^2 + z^2}$.

20. Find the mass of the solid enclosed between the spheres $x^2 + y^2 + z^2 = 1$ and $x^2 + y^2 + z^2 = 4$ if the density is $\delta(x, y, z) = (x^2 + y^2 + z^2)^{-1/2}$.

21. Find the mass of a spherical solid of radius a if the density is proportional to the distance from the center. (Let k be the constant of proportionality.)

In Exercises 22–24, use cylindrical coordinates to find the centroid of the solid.

22. The solid bounded by the cone $z = \sqrt{x^2 + y^2}$ and the plane $z = 2$.

23. The solid that is bounded above by the sphere
$$x^2 + y^2 + z^2 = 2$$
and below by the paraboloid $z = x^2 + y^2$.

24. The solid bounded above by the paraboloid $z = x^2 + y^2$, below by the plane $z = 0$, and laterally by the cylinder $(x - 1)^2 + y^2 = 1$. [*Hint:* Use the Wallis formulas in Exercises 34 and 36 of Section 9.3 for the integration.]

In Exercises 25–27, use spherical coordinates to find the centroid.

25. The solid in the first octant bounded by the coordinate planes and the sphere $x^2 + y^2 + z^2 = a^2$.

26. The solid bounded above by the sphere $\rho = 4$ and below by the cone $\phi = \pi/3$.

27. The solid that is enclosed by the hemispheres $y = \sqrt{9 - x^2 - z^2}$, $y = \sqrt{4 - x^2 - z^2}$, and the plane $y = 0$.

In Exercises 28–31, use cylindrical or spherical coordinates to evaluate the integral.

28. $\displaystyle\int_0^a \int_0^{\sqrt{a^2-x^2}} \int_0^{a^2-x^2-y^2} x^2 \, dz \, dy \, dx \quad (a > 0).$

29. $\displaystyle\int_{-1}^1 \int_0^{\sqrt{1-x^2}} \int_0^{\sqrt{1-x^2-y^2}} e^{-(x^2+y^2+z^2)^{3/2}} \, dz \, dy \, dx.$

30. $\displaystyle\int_0^2 \int_0^{\sqrt{4-y^2}} \int_{\sqrt{x^2+y^2}}^{\sqrt{8-x^2-y^2}} z^2 \, dz \, dx \, dy.$

31. $\displaystyle\int_{-3}^3 \int_{-\sqrt{9-y^2}}^{\sqrt{9-y^2}} \int_{-\sqrt{9-x^2-y^2}}^{\sqrt{9-x^2-y^2}} \sqrt{x^2 + y^2 + z^2} \, dz \, dx \, dy.$

32. Let G be the solid in the first octant bounded by the sphere $x^2 + y^2 + z^2 = 4$ and the coordinate planes. In each part evaluate
$$\iiint_G xyz \, dV$$

(a) using rectangular coordinates

(b) using cylindrical coordinates

(c) using spherical coordinates.

Solve Exercises 33–38 using either cylindrical or spherical coordinates, whichever seems appropriate.

33. Find the center of gravity of the solid hemisphere bounded by $z = \sqrt{a^2 - x^2 - y^2}$ and $z = 0$ if the density is proportional to the distance from the origin.

34. Find the center of gravity of the solid in the first octant bounded by the cylinder $x^2 + y^2 = a^2$, the coordinate planes, and the plane $z = a$ if the density of the solid is $\delta(x, y, z) = xyz$.

35. Find the center of gravity of the solid bounded by the paraboloid $z = 1 - x^2 - y^2$ and the xy-plane if the density is $\delta(x, y, z) = x^2 + y^2 + z^2$.

36. Find the center of gravity of the solid that is bounded by the cylinder $x^2 + y^2 = 1$, the cone $z = \sqrt{x^2 + y^2}$, and the xy-plane if the density is $\delta(x, y, z) = z$.

37. Find the volume that is enclosed by the spheres
$$x^2 + y^2 + z^2 = 9 \quad \text{and} \quad x^2 + y^2 + (z - 2)^2 = 4$$

38. Suppose that the density at a point on a spherical planet is assumed to be
$$\delta = \delta_0 e^{[(\rho/R)^3 - 1]}$$
where δ_0 is a positive constant, R is the radius of the planet, and ρ is the distance from the point to the planet's center. Calculate the mass of the planet.

39. In this exercise we shall obtain a formula for the volume of the spherical wedge in Figure 17.7.7.

(a) Use a triple integral in cylindrical coordinates to show that the volume of the solid bounded above by a sphere $\rho = \rho_0$, below by a cone $\phi = \phi_0$, and on the sides by $\theta = \theta_1$ and $\theta = \theta_2$ $(\theta_1 < \theta_2)$ is
$$V = \tfrac{1}{3} \rho_0^3 (1 - \cos \phi_0)(\theta_2 - \theta_1)$$
[*Hint:* In cylindrical coordinates, the sphere has the equation $r^2 + z^2 = \rho_0^2$ and the cone has the equation $z = r \cot \phi_0$. For simplicity, consider only the case $0 < \phi_0 < \pi/2$.]

(b) Subtract appropriate volumes and use the result in part (a) to deduce that the volume ΔV of the spherical wedge is
$$\Delta V = \frac{\rho_2^3 - \rho_1^3}{3} (\cos \phi_1 - \cos \phi_2)(\theta_2 - \theta_1)$$

(c) Apply the Mean-Value Theorem to the functions $\cos \phi$ and ρ^3 to deduce that the formula in part (b) can be written as
$$\Delta V = \rho^{*2} \sin \phi^* \, \Delta \rho \, \Delta \phi \, \Delta \theta$$
where ρ^* is between ρ_1 and ρ_2, ϕ^* is between ϕ_1 and ϕ_2, and $\Delta \rho = \rho_2 - \rho_1$, $\Delta \phi = \phi_2 - \phi_1$, $\Delta \theta = \theta_2 - \theta_1$.

In Exercises 40–44, use the formulas in Exercise 36, Section 17.6, to find the indicated moment of inertia for the given homogeneous solid. Express your answer in terms of the mass M of the solid.

40. I_z for the solid cylinder $x^2 + y^2 \leq R^2$, $0 \leq z \leq h$.

41. I_y for the solid cylinder $x^2 + y^2 \leq R^2$, $0 \leq z \leq h$.

42. I_z for the hollow cylinder $R_1^2 \leq x^2 + y^2 \leq R_2^2$, $0 \leq z \leq h$.

43. I_z for the solid sphere $x^2 + y^2 + z^2 \leq R^2$.

44. I_z for the solid cone $\dfrac{h}{R}\sqrt{x^2 + y^2} \leq z \leq h$.

45. If d is the distance between two point masses m and m', then according to Newton's *inverse-square law of gravitational attraction* each point mass attracts the other with a force of magnitude kmm'/d^2, where k is a constant. Because force is a vector, the attraction between a lamina or solid and a point mass can be found by considering the components of the force. For example, assume that a solid G with mass density δ attracts a unit point mass at the origin (see the figure). The component in the z-direction of the force that G exerts on the unit mass can be found by approximating the corresponding components exerted by small subregions, summing, and taking the limit. This leads to the formula, using the spherical coordinates ρ and ϕ,

$$\iiint_G \frac{k\delta \cos \phi}{\rho^2}\, dV$$

for the component of force in the z-direction.

(a) Use cylindrical coordinates to find the component in the z-direction of the force of attraction on a unit mass at the origin by the homogeneous solid cylinder $x^2 + y^2 \leq R^2$, $0 < a \leq z \leq a + h$.

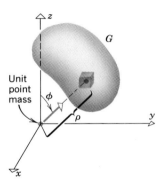

(b) Explain why the force in part (a) is the total force that acts on the unit mass.

In Exercises 46–48, use the formula in Exercise 45 to find the magnitude of the force of attraction on a unit mass at the origin by the given homogeneous lamina or solid. (The regions are all such that the force in the z-direction is the only force of attraction.)

46. The circular lamina $x^2 + y^2 \leq R^2$, $z = a > 0$. [*Hint:* If S is a lamina with area A, and δ is the *surface* density, then the formula is

$$\iint_S \frac{k\delta \cos \phi}{\rho^2}\, dA$$

Use polar coordinates to do the integration.]

47. The solid cone $\dfrac{h}{R}r \leq z \leq h$. (Use spherical coordinates.)

48. The solid sphere of radius R with its center on the z-axis at $z = a > R$. (Use spherical coordinates.) Show that the force is the same as if the total mass of the sphere were concentrated at its center.

17.8 CHANGE OF VARIABLES IN MULTIPLE INTEGRALS; JACOBIANS

In this section we shall discuss a general method for evaluating double and triple integrals by substitution. Most of the results in this section are very difficult to prove, so our approach will be informal and motivational. Our goal is to provide a geometric understanding of the basic principles and an exposure to computational techniques.

☐ **CHANGE OF VARIABLE IN A SINGLE INTEGRAL**

To motivate techniques for evaluating double and triple integrals by substitution, it will be helpful to consider the effect of a substitution $x = g(u)$ on a single definite integral over an interval $[a, b]$, where g is differentiable and either increasing $[g'(u) > 0]$ or decreasing $[g'(u) < 0]$. In either case, g is one-to-one and

$$\int_a^b f(x)\, dx = \int_{g^{-1}(a)}^{g^{-1}(b)} f(g(u))g'(u)\, du$$

In this relationship $f(x)$ and dx are expressed in terms of u, and the u-limits of integration result from solving the equations

$$a = g(u) \quad \text{and} \quad b = g(u)$$

In the case where g is decreasing we have $g^{-1}(b) < g^{-1}(a)$, which is contrary to our usual convention of writing definite integrals with the larger limit of integration at the top. We can remedy this by reversing the limits of integration and writing

$$\int_a^b f(x)\, dx = -\int_{g^{-1}(b)}^{g^{-1}(a)} f(g(u))g'(u)\, du = \int_{g^{-1}(b)}^{g^{-1}(a)} f(g(u))|g'(u)|\, du$$

where the absolute value results from the fact that $g'(u)$ is negative. Thus, regardless of whether g is increasing or decreasing we can write

$$\int_a^b f(x)\, dx = \int_\alpha^\beta f(g(u))|g'(u)|\, du \tag{1}$$

where α and β are the u-limits of integration and $\alpha < \beta$.

The expression $g'(u)$ that appears in (1) is called the **_Jacobian_** of the change of variable $x = g(u)$ in honor of C. G. J. Jacobi,* who made the first serious study of change of variables in multiple integrals in the mid 1800s. Formula (1) reveals three effects of the change of variable $x = g(u)$:

- The new integrand becomes $f(g(u))$ times the absolute value of the Jacobian.
- dx becomes du.
- The x-interval of integration is transformed into a u-interval of integration.

Our goal in this section is to show that analogous results hold for changing variables in double and triple integrals.

☐ **TRANSFORMATIONS OF THE PLANE**

Equations of the form

$$x = x(u, v), \quad y = y(u, v) \tag{2}$$

associate a point (x, y) with a point (u, v), and hence the two equations define a function T that associates points in the xy-plane with points in the uv-plane according to the formula

$$T(u, v) = (x(u, v), y(u, v))$$

---

*CARL GUSTAV JACOB JACOBI (1804–1851). German mathematician. Jacobi, the son of a banker, grew up in a background of wealth and culture and showed brilliance in mathematics early. He resisted studying mathematics by rote, preferring instead to learn general principles from the works of the masters, Euler and Lagrange. He entered the University of Berlin at age 16 as a student of mathematics and classical studies. However, he soon realized that he could not do both and turned fully to mathematics with a blazing intensity that he would maintain throughout his life. He received his Ph.D. in 1825 and was able to secure a position as a lecturer at the University of Berlin by giving up Judaism and becoming a Christian. However, his promotion opportunities remained limited and he moved on to the University of Königsberg. Jacobi was born to teach—he had a dynamic personality and delivered his lectures with a clarity and enthusiasm that frequently left his audience spellbound. However, in spite of extensive teaching commitments, he was able to publish volumes of revolutionary mathematical research that eventually made him the leading European mathematician after Gauss. His main body of research was in the area of elliptic functions, a branch of mathematics with important applications in astronomy and physics as well as in other fields of mathematics. Because of his family wealth, Jacobi was not dependent on his teaching salary in his early years. However, his comfortable world eventually collapsed. In 1840 his family went bankrupt and he was personally wiped out financially. In 1842 he had a nervous breakdown from overwork. In 1843 he became seriously ill with diabetes and moved to Berlin with the help of a government grant to defray his medical expenses. In 1848 he made a stupid political remark that caused the government to withdraw the grant, eventually resulting in the loss of his home. His health continued to decline and in 1851 he finally succumbed to successive bouts of influenza and smallpox. In spite of all his problems, Jacobi was a tireless worker to the end. When a friend expressed concern about the effect of the hard work on his health, Jacobi replied, "Certainly, I have sometimes endangered my health by overwork, but what of it? Only cabbages have no nerves, no worries. And what do they get out of their perfect well-being?"

We call T a **transformation** from the uv-plane to the xy-plane and (x, y) the **image** of (u, v) under the transformation T. We also say that T **maps** (x, y) into (u, v). The set R of all images in the xy-plane of a set S in the uv-plane is called the **image of S under T**. If distinct points in the uv-plane have distinct images in the xy-plane, then T is said to be **one-to-one**. In this case the equations in (2) define u and v as functions of x and y, say

$$u = u(x, y), \quad v = v(x, y)$$

These equations, which can often be obtained by solving (2) for u and v in terms of x and y, define a transformation from the xy-plane to the uv-plane that maps the image of (u, v) under T back into (u, v). This transformation is denoted by T^{-1} and is called the **inverse of T** (Figure 17.8.1).

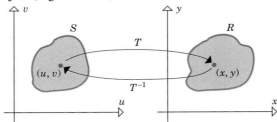

Figure 17.8.1

One way to visualize the geometric effect of a transformation T is to determine the images in the xy-plane of the vertical and horizontal lines in the uv-plane. Sets of points in the xy-plane that are images of horizontal lines (v constant) are called **u-curves**, and sets of points that are images of vertical lines (u constant) are called **v-curves** (Figure 17.8.2).

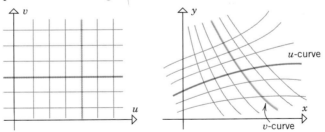

Figure 17.8.2

Example 1 Let T be the transformation from the uv-plane to the xy-plane defined by the equations

$$x = \tfrac{1}{4}(u + v), \quad y = \tfrac{1}{2}(u - v) \tag{3}$$

(a) Find $T(1, 3)$.

(b) Sketch the u-curves corresponding to $v = -2, -1, 0, 1, 2$.

(c) Sketch the v-curves corresponding to $u = -2, -1, 0, 1, 2$.

(d) Sketch the image under T of the square region in the uv-plane bounded by the lines $u = -2, u = 2, v = -2,$ and $v = 2$.

Solution (a). Substituting $u = 1$ and $v = 3$ in (3) yields $T(1, 3) = (1, -1)$.

Solutions (b and c). In these parts it will be convenient to express the transformation equations with u and v as functions of x and y. We leave it for the reader to show that

$$u = 2x + y, \quad v = 2x - y$$

Thus, the u-curves corresponding to $v = -2, -1, 0, 1,$ and 2 are

$$2x - y = -2, \quad 2x - y = -1, \quad 2x - y = 0, \quad 2x - y = 1, \quad 2x - y = 2$$

and the v-curves corresponding to $u = -2, -1, 0, 1,$ and 2 are

$$2x + y = -2, \quad 2x + y = -1, \quad 2x + y = 0, \quad 2x + y = 1, \quad 2x + y = 2$$

In Figure 17.8.3 the u-curves are shown in green and the v-curves in blue.

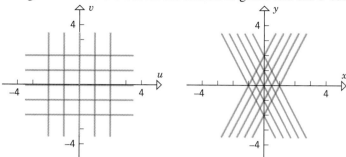

Figure 17.8.3

Solution (d).   The image of a region can often be found by finding the image of its boundary. In this case the images of the boundary lines $u = -2,$ $u = 2,$ $v = -2,$ and $v = 2$ enclose the diamond-shaped region in the xy-plane shown in Figure 17.8.4.

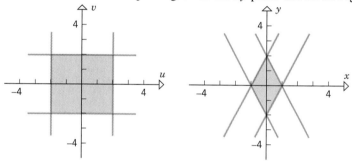

Figure 17.8.4

☐ **JACOBIANS IN TWO VARIABLES**

To derive the change-of-variable formula for double integrals, we will need to understand the relationship between the area of a *small* rectangular region in the uv-plane and the area of its image in the xy-plane under a transformation T given by the equations

$$x = x(u, v), \quad y = y(u, v)$$

For this purpose, suppose that Δu and Δv are positive, and consider a rectangular region S in the uv-plane enclosed by the lines

$$u = u_0, \quad u = u_0 + \Delta u, \quad v = v_0, \quad v = v_0 + \Delta v$$

If the functions $x(u, v)$ and $y(u, v)$ are continuous, and if Δu and Δv are not too large, then the image of S in the xy-plane is a "curvilinear" rectangular region R enclosed by the u-curves and v-curves corresponding to these lines (Figure 17.8.5).

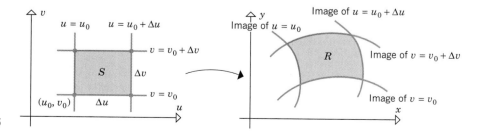

Figure 17.8.5

If we let

$$\mathbf{r} = \mathbf{r}(u, v) = x(u, v)\mathbf{i} + y(u, v)\mathbf{j}$$

be the position vector to the point in the xy-plane that corresponds to the point (u, v) in the uv-plane, then the u-curve corresponding to $v = v_0$ and the v-curve corresponding to $u = u_0$ can be represented in vector form as

$$\mathbf{r}(u, v_0) = x(u, v_0)\mathbf{i} + y(u, v_0)\mathbf{j} \qquad \boxed{u\text{-curve}}$$

$$\mathbf{r}(u_0, v) = x(u_0, v)\mathbf{i} + y(u_0, v)\mathbf{j} \qquad \boxed{v\text{-curve}}$$

Since we are assuming Δu and Δv to be small, the region R can be approximated by a parallelogram determined by the "secant vectors"

$$\mathbf{a} = \mathbf{r}(u_0 + \Delta u, v_0) - \mathbf{r}(u_0, v_0) \tag{4}$$

$$\mathbf{b} = \mathbf{r}(u_0, v_0 + \Delta v) - \mathbf{r}(u_0, v_0) \tag{5}$$

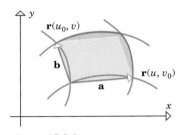

$\mathbf{r}(u_0, v)$

$\mathbf{b}$

$\mathbf{a}$

$\mathbf{r}(u, v_0)$

Figure 17.8.6

shown in Figure 17.8.6. A more useful approximation of R can be obtained by using Formulas (4) and (5) of Section 16.3 to approximate these secant vectors by tangent vectors as follows:

$$\mathbf{a} = \frac{\mathbf{r}(u_0 + \Delta u, v_0) - \mathbf{r}(u_0, v_0)}{\Delta u} \Delta u$$

$$\approx \frac{\partial \mathbf{r}}{\partial u} \Delta u = \left(\frac{\partial x}{\partial u}\mathbf{i} + \frac{\partial y}{\partial u}\mathbf{j} \right) \Delta u$$

$$\mathbf{b} = \frac{\mathbf{r}(u_0, v_0 + \Delta v) - \mathbf{r}(u_0, v_0)}{\Delta v} \Delta v$$

$$\approx \frac{\partial \mathbf{r}}{\partial v} \Delta v = \left(\frac{\partial x}{\partial v}\mathbf{i} + \frac{\partial y}{\partial v}\mathbf{j} \right) \Delta v$$

where the partial derivatives are evaluated at (u_0, v_0). But at $T(u_0, v_0)$ the vectors $\partial \mathbf{r}/\partial u$ and $\partial \mathbf{r}/\partial v$ are tangent to the u-curve and v-curve, respectively, and consequently so are $(\partial \mathbf{r}/\partial u)\,\Delta u$ and $(\partial \mathbf{r}/\partial v)\,\Delta v$. Thus, the region R can be approximated by the parallelogram determined by the tangent vectors

$$\frac{\partial \mathbf{r}}{\partial u} \Delta u = \left(\frac{\partial x}{\partial u}\mathbf{i} + \frac{\partial y}{\partial u}\mathbf{j} \right) \Delta u$$

$$\frac{\partial \mathbf{r}}{\partial v} \Delta v = \left(\frac{\partial x}{\partial v}\mathbf{i} + \frac{\partial y}{\partial v}\mathbf{j} \right) \Delta v$$

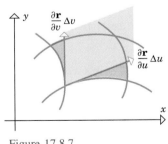

$\dfrac{\partial \mathbf{r}}{\partial v} \Delta v$

$\dfrac{\partial \mathbf{r}}{\partial u} \Delta u$

Figure 17.8.7

(Figure 17.8.7). Hence, it follows that the area of the region R, which we shall denote by ΔA, can be approximated by the area of the parallelogram determined by these vectors. Thus, from Formula (7) of Section 14.4 we have

$$\Delta A \approx \left\| \frac{\partial \mathbf{r}}{\partial u} \Delta u \times \frac{\partial \mathbf{r}}{\partial v} \Delta v \right\| = \left\| \frac{\partial \mathbf{r}}{\partial u} \times \frac{\partial \mathbf{r}}{\partial v} \right\| \Delta u\, \Delta v \tag{6}$$

where the derivatives are evaluated at (u_0, v_0). Computing the cross product, we obtain

$$\frac{\partial \mathbf{r}}{\partial u} \times \frac{\partial \mathbf{r}}{\partial v} = \begin{vmatrix} \mathbf{i} & \mathbf{j} & \mathbf{k} \\ \dfrac{\partial x}{\partial u} & \dfrac{\partial y}{\partial u} & 0 \\ \dfrac{\partial x}{\partial v} & \dfrac{\partial y}{\partial v} & 0 \end{vmatrix} = \begin{vmatrix} \dfrac{\partial x}{\partial u} & \dfrac{\partial y}{\partial u} \\ \dfrac{\partial x}{\partial v} & \dfrac{\partial y}{\partial v} \end{vmatrix} \mathbf{k} = \begin{vmatrix} \dfrac{\partial x}{\partial u} & \dfrac{\partial x}{\partial v} \\ \dfrac{\partial y}{\partial u} & \dfrac{\partial y}{\partial v} \end{vmatrix} \mathbf{k} \tag{7}$$

The determinant in (7) is sufficiently important that it has its own terminology and notation.

17.8.1   DEFINITION.   If T is the transformation from the uv-plane to the xy-plane defined by the equations $x = x(u, v)$, $y = y(u, v)$, then the **Jacobian of T** is denoted by $J(u, v)$ or by $\partial(x, y)/\partial(u, v)$ and is defined by

$$J(u, v) = \frac{\partial(x, y)}{\partial(u, v)} = \begin{vmatrix} \dfrac{\partial x}{\partial u} & \dfrac{\partial x}{\partial v} \\ \dfrac{\partial y}{\partial u} & \dfrac{\partial y}{\partial v} \end{vmatrix} = \frac{\partial x}{\partial u}\frac{\partial y}{\partial v} - \frac{\partial y}{\partial u}\frac{\partial x}{\partial v}.$$

Using the notation in this definition, it follows from (6) and (7) that

$$\Delta A \approx \left\| \frac{\partial(x, y)}{\partial(u, v)}\mathbf{k} \right\| \Delta u\, \Delta v$$

or since $\mathbf{k}$ is a unit vector,

$$\Delta A \approx \left| \frac{\partial(x, y)}{\partial(u, v)} \right| \Delta u\, \Delta v \tag{8}$$

At the point (u_0, v_0) this important formula relates the areas of the regions R and S in Figure 17.8.5: It tells us that *for small values of Δu and Δv, the area of R is approximately the absolute value of the Jacobian times the area of S.* Moreover, it is proved in advanced calculus courses that the error in the approximation approaches zero as $\Delta u \to 0$ and $\Delta v \to 0$.

☐ **CHANGE OF VARIABLES IN DOUBLE INTEGRALS**

Our next objective is to provide a geometric motivation for the following result.

17.8.2   CHANGE-OF-VARIABLE FORMULA FOR DOUBLE INTEGRALS.   If the transformation $x = x(u, v)$, $y = y(u, v)$ maps the region S in the uv-plane into the region R in the xy-plane, and if the Jacobian $\partial(x, y)/\partial(u, v)$ is nonzero and does not change sign on S, then with appropriate restrictions on the transformation and the regions it follows that

$$\iint\limits_{R} f(x, y)\, dA_{xy} = \iint\limits_{S} f(x(u, v), y(u, v)) \left| \frac{\partial(x, y)}{\partial(u, v)} \right| dA_{uv} \tag{9}$$

where we have attached subscripts to the dA's to help identify the associated variables.

REMARK.   A precise statement of conditions under which Formula (9) holds would take us beyond the scope of this course. Suffice it to say that the formula holds if T is a one-to-one transformation, $f(x, y)$ is continuous on R, the partial derivatives of $x(u, v)$ and $y(u, v)$ exist and are continuous on S, and the regions R and S are not too complicated.

To motivate Formula (9), we proceed as follows:

• Subdivide the region S in the uv-plane into pieces by lines parallel to the coordinate axes, and exclude from consideration any pieces that contain points outside of S. This leaves only rectangular regions that are subsets of S. Assume that there are n such regions and denote the kth such region by S_k. Assume that S_k has dimensions Δu_k by Δv_k and, as shown in Figure 17.8.8a, let (u_k^*, v_k^*) be its "lower left corner."

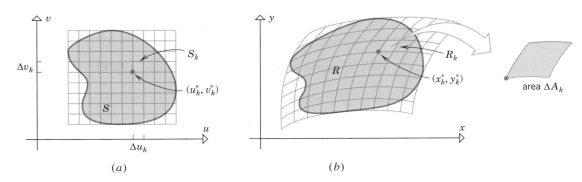

Figure 17.8.8

(a)

(b)

- As shown in Figure 17.8.8*b*, the transformation *T* defined by the equations $x = x(u, v)$, $y = y(u, v)$ maps S_k into a curvilinear rectangular region R_k in the *xy*-plane and maps the point (u_k^*, v_k^*) into the point $(x_k^*, y_k^*) = (x(u_k^*, v_k^*), y(u_k^*, v_k^*))$ in R_k. Denote the area of R_k by ΔA_k.

- In rectangular coordinates the double integral of $f(x, y)$ over a region *R* is defined as a limit of Riemann sums in which *R* is subdivided into *rectangular* subregions. It is proved in advanced calculus courses that under appropriate conditions subdivisions into *curvilinear* rectangular subregions can be used instead. Accepting this to be so, we can approximate the double integral of $f(x, y)$ over *R* as

$$\iint_R f(x, y)\, dA_{xy} \approx \sum_{k=1}^{n} f(x_k^*, y_k^*)\, \Delta A_k$$

$$\approx \sum_{k=1}^{n} f(x(u_k^*, v_k^*), y(u_k^*, v_k^*)) \left| \frac{\partial(x, y)}{\partial(u, v)} \right| \Delta u_k\, \Delta v_k$$

where the Jacobian is evaluated at (u_k^*, v_k^*). But the last expression is a Riemann sum for the integral

$$\iint_S f(x(u, v), y(u, v)) \left| \frac{\partial(x, y)}{\partial(u, v)} \right| dA_{uv}$$

so Formula (9) follows if we assume that the errors in the approximations approach zero as $n \to +\infty$.

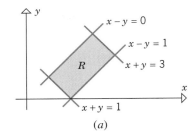

(a)

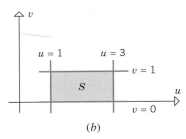

(b)

Figure 17.8.9

Example 2 Evaluate

$$\iint_R \frac{x - y}{x + y}\, dA$$

where *R* is the region enclosed by the lines $x - y = 0$, $x - y = 1$, $x + y = 1$, and $x + y = 3$ (Figure 17.8.9*a*).

Solution. This integral would be tedious to evaluate directly because the region *R* is oriented in such a way that we would have to subdivide it and integrate over each part separately. However, the occurrence of the expressions $x - y$ and $x + y$ in the equations of the boundary suggests that the transformation

$$u = x + y, \quad v = x - y \tag{10}$$

would be helpful, since with this transformation the boundary lines

$$x + y = 1, \quad x + y = 3, \quad x - y = 0, \quad x - y = 1$$

are u-curves and v-curves corresponding to the lines

$$u = 1, \quad u = 3, \quad v = 0, \quad v = 1$$

in the uv-plane. These lines enclose the region S shown in Figure 17.8.9b. To find the Jacobian $\partial(x, y)/\partial(u, v)$ of this transformation, we first solve (10) for x and y in terms of u and v. This yields

$$x = \tfrac{1}{2}(u + v), \quad y = \tfrac{1}{2}(u - v)$$

from which we obtain

$$\frac{\partial(x, y)}{\partial(u, v)} = \begin{vmatrix} \dfrac{\partial x}{\partial u} & \dfrac{\partial x}{\partial v} \\[2mm] \dfrac{\partial y}{\partial u} & \dfrac{\partial y}{\partial v} \end{vmatrix} = \begin{vmatrix} \tfrac{1}{2} & \tfrac{1}{2} \\[1mm] \tfrac{1}{2} & -\tfrac{1}{2} \end{vmatrix} = -\tfrac{1}{4} - \tfrac{1}{4} = -\tfrac{1}{2}$$

Thus, from Formula (9), but with the notation dA rather than dA_{xy},

$$\iint\limits_{R} \frac{x - y}{x + y}\, dA = \iint\limits_{S} \frac{v}{u} \left| \frac{\partial(x, y)}{\partial(u, v)} \right| dA_{uv}$$

$$= \iint\limits_{S} \frac{v}{u} \left| -\frac{1}{2} \right| dA_{uv} = \frac{1}{2} \int_{0}^{1} \int_{1}^{3} \frac{v}{u}\, du\, dv$$

$$= \frac{1}{2} \int_{0}^{1} v \ln |u| \Big]_{u=1}^{3} dv$$

$$= \frac{1}{2} \ln 3 \int_{0}^{1} v\, dv = \frac{1}{4} \ln 3 \quad \blacktriangleleft$$

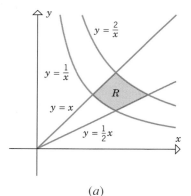

(a)

(b)

Figure 17.8.10

Example 3 Evaluate

$$\iint\limits_{R} e^{xy}\, dA$$

where R is the region enclosed by the lines $y = \tfrac{1}{2}x$ and $y = x$ and the hyperbolas $y = 1/x$ and $y = 2/x$ (Figure 17.8.10a).

Solution. As in the preceding example, we look for a transformation in which the boundary curves in the xy-plane become u-curves and v-curves. For this purpose we rewrite the four boundary curves as

$$\frac{y}{x} = \tfrac{1}{2}, \quad \frac{y}{x} = 1, \quad xy = 1, \quad xy = 2$$

which suggests the transformation

$$u = \frac{y}{x}, \quad v = xy \tag{11}$$

With this transformation the boundary curves in the xy-plane are u-curves and v-curves corresponding to the lines

$$u = \tfrac{1}{2}, \quad u = 1, \quad v = 1, \quad v = 2$$

in the uv-plane. These lines enclose the region S shown in Figure 17.8.10b. To find the Jacobian $\partial(x, y)/\partial(u, v)$ of this transformation, we first solve (11) for x and y in terms of u and v. This yields

$$x = \sqrt{v/u}, \quad y = \sqrt{uv}$$

from which we obtain

$$\frac{\partial(x, y)}{\partial(u, v)} = \begin{vmatrix} \dfrac{\partial x}{\partial u} & \dfrac{\partial x}{\partial v} \\[2mm] \dfrac{\partial y}{\partial u} & \dfrac{\partial y}{\partial v} \end{vmatrix} = \begin{vmatrix} -\dfrac{1}{2u}\sqrt{\dfrac{v}{u}} & \dfrac{1}{2\sqrt{uv}} \\[3mm] \dfrac{1}{2}\sqrt{\dfrac{v}{u}} & \dfrac{1}{2}\sqrt{\dfrac{u}{v}} \end{vmatrix} = -\frac{1}{4u} - \frac{1}{4u} = -\frac{1}{2u}$$

Thus, from Formula (9), but with the notation dA rather than dA_{xy},

$$\iint_R e^{xy}\, dA = \iint_S e^v \left| -\frac{1}{2u} \right| dA_{uv} = \frac{1}{2} \iint_S \frac{1}{u} e^v\, dA_{uv}$$

$$= \frac{1}{2} \int_1^2 \int_{1/2}^1 \frac{1}{u} e^v\, du\, dv = \frac{1}{2} \int_1^2 e^v \ln|u| \Big]_{u=1/2}^1 dv$$

$$= \frac{1}{2} \ln 2 \int_1^2 e^v\, dv = \frac{1}{2}(e^2 - e) \ln 2 \quad \blacktriangleleft$$

☐ **CHANGE OF VARIABLES IN TRIPLE INTEGRALS**

Equations of the form

$$x = x(u, v, w), \quad y = y(u, v, w), \quad z = z(u, v, w) \tag{12}$$

define a **transformation** T from uvw-space to xyz-space. Just as a transformation $x = x(u, v)$, $y = y(u, v)$ in two variables maps small rectangles in the uv-plane into curvilinear rectangles in the xy-plane, so (12) maps small rectangular parallelepipeds in uvw-space into curvilinear parallelepipeds in xyz-space (Figure 17.8.11). The definition of the Jacobian of (12) is similar to Definition 17.8.1.

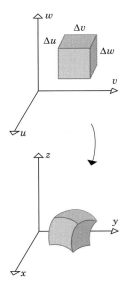
Figure 17.8.11

17.8.3 DEFINITION. If T is the transformation from uvw-space to xyz-space defined by the equations $x = x(u, v, w)$, $y = y(u, v, w)$, $z = z(u, v, w)$, then the **Jacobian of T** is denoted by $J(u, v, w)$ or $\partial(x, y, z)/\partial(u, v, w)$ and is defined by

$$J(u, v, w) = \frac{\partial(x, y, z)}{\partial(u, v, w)} = \begin{vmatrix} \dfrac{\partial x}{\partial u} & \dfrac{\partial x}{\partial v} & \dfrac{\partial x}{\partial w} \\[2mm] \dfrac{\partial y}{\partial u} & \dfrac{\partial y}{\partial v} & \dfrac{\partial y}{\partial w} \\[2mm] \dfrac{\partial z}{\partial u} & \dfrac{\partial z}{\partial v} & \dfrac{\partial z}{\partial w} \end{vmatrix}$$

For small values of Δu, Δv, and Δw, the volume ΔV of the curvilinear parallelepiped in Figure 17.8.11 is related to the volume $\Delta u\, \Delta v\, \Delta w$ of the rectangular parallelepiped by

$$\Delta V \approx \left| \frac{\partial(x, y, z)}{\partial(u, v, w)} \right| \Delta u\, \Delta v\, \Delta w \tag{13}$$

which is the analog of Formula (8). Using this relationship and an argument similar to the one that led to Formula (9), one can obtain the following result.

17.8.4 CHANGE-OF-VARIABLE FORMULA FOR TRIPLE INTEGRALS. If the transformation $x = x(u, v, w)$, $y = y(u, v, w)$, $z = z(u, v, w)$ maps the region S in uvw-space into the region R in xyz-space, and if the Jacobian $\partial(x, y, z)/\partial(u, v, w)$ is nonzero and does not change sign on S, then with appropriate restrictions on the transformation and the regions it follows that

$$\iiint_R f(x, y, z)\, dV_{xyz} = \iiint_S f(x(u, v, w), y(u, v, w), z(u, v, w)) \left| \frac{\partial(x, y, z)}{\partial(u, v, w)} \right| dV_{uvw}$$

$$\tag{14}$$

Example 4   Find the volume of the region G enclosed by the ellipsoid

$$\frac{x^2}{a^2} + \frac{y^2}{b^2} + \frac{z^2}{c^2} = 1$$

Solution.   The volume V is given by the triple integral

$$V = \iiint\limits_{G} dV$$

To evaluate this integral, we make the change of variables

$$x = au, \quad y = bv, \quad z = cw \tag{15}$$

which maps the region G in xyz-space into the region S in uvw-space enclosed by a sphere of radius 1. This can be seen from (15) by noting that

$$\frac{x^2}{a^2} + \frac{y^2}{b^2} + \frac{z^2}{c^2} = 1 \quad \text{becomes} \quad u^2 + v^2 + w^2 = 1$$

The Jacobian of (15) is

$$\frac{\partial(x, y, z)}{\partial(u, v, w)} = \begin{vmatrix} \dfrac{\partial x}{\partial u} & \dfrac{\partial x}{\partial v} & \dfrac{\partial x}{\partial w} \\ \dfrac{\partial y}{\partial u} & \dfrac{\partial y}{\partial v} & \dfrac{\partial y}{\partial w} \\ \dfrac{\partial z}{\partial u} & \dfrac{\partial z}{\partial v} & \dfrac{\partial z}{\partial w} \end{vmatrix} = \begin{vmatrix} a & 0 & 0 \\ 0 & b & 0 \\ 0 & 0 & c \end{vmatrix} = abc$$

Thus, from Formula (14), but with the notation dV rather than dV_{xyz},

$$V = \iiint\limits_{G} dV = \iiint\limits_{S} \left| \frac{\partial(x, y, z)}{\partial(u, v, w)} \right| dV_{uvw} = abc \iiint\limits_{S} dV_{uvw}$$

The last integral is the volume enclosed by a sphere of radius 1, which we know to be $\frac{4}{3}\pi$. Thus, the volume enclosed by the ellipsoid is $V = \frac{4}{3}\pi abc$.   ◄

Jacobians also relate to the problems of converting double and triple integrals in rectangular coordinates to iterated integrals in polar, cylindrical, or spherical coordinates. For example, from Formula (14) the relationship between a triple integral in rectangular coordinates and the corresponding iterated integral in cylindrical coordinates is

$$\iiint\limits_{G} f(x, y, z)\, dV_{xyz} = \iiint\limits_{S} f(r \cos\theta, r \sin\theta, z) \left| \frac{\partial(x, y, z)}{\partial(r, \theta, z)} \right| dV_{r\theta z}$$

$$= \iiint\limits_{\substack{\text{appropriate} \\ r\theta z\text{-limits}}} f(r \cos\theta, r \sin\theta, z)\, r\, dr\, d\theta\, dz \tag{16}$$

The factor r that appears in the final integrand is precisely the absolute value of the Jacobian of the transformation

$$x = r \cos\theta, \quad y = r \sin\theta, \quad z = z \tag{17}$$

from rectangular to cylindrical coordinates, since

$$
\frac{\partial(x, y, z)}{\partial(r, \theta, z)} = \begin{vmatrix} \dfrac{\partial x}{\partial r} & \dfrac{\partial x}{\partial \theta} & \dfrac{\partial x}{\partial z} \\[2mm] \dfrac{\partial y}{\partial r} & \dfrac{\partial y}{\partial \theta} & \dfrac{\partial y}{\partial z} \\[2mm] \dfrac{\partial z}{\partial r} & \dfrac{\partial z}{\partial \theta} & \dfrac{\partial z}{\partial z} \end{vmatrix} = \begin{vmatrix} \cos\theta & -r\sin\theta & 0 \\ \sin\theta & r\cos\theta & 0 \\ 0 & 0 & 1 \end{vmatrix}
$$

$$
= r\cos^2\theta + r\sin^2\theta = r
$$

and $r \geq 0$ in (17).

▶ Exercise Set 17.8

In Exercises 1–4, find the Jacobian $\partial(x, y)/\partial(u, v)$.

1. $x = u + 4v, \; y = 3u - 5v$.

2. $x = u + 2v^2, \; y = 2u^2 - v$.

3. $x = \sin u + \cos v, \; y = -\cos u + \sin v$.

4. $x = \dfrac{2u}{u^2 + v^2}, \; y = -\dfrac{2v}{u^2 + v^2}$.

In Exercises 5–8, solve for x and y in terms of u and v, and then find the Jacobian $\partial(x, y)/\partial(u, v)$.

5. $u = 2x - 5y, \; v = x + 2y$.

6. $u = e^x, \; v = ye^{-x}$.

7. $u = x^2 - y^2, \; v = x^2 + y^2 \; (x > 0, y > 0)$.

8. $u = xy, \; v = xy^3 \; (x > 0, y > 0)$.

In Exercises 9–11, find the Jacobian $\partial(x, y, z)/\partial(u, v, w)$.

9. $x = 3u + v, \; y = u - 2w, \; z = v + w$.

10. $x = u - uv, \; y = uv - uvw, \; z = uvw$.

11. $u = xy, \; v = y, \; w = x + z$.

12. (a) Consider the transformation $x = r\cos\theta, \; y = r\sin\theta$ from rectangular to polar coordinates, where $r \geq 0$. Show that

$$
\left| \frac{\partial(x, y)}{\partial(r, \theta)} \right| = r
$$

(b) Consider the transformation $x = \rho\sin\phi\cos\theta,$ $y = \rho\sin\phi\sin\theta, \; z = \rho\cos\phi$ from rectangular to spherical coordinates, where $0 \leq \phi \leq \pi$. Show that

$$
\left| \frac{\partial(x, y, z)}{\partial(\rho, \theta, \phi)} \right| = \rho^2 \sin\phi
$$

13. Use the transformation $u = x - 2y, \; v = 2x + y$ to find

$$
\iint\limits_R \frac{x - 2y}{2x + y} \, dA
$$

where R is the rectangular region enclosed by the lines $x - 2y = 1, \; x - 2y = 4, \; 2x + y = 1, \; 2x + y = 3$.

14. Use the transformation $u = x + y, \; v = x - y$ to find

$$
\iint\limits_R (x - y)e^{x^2 - y^2} \, dA
$$

where R is the rectangular region enclosed by the lines $x + y = 0, \; x + y = 1, \; x - y = 1, \; x - y = 4$.

15. Use the transformation $u = \frac{1}{2}(x + y), \; v = \frac{1}{2}(x - y)$ to find

$$
\iint\limits_R \sin\tfrac{1}{2}(x + y) \cos\tfrac{1}{2}(x - y) \, dA
$$

where R is the triangular region with vertices $(0, 0), (2, 0), (1, 1)$.

16. Use the transformation $u = y/x, \; v = xy$ to find

$$
\iint\limits_R xy^3 \, dA
$$

where R is the region in the first quadrant enclosed by $y = x, \; y = 3x, \; xy = 1, \; xy = 4$.

17. Use the transformation $x = 3u, \; y = 4v$ to find

$$
\iint\limits_R \sqrt{16x^2 + 9y^2} \, dA
$$

where R is the region enclosed by the ellipse $x^2/9 + y^2/16 = 1$. [Hint: Use polar coordinates to evaluate the transformed integral.]

18. Use the transformation $x = 2u, \; y = v$ to find

$$
\iint\limits_R e^{-(x^2 + 4y^2)} \, dA
$$

where R is the region enclosed by the ellipse $x^2/4 + y^2 = 1$. [Hint: Use polar coordinates to evaluate the transformed integral.]

It will be shown in Exercise 38 that the following relationship holds for a one-to-one transformation:

$$\frac{\partial(x, y)}{\partial(u, v)} \cdot \frac{\partial(u, v)}{\partial(x, y)} = 1 \qquad (A)$$

In Exercises 19–22, verify this result by finding the two factors on the left of this equation.

19. $x = u - uv$, $y = uv$.

20. $x = uv$, $y = v^2$.

21. $x = v^2/u$, $y = v/u$.

22. $x = \frac{1}{2}(u^2 + v^2)$, $y = \frac{1}{2}(u^2 - v^2)$ $(u > 0, v > 0)$.

23. Evaluate the integral in Exercise 16 by using Formula (A) above to find $\partial(x, y)/\partial(u, v)$ from $\partial(u, v)/\partial(x, y)$ without solving for x and y in terms of u and v.

In Exercises 24–26, use the method of Exercise 23 to find the Jacobian $\partial(x, y)/\partial(u, v)$ required for the integration.

24. Use the transformation $u = x^2 - y^2$, $v = x^2 + y^2$ to find

$$\iint\limits_{R} xy \, dA$$

where R is the region in the first quadrant enclosed by the hyperbolas $x^2 - y^2 = 1$, $x^2 - y^2 = 4$ and the circles $x^2 + y^2 = 9$, $x^2 + y^2 = 16$.

25. Use the transformation $u = xy$, $v = xy^4$ to find

$$\iint\limits_{R} \sin(xy) \, dA$$

where R is the region enclosed by the curves $xy = \pi$, $xy = 2\pi$, $xy^4 = 1$, $xy^4 = 2$.

26. Use the transformation $u = xy$, $v = x^2 - y^2$ to find

$$\iint\limits_{R} (x^4 - y^4)e^{xy} \, dA$$

where R is the region in the first quadrant enclosed by the hyperbolas $xy = 1$, $xy = 3$, $x^2 - y^2 = 3$, $x^2 - y^2 = 4$.

In Exercises 27–30, evaluate the integral by making an appropriate change of variables.

27. $\displaystyle\iint\limits_{R} \frac{y - 4x}{y + 4x} \, dA$, where R is the region enclosed by the lines $y = 4x$, $y = 4x + 2$, $y = 2 - 4x$, $y = 5 - 4x$.

28. $\displaystyle\iint\limits_{R} (x^2 - y^2) \, dA$, where R is the rectangular region enclosed by the lines $y = -x$, $y = 1 - x$, $y = x$, $y = x + 2$.

29. $\displaystyle\iint\limits_{R} \frac{\sin(x - y)}{\cos(x + y)} \, dA$, where R is the triangular region enclosed by the lines $y = 0$, $y = x$, $x + y = \pi/4$.

30. $\displaystyle\iint\limits_{R} e^{(y-x)/(y+x)} \, dA$, where R is the region in the first quadrant enclosed by the trapezoid with vertices $(0, 1)$, $(1, 0)$, $(0, 4)$, $(4, 0)$.

31. Use an appropriate change of variables to find the area of the region in the first quadrant enclosed by the curves $y = x$, $y = 2x$, $x = y^2$, $x = 4y^2$.

32. Use an appropriate change of variables to find the volume of the solid bounded above by the plane $x + y + z = 9$, below by the xy-plane, and laterally by the elliptic cylinder $4x^2 + 9y^2 = 36$. [*Hint:* Express the volume as a double integral in xy-coordinates, then use polar coordinates to evaluate the transformed integral.]

33. Use the transformation $u = x$, $v = z - y$, $w = xy$ to find

$$\iiint\limits_{G} (z - y)^2 xy \, dV$$

where G is the region enclosed by the surfaces $x = 1$, $x = 3$, $z = y$, $z = y + 1$, $xy = 2$, $xy = 4$.

34. Use the transformation $u = xy$, $v = yz$, $w = xz$ to find

$$\iiint\limits_{G} \sqrt{xyz} \, dV$$

where G is the region in the first octant enclosed by the hyperbolic cylinders $xy = 1$, $xy = 2$, $yz = 1$, $yz = 3$, $xz = 1$, $xz = 4$.

35. Use the transformation $x = au$, $y = bv$, $z = cw$ to find

$$\iiint\limits_{G} x^2 \, dV$$

where G is the region enclosed by the ellipsoid $x^2/a^2 + y^2/b^2 + z^2/c^2 = 1$. [*Hint:* Use spherical coordinates to evaluate the transformed integral.]

36. Find the volume of the region G in Exercise 34.

37. Use the transformation $u = y/z$, $v = 4x - y$, $w = y/z^2$ to find the volume of the region enclosed by the surfaces $y = z$, $y = 2z$, $y = 4x$, $y = 4x - 12$, $y = z^2$, $y = 4z^2$.

38. (a) Verify that

$$\begin{vmatrix} a_1 & b_1 \\ c_1 & d_1 \end{vmatrix} \begin{vmatrix} a_2 & b_2 \\ c_2 & d_2 \end{vmatrix} = \begin{vmatrix} a_1a_2 + b_1c_2 & a_1b_2 + b_1d_2 \\ c_1a_2 + d_1c_2 & c_1b_2 + d_1d_2 \end{vmatrix}$$

(b) If $x = x(u, v)$, $y = y(u, v)$ is a one-to-one transformation, then $u = u(x, y)$, $v = v(x, y)$. Assuming differentiability, use the result in part (a) and the chain rule to show that

$$\frac{\partial(x, y)}{\partial(u, v)} \cdot \frac{\partial(u, v)}{\partial(x, y)} = 1$$

◆ TECHNOLOGY EXERCISES   *Chapter 17*

Most of these exercises require access to a graphing calculator or a computer algebra system (CAS) such as *Mathematica*, *Maple*, or *Derive*. When you are asked to *find* an answer or to *solve* an equation, you may choose to find an exact result or a numerical approximation, depending on the particular technology you are using and on your own imagination. The form of your answers may differ from those of other students or from those in the answer section of the text, depending on how you solve the problems and the accuracy you use in your numerical approximations. Those exercises that are more appropriate for a CAS than a graphing calculator are labeled with the icon ◆.

◆ **1. Volume:** Find the volume of the solid in the first octant beneath the surface $z = \sqrt{1 + x + y}$ and above the region in the xy-plane enclosed by $y = \sin x$ and $y = x/2$.

◆ **2. Volume in polar coordinates:** Use polar coordinates to find the volume of the solid that lies beneath the surface $z = x^2\sqrt{1 + x^2 + y^2}$ and above the region in the xy-plane enclosed by $r = 1 + \cos\theta$.

◆ **3. Surface area:** Find the area of the portion of the surface $z = e^{-x^2 - y^2}$ that is inside the cylinder $x^2 + y^2 = 1$.

◆ **4. Spherical coordinates:** Use spherical coordinates to find the exact value of

$$\iiint\limits_{G} \frac{1}{x^2 + y^2 + (z - 2)^2}\, dV$$

where G is the solid that is enclosed by the sphere $x^2 + y^2 + z^2 = 1$.

◆ **5. The Theorem of Pappus:** Let R be the region in the xy-plane enclosed by the graphs of $y = 3 + x - x^2$ and $y = 1 + \cos x$.

(a) Find the area of R.

(b) Find the y-coordinate of the centroid of R.

(c) Use the results in parts (a) and (b) and the Theorem of Pappus to find the volume of the solid generated by revolving R about the x-axis. Check your result by using the method of washers to find the volume.

◆ **6. Centroid:** Find the centroid of the solid beneath the surface $z = 1/(1 + x^2 + y^2)$ and above the region in the xy-plane enclosed by $y = \sin x$ and the x-axis from $x = 0$ to $x = \pi$.

7. Centroid: Let G be the solid that is enclosed by the surface $z = 1/(1 + x^2 + y^2)$, the xy-plane, and the cylinder $x^2 + y^2 = a^2$. Find the value of a for which the centroid of G is $(0, 0, \frac{1}{4})$.

◆ **8. Centroid of a solid of infinite extent:** Find the centroid of the solid beneath the surface $z = e^{2x + y - x^2 - y^2}$ and above the xy-plane.

C. F. Gauss.

Carl Friedrich Gauss (1777–1855)

18 TOPICS IN VECTOR CALCULUS

18.1 VECTOR FIELDS

In this section we consider functions that associate vectors with points in 2-space or 3-space. We shall see that such functions play an important role in the study of fluid flow, gravitational force fields, electromagnetic force fields, and a wide range of other applied problems.

□ **VECTOR FIELDS**

To motivate the mathematical ideas in this section, consider a unit point mass located at any point in the universe. According to Newton's Universal Law of Gravitation, the earth exerts an attractive force on the mass that is directed toward the earth's center and has a magnitude that is inversely proportional to the square of the distance from the mass to the earth's center (Figure 18.1.1). This association of force vectors with points in space is called the earth's *gravitational field*. A similar idea arises in fluid flow. Imagine a stream in which the water flows horizontally at every level, and consider the layer of water at a

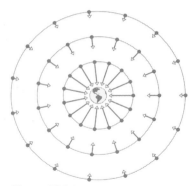

Figure 18.1.1

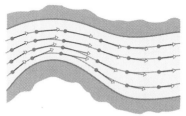

Figure 18.1.2

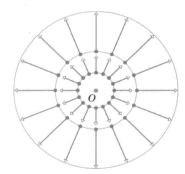

Figure 18.1.3

specific depth. At each point of the layer, the water has a certain velocity, which we can represent by a vector at that point (Figure 18.1.2). This association of velocity vectors with points in the two-dimensional layer is called the *flow field* at that layer. These ideas are captured in the following definition.

> **18.1.1**   DEFINITION.   A *vector field* is a function that associates a unique vector $\mathbf{F}(P)$ with each point P in a region of 2-space or 3-space.

Example 1   Let O be a fixed point in 2-space, and for each point P in 2-space define the vector field $\mathbf{F}(P)$ by $\mathbf{F}(P) = \overrightarrow{OP}$. Some typical vectors in this vector field are shown in Figure 18.1.3. In that figure we have followed the standard convention of positioning the vector $\mathbf{F}(P)$ with its initial point at P.   ◄

Observe that the concept of a vector field has been defined without reference to a coordinate system; it is said to be a *coordinate-free* definition. However, for computational purposes it is often desirable to work with vector fields in coordinate systems. If $\mathbf{F}(P)$ is a vector field in 2-space with an xy-coordinate system, then the point P has coordinates (x, y), and the components of the vector $\mathbf{F}(P)$ are functions of x and y. Thus, $\mathbf{F}(P)$ can be expressed as

$$\mathbf{F}(x, y) = f(x, y)\mathbf{i} + g(x, y)\mathbf{j}$$

Similarly, in 3-space with an xyz-coordinate system, a vector field $\mathbf{F}(P)$ can be expressed as

$$\mathbf{F}(x, y, z) = f(x, y, z)\mathbf{i} + g(x, y, z)\mathbf{j} + h(x, y, z)\mathbf{k}$$

Just as it is impossible to describe a curve completely by plotting finitely many points, so it is impossible to describe a vector field completely by drawing finitely many vectors. Nevertheless, it is often possible to get a useful picture of a vector field by sketching a finite number of vectors that are well chosen.

Example 2   Figure 18.1.4 shows sketches of three vector fields in 2-space. For simplicity, we have omitted the scales and selected vectors that do not overlap; nevertheless, the sketches still provide some useful geometric insight into the behavior of the fields.   ◄

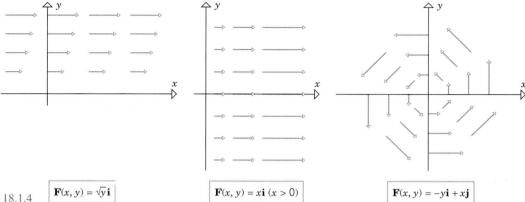

Figure 18.1.4

| $\mathbf{F}(x, y) = \sqrt{y}\,\mathbf{i}$ | $\mathbf{F}(x, y) = x\mathbf{i}\ (x > 0)$ | $\mathbf{F}(x, y) = -y\mathbf{i} + x\mathbf{j}$ |

REMARK.   Sometimes it is helpful to denote the vector fields $\mathbf{F}(x, y)$ and $\mathbf{F}(x, y, z)$ entirely in vector notation by identifying (x, y) with the radius vector $\mathbf{r} = x\mathbf{i} + y\mathbf{j}$ and (x, y, z) with the radius vector $\mathbf{r} = x\mathbf{i} + y\mathbf{j} + z\mathbf{k}$. With this notation a vector field in either 2-space or 3-space can be written as $\mathbf{F}(\mathbf{r})$. When no confusion is likely to arise, we shall sometimes omit the $\mathbf{r}$ altogether and denote the vector field as $\mathbf{F}$.

☐ **INVERSE-SQUARE FIELDS**

According to Newton's Universal Law of Gravitation, objects with masses m and M attract each other with a force $\mathbf{F}$ of magnitude

$$\|\mathbf{F}\| = \frac{GmM}{r^2} \tag{1}$$

where r is the distance between the objects (treated as point masses) and G is a constant. If we assume that the object of mass M is located at the origin of an xyz-coordinate system and $\mathbf{r}$ is the radius vector to the object of mass m, then $r = \|\mathbf{r}\|$, and the force $\mathbf{F}(\mathbf{r})$ exerted by the object of mass M on the object of mass m is in the direction of the unit vector $-\mathbf{r}/\|\mathbf{r}\|$. Thus, from (1)

$$\mathbf{F}(\mathbf{r}) = -\frac{GmM}{\|\mathbf{r}\|^2} \frac{\mathbf{r}}{\|\mathbf{r}\|} = -\frac{GmM}{\|\mathbf{r}\|^3} \mathbf{r}$$

If m and M are constant, and we let $c = -GmM$, then this formula can be expressed as

$$\mathbf{F}(\mathbf{r}) = \frac{c}{\|\mathbf{r}\|^3} \mathbf{r}$$

Vector fields of this form arise in electromagnetic as well as gravitational problems. Such fields are so important that they have their own terminology.

18.1.2 DEFINITION. If $\mathbf{r}$ is a radius vector in 2-space or 3-space, and if c is a constant, then a vector field of the form

$$\mathbf{F}(\mathbf{r}) = \frac{c}{\|\mathbf{r}\|^3} \mathbf{r} \tag{2}$$

is called an ***inverse-square field***.

Observe that if $c > 0$ in (2), then $\mathbf{F}(\mathbf{r})$ has the same direction as $\mathbf{r}$, so each vector in the field is directed away from the origin; and if $c < 0$, then $\mathbf{F}(\mathbf{r})$ is oppositely directed to $\mathbf{r}$, so each vector in the field is directed toward the origin. In either case the magnitude of $\mathbf{F}(\mathbf{r})$ is inversely proportional to the square of distance from the tip of $\mathbf{r}$ to the origin, since

$$\|\mathbf{F}(\mathbf{r})\| = \frac{|c|}{\|\mathbf{r}\|^3} \|\mathbf{r}\| = \frac{|c|}{\|\mathbf{r}\|^2}$$

Example 3 ***Coulomb's law*** states that *the electrostatic force exerted by one charged particle on another is directly proportional to the product of the charges and inversely proportional to the square of the distance between them.* This has the same form as Newton's Universal Law of Gravitation, so the electrostatic force field exerted by a charged particle is an inverse-square field. Specifically, if a particle of charge Q is at the origin of a coordinate system, and if $\mathbf{r}$ is the radius vector to a particle of charge q, then the force $\mathbf{F}(\mathbf{r})$ that the particle of charge Q exerts on the particle of charge q is of the form

$$\mathbf{F}(\mathbf{r}) = \frac{qQ}{4\pi\epsilon_0 \|\mathbf{r}\|^3} \mathbf{r}$$

where ϵ_0 is a positive constant (called the ***permittivity constant***). This formula is of form (2) with $c = qQ/4\pi\epsilon_0$. ◄

In our later work it will sometimes be convenient to express (2) in terms of (x, y) or (x, y, z) rather than $\mathbf{r}$. In the three-dimensional case $\|\mathbf{r}\| = \sqrt{x^2 + y^2 + z^2}$, so (2) can be written as

$$\mathbf{F}(x, y, z) = \frac{c}{(x^2 + y^2 + z^2)^{3/2}} (x\mathbf{i} + y\mathbf{j} + z\mathbf{k}) \tag{3}$$

Similarly, in the two-dimensional case

$$\mathbf{F}(x, y) = \frac{c}{(x^2 + y^2)^{3/2}} (x\mathbf{i} + y\mathbf{j}) \tag{4}$$

☐ **GRADIENT FIELDS**

An important class of vector fields arises from the process of finding gradients. Recall that if ϕ is a function of three variables, then the gradient of ϕ is defined as

$$\nabla\phi = \frac{\partial\phi}{\partial x}\mathbf{i} + \frac{\partial\phi}{\partial y}\mathbf{j} + \frac{\partial\phi}{\partial z}\mathbf{k}$$

This formula defines a vector field in 3-space called the ***gradient field of ϕ***. Similarly, the gradient of a function of two variables defines a gradient field in 2-space. At each point in a gradient field where the gradient is nonzero, the vector points in the direction in which the rate of increase of ϕ is maximum.

Example 4   Sketch the gradient field of $\phi(x, y) = x + y$.

Solution.   The gradient of ϕ is

$$\nabla\phi = \frac{\partial\phi}{\partial x}\mathbf{i} + \frac{\partial\phi}{\partial y}\mathbf{j} = \mathbf{i} + \mathbf{j}$$

which is the same at each point. A portion of the vector field is sketched in Figure 18.1.5.   ◄

If $\mathbf{F}$ is an arbitrary vector field, one can ask whether it is the gradient field for some *scalar* function ϕ. We shall see later that not every vector field is a gradient field. However, those vector fields that are gradient fields are sufficiently important that they have a special name.

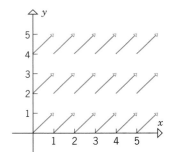

Figure 18.1.5

> **18.1.3**   DEFINITION.   A vector field $\mathbf{F}$ is said to be ***conservative*** in a region if it is the gradient field for some function ϕ in that region. The function ϕ is called a ***potential function*** for $\mathbf{F}$ in the region.

Example 5   Inverse-square fields are conservative in any region that does not contain the origin. For example, in the three-dimensional case the function

$$\phi(x, y, z) = -\frac{c}{(x^2 + y^2 + z^2)^{1/2}} \tag{5}$$

is a potential function for (3) in any region not containing the origin, since

$$\nabla\phi(x, y, z) = \frac{\partial\phi}{\partial x}\mathbf{i} + \frac{\partial\phi}{\partial y}\mathbf{j} + \frac{\partial\phi}{\partial z}\mathbf{k}$$

$$= \frac{cx}{(x^2 + y^2 + z^2)^{3/2}}\mathbf{i} + \frac{cy}{(x^2 + y^2 + z^2)^{3/2}}\mathbf{j} + \frac{cz}{(x^2 + y^2 + z^2)^{3/2}}\mathbf{k}$$

$$= \frac{c}{(x^2 + y^2 + z^2)^{3/2}} (x\mathbf{i} + y\mathbf{j} + z\mathbf{k})$$

$$= \mathbf{F}(x, y, z)$$

In a later section we shall discuss methods for finding potential functions for conservative vector fields.   ◄

☐ **DIVERGENCE AND CURL**

We shall now define two important operations on vector fields in 3-space—the *divergence* and the *curl* of the field. These names originate in the study of fluid flow, in which case the divergence relates to the way in which fluid flows toward or away from a point and the curl relates to the rotational properties of the fluid at a point. We will investigate the physical interpretations of these operations in more detail later, but for now we shall focus only on their computation.

18.1.4 DEFINITION. If $\mathbf{F}(x, y, z) = f(x, y, z)\mathbf{i} + g(x, y, z)\mathbf{j} + h(x, y, z)\mathbf{k}$, then we define the **divergence of F**, written div **F**, by

$$\operatorname{div} \mathbf{F} = \frac{\partial f}{\partial x} + \frac{\partial g}{\partial y} + \frac{\partial h}{\partial z} \tag{6}$$

18.1.5 DEFINITION. If $\mathbf{F}(x, y, z) = f(x, y, z)\mathbf{i} + g(x, y, z)\mathbf{j} + h(x, y, z)\mathbf{k}$, then we define the **curl of F**, written curl **F**, by

$$\operatorname{curl} \mathbf{F} = \left(\frac{\partial h}{\partial y} - \frac{\partial g}{\partial z} \right)\mathbf{i} + \left(\frac{\partial f}{\partial z} - \frac{\partial h}{\partial x} \right)\mathbf{j} + \left(\frac{\partial g}{\partial x} - \frac{\partial f}{\partial y} \right)\mathbf{k} \tag{7}$$

REMARK. Observe that div **F** and curl **F** depend on the point at which they are computed, and hence are more properly written as div $\mathbf{F}(x, y, z)$ and curl $\mathbf{F}(x, y, z)$. However, even though these functions are expressed in terms of x, y, and z, it can be proved that their values at a fixed point depend on the point but not on the coordinate system selected. This is important in applications, since it allows physicists and engineers to compute the curl and divergence in any convenient coordinate system.

Before proceeding to some examples, we note that div **F** has scalar values, whereas curl **F** has vector values (i.e., curl **F** is itself a vector field). Moreover, for computational purposes it is useful to note that the formula for the curl can be expressed in the determinant form

$$\operatorname{curl} \mathbf{F} = \begin{vmatrix} \mathbf{i} & \mathbf{j} & \mathbf{k} \\ \dfrac{\partial}{\partial x} & \dfrac{\partial}{\partial y} & \dfrac{\partial}{\partial z} \\ f & g & h \end{vmatrix} \tag{8}$$

The reader should verify that Formula (7) results if we compute this determinant by interpreting a ''product'' such as $(\partial/\partial x)(g)$ to mean $\partial g/\partial x$. Keep in mind, however, that (8) is just a mnemonic device and not a true determinant, since the entries in a determinant must be numbers, not vectors and partial derivative symbols.

Example 6 Find the divergence and the curl of the vector field

$$\mathbf{F}(x, y, z) = x^2 y \mathbf{i} + 2y^3 z \mathbf{j} + 3z \mathbf{k}$$

Solution. From (6)

$$\operatorname{div} \mathbf{F} = \frac{\partial}{\partial x}(x^2 y) + \frac{\partial}{\partial y}(2y^3 z) + \frac{\partial}{\partial z}(3z)$$

$$= 2xy + 6y^2 z + 3$$

and from (8)

$$\text{curl }\mathbf{F} = \begin{vmatrix} \mathbf{i} & \mathbf{j} & \mathbf{k} \\ \dfrac{\partial}{\partial x} & \dfrac{\partial}{\partial y} & \dfrac{\partial}{\partial z} \\ x^2 y & 2y^3 z & 3z \end{vmatrix}$$

$$= \left[\frac{\partial}{\partial y}(3z) - \frac{\partial}{\partial z}(2y^3 z) \right] \mathbf{i} + \left[\frac{\partial}{\partial z}(x^2 y) - \frac{\partial}{\partial x}(3z) \right] \mathbf{j}$$

$$+ \left[\frac{\partial}{\partial x}(2y^3 z) - \frac{\partial}{\partial y}(x^2 y) \right] \mathbf{k}$$

$$= -2y^3 \mathbf{i} - x^2 \mathbf{k} \qquad \blacktriangleleft$$

Example 7 Show that the divergence of the inverse-square field

$$\mathbf{F}(x, y, z) = \frac{c}{(x^2 + y^2 + z^2)^{3/2}}(x\mathbf{i} + y\mathbf{j} + z\mathbf{k})$$

is zero.

Solution. The computations can be simplified by letting $r = (x^2 + y^2 + z^2)^{1/2}$, in which case $\mathbf{F}$ can be expressed as

$$\mathbf{F}(x, y, z) = \frac{cx\mathbf{i} + cy\mathbf{j} + cz\mathbf{k}}{r^3} = \frac{cx}{r^3}\mathbf{i} + \frac{cy}{r^3}\mathbf{j} + \frac{cz}{r^3}\mathbf{k}$$

We leave it for the reader to show that

$$\frac{\partial r}{\partial x} = \frac{x}{r}, \quad \frac{\partial r}{\partial y} = \frac{y}{r}, \quad \frac{\partial r}{\partial z} = \frac{z}{r}$$

Thus,

$$\text{div }\mathbf{F} = c\left[\frac{\partial}{\partial x}\left(\frac{x}{r^3}\right) + \frac{\partial}{\partial y}\left(\frac{y}{r^3}\right) + \frac{\partial}{\partial z}\left(\frac{z}{r^3}\right) \right] \tag{9}$$

But

$$\frac{\partial}{\partial x}\left(\frac{x}{r^3}\right) = \frac{r^3 - x(3r^2)(x/r)}{(r^3)^2} = \frac{1}{r^3} - \frac{3x^2}{r^5}$$

$$\frac{\partial}{\partial y}\left(\frac{y}{r^3}\right) = \frac{1}{r^3} - \frac{3y^2}{r^5}$$

$$\frac{\partial}{\partial z}\left(\frac{z}{r^3}\right) = \frac{1}{r^3} - \frac{3z^2}{r^5}$$

Substituting these expressions in (9) yields

$$\text{div }\mathbf{F} = c\left[\frac{3}{r^3} - \frac{3x^2 + 3y^2 + 3z^2}{r^5} \right] = c\left[\frac{3}{r^3} - \frac{3r^2}{r^5} \right] = 0 \qquad \blacktriangleleft$$

☐ **THE ∇ OPERATOR**

Thus far, the symbol ∇ that appears in the gradient expression $\nabla\phi$ has not been given a meaning of its own. However, it is often convenient to view ∇ as an operator

$$\nabla = \frac{\partial}{\partial x}\mathbf{i} + \frac{\partial}{\partial y}\mathbf{j} + \frac{\partial}{\partial z}\mathbf{k} \tag{10}$$

which when applied to $\phi(x, y, z)$ produces the gradient

$$\nabla\phi = \frac{\partial\phi}{\partial x}\mathbf{i} + \frac{\partial\phi}{\partial y}\mathbf{j} + \frac{\partial\phi}{\partial z}\mathbf{k}$$

We call (10) the **del operator**. This is analogous to the derivative operator d/dx, which when applied to $f(x)$ produces the derivative $f'(x)$.

The del operator allows us to express the divergence of a vector field

$$\mathbf{F} = f(x, y, z)\mathbf{i} + g(x, y, z)\mathbf{j} + h(x, y, z)\mathbf{k}$$

in dot product notation as

$$\text{div } \mathbf{F} = \nabla \cdot \mathbf{F} = \frac{\partial f}{\partial x} + \frac{\partial g}{\partial y} + \frac{\partial h}{\partial z} \tag{11}$$

and the curl of this field in cross-product notation as

$$\text{curl } \mathbf{F} = \nabla \times \mathbf{F} = \begin{vmatrix} \mathbf{i} & \mathbf{j} & \mathbf{k} \\ \dfrac{\partial}{\partial x} & \dfrac{\partial}{\partial y} & \dfrac{\partial}{\partial z} \\ f & g & h \end{vmatrix} \tag{12}$$

☐ **THE LAPLACIAN ∇^2**

The operator that results by taking the dot product of the del operator with itself is denoted by ∇^2 and is called the **Laplacian* operator**. This operator has the form

$$\nabla^2 = \nabla \cdot \nabla = \frac{\partial^2}{\partial x^2} + \frac{\partial^2}{\partial y^2} + \frac{\partial^2}{\partial z^2} \tag{13}$$

When applied to $\phi(x, y, z)$ the Laplacian operator produces the function

$$\nabla^2 \phi = \frac{\partial^2 \phi}{\partial x^2} + \frac{\partial^2 \phi}{\partial y^2} + \frac{\partial^2 \phi}{\partial z^2}$$

Note that $\nabla^2 \phi$ can also be expressed as div $(\nabla\phi)$. The equation $\nabla^2 \phi = 0$ or, equivalently,

$$\frac{\partial^2 \phi}{\partial x^2} + \frac{\partial^2 \phi}{\partial y^2} + \frac{\partial^2 \phi}{\partial z^2} = 0$$

---

*PIERRE-SIMON DE LAPLACE (1749–1827). French mathematician and physicist. Laplace is sometimes referred to as the French Isaac Newton because of his work in celestial mechanics. In a five-volume treatise entitled *Traité de Mécanique Céleste*, he solved extremely difficult problems involving gravitational interactions between the planets. In particular, he was able to show that our solar system is stable and not prone to catastrophic collapse as a result of these interactions. This was an issue of major concern at the time because Jupiter's orbit appeared to be shrinking and Saturn's expanding; Laplace showed that these were expected periodic anomalies. In addition to his work in celestial mechanics, he founded modern probability theory, showed with Lavoisier that respiration is a form of combustion, and developed methods that fostered many new branches of pure mathematics.

Laplace was born to moderately successful parents in Normandy, his father being a farmer and cider merchant. He matriculated in the theology program at the University of Caen at age 16 but left for Paris at age 18 with a letter of introduction to the influential mathematician d'Alembert, who eventually helped him undertake a career in mathematics. Laplace was a prolific writer, and after his election to the Academy of Sciences in 1773, the secretary wrote that the Academy had never received so many important research papers by so young a person in such a short time. Laplace had little interest in pure mathematics—he regarded mathematics merely as a tool for solving applied problems. In his impatience with mathematical detail, he frequently omitted complicated arguments with the statement, ''It is easy to show that. . . .'' He admitted, however, that as time passed he often had trouble reconstructing the omitted details himself!

At the height of his fame, Laplace served on many government committees and held the posts of Minister of the Interior and chancellor of the Senate. He barely escaped imprisonment and execution during the period of the Revolution, probably because he was able to convince each opposing party that he sided with them. Napoleon described him as a great mathematician but a poor administrator who ''sought subtleties everywhere, had only doubtful ideas, and . . . carried the spirit of the infinitely small into administration.'' In spite of his genius, Laplace was both egotistic and insecure, attempting to ensure his place in history by conveniently failing to credit mathematicians whose work he used—an unnecessary pettiness since his own work was so brilliant. However, on the positive side he was supportive of young mathematicians, often treating them as his own children. Laplace ranks as one of the most influential mathematicians in history.

is known as **Laplace's equation**. This equation plays an important role in a wide variety of applications, resulting from the fact that it is satisfied by the potential function (5) for the inverse-square field (3).

▶ Exercise Set 18.1

In Exercises 1–4, sketch the vector field by drawing some typical nonintersecting vectors. The vectors need not be drawn to the same scale as the coordinate axes, but they should be in the correct proportions relative to each other.

1. $\mathbf{F}(x, y) = 2\mathbf{i} - \mathbf{j}$.   **2.** $\mathbf{F}(x, y) = y\mathbf{j}$, $y > 0$.

3. $\mathbf{F}(x, y) = y\mathbf{i} - x\mathbf{j}$. [*Note:* Each vector in the field is perpendicular to the position vector $\mathbf{r} = x\mathbf{i} + y\mathbf{j}$.]

4. $\mathbf{F}(x, y) = \dfrac{x\mathbf{i} + y\mathbf{j}}{\sqrt{x^2 + y^2}}$. [*Note:* Each vector in the field is a unit vector in the same direction as the position vector $\mathbf{r} = x\mathbf{i} + y\mathbf{j}$.]

In Exercises 5–10, find div $\mathbf{F}$ and curl $\mathbf{F}$.

5. $\mathbf{F}(x, y, z) = x^2\mathbf{i} - 2\mathbf{j} + yz\mathbf{k}$.

6. $\mathbf{F}(x, y, z) = xz^3\mathbf{i} + 2y^4x^2\mathbf{j} + 5z^2y\mathbf{k}$.

7. $\mathbf{F}(x, y, z) = 7y^3z^2\mathbf{i} - 8x^2z^5\mathbf{j} - 3xy^4\mathbf{k}$.

8. $\mathbf{F}(x, y, z) = e^{xy}\mathbf{i} - \cos y\mathbf{j} + \sin^2 z\mathbf{k}$.

9. $\mathbf{F}(x, y, z) = \dfrac{1}{\sqrt{x^2 + y^2 + z^2}}(x\mathbf{i} + y\mathbf{j} + z\mathbf{k})$.

10. $\mathbf{F}(x, y, z) = \ln x\mathbf{i} + e^{xyz}\mathbf{j} + \tan^{-1}(z/x)\mathbf{k}$.

In Exercises 11–18, let k be a constant, and let $\mathbf{F} = \mathbf{F}(x, y, z)$, $\mathbf{G} = \mathbf{G}(x, y, z)$, and $\phi = \phi(x, y, z)$. Prove the following identities, assuming that all derivatives involved exist and are continuous.

11. div $(k\mathbf{F}) = k$ div $\mathbf{F}$.   **12.** curl $(k\mathbf{F}) = k$ curl $\mathbf{F}$.

13. div $(\mathbf{F} + \mathbf{G}) = $ div $\mathbf{F} + $ div $\mathbf{G}$.

14. curl $(\mathbf{F} + \mathbf{G}) = $ curl $\mathbf{F} + $ curl $\mathbf{G}$.

15. div $(\phi\mathbf{F}) = \phi$ div $\mathbf{F} + \nabla\phi \cdot \mathbf{F}$.

16. curl $(\phi\mathbf{F}) = \phi$ curl $\mathbf{F} + \nabla\phi \times \mathbf{F}$.

17. div (curl $\mathbf{F}$) $= 0$.   **18.** curl $(\nabla\phi) = \mathbf{0}$.

In Exercises 19 and 20, let $\mathbf{r} = x\mathbf{i} + y\mathbf{j} + z\mathbf{k}$, and let $\mathbf{u} = a\mathbf{i} + b\mathbf{j} + c\mathbf{k}$, where a, b, and c are constants.

19. Show that div $(\mathbf{u} \times \mathbf{r}) = 0$.

20. Show that curl $(\mathbf{u} \times \mathbf{r}) = 2\mathbf{u}$.

In Exercises 21 and 22, let $\mathbf{F} = f(x)\mathbf{i} + g(x)\mathbf{j} + h(x)\mathbf{k}$, where f, g, and h are differentiable functions.

21. Find div $\mathbf{F}$.   **22.** Find curl $\mathbf{F}$.

In Exercises 23–29, prove the result, assuming that $\mathbf{r} = x\mathbf{i} + y\mathbf{j} + z\mathbf{k}$, $r = \|\mathbf{r}\|$, and $\mathbf{F} = f(r)\mathbf{r}$, where f is a differentiable function of r.

23. div $\mathbf{r} = 3$.   **24.** curl $\mathbf{r} = \mathbf{0}$.

25. $\nabla r = \dfrac{1}{r}\mathbf{r}$.

26. $\nabla f(r) = \dfrac{f'(r)}{r}\mathbf{r}$. [Use the chain rule and Exercise 25.]

27. div $\mathbf{F} = 3f(r) + rf'(r)$. [Use Exercises 15, 23, and 26.]

28. curl $\mathbf{F} = \mathbf{0}$. [Use Exercises 16, 24, and 26.]

29. $\nabla^2 f(r) = 2\dfrac{f'(r)}{r} + f''(r)$. [Use Exercises 15, 23, and 26.]

30. (a) Use the result in Exercise 27 to show that the divergence of the inverse-square field $\mathbf{F} = \mathbf{r}/r^3$ is zero.

    (b) Use the result of Exercise 27 to show that if $\mathbf{F}$ is a vector field of the form $\mathbf{F} = f(r)\mathbf{r}$ and if div $\mathbf{F} = 0$, then $\mathbf{F}$ is an inverse-square field. [*Suggestion:* Multiply $3f(r) + rf'(r) = 0$ through by r^2, and write the result as a derivative of a product.]

31. A curve C is called a **flow line** of a vector field $\mathbf{F}$ if $\mathbf{F}$ is a tangent vector to C at each point along C (Figure 18.1.6).

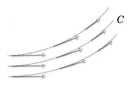

Flow lines of a vector field

Figure 18.1.6

    (a) Let C be a flow line for the vector field $\mathbf{F}(x, y) = -y\mathbf{i} + x\mathbf{j}$, and let (x, y) be a point on C for which $y \neq 0$. Show that the flow lines satisfy the differential equation

$$\frac{dy}{dx} = -\frac{x}{y}$$

    (b) Solve the differential equation in part (a) by separation of variables, and show that the flow lines are concentric circles centered at the origin. [*Note:* See the third part of Figure 18.1.4.]

In Exercises 32–34, find a differential equation satisfied by the flow lines of **F** (see Exercise 31), and solve it to find equations for the flow lines of **F**. Sketch some typical flow lines and tangent vectors.

32. $\mathbf{F}(x, y) = \mathbf{i} + x\mathbf{j}$.

33. $\mathbf{F}(x, y) = x\mathbf{i} + \mathbf{j}, \ x > 0$.

34. $\mathbf{F}(x, y) = x\mathbf{i} - y\mathbf{j}, \ x > 0$ and $y > 0$.

■ **18.2** LINE INTEGRALS

> *In previous chapters we considered three kinds of integrals in rectangular coordinates: single integrals over intervals, double integrals over two-dimensional regions, and triple integrals over three-dimensional regions. In this section we shall discuss line integrals, which are integrals over curves in two- or three-dimensional space.*

☐ **SURFACE AREA AS A MOTIVATION FOR LINE INTEGRALS**

Integrals over curves arise in a variety of problems. One such problem can be stated as follows:

> **18.2.1** PROBLEM. *Let C be the graph in the xy-plane of a smooth vector-valued function* $\mathbf{r}(t) = x(t)\mathbf{i} + y(t)\mathbf{j}$, *and let* $f(x, y) = f(x(t), y(t))$ *be continuous and nonnegative for* $a \leq t \leq b$. *Find the area of the surface swept out by the vertical line segment from the point* $P(x(t), y(t), 0)$ *to the point* $Q\big(x(t), y(t), f(x(t), y(t))\big)$ *as t varies from a to b (Figure 18.2.1).*

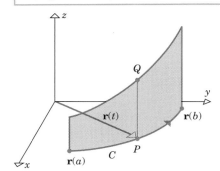

Figure 18.2.1

- Divide the curve C into n arcs by choosing a succession of distinct points $P_1, P_2, \ldots, P_{n-1}$ on C between $\mathbf{r}(a)$ and $\mathbf{r}(b)$ in the direction of increasing t. As shown in Figure 18.2.2, these points divide the surface into n strips. If we denote the area of the kth strip by ΔA_k, then the total area A can be expressed as

$$A = \Delta A_1 + \Delta A_2 + \cdots + \Delta A_n = \sum_{k=1}^{n} \Delta A_k$$

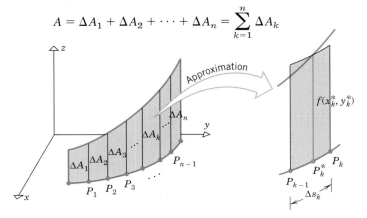

Figure 18.2.2

- Let us now approximate the area ΔA_k as follows: Denote the length of the kth arc by Δs_k, and choose an arbitrary point $P_k^*(x_k^*, y_k^*)$ on this arc. If the kth arc is small, then the value of f cannot change much over the arc, since f is continuous on C. Thus, we can reasonably assume that f has a constant value of $f(x_k^*, y_k^*)$ over the kth arc and that ΔA_k can be approximated by the area of a rectangle of base Δs_k and height $f(x_k^*, y_k^*)$ (Figure 18.2.2); that is,

$$\Delta A_k \approx f(x_k^*, y_k^*)\, \Delta s_k$$

from which it follows that

$$A \approx \sum_{k=1}^{n} f(x_k^*, y_k^*)\, \Delta s_k$$

- If we now increase n in such a way that the length of each arc approaches zero, then it is plausible that the error in this approximation approaches zero, and the exact surface area is

$$A = \lim_{n \to +\infty} \sum_{k=1}^{n} f(x_k^*, y_k^*)\, \Delta s_k \tag{1}$$

The limit in (1) is called the **line integral of f over (or along) C** and is denoted by

$$\int_C f(x, y)\, ds = \lim_{n \to +\infty} \sum_{k=1}^{n} f(x_k^*, y_k^*)\, \Delta s_k \tag{2}$$

With this notation, the area of the surface in Problem 18.2.1 can be expressed as

$$A = \int_C f(x, y)\, ds \tag{3}$$

This result was derived under the assumption that f is nonnegative on C. In the case where $f(x, y)$ can have both positive and negative values on C, the line integral over C represents a difference of two areas, the area above C and below $z = f(x, y)$ minus the area below C and above $z = f(x, y)$.

☐ **EVALUATING LINE INTEGRALS**

Except in the simplest cases, it is impractical to evaluate a line integral directly from (2), so we shall now show how to express a line integral as an ordinary definite integral. For this purpose, assume that the points P_{k-1} and P_k in Figure 18.2.3 correspond to parameter values of t_{k-1} and t_k, respectively, and that $P_k^*(x_k^*, y_k^*)$ corresponds to the parameter value t_k^*. If we let $\Delta t_k = t_k - t_{k-1}$, then we can approximate Δs_k as

$$\Delta s_k \approx \sqrt{(\Delta x_k)^2 + (\Delta y_k)^2} = \sqrt{\left(\frac{\Delta x_k}{\Delta t_k}\right)^2 + \left(\frac{\Delta y_k}{\Delta t_k}\right)^2}\, \Delta t_k \tag{4}$$

from which it follows that (2) can be expressed as

$$\int_C f(x, y)\, ds = \lim_{n \to +\infty} \sum_{k=1}^{n} f(x(t_k^*), y(t_k^*)) \sqrt{\left(\frac{\Delta x_k}{\Delta t_k}\right)^2 + \left(\frac{\Delta y_k}{\Delta t_k}\right)^2}\, \Delta t_k$$

which suggests that

$$\int_C f(x, y)\, ds = \int_a^b f(x(t), y(t)) \sqrt{\left(\frac{dx}{dt}\right)^2 + \left(\frac{dy}{dt}\right)^2}\, dt \tag{5}$$

Recall from Formula (16) of Section 15.3 that if s is an arc-length parameter for C, then

$$\frac{ds}{dt} = \sqrt{\left(\frac{dx}{dt}\right)^2 + \left(\frac{dy}{dt}\right)^2}$$

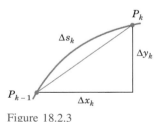

Figure 18.2.3

which is commonly expressed as

$$ds = \sqrt{\left(\frac{dx}{dt}\right)^2 + \left(\frac{dy}{dt}\right)^2} \, dt \tag{6}$$

Thus, the definite integral on the right side of (5) can be obtained from the line integral on the left side by expressing $f(x, y)$ in terms of t and using (6) to express ds in terms of dt. Because ds is so closely related to arc length, (5) is sometimes called the **integral of f over C with respect to arc length**. We leave it as an exercise to show that in the special case where C is expressed in terms of an arc-length parameter s (i.e., $t = s$), then (5) simplifies to

$$\int_C f(x, y) \, ds = \int_a^b f(x(s), y(s)) \, ds \tag{7}$$

Example 1 Find the area of the surface extending upward from the circle $x^2 + y^2 = 1$ to the parabolic cylinder $z = 1 - x^2$ (Figure 18.2.4).

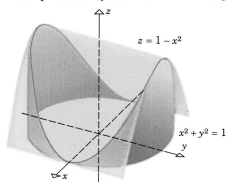

Figure 18.2.4

Solution. Denote the circle by C and represent it as

$$\mathbf{r}(t) = \cos t \, \mathbf{i} + \sin t \, \mathbf{j} \quad (0 \leq t \leq 2\pi)$$

From (3), the area A of the surface is

$$A = \int_C (1 - x^2) \, ds = \int_0^{2\pi} (1 - \cos^2 t) \sqrt{(-\sin t)^2 + (\cos t)^2} \, dt$$

$$= \int_0^{2\pi} \sin^2 t \, dt = \frac{1}{2} \int_0^{2\pi} (1 - \cos 2t) \, dt = \pi \qquad \blacktriangleleft$$

☐ **LINE INTEGRALS WITH RESPECT TO x AND y**

In addition to line integrals with respect to arc length, there are two other important types of line integrals; these result from replacing Δs_k in (2) by $\Delta x_k = x(t_k) - x(t_{k-1})$ and $\Delta y_k = y(t_k) - y(t_{k-1})$, respectively. We define

$$\int_C f(x, y) \, dx = \lim_{n \to +\infty} \sum_{k=1}^{n} f(x(t_k^*), y(t_k^*)) \, \Delta x_k$$

$$= \lim_{n \to +\infty} \sum_{k=1}^{n} f(x(t_k^*), y(t_k^*)) \frac{\Delta x_k}{\Delta t_k} \Delta t_k \tag{8}$$

$$\int_C f(x, y) \, dy = \lim_{n \to +\infty} \sum_{k=1}^{n} f(x(t_k^*), y(t_k^*)) \, \Delta y_k$$

$$= \lim_{n \to +\infty} \sum_{k=1}^{n} f(x(t_k^*), y(t_k^*)) \frac{\Delta y_k}{\Delta t_k} \Delta t_k \tag{9}$$

from which it follows that

$$\int_C f(x, y)\, dx = \int_a^b f(x(t), y(t)) x'(t)\, dt \tag{10}$$

$$\int_C f(x, y)\, dy = \int_a^b f(x(t), y(t)) y'(t)\, dt \tag{11}$$

We call (10) the **line integral of f over C with respect to x** and (11) the **line integral of f over C with respect to y**.

REMARK.   In words, the line integrals with respect to x and y can be written as definite integrals by expressing $f(x, y)$ in terms of t and expressing dx and dy in terms of t as $dx = x'(t)\, dt$ and $dy = y'(t)\, dt$. Note that a line integral with respect to x is zero over a vertical line segment [since $x'(t) = 0$], and a line integral with respect to y is zero over a horizontal line segment [since $y'(t) = 0$].

Frequently, the line integrals in (10) and (11) occur in combination, in which case we dispense with one of the integral signs and write

$$\int_C f(x, y)\, dx + g(x, y)\cdot dy = \int_C f(x, y)\, dx + \int_C g(x, y)\, dy \tag{12}$$

Example 2   Evaluate

$$\int_C 2xy\, dx + (x^2 + y^2)\, dy$$

over the circular arc C given by $x = \cos t$, $y = \sin t$ $(0 \le t \le \pi/2)$ (Figure 18.2.5).

Solution.   From (10) and (11)

$$\int_C 2xy\, dx = \int_0^{\pi/2} (2 \cos t \sin t)\left[\frac{d}{dt}(\cos t)\right] dt$$

$$= -2 \int_0^{\pi/2} \sin^2 t \cos t\, dt = -\frac{2}{3} \sin^3 t \Big]_0^{\pi/2} = -\frac{2}{3}$$

$$\int_C (x^2 + y^2)\, dy = \int_0^{\pi/2} (\cos^2 t + \sin^2 t)\left[\frac{d}{dt}(\sin t)\right] dt$$

$$= \int_0^{\pi/2} \cos t\, dt = \sin t \Big]_0^{\pi/2} = 1$$

Thus, from (12)

$$\int_C 2xy\, dx + (x^2 + y^2)\, dy = \int_C 2xy\, dx + \int_C (x^2 + y^2)\, dy$$

$$= -\frac{2}{3} + 1 = \frac{1}{3} \quad \blacktriangleleft$$

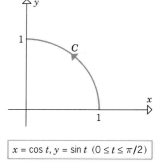

$x = \cos t, y = \sin t \ (0 \le t \le \pi/2)$

Figure 18.2.5

☐ **LINE INTEGRALS IN 3-SPACE**

The notion of a line integral can be extended to 3-space. If C is a curve in 3-space represented by a smooth vector-valued function

$$\mathbf{r}(t) = x(t)\mathbf{i} + y(t)\mathbf{j} + z(t)\mathbf{k} \quad (a \le t \le b)$$

and if $f(x, y, z)$, $g(x, y, z)$, and $h(x, y, z)$ are continuous functions of t on C, then

$$\int_C f(x, y, z)\, ds = \int_a^b f(x(t), y(t), z(t)) \sqrt{\left(\frac{dx}{dt}\right)^2 + \left(\frac{dy}{dt}\right)^2 + \left(\frac{dz}{dt}\right)^2}\, dt \tag{13}$$

$$\int_C f(x, y, z)\, dx = \int_a^b f(x(t), y(t), z(t))x'(t)\, dt \tag{14a}$$

$$\int_C g(x, y, z)\, dy = \int_a^b g(x(t), y(t), z(t))y'(t)\, dt \tag{14b}$$

$$\int_C h(x, y, z)\, dz = \int_a^b h(x(t), y(t), z(t))z'(t)\, dt \tag{14c}$$

$$\int_C f(x, y, z)\, dx + g(x, y, z)\, dy + h(x, y, z)\, dz$$
$$= \int_C f(x, y, z)\, dx + \int_C g(x, y, z)\, dy + \int_C h(x, y, z)\, dz \tag{15}$$

Example 3   Evaluate

$$\int_C (xy + z^3)\, ds$$

where C is the portion of the helix given by the parametric equations

$$x = \cos t, \quad y = \sin t, \quad z = t \quad (0 \le t \le \pi)$$

Solution.   From Formula (13) we can evaluate the given line integral as a definite integral by using the parametric equations to express x, y, and z in terms of t and replacing ds by

$$ds = \sqrt{\left(\frac{dx}{dt}\right)^2 + \left(\frac{dy}{dt}\right)^2 + \left(\frac{dz}{dt}\right)^2}\, dt = \sqrt{(-\sin t)^2 + (\cos t)^2 + 1}\, dt$$
$$= \sqrt{2}\, dt$$

which yields

$$\int_C (xy + z^3)\, ds = \int_0^\pi (\cos t \sin t + t^3) \sqrt{2}\, dt = \sqrt{2} \left[\frac{\sin^2 t}{2} + \frac{t^4}{4} \right]_0^\pi$$
$$= \frac{\sqrt{2}\, \pi^4}{4} \quad \blacktriangleleft$$

☐ **LINE INTEGRALS OVER PIECEWISE SMOOTH CURVES**

Thus far, we have considered only line integrals over smooth curves. However, the notion of a line integral can be extended to curves formed from finitely many smooth curves $C_1, C_2, \ldots, C_n$ joined end to end. Such a curve is called *piecewise smooth* (Figure 18.2.6). We define a line integral over a piecewise smooth curve C to be the sum of the integrals over the pieces:

$$\int_C = \int_{C_1} + \int_{C_2} + \cdots + \int_{C_n}$$

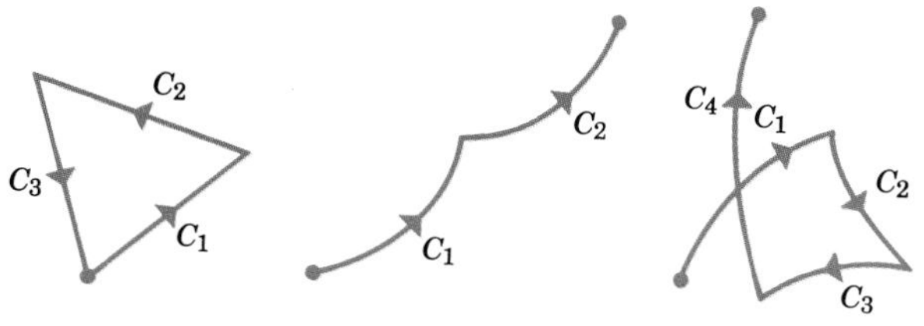

Figure 18.2.6

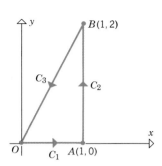

Figure 18.2.7

Example 4   Evaluate

$$\int_C x^2 y \, dx + x \, dy$$

in a counterclockwise direction around the triangular path shown in Figure 18.2.7.

Solution.   We shall integrate over C_1, C_2, and C_3 separately and add the results. For each of the three integrals we must find parametric equations that trace the path of integration in the correct direction. For this purpose recall from Example 4 of Section 15.1 that the graph of the vector-valued function

$$\mathbf{r}(t) = (1 - t)\mathbf{r}_0 + t\mathbf{r}_1 \quad (0 \le t \le 1)$$

is the line segment joining $\mathbf{r}_0$ and $\mathbf{r}_1$, oriented in the direction from $\mathbf{r}_0$ to $\mathbf{r}_1$. Thus, the line segments C_1, C_2, and C_3 can be represented in vector notation as

$$C_1: \mathbf{r}(t) = (1 - t)\langle 0, 0\rangle + t\langle 1, 0\rangle = \langle t, 0\rangle$$
$$C_2: \mathbf{r}(t) = (1 - t)\langle 1, 0\rangle + t\langle 1, 2\rangle = \langle 1, 2t\rangle$$
$$C_3: \mathbf{r}(t) = (1 - t)\langle 1, 2\rangle + t\langle 0, 0\rangle = \langle 1 - t, 2 - 2t\rangle$$

where t varies from 0 to 1 in each case. From these equations we obtain

$$\int_{C_1} x^2 y \, dx + x \, dy = \int_0^1 (t^2)(0)\frac{d}{dt}[t]\,dt + \int_0^1 (t)\frac{d}{dt}[0]\,dt = 0$$

$$\int_{C_2} x^2 y \, dx + x \, dy = \int_0^1 (1^2)(2t)\frac{d}{dt}[1]\,dt + \int_0^1 (1)\frac{d}{dt}[2t]\,dt = 0 + 2 = 2$$

$$\int_{C_3} x^2 y \, dx + x \, dy = \int_0^1 (1 - t)^2(2 - 2t)\frac{d}{dt}[1 - t]\,dt + \int_0^1 (1 - t)\frac{d}{dt}[2 - 2t]\,dt$$

$$= 2\int_0^1 (t - 1)^3\,dt + 2\int_0^1 (t - 1)\,dt = -\tfrac{1}{2} - 1 = -\tfrac{3}{2}$$

Thus,

$$\int_C x^2 y \, dx + x \, dy = 0 + 2 + \left(-\tfrac{3}{2}\right) = \tfrac{1}{2} \quad \blacktriangleleft$$

☐ **CHANGE OF PARAMETER
IN LINE INTEGRALS**

Since the parametric equations of a curve are used to evaluate line integrals over that curve, it seems possible that two different parametrizations of a curve C might produce different values for the same line integral over C. The following theorems deal with this question.

18.2.2   THEOREM (*Independence of Parametrization*).   *If C is a smooth parametric curve, then the value of any line integral over C is unchanged by a smooth change of parameter that preserves the orientation of C.*

18.2.3   THEOREM (*Reversal of Orientation*).   *If C is a smooth parametric curve, then a smooth change of parameter that reverses the orientation of C changes the sign of a line integral over C with respect to x, y, or z, but leaves the value of a line integral over C with respect to arc length unchanged.*

REMARK.   We will not prove these results, but some insight into the second theorem can be obtained by noting, for example, that reversing the orientation of C changes the sign of Δx_k in (8) and of Δy_k in (9) and hence changes the sign of the integrals; but reversing the orientation has no effect on Δs_k in (2), since Δs_k is an arc length and hence is positive, regardless of the orientation.

If a curve C is traced in a certain direction, then the same curve traced in the opposite direction is often denoted by the symbol $-C$. Thus, it follows from Theorem 18.2.3 that

$$\int_{-C} f(x, y)\, dx + g(x, y)\, dy = -\int_{C} f(x, y)\, dx + g(x, y)\, dy \tag{16}$$

$$\int_{-C} f(x, y)\, ds = \int_{C} f(x, y)\, ds \tag{17}$$

and similarly for line integrals in 3-space.

☐ **ARC LENGTH AS A LINE INTEGRAL**

If C is the graph in 3-space of a smooth vector-valued function $\mathbf{r}(t) = x(t)\mathbf{i} + y(t)\mathbf{j} + z(t)\mathbf{k}$, where $a \leq t \leq b$, then it follows from (13) that

$$\int_{C} ds = \int_{a}^{b} \sqrt{\left(\frac{dx}{dt}\right)^2 + \left(\frac{dy}{dt}\right)^2 + \left(\frac{dz}{dt}\right)^2}\, dt$$

which by Formula (10) of Section 15.3 is the arc length L of the curve C. Thus, the arc length L of a smooth parametric curve C can be expressed as

$$L = \int_{C} ds \tag{18}$$

The same result holds in 2-space (verify).

☐ **MASS OF A WIRE AS A LINE INTEGRAL**

We consider an idealized bent wire in 3-space that is sufficiently thin that it can be represented as a curve C. If the composition of the wire is uniform throughout, then the wire is called **homogeneous** and its **linear mass density** is defined to be the total mass divided by the total length. However, if the wire is not homogeneous, then the composition may vary from point to point, and the density is specified by a **density function** $\delta(x, y, z)$, whose value at a point (x, y, z) is viewed as a limit

$$\delta(x, y, z) = \lim_{\Delta s \to 0} \frac{\Delta M}{\Delta s} \tag{19}$$

in which ΔM and Δs represent the mass and length of a small section of wire centered at (x, y, z). Thus, if the wire is subdivided into n small pieces, and if (x_k^*, y_k^*, z_k^*) is the center of the kth piece and Δs_k is the length of the kth piece, then from (19) the mass ΔM_k of that piece can be approximated as

$$\Delta M_k \approx \delta(x_k^*, y_k^*, z_k^*)\, \Delta s_k$$

and hence the mass M of the entire wire can be approximated as

$$M = \sum_{k=1}^{n} \Delta M_k \approx \sum_{k=1}^{n} \delta(x_k^*, y_k^*, z_k^*)\, \Delta s_k \tag{20}$$

If we now increase n in such a way that the lengths of the pieces approach zero, then it is plausible that the error in (20) will approach zero, and the exact value of M will be given by the line integral

$$M = \int_{C} \delta(x, y, z)\, ds \tag{21}$$

The same result holds for wires in 2-space, but with two variables rather than three.

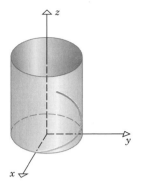

Figure 18.2.8

Example 5   Find the mass of a thin wire shaped in the form of a helix with the parametric equations

$$x = \cos t, \quad y = \sin t, \quad z = 2t \quad (0 \le t \le \pi)$$

if the density function is $\delta(x, y, z) = kz \ (k > 0)$ (Figure 18.2.8).

Solution.   From (21) with C denoting the helix, the mass M is

$$M = \int_C kz \, ds = k \int_0^\pi 2t \sqrt{(-\sin t)^2 + (\cos t)^2 + 2^2} \, dt$$

$$= 2\sqrt{5} \, k \int_0^\pi t \, dt = \sqrt{5} \, k\pi^2 \quad \blacktriangleleft$$

☐ **WORK AS A LINE INTEGRAL**

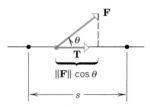

Figure 18.2.9

In everyday language we use the word *work* to denote the exertion of muscular or intellectual effort. However, in physics and engineering the term has a more definite and technical meaning—work is done whenever an object undergoes a displacement while subjected to a force that has a component in the direction of motion. More precisely:

If a constant force $\mathbf{F}$ acts on a particle that moves a distance s along a straight line, and if the force vector $\mathbf{F}$ makes an angle θ with a unit vector $\mathbf{T}$ in the direction of motion, then the work W done by $\mathbf{F}$ on the particle is defined to be

$$W = (\|\mathbf{F}\| \cos \theta)s = (\mathbf{F} \cdot \mathbf{T})s \tag{22}$$

that is, the work W is the scalar component of force in the direction of motion times the distance traveled by the particle (Figure 18.2.9).

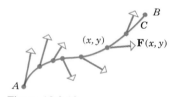

Figure 18.2.10

Although this notion of work is adequate for certain simple applications, it does not apply to problems in which the particle is subjected to a variable force as it moves along a curved path. To deal with such problems we must extend the notion of work to allow for a particle that moves from a point A to a point B along a curve C in 2-space or 3-space, while subjected to a force $\mathbf{F}$ that varies from point to point along the curve (Figure 18.2.10). Focusing on 3-space for the moment, suppose that a particle with position vector $\mathbf{r}(t)$ moves from A to B along a curve C as t increases from a to b. At each point (x, y, z) on C we denote the unit tangent vector to C by $\mathbf{T}$ [or $\mathbf{T}(x, y, z)$] and the force on the particle by $\mathbf{F}$ [or $\mathbf{F}(x, y, z)$] (Figure 18.2.11). We shall assume that $\mathbf{T}$ and $\mathbf{F}$ are continuous on C. To motivate an appropriate definition of work in this case we shall use a limiting process:

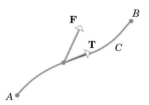

Figure 18.2.11

- As shown in Figure 18.2.12, subdivide C into n arcs by successively inserting distinct points $P_1, P_2, P_3, \ldots, P_{n-1}$ between A and B in the direction of increasing t, and denote the length of the kth arc by Δs_k. Let (x_k^*, y_k^*, z_k^*) be any point on this arc, and let $\mathbf{F}_k^* = \mathbf{F}(x_k^*, y_k^*, z_k^*)$ and $\mathbf{T}_k^* = \mathbf{T}(x_k^*, y_k^*, z_k^*)$ be the force vector and the unit tangent vector at this point.

- If P_{k-1} and P_k are close together, then the kth arc is nearly straight, and the vectors $\mathbf{F}$ and $\mathbf{T}$ cannot vary much from $\mathbf{F}_k^*$ and $\mathbf{T}_k^*$ over the small arc. Thus from (22), the work W_k performed as the particle moves along C from P_{k-1} to P_k can be approximated by

$$W_k \approx (\mathbf{F}_k^* \cdot \mathbf{T}_k^*) \Delta s_k$$

and the total work W done as the particle moves along C from A to B can be approximated by

$$W \approx \sum_{k=1}^{n} (\mathbf{F}_k^* \cdot \mathbf{T}_k^*) \Delta s_k$$

- There are two sources of error in this approximation, the error in assuming the kth arc to be straight and the error in assuming the force and unit tangent vectors to be constant over the kth arc. However, it seems plausible that if we increase n in such a way that the

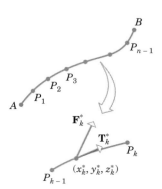

Figure 18.2.12

lengths of the arcs approach zero, then the errors will diminish and the exact value of the work will be given by

$$W = \lim_{n \to +\infty} \sum_{k=1}^{n} (\mathbf{F}_k^* \cdot \mathbf{T}_k^*) \, \Delta s_k$$

- If we now assume that the point (x_k^*, y_k^*, z_k^*) corresponds to the parameter value t_k^*, then

$$\mathbf{F}_k^* = \mathbf{F}(x(t_k^*), y(t_k^*), z(t_k^*)) \quad \text{and} \quad \mathbf{T}_k^* = \mathbf{T}(x(t_k^*), y(t_k^*), z(t_k^*))$$

so from (2), W can be expressed as the line integral

$$W = \int_C \mathbf{F}(x, y, z) \cdot \mathbf{T}(x, y, z) \, ds$$

or more briefly,

$$W = \int_C \mathbf{F} \cdot \mathbf{T} \, ds \tag{23}$$

We take this to be the definition of the **work done by F along C**. In words, *the work done by the force* **F** *along the curve* C *is obtained by integrating the tangential component of force along the curve.* The definition in 2-space is the same, except for the number of variables.

☐ **A METHOD FOR CALCULATING WORK**

For calculations it is helpful to write (23) in a different form: Recall from Formula (2) of Section 15.4 that the unit tangent vector **T** can be expressed as

$$\mathbf{T} = \frac{d\mathbf{r}}{ds}$$

which suggests that (23) can be expressed as

$$W = \int_C \mathbf{F} \cdot d\mathbf{r} \tag{24}$$

in which $d\mathbf{r}$ is interpreted as

$$d\mathbf{r} = dx\,\mathbf{i} + dy\,\mathbf{j} \quad \text{or} \quad d\mathbf{r} = dx\,\mathbf{i} + dy\,\mathbf{j} + dz\,\mathbf{k} \tag{25}$$

depending on whether C is in 2-space or 3-space. Alternatively, if

$$\mathbf{F} = f(x, y)\mathbf{i} + g(x, y)\mathbf{j} \quad \text{or} \quad \mathbf{F} = f(x, y, z)\mathbf{i} + g(x, y, z)\mathbf{j} + h(x, y, z)\mathbf{k}$$

then (24) can be expressed in scalar form as

$$W = \int_C f(x, y) \, dx + g(x, y) \, dy \tag{26}$$

$$W = \int_C f(x, y, z) \, dx + g(x, y, z) \, dy + h(x, y, z) \, dz \tag{27}$$

Example 6 Find the work done by the force field

$$\mathbf{F}(x, y) = x^3 y\mathbf{i} + (x - y)\mathbf{j}$$

on a particle that moves along the parabola $y = x^2$ from $(-2, 4)$ to $(1, 1)$.

Solution. If we use $x = t$ as the parameter, the path C of the particle is represented by

$$x = t, \quad y = t^2 \quad (-2 \le t \le 1)$$

or in vector notation as

$$\mathbf{r}(t) = t\mathbf{i} + t^2\mathbf{j} \quad (-2 \le t \le 1)$$

Thus, from (24) the work W done by $\mathbf{F}$ is

$$W = \int_C \mathbf{F} \cdot d\mathbf{r} = \int_C (x^3 y\mathbf{i} + (x - y)\mathbf{j}) \cdot (dx\mathbf{i} + dy\mathbf{j})$$

$$= \int_C x^3 y \, dx + (x - y) \, dy = \int_{-2}^{1} (t^5 + (t - t^2)(2t)) \, dt$$

$$= \frac{1}{6}t^6 + \frac{2}{3}t^3 - \frac{1}{2}t^4 \Big]_{-2}^{1} = 3$$

where the units for W depend on the units chosen for force and distance. ◀

REMARK.   Reversing the orientation of C in (23) reverses the direction of the vector $\mathbf{T}$ in the integrand, and hence reverses the sign of W. Thus, for the force field in the preceding example, the work performed on a particle moving from $(1, 1)$ to $(-2, 4)$ along the parabola is -3.

Example 7   Find the work done by the force field $\mathbf{F}(x, y, z) = yz\mathbf{i} + xz\mathbf{j} + xy\mathbf{k}$ on a particle that moves along the twisted cubic $\mathbf{r}(t) = t\mathbf{i} + t^2\mathbf{j} + t^3\mathbf{k}$ $(0 \le t \le 1)$.

Solution.   From (24)

$$W = \int_C \mathbf{F} \cdot d\mathbf{r} = \int_C yz \, dx + xz \, dy + xy \, dz$$

$$= \int_0^1 (t^2 t^3 + (t t^3)(2t) + (t t^2)(3t^2)) \, dt$$

$$= \int_0^1 (t^5 + 2t^5 + 3t^5) \, dt = \int_0^1 6t^5 \, dt = 1 \quad ◀$$

▶ Exercise Set 18.2

---

1. Find the area of the surface extending upward from the parabola $y = x^2$ $(0 \le x \le 2)$ to the plane $z = 3x$.

2. Find the area of the surface extending upward from the semicircle $y = \sqrt{4 - x^2}$ to the surface $z = x^2 y$.

In Exercises 3–6, evaluate the line integral.

3. $\int_C \dfrac{1}{1 + x} ds$, where C is the curve $x = t$, $y = \frac{2}{3}t^{3/2}$, $0 \le t \le 3$.

4. $\int_C \dfrac{x}{1 + y^2} ds$, where C is the line $x = 1 + 2t$, $y = t$, $0 \le t \le 1$.

5. $\int_C 3x^2 yz \, ds$, where C is the curve $x = t$, $y = t^2$, $z = \frac{2}{3}t^3$, $0 \le t \le 1$.

6. $\int_C \dfrac{e^{-z}}{x^2 + y^2} ds$, where C is the helix $x = 2\cos t$, $y = 2\sin t$, $z = t$, $0 \le t \le 2\pi$.

7. Let C be the curve $x = t$, $y = t^2$, $0 \le t \le 1$. Find

(a) $\int_C (2x + y) \, dx$ (b) $\int_C (x^2 - y) \, dy$

(c) $\int_C (2x + y) \, dx + (x^2 - y) \, dy$.

8. Find $\int_C (3x + 2y) \, dx + (2x - y) \, dy$, where C is

(a) the line segment from $(0, 0)$ to $(1, 1)$

(b) the parabolic arc $y = x^2$ from $(0, 0)$ to $(1, 1)$

(c) the curve $y = \sin (\pi x/2)$ from $(0, 0)$ to $(1, 1)$

(d) the curve $x = y^3$ from $(0, 0)$ to $(1, 1)$.

In Exercises 9–16, evaluate the line integral.

9. $\int_C y \, dx - x^2 \, dy$, where C is the curve $x = t$, $y = \frac{1}{2}t^2$, $0 \le t \le 2$.

10. $\int_C e^{y-x} \, dx + xy \, dy$, where C is the curve $x = 2 + t$, $y = 3 - t$, $0 \le t \le 1$.

11. $\int_C (x + 2y) \, dx + (x - y) \, dy$, where C is the curve $x = 2 \cos t$, $y = 4 \sin t$, $0 \le t \le \pi/4$.

12. $\int_C (x^2 - y^2) \, dx + x \, dy$, where C is the curve $x = t^{2/3}$, $y = t$, $-1 \le t \le 1$.

13. $\int_C -y \, dx + x \, dy$ along $y^2 = 3x$ from the point $(3, 3)$ to the point $(0, 0)$.

14. $\int_C (y - x) \, dx + x^2 y \, dy$ along $y^2 = x^3$ from the point $(1, -1)$ to the point $(1, 1)$.

15. $\int_C (x^2 + y^2) \, dx - x \, dy$ along the quarter-circle $x^2 + y^2 = 1$ from $(1, 0)$ to $(0, 1)$.

16. $\int_C (y - x) \, dx + xy \, dy$ along the line segment from $(3, 4)$ to $(2, 1)$.

17. Find $\int_C \mathbf{F} \cdot d\mathbf{r}$, where $\mathbf{F}(x, y) = x^2 \mathbf{i} + xy \mathbf{j}$ and C is the semicircle given by $\mathbf{r}(t) = 2 \cos t \, \mathbf{i} + 2 \sin t \, \mathbf{j}$, $0 \le t \le \pi$.

18. Find $\int_C \mathbf{F} \cdot d\mathbf{r}$, where $\mathbf{F}(x, y) = x^2 y \mathbf{i} + 4 \mathbf{j}$ and C is the curve $\mathbf{r}(t) = e^t \mathbf{i} + e^{-t} \mathbf{j}$, $0 \le t \le 1$.

19. Find $\int_C \mathbf{F} \cdot d\mathbf{r}$, where $\mathbf{F}(x, y) = (x^2 + y^2)^{-3/2}(x\mathbf{i} + y\mathbf{j})$ and C is the curve $\mathbf{r}(t) = e^t \sin t \, \mathbf{i} + e^t \cos t \, \mathbf{j}$, $0 \le t \le 1$.

20. Evaluate $\int_C y \, dx - x \, dy$ over each of the curves:

(a)
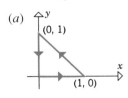
$(0, 1)$
$(1, 0)$

(b)
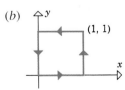
$(1, 1)$

(c)
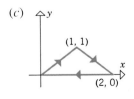
$(1, 1)$
$(2, 0)$

(d)
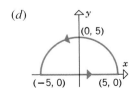
$(0, 5)$
$(-5, 0)$ $(5, 0)$

In Exercises 21–24, evaluate the line integral.

21. $\int_C yz \, dx - xz \, dy + xy \, dz$, where C is the curve $x = e^t$, $y = e^{3t}$, $z = e^{-t}$, $0 \le t \le 1$.

22. $\int_C y \, dx + z \, dy - x \, dz$, where C is the line segment from $(1, 1, 1)$ to $(-3, 2, 0)$.

23. $\int_C \mathbf{F} \cdot d\mathbf{r}$, where $\mathbf{F}(x, y, z) = z\mathbf{i} + x\mathbf{j} + y\mathbf{k}$ and C is the curve $\mathbf{r}(t) = \sin t \, \mathbf{i} + 3 \sin t \, \mathbf{j} + \sin^2 t \, \mathbf{k}$, $0 \le t \le \pi/2$.

24. $\int_C x^2 z \, dx - yx^2 \, dy + 3xz \, dz$, where C is the triangular path from $(0, 0, 0)$ to $(1, 1, 0)$ to $(1, 1, 1)$ to $(0, 0, 0)$.

25. Find the mass of a thin wire shaped in the form of the circular arc $y = \sqrt{9 - x^2}$ $(0 \le x \le 3)$ if the density function is $\delta(x, y) = kx\sqrt{y}$ $(k > 0)$.

26. Find the mass of a thin wire shaped in the form of the curve $x = e^t \cos t$, $y = e^t \sin t$ $(0 \le t \le 1)$ if the density function δ is proportional to the distance from the origin.

27. Find the mass of a thin wire shaped in the form of the helix $x = 3 \cos t$, $y = 3 \sin t$, $z = 4t$ $(0 \le t \le \pi/2)$ if the density function is $\delta = kx/(1 + y^2)$ $(k > 0)$.

28. Find the mass of a thin wire shaped in the form of the curve $x = 2t$, $y = \ln t$, $z = 4\sqrt{t}$ $(1 \le t \le 4)$ if the density function is proportional to the distance above the xy-plane.

29. Find the work done if a particle moves along the parabolic arc $x = y^2$ from $(0, 0)$ to $(1, 1)$ while subject to the force $\mathbf{F}(x, y) = xy\mathbf{i} + x^2\mathbf{j}$.

30. Find the work done if a particle moves along the curve $x = t$, $y = 1/t$, $1 \le t \le 3$ while subject to the force $\mathbf{F}(x, y) = (x^2 + xy)\mathbf{i} + (y - x^2 y)\mathbf{j}$.

31. Find the work done by the force
$$\mathbf{F}(x, y) = \frac{1}{x^2 + y^2}\mathbf{i} + \frac{4}{x^2 + y^2}\mathbf{j}$$
acting on a particle that moves along the curve C shown.

(a)
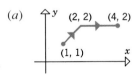
$(2, 2)$ $(4, 2)$
$(1, 1)$

(b)
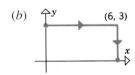
$(6, 3)$

(c)
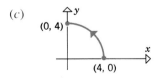
$(0, 4)$
$(4, 0)$

32. Find the work that is done by the force
$$\mathbf{F}(x, y, z) = xy\mathbf{i} + yz\mathbf{j} + xz\mathbf{k}$$
acting on a particle that moves along the curve given by $\mathbf{r}(t) = t\mathbf{i} + t^2\mathbf{j} + t^3\mathbf{k}$, $0 \le t \le 1$.

33. Find the work that is done by a force
$$\mathbf{F}(x, y, z) = (x + y)\mathbf{i} + xy\mathbf{j} - z^2\mathbf{k}$$
acting on a particle that moves along the line segment

from $(0, 0, 0)$ to $(1, 3, 1)$ and then along the line segment from $(1, 3, 1)$ to $(2, -1, 4)$.

34. Evaluate the integral $\displaystyle\int_{-C} \frac{x\,dy - y\,dx}{x^2 + y^2}$, where C is the circle $x^2 + y^2 = a^2$ traversed counterclockwise.

35. A particle moves from the point $(0, 0)$ to the point $(1, 0)$ along the curve $x = t$, $y = \lambda t(1 - t)$ while subject to the force $\mathbf{F}(x, y) = xy\mathbf{i} + (x - y)\mathbf{j}$. For what value of λ is the work done equal to 1?

36. A farmer weighing 150 lb carries a sack of grain weighing 20 lb up a circular helical staircase around a silo of radius 25 ft. As the farmer climbs, grain leaks from the sack at a rate of 1 lb per 10 ft of ascent. How much work is performed by the farmer in climbing through a vertical distance of 60 ft in exactly four revolutions?

■ **18.3** INDEPENDENCE OF PATH; CONSERVATIVE VECTOR FIELDS

> *In general, the value of a line integral $\int_C \mathbf{F} \cdot d\mathbf{r}$ depends on the curve C. However, we shall show in this section that when the integrand satisfies appropriate conditions, the value of the integral depends only on the location of the endpoints of the curve C and not on the shape of the curve, in which case the evaluation of the line integral is greatly simplified. Vector fields in which line integrals depend only on the endpoints of the curve of integration are of special importance in physics and engineering, and we shall discuss such vector fields in this section.*

☐ **INDEPENDENCE OF PATH**

The curve C in a line integral is often called the **path of integration**. The following example illustrates how different paths of integration from one point to another can result in the same value for a line integral.

Example 1   Let $\mathbf{F}(x, y) = y\mathbf{i} + x\mathbf{j}$. Evaluate the line integral

$$\int_C \mathbf{F} \cdot d\mathbf{r}$$

over the following curves (Figure 18.3.1):

(a)   The line segment $y = x$ from $(0, 0)$ to $(1, 1)$.
(b)   The parabola $y = x^2$ from $(0, 0)$ to $(1, 1)$.
(c)   The cubic $y = x^3$ from $(0, 0)$ to $(1, 1)$.

Solution (a).   With $x = t$ as the parameter, the path of integration is given by

$$x = t, \quad y = t \qquad (0 \le t \le 1)$$

Thus,

$$\int_C \mathbf{F} \cdot d\mathbf{r} = \int_C (y\mathbf{i} + x\mathbf{j}) \cdot (dx\mathbf{i} + dy\mathbf{j}) = \int_C y\,dx + x\,dy$$

$$= \int_0^1 2t\,dt = 1$$

Solution (b).   With $x = t$ as the parameter, the path of integration is given by

$$x = t, \quad y = t^2 \qquad (0 \le t \le 1)$$

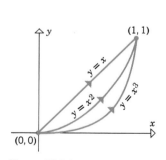

Figure 18.3.1

Thus,

$$\int_C \mathbf{F} \cdot d\mathbf{r} = \int_C y\,dx + x\,dy = \int_0^1 3t^2\,dt = 1$$

Solution (c). With $x = t$ as the parameter, the path of integration is given by

$$x = t, \quad y = t^3 \quad (0 \le t \le 1)$$

Thus,

$$\int_C \mathbf{F} \cdot d\mathbf{r} = \int_C y\,dx + x\,dy = \int_0^1 4t^3\,dt = 1 \quad \blacktriangleleft$$

The results in this example are not accidental; we shall soon see that the value of this line integral is the same over *all* piecewise smooth paths from $(0, 0)$ to $(1, 1)$. A line integral whose value is the same over all piecewise smooth paths from one endpoint to the other is said to be ***independent of path***.

Recall from the First Fundamental Theorem of Calculus (Theorem 5.7.1) that if F is an antiderivative of f, then

$$\int_a^b f(x)\,dx = F(b) - F(a)$$

The following result is the analog of that theorem for line integrals.

18.3.1   THEOREM (***The Fundamental Theorem of Line Integrals***).   *Suppose that* $\mathbf{F}(x, y) = f(x, y)\mathbf{i} + g(x, y)\mathbf{j}$, *where f and g are continuous in some open region containing the points (x_0, y_0) and (x_1, y_1). If*

$$\mathbf{F}(x, y) = \nabla\phi(x, y)$$

at each point of this region, then for any piecewise smooth curve C that starts at (x_0, y_0), ends at (x_1, y_1), and lies entirely in the region, we have

$$\int_C \mathbf{F}(x, y) \cdot d\mathbf{r} = \phi(x_1, y_1) - \phi(x_0, y_0) \tag{1}$$

We shall give the proof for a smooth curve C. The proof for piecewise smooth curves is obtained by considering each piece separately. The details are left for the exercises.

Proof. If C is given parametrically by $x = x(t)$, $y = y(t)$ $(a \le t \le b)$, then the initial and final points of the curve C are

$$(x_0, y_0) = (x(a), y(a)) \quad \text{and} \quad (x_1, y_1) = (x(b), y(b))$$

Since $\mathbf{F}(x, y) = \nabla\phi$, it follows that

$$\mathbf{F}(x, y) = \frac{\partial\phi}{\partial x}\mathbf{i} + \frac{\partial\phi}{\partial y}\mathbf{j}$$

so

$$\int_C \mathbf{F}(x, y) \cdot d\mathbf{r} = \int_C \frac{\partial\phi}{\partial x}\,dx + \frac{\partial\phi}{\partial y}\,dy = \int_a^b \left[\frac{\partial\phi}{\partial x}\frac{dx}{dt} + \frac{\partial\phi}{\partial y}\frac{dy}{dt} \right] dt$$

$$= \int_a^b \frac{d}{dt}[\phi(x(t), y(t))]\,dt = \phi[x(t), y(t)] \Big]_{t=a}^{b}$$

$$= \phi(x(b), y(b)) - \phi(x(a), y(a))$$

$$= \phi(x_1, y_1) - \phi(x_0, y_0) \quad \blacksquare$$

For line integrals that are independent of path, it is common to write (1) as

$$\int_{(x_0, y_0)}^{(x_1, y_1)} \mathbf{F} \cdot d\mathbf{r} = \phi(x_1, y_1) - \phi(x_0, y_0) \tag{2}$$

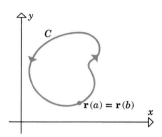

Figure 18.3.2

Example 2   The Fundamental Theorem of Line Integrals explains the results in Example 1, since $\mathbf{F}(x, y) = y\mathbf{i} + x\mathbf{j}$ is the gradient of the function $\phi(x, y) = xy$ (verify). Thus, for any piecewise smooth curve C from $(0, 0)$ to $(1, 1)$, it follows from (2) that

$$\int_C \mathbf{F} \cdot d\mathbf{r} = \int_{(0,0)}^{(1,1)} \mathbf{F} \cdot d\mathbf{r} = \phi(1, 1) - \phi(0, 0) = 1 - 0 = 1 \quad \blacktriangleleft$$

A curve C represented by a vector-valued function $\mathbf{r}(t)$ ($a \le t \le b$) is said to be **closed** if the initial point $\mathbf{r}(a)$ and the terminal point $\mathbf{r}(b)$ coincide (Figure 18.3.2). If C is a piecewise smooth closed curve in 2-space, and if $\mathbf{F}$ satisfies the conditions of Theorem 18.3.1, then the points (x_0, y_0) and (x_1, y_1) in the theorem are the same and hence

$$\int_C \mathbf{F} \cdot d\mathbf{r} = \phi(x_1, y_1) - \phi(x_0, y_0) = 0 \tag{3}$$

☐ **LINE INTEGRALS OF CONSERVATIVE VECTOR FIELDS**

Recall from Definition 18.1.3 that $\mathbf{F}(x, y)$ is called a *conservative vector field* if $\mathbf{F}$ is the gradient of some potential function ϕ, that is, $\mathbf{F}(x, y) = \nabla\phi(x, y)$. In this case the Fundamental Theorem of Line Integrals implies that the line integral $\int_C \mathbf{F} \cdot d\mathbf{r}$ is independent of the path and has a value of zero if the path is closed. We want to show that the converse results are also true; that is, a vector field in which all line integrals are independent of path or in which all line integrals around closed paths are zero must be conservative. However, to do so we will need to assume that the domain D of $\mathbf{F}(x, y)$ is **connected**; that is, any two points in D can be connected by a piecewise smooth curve that lies entirely in D (Figure 18.3.3).

Connected

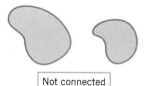

Not connected

Figure 18.3.3

18.3.2   THEOREM.   *If* $\mathbf{F}(x, y) = f(x, y)\mathbf{i} + g(x, y)\mathbf{j}$ *is continuous on an open connected region, then the following are equivalent (i.e., all are true or all are false):*

(*a*) $\mathbf{F}(x, y)$ *is a conservative vector field in the region.*

(*b*) $\displaystyle\int_C \mathbf{F} \cdot d\mathbf{r} = 0$ *for every piecewise smooth closed curve C in the region.*

(*c*) $\displaystyle\int_C \mathbf{F} \cdot d\mathbf{r}$ *is independent of path for every piecewise smooth curve C in the region.*

Proof.   To prove a theorem of this form, it suffices to show that (*a*) implies (*b*), (*b*) implies (*c*), and (*c*) implies (*a*). We have already seen that (*a*) implies (*b*) [see (3)], and we shall leave it as an exercise to show that (*b*) implies (*c*). Thus, it remains to show that if $\int_C \mathbf{F} \cdot d\mathbf{r}$ is independent of path for every piecewise smooth curve C in the region, then $\mathbf{F}(x, y)$ is a conservative vector field in the region. To prove this, assume that

$$\mathbf{F} = f(x, y)\mathbf{i} + g(x, y)\mathbf{j}$$

We must find a function ϕ such that $\nabla\phi = \mathbf{F}$, that is,

$$\frac{\partial\phi}{\partial x} = f(x, y) \quad \text{and} \quad \frac{\partial\phi}{\partial y} = g(x, y)$$

For this purpose, let D denote the domain of $\mathbf{F}$, and let (a, b) be any fixed point in D. For any point (x, y) in D define

$$\phi(x, y) = \int_{(a, b)}^{(x, y)} \mathbf{F} \cdot d\mathbf{r} \tag{4}$$

(which is possible because the integral is assumed to be independent of path in D). We will show that $\nabla \phi = \mathbf{F}$. Since D is open, we can find a circular disk centered at (x, y) whose points lie entirely in D. As shown in Figure 18.3.4, choose any point (x_1, y) in this disk that lies on the same horizontal line as (x, y) but which is different from (x, y). Because the integral in (4) is independent of path, we can evaluate it by first integrating from (a, b) to (x_1, y) along an arbitrary piecewise smooth curve C_1 in D, and then continuing along the horizontal line segment C_2 from (x_1, y) to (x, y). This yields

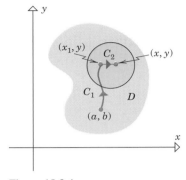

Figure 18.3.4

$$\phi(x, y) = \int_{C_1} \mathbf{F} \cdot d\mathbf{r} + \int_{C_2} \mathbf{F} \cdot d\mathbf{r} = \int_{(a, b)}^{(x_1, y)} \mathbf{F} \cdot d\mathbf{r} + \int_{C_2} \mathbf{F} \cdot d\mathbf{r}$$

Since the first term does not depend on x, its partial derivative with respect to x is zero and hence

$$\frac{\partial \phi}{\partial x} = \frac{\partial}{\partial x} \int_{C_2} \mathbf{F} \cdot d\mathbf{r} = \frac{\partial}{\partial x} \int_{C_2} f(x, y) \, dx + g(x, y) \, dy$$

However, the line integral with respect to y is zero along the horizontal line segment C_2, so this equation simplifies to

$$\frac{\partial \phi}{\partial x} = \frac{\partial}{\partial x} \int_{C_2} f(x, y) \, dx \tag{5}$$

But the line segment C_2 from (x_1, y) to (x, y) can be expressed parametrically as

$$x = t, \quad y = y$$

where t varies from x_1 to x. Thus, (5) can be written as

$$\frac{\partial \phi}{\partial x} = \frac{\partial}{\partial x} \int_{x_1}^{x} f(t, y) \, dt = f(x, y)$$

where the last equality was obtained from the Second Fundamental Theorem of Calculus (Theorem 5.9.1) by treating y as a constant and the integrand as a function of t alone. The proof that $\partial \phi / \partial y = g(x, y)$ can be obtained in a similar manner by joining (x, y) to a point (x, y_1) with a vertical line segment (Exercise 26). ∎

□ **A TEST FOR CONSERVATIVE VECTOR FIELDS**

Although Theorem 18.3.2 is an important characterization of conservative vector fields, it is not an effective computational tool for determining if a vector field is conservative. To find a useful way of doing this we will need some terminology. A plane curve $\mathbf{r}(t)$, where $a \leq t \leq b$, is said to be *simple* if it does not intersect itself anywhere between its endpoints. As shown in Figure 18.3.5, a simple curve may or may not be closed.

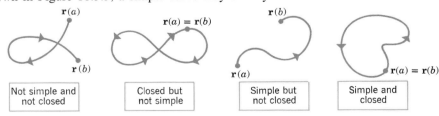

Figure 18.3.5

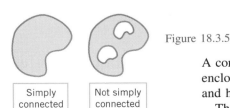

Figure 18.3.6

A connected set D in 2-space is called *simply connected* if no simple closed curve in D encloses points not in D. Informally stated, a simply connected set in 2-space is connected and has no holes (Figure 18.3.6).

   The following theorem is the primary tool for determining whether a given vector field in 2-space is conservative.

> **18.3.3   THEOREM** (*The Conservative Field Test*).   *Let* $\mathbf{F}(x, y) = f(x, y)\mathbf{i} + g(x, y)\mathbf{j}$, *where* f *and* g *have continuous first partial derivatives in an open simply connected region. Then* $\mathbf{F}$ *is a conservative vector field on that region if and only if*
>
> $$\frac{\partial f}{\partial y} = \frac{\partial g}{\partial x} \tag{6}$$
>
> *at each point of the region.*

The complete proof of this theorem requires results from advanced calculus and will be omitted. However, it is not hard to see why Formula (6) holds when $\mathbf{F}$ is conservative, that is, when $\mathbf{F}$ is the gradient of a potential function ϕ. In this case we have

$$\frac{\partial \phi}{\partial x} = f \quad \text{and} \quad \frac{\partial \phi}{\partial y} = g \tag{7}$$

so that

$$\frac{\partial f}{\partial y} = \frac{\partial}{\partial y}\left(\frac{\partial \phi}{\partial x}\right) = \frac{\partial^2 \phi}{\partial y\, \partial x} \quad \text{and} \quad \frac{\partial g}{\partial x} = \frac{\partial}{\partial x}\left(\frac{\partial \phi}{\partial y}\right) = \frac{\partial^2 \phi}{\partial x\, \partial y}$$

But the mixed partial derivatives in these equations are equal (Theorem 16.4.6), so (6) follows.

WARNING.   In (6), the **i**-component of $\mathbf{F}$ is differentiated with respect to y and the **j**-component with respect to x. It is easy to get this backwards by mistake.

Once it is established that a vector field is conservative, a potential function for the field can be obtained by first integrating either of the equations in (7). This is illustrated in the following example.

Example 3   Let $\mathbf{F}(x, y) = 2xy^3\mathbf{i} + (1 + 3x^2y^2)\mathbf{j}$.

(a)   Show that $\mathbf{F}$ is a conservative vector field on the entire xy-plane.
(b)   Find ϕ by first integrating $\partial\phi/\partial x$.
(c)   Find ϕ by first integrating $\partial\phi/\partial y$.

Solution (*a*).   Since $f(x, y) = 2xy^3$ and $g(x, y) = 1 + 3x^2y^2$, we have

$$\frac{\partial f}{\partial y} = 6xy^2 = \frac{\partial g}{\partial x}$$

so (6) holds for all (x, y).

Solution (*b*).   Since the field $\mathbf{F}$ is conservative, there is a potential function ϕ such that

$$\frac{\partial \phi}{\partial x} = 2xy^3 \quad \text{and} \quad \frac{\partial \phi}{\partial y} = 1 + 3x^2y^2 \tag{8}$$

Integrating the first of these equations with respect to x (and treating y as a constant) yields

$$\phi = \int 2xy^3\, dx = x^2y^3 + k(y) \tag{9}$$

where $k(y)$ represents the "constant" of integration. We are justified in treating the constant of integration as a function of y, since y is held constant in the integration process. To find $k(y)$ we differentiate (9) with respect to y and use the second equation in (8) to obtain

$$\frac{\partial \phi}{\partial y} = 3x^2y^2 + k'(y) = 1 + 3x^2y^2$$

from which it follows that $k'(y) = 1$. Thus,

$$k(y) = \int k'(y)\, dy = \int 1\, dy = y + K$$

where K is a (numerical) constant of integration. Substituting in (9) we obtain

$$\phi = x^2 y^3 + y + K$$

The appearance of the arbitrary constant K tells us that ϕ is not unique. The reader may want to verify that $\nabla\phi = \mathbf{F}$.

Solution (c).   Integrating the second equation in (8) with respect to y (and treating x as a constant) yields

$$\phi = \int (1 + 3x^2 y^2)\, dy = y + x^2 y^3 + k(x) \tag{10}$$

where $k(x)$ is the "constant" of integration. Differentiating (10) with respect to x and using the first equation in (8) yields

$$\frac{\partial \phi}{\partial x} = 2xy^3 + k'(x) = 2xy^3$$

from which it follows that $k'(x) = 0$ and consequently that $k(x) = K$, where K is a numerical constant of integration. Substituting this in (10) yields

$$\phi = y + x^2 y^3 + K$$

which agrees with the solution in part (b).   ◀

Example 4   Use the potential function obtained in Example 3 to evaluate the integral

$$\int_{(1,4)}^{(3,1)} 2xy^3\, dx + (1 + 3x^2 y^2)\, dy$$

Solution.   The integrand can be expressed as $\mathbf{F} \cdot d\mathbf{r}$, where $\mathbf{F}$ is the vector field in Example 3. Thus, using Formula (2) and the potential function $\phi = y + x^2 y^3 + K$ for $\mathbf{F}$ we obtain

$$\int_{(1,4)}^{(3,1)} 2xy^3\, dx + (1 + 3x^2 y^2)\, dy = \int_{(1,4)}^{(3,1)} \mathbf{F} \cdot d\mathbf{r} = \phi(3, 1) - \phi(1, 4)$$
$$= (10 + K) - (68 + K) = -58 \qquad ◀$$

REMARK.   Note that the constant K drops out. In future integration problems we shall omit K from the computations.

Example 5   Let $\mathbf{F}(x, y) = e^y \mathbf{i} + xe^y \mathbf{j}$.

(a)   Verify that the vector field $\mathbf{F}$ is conservative on the entire xy-plane.

(b)   Find the work done by the field on a particle that moves from $(1, 0)$ to $(-1, 0)$ over the semicircular path C shown in Figure 18.3.7.

Solution (a).   For the given field we have $f(x, y) = e^y$ and $g(x, y) = xe^y$. Thus,

$$\frac{\partial}{\partial y}(e^y) = e^y = \frac{\partial}{\partial x}(xe^y)$$

so (6) holds for all (x, y) and hence $\mathbf{F}$ is conservative on the entire xy-plane.

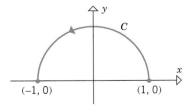

Figure 18.3.7

Solution (b).   From Formula (24) of Section 18.2, the work done by the field is

$$W = \int_C \mathbf{F} \cdot d\mathbf{r} = \int_C e^y\, dx + x e^y\, dy \tag{11}$$

However, the calculations involved in integrating along C are tedious, so it is preferable to apply Theorem 18.3.1, taking advantage of the fact that the field is conservative and the integral is independent of path. Thus, we write (11) as

$$W = \int_{(1,0)}^{(-1,0)} e^y\, dx + x e^y\, dy = \phi(-1,0) - \phi(1,0) \tag{12}$$

As illustrated in Example 3, we can find ϕ by integrating either of the equations

$$\frac{\partial \phi}{\partial x} = e^y \quad \text{and} \quad \frac{\partial \phi}{\partial y} = x e^y \tag{13}$$

We shall integrate the first. We obtain

$$\phi = \int e^y\, dx = x e^y + k(y) \tag{14}$$

Differentiating this equation with respect to y and using the second equation in (13) yields

$$\frac{\partial \phi}{\partial y} = x e^y + k'(y) = x e^y$$

from which it follows that $k'(y) = 0$ or $k(y) = K$. Thus, from (14)

$$\phi = x e^y + K$$

and hence from (12)

$$W = \phi(-1,0) - \phi(1,0) = (-1)e^0 - 1 e^0 = -2 \quad \blacktriangleleft$$

☐ **CONSERVATIVE VECTOR FIELDS IN 3-SPACE**

All of the results in this section have analogs in 3-space: Theorems 18.3.1 and 18.3.2 can be extended to vector fields in 3-space simply by adding a third variable and modifying the hypotheses appropriately. For example, in 3-space, Formula (1) becomes

$$\int_C \mathbf{F}(x, y, z) \cdot d\mathbf{r} = \phi(x_1, y_1, z_1) - \phi(x_0, y_0, z_0) \tag{15}$$

Theorem 18.3.3 can also be extended to vector fields in 3-space. We leave it for the exercises to show that if $\mathbf{F}(x, y, z) = f(x, y, z)\mathbf{i} + g(x, y, z)\mathbf{j} + h(x, y, z)\mathbf{k}$ is a conservative field, then

$$\frac{\partial f}{\partial y} = \frac{\partial g}{\partial x}, \quad \frac{\partial f}{\partial z} = \frac{\partial h}{\partial x}, \quad \frac{\partial g}{\partial z} = \frac{\partial h}{\partial y} \tag{16}$$

that is, curl $\mathbf{F} = \mathbf{0}$. Conversely, a vector field satisfying these conditions on a suitably restricted region is conservative on that region if f, g, and h are continuous and have continuous first partial derivatives in the region. Some problems involving Formulas (15) and (16) are given in the exercises.

☐ **POTENTIAL ENERGY**

If $\mathbf{F}(x, y, z)$ is a conservative force field with a potential function $\phi(x, y, z)$, then physicists call $V(x, y, z) = -\phi(x, y, z)$ the **potential energy** of the field at the point (x, y, z). Thus, it follows from the 3-space version of Theorem 18.3.1 that the work W done by $\mathbf{F}$ on a particle that moves along any path C from a point (x_0, y_0, z_0) to a point (x_1, y_1, z_1) is related to the potential energy by the equation

$$W = \int_C \mathbf{F} \cdot d\mathbf{r} = \phi(x_1, y_1, z_1) - \phi(x_0, y_0, z_0)$$
$$= -[V(x_1, y_1, z_1) - V(x_0, y_0, z_0)]$$

That is, the work done by the force field is the negative of the change in potential energy. In particular, it follows from the 3-space analog of Theorem 18.3.2(b) that if a particle traverses a closed path, then the work done by the field is zero, and there is no change in potential energy.

▶ Exercise Set 18.3

In Exercises 1–6, determine whether **F** is conservative. If it is, find a potential function for it.

1. $\mathbf{F}(x, y) = x\mathbf{i} + y\mathbf{j}$.

2. $\mathbf{F}(x, y) = 3y^2\mathbf{i} + 6xy\mathbf{j}$.

3. $\mathbf{F}(x, y) = x^2y\mathbf{i} + 5xy^2\mathbf{j}$.

4. $\mathbf{F}(x, y) = e^x \cos y\mathbf{i} - e^x \sin y\mathbf{j}$.

5. $\mathbf{F}(x, y) = (\cos y + y \cos x)\mathbf{i} + (\sin x - x \sin y)\mathbf{j}$.

6. $\mathbf{F}(x, y) = x \ln y\mathbf{i} + y \ln x\mathbf{j}$.

7. Show that $\displaystyle\int_{(-1, 2)}^{(1, 3)} y^2\, dx + 2xy\, dy$ is independent of the path and evaluate the integral by

   (a) using Theorem 18.3.1

   (b) integrating along the line segment from $(-1, 2)$ to $(1, 3)$.

8. Show that $\displaystyle\int_{(0, 1)}^{(\pi, -1)} y \sin x\, dx - \cos x\, dy$ is independent of the path, and evaluate the integral by

   (a) using Theorem 18.3.1

   (b) integrating along the line segment from $(0, 1)$ to $(\pi, -1)$.

In Exercises 9–14, show that the integral is independent of the path, and find its value.

9. $\displaystyle\int_{(1, 2)}^{(4, 0)} 3y\, dx + 3x\, dy$.

10. $\displaystyle\int_{(0, 0)}^{(1, \pi/2)} e^x \sin y\, dx + e^x \cos y\, dy$.

11. $\displaystyle\int_{(0, 0)}^{(3, 2)} 2xe^y\, dx + x^2e^y\, dy$.

12. $\displaystyle\int_{(-1, 2)}^{(0, 1)} (3x - y + 1)\, dx - (x + 4y + 2)\, dy$.

13. $\displaystyle\int_{(2, -2)}^{(-1, 0)} 2xy^3\, dx + 3y^2x^2\, dy$.

14. $\displaystyle\int_{(1, 1)}^{(3, 3)} \left(e^x \ln y - \frac{e^y}{x}\right) dx + \left(\frac{e^x}{y} - e^y \ln x\right) dy$, where x and y are positive.

In Exercises 15–18, find the work done by the conservative force **F** as it acts on a particle moving from P to Q.

15. $\mathbf{F}(x, y) = xy^2\mathbf{i} + x^2y\mathbf{j}$; $P(1, 1)$, $Q(0, 0)$.

16. $\mathbf{F}(x, y) = ye^{xy}\mathbf{i} + xe^{xy}\mathbf{j}$; $P(-1, 1)$, $Q(2, 0)$.

17. $\mathbf{F}(x, y) = \dfrac{y}{x^2 + y^2}\mathbf{i} - \dfrac{x}{x^2 + y^2}\mathbf{j}$; $P(0, 2)$, $Q(3, 3)$.
    (Assume that $y > 0$.)

18. $\mathbf{F}(x, y) = e^{-y} \cos x\mathbf{i} - e^{-y} \sin x\mathbf{j}$; $P(\pi/2, 1)$, $Q(-\pi/2, 0)$.

19. Find $\displaystyle\int_C \mathbf{F} \cdot d\mathbf{r}$, where
    $$\mathbf{F}(x, y) = (e^y + ye^x)\mathbf{i} + (xe^y + e^x)\mathbf{j}$$
    and C is the curve
    $$\mathbf{r}(t) = \sin(\pi t/2)\mathbf{i} + \ln t\mathbf{j}, \quad 1 \le t \le 2$$
    [*Hint:* First determine whether **F** is conservative.]

20. Find $\displaystyle\int_C \mathbf{F} \cdot d\mathbf{r}$, where
    $$\mathbf{F}(x, y) = 2xy\mathbf{i} + (x^2 + \cos y)\mathbf{j}$$
    and C is the curve
    $$\mathbf{r}(t) = t\mathbf{i} + t \cos(t/3)\mathbf{j}, \quad 0 \le t \le \pi$$

21. (a) Show that the inverse-square field
    $$\mathbf{F}(\mathbf{r}) = \frac{1}{\|\mathbf{r}\|^3}\mathbf{r} \quad (\mathbf{r} = x\mathbf{i} + y\mathbf{j})$$
    is conservative everywhere, except at the origin, by finding a potential function for it.

    (b) Find $\displaystyle\int_C \mathbf{F} \cdot d\mathbf{r}$, where C is the straight line segment from $P(0, -1)$ to $Q(-3, 4)$.

    (c) Find $\displaystyle\int_C \mathbf{F} \cdot d\mathbf{r}$, where C is any piecewise smooth closed curve that does not go through the origin.

22. Show that the vector field
    $$\mathbf{F}(x, y) = \frac{2xy}{(x^2 + y^2)^2}\mathbf{i} + \frac{y^2 - x^2}{(x^2 + y^2)^2}\mathbf{j}$$
    is conservative, except at the origin, by finding a potential function for it.

23. Find a function h for which
    $$\mathbf{F}(x, y) = h(x)[x \sin y + y \cos y]\mathbf{i}$$
    $$+ h(x)[x \cos y - y \sin y]\mathbf{j}$$
    is conservative.

24. Prove Theorem 18.3.1 if C is a piecewise smooth curve composed of smooth curves $C_1, C_2, \ldots, C_n$.

25. Prove that (b) implies (c) in Theorem 18.3.2. [*Hint:* Consider any two piecewise smooth oriented curves C_1 and C_2 in the region from a point P to a point Q, and integrate around the closed curve consisting of C_1 and $-C_2$.]

26. Complete the proof of Theorem 18.3.2 by showing that $\partial\phi/\partial y = g(x, y)$, where $\phi(x, y)$ is the function in (4).

27. Prove: If $\mathbf{F}(x, y, z) = f(x, y, z)\mathbf{i} + g(x, y, z)\mathbf{j} + h(x, y, z)\mathbf{k}$ is a conservative field and f, g, and h have continuous first partial derivatives in a region, then

$$\frac{\partial f}{\partial y} = \frac{\partial g}{\partial x}, \quad \frac{\partial f}{\partial z} = \frac{\partial h}{\partial x}, \quad \frac{\partial g}{\partial z} = \frac{\partial h}{\partial y}$$

in the region.

28. Use the result of Exercise 27 to show that curl $\mathbf{F} = \mathbf{0}$ in the region if $\mathbf{F}$ satisfies the stated conditions.

It can be shown that if the components of $\mathbf{F}(x, y, z)$ have continuous first partial derivatives everywhere, except possibly at finitely many points, then $\mathbf{F}$ is conservative (except at those points) if and only if curl $\mathbf{F} = \mathbf{0}$. Use this result in Exercises 29–34 to determine whether $\mathbf{F}$ is conservative; if so, find a potential function for it.

29. $\mathbf{F}(x, y, z) = z^2\mathbf{i} + e^{-y}\mathbf{j} + 2xz\mathbf{k}$.

30. $\mathbf{F}(x, y, z) = xy\mathbf{i} + x^2\mathbf{j} + \sin z\mathbf{k}$.

31. $\mathbf{F}(x, y, z) = yz\mathbf{i} + xz\mathbf{j} + (xy + 3z^2)\mathbf{k}$.

32. $\mathbf{F}(x, y, z) = \sin x\mathbf{i} + z\mathbf{j} + y\mathbf{k}$.

33. $\mathbf{F}(x, y, z) = z\mathbf{i} + 2yz\mathbf{j} + y^2\mathbf{k}$.

34. $\mathbf{F}(x, y, z) = 2xz\mathbf{i} + \cos y\mathbf{j} + (x^2 + 2z)\mathbf{k}$.

35. In Formula (5) of Section 18.1 a potential function was given for the inverse-square field

$$\mathbf{F}(\mathbf{r}) = \frac{1}{\|\mathbf{r}\|^3}\mathbf{r}$$

Use this potential function to evaluate the integrals:

(a) $\displaystyle\int_C \mathbf{F} \cdot d\mathbf{r}$, where C is the line segment from $P(1, 1, 2)$ to $Q(3, 2, 1)$

(b) $\displaystyle\int_C \mathbf{F} \cdot d\mathbf{r}$, where C is the curve
$\mathbf{r}(t) = (2t^2 + 1)\mathbf{i} + (t^3 + 1)\mathbf{j} + (2 - \sqrt{t})\mathbf{k}, 0 \leq t \leq 1$

(c) $\displaystyle\int_C \mathbf{F} \cdot d\mathbf{r}$, where C is any piecewise smooth closed curve that does not pass through the origin.

36. (a) Show that the vector field

$$\mathbf{F}(\mathbf{r}) = \frac{1}{\|\mathbf{r}\|^2}\mathbf{r} \quad (\mathbf{r} = x\mathbf{i} + y\mathbf{j} + z\mathbf{k})$$

is conservative everywhere, except at the origin, by finding a potential function for it.

(b) Find $\displaystyle\int_C \mathbf{F} \cdot d\mathbf{r}$, where C is any piecewise smooth curve from $P(2, 2, 1)$ to $Q(a, b, c)$ that does not go through the origin.

(c) Find $\displaystyle\int_C \mathbf{F} \cdot d\mathbf{r}$, where C is any piecewise smooth closed curve that does not go through the origin.

■ 18.4 GREEN'S THEOREM

In this section we shall discuss a remarkable and beautiful theorem that expresses the double integral over a plane region in terms of a line integral over its boundary.

☐ **GREEN'S THEOREM**

18.4.1 THEOREM (*Green's* * *Theorem*). *Let R be a simply connected plane region whose boundary is a simple, closed, piecewise smooth curve C oriented counterclockwise. If $f(x, y)$ and $g(x, y)$ have continuous first partial derivatives on some open set containing R, then*

$$\int_C f(x, y)\, dx + g(x, y)\, dy = \iint_R \left(\frac{\partial g}{\partial x} - \frac{\partial f}{\partial y} \right) dA \tag{1}$$

Proof. For simplicity, we shall prove the theorem only for regions that are simultaneously type I and type II (see Section 17.2). Such a region is shown in Figure 18.4.1. The crux of the proof is to show that

$$\int_C f(x, y)\, dx = -\iint_R \frac{\partial f}{\partial y}\, dA \quad \text{and} \quad \int_C g(x, y)\, dy = \iint_R \frac{\partial g}{\partial x}\, dA \tag{2a–b}$$

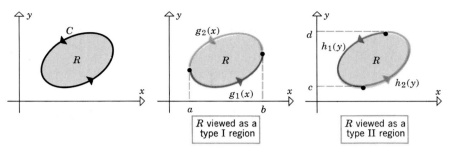

Figure 18.4.1

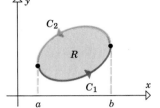

Figure 18.4.2

To prove (2a), view R as a type I region and let C_1 and C_2 be the lower and upper boundary curves, oriented as in Figure 18.4.2. Then

$$\int_C f(x, y)\, dx = \int_{C_1} f(x, y)\, dx + \int_{C_2} f(x, y)\, dx$$

or, equivalently,

$$\int_C f(x, y)\, dx = \int_{C_1} f(x, y)\, dx - \int_{-C_2} f(x, y)\, dx \tag{3}$$

(This step will help simplify our calculations since C_1 and $-C_2$ are then both oriented left to right.) With $x = t$ as the parameter, C_1 and $-C_2$ are represented by

$$C_1 : x = t,\, y = g_1(t) \quad (a \le t \le b)$$
$$-C_2 : x = t,\, y = g_2(t) \quad (a \le t \le b)$$

Thus, (3) yields

$$\int_C f(x, y)\, dx = \int_a^b f(t, g_1(t)) \left(\frac{dx}{dt}\right) dt - \int_a^b f(t, g_2(t)) \left(\frac{dx}{dt}\right) dt$$

$$= \int_a^b f(t, g_1(t))\, dt - \int_a^b f(t, g_2(t))\, dt$$

$$= -\int_a^b [f(t, g_2(t)) - f(t, g_1(t))]\, dt$$

$$= -\int_a^b \left[f(t, y) \right]_{y=g_1(t)}^{y=g_2(t)} dt = -\int_a^b \left[\int_{g_1(t)}^{g_2(t)} \frac{\partial f}{\partial y}\, dy \right] dt$$

$$= -\int_a^b \int_{g_1(x)}^{g_2(x)} \frac{\partial f}{\partial y}\, dy\, dx = -\iint_R \frac{\partial f}{\partial y}\, dA$$

Since $x = t$

* GEORGE GREEN (1793–1841). English mathematician and physicist. Green left school at an early age to work in his father's bakery and consequently had little early formal education. When his father opened a mill, the boy used the top room as a study in which he taught himself physics and mathematics from library books. In 1828 Green published his most important work, *An Essay on the Application of Mathematical Analysis to the Theories of Electricity and Magnetism.* Although Green's Theorem appeared in that paper, the result went virtually unnoticed because of the small pressrun and local distribution. Following the death of his father in 1829, Green was urged by friends to seek a college education. In 1833, after four years of self-study to close the gaps in his elementary education, Green was admitted to Caius College, Cambridge. He graduated four years later, but with a disappointing performance on his final examinations—possibly because he was more interested in his own research. After a succession of works on light and sound, he was named to be Perse Fellow at Caius College. Two years later he died. In 1845, four years after his death, his paper of 1828 was published and the theories developed therein by this obscure, self-taught baker's son helped pave the way to the modern theories of electricity and magnetism.

The proof of (2b) is obtained similarly by treating R as a type II region. The details are omitted.   ∎

Example 1   Use Green's Theorem to evaluate

$$\int_C x^2 y \, dx + x \, dy$$

over the triangular path shown in Figure 18.4.3.

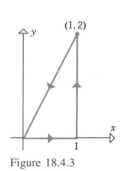

(1, 2)

Figure 18.4.3

Solution.   Since $f(x, y) = x^2 y$ and $g(x, y) = x$, it follows from (1) that

$$\int_C x^2 y \, dx + x \, dy = \iint_R \left[\frac{\partial}{\partial x}(x) - \frac{\partial}{\partial y}(x^2 y) \right] dA = \int_0^1 \int_0^{2x} (1 - x^2) \, dy \, dx$$

$$= \int_0^1 (2x - 2x^3) \, dx = \left[x^2 - \frac{x^4}{2} \right]_0^1 = \frac{1}{2}$$

This agrees with the result obtained in Example 4 of Section 18.2, where we evaluated the line integral directly. Note how much simpler this solution is.   ◀

☐ **FINDING WORK USING GREEN'S THEOREM**

Because Formula (1) requires the curve C to be closed, it is desirable to have a notation for indicating this. One way of doing this is to write (1) as

$$\oint_C f(x, y) \, dx + g(x, y) \, dy = \iint_R \left(\frac{\partial g}{\partial x} - \frac{\partial f}{\partial y} \right) dA$$

where the symbol ○ on the integral sign indicates that C is a simple closed curve.

Example 2   Find the work done by the force field $\mathbf{F}(x, y) = (e^x - y^3)\mathbf{i} + (\cos y + x^3)\mathbf{j}$ on a particle that travels once around the unit circle $x^2 + y^2 = 1$ in the counterclockwise direction.

Solution.   From Formula (24) of Section 18.2, the work W done by the field is

$$W = \oint_C \mathbf{F} \cdot d\mathbf{r} = \oint_C (e^x - y^3) \, dx + (\cos y + x^3) \, dy \tag{4}$$

However, the force field $\mathbf{F}$ is not conservative, since Equation (6) in Theorem 18.3.3 does not hold (verify), and consequently the work done by $\mathbf{F}$ need not be zero, even though C is a simple closed curve. Moreover, evaluating (4) directly by parametrizing the circle leads to a complicated line integral in terms of t, so the method of choice for solving this problem is Green's Theorem. This yields

$$W = \oint_C (e^x - y^3) \, dx + (\cos y + x^3) \, dy$$

$$= \iint_R \left[\frac{\partial}{\partial x}(\cos y + x^3) - \frac{\partial}{\partial y}(e^x - y^3) \right] dA$$

$$= \iint_R (3x^2 + 3y^2) \, dA = 3 \iint_R (x^2 + y^2) \, dA$$

$$= 3 \int_0^{2\pi} \int_0^1 (r^2) r \, dr \, d\theta = \frac{3}{4} \int_0^{2\pi} d\theta = \frac{3\pi}{2}   ◀$$

We converted to polar coordinates.

REMARK. Green's Theorem provides another viewpoint about work in conservative vector fields. In a force field $\mathbf{F}(x, y) = f(x, y)\mathbf{i} + g(x, y)\mathbf{j}$, the line integral in (1) represents the work done by the field on a particle moving over the closed curve C. However, if the field is conservative, then from Theorem 18.3.3 the integrand of the double integral in (1) is zero, and hence Green's Theorem implies that the work done by the field is zero, which agrees with results obtained in the preceding section.

□ **FINDING AREAS USING GREEN'S THEOREM**

Green's Theorem leads to some new formulas for the area A of a region R that satisfies the conditions of the theorem. Two such formulas can be obtained as follows:

$$A = \iint_R dA = \oint_C x\, dy \quad \text{and} \quad A = \iint_R dA = \oint_C (-y)\, dx$$

| Set $f(x, y) = 0$ and $g(x, y) = x$ in (1). |

| Set $f(x, y) = -y$ and $g(x, y) = 0$ in (1). |

A third formula can be obtained by adding these two equations together. Thus, we have the following three formulas that express the area A of a region R in terms of line integrals over the boundary:

$$A = \oint_C x\, dy = -\oint_C y\, dx = \frac{1}{2}\oint_C -y\, dx + x\, dy \tag{5}$$

REMARK. Although the third formula in (5) looks more complicated than the other two, it often leads to simpler integrations; but each has its use.

Example 3 Use the third formula in (5) to find the area enclosed by the ellipse $x^2/a^2 + y^2/b^2 = 1$.

Solution. The ellipse, with counterclockwise orientation, can be represented parametrically by

$$x = a\cos t, \quad y = b\sin t \quad (0 \le t \le 2\pi)$$

If we denote this curve by C, then the area A enclosed by the ellipse is

$$A = \frac{1}{2}\oint_C -y\, dx + x\, dy$$

$$= \frac{1}{2}\int_0^{2\pi} [(-b\sin t)(-a\sin t) + (a\cos t)(b\cos t)]\, dt$$

$$= \frac{1}{2}ab\int_0^{2\pi} (\sin^2 t + \cos^2 t)\, dt = \frac{1}{2}ab\int_0^{2\pi} dt = \pi ab \quad \blacktriangleleft$$

□ **GREEN'S THEOREM FOR MULTIPLY CONNECTED REGIONS**

Recall that a region in 2-space is called *simply connected* if it is connected and has no holes. Regions of 2-space that are connected but have finitely many holes are sometimes called ***multiply connected***. Although Green's Theorem (18.4.1) was stated for simply connected regions, it can be extended to multiply connected regions with some appropriate modifications. For example, let R be the multiply connected region shown in Figure 18.4.4a. The boundary of R consists of two disjoint curves C_1 and C_2, where C_1 is the "outside boundary" and C_2 is the boundary of the hole in R. As shown in the figure, suppose that the boundary curves are oriented so that the region R is on the left as the curves are traversed in the direction of orientation. Thus, the outside boundary is oriented counterclockwise, and the boundary of the hole is oriented clockwise. As shown in Figure

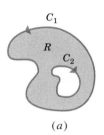

(a)

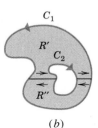

(b)

Figure 18.4.4

18.4.4b, let us divide R into two regions R' and R'' by introducing two "cuts" in R. The cuts are shown as line segments, but any piecewise smooth curves will suffice. If we assume that f and g satisfy the hypotheses of Green's Theorem on R (and hence on R' and R''), then we can apply this theorem to both R' and R'' to obtain

$$\iint_R \left(\frac{\partial g}{\partial x} - \frac{\partial f}{\partial y}\right) dA = \iint_{R'} \left(\frac{\partial g}{\partial x} - \frac{\partial f}{\partial y}\right) dA + \iint_{R''} \left(\frac{\partial g}{\partial x} - \frac{\partial f}{\partial y}\right) dA$$

$$= \underbrace{\oint f(x, y)\, dx + g(x, y)\, dy}_{\substack{\text{Boundary} \\ \text{of } R'}} + \underbrace{\oint f(x, y)\, dx + g(x, y)\, dy}_{\substack{\text{Boundary} \\ \text{of } R''}}$$

However, the two line integrals are taken in opposite directions along the cuts, and hence cancel there, leaving only the contributions along C_1 and C_2. Thus,

$$\iint_R \left(\frac{\partial g}{\partial x} - \frac{\partial f}{\partial y}\right) dA = \oint_{C_1} f(x, y)\, dx + g(x, y)\, dy + \oint_{C_2} f(x, y)\, dx + g(x, y)\, dy \quad (6)$$

In general, Green's Theorem applies to a multiply connected region R, provided f and g satisfy the hypotheses of the theorem, and we integrate over the *entire* boundary with the boundary curves oriented so that the region is on the left as we travel in the direction of orientation. For a region with n holes, the analog of Formula (6) would involve $n + 1$ line integrals, one for the outside boundary and one for the boundary of each hole.

Example 4 Evaluate the integral

$$\oint_C \frac{-y\, dx + x\, dy}{x^2 + y^2}$$

if C is a piecewise smooth simple closed curve oriented counterclockwise such that:

(a) C does not enclose the origin;

(b) C encloses the origin.

Solution (a). Let

$$f(x, y) = -\frac{y}{x^2 + y^2}, \quad g(x, y) = \frac{x}{x^2 + y^2}$$

so that

$$\frac{\partial g}{\partial x} = \frac{y^2 - x^2}{(x^2 + y^2)^2} = \frac{\partial f}{\partial y}$$

if x and y are not both zero. Thus, if C does not enclose the origin, we have

$$\frac{\partial g}{\partial x} - \frac{\partial f}{\partial y} = 0$$

on the simply connected region enclosed by C, and hence the given integral is zero by Green's Theorem.

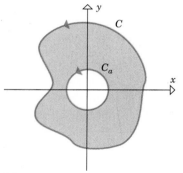

Figure 18.4.5

Solution (b). Because f and g are undefined at the origin, the hypotheses of Green's Theorem are not satisfied on the region enclosed by C. Moreover, we cannot evaluate the integral directly since we are not given a specific curve C. To circumvent these problems, we construct a circle C_a with *counterclockwise* orientation, centered at the origin, and of sufficiently small radius a that it lies inside the region enclosed by C (Figure 18.4.5). The functions f and g satisfy the hypotheses of Green's Theorem on the *multiply connected* region R enclosed between C and C_a, and hence

$$\oint_C \frac{-y\, dx + x\, dy}{x^2 + y^2} + \oint_{-C_a} \frac{-y\, dx + x\, dy}{x^2 + y^2} = \iint_R 0\, dA = 0 \quad (7)$$

(where the integration is over $-C_a$ to keep the region R to the left). It follows from (7) that

$$\oint_C \frac{-y\,dx + x\,dy}{x^2 + y^2} = \oint_{C_a} \frac{-y\,dx + x\,dy}{x^2 + y^2}$$

This shows that the integral over the arbitrary curve C is the same as the integral over the circle. Therefore, parametrizing the circle as $x = a\cos t$, $y = a\sin t$ ($0 \le t \le 2\pi$) yields

$$\oint_C \frac{-y\,dx + x\,dy}{x^2 + y^2} = \int_0^{2\pi} \frac{(-a\sin t)(-a\sin t)\,dt + (a\cos t)(a\cos t)\,dt}{(a\cos t)^2 + (a\sin t)^2}$$

$$= \int_0^{2\pi} 1\,dt = 2\pi \quad \blacktriangleleft$$

▶ Exercise Set 18.4

In Exercises 1 and 2, evaluate the line integral using Green's Theorem and check the answer by evaluating directly.

1. $\oint_C y^2\,dx + x^2\,dy$, where C is the square with vertices $(0, 0)$, $(1, 0)$, $(1, 1)$, and $(0, 1)$ oriented counterclockwise.

2. $\oint_C y\,dx + x\,dy$, where C is the unit circle oriented counterclockwise.

In Exercises 3–13, use Green's Theorem to evaluate the integral. In each exercise, assume that the curve C is oriented counterclockwise.

3. $\oint_C 3xy\,dx + 2xy\,dy$, where C is the rectangle bounded by $x = -2$, $x = 4$, $y = 1$, and $y = 2$.

4. $\oint_C (x^2 - y^2)\,dx + x\,dy$, where C is the circle $x^2 + y^2 = 9$.

5. $\oint_C x\cos y\,dx - y\sin x\,dy$, where C is the square with vertices $(0, 0)$, $(0, \pi/2)$, $(\pi/2, \pi/2)$, and $(\pi/2, 0)$.

6. $\oint_C y\tan^2 x\,dx + \tan x\,dy$, where C is the circle $x^2 + (y + 1)^2 = 1$.

7. $\oint_C (x^2 - y)\,dx + x\,dy$, where C is the circle $x^2 + y^2 = 4$.

8. $\oint_C (e^x + y^2)\,dx + (e^y + x^2)\,dy$, where C is the boundary of the region between $y = x^2$ and $y = x$.

9. $\oint_C \ln(1 + y)\,dx - \frac{xy}{1 + y}\,dy$, where C is the triangle with vertices $(0, 0)$, $(2, 0)$, and $(0, 4)$.

10. $\oint_C x^2 y\,dx - y^2 x\,dy$, where C is the boundary of the region in the first quadrant, enclosed between the coordinate axes and the circle $x^2 + y^2 = 16$.

11. $\oint_C \tan^{-1} y\,dx - \frac{y^2 x}{1 + y^2}\,dy$, where C is the square with vertices $(0, 0)$, $(0, 1)$, $(1, 1)$, and $(1, 0)$.

12. $\oint_C \cos x\sin y\,dx + \sin x\cos y\,dy$, where C is the triangle with vertices $(0, 0)$, $(3, 3)$, and $(0, 3)$.

13. $\oint_C x^2 y\,dx + (y + xy^2)\,dy$, where C is the boundary of the region enclosed by $y = x^2$ and $x = y^2$.

14. Let C be the boundary of the region enclosed between $y = x^2$ and $y = 2x$. Assuming that C is oriented counterclockwise, evaluate the following integrals by Green's Theorem:

(a) $\oint_C (6xy - y^2)\,dx$ (b) $\oint_C (6xy - y^2)\,dy$.

15. Find the area of the ellipse in Example 3 using

(a) the first formula in (5)

(b) the second formula in (5).

16. Use a line integral to find the area of the region enclosed by

$$x = a\cos^3 t, \quad y = a\sin^3 t \quad (0 \le t \le 2\pi)$$

17. Use a line integral to find the area of the triangle with vertices $(0, 0)$, $(a, 0)$, and $(0, b)$, where $a > 0$ and $b > 0$.

18. Use a line integral to find the area of the region in the first quadrant enclosed by $y = x$, $y = 1/x$, and $y = x/9$.

19. Use the formula

$$A = \frac{1}{2}\oint_C -y\,dx + x\,dy$$

to find the area of the region swept out by the line from the origin to the ellipse $x = a\cos t$, $y = b\sin t$ if t varies from $t = 0$ to $t = t_0$ ($0 \le t_0 \le 2\pi$).

20. Use the formula

$$A = \frac{1}{2}\oint_C -y\,dx + x\,dy$$

to find the area of the region swept out by the line from the origin to the hyperbola $x = a\cosh t$, $y = b\sinh t$ if t varies from $t = 0$ to $t = t_0$ ($t_0 \ge 0$).

21. A particle, starting at $(5, 0)$, traverses the upper semicircle $x^2 + y^2 = 25$ and returns to its starting point along the x-axis. Use Green's Theorem to find the work done on the particle by a force $\mathbf{F}(x, y) = xy\mathbf{i} + (\frac{1}{2}x^2 + yx)\mathbf{j}$.

22. A particle moves counterclockwise one time around the closed curve formed by $y = 0$, $x = 2$, and $y = x^3/4$. Use Green's Theorem to find the work done by a force $\mathbf{F}(x, y) = \sqrt{y}\mathbf{i} + \sqrt{x}\mathbf{j}$.

23. (a) Let R be a plane region with area A whose boundary is a piecewise smooth simple closed curve C. Use Green's Theorem to prove that the centroid $(\bar{x}, \bar{y})$ of R is given by
$$\bar{x} = \frac{1}{2A} \oint_C x^2 \, dy, \quad \bar{y} = -\frac{1}{2A} \oint_C y^2 \, dx$$

    (b) Use the result in part (a) to find the centroid of the region enclosed between the x-axis and the upper half of the circle $x^2 + y^2 = a^2$.

24. Evaluate $\oint_C y \, dx - x \, dy$, where C is the cardioid
$$r = a(1 + \cos \theta), \quad 0 \le \theta \le 2\pi$$

25. (a) Let C be the line segment from a point (a, b) to a point (c, d). Show that
$$\int_C -y \, dx + x \, dy = ad - bc$$

(b) Use the result in part (a) to show that the area A of a triangle with successive vertices (x_1, y_1), (x_2, y_2), and (x_3, y_3) going counterclockwise is
$$A = \tfrac{1}{2} [(x_1 y_2 - x_2 y_1) \\ + (x_2 y_3 - x_3 y_2) + (x_3 y_1 - x_1 y_3)]$$

(c) Find a formula for the area of a polygon with successive vertices (x_1, y_1), $(x_2, y_2), \ldots, (x_n, y_n)$ going counterclockwise.

(d) Use the result in part (c) to find the area of a quadrilateral with vertices $(0, 0)$, $(3, 4)$, $(-2, 2)$, $(-1, 0)$.

26. Find a simple closed curve C that maximizes the value of
$$\oint_C \tfrac{1}{3}y^3 \, dx + (x - \tfrac{1}{3}x^3) \, dy$$

In Exercises 27 and 28, assume that the boundary curves of the regions are oriented so that the region is on the left traveling in the direction of orientation.

27. Find $\oint_C \mathbf{F} \cdot d\mathbf{r}$, where $\mathbf{F}(x, y) = (x^2 + y)\mathbf{i} + (4x - \cos y)\mathbf{j}$ and C is the boundary of the region R that is inside the square with vertices $(0, 0)$, $(5, 0)$, $(5, 5)$, $(0, 5)$ but is outside the rectangle with vertices $(1, 1)$, $(3, 1)$, $(3, 2)$, $(1, 2)$.

28. Find $\oint_C \mathbf{F} \cdot d\mathbf{r}$, where $\mathbf{F}(x, y) = (e^{-x} + 3y)\mathbf{i} + x\mathbf{j}$ and C is the boundary of the region R between the circles $x^2 + y^2 = 16$ and $x^2 - 2x + y^2 = 3$.

18.5 INTRODUCTION TO SURFACE INTEGRALS

In previous sections we considered four kinds of integrals—integrals over intervals, double integrals over two-dimensional regions, triple integrals over three-dimensional solids, and line integrals over curves in two- or three-dimensional space. In this section we shall discuss integrals over surfaces in three-dimensional space. Such integrals occur in problems involving fluid and heat flow, electricity, magnetism, mass, and center of gravity.

☐ **DEFINITION OF A SURFACE INTEGRAL**

Recall that if C is a smooth curve in 3-space, and $f(x, y, z)$ is continuous on C, then the line integral of f over C with respect to arc length is defined by subdividing C into n arcs and defining the line integral as the limit

$$\int_C f(x, y, z) \, ds = \lim_{n \to +\infty} \sum_{k=1}^n f(x_k^*, y_k^*, z_k^*) \, \Delta s_k$$

where (x_k^*, y_k^*, z_k^*) is a point on the kth arc and Δs_k is the length of the kth arc. We will define *surface integrals* in an analogous manner.

Let σ be a surface in 3-space with finite surface area, and let $f(x, y, z)$ be a continuous function defined on σ. As shown in Figure 18.5.1, subdivide σ into parts, $\sigma_1, \sigma_2, \ldots, \sigma_n$ with areas $\Delta S_1, \Delta S_2, \ldots, \Delta S_n$, and form the sum

$$\sum_{k=1}^n f(x_k^*, y_k^*, z_k^*) \, \Delta S_k \tag{1}$$

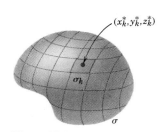

Figure 18.5.1

where (x_k^*, y_k^*, z_k^*) is an arbitrary point on σ_k. Now repeat the subdivision process, dividing σ into more and more parts in such a way that the maximum dimension of each part approaches zero as $n \to +\infty$. If (1) approaches a limit that does not depend on the way the subdivisions are made or how the points (x_k^*, y_k^*, z_k^*) are chosen, then this limit is called the *surface integral* of $f(x, y, z)$ over σ and is denoted by

$$\iint_\sigma f(x, y, z) \, dS = \lim_{n \to +\infty} \sum_{k=1}^{n} f(x_k^*, y_k^*, z_k^*) \, \Delta S_k \tag{2}$$

☐ **EVALUATING SURFACE INTEGRALS**

There are various procedures for evaluating surface integrals, depending on how the surface σ is represented. The following theorem provides a method for evaluating a surface integral when σ is of the form $z = g(x, y)$, or of the form $y = g(x, z)$ or $x = g(y, z)$.

18.5.1   THEOREM.

(a) *Let σ be a surface with equation $z = g(x, y)$ and let R be its projection on the xy-plane. If g has continuous first partial derivatives on R and $f(x, y, z)$ is continuous on σ, then*

$$\iint_\sigma f(x, y, z) \, dS = \iint_R f[x, y, g(x, y)] \sqrt{\left(\frac{\partial z}{\partial x}\right)^2 + \left(\frac{\partial z}{\partial y}\right)^2 + 1} \, dA \tag{3}$$

(b) *Let σ be a surface with equation $y = g(x, z)$ and let R be its projection on the xz-plane. If g has continuous first partial derivatives on R and $f(x, y, z)$ is continuous on σ, then*

$$\iint_\sigma f(x, y, z) \, dS = \iint_R f[x, g(x, z), z] \sqrt{\left(\frac{\partial y}{\partial x}\right)^2 + \left(\frac{\partial y}{\partial z}\right)^2 + 1} \, dA \tag{4}$$

(c) *Let σ be a surface with equation $x = g(y, z)$ and let R be its projection on the yz-plane. If g has continuous first partial derivatives on R and $f(x, y, z)$ is continuous on σ, then*

$$\iint_\sigma f(x, y, z) \, dS = \iint_R f[g(y, z), y, z] \sqrt{\left(\frac{\partial x}{\partial y}\right)^2 + \left(\frac{\partial x}{\partial z}\right)^2 + 1} \, dA \tag{5}$$

To illustrate how the formulas in this theorem can be obtained, consider the case where σ is represented by an equation of the form $z = g(x, y)$ over a region R of the xy-plane. In this case it follows from Formula (3) of Section 17.4 (with g instead of f) that the surface area ΔS_k in Formula (2) can be approximated by

$$\Delta S_k \approx \sqrt{g_x(x_k^*, y_k^*)^2 + g_y(x_k^*, y_k^*)^2 + 1} \, \Delta A_k$$

in which case (2) becomes

$$\iint_\sigma f(x, y, z) \, dS = \lim_{n \to +\infty} \sum_{k=1}^{n} f(x_k^*, y_k^*, z_k^*) \sqrt{g_x(x_k^*, y_k^*)^2 + g_y(x_k^*, y_k^*)^2 + 1} \, \Delta A_k$$

$$= \lim_{n \to +\infty} \sum_{k=1}^{n} f(x_k^*, y_k^*, g(x_k^*, y_k^*)) \sqrt{g_x(x_k^*, y_k^*)^2 + g_y(x_k^*, y_k^*)^2 + 1} \, \Delta A_k$$

Noting that the limit in this equation is a double integral over the region R, we can rewrite the equation as

$$\iint_\sigma f(x, y, z)\, dS = \iint_R f(x, y, g(x, y)) \sqrt{g_x(x, y)^2 + g_y(x, y)^2 + 1}\, dA$$

or, equivalently,

$$\iint_\sigma f(x, y, z)\, dS = \iint_R f(x, y, g(x, y)) \sqrt{\left(\frac{\partial z}{\partial x}\right)^2 + \left(\frac{\partial z}{\partial y}\right)^2 + 1}\, dA$$

Example 1   Evaluate the surface integral

$$\iint_\sigma xz\, dS$$

where σ is the part of the plane $x + y + z = 1$ that lies in the first octant.

Solution.   The equation of the plane can be written as

$$z = 1 - x - y$$

which is of the form $z = g(x, y)$. Consequently, we can apply Formula (3) with $z = g(x, y) = 1 - x - y$ and $f(x, y, z) = xz$. Thus,

$$\frac{\partial z}{\partial x} = -1 \quad \text{and} \quad \frac{\partial z}{\partial y} = -1$$

so (3) becomes

$$\iint_\sigma xz\, dS = \iint_R x(1 - x - y)\sqrt{(-1)^2 + (-1)^2 + 1}\, dA \tag{6}$$

where R is the projection of σ on the xy-plane (Figure 18.5.2). Rewriting the double integral in (6) as an iterated integral yields

$$\iint_\sigma xz\, dS = \sqrt{3} \int_0^1 \int_0^{1-x} (x - x^2 - xy)\, dy\, dx$$

$$= \sqrt{3} \int_0^1 \left[xy - x^2 y - \frac{xy^2}{2} \right]_{y=0}^{1-x} dx$$

$$= \sqrt{3} \int_0^1 \left(\frac{x}{2} - x^2 + \frac{x^3}{2} \right) dx$$

$$= \sqrt{3} \left[\frac{x^2}{4} - \frac{x^3}{3} + \frac{x^4}{8} \right]_0^1 = \frac{\sqrt{3}}{24} \quad \blacktriangleleft$$

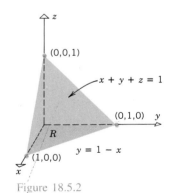

Figure 18.5.2

Example 2   Evaluate the surface integral

$$\iint_\sigma y^2 z^2\, dS$$

where σ is the part of the cone $z = \sqrt{x^2 + y^2}$ that lies between the planes $z = 1$ and $z = 2$ (Figure 18.5.3)

Solution.   We shall apply Formula (3) with $z = g(x, y) = \sqrt{x^2 + y^2}$ and $f(x, y, z) = y^2 z^2$. Thus,

$$\frac{\partial z}{\partial x} = \frac{x}{\sqrt{x^2 + y^2}} \quad \text{and} \quad \frac{\partial z}{\partial y} = \frac{y}{\sqrt{x^2 + y^2}}$$

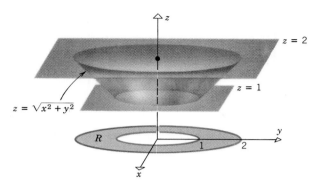

Figure 18.5.3

so

$$\sqrt{\left(\frac{\partial z}{\partial x}\right)^2 + \left(\frac{\partial z}{\partial y}\right)^2 + 1} = \sqrt{2}$$

(verify), and (3) yields

$$\iint_\sigma y^2 z^2 \, dS = \iint_R y^2 (\sqrt{x^2 + y^2})^2 \sqrt{2} \, dA = \sqrt{2} \iint_R y^2 (x^2 + y^2) \, dA$$

where R is the annulus enclosed between $x^2 + y^2 = 1$ and $x^2 + y^2 = 4$ (Figure 18.5.3). Using polar coordinates to evaluate this double integral over the annulus R yields

$$\iint_\sigma y^2 z^2 \, dS = \sqrt{2} \int_0^{2\pi} \int_1^2 (r \sin \theta)^2 (r^2) r \, dr \, d\theta$$

$$= \sqrt{2} \int_0^{2\pi} \int_1^2 r^5 \sin^2 \theta \, dr \, d\theta$$

$$= \sqrt{2} \int_0^{2\pi} \frac{r^6}{6} \sin^2 \theta \Big]_{r=1}^2 \, d\theta = \frac{21}{\sqrt{2}} \int_0^{2\pi} \sin^2 \theta \, d\theta$$

$$= \frac{21}{\sqrt{2}} \left[\frac{1}{2} \theta - \frac{1}{4} \sin 2\theta \right]_0^{2\pi} = \frac{21\pi}{\sqrt{2}} \quad \boxed{\text{See Example 1, Section 9.3.}} \quad \blacktriangleleft$$

☐ **SURFACE AREA AS A SURFACE INTEGRAL**

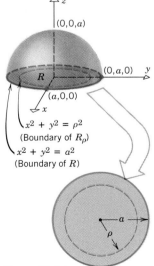

Figure 18.5.4

In the special case where $f(x, y, z)$ is 1, Formula (2) becomes

$$\iint_\sigma dS = \lim_{n \to +\infty} \sum_{k=1}^n \Delta S_k$$

But for all values of n the sum is the surface area of the surface σ; and hence so is the limit. Thus, the surface area S of the surface σ can be expressed as

$$S = \iint_\sigma dS \qquad (7)$$

Example 3 Use Formula (3) to show that the surface area of the upper hemisphere of radius a given by the equation $z = \sqrt{a^2 - x^2 - y^2}$ is $2\pi a^2$.

Solution. As shown in Figure 18.5.4, the projection of the hemisphere on the xy-plane is the circular region R of radius a centered at the origin. However, we cannot apply Theorem 18.5.1 directly because the partial derivatives

$$\frac{\partial z}{\partial x} = \frac{-x}{\sqrt{a^2 - x^2 - y^2}} \quad \text{and} \quad \frac{\partial z}{\partial y} = \frac{-y}{\sqrt{a^2 - x^2 - y^2}}$$

do not exist on the boundary of the region R since $x^2 + y^2 = a^2$ there. In order to overcome this problem we evaluate the double integral over a slightly smaller circular region R_ρ of radius ρ, and then let ρ approach a. The computations are as follows:

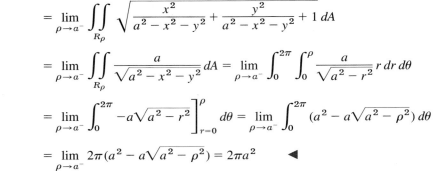

$$\iint_\sigma dS = \lim_{\rho \to a^-} \iint_{R_\rho} \sqrt{\left(\frac{\partial z}{\partial x}\right)^2 + \left(\frac{\partial z}{\partial y}\right)^2 + 1}\; dA$$

$$= \lim_{\rho \to a^-} \iint_{R_\rho} \sqrt{\frac{x^2}{a^2 - x^2 - y^2} + \frac{y^2}{a^2 - x^2 - y^2} + 1}\; dA$$

$$= \lim_{\rho \to a^-} \iint_{R_\rho} \frac{a}{\sqrt{a^2 - x^2 - y^2}}\, dA = \lim_{\rho \to a^-} \int_0^{2\pi} \int_0^{\rho} \frac{a}{\sqrt{a^2 - r^2}}\, r\, dr\, d\theta$$

$$= \lim_{\rho \to a^-} \int_0^{2\pi} \left. -a\sqrt{a^2 - r^2}\, \right]_{r=0}^{\rho} d\theta = \lim_{\rho \to a^-} \int_0^{2\pi} (a^2 - a\sqrt{a^2 - \rho^2})\, d\theta$$

$$= \lim_{\rho \to a^-} 2\pi(a^2 - a\sqrt{a^2 - \rho^2}) = 2\pi a^2 \quad \blacktriangleleft$$

MASS OF A CURVED LAMINA AS A SURFACE INTEGRAL

The thickness of a curved lamina is negligible.

Figure 18.5.5

In Section 17.6 we defined a *lamina* to be an idealized thin object in 2-space. There is an analogous concept in 3-space: We define a **curved lamina** to be an idealized object in 3-space that is sufficiently thin that it can be regarded as a surface. A curved lamina may look like a bent plate, as in Figure 18.5.5, or it may enclose a region of 3-space, like the shell of an egg. If the composition of a curved lamina is uniform throughout, then the lamina is called **homogeneous** and its **density** is defined to be the total mass divided by the total surface area. However, if the lamina is not homogeneous, then the density varies from point to point, in which case the density of a curved lamina σ is described by a **density function** $\delta(x, y, z)$, whose value at point (x, y, z) is viewed as a limit

$$\delta(x, y, z) = \lim_{\Delta S \to 0} \frac{\Delta M}{\Delta S} \tag{8}$$

in which ΔM and ΔS represent the mass and surface area of a small section of lamina containing the point (x, y, z) (Figure 18.5.6). Thus, if the lamina is subdivided into n small pieces, and if (x_k^*, y_k^*, z_k^*) is a point in the kth piece and ΔS_k is the surface area of the kth piece, then from (8) the mass ΔM_k of that piece can be approximated as

$$\Delta M_k \approx \delta(x_k^*, y_k^*, z_k^*)\, \Delta S_k$$

and hence the mass M of the entire lamina can be approximated as

$$M = \sum_{k=1}^n \Delta M_k \approx \sum_{k=1}^n \delta(x_k^*, y_k^*, z_k^*)\, \Delta S_k \tag{9}$$

If we now increase n in such a way that the dimensions of the pieces approach zero, then it is plausible that the error in (9) will approach zero, and the exact value of M will be given by the surface integral

$$M = \iint_\sigma \delta(x, y, z)\, dS \tag{10}$$

Example 4 A curved lamina σ is the portion of the paraboloid $z = x^2 + y^2$ below the plane $z = 1$ (Figure 18.5.7) and has constant density $\delta(x, y, z) = \delta_0$. Find the mass of the lamina.

Solution. Since $z = g(x, y) = x^2 + y^2$, it follows that

$$\frac{\partial z}{\partial x} = 2x \quad \text{and} \quad \frac{\partial z}{\partial y} = 2y$$

Figure 18.5.6 appears at left with label (x, y, z).

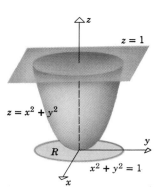

Figure 18.5.7

Substituting these expressions and $\delta(x, y, z) = \delta(x, y, g(x, y)) = \delta_0$ into (10) yields

$$M = \iint_\sigma \delta_0 \, dS = \iint_R \delta_0 \sqrt{(2x)^2 + (2y)^2 + 1} \, dA = \delta_0 \iint_R \sqrt{4x^2 + 4y^2 + 1} \, dA \qquad (11)$$

where R is the circular region enclosed by $x^2 + y^2 = 1$. To evaluate (11) we use polar coordinates:

$$M = \delta_0 \int_0^{2\pi} \int_0^1 \sqrt{4r^2 + 1} \, r \, dr \, d\theta = \frac{\delta_0}{12} \int_0^{2\pi} (4r^2 + 1)^{3/2} \Big]_{r=0}^1 d\theta$$

$$= \frac{\delta_0}{12} \int_0^{2\pi} (5^{3/2} - 1) \, d\theta = \frac{\pi \delta_0}{6} (5\sqrt{5} - 1) \qquad \blacktriangleleft$$

☐ **SURFACE INTEGRALS IN PARAMETRIC FORM**

In Section 17.4 we showed that if a surface σ is represented by a vector-valued function

$$\mathbf{r}(u, v) = x(u, v)\mathbf{i} + y(u, v)\mathbf{j} + z(u, v)\mathbf{k}$$

where (u, v) varies over a region R in the uv-plane, then the surface area ΔS_k of a small section σ_k of the surface σ can be approximated by

$$\Delta S_k \approx \left\| \frac{\partial \mathbf{r}}{\partial u} \times \frac{\partial \mathbf{r}}{\partial v} \right\| \Delta A_k$$

[see Formula (4) of Section 17.4]. If we use this formula to approximate ΔS_k in (2), and if we express the x, y, and z coordinates in terms of u and v, then the resulting formula for the surface integral is

$$\iint_\sigma f(x, y, z) \, dS = \iint_R f(x(u, v), y(u, v), z(u, v)) \left\| \frac{\partial \mathbf{r}}{\partial u} \times \frac{\partial \mathbf{r}}{\partial v} \right\| dA \qquad (12)$$

Some problems that use this formula are given in the exercises.

▶ Exercise Set 18.5

In Exercises 1–6, evaluate the surface integrals.

1. $\iint_\sigma z^2 \, dS$, where σ is the portion of the cone

$z = \sqrt{x^2 + y^2}$ between the planes $z = 1$ and $z = 2$.

2. $\iint_\sigma xy \, dS$, where σ is the portion of the plane

$x + y + z = 1$ lying in the first octant.

3. $\iint_\sigma x^2 y \, dS$, where σ is the portion of the cylinder

$x^2 + z^2 = 1$ between the planes $y = 0$, $y = 1$, and above the xy-plane.

4. $\iint_\sigma (x^2 + y^2) z \, dS$, where σ is the portion of the sphere

$x^2 + y^2 + z^2 = 4$ above the plane $z = 1$.

5. $\iint_\sigma (x + y + z) \, dS$, where σ is the portion of the plane

$x + y = 1$ in the first octant between $z = 0$ and $z = 1$.

6. $\iint_\sigma (x + y) \, dS$, where σ is the portion of the plane

$z = 6 - 2x - 3y$ in the first octant.

7. Evaluate

$$\iint_\sigma (x + y + z) \, dS$$

over the surface of the cube defined by the inequalities $0 \le x \le 1, 0 \le y \le 1, 0 \le z \le 1$. [*Hint:* Integrate over each face separately.]

8. Evaluate

$$\iint_\sigma (z + 1) \, dS$$

where σ is the upper hemisphere

$$z = \sqrt{1 - x^2 - y^2}$$

9. Evaluate

$$\iint_\sigma \sqrt{x^2 + y^2 + z^2} \, dS$$

over the portion of the cone $z = \sqrt{x^2 + y^2}$ below the plane $z = 1$.

10. Evaluate

$$\iint_{\sigma} (x^2 + y^2)\, dS$$

over the surface of the sphere $x^2 + y^2 + z^2 = a^2$. [*Hint:* Divide the sphere into two parts and integrate over each part separately.]

In Exercises 11 and 12, set up, but do not evaluate, an iterated integral equal to the given surface integral by projecting σ on (a) the xy-plane, (b) the yz-plane, and (c) the xz-plane.

11. $\iint_{\sigma} xyz\, dS$, where σ is the portion of the plane

$2x + 3y + 4z = 12$ in the first octant.

12. $\iint_{\sigma} xz\, dS$, where σ is the portion of the sphere

$x^2 + y^2 + z^2 = a^2$ in the first octant.

In Exercises 13 and 14, set up, but do not evaluate, two different iterated integrals equal to the given integral.

13. $\iint_{\sigma} xyz\, dS$, where σ is the portion of the surface $y^2 = x$

between the planes $z = 0$, $z = 4$, $y = 1$, and $y = 2$.

14. $\iint_{\sigma} x^2 y\, dS$, where σ is the portion of the cylinder

$y^2 + z^2 = a^2$ in the first octant between the planes $x = 0$, $x = 9$, $z = y$, and $z = 2y$.

In Exercises 15–18, find the mass of the given lamina assuming the density to be a constant δ_0.

15. The lamina that is the portion of the paraboloid $z = 1 - x^2 - y^2$ above the xy-plane.

16. The lamina that is the portion of the plane $2x + 2y + z = 8$ in the first octant.

17. The lamina that is the portion of the circular cylinder $x^2 + z^2 = 4$ that lies directly above the rectangle $R = \{(x, y): 0 \le x \le 1, 0 \le y \le 4\}$ in the xy-plane.

18. The lamina that is the portion of the paraboloid $2z = x^2 + y^2$ inside the cylinder $x^2 + y^2 = 8$.

19. Find the mass of the lamina that is the portion of the surface $y^2 = 4 - z$ between the planes $x = 0$, $x = 3$, $y = 0$, and $y = 3$ if the density is $\delta(x, y, z) = y$.

20. Find the mass of the lamina that is the portion of the cone $z = \sqrt{x^2 + y^2}$ between $z = 1$ and $z = 4$ if the density is $\delta(x, y, z) = x^2 z$.

21. Use (10) to verify that if a curved lamina has constant density δ_0, then its mass is the density times the surface area.

22. Show that the mass of the spherical lamina given by $x^2 + y^2 + z^2 = a^2$ is $2\pi a^3$ if the density at any point is equal to the distance from the point to the xy-plane.

The centroid of a curved lamina σ is defined by

$$\bar{x} = \frac{\displaystyle\iint_{\sigma} x\, dS}{\text{area of } \sigma}, \quad \bar{y} = \frac{\displaystyle\iint_{\sigma} y\, dS}{\text{area of } \sigma}, \quad \bar{z} = \frac{\displaystyle\iint_{\sigma} z\, dS}{\text{area of } \sigma}$$

In Exercises 23 and 24, find the centroid of the lamina.

23. The lamina that is the portion of the paraboloid $z = \frac{1}{2}(x^2 + y^2)$ below the plane $z = 4$.

24. The lamina that is the portion of the sphere $x^2 + y^2 + z^2 = 4$ above the plane $z = 1$.

In Exercises 25–28, evaluate the surface integrals.

25. $\iint_{\sigma} xyz\, dS$, where σ is the portion of the cone

$$\mathbf{r}(u, v) = u \cos v\mathbf{i} + u \sin v\mathbf{j} + 3u\mathbf{k}$$

for which $1 \le u \le 2$, $0 \le v \le \pi/2$.

26. $\iint_{\sigma} \dfrac{x^2 + z^2}{y}\, dS$, where σ is the portion of the cylinder

$$\mathbf{r}(u, v) = 2 \cos v\mathbf{i} + u\mathbf{j} + 2 \sin v\mathbf{k}$$

for which $1 \le u \le 3$, $0 \le v \le 2\pi$.

27. $\iint_{\sigma} \dfrac{1}{\sqrt{1 + 4x^2 + 4y^2}}\, dS$, where σ is the portion of the paraboloid

$$\mathbf{r}(u, v) = u \cos v\mathbf{i} + u \sin v\mathbf{j} + u^2\mathbf{k}$$

for which $0 \le u \le \sin v$, $0 \le v \le \pi$.

28. $\iint_{\sigma} e^{-z}\, dS$, where σ is the hemisphere

$$\mathbf{r}(u, v) = 2 \sin u \cos v\mathbf{i} + 2 \sin u \sin v\mathbf{j} + 2 \cos u\mathbf{k}$$

for which $0 \le u \le \pi/2$, $0 \le v \le 2\pi$.

■ **18.6** SURFACE INTEGRALS OF VECTOR FIELDS; FLUX

> *In this section we shall focus on applications of surface integrals to vector fields such as those associated with fluid flow and electrostatic forces. However, the ideas that we shall develop are quite general and are applicable to almost all vector fields.*

□ **FLOW FIELDS**

Throughout this section

$$\mathbf{F}(x, y, z) = f(x, y, z)\mathbf{i} + g(x, y, z)\mathbf{j} + h(x, y, z)\mathbf{k}$$

can be any vector field in 3-space. However, we will motivate most of our ideas by focusing on vector fields arising in fluid flow and electrostatics. In the case of fluid flow, **F** will represent the velocity of a fluid particle at the point (x, y, z) (Figure 18.6.1a). In the case of electrostatics, we will assume that charged particles are distributed at various fixed points in 3-space and that $\mathbf{F}(x, y, z)$ is the combined force that these charges exert on a *test particle* of unit positive charge located at the point (x, y, z) (Figures 18.6.1b and 18.6.1c). In both the fluid flow and electrostatic cases it will be convenient to call **F** a *flow field*, even though nothing is actually flowing in the electrostatic case.

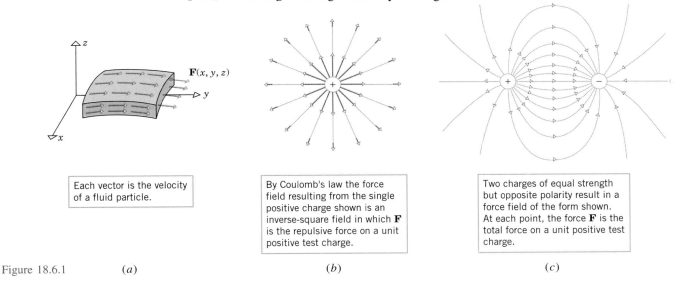

Each vector is the velocity of a fluid particle.

By Coulomb's law the force field resulting from the single positive charge shown is an inverse-square field in which **F** is the repulsive force on a unit positive test charge.

Two charges of equal strength but opposite polarity result in a force field of the form shown. At each point, the force **F** is the total force on a unit positive test charge.

Figure 18.6.1 (a) (b) (c)

□ **ORIENTED SURFACES**

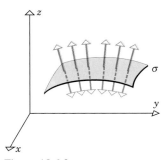

Figure 18.6.2

Suppose that a surface σ is constructed from permeable material through which fluid can flow freely. We shall be concerned with the flow of fluid in directions *normal* to the surface σ. However, because surfaces in 3-space can be extremely complex, it will be necessary to impose certain restrictions on σ to keep the mathematics from getting out of hand.

We will call a surface σ *smooth* if it has a tangent plane (and hence a normal vector) at every point that is not on the boundary of σ. In general, if (x, y, z) is a point on a smooth surface σ, then there are two oppositely directed unit normal vectors at the point, each of which defines a possible direction of fluid flow normal to the surface. For example, in Figure 18.6.2 the fluid can flow roughly "upward" or "downward" through the surface, and in Figure 18.6.3 the fluid can flow "inward" or "outward" through the sphere.

A smooth surface σ is said to be *orientable* if it is possible to construct a unit normal vector at each point of the surface in such a way that the vectors vary continuously (i.e., have no abrupt changes in direction) as we traverse any smooth curve on the surface. The unit normal vectors are then said to form an *orientation* of the surface. It is proved in

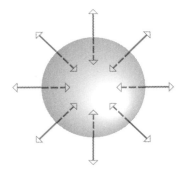

Figure 18.6.3

advanced mathematics courses that an orientable surface has only two possible orientations. For example, the surface in Figure 18.6.4 can be oriented upward, as in part (*a*), or downward, as in part (*b*). However, the vectors in part (*c*) do not define an orientation because the directions of those vectors change abruptly as we cross the curve drawn on the surface. As shown in Figure 18.6.3 a sphere can be oriented by *inward normals* or by *outward normals*.

Figure 18.6.4

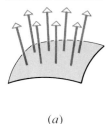 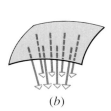 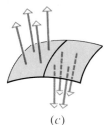

(*a*)   (*b*)   (*c*)

REMARK.   Observe that the notion of orientability applies only to smooth surfaces. However, many important surfaces are not themselves smooth, but are unions of finitely many smooth orientable surfaces. A box, for example, is not smooth, since no tangent planes exist at the edges. However, each face is a smooth orientable surface. It should also be noted that not every smooth surface is orientable. In the exercises we discuss the **Möbius strip**, which is a famous nonorientable surface.

☐ **FLUX**

In physics, the term *fluid* is used to describe both liquids and gases. Liquids (such as water) are usually regarded to be **incompressible**, meaning that the liquid has a uniform density (mass per unit volume) that cannot be altered by compressive forces. Gases (such as air) are regarded to be **compressible**, meaning that the density may vary from point to point and can be altered by compressive forces. In all of our discussions we shall be concerned only with incompressible fluids. Moreover, we shall assume that the velocity of the fluid at a fixed point does not vary with time. Fluid flows with this property are said to be in a **steady state**.

Our first goal in this section is to define a fundamental concept of physics known as *flux* (from the Latin word *fluxus*, meaning ''flow''). This concept is applicable in any vector field, but we shall motivate it in the context of steady-state flow of an incompressible fluid. We consider the following problem:

> **18.6.1**   PROBLEM.   *Let* $\mathbf{F}(x, y, z) = f(x, y, z)\mathbf{i} + g(x, y, z)\mathbf{j} + h(x, y, z)\mathbf{k}$ *be a fluid flow field for which f, g, and h are continuous and have continuous first partial derivatives, and let σ be an orientable surface in the fluid with an orientation defined by the unit normal vectors* $\mathbf{n} = \mathbf{n}(x, y, z)$. *Find the net mass* Φ *of fluid that passes through the surface per unit time, where the net mass is interpreted to mean the mass passing through the surface in the direction of orientation minus the mass passing through the surface in the direction opposite to the orientation.*

REMARK.   For simplicity, we shall assume that the fluid in this problem has density 1. Since density is mass per unit volume, it follows from this assumption that the net mass of fluid that passes through the surface per unit time is numerically the same as the net volume of fluid that passes through the surface per unit time (although the units differ). Thus, we can replace the word ''mass'' by ''volume'' in this problem.

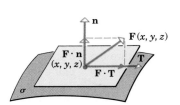

Figure 18.6.5

Before attempting to solve this problem, we note that at each point (x, y, z) on the surface the velocity vector $\mathbf{F}(x, y, z)$ can be resolved into two components, a component $\mathbf{F} \cdot \mathbf{n}$ along the unit normal vector $\mathbf{n}$ in the orientation of σ and a component $\mathbf{F} \cdot \mathbf{T}$ tangent to the surface (Figure 18.6.5). The component $\mathbf{F} \cdot \mathbf{T}$ is the component of flow *along* the surface, and $\mathbf{F} \cdot \mathbf{n}$ is the component *across* the surface. Fluid crosses σ in the direction of orientation where $\mathbf{F} \cdot \mathbf{n}$ is positive and opposite to that direction where $\mathbf{F} \cdot \mathbf{n}$ is negative.

To solve Problem 18.6.1, we subdivide σ into n parts $\sigma_1, \sigma_2, \ldots, \sigma_n$ with areas

$$\Delta S_1, \Delta S_2, \ldots, \Delta S_n$$

If the parts are small and the flow is not too erratic, it is reasonable to assume that the velocity does not vary much on each part. Thus, if (x_k^*, y_k^*, z_k^*) is any point in the kth part, we can assume that $\mathbf{F}(x, y, z)$ is constant and equal to $\mathbf{F}(x_k^*, y_k^*, z_k^*)$ throughout the part, and that the component of flow across the surface σ_k is

$$\mathbf{F}(x_k^*, y_k^*, z_k^*) \cdot \mathbf{n}(x_k^*, y_k^*, z_k^*) \tag{1}$$

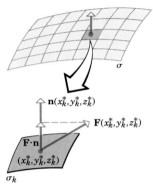

(Figure 18.6.6). If we observe the flow across σ_k for one unit of time, then the section of fluid initially on the surface will move to some new position, sweeping out a solid whose volume ΔV_k can be approximated by

$$\Delta V_k \approx \begin{bmatrix} \text{distance traveled by} \\ \text{the fluid in one unit} \\ \text{of time} \end{bmatrix} \Delta S_k \tag{2}$$

(Figure 18.6.7). But (1) is the component of fluid velocity across the surface and therefore represents either the distance traveled by the fluid in one unit of time or the negative of that distance, depending on whether $\mathbf{F} \cdot \mathbf{n}$ is positive or negative. Thus, from (2)

$$\mathbf{F}(x_k^*, y_k^*, z_k^*) \cdot \mathbf{n}(x_k^*, y_k^*, z_k^*) \, \Delta S_k$$

Figure 18.6.6

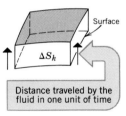

is either ΔV_k [if the fluid is moving in the direction of $\mathbf{n}(x_k^*, y_k^*, z_k^*)$] or $-\Delta V_k$ [if the fluid is moving in the direction opposite to $\mathbf{n}(x_k^*, y_k^*, z_k^*)$]. Thus, the net volume of fluid passing through σ in one unit of time can be approximated as

$$\Phi \approx \sum_{k=1}^{n} \mathbf{F}(x_k^*, y_k^*, z_k^*) \cdot \mathbf{n}(x_k^*, y_k^*, z_k^*) \, \Delta S_k$$

If we now increase n in such a way that dimensions of the surface parts approach zero, then it is plausible that the errors in the approximations approach zero, and the exact value of Φ is

$$\Phi = \lim_{n \to +\infty} \sum_{k=1}^{n} \mathbf{F}(x_k^*, y_k^*, z_k^*) \cdot \mathbf{n}(x_k^*, y_k^*, z_k^*) \, \Delta S_k$$

Figure 18.6.7

which can be expressed as the surface integral

$$\Phi = \iint_{\sigma} \mathbf{F}(x, y, z) \cdot \mathbf{n}(x, y, z) \, dS \tag{3}$$

The quantity Φ defined by this integral is called the *__flux of F across σ__*. The concept of flux is applicable to any vector field in 3-space. However, in the case of incompressible steady-state fluid flow with a fluid of mass 1, we have seen that Φ represents the net mass of fluid passing through σ per unit time.

☐ **CALCULATING FLUX**

To calculate the flux of $\mathbf{F}$ across an oriented surface σ from (3), it is first necessary to find a formula for the normal vector $\mathbf{n}(x, y, z)$ that appears in the integrand. Such formulas depend on the form in which the surface σ is expressed. For example, Table 18.6.1 lists the two possible orientations and the formulas for $\mathbf{n} = \mathbf{n}(x, y, z)$ for surfaces of the form $z = g(x, y)$, $y = g(x, z)$, and $x = g(y, z)$.

The formulas in Table 18.6.1 can be derived by noting that the surfaces $z = g(x, y)$, $y = g(x, z)$, and $x = g(y, z)$ can all be expressed in the form $G(x, y, z) = 0$ by taking the function g to the left side of the equation. In all three cases the surface σ is a level surface

Table 18.6.1

$z = g(x, y)$		$y = g(x, z)$		$x = g(y, z)$	
Positive **k** component	$\mathbf{n} = \dfrac{-\frac{\partial z}{\partial x}\mathbf{i} - \frac{\partial z}{\partial y}\mathbf{j} + \mathbf{k}}{\sqrt{\left(\frac{\partial z}{\partial x}\right)^2 + \left(\frac{\partial z}{\partial y}\right)^2 + 1}}$ Positive orientation (up)	Positive **j** component	$\mathbf{n} = \dfrac{-\frac{\partial y}{\partial x}\mathbf{i} + \mathbf{j} - \frac{\partial y}{\partial z}\mathbf{k}}{\sqrt{\left(\frac{\partial y}{\partial x}\right)^2 + \left(\frac{\partial y}{\partial z}\right)^2 + 1}}$ Positive orientation (right)	Positive **i** component	$\mathbf{n} = \dfrac{\mathbf{i} - \frac{\partial x}{\partial y}\mathbf{j} - \frac{\partial x}{\partial z}\mathbf{k}}{\sqrt{\left(\frac{\partial x}{\partial y}\right)^2 + \left(\frac{\partial x}{\partial z}\right)^2 + 1}}$ Positive orientation (front)
Negative **k** component	$\mathbf{n} = \dfrac{\frac{\partial z}{\partial x}\mathbf{i} + \frac{\partial z}{\partial y}\mathbf{j} - \mathbf{k}}{\sqrt{\left(\frac{\partial z}{\partial x}\right)^2 + \left(\frac{\partial z}{\partial y}\right)^2 + 1}}$ Negative orientation (down)	Negative **j** component	$\mathbf{n} = \dfrac{\frac{\partial y}{\partial x}\mathbf{i} - \mathbf{j} + \frac{\partial y}{\partial z}\mathbf{k}}{\sqrt{\left(\frac{\partial y}{\partial x}\right)^2 + \left(\frac{\partial y}{\partial z}\right)^2 + 1}}$ Negative orientation (left)	Negative **i** component	$\mathbf{n} = \dfrac{-\mathbf{i} + \frac{\partial x}{\partial y}\mathbf{j} + \frac{\partial x}{\partial z}\mathbf{k}}{\sqrt{\left(\frac{\partial x}{\partial y}\right)^2 + \left(\frac{\partial x}{\partial z}\right)^2 + 1}}$ Negative orientation (back)

for $G(x, y, z)$, so it follows from Theorem 16.7.6 that ∇G is normal to σ at (x, y, z) and hence that

$$\frac{\nabla G}{\|\nabla G\|} \quad \text{and} \quad -\frac{\nabla G}{\|\nabla G\|} \tag{4}$$

are unit normal vectors to σ at (x, y, z). These are the vectors listed in Table 18.6.1. For example, if $z = g(x, y)$, then

$$G(x, y, z) = z - g(x, y)$$

so

$$\nabla G = -\frac{\partial g}{\partial x}\mathbf{i} - \frac{\partial g}{\partial y}\mathbf{j} + \mathbf{k} = -\frac{\partial z}{\partial x}\mathbf{i} - \frac{\partial z}{\partial y}\mathbf{j} + \mathbf{k} \tag{5}$$

and hence

$$\frac{\nabla G}{\|\nabla G\|} = \frac{-\frac{\partial z}{\partial x}\mathbf{i} - \frac{\partial z}{\partial y}\mathbf{j} + \mathbf{k}}{\sqrt{\left(\frac{\partial z}{\partial x}\right)^2 + \left(\frac{\partial z}{\partial y}\right)^2 + 1}}$$

which is the normal for the positive orientation of $z = g(x, y)$ shown in Table 18.6.1. Multiplying by -1 produces the negative orientation.

Example 1 Let σ be the portion of the surface $z = 1 - x^2 - y^2$ that lies above the xy-plane, and suppose that σ is oriented upward (Figure 18.6.8). Find the flux Φ of the flow field $\mathbf{F}(x, y, z) = x\mathbf{i} + y\mathbf{j} + z\mathbf{k}$ across σ.

Solution. Because the surface σ is of the form $z = g(x, y)$, we use Formula (3) of Section 18.5 to express (3) as a double integral. This yields

$$\Phi = \iint_\sigma \mathbf{F} \cdot \mathbf{n} \, dS = \iint_R (\mathbf{F} \cdot \mathbf{n}) \sqrt{\left(\frac{\partial z}{\partial x}\right)^2 + \left(\frac{\partial z}{\partial y}\right)^2 + 1} \, dA$$

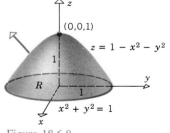

Figure 18.6.8

where R is the projection of the surface σ on the xy-plane (Figure 18.6.8). Since σ is oriented upward, we substitute the formula for $\mathbf{n}$ given in Table 18.6.1 for the positive (up) orientation of $z = g(x, y)$; this yields

$$\Phi = \iint_R \mathbf{F} \cdot \left[\frac{-\dfrac{\partial z}{\partial x}\mathbf{i} - \dfrac{\partial z}{\partial y}\mathbf{j} + \mathbf{k}}{\sqrt{\left(\dfrac{\partial z}{\partial x}\right)^2 + \left(\dfrac{\partial z}{\partial y}\right)^2 + 1}} \right] \sqrt{\left(\frac{\partial z}{\partial x}\right)^2 + \left(\frac{\partial z}{\partial y}\right)^2 + 1}\, dA$$

$$= \iint_R \mathbf{F} \cdot \left(-\frac{\partial z}{\partial x}\mathbf{i} - \frac{\partial z}{\partial y}\mathbf{j} + \mathbf{k} \right) dA$$

$$= \iint_R (x\mathbf{i} + y\mathbf{j} + z\mathbf{k}) \cdot (2x\mathbf{i} + 2y\mathbf{j} + \mathbf{k})\, dA$$

$$= \iint_R (x^2 + y^2 + 1)\, dA \qquad \boxed{\text{Since } z = 1 - x^2 - y^2 \text{ on the surface.}}$$

$$= \int_0^{2\pi} \int_0^1 (r^2 + 1)r\, dr\, d\theta \qquad \boxed{\text{Using polar coordinates to evaluate the integral}}$$

$$= \int_0^{2\pi} \left(\frac{3}{4}\right) d\theta = \frac{3\pi}{2} \qquad \blacktriangleleft$$

The cancelation of the radicals in the preceding example was not accidental; it is a consequence of the following theorem.

18.6.2 THEOREM. *Let σ be a smooth surface of the form $z = g(x, y)$, $y = g(x, z)$, or $x = g(y, z)$, and suppose that the equation is rewritten as $G(x, y, z) = 0$ by taking g to the left side. Let R be the projection of σ on the xy-plane if $z = g(x, y)$, on the xz-plane if $y = g(x, z)$, and on the yz-plane if $x = g(y, z)$. If g is continuous, and has continuous first partial derivatives on R, then*

$$\iint_\sigma \mathbf{F} \cdot \mathbf{n}\, dS = \pm \iint_R \mathbf{F} \cdot \nabla G\, dA \tag{6}$$

where the $+$ sign is used if σ has positive orientation and the $-$ sign if it has negative orientation.

Although we omit the formal proof, the basic idea is to apply Theorem 18.5.1. In all three parts of that theorem the radical can be expressed in terms of the function G as $\|\nabla G\|$ (verify). Thus, from Theorem 18.5.1 and the formulas in (4) for the normal it follows that

$$\iint_\sigma \mathbf{F} \cdot \mathbf{n}\, dS = \iint_R \mathbf{F} \cdot \left(\pm \frac{\nabla G}{\|\nabla G\|} \right) \|\nabla G\|\, dA$$

$$= \pm \iint_R \mathbf{F} \cdot \nabla G\, dA$$

In the case where $z = g(x, y)$, we have $G(x, y, z) = z - g(x, y)$, so that

$$\nabla G = -\frac{\partial g}{\partial x}\mathbf{i} - \frac{\partial g}{\partial y}\mathbf{j} + \mathbf{k} = -\frac{\partial z}{\partial x}\mathbf{i} - \frac{\partial z}{\partial y}\mathbf{j} + \mathbf{k}$$

Substituting this expression for ∇G in (6) and taking R to be the projection of the surface $z = g(x, y)$ on the xy-plane yields the following formulas:

$$\iint_\sigma \mathbf{F} \cdot \mathbf{n}\, dS = \iint_R \mathbf{F} \cdot \left(-\frac{\partial z}{\partial x}\mathbf{i} - \frac{\partial z}{\partial y}\mathbf{j} + \mathbf{k}\right) dA \qquad \boxed{\sigma \text{ oriented up}} \qquad (7)$$

$$\iint_\sigma \mathbf{F} \cdot \mathbf{n}\, dS = \iint_R \mathbf{F} \cdot \left(\frac{\partial z}{\partial x}\mathbf{i} + \frac{\partial z}{\partial y}\mathbf{j} - \mathbf{k}\right) dA \qquad \boxed{\sigma \text{ oriented down}} \qquad (8)$$

The derivation of the corresponding formulas for the cases where $y = g(x, z)$ and $x = g(y, z)$ are left as exercises.

Example 2   Let σ be the sphere $x^2 + y^2 + z^2 = a^2$ oriented by outward normals (Figure 18.6.9), and let $\mathbf{F}(x, y, z) = z\mathbf{k}$. Evaluate

$$\iint_\sigma \mathbf{F} \cdot \mathbf{n}\, dS$$

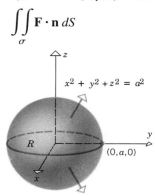

Figure 18.6.9

Solution.   On the upper hemisphere the outward unit normal is upward, while on the lower hemisphere it is downward. Since different formulas for these normals apply on the two hemispheres, it is desirable to write

$$\iint_\sigma \mathbf{F} \cdot \mathbf{n}\, dS = \iint_{\sigma_1} \mathbf{F} \cdot \mathbf{n}\, dS + \iint_{\sigma_2} \mathbf{F} \cdot \mathbf{n}\, dS \qquad (9)$$

where σ_1 is the upper hemisphere and σ_2 is the lower hemisphere.

The upper hemisphere σ_1 has the equation

$$z = \sqrt{a^2 - x^2 - y^2} \qquad (10)$$

so that (7) yields

$$\iint_{\sigma_1} \mathbf{F} \cdot \mathbf{n}\, dS = \iint_R (z\mathbf{k}) \cdot \left(\frac{x}{\sqrt{a^2 - x^2 - y^2}}\mathbf{i} + \frac{y}{\sqrt{a^2 - x^2 - y^2}}\mathbf{j} + \mathbf{k}\right) dA$$

$$= \iint_R z\, dA = \iint_R \sqrt{a^2 - x^2 - y^2}\, dA \qquad \boxed{\begin{array}{l} R \text{ is the region shown} \\ \text{in Figure 18.6.9.} \end{array}}$$

$$= \int_0^{2\pi} \int_0^a \sqrt{a^2 - r^2}\, r\, dr\, d\theta$$

$$= \int_0^{2\pi} -\frac{1}{3}(a^2 - r^2)^{3/2}\bigg]_0^a d\theta$$

$$= \int_0^{2\pi} \frac{1}{3}a^3\, d\theta = \frac{2\pi a^3}{3}$$

The lower hemisphere σ_2 has the equation

$$z = -\sqrt{a^2 - x^2 - y^2}$$

so that (8) yields

$$\iint\limits_{\sigma_2} \mathbf{F} \cdot \mathbf{n} \, dS = \iint\limits_{R} (z\mathbf{k}) \cdot \left(\frac{x}{\sqrt{a^2 - x^2 - y^2}} \mathbf{i} + \frac{y}{\sqrt{a^2 - x^2 - y^2}} \mathbf{j} - \mathbf{k} \right) dA$$

$$= \iint\limits_{R} -z \, dA = \iint\limits_{R} \sqrt{a^2 - x^2 - y^2} \, dA$$

$$= \frac{2\pi a^3}{3} \quad \boxed{\begin{array}{l} \text{The computations} \\ \text{are identical to} \\ \text{those above.} \end{array}}$$

Thus, from (9)

$$\iint\limits_{\sigma} \mathbf{F} \cdot \mathbf{n} \, dS = \frac{2\pi a^3}{3} + \frac{2\pi a^3}{3} = \frac{4\pi a^3}{3} \quad \blacktriangleleft$$

▶ Exercise Set 18.6

1. Use three different formulas in Table 18.6.1 to calculate the unit normal to $2x + 3y + 4z = 9$ at $(1, 1, 1)$ that has positive components.

2. Use three different formulas in Table 18.6.1 to calculate the unit normal to $x^2 + y^2 + z^2 = 9$ at $(2, 1, -2)$ that points below the xy-plane.

3. In parts (a)–(c), use any appropriate formula to calculate the indicated unit normal.

    (a) The unit normal to $z = x^2 + y^2$ at $(1, 2, 5)$ that points toward the z-axis

    (b) The unit normal to $z = \sqrt{x^2 + y^2}$ at $(-3, 4, 5)$ that points toward the xz-plane

    (c) The unit normal to the cylinder $x^2 + z^2 = 25$ at $(3, 2, -4)$ that points away from the xy-plane.

4. In each part, use any appropriate formula to calculate the indicated unit normal.

    (a) The unit normal to the surface $y^2 = x$ at $(1, 1, 2)$ that points toward the xz-plane

    (b) The unit normal to the hyperbolic paraboloid $y = z^2 - x^2$ at $(1, 3, 2)$ that points toward the yz-plane

    (c) The unit normal to the cone $x^2 = y^2 + z^2$ at $(\sqrt{2}, -1, -1)$ that points away from the xy-plane.

In Exercises 5–15, evaluate $\iint\limits_{\sigma} \mathbf{F} \cdot \mathbf{n} \, dS$.

5. $\mathbf{F}(x, y, z) = x\mathbf{i} + y\mathbf{j} + 2z\mathbf{k}$; σ is the portion of the surface $z = 1 - x^2 - y^2$ above the xy-plane, oriented by upward normals.

6. $\mathbf{F}(x, y, z) = (x + y)\mathbf{i} + (y + z)\mathbf{j} + (z + x)\mathbf{k}$; σ is the portion of the plane $x + y + z = 1$ in the first octant, oriented by unit normals with positive components.

7. $\mathbf{F}(x, y, z) = z^2\mathbf{k}$; σ is the upper hemisphere given by $z = \sqrt{1 - x^2 - y^2}$, oriented by upward unit normals.

8. $\mathbf{F}(x, y, z) = x^2\mathbf{i} + yx\mathbf{j} + zx\mathbf{k}$; σ is the portion of the plane $6x + 3y + 2z = 6$ in the first octant, oriented by unit normals with positive components.

9. $\mathbf{F}(x, y, z) = x\mathbf{i} + y\mathbf{j} + z\mathbf{k}$; σ is the upper hemisphere $z = \sqrt{9 - x^2 - y^2}$, oriented by upward unit normals.

10. $\mathbf{F}(x, y, z) = \mathbf{i} + \mathbf{j} + \mathbf{k}$; σ is the portion of the cone $z = \sqrt{x^2 + y^2}$ below the plane $z = 1$, oriented by downward unit normals.

11. $\mathbf{F}(x, y, z) = x\mathbf{i} + y\mathbf{j} + 2z\mathbf{k}$; σ is the portion of the cone $z^2 = x^2 + y^2$ between the planes $z = 1$ and $z = 2$, oriented by upward unit normals.

12. $\mathbf{F}(x, y, z) = y\mathbf{j} + \mathbf{k}$; σ is the portion of the paraboloid $z = x^2 + y^2$ below the plane $z = 4$, oriented by downward unit normals.

13. $\mathbf{F}(x, y, z) = x\mathbf{k}$; σ is the portion of the paraboloid $z = x^2 + y^2$ below the plane $z = y$, oriented by downward unit normals.

14. $\mathbf{F}(x, y, z) = x\mathbf{i} + y\mathbf{j} + z\mathbf{k}$; σ is the portion of the cylinder $z^2 = 1 - x^2$ between the planes $y = 1$ and $y = -2$, oriented by outward unit normals.

15. $\mathbf{F}(x, y, z) = x\mathbf{i} + y\mathbf{j} + z\mathbf{k}$, where σ is the sphere $x^2 + y^2 + z^2 = a^2$ oriented by outward unit normals.

16. Let σ be the surface of the cube bounded by the planes $x = \pm 1, y = \pm 1, z = \pm 1$, oriented by outward unit normals. In each part, find the flux of $\mathbf{F}$ over σ.

    (a) $\mathbf{F}(x, y, z) = x\mathbf{i}$

    (b) $\mathbf{F}(x, y, z) = x\mathbf{i} + y\mathbf{j} + z\mathbf{k}$

    (c) $\mathbf{F}(x, y, z) = x^2\mathbf{i} + y^2\mathbf{j} + z^2\mathbf{k}$.

17. Show that reversing the orientation of σ reverses the sign of

$$\iint\limits_{\sigma} \mathbf{F} \cdot \mathbf{n} \, dS$$

18. Let $\mathbf{F} = \|\mathbf{r}\|^k \mathbf{r}$, where $\mathbf{r} = x\mathbf{i} + y\mathbf{j} + z\mathbf{k}$ and k is a constant. (If $k = -3$, this is an inverse-square field.) Let σ be the sphere of radius a centered at the origin and oriented by the outward normal $\mathbf{n} = \mathbf{r}/\|\mathbf{r}\| = \mathbf{r}/a$.

   (a) Evaluate $\iint\limits_{\sigma} \mathbf{F} \cdot \mathbf{n} \, dS$ without performing any integrations. [*Hint:* The surface area of a sphere of radius a is $4\pi a^2$.]

   (b) For what value of k is the integral in part (a) independent of the radius of the sphere?

19. Obtain the analogs of Formulas (7) and (8) for

   (a) surfaces of the form $x = g(y, z)$

   (b) surfaces of the form $y = g(x, z)$.

20. Evaluate $\iint\limits_{\sigma} \mathbf{F} \cdot \mathbf{n} \, dS$ where σ is the portion of the paraboloid $x = y^2 + z^2$ with $x \le 1$ and $z \ge 0$ oriented by backward unit normals and

$$\mathbf{F}(x, y, z) = y\mathbf{i} - z\mathbf{j} + 8\mathbf{k}$$

   [*Hint:* Exercise 19(a).]

21. Evaluate $\iint\limits_{\sigma} \mathbf{F} \cdot \mathbf{n} \, dS$ if σ is the hemisphere $y = \sqrt{1 - x^2 - z^2}$ oriented by right unit normals and $\mathbf{F}(x, y, z) = x\mathbf{i} + y\mathbf{j} + z\mathbf{k}$. [*Hint:* Exercise 19(b).]

22. The best-known example of a nonorientable surface is the **Möbius strip**, which can be visualized by taking a band of paper, twisting it once, and gluing the ends together (Figures 18.6.10*a* and 18.6.10*b*). In Figure 18.6.10*c*, we have tried to construct unit normal vectors whose directions vary continuously moving counterclockwise around the dashed curve starting and finishing on the vertical line.

   (a) Explain why these vectors are not part of an orientation of the surface.

   (b) Explain why the surface is not orientable.

If a surface is represented parametrically by

$$\mathbf{r}(u, v) = x(u, v)\mathbf{i} + y(u, v)\mathbf{j} + z(u, v)\mathbf{k}$$

where (u, v) varies over a region R in the uv-plane, then from (6) in Section 16.3 a unit normal vector to the surface is given by

$$\mathbf{n} = \frac{\dfrac{\partial \mathbf{r}}{\partial u} \times \dfrac{\partial \mathbf{r}}{\partial v}}{\left\| \dfrac{\partial \mathbf{r}}{\partial u} \times \dfrac{\partial \mathbf{r}}{\partial v} \right\|}$$

If a surface σ has the orientation determined by this $\mathbf{n}$, then from (12) in Section 18.5 with $f(x, y, z) = \mathbf{F} \cdot \mathbf{n}$ we obtain

$$\iint\limits_{\sigma} \mathbf{F} \cdot \mathbf{n} \, dS = \iint\limits_{R} \mathbf{F} \cdot \left(\frac{\partial \mathbf{r}}{\partial u} \times \frac{\partial \mathbf{r}}{\partial v} \right) dA$$

where the integrand on the right is expressed in terms of u and v. In Exercises 23–26, use this to evaluate $\iint\limits_{\sigma} \mathbf{F} \cdot \mathbf{n} \, dS$.

23. $\mathbf{F}(x, y, z) = x\mathbf{i} + y\mathbf{j} + \mathbf{k}$; σ is the portion of the paraboloid

$$\mathbf{r}(u, v) = u \cos v\mathbf{i} + u \sin v\mathbf{j} + (1 - u^2)\mathbf{k}$$

for which $1 \le u \le 2, \, 0 \le v \le 2\pi$.

24. $\mathbf{F}(x, y, z) = e^{-y}\mathbf{i} - y\mathbf{j} + x \sin z\mathbf{k}$; σ is the portion of the elliptic cylinder

$$\mathbf{r}(u, v) = 2 \cos v\mathbf{i} + \sin v\mathbf{j} + u\mathbf{k}$$

for which $0 \le u \le 5, \, 0 \le v \le 2\pi$.

25. $\mathbf{F}(x, y, z) = \sqrt{x^2 + y^2}\mathbf{k}$; σ is the portion of the cone

$$\mathbf{r}(u, v) = u \cos v\mathbf{i} + u \sin v\mathbf{j} + 2u\mathbf{k}$$

for which $0 \le u \le \sin v, \, 0 \le v \le \pi$.

26. $\mathbf{F}(x, y, z) = z^2\mathbf{k}$; σ is the portion of the sphere

$$\mathbf{r}(u, v) = 2 \sin u \cos v\mathbf{i} + 2 \sin u \sin v\mathbf{j} + 2 \cos u\mathbf{k}$$

for which $0 \le u \le \pi/3, \, 0 \le v \le 2\pi$.

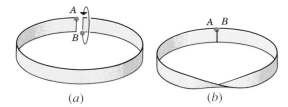

(*a*)   (*b*)   (*c*)   Figure 18.6.10

■ 18.7   THE DIVERGENCE THEOREM

Recall that Green's Theorem establishes a relationship between a double integral over a plane region and a line integral over the boundary of the region. In this section we shall establish a relationship between a triple integral over a solid and a surface integral over the boundary of the solid. This relationship is the cornerstone of many important principles in physics.

□ **THE DIVERGENCE THEOREM**

In this section we will be concerned with surfaces that are boundaries of solids in 3-space—for example, the surface of a solid sphere, the surface of a solid box, or the surface of a solid cylinder (the wall and the end caps). Such surfaces are said to be ***closed***. If a closed surface is orientable or piecewise orientable, then the two possible orientations are called ***inward*** (toward the solid) and ***outward*** (away from the solid).

In Section 18.1 we defined the *divergence* of a vector field

$$\mathbf{F}(x, y, z) = f(x, y, z)\mathbf{i} + g(x, y, z)\mathbf{j} + h(x, y, z)\mathbf{k}$$

as

$$\operatorname{div} \mathbf{F} = \frac{\partial f}{\partial x} + \frac{\partial g}{\partial y} + \frac{\partial h}{\partial z}$$

The following result, known as the ***Divergence Theorem*** or ***Gauss'* Theorem***, shows that under appropriate conditions the flux of a vector field across a closed surface with outward

---

* CARL FRIEDRICH GAUSS (1777–1855). German mathematician and scientist. Sometimes called the "prince of mathematicians," Gauss ranks with Newton and Archimedes as one of the three greatest mathematicians who ever lived. His father, a laborer, was an uncouth but honest man who would have liked Gauss to take up a trade such as gardening or bricklaying; but the boy's genius for mathematics was not to be denied. In the entire history of mathematics there may never have been a child so precocious as Gauss—by his own account he worked out the rudiments of arithmetic before he could talk. One day, before he was even three years old, his genius became apparent to his parents in a very dramatic way. His father was preparing the weekly payroll for the laborers under his charge while the boy watched quietly from a corner. At the end of the long and tedious calculation, Gauss informed his father that there was an error in the result and stated the answer, which he had worked out in his head. To the astonishment of his parents, a check of the computations showed Gauss to be correct!

For his elementary education Gauss was enrolled in a squalid school run by a man named Büttner whose main teaching technique was thrashing. Büttner was in the habit of assigning long addition problems which, unknown to his students, were arithmetic progressions that he could sum up using formulas. On the first day that Gauss entered the arithmetic class, the students were asked to sum the numbers from 1 to 100. But no sooner had Büttner stated the problem than Gauss turned over his slate and exclaimed in his peasant dialect, "Ligget se'." (Here it lies.) For nearly an hour Büttner glared at Gauss, who sat with folded hands while his classmates toiled away. When Büttner examined the slates at the end of the period, Gauss' slate contained a single number, 5050—the only correct solution in the class.

To his credit, Büttner recognized the genius of Gauss and with the help of his assistant, John Bartels, had him brought to the attention of Karl Wilhelm Ferdinand, Duke of Brunswick. The shy and awkward boy, who was then fourteen, so captivated the Duke that he subsidized him through preparatory school, college, and the early part of his career.

From 1795 to 1798 Gauss studied mathematics at the University of Göttingen, receiving his degree in absentia from the University of Helmstadt. For his dissertation, he gave the first complete proof of the fundamental theorem of algebra, which states that every polynomial equation has as many solutions as its degree. At age 19 he solved a problem that baffled Euclid, inscribing a regular polygon of seventeen sides in a circle using straightedge and compass; and in 1801, at age 24, he published his first masterpiece, *Disquisitiones Arithmeticae,* considered by many to be one of the most brilliant achievements in mathematics. In that book Gauss systematized the study of number theory (properties of the integers) and formulated the basic concepts that form the foundation of that subject.

(continued on next page)

orientation is equal to the triple integral of the divergence of the field over the solid region enclosed by the surface.

18.7.1   THEOREM (*Divergence Theorem*).   *Let G be a solid with surface σ oriented outward. If*

$$\mathbf{F}(x, y, z) = f(x, y, z)\mathbf{i} + g(x, y, z)\mathbf{j} + h(x, y, z)\mathbf{k}$$

where f, g, and h have continuous first partial derivatives on some open set containing G, then

$$\iint_{\sigma} \mathbf{F} \cdot \mathbf{n}\, dS = \iiint_{G} \operatorname{div} \mathbf{F}\, dV \tag{1}$$

The proof of this theorem for a general solid G is too difficult to present here. However, we can give a proof for the special case where G is a simple solid (see Figure 17.5.3 and the discussion preceding it).

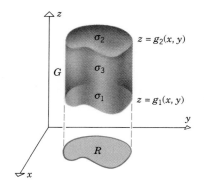

Proof (for simple solids).   Let G be a simple solid with upper surface $z = g_2(x, y)$, lower surface $z = g_1(x, y)$, and projection R on the xy-plane. Let σ_1 denote the lower surface, σ_2 the upper surface, and σ_3 the lateral surface (Figure 18.7.1*a*). If the upper surface and lower surface meet as in Figure 18.7.1*b*, then there is no lateral surface σ_3. Our proof will allow for both cases shown in those figures.

We want to show that

$$\iint_{\sigma} \mathbf{F} \cdot \mathbf{n}\, dS = \iiint_{G} \operatorname{div} \mathbf{F}\, dV$$

Figure 18.7.1

(continued)

   In the same year that the *Disquisitiones* was published, Gauss again applied his phenomenal computational skills in a dramatic way. The astronomer Giuseppi Piazzi had observed the asteroid Ceres for $\frac{1}{40}$ of its orbit, but lost it in the sun. Using only three observations and the "method of least squares" that he had developed in 1795, Gauss computed the orbit with such accuracy that astronomers had no trouble relocating it the following year. This achievement brought him instant recognition as the premier mathematician in Europe, and in 1807 he was made Professor of Astronomy and head of the astronomical observatory at Göttingen.

   In the years that followed, Gauss revolutionized mathematics by bringing to it standards of precision and rigor undreamed of by his predecessors. He had a passion for perfection that drove him to polish and rework his papers rather than publish less finished work in greater numbers—his favorite saying was "Pauca, sed matura" (Few, but ripe). As a result, many of his important discoveries were squirreled away in diaries that remained unpublished until years after his death.

   Among his myriad achievements, Gauss discovered the Gaussian or "bell-shaped" error curve fundamental in probability, gave the first geometric interpretation of complex numbers and established their fundamental role in mathematics, developed methods of characterizing surfaces intrinsically by means of the curves that they contain, developed the theory of conformal (angle-preserving) maps, and discovered non-Euclidean geometry 30 years before the ideas were published by others. In physics he made major contributions to the theory of lenses and capillary action, and with Wilhelm Weber he did fundamental work in electromagnetism. Gauss invented the heliotrope, bifilar magnetometer, and an electrotelegraph.

   Gauss was deeply religious and aristocratic in demeanor. He mastered foreign languages with ease, read extensively, and enjoyed minerology and botany as hobbies. He disliked teaching and was usually cool and discouraging to other mathematicians, possibly because he had already anticipated their work. It has been said that if Gauss had published all of his discoveries, the current state of mathematics would be advanced by 50 years. He was without a doubt the greatest mathematician of the modern era.

or equivalently,

$$\iint_\sigma [f(x, y, z)\mathbf{i} + g(x, y, z)\mathbf{j} + h(x, y, z)\mathbf{k}] \cdot \mathbf{n}\, dS = \iiint_G \left(\frac{\partial f}{\partial x} + \frac{\partial g}{\partial y} + \frac{\partial h}{\partial z} \right) dV$$

To prove this, it suffices to prove the following three equalities:

$$\iint_\sigma [f(x, y, z)\mathbf{i} \cdot \mathbf{n}]\, dS = \iiint_G \frac{\partial f}{\partial x}\, dV \tag{2a}$$

$$\iint_\sigma [g(x, y, z)\mathbf{j} \cdot \mathbf{n}]\, dS = \iiint_G \frac{\partial g}{\partial y}\, dV \tag{2b}$$

$$\iint_\sigma [h(x, y, z)\mathbf{k} \cdot \mathbf{n}]\, dS = \iiint_G \frac{\partial h}{\partial z}\, dV \tag{2c}$$

Since the proofs of all three formulas are similar, we shall prove only the third.
It follows from Theorem 17.5.2 that

$$\iiint_G \frac{\partial h}{\partial z}\, dV = \iint_R \left[\int_{g_1(x, y)}^{g_2(x, y)} \frac{\partial h}{\partial z}\, dz \right] dA = \iint_R \Big[h(x, y, z) \Big]_{z=g_1(x, y)}^{g_2(x, y)} dA$$

so

$$\iiint_G \frac{\partial h}{\partial z}\, dV = \iint_R [h(x, y, g_2(x, y)) - h(x, y, g_1(x, y))]\, dA \tag{3}$$

We shall evaluate the surface integral in (2c) by integrating over each surface of G separately. If there is a lateral surface σ_3, then at each point of this surface $\mathbf{n} \cdot \mathbf{k} = 0$ since $\mathbf{n}$ is horizontal and $\mathbf{k}$ is vertical. Thus,

$$\iint_{\sigma_3} [h(x, y, z)\mathbf{k} \cdot \mathbf{n}]\, dS = 0$$

Therefore, regardless of whether or not G has a lateral surface, we can write

$$\iint_\sigma [h(x, y, z)\mathbf{k} \cdot \mathbf{n}]\, dS = \iint_{\sigma_1} [h(x, y, z)\mathbf{k} \cdot \mathbf{n}]\, dS + \iint_{\sigma_2} [h(x, y, z)\mathbf{k} \cdot \mathbf{n}]\, dS \tag{4}$$

On the upper surface σ_2, the outer normal is an upward normal, and on the lower surface σ_1, the outer normal is a downward normal. Thus, Formulas (7) and (8) of Section 18.6 imply that

$$\iint_{\sigma_2} [h(x, y, z)\mathbf{k} \cdot \mathbf{n}]\, dS = \iint_R \left[h(x, y, g_2(x, y))\mathbf{k} \cdot \left(-\frac{\partial z}{\partial x}\mathbf{i} - \frac{\partial z}{\partial y}\mathbf{j} + \mathbf{k} \right) \right] dA$$

$$= \iint_R [h(x, y, g_2(x, y))]\, dA \tag{5}$$

and

$$\iint_{\sigma_1} [h(x, y, z)\mathbf{k} \cdot \mathbf{n}]\, dS = \iint_R \left[h(x, y, g_1(x, y))\mathbf{k} \cdot \left(\frac{\partial z}{\partial x}\mathbf{i} + \frac{\partial z}{\partial y}\mathbf{j} - \mathbf{k} \right) \right] dA$$

$$= -\iint_R [h(x, y, g_1(x, y))]\, dA \tag{6}$$

Substituting (5) and (6) into (4) and combining the terms into a single integral yields

$$\iint_\sigma [h(x, y, z)\mathbf{k} \cdot \mathbf{n}] \, dS = \iint_R [h(x, y, g_2(x, y)) - h(x, y, g_1(x, y))] \, dA \qquad (7)$$

Equation (2c) now follows from (3) and (7).   ∎

Example 1   Let σ be the sphere $x^2 + y^2 + z^2 = a^2$ oriented outward. Use the Divergence Theorem to find the flux of the vector field $\mathbf{F}(x, y, z) = z\mathbf{k}$ across σ.

Solution.   The divergence of the vector field is

$$\operatorname{div} \mathbf{F} = \frac{\partial z}{\partial z} = 1$$

Thus, if G denotes the spherical solid enclosed by σ, then it follows from (1) that the flux Φ across σ is

$$\Phi = \iint_\sigma \mathbf{F} \cdot \mathbf{n} \, dS = \iiint_G dV = \text{volume of } G = \frac{4\pi a^3}{3} \qquad \blacktriangleleft$$

REMARK.   We solved this same problem in Example 2 of Section 18.6 by evaluating the integral for Φ directly. Note how much simpler this solution is.

Example 2   Let σ be the surface of the cube shown in Figure 18.7.2 oriented outward. Use the Divergence Theorem to find the flux of the vector field

$$\mathbf{F}(x, y, z) = 2x\mathbf{i} + 3y\mathbf{j} + z^2\mathbf{k}$$

across σ.

Solution.   The divergence of the vector field is

$$\operatorname{div} \mathbf{F} = \frac{\partial}{\partial x}(2x) + \frac{\partial}{\partial y}(3y) + \frac{\partial}{\partial z}(z^2) = 5 + 2z$$

Thus, if G denotes the solid cube enclosed by σ, then it follows from (1) that the flux Φ across σ is

$$\Phi = \iint_\sigma \mathbf{F} \cdot \mathbf{n} \, dS = \iiint_G (5 + 2z) \, dV = \int_0^1 \int_0^1 \int_0^1 (5 + 2z) \, dz \, dy \, dx$$

$$= \int_0^1 \int_0^1 [5z + z^2]_{z=0}^1 \, dy \, dx = \int_0^1 \int_0^1 6 \, dy \, dx = 6 \qquad \blacktriangleleft$$

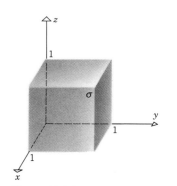

Figure 18.7.2

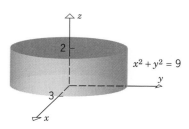

Figure 18.7.3

Example 3   Let σ be the surface of the solid enclosed by the circular cylinder $x^2 + y^2 = 9$ and the planes $z = 0$ and $z = 2$, oriented outward (Figure 18.7.3). Use the Divergence Theorem to find the flux of the vector field $\mathbf{F}(x, y, z) = x^3\mathbf{i} + y^3\mathbf{j} + z^2\mathbf{k}$ across σ.

Solution.   The divergence of the vector field is

$$\operatorname{div} \mathbf{F} = \frac{\partial}{\partial x}(x^3) + \frac{\partial}{\partial y}(y^3) + \frac{\partial}{\partial z}(z^2) = 3x^2 + 3y^2 + 2z$$

Thus, if G denotes the solid circular cylinder enclosed by σ, then it follows from (1) that the flux Φ across σ is

$$\Phi = \iint_{\sigma} \mathbf{F} \cdot \mathbf{n}\, dS = \iiint_{G} (3x^2 + 3y^2 + 2z)\, dV$$

$$= \int_{0}^{2\pi} \int_{0}^{3} \int_{0}^{2} (3r^2 + 2z) r \, dz \, dr \, d\theta \quad \boxed{\text{Using cylindrical coordinates}}$$

$$= \int_{0}^{2\pi} \int_{0}^{3} \left[3r^3 z + z^2 r \right]_{z=0}^{2} dr \, d\theta$$

$$= \int_{0}^{2\pi} \int_{0}^{3} (6r^3 + 4r) \, dr \, d\theta$$

$$= \int_{0}^{2\pi} \left[\frac{3r^4}{2} + 2r^2 \right]_{0}^{3} d\theta$$

$$= \int_{0}^{2\pi} \frac{279}{2} \, d\theta = 279\pi \quad \blacktriangleleft$$

Example 4 Let σ be the surface of the solid enclosed by the hemisphere $z = \sqrt{a^2 - x^2 - y^2}$ and the plane $z = 0$, oriented outward (Figure 18.7.4). Use the Divergence Theorem to find the flux of the vector field $\mathbf{F}(x, y, z) = x^3\mathbf{i} + y^3\mathbf{j} + z^3\mathbf{k}$ across σ.

Solution. The divergence of the vector field is

$$\operatorname{div} \mathbf{F} = \frac{\partial}{\partial x}(x^3) + \frac{\partial}{\partial y}(y^3) + \frac{\partial}{\partial z}(z^3) = 3x^2 + 3y^2 + 3z^2$$

Thus, if G denotes the hemispheric solid enclosed by σ, then it follows from (1) that the flux Φ across σ is

$$\Phi = \iint_{\sigma} \mathbf{F} \cdot \mathbf{n}\, dS = \iiint_{G} (3x^2 + 3y^2 + 3z^2)\, dV$$

$$= \int_{0}^{2\pi} \int_{0}^{\pi/2} \int_{0}^{a} (3\rho^2)\rho^2 \sin\phi \, d\rho \, d\phi \, d\theta \quad \boxed{\text{Using spherical coordinates}}$$

$$= 3 \int_{0}^{2\pi} \int_{0}^{\pi/2} \int_{0}^{a} \rho^4 \sin\phi \, d\rho \, d\phi \, d\theta$$

$$= 3 \int_{0}^{2\pi} \int_{0}^{\pi/2} \left[\frac{\rho^5}{5} \sin\phi \right]_{\rho=0}^{a} d\phi \, d\theta$$

$$= \frac{3a^5}{5} \int_{0}^{2\pi} \int_{0}^{\pi/2} \sin\phi \, d\phi \, d\theta$$

$$= \frac{3a^5}{5} \int_{0}^{2\pi} \left[-\cos\phi \right]_{0}^{\pi/2} d\theta$$

$$= \frac{3a^5}{5} \int_{0}^{2\pi} d\theta = \frac{6\pi a^5}{5} \quad \blacktriangleleft$$

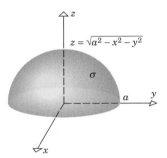

z = \sqrt{a^2 - x^2 - y^2}

Figure 18.7.4

□ **DIVERGENCE VIEWED AS FLUX DENSITY**

The Divergence Theorem provides a useful way of interpreting the divergence of a vector field $\mathbf{F}$ in 3-space. Suppose that G is a *small* spherical region centered at the point P_0 and that its surface $\sigma(G)$ is oriented outward. Denote the volume of the region by $\operatorname{vol}(G)$ and the flux of $\mathbf{F}$ across $\sigma(G)$ by $\Phi(G)$. If $\operatorname{div} \mathbf{F}$ is continuous on G, then over the small region G

the value of div $\mathbf{F}$ will not vary much from its value div $\mathbf{F}(P_0)$ at the center, and we can reasonably approximate div $\mathbf{F}$ by the constant div $\mathbf{F}(P_0)$ on G. Thus, the Divergence Theorem implies that the flux $\Phi(G)$ of $\mathbf{F}$ across $\sigma(G)$ can be approximated as

$$\Phi(G) = \iint\limits_{\sigma(G)} \mathbf{F} \cdot \mathbf{n}\, dS = \iiint\limits_{G} \text{div } \mathbf{F}\, dA \approx \text{div } \mathbf{F}(P_0) \iiint\limits_{G} dV = \text{div } \mathbf{F}(P_0)\, \text{vol}(G)$$

from which we obtain the following approximation of div $\mathbf{F}(P_0)$:

$$\text{div } \mathbf{F}(P_0) \approx \frac{\Phi(G)}{\text{vol}(G)} \tag{8}$$

The expression on the right side of (8) is called the ***flux density of*** $\mathbf{F}$ ***over*** $\mathbf{G}$. If we now let the radius of the sphere approach zero [so that vol(G) approaches zero], then it is plausible that the error in this approximation approaches zero, and the divergence of $\mathbf{F}$ at the point P_0 is given exactly by

$$\text{div } \mathbf{F}(P_0) = \lim_{\text{vol}(G) \to 0} \frac{\Phi(G)}{\text{vol}(G)}$$

or equivalently,

$$\text{div } \mathbf{F}(P_0) = \lim_{\text{vol}(G) \to 0} \frac{1}{\text{vol}(G)} \iint\limits_{\sigma(G)} \mathbf{F} \cdot \mathbf{n}\, dS \tag{9}$$

This limit, called the ***flux density of*** $\mathbf{F}$ ***at the point*** P_0, is sometimes taken as the definition of divergence. This results in a definition of divergence that does not require the introduction of a coordinate system, unlike the definition of divergence in Section 18.1.

☐ **INTERPRETATION OF DIVERGENCE IN FLUID FLOW**

If P_0 is a point in an incompressible fluid at which div $\mathbf{F}(P_0) > 0$, then it follows from (8) that $\Phi(G) > 0$ for a sufficiently small sphere G centered at P_0. Thus, there is a greater volume of fluid going out through the surface of G than coming in. But this can only happen if there is some point *inside* the sphere at which fluid is flowing in; otherwise the net outward flow through the surface would result in a decrease in density within the sphere, contradicting the incompressibility assumption. Similarly, if div $\mathbf{F}(P_0) < 0$, there would have to be a point *inside* the sphere at which fluid is draining out; otherwise the net inward flow through the surface would result in an increase in density within the sphere. In an incompressible fluid, points at which div $\mathbf{F}(P_0) > 0$ are called ***sources*** and points at which div $\mathbf{F}(P_0) < 0$ are called ***sinks***. Fluid enters the flow at a source and drains out at a sink. In an incompressible fluid without sources or sinks we must have

$$\text{div } \mathbf{F}(P) = 0$$

at every point P. In hydrodynamics this is called the ***continuity equation for incompressible fluids*** and is sometimes taken as the defining characteristic of an incompressible fluid.

☐ **GAUSS' LAW FOR INVERSE-SQUARE FIELDS**

Some of the major principles of physics are consequences of the following result, which we shall obtain by applying the Divergence Theorem to inverse-square fields (see Definition 18.1.2).

18.7.2 GAUSS' LAW FOR INVERSE-SQUARE FIELDS. If

$$\mathbf{F}(\mathbf{r}) = \frac{c}{\|\mathbf{r}\|^3} \mathbf{r}$$

is an inverse-square field in 3-space, and if σ is a closed orientable surface that surrounds the origin and has outward orientation, then the flux Φ of $\mathbf{F}$ across σ is

$$\Phi = \iint\limits_{\sigma} \mathbf{F} \cdot \mathbf{n}\, dS = 4\pi c \tag{10}$$

Recall from Formula (3) of Section 18.1 that **F** can be expressed in component form as

$$\mathbf{F}(x, y, z) = \frac{c}{(x^2 + y^2 + z^2)^{3/2}} (x\mathbf{i} + y\mathbf{j} + z\mathbf{k}) \tag{11}$$

Since the components of **F** are not continuous at the origin, we cannot apply the Divergence Theorem over the solid enclosed by σ. However, we can circumvent this difficulty by constructing a sphere of radius a centered at the origin, where the radius is sufficiently small that the sphere lies entirely within the region enclosed by σ (Figure 18.7.5). We shall denote the surface of this sphere by σ_a. The solid G enclosed between σ_a and σ is a three-dimensional *simply connected solid* in which σ is the outer boundary and σ_a is the boundary of a cavity inside the solid. The components of **F** satisfy the hypotheses of the Divergence Theorem on G.

Just as we were able to extend Green's Theorem to multiply connected regions in the plane, so it is possible to extend the Divergence Theorem to multiply connected solids in 3-space, provided the surface integral in the theorem is taken over the *entire* boundary with the outside boundary oriented outward (away from G) and the boundaries of the cavities oriented inward (toward the cavities). Thus, if **F** is the inverse-square field in (11), and if σ_a is oriented inward, then the Divergence Theorem yields

$$\iiint\limits_{G} \operatorname{div} \mathbf{F}\, dV = \iint\limits_{\sigma} \mathbf{F} \cdot \mathbf{n}\, dS + \iint\limits_{\sigma_a} \mathbf{F} \cdot \mathbf{n}\, dS \tag{12}$$

But we showed in Example 7 of Section 18.1 that $\operatorname{div} \mathbf{F} = 0$, so (12) yields

$$\iint\limits_{\sigma} \mathbf{F} \cdot \mathbf{n}\, dS = - \iint\limits_{\sigma_a} \mathbf{F} \cdot \mathbf{n}\, dS \tag{13}$$

We can evaluate the surface integral over σ_a by expressing the integrand in terms of components; however, it is easier to leave it in vector form. At each point on the sphere the unit normal **n** points inward along a radius from the origin, and hence $\mathbf{n} = -\mathbf{r}/\|\mathbf{r}\|$. Thus, (13) yields

$$\iint\limits_{\sigma} \mathbf{F} \cdot \mathbf{n}\, dS = - \iint\limits_{\sigma_a} \frac{c}{\|\mathbf{r}\|^3} \mathbf{r} \cdot \left(-\frac{\mathbf{r}}{\|\mathbf{r}\|} \right) dS$$

$$= \iint\limits_{\sigma_a} \frac{c}{\|\mathbf{r}\|^4} (\mathbf{r} \cdot \mathbf{r})\, dS$$

$$= \iint\limits_{\sigma_a} \frac{c}{\|\mathbf{r}\|^2}\, dS$$

$$= \frac{c}{a^2} \iint\limits_{\sigma_a} dS \qquad \boxed{\|\mathbf{r}\| = a \text{ on } \sigma_a}$$

$$= \frac{c}{a^2} (4\pi a^2) \qquad \boxed{\begin{array}{l}\text{The integral is the surface}\\\text{area of the sphere.}\end{array}}$$

$$= 4\pi c$$

which establishes (10).

□ **GAUSS' LAW IN ELECTROSTATICS**

It follows from Example 3 of Section 18.1 with $q = 1$ that a single charged particle of charge Q located at the origin creates an inverse-square field

$$\mathbf{F}(\mathbf{r}) = \frac{Q}{4\pi\epsilon_0 \|\mathbf{r}\|^3} \mathbf{r}$$

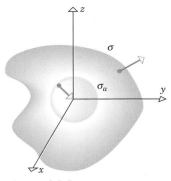

Figure 18.7.5

in which $\mathbf{F}(\mathbf{r})$ is the electrical force exerted by Q on a unit positive charge ($q = 1$) located at the point with position vector $\mathbf{r}$. In this case Gauss' law (18.7.2) states that the outward flux Φ across any closed orientable surface σ that surrounds Q is

$$\Phi = \iint_{\sigma} \mathbf{F} \cdot \mathbf{n}\, dS = 4\pi \left(\frac{Q}{4\pi\epsilon_0} \right) = \frac{Q}{\epsilon_0}$$

This result can be extended to fields resulting from more than one charge. It is one of the fundamental laws in the study of electricity and magnetism.

▶ Exercise Set 18.7

In Exercises 1–13, use the Divergence Theorem to evaluate $\iint_{\sigma} \mathbf{F} \cdot \mathbf{n}\, dS$, where $\mathbf{n}$ is the outer unit normal to σ.

1. $\mathbf{F}(x, y, z) = 4x\mathbf{i} - 3y\mathbf{j} + 7z\mathbf{k}$; σ is the surface of the cube bounded by the coordinate planes and the planes $x = 1$, $y = 1$, and $z = 1$.

2. $\mathbf{F}(x, y, z) = (x^2 + y)\mathbf{i} + z^2\mathbf{j} + (e^y - z)\mathbf{k}$; σ is the surface of the rectangular solid bounded by the coordinate planes and the planes $x = 3$, $y = 1$, and $z = 2$.

3. $\mathbf{F}(x, y, z) = 2x\mathbf{i} + 2y\mathbf{j} + 2z\mathbf{k}$, where σ is the sphere $x^2 + y^2 + z^2 = 9$.

4. $\mathbf{F}(x, y, z) = z^3\mathbf{i} - x^3\mathbf{j} + y^3\mathbf{k}$, where σ is the sphere $x^2 + y^2 + z^2 = a^2$.

5. $\mathbf{F}(x, y, z) = (x - z)\mathbf{i} + (y - x)\mathbf{j} + (z - y)\mathbf{k}$; σ is the surface of the cylindrical solid bounded by $x^2 + y^2 = a^2$, $z = 0$, and $z = 1$.

6. $\mathbf{F}(x, y, z) = x\mathbf{i} + y\mathbf{j} + z\mathbf{k}$; σ is the surface of the solid bounded by the paraboloid $z = 1 - x^2 - y^2$ and the xy-plane.

7. $\mathbf{F}(x, y, z) = x^3\mathbf{i} + y^3\mathbf{j} + z^3\mathbf{k}$; σ is the surface of the cylindrical solid bounded by $x^2 + y^2 = 4$, $z = 0$, and $z = 3$.

8. $\mathbf{F}(x, y, z) = (x^3 - e^y)\mathbf{i} + (y^3 + \sin z)\mathbf{j} + (z^3 - xy)\mathbf{k}$, where σ is the surface of the solid bounded by $z = \sqrt{4 - x^2 - y^2}$ and the xy-plane. [*Hint:* Use spherical coordinates.]

9. $\mathbf{F}(x, y, z) = (x^2 + y)\mathbf{i} + xy\mathbf{j} - (2xz + y)\mathbf{k}$; σ is the surface of the tetrahedron in the first octant bounded by $x + y + z = 1$ and the coordinate planes.

10. $\mathbf{F}(x, y, z) = 2xz\mathbf{i} + yz\mathbf{j} + z^2\mathbf{k}$, where σ is the surface of the hemispherical solid bounded above by $z = \sqrt{a^2 - x^2 - y^2}$ and below by the xy-plane.

11. $\mathbf{F}(x, y, z) = x^2\mathbf{i} + y^2\mathbf{j} + z^2\mathbf{k}$; σ is the surface of the conical solid bounded by $z = \sqrt{x^2 + y^2}$ and $z = 1$.

12. $\mathbf{F}(x, y, z) = x^2 y\mathbf{i} - xy^2\mathbf{j} + (z + 2)\mathbf{k}$; σ is the surface of the solid bounded above by the plane $z = 2x$ and below by the paraboloid $z = x^2 + y^2$.

13. $\mathbf{F}(x, y, z) = x^3\mathbf{i} + x^2 y\mathbf{j} + xy\mathbf{k}$; σ is the surface of the solid bounded by $z = 4 - x^2$, $y + z = 5$, $z = 0$, and $y = 0$.

14. Find $\iint_{\sigma} \mathbf{F} \cdot \mathbf{n}\, dS$, where $\mathbf{F}(x, y, z) = a\mathbf{i} + b\mathbf{j} + c\mathbf{k}$, σ is the surface of a solid G, and $\mathbf{n}$ is an outward unit normal (a, b, and c constants).

15. Prove that if $\mathbf{F}(x, y, z) = x\mathbf{i} + y\mathbf{j} + z\mathbf{k}$ and σ is the surface of a solid G oriented by outward unit normals, then

$$\iint_{\sigma} \mathbf{F} \cdot \mathbf{n}\, dS = 3V$$

where V is the volume of G.

16. Use the result in Exercise 15 to find $\iint_{\sigma} \mathbf{F} \cdot \mathbf{n}\, dS$, where $\mathbf{F}(x, y, z) = x\mathbf{i} + y\mathbf{j} + z\mathbf{k}$ and S is the surface of the cylindrical solid bounded by $x^2 + 4x + y^2 = 5$, $z = -1$, and $z = 4$.

In Exercises 17–20, determine whether the flow field $\mathbf{F}(x, y, z)$ is free of sources and sinks. If it is not, find the location of all sources and sinks.

17. $\mathbf{F}(x, y, z) = (y + z)\mathbf{i} - xz^3\mathbf{j} + (x^2 \sin y)\mathbf{k}$.

18. $\mathbf{F}(x, y, z) = xy\mathbf{i} - xyj + y^2\mathbf{k}$.

19. $\mathbf{F}(x, y, z) = x^3\mathbf{i} + y^3\mathbf{j} + z^3\mathbf{k}$.

20. $\mathbf{F}(x, y, z) = (x^3 - x)\mathbf{i} + (y^3 - y)\mathbf{j} + (z^3 - z)\mathbf{k}$.

■ 18.8 STOKES' THEOREM

*In this section we shall discuss a generalization of Green's Theorem to three dimensions, called **Stokes'* Theorem**. This theorem has applications in various branches of physics and plays an important role in analyzing the rotational motion of fluids. It also provides a physical interpretation of the curl of a vector field.*

☐ **RELATIVE ORIENTATION OF CURVES AND SURFACES**

In this section we shall be concerned with oriented surfaces in 3-space that are bounded by simple closed curves. If an oriented surface σ is bounded by a curve C (Figure 18.8.1a), then there are two possible orientations for C, which can be described as follows: Imagine a person walking along the curve C so the person's head is in the direction of orientation of σ. The person is said to be walking in the **positive direction** of C if the surface σ is on the left (Figure 18.8.1b) and is said to be walking in the **negative direction** of C if the surface σ is on the right (Figure 18.8.1c). This establishes a right-hand relationship between the orientations of σ and C in the sense that if the fingers of the right hand are cupped in the positive direction of C, then the thumb points (roughly) in the direction of orientation of σ.

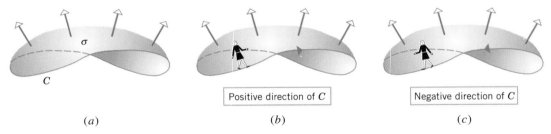

	Positive direction of C	Negative direction of C
(a)	(b)	(c)

Figure 18.8.1

---

*GEORGE GABRIEL STOKES (1819–1903). Irish mathematician and physicist. Born in Skreen, Ireland, Stokes came from a family deeply rooted in the Church of Ireland. His father was a rector, his mother the daughter of a rector, and three of his brothers took holy orders. He received his early education from his father and a local parish clerk. In 1837, he entered Pembroke College and after graduating with top honors accepted a fellowship at the college. In 1847 he was appointed Lucasian professor of mathematics at Cambridge, a position once held by Isaac Newton, but one that had lost its esteem through the years. By virtue of his accomplishments, Stokes ultimately restored the position to the eminence it once held. Unfortunately, the position paid very little and Stokes was forced to teach at the Government School of Mines during the 1850s to supplement his income.

Stokes was one of several outstanding nineteenth century scientists who helped turn the physical sciences in a more empirical direction. He systematically studied hydrodynamics, elasticity of solids, behavior of waves in elastic solids, and diffraction of light. For Stokes, mathematics was a tool for his physical studies. He wrote classic papers on the motion of viscous fluids that laid the foundation for modern hydrodynamics; he elaborated on the wave theory of light; and he wrote papers on gravitational variation that established him as a founder of the modern science of geodesy.

Stokes was honored in his later years with degrees, medals, and memberships in foreign societies. He was knighted in 1889. Throughout his life, Stokes gave generously of his time to learned societies and readily assisted those who sought his help in solving problems. He was deeply religious and vitally concerned with the relationship between science and religion.

Before proceeding to the main theorem, it will be helpful to recall from Section 18.1 that the curl of a vector field

$$\mathbf{F}(x, y, z) = f(x, y, z)\mathbf{i} + g(x, y, z)\mathbf{j} + h(x, y, z)\mathbf{k}$$

is defined as

$$\operatorname{curl} \mathbf{F} = \left(\frac{\partial h}{\partial y} - \frac{\partial g}{\partial z}\right)\mathbf{i} + \left(\frac{\partial f}{\partial z} - \frac{\partial h}{\partial x}\right)\mathbf{j} + \left(\frac{\partial g}{\partial x} - \frac{\partial f}{\partial y}\right)\mathbf{k} = \begin{vmatrix} \mathbf{i} & \mathbf{j} & \mathbf{k} \\ \dfrac{\partial}{\partial x} & \dfrac{\partial}{\partial y} & \dfrac{\partial}{\partial z} \\ f & g & h \end{vmatrix} \tag{1}$$

☐ STOKES' THEOREM

The following result, known as **Stokes' Theorem**, shows that under appropriate conditions the surface integral of the normal component of curl **F** is equal to the line integral over the boundary of the tangential component of **F**. The proof is difficult and is left for advanced courses.

18.8.1   THEOREM (**Stokes' Theorem**).   *Let σ be a piecewise smooth orientable surface that is bounded by a simple, closed, piecewise smooth curve C with positive orientation. If the components of the vector field*

$$\mathbf{F}(x, y, z) = f(x, y, z)\mathbf{i} + g(x, y, z)\mathbf{j} + h(x, y, z)\mathbf{k}$$

are continuous and have continuous first partial derivatives on some open set containing σ, and if **T** *is the unit tangent vector to C, then*

$$\oint_C \mathbf{F} \cdot \mathbf{T}\, ds = \iint_\sigma (\operatorname{curl} \mathbf{F}) \cdot \mathbf{n}\, dS \tag{2}$$

For computational purposes, it is desirable to express (2) in a different form. Recall from Formula (2) of Section 15.4 that

$$\frac{d\mathbf{r}}{ds} = \mathbf{T}$$

from which we obtain the following alternative form of (2):

$$\oint_C \mathbf{F} \cdot d\mathbf{r} = \iint_\sigma (\operatorname{curl} \mathbf{F}) \cdot \mathbf{n}\, dS \tag{3}$$

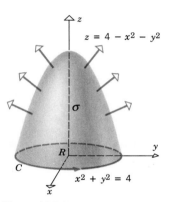

Figure 18.8.2

Example 1   Let σ be the portion of the paraboloid $z = 4 - x^2 - y^2$ for which $z \geq 0$, and let C be the circle $x^2 + y^2 = 4$ that forms the boundary of σ in the xy-plane (Figure 18.8.2). Verify Stokes' Theorem for the vector field $\mathbf{F}(x, y, z) = 2z\mathbf{i} + 3x\mathbf{j} + 5y\mathbf{k}$ if σ is oriented upward.

Solution.   We will verify (3). Since σ is oriented upward, the positive orientation of C is counterclockwise looking down the positive z-axis. Thus, C can be represented parametrically (with positive orientation) by

$$x = 2\cos t, \quad y = 2\sin t, \quad z = 0 \qquad (0 \leq t \leq 2\pi) \tag{4}$$

Therefore

$$\oint_C \mathbf{F} \cdot d\mathbf{r} = \oint_C 2z\, dx + 3x\, dy + 5y\, dz$$

$$= \int_0^{2\pi} [0 + (6\cos t)(2\cos t) + 0]\, dt$$

$$= \int_0^{2\pi} 12\cos^2 t\, dt = 12 \left[\frac{1}{2} t + \frac{1}{4} \sin 2t \right]_0^{2\pi} = 12\pi$$

To evaluate the right side of (3), we note that

$$\text{curl } \mathbf{F} = \begin{vmatrix} \mathbf{i} & \mathbf{j} & \mathbf{k} \\ \dfrac{\partial}{\partial x} & \dfrac{\partial}{\partial y} & \dfrac{\partial}{\partial z} \\ 2z & 3x & 5y \end{vmatrix} = 5\mathbf{i} + 2\mathbf{j} + 3\mathbf{k}$$

Since σ is oriented up and is expressed in the form $z = g(x, y) = 4 - x^2 - y^2$, it follows from Formula (7) of Section 18.6 with curl $\mathbf{F}$ replacing $\mathbf{F}$ that

$$\iint_\sigma (\text{curl } \mathbf{F}) \cdot \mathbf{n}\, dS = \iint_R (\text{curl } \mathbf{F}) \cdot \left(-\frac{\partial z}{\partial x}\mathbf{i} - \frac{\partial z}{\partial y}\mathbf{j} + \mathbf{k} \right) dA$$

$$= \iint_R (5\mathbf{i} + 2\mathbf{j} + 3\mathbf{k}) \cdot (2x\mathbf{i} + 2y\mathbf{j} + \mathbf{k})\, dA$$

$$= \iint_R (10x + 4y + 3)\, dA$$

$$= \int_0^{2\pi} \int_0^2 (10r\cos\theta + 4r\sin\theta + 3)r\, dr\, d\theta$$

$$= \int_0^{2\pi} \left[\frac{10r^3}{3}\cos\theta + \frac{4r^3}{3}\sin\theta + \frac{3r^2}{2} \right]_{r=0}^{2} d\theta$$

$$= \int_0^{2\pi} \left(\frac{80}{3}\cos\theta + \frac{32}{3}\sin\theta + 6 \right) d\theta$$

$$= \left[\frac{80}{3}\sin\theta - \frac{32}{3}\cos\theta + 6\theta \right]_0^{2\pi} = 12\pi$$

which agrees with the value of the line integral obtained above. Note, however, that the line integral was easier to evaluate, and hence would be the computational method of choice in this case. ◀

REMARK.   Note that if σ_1 and σ_2 are surfaces with the same positively oriented boundary C, then for any vector field $\mathbf{F}$ that satisfies the hypotheses of Stokes' Theorem, it must be the case that

$$\iint_{\sigma_1} \text{curl } \mathbf{F} \cdot \mathbf{n}\, dS = \iint_{\sigma_2} \text{curl } \mathbf{F} \cdot \mathbf{n}\, dS$$

since (2) implies that the two surface integrals are equal to the same line integral over C. For example, the parabolic surface in Example 1 can be replaced by the upper hemisphere of radius 2 (with upward orientation), or even by the circular region R in Figure 18.8.2 (with upward orientation), without altering the value of the surface integral, since the parabolic surface, the upper hemisphere, and the region R have the same oriented boundary C in the xy-plane.

Example 2 Let C be the rectangle in the plane $z = y$ oriented as in Figure 18.8.3, and let $\mathbf{F}(x, y, z) = x^2\mathbf{i} + 4xy^3\mathbf{j} + y^2x\mathbf{k}$. Find

$$\oint_C \mathbf{F} \cdot d\mathbf{r}$$

Solution. To evaluate the integral directly would require four separate integrations, one over each side of the rectangle. Instead, we shall apply Stokes' Theorem, so we need only evaluate a single surface integral over the rectangular surface σ bounded by C. For the positive direction of C to be as shown in Figure 18.8.3, the surface σ must be oriented by downward normals.

Since the surface σ has equation $z = y$ and

$$\text{curl } \mathbf{F} = \begin{vmatrix} \mathbf{i} & \mathbf{j} & \mathbf{k} \\ \dfrac{\partial}{\partial x} & \dfrac{\partial}{\partial y} & \dfrac{\partial}{\partial z} \\ x^2 & 4xy^3 & y^2x \end{vmatrix} = 2yx\mathbf{i} - y^2\mathbf{j} + 4y^3\mathbf{k}$$

it follows from Formula (8) of Section 18.6 with curl $\mathbf{F}$ replacing $\mathbf{F}$ that

$$\iint_\sigma (\text{curl } \mathbf{F}) \cdot \mathbf{n} \, dS = \iint_R (\text{curl } \mathbf{F}) \cdot \left(\frac{\partial z}{\partial x}\mathbf{i} + \frac{\partial z}{\partial y}\mathbf{j} - \mathbf{k} \right) dA$$

$$= \iint_R (2yx\mathbf{i} - y^2\mathbf{j} + 4y^3\mathbf{k}) \cdot (0\mathbf{i} + \mathbf{j} - \mathbf{k}) \, dA$$

$$= \int_0^1 \int_0^3 (-y^2 - 4y^3) \, dy \, dx$$

$$= -\int_0^1 \left[\frac{y^3}{3} + y^4 \right]_{y=0}^3 dx$$

$$= -\int_0^1 90 \, dx = -90 \qquad \blacktriangleleft$$

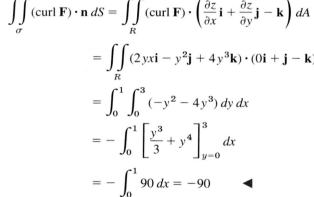

Figure 18.8.3

☐ **RELATIONSHIP BETWEEN GREEN'S THEOREM AND STOKES' THEOREM**

It is sometimes convenient to regard a vector field

$$\mathbf{F}(x, y) = f(x, y)\mathbf{i} + g(x, y)\mathbf{j}$$

in 2-space as a vector field in 3-space by expressing it as

$$\mathbf{F}(x, y) = f(x, y)\mathbf{i} + g(x, y)\mathbf{j} + 0\mathbf{k} \tag{5}$$

If R is a region in the xy-plane enclosed by a curve C, then we can treat R as a *flat* surface, and we can treat a surface integral over R as an ordinary double integral over R. Thus, if we orient R upward and C counterclockwise looking down the positive z-axis, then Formula (3) applied to (5) yields

$$\oint_C \mathbf{F} \cdot d\mathbf{r} = \iint_R \text{curl } \mathbf{F} \cdot \mathbf{k} \, dA \tag{6}$$

But

$$\text{curl } \mathbf{F} = \begin{vmatrix} \mathbf{i} & \mathbf{j} & \mathbf{k} \\ \dfrac{\partial}{\partial x} & \dfrac{\partial}{\partial y} & \dfrac{\partial}{\partial z} \\ f & g & 0 \end{vmatrix} = -\frac{\partial g}{\partial z}\mathbf{i} + \frac{\partial f}{\partial z}\mathbf{j} + \left(\frac{\partial g}{\partial x} - \frac{\partial f}{\partial y}\right)\mathbf{k} = \left(\frac{\partial g}{\partial x} - \frac{\partial f}{\partial y}\right)\mathbf{k}$$

since $\partial g/\partial z = \partial f/\partial z = 0$ (why?) Substituting this expression in (6) and expressing the integrals in terms of components yields

$$\oint_C f\,dx + g\,dy = \iint_R \left(\frac{\partial g}{\partial x} - \frac{\partial f}{\partial y}\right) dA$$

(verify) which is Green's Theorem (18.4.1). Thus, Green's Theorem is a special case of Stokes' Theorem.

☐ **RELATIONSHIP BETWEEN CURL AND CIRCULATION**

Stokes' Theorem provides a useful way of interpreting the curl of a vector field $\mathbf{F}$ in 3-space. The ideas that we shall develop here are applicable to most vector fields, but we shall motivate them in the context of incompressible steady-state fluid flow.

If we imagine a small piece of straw floating on the surface of a stream, the straw will be carried downstream by the current, but at the same time it may spin due to whirlpools or eddies in the flow. It is the curl of the flow field that relates to the spinning motion of the straw. To see why this is so, let σ_a be a *small* oriented disk-shaped region of radius a and centered at P_0 (Figure 18.8.4). Denote the circular boundary of the region by C_a and the area of σ_a by $A(\sigma_a)$. At each point of C_a the flow field $\mathbf{F}$ has a component $\mathbf{F} \cdot \mathbf{u}$ along the outward unit normal to C_a and a component $\mathbf{F} \cdot \mathbf{T}$ along the unit tangent to C_a (Figure 18.8.5). Fluid moving in the direction of $\mathbf{u}$ flows *through* the circle, and fluid moving in the direction of $\mathbf{T}$ moves *along* the circle. Physicists call the integral

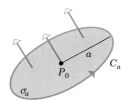

Figure 18.8.4

$$\oint_{C_a} \mathbf{F} \cdot \mathbf{T}\,ds$$

the *circulation of* $\mathbf{F}$ *around* C_a and use it as a measure of the tendency for fluid to flow in the positive direction around the circle C_a. For example, in the extreme case where the flow is normal to the circle at each point, the circulation around C_a is zero, since $\mathbf{F} \cdot \mathbf{T} = 0$ at each point of the circle (Figure 18.8.6a). The more closely that $\mathbf{F}$ aligns with $\mathbf{T}$ along the circle, the larger the value of $\mathbf{F} \cdot \mathbf{T}$, and the larger the value of the circulation. The maximum circulation occurs when $\mathbf{F}$ is aligned with $\mathbf{T}$ at each point of the circle (Figure 18.8.6b).

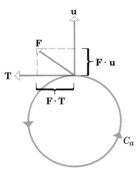

Figure 18.8.5

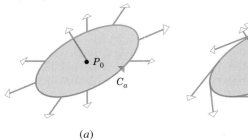

Figure 18.8.6   (a)   (b)

To see the relationship between circulation and curl, suppose that curl $\mathbf{F}$ is continuous on σ_a, so that when σ_a is small the value of curl $\mathbf{F}$ at any point of σ_a will not vary much from the value of curl $\mathbf{F}(P_0)$ at the center. Thus, for a small disk σ_a we can reasonably assume that curl $\mathbf{F}$ has a constant value of curl $\mathbf{F}(P_0)$ on σ_a. Moreover, because the surface σ_a is flat, the unit normal vectors that orient σ_a are all equal; that is, the orienting vector **n** is constant on σ_a. Thus, version (2) of Stokes' Theorem implies that

$$\oint_{C_a} \mathbf{F} \cdot \mathbf{T} \, ds = \iint_{\sigma_a} (\text{curl } \mathbf{F}) \cdot \mathbf{n} \, dS \approx \text{curl } \mathbf{F}(P_0) \cdot \mathbf{n} \iint_{\sigma_a} dS$$

where the line integral is taken in the positive direction of C_a. But the double integral in this equation represents the surface area of σ_a, so it follows that

$$\oint_{C_a} \mathbf{F} \cdot \mathbf{T} \, ds \approx [\text{curl } \mathbf{F}(P_0) \cdot \mathbf{n}] \, A(\sigma_a)$$

from which we obtain the following relationship between curl and circulation:

$$\text{curl } \mathbf{F}(P_0) \cdot \mathbf{n} \approx \frac{1}{A(\sigma_a)} \oint_{C_a} \mathbf{F} \cdot \mathbf{T} \, ds \tag{7}$$

The quantity on the right side of (7) is called the ***circulation density of F around*** C_a. If we now let the radius a of the disk approach zero (with **n** fixed), then it is plausible that the error in this approximation will approach zero and the exact value of curl $\mathbf{F}(P_0) \cdot \mathbf{n}$ will be given by

$$\text{curl } \mathbf{F}(P_0) \cdot \mathbf{n} = \lim_{a \to 0} \frac{1}{A(\sigma_a)} \oint_{C_a} \mathbf{F} \cdot \mathbf{T} \, ds \tag{8}$$

Thus, we call curl $\mathbf{F}(P_0) \cdot \mathbf{n}$ the ***circulation density of F at P_0 in the direction of*** **n**. Since curl $\mathbf{F}(P_0) \cdot \mathbf{n}$ has its maximum value when **n** is in the same direction as curl $\mathbf{F}(P_0)$, it follows that in the vicinity of the point P_0 the maximum circulation occurs around small circles in the plane normal to curl $\mathbf{F}(P_0)$. Physically, if a small paddle wheel is immersed in the fluid so that the pivot point is at P_0, then the paddles will turn most rapidly when the spindle is aligned with curl $\mathbf{F}(P_0)$ (Figure 18.8.7). If curl $\mathbf{F} = \mathbf{0}$ at each point of a region, then **F** is said to be ***irrotational*** in that region, since no circulation occurs about any point of the region.

We conclude by noting that if mutually orthogonal unit vectors **i**, **j**, and **k** are substituted for **n** in (8), then the resulting equations produce the components of curl $\mathbf{F}(P_0)$ in the directions of those vectors. Thus (8) is sometimes taken as the definition of curl. This definition of curl does not require the introduction of a coordinate system, unlike the definition in Section 18.1.

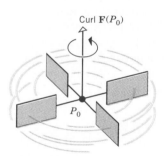

Curl $\mathbf{F}(P_0)$

P_0

Figure 18.8.7

► Exercise Set 18.8

> In Exercises 1–8, use Stokes' Theorem to evaluate the integral $\oint_C \mathbf{F} \cdot d\mathbf{r}$.

1. $\mathbf{F}(x, y, z) = z^2\mathbf{i} + 2x\mathbf{j} - y^3\mathbf{k}$; C is the circle $x^2 + y^2 = 1$ in the xy-plane with counterclockwise orientation looking down the positive z-axis.

2. $\mathbf{F}(x, y, z) = xz\mathbf{i} + 3x^2y^2\mathbf{j} + yx\mathbf{k}$; C is the rectangle in the plane $z = y$ shown in Figure 18.8.3.

3. $\mathbf{F}(x, y, z) = 3z\mathbf{i} + 4x\mathbf{j} + 2y\mathbf{k}$; C is the boundary of the paraboloid shown in Figure 18.8.2.

4. $\mathbf{F}(x, y, z) = -3y^2\mathbf{i} + 4z\mathbf{j} + 6x\mathbf{k}$; C is the triangle in the plane $z = \frac{1}{2}y$ with vertices $(2, 0, 0)$, $(0, 2, 1)$, and $(0, 0, 0)$ with a counterclockwise orientation looking down the positive z-axis.

5. $\mathbf{F}(x, y, z) = xy\mathbf{i} + x^2\mathbf{j} + z^2\mathbf{k}$; C is the intersection of the paraboloid $z = x^2 + y^2$ and the plane $z = y$ with a counterclockwise orientation looking down the positive z-axis.

6. $\mathbf{F}(x, y, z) = xy\mathbf{i} + yz\mathbf{j} + zx\mathbf{k}$; C is the triangle in the plane $x + y + z = 1$ with vertices $(1, 0, 0)$, $(0, 1, 0)$, and $(0, 0, 1)$ with a counterclockwise orientation looking from the first octant toward the origin.

7. $\mathbf{F}(x, y, z) = (x - y)\mathbf{i} + (y - z)\mathbf{j} + (z - x)\mathbf{k}$; C is the circle $x^2 + y^2 = a^2$ in the xy-plane with counterclockwise orientation looking down the positive z-axis.

8. $\mathbf{F}(x, y, z) = (z + \sin x)\mathbf{i} + (x + y^2)\mathbf{j} + (y + e^z)\mathbf{k}$; C is the intersection of the sphere $x^2 + y^2 + z^2 = 1$ and the cone $z = \sqrt{x^2 + y^2}$ with counterclockwise orientation looking down the positive z-axis.

In Exercises 9–12, verify Stokes' Theorem by computing the line and surface integrals in (3) and showing that they are equal.

9. $\mathbf{F}(x, y, z) = (x - y)\mathbf{i} + (y - z)\mathbf{j} + (z - x)\mathbf{k}$; σ is the portion of the plane $x + y + z = 1$ in the first octant.

10. $\mathbf{F}(x, y, z) = x^2\mathbf{i} + y^2\mathbf{j} + z^2\mathbf{k}$; σ is the portion of the cone $z = \sqrt{x^2 + y^2}$ below the plane $z = 1$.

11. $\mathbf{F}(x, y, z) = x\mathbf{i} + y\mathbf{j} + z\mathbf{k}$; σ is the upper hemisphere $z = \sqrt{a^2 - x^2 - y^2}$.

12. $\mathbf{F}(x, y, z) = (z - y)\mathbf{i} + (z + x)\mathbf{j} - (x + y)\mathbf{k}$; σ is the portion of the paraboloid $z = 9 - x^2 - y^2$ above the xy-plane.

13. Use the result in Exercise 17 of Section 18.1 to help prove that

$$\iint_{\sigma} (\text{curl } \mathbf{F}) \cdot \mathbf{n} \, dS = 0$$

where σ is the surface of a solid G, $\mathbf{n}$ is the unit outer normal to σ, and the components of $\mathbf{F}$ have continuous second partial derivatives on and within σ.

14. Consider the flow field given by the formula

$$\mathbf{F}(x, y, z) = (x - z)\mathbf{i} + (y - x)\mathbf{j} + (z - xy)\mathbf{k}$$

(a) Use Stokes' Theorem to find the circulation around the triangle with vertices $A(1, 0, 0)$, $B(0, 2, 0)$, and $C(0, 0, 1)$ oriented counterclockwise looking from the origin toward the first octant.

(b) Find the circulation density of $\mathbf{F}$ at the origin in the direction of $\mathbf{k}$.

(c) Find the unit vector $\mathbf{n}$ such that the circulation density of $\mathbf{F}$ at the origin is maximum in the direction of $\mathbf{n}$.

◆ TECHNOLOGY EXERCISES   Chapter 18

Most of these exercises require access to a graphing calculator or a computer algebra system (CAS) such as *Mathematica*, *Maple*, or *Derive*. When you are asked to *find* an answer or to *solve* an equation, you may choose to find an exact result or a numerical approximation, depending on the particular technology you are using and on your own imagination. The form of your answers may differ from those of other students or from those in the answer section of the text, depending on how you solve the problems and the accuracy you use in your numerical approximations. Those exercises that are more appropriate for a CAS than a graphing calculator are labeled with the icon ◆.

◆ **1. Work:** For any constant c, the curve $y = c(x - x^2)$ passes through the origin and the point $(1, 0)$. Suppose that a particle moves along such a curve from the origin to the point $(1, 0)$ while subject to the force

$$\mathbf{F}(x, y) = (3x^3 + y^2 - y^4)\mathbf{i} + (2y^3 - x^2)\mathbf{j}$$

Find the values of c such that the work done by $\mathbf{F}$ is zero.

◆ **2. Maximum work:** Find the value of c in Exercise 1 so that the work done by $\mathbf{F}$ is maximum.

◆ **3. Work and kinetic energy:** The *kinetic energy* of a particle is defined as $K = \frac{1}{2}mv^2$, where m is the mass of the particle and v is its speed. If m is measured in kilograms (kg) and v is measured in meters per second (m/sec), then K is in joules (J). The **work–energy principle** of physics states that if $\mathbf{F}$ is the only force acting on a particle, then the change in kinetic energy is equal to the work done on the particle by $\mathbf{F}$, where $\mathbf{F}$ is measured in newtons and distance is measured in meters. Assume that

$$\mathbf{F}(x, y, z) = y^2\mathbf{i} + xz\mathbf{j} + x\mathbf{k}$$

is the only force acting on a particle of mass 0.13 kg that moves along the curve C described parametrically by

$$\mathbf{r}(t) = t\mathbf{i} + \sin t\mathbf{j} + e^t\mathbf{k}$$

where **F** is in newtons and **r** is in meters. Suppose that the particle is moving with a speed of 1.5 m/sec in the positive direction along C if $t = 0$. Find its speed when it reaches the point at which $t = 2$.

◆ 4. **Conservative vector fields:** Suppose that a particle moves along a curve C from the point $(-1, 0)$ to the point $(1, 3)$ while subject to the force

$$\mathbf{F}(x, y) = \frac{2xe^y}{x^2 e^y + 1}\mathbf{i} + \frac{x^2 e^y}{x^2 e^y + 1}\mathbf{j}$$

(a) Find the work done by **F** along the line segment C from $(-1, 0)$ to $(1, 3)$ by evaluating $\int_C \mathbf{F} \cdot d\mathbf{r}$.

(b) Show that **F** is conservative by finding a potential function $\phi(x, y)$.

(c) Use the function $\phi(x, y)$ obtained in part (b) to check your result in part (a).

◆ 5. **Independence of path:** Suppose that a particle moves along a curve C from the origin to the point $(1, 1, 1)$ while subject to the force

$$\mathbf{F}(x, y, z) = 3x^2 y^2 z^4 \mathbf{i} + 2x^3 yz^4 \mathbf{j} + 4x^3 y^2 z^3 \mathbf{k}$$

(a) The parametric curves

$$C_1 : \mathbf{r}(t) = t\mathbf{i} + t^2\mathbf{j} + t^3\mathbf{k}$$
$$C_2 : \mathbf{r}(t) = \sin\tfrac{1}{2}\pi t\,\mathbf{i} + t^3\mathbf{j} + \sin\tfrac{5}{2}\pi t\,\mathbf{k}$$

are traced from the origin to the point $(1, 1, 1)$ as t varies from 0 to 1. Show that the work done by **F** along C_1 is equal to the work done by **F** along C_2 by evaluating $\int_{C_1} \mathbf{F} \cdot d\mathbf{r}$ and $\int_{C_2} \mathbf{F} \cdot d\mathbf{r}$.

(b) Show that **F** is conservative by finding a potential function $\phi(x, y, z)$.

(c) Use the function $\phi(x, y, z)$ obtained in part (b) to check your result in part (a).

◆ 6. **Green's Theorem:** Suppose that C is the ellipse $x^2/3 + y^2/5 = 1$ oriented counterclockwise, and R is the region enclosed by C. Verify Green's Theorem for $f(x, y) = x^3 y^2$ and $g(x, y) = xy^2$ by evaluating

$$\oint_C f(x, y)\, dx + g(x, y)\, dy$$

and

$$\iint_R \left(\frac{\partial g}{\partial x} - \frac{\partial f}{\partial y} \right) dA$$

[*Suggestion:* Parametrize C.]

◆ 7. **Mass of a lamina:** Find the mass of the curved lamina $z = e^{-x^2 - y^2}$ that lies above the region in the xy-plane enclosed by $x^2 + y^2 = 9$ given that the density function is $\delta(x, y, z) = \sqrt{x^2 + y^2}$.

◆ 8. **The Divergence Theorem:** Let σ be the surface of the solid G enclosed by the paraboloid $z = 1 - x^2 - y^2$ and the plane $z = 0$. Assuming that σ is oriented outward, verify the Divergence Theorem for the vector field

$$\mathbf{F} = (x^2 y - z^2)\mathbf{i} + (y^3 - x)\mathbf{j} + (2x + 3z - 1)\mathbf{k}$$

by evaluating

$$\iint_\sigma \mathbf{F} \cdot \mathbf{n}\, dS \quad \text{and} \quad \iiint_G \text{div}\, \mathbf{F}\, dV$$

◆ 9. **Stokes' Theorem:** Let σ be the portion of the paraboloid $z = 1 - x^2 - y^2$ for which $z \geq 0$, and let C be the circle $x^2 + y^2 = 1$ that forms the boundary of σ in the xy-plane. Assuming that σ is oriented upward, verify Stokes' Theorem for the vector field

$$\mathbf{F} = (x^2 y - z^2)\mathbf{i} + (y^3 - x)\mathbf{j} + (2x + 3z - 1)\mathbf{k}$$

by evaluating

$$\oint_C \mathbf{F} \cdot d\mathbf{r} \quad \text{and} \quad \iint_\sigma (\text{curl}\, \mathbf{F}) \cdot \mathbf{n}\, dS$$

10. **Flux across a sphere:** Let

$$\mathbf{F}(x, y, z) = a^2 x\mathbf{i} + (y/a)\mathbf{j} + az^2\mathbf{k}$$

and let σ be the sphere of radius 1, centered at the origin and oriented outward. Find all values of a such that the flux of **F** across σ is 10.

Gabriel Cramer (1704–1752)

19 SECOND-ORDER DIFFERENTIAL EQUATIONS

In Section 7.7 we showed how to solve first-order linear differential equations. In this section we shall show how to solve certain second-order linear differential equations.

☐ **SECOND-ORDER LINEAR DIFFERENTIAL EQUATIONS**

In this section we shall be concerned with the general *second-order linear differential equation*

$$\frac{d^2y}{dx^2} + p(x)\frac{dy}{dx} + q(x)y = r(x) \tag{1}$$

or in an alternative notation,

$$y'' + p(x)y' + q(x)y = r(x)$$

where $p(x)$, $q(x)$, and $r(x)$ are continuous functions. If $r(x) = 0$ for all x, then (1) reduces to

$$\frac{d^2y}{dx^2} + p(x)\frac{dy}{dx} + q(x)y = 0$$

which is called the general second-order linear ***homogeneous*** differential equation. If $r(x)$ is not identically zero, then (1) is said to be ***nonhomogeneous***. Some examples of second-order linear differential equations are

$$\frac{d^2y}{dx^2} + x^2\frac{dy}{dx} - xy = e^x \qquad \boxed{p(x) = x^2, \quad q(x) = -x, \quad r(x) = e^x}$$

$$y'' + y' - 3y = \sin x \qquad \boxed{p(x) = 1, \quad q(x) = -3, \quad r(x) = \sin x}$$

$$y'' + e^x y = 3 \qquad \boxed{p(x) = 0, \quad q(x) = e^x, \quad r(x) = 3}$$

$$\frac{d^2y}{dx^2} - \frac{dy}{dx} + 2y = 0 \qquad \boxed{p(x) = -1, \quad q(x) = 2, \quad r(x) = 0}$$

The last equation is homogeneous and the first three are nonhomogeneous.

In Section 7.7 we gave a general procedure for solving first-order linear differential equations by integration. For second-order linear differential equations the situation is more complicated and simple general procedures for solving such equations can only be given in special cases. Before we can pursue this matter further, we shall need some preliminary results.

☐ **LINEAR INDEPENDENCE**

Two continuous functions f and g are said to be ***linearly dependent*** if one is a constant multiple of the other. If neither is a constant multiple of the other, then they are called ***linearly independent***. Thus,

$$f(x) = \sin x \quad \text{and} \quad g(x) = 3\sin x$$

are linearly dependent, but

$$f(x) = x \quad \text{and} \quad g(x) = x^2$$

are linearly independent.

The following theorem is central to the study of second-order linear differential equations.

19.1.1   THEOREM.   *If $y_1 = y_1(x)$ and $y_2 = y_2(x)$ are linearly independent solutions of the homogeneous equation*

$$\frac{d^2y}{dx^2} + p(x)\frac{dy}{dx} + q(x)y = 0 \tag{2}$$

then

$$y(x) = c_1 y_1(x) + c_2 y_2(x) \tag{3}$$

is the general solution of (2) in the sense that every solution of (2) can be obtained from (3) by choosing appropriate values for the arbitrary constants c_1 and c_2; conversely, (3) is a solution of (2) for all choices of c_1 and c_2.

(Readers interested in a proof of this theorem are referred to *Elementary Differential Equations and Boundary Value Problems*, John Wiley & Sons, New York, 1992, by William E. Boyce and Richard C. DiPrima.)

REMARK.   The expression on the right side of (3) is called a ***linear combination*** of $y_1(x)$ and $y_2(x)$. Thus, Theorem 19.1.1 tells us that once we find two linearly independent solutions of (2), we know *all* the solutions because every other solution can be expressed as a linear combination of those two.

☐ **CONSTANT**
COEFFICIENTS

For the remainder of this section we shall restrict our attention to second-order linear homogeneous equations of the form

$$\frac{d^2y}{dx^2} + p\frac{dy}{dx} + qy = 0 \tag{4}$$

where p and q are *constants*. Our objective is to find two linearly independent solutions of this equation. We shall start by looking for solutions of the form $y = e^{mx}$. This is motivated by the fact that the first and second derivatives of this function are multiples of y, suggesting that a solution of (4) might result by choosing m appropriately. To find such an m, we substitute

$$y = e^{mx}, \quad \frac{dy}{dx} = me^{mx}, \quad \frac{d^2y}{dx^2} = m^2e^{mx} \tag{5}$$

into (4) to obtain

$$(m^2 + pm + q)e^{mx} = 0 \tag{6}$$

which is satisfied if and only if

$$m^2 + pm + q = 0 \tag{7}$$

since $e^{mx} \neq 0$ for every x.

Equation (7), which is called the **auxiliary equation** for (4), can be obtained from (4) by replacing d^2y/dx^2 by m^2, dy/dx by m $(= m^1)$, and y by 1 $(= m^0)$. The solutions, m_1 and m_2, of the auxiliary equation can be obtained by factoring or by the quadratic formula. These solutions are

$$m_1 = \frac{-p + \sqrt{p^2 - 4q}}{2}, \quad m_2 = \frac{-p - \sqrt{p^2 - 4q}}{2} \tag{8}$$

Depending on whether $p^2 - 4q$ is positive, zero, or negative, these roots will be distinct and real, equal and real, or complex conjugates.* We shall consider each of these cases separately.

☐ **DISTINCT REAL ROOTS**

If m_1 and m_2 are distinct real roots, then (4) has the two solutions

$$y_1 = e^{m_1x}, \quad y_2 = e^{m_2x}$$

Neither of the functions e^{m_1x} and e^{m_2x} is a constant multiple of the other (Exercise 29), so the general solution of (4) in this case is

$$y(x) = c_1e^{m_1x} + c_2e^{m_2x} \tag{9}$$

Example 1   Find the general solution of $y'' - y' - 6y = 0$.

Solution.   The auxiliary equation is

$$m^2 - m - 6 = 0 \quad \text{or equivalently,} \quad (m + 2)(m - 3) = 0$$

so its roots are $m = -2$, $m = 3$. Thus, from (9) the general solution of the differential equation is

$$y = c_1e^{-2x} + c_2e^{3x}$$

where c_1 and c_2 are arbitrary constants.   ◀

---

*Recall that the complex solutions of a polynomial equation, and in particular of a quadratic equation, occur as conjugate pairs $a + bi$ and $a - bi$.

☐ **EQUAL REAL ROOTS**

If m_1 and m_2 are equal real roots, say $m_1 = m_2 \ (= m)$, then the auxiliary equation yields only one solution of (4):

$$y_1(x) = e^{mx}$$

We shall now show that

$$y_2(x) = xe^{mx} \tag{10}$$

is a second linearly independent solution. To see that this is so, note that $p^2 - 4q = 0$ in (8) since the roots are equal. Thus,

$$m = m_1 = m_2 = -p/2$$

and (10) becomes

$$y_2(x) = xe^{(-p/2)x}$$

Differentiating yields

$$y_2'(x) = \left(1 - \frac{p}{2}x\right)e^{(-p/2)x} \quad \text{and} \quad y_2''(x) = \left(\frac{p^2}{4}x - p\right)e^{-(p/2)x}$$

so

$$y_2''(x) + py_2'(x) + qy_2(x) = \left[\left(\frac{p^2}{4}x - p\right) + p\left(1 - \frac{p}{2}x\right) + qx\right]e^{(-p/2)x}$$

$$= \left[-\frac{p^2}{4} + q\right]xe^{(-p/2)x} \tag{11}$$

But $p^2 - 4q = 0$ implies that $(-p^2/4) + q = 0$, so (11) becomes

$$y_2''(x) + py_2'(x) + qy_2(x) = 0$$

which tells us that $y_2(x)$ is a solution of (4). It can be shown that

$$y_1(x) = e^{mx} \quad \text{and} \quad y_2(x) = xe^{mx}$$

are linearly independent (Exercise 29), so the general solution of (4) in this case is

$$y = c_1 e^{mx} + c_2 xe^{mx} \tag{12}$$

Example 2   Find the general solution of $y'' - 8y' + 16y = 0$.

Solution.   The auxiliary equation is

$$m^2 - 8m + 16 = 0 \quad \text{or equivalently,} \quad (m - 4)^2 = 0$$

so $m = 4$ is the only root. Thus, from (12) the general solution of the differential equation is

$$y = c_1 e^{4x} + c_2 xe^{4x} \qquad \blacktriangleleft$$

☐ **COMPLEX ROOTS**

If the auxiliary equation has complex roots $m_1 = a + bi$ and $m_2 = a - bi$, then $y_1(x) = e^{ax}\cos bx$ and $y_2(x) = e^{ax}\sin bx$ are linearly independent solutions of (4) and

$$y = e^{ax}(c_1 \cos bx + c_2 \sin bx) \tag{13}$$

is the general solution. The proof is discussed in the exercises (Exercise 30).

Example 3   Find the general solution of $y'' + y' + y = 0$.

Solution. The auxiliary equation $m^2 + m + 1 = 0$ has roots

$$m_1 = \frac{-1 + \sqrt{1 - 4}}{2} = -\frac{1}{2} + \frac{\sqrt{3}}{2}i$$

$$m_2 = \frac{-1 - \sqrt{1 - 4}}{2} = -\frac{1}{2} - \frac{\sqrt{3}}{2}i$$

Thus, from (13) with $a = -1/2$ and $b = \sqrt{3}/2$, the general solution of the differential equation is

$$y = e^{-x/2}\left(c_1 \cos \frac{\sqrt{3}}{2}x + c_2 \sin \frac{\sqrt{3}}{2}x\right) \quad \blacktriangleleft$$

☐ **INITIAL-VALUE PROBLEMS**

When a physical problem leads to a second-order differential equation, there are usually two conditions in the problem that determine specific values for the two arbitrary constants in the general solution of the equation. Conditions that specify the value of the solution $y(x)$ and its derivative $y'(x)$ at some point $x = x_0$ are called *initial conditions*. A second-order differential equation with initial conditions is called a *second-order initial-value problem*.

Example 4 Solve the initial-value problem

$$y'' - y = 0, \quad y(0) = 1, \quad y'(0) = 0$$

Solution. We must first solve the differential equation. The auxiliary equation

$$m^2 - 1 = 0$$

has distinct real roots $m_1 = 1$, $m_2 = -1$, so from (9) the general solution is

$$y(x) = c_1 e^x + c_2 e^{-x} \tag{14}$$

and the derivative of this solution is

$$y'(x) = c_1 e^x - c_2 e^{-x} \tag{15}$$

Substituting $x = 0$ in (14) and (15) and using the initial conditions $y(0) = 1$ and $y'(0) = 0$ yields the system of equations

$$c_1 + c_2 = 1$$

$$c_1 - c_2 = 0$$

Solving this system yields $c_1 = \frac{1}{2}$, $c_2 = \frac{1}{2}$, so from (14) the solution of the initial-value problem is

$$y(x) = \tfrac{1}{2}e^x + \tfrac{1}{2}e^{-x} = \cosh x \quad \blacktriangleleft$$

The following summary is included as a ready reference for the solution of second-order homogeneous linear differential equations with constant coefficients.

Summary

EQUATION: $\quad y'' + py' + qy = 0$
AUXILIARY EQUATION: $\quad m^2 + pm + q = 0$

CASE	GENERAL SOLUTION
Distinct real roots m_1, m_2 of the auxiliary equation	$y = c_1 e^{m_1 x} + c_2 e^{m_2 x}$
Equal real roots $m_1 = m_2 \ (= m)$ of the auxiliary equation	$y = c_1 e^{mx} + c_2 x e^{mx}$
Complex roots $m_1 = a + bi$, $m_2 = a - bi$ of the auxiliary equation	$y = e^{ax}(c_1 \cos bx + c_2 \sin bx)$

▶ Exercise Set 19.1

1. Verify that the following are solutions of the differential equation $y'' - y' - 2y = 0$ by substituting these functions into the equation:
   (a) e^{2x} and e^{-x}
   (b) $c_1 e^{2x} + c_2 e^{-x}$ (c_1, c_2 constants).

2. Verify that the following are solutions of the differential equation $y'' + 4y' + 4y = 0$ by substituting these functions into the equation:
   (a) e^{-2x} and xe^{-2x}
   (b) $c_1 e^{-2x} + c_2 xe^{-2x}$ (c_1, c_2 constants).

In Exercises 3–16, find the general solution of the differential equation.

3. $y'' + 3y' - 4y = 0$.   4. $y'' + 6y' + 5y = 0$.

5. $y'' - 2y' + y = 0$.   6. $y'' + 6y' + 9y = 0$.

7. $y'' + 5y = 0$.   8. $y'' + y = 0$.

9. $\dfrac{d^2y}{dx^2} - \dfrac{dy}{dx} = 0$.   10. $\dfrac{d^2y}{dx^2} + 3\dfrac{dy}{dx} = 0$.

11. $\dfrac{d^2y}{dt^2} + 4\dfrac{dy}{dt} + 4y = 0$.   12. $\dfrac{d^2y}{dt^2} - 10\dfrac{dy}{dt} + 25y = 0$.

13. $\dfrac{d^2y}{dx^2} - 4\dfrac{dy}{dx} + 13y = 0$.   14. $\dfrac{d^2y}{dx^2} - 6\dfrac{dy}{dx} + 25y = 0$.

15. $8y'' - 2y' - y = 0$.   16. $9y'' - 6y' + y = 0$.

In Exercises 17–22, solve the initial-value problem.

17. $y'' + 2y' - 3y = 0$; $y(0) = 1, y'(0) = 5$.

18. $y'' - 6y' - 7y = 0$; $y(0) = 5, y'(0) = 3$.

19. $y'' - 6y' + 9y = 0$; $y(0) = 2, y'(0) = 1$.

20. $y'' + 4y' + y = 0$; $y(0) = 5, y'(0) = 4$.

21. $y'' + 4y' + 5y = 0$; $y(0) = -3, y'(0) = 0$.

22. $y'' - 6y' + 13y = 0$; $y(0) = -1, y'(0) = 1$.

23. In each part find a second-order linear homogeneous differential equation with constant coefficients that has the given functions as solutions.
    (a) $y_1 = e^{5x}$, $y_2 = e^{-2x}$
    (b) $y_1 = e^{4x}$, $y_2 = xe^{4x}$
    (c) $y_1 = e^{-x}\cos 4x$, $y_2 = e^{-x}\sin 4x$.

24. Show that if e^x and e^{-x} are solutions of a second-order linear homogeneous differential equation, then so are $\cosh x$ and $\sinh x$.

25. Find all values of k for which the differential equation $y'' + ky' + ky = 0$ has a general solution of the given form.
    (a) $y = c_1 e^{ax} + c_2 e^{bx}$ (b) $y = c_1 e^{ax} + c_2 xe^{ax}$
    (c) $y = c_1 e^{ax}\cos bx + c_2 e^{ax}\sin bx$.

26. The equation
$$x^2\frac{d^2y}{dx^2} + px\frac{dy}{dx} + qy = 0 \quad (x > 0)$$
where p and q are constants, is called *Euler's equidimensional equation*. Show that the substitution $x = e^z$ transforms this equation into the equation
$$\frac{d^2y}{dz^2} + (p - 1)\frac{dy}{dz} + qy = 0$$

27. Use the result in Exercise 26 to find the general solution of
    (a) $x^2\dfrac{d^2y}{dx^2} + 3x\dfrac{dy}{dx} + 2y = 0 \quad (x > 0)$
    (b) $x^2\dfrac{d^2y}{dx^2} - x\dfrac{dy}{dx} - 2y = 0 \quad (x > 0)$.

28. Let $y(x)$ be a solution of $y'' + py' + qy = 0$. Prove: If p and q are positive constants, then $\lim\limits_{x \to +\infty} y(x) = 0$.

29. The *Wronskian* of two differentiable functions y_1 and y_2 is denoted by $W(y_1, y_2)$ and is defined to be the function
$$W(y_1, y_2) = y_1 y_2' - y_1' y_2 = \begin{vmatrix} y_1 & y_2 \\ y_1' & y_2' \end{vmatrix}$$
The value of $W(y_1, y_2)$ at a point x is denoted by $W(y_1, y_2)(x)$ or often more simply by $W(x)$. It can be proved that two solutions, $y_1 = y_1(x)$ and $y_2 = y_2(x)$, of Equation (2) are linearly dependent if and only if $W(x) = 0$ for all x. Equivalently, the functions are linearly independent if and only if $W(x) \neq 0$ for at least one value of x. Use this result to prove that the following solutions of Equation (4) are linearly independent:
    (a) $y_1 = e^{m_1 x}$, $y_2 = e^{m_2 x}$ ($m_1 \neq m_2$)
    (b) $y_1 = e^{mx}$, $y_2 = xe^{mx}$.

30. Prove: If the auxiliary equation of
$$y'' + py' + qy = 0$$
has complex roots $a + bi$ and $a - bi$, then the general solution of this differential equation is
$$y(x) = e^{ax}(c_1\cos bx + c_2\sin bx)$$
[*Hint:* By substitution, verify that $y_1 = e^{ax}\cos bx$ and $y_2 = e^{ax}\sin bx$ are solutions of the differential equation. Then use Exercise 29 to prove that y_1 and y_2 are linearly independent.]

31. Suppose that the auxiliary equation of the equation $y'' + py' + qy = 0$ has distinct real roots μ and m.
    (a) Show that the function
$$g_\mu(x) = \frac{e^{\mu x} - e^{mx}}{\mu - m}$$
is a solution of the differential equation.

(b)   Use L'Hôpital's rule to show that

$$\lim_{\mu \to m} g_{\mu}(x) = xe^{mx}$$

[*Note:* Can you see how the result in part (b) makes it plausible that the function $y(x) = xe^{mx}$ is a solution of $y'' + py' + qy = 0$ when m is a repeated root of the auxiliary equation?]

32. Consider the problem of solving the differential equation

$$y'' + \lambda y = 0$$

subject to the conditions $y(0) = 0$, $y(\pi) = 0$.

(a)   Show that if $\lambda \le 0$, then $y = 0$ is the only solution.

(b)   Show that if $\lambda > 0$, then the solution is

$$y = c \sin \sqrt{\lambda} x$$

where c is an arbitrary constant, if

$$\lambda = 1, 2^2, 3^2, 4^2, \ldots$$

and the only solution is $y = 0$ otherwise.

■ 19.2 SECOND-ORDER LINEAR NONHOMOGENEOUS DIFFERENTIAL EQUATIONS WITH CONSTANT COEFFICIENTS; UNDETERMINED COEFFICIENTS

In this section we shall study techniques for solving second-order linear differential equations that are not homogeneous.

□ **THE COMPLEMENTARY EQUATION**

The following theorem is the key result for solving a second-order *nonhomogeneous* linear differential equation with constant coefficients

$$y'' + py' + qy = r(x) \tag{1}$$

where p and q are constants and $r(x)$ is a continuous function of x.

19.2.1   THEOREM.   *The general solution of* (1) *is*

$$y(x) = c_1 y_1(x) + c_2 y_2(x) + y_p(x)$$

where $c_1 y_1(x) + c_2 y_2(x)$ *is the general solution of the homogeneous equation*

$$y'' + py' + qy = 0 \tag{2}$$

and $y_p(x)$ *is any solution of* (1).

The proof is deferred to the end of the section.

REMARK.   Equation (2) is called the ***complementary equation*** to (1) and $y_p(x)$ is called a ***particular solution*** of (1). Thus, Theorem 19.2.1 states that *the general solution of* (1) *is obtained by adding a particular solution of* (1) *to the general solution of the complementary equation.*

□ **THE METHOD OF UNDETERMINED COEFFICIENTS**

Since we already know how to obtain the general solution of (2), we shall focus our attention on the problem of obtaining a particular solution of (1). In this section we shall discuss a procedure for doing this called the method of ***undetermined coefficients***, and in the next section we shall discuss a second procedure called *variation of parameters.*

The first step in the method of undetermined coefficients is to make a reasonable guess about the form of the particular solution. This guess will involve one or more unknown coefficients that we shall then determine from conditions obtained by substituting the proposed solution into the differential equation. We shall begin with some examples that illustrate the basic idea.

Example 1   Find a particular solution of the differential equation

$$y'' + 2y' - 8y = e^{3x} \tag{3}$$

Solution.   It should be clear that such functions as $\sin x$, $\cos x$, $\ln x$, or x^3 are not reasonable possibilities for $y_p(x)$ because the derivatives of such functions do not yield expressions involving e^{3x}. Since the obvious choice for $y_p(x)$ is an expression involving e^{3x}, we shall *guess* that $y_p(x)$ has the form

$$y_p(x) = Ae^{3x} \tag{4}$$

where A is an unknown constant to be determined. It follows that

$$y_p'(x) = 3Ae^{3x} \tag{5}$$

$$y_p''(x) = 9Ae^{3x} \tag{6}$$

Substituting expressions (4), (5), and (6) for y, y', and y'' in (3) yields

$$9Ae^{3x} + 2(3Ae^{3x}) - 8(Ae^{3x}) = e^{3x}$$

or

$$7Ae^{3x} = e^{3x}$$

Thus, $A = \frac{1}{7}$. From this result and (4), a particular solution of (3) is

$$y_p(x) = \tfrac{1}{7}e^{3x} \qquad \blacktriangleleft$$

Example 2   Find a particular solution of the differential equation

$$y'' - y' - 6y = e^{3x} \tag{7}$$

Solution.   As in Example 1, a reasonable guess is $y_p(x) = Ae^{3x}$. However, if we substitute (4), (5), and (6) in (7), we obtain

$$9Ae^{3x} - 3Ae^{3x} - 6Ae^{3x} = e^{3x} \quad \text{or} \quad 0 = e^{3x}$$

which is a contradiction. The problem here is that $y_p(x) = Ae^{3x}$ is a solution of the complementary equation

$$y'' - y' - 6y = 0$$

which makes it impossible for y_p to satisfy (7), no matter how A is selected. How should we proceed in this case? Experience has shown that multiplying Ae^{3x} by x produces the correct form for the solution. Thus, we shall try

$$y_p(x) = Axe^{3x} \tag{8}$$

It follows that

$$y_p'(x) = 3Axe^{3x} + Ae^{3x} \tag{9}$$

$$y_p''(x) = 9Axe^{3x} + 6Ae^{3x} \tag{10}$$

Substituting expressions (8), (9), and (10) for y, y', and y'' in (7) yields

$$(9Axe^{3x} + 6Ae^{3x}) - (3Axe^{3x} + Ae^{3x}) - 6(Axe^{3x}) = e^{3x}$$

or

$$5Ae^{3x} = e^{3x}$$

Thus, $A = \frac{1}{5}$, so (8) implies that a particular solution of (7) is

$$y_p(x) = \tfrac{1}{5}xe^{3x} \qquad \blacktriangleleft$$

REMARK. Had it turned out that Axe^{3x} was also a solution of the complementary equation, then we would have tried $y_p(x) = Ax^2 e^{3x}$ as our candidate for a particular solution.

□ SUMMARY

In summary, a particular solution of an equation of the form

$$y'' + py' + qy = ke^{ax}$$

can be obtained as follows:

> **Step 1.** Start with $y_p = Ae^{ax}$ as an initial guess.
>
> **Step 2.** Determine if the initial guess is a solution of the complementary equation $y'' + py' + qy = 0$.
>
> **Step 3.** If the initial guess is not a solution of the complementary equation, then $y_p = Ae^{ax}$ is the correct form of a particular solution.
>
> **Step 4.** If the initial guess is a solution of the complementary equation, then multiply it by the smallest positive integer power of x required to produce a function that is not a solution of the complementary equation. This will yield either $y_p = Axe^{ax}$ or $y_p = Ax^2 e^{ax}$.

Table 19.2.1, which we provide without proof, restates the preceding procedure more compactly and also explains how to find a particular solution of (1) when $r(x)$ is a polynomial or a combination of sine and cosine functions.

Table 19.2.1

EQUATION	INITIAL GUESS FOR y_p
$y'' + py' + qy = ke^{ax}$	$y_p = Ae^{ax}$
$y'' + py' + qy = a_0 + a_1 x + \cdots + a_n x^n$	$y_p = A_0 + A_1 x + \cdots + A_n x^n$
$y'' + py' + qy = a_1 \cos bx + a_2 \sin bx$	$y_p = A_1 \cos bx + A_2 \sin bx$

MODIFICATION RULE

If any term in the initial guess is a solution of the complementary equation, then the correct form for y_p is obtained by multiplying the initial guess by the smallest positive integer power of x required so that no term is a solution of the complementary equation.

Example 3 Find the general solution of

$$y'' + y' = 4x^2 \tag{11}$$

Solution. We begin by finding the general solution of the complementary equation

$$y'' + y' = 0$$

The auxiliary equation for this homogeneous equation is

$$m^2 + m = 0$$

which has roots $m_1 = 0$, $m_2 = -1$. Thus, the general solution, $y_c(x)$, of the complementary equation is

$$y_c(x) = c_1 e^{0x} + c_2 e^{-x} = c_1 + c_2 e^{-x}$$

Since the right-hand side of (11) is a second-degree polynomial, our initial guess for y_p is

$$y_p = A_0 + A_1 x + A_2 x^2$$

But A_0 is a solution of the complementary equation (take $c_1 = A_0$, $c_2 = 0$), so our second guess for y_p is

$$y_p = x(A_0 + A_1 x + A_2 x^2) = A_0 x + A_1 x^2 + A_2 x^3 \tag{12}$$

Since no term of this function is a solution of the complementary equation, this is the correct form for y_p. Differentiating (12) we obtain

$$y_p'(x) = A_0 + 2A_1 x + 3A_2 x^2 \tag{13}$$

$$y_p''(x) = 2A_1 + 6A_2 x \tag{14}$$

Substituting (13) and (14) in (11) yields

$$(2A_1 + 6A_2 x) + (A_0 + 2A_1 x + 3A_2 x^2) = 4x^2$$

or

$$(A_0 + 2A_1) + (2A_1 + 6A_2)x + 3A_2 x^2 = 4x^2$$

To solve for A_0, A_1, and A_2, we equate corresponding coefficients on the two sides of this equation; this yields

$$A_0 + 2A_1 = 0$$

$$2A_1 + 6A_2 = 0$$

$$3A_2 = 4$$

from which we obtain

$$A_2 = \tfrac{4}{3}, \quad A_1 = -4, \quad A_0 = 8$$

Substituting these values in (12) yields the particular solution

$$y_p(x) = 8x - 4x^2 + \tfrac{4}{3}x^3$$

Thus, the general solution of (11) is

$$y(x) = y_c(x) + y_p(x) = c_1 + c_2 e^{-x} + 8x - 4x^2 + \tfrac{4}{3}x^3 \quad \blacktriangleleft$$

Example 4   Find the general solution of

$$y'' - 2y' + y = \sin 2x \tag{15}$$

Solution.   We begin by finding the general solution of the complementary equation

$$y'' - 2y' + y = 0$$

The auxiliary equation for this homogeneous equation is

$$m^2 - 2m + 1 = 0$$

which has the repeated root $m_1 = 1$, $m_2 = 1$. Thus, the general solution of the complementary equation is

$$y_c(x) = c_1 e^x + c_2 x e^x \tag{16}$$

Since $\sin 2x$ has the form $a_1 \cos bx + a_2 \sin bx$ ($a_1 = 0$, $a_2 = 1$, $b = 2$), our initial guess for y_p is

$$y_p(x) = A_1 \cos 2x + A_2 \sin 2x \tag{17}$$

Since no term of y_p is a solution of the complementary equation [see (16)], this is the correct form for a particular solution. Differentiating (17) we obtain

$$y_p'(x) = -2A_1 \sin 2x + 2A_2 \cos 2x \tag{18}$$

$$y_p''(x) = -4A_1 \cos 2x - 4A_2 \sin 2x \tag{19}$$

Substituting (17), (18), and (19) in (15) yields

$$(-4A_1 \cos 2x - 4A_2 \sin 2x) - 2(-2A_1 \sin 2x + 2A_2 \cos 2x)$$
$$+ (A_1 \cos 2x + A_2 \sin 2x) = \sin 2x$$

or

$$(-3A_1 - 4A_2) \cos 2x + (4A_1 - 3A_2) \sin 2x = \sin 2x$$

To solve for A_1 and A_2 we equate the coefficients of $\sin 2x$ and $\cos 2x$ on the two sides of this equation; this yields

$$-3A_1 - 4A_2 = 0$$
$$4A_1 - 3A_2 = 1$$

from which we obtain

$$A_1 = \tfrac{4}{25}, \quad A_2 = -\tfrac{3}{25}$$

(Verify.) From this result and (17), a particular solution of (15) is

$$y_p(x) = \tfrac{4}{25} \cos 2x - \tfrac{3}{25} \sin 2x$$

Thus, the general solution of (15) is

$$y(x) = y_c(x) + y_p(x) = c_1 e^x + c_2 x e^x + \tfrac{4}{25} \cos 2x - \tfrac{3}{25} \sin 2x \qquad \blacktriangleleft$$

We conclude this section with a proof of Theorem 19.2.1.

■ PROOF

Proof of Theorem 19.2.1. To prove that

$$y(x) = c_1 y_1(x) + c_2 y_2(x) + y_p(x) \tag{20}$$

is the general solution of (1) we must prove two results: first, that for all choices of c_1 and c_2 the function $y(x)$ defined by (20) satisfies (1); and second, that every solution of (1) can be obtained from (20) by choosing appropriate values for the constants c_1 and c_2.

To prove the first statement, let c_1 and c_2 have any real values. Then differentiating (20) yields

$$y'(x) = c_1 y_1'(x) + c_2 y_2'(x) + y_p'(x)$$
$$y''(x) = c_1 y_1''(x) + c_2 y_2''(x) + y_p''(x)$$

so

$$y''(x) + py'(x) + qy(x) = c_1(y_1''(x) + py_1'(x) + qy_1(x))$$
$$+ c_2(y_2''(x) + py_2'(x) + qy_2(x))$$
$$+ y_p''(x) + py_p'(x) + qy_p(x) \tag{21}$$

But $y_1(x)$ and $y_2(x)$ satisfy (2) and $y_p(x)$ satisfies (1), so (21) reduces to

$$y''(x) + py'(x) + qy(x) = 0 + 0 + r(x) = r(x)$$

which proves that $y(x)$ satisfies (1).

To prove the second statement, let $y(x)$ be any solution of (1). We must find values of c_1 and c_2 such that

$$y(x) = c_1 y_1(x) + c_2 y_2(x) + y_p(x) \tag{22}$$

But $y(x) - y_p(x)$ satisfies (2) since

$$(y(x) - y_p(x))'' + p(y(x) - y_p(x))' + q(y(x) - y_p(x))$$
$$= [y''(x) + py'(x) + qy(x)] - [y_p''(x) + py_p'(x) + qy_p(x)]$$
$$= r(x) - r(x) = 0$$

Thus, since $c_1 y_1(x) + c_2 y_2(x)$ is the general solution of (2), there exist values of c_1 and c_2 such that the solution $y(x) - y_p(x)$ can be written as

$$y(x) - y_p(x) = c_1 y_1(x) + c_2 y_2(x)$$

from which (22) follows. ∎

▶ Exercise Set 19.2

In Exercises 1–24, use the method of undetermined coefficients to find the general solution of the equation.

1. $y'' + 6y' + 5y = 2e^{3x}$.   **2.** $y'' + 3y' - 4y = 5e^{7x}$.

3. $y'' - 9y' + 20y = -3e^{5x}$.

4. $y'' + 7y' - 8y = 7e^{x}$.   **5.** $y'' + 2y' + y = e^{-x}$.

6. $y'' + 4y' + 4y = 4e^{-2x}$.   **7.** $y'' + y' - 12y = 4x^2$.

8. $y'' - 4y' - 5y = -6x^2$.   **9.** $y'' - 6y' = x - 1$.

10. $y'' + 3y' = 2x + 2$.   **11.** $y'' - x^3 + 1 = 0$.

12. $y'' + 3x^3 + x = 0$.   **13.** $y'' - y' - 2y = 10\cos x$.

14. $y'' - 3y' - 4y = 2\sin x$.

15. $y'' - 4y = 2\sin 2x + 3\cos 2x$.

16. $y'' - 9y = \cos 3x - \sin 3x$.

17. $y'' + y = \sin x$.   **18.** $y'' + 4y = \cos 2x$.

19. $y'' - 3y' + 2y = x$.   **20.** $y'' + 4y' + 4y = 3x + 3$.

21. $y'' + 4y' + 9y = x^2 + 3x$.

22. $y'' - y = 1 + x + x^2$.   **23.** $y'' + 4y = \sin x \cos x$.

24. $y'' + 4y = \cos^2 x - \sin^2 x$.

25. (a) Prove: If $y_1(x)$ is a solution of
$$y'' + p(x)y' + q(x)y = r_1(x)$$
and $y_2(x)$ is a solution of
$$y'' + p(x)y' + q(x)y = r_2(x)$$
then $y_1(x) + y_2(x)$ is a solution of
$$y'' + p(x)y' + q(x)y = r_1(x) + r_2(x)$$

  (b) Use the result in part (a) to find a particular solution of the equation
$$y'' + 3y' - 4y = x + e^x$$

  (c) State a generalization of the result in part (a) that is applicable to the equation
$$y'' + p(x)y' + q(x)y = r_1(x) + r_2(x) + \cdots + r_n(x)$$

In Exercises 26–34, use the results in Exercise 25 to find the general solution of the differential equation.

26. $y'' - y' - 2y = x + e^{-x}$.   **27.** $y'' - y = 1 + e^x$.

28. $y'' - 4y' + 3y = 2\cos x + 4\sin x$.

29. $y'' + 4y = 1 + x + \sin x$.

30. $y'' + 2y' + y = 2 + 3x + 3e^x + 2\cos 2x$.

31. $y'' - 2y' + y = \sinh x$. [*Hint:* $\sinh x = \frac{1}{2}(e^x - e^{-x})$.]

32. $y'' + 4y' - 5y = \cosh x$. [*Hint:* $\cosh x = \frac{1}{2}(e^x + e^{-x})$.]

33. $y'' + y = 12\cos^2 x$. [*Hint:* $\cos^2 x = \frac{1}{2}(1 + \cos 2x)$.]

34. $y'' + 2y' + y = \sin^2 x$. [*Hint:* $\sin^2 x = \frac{1}{2}(1 - \cos 2x)$.]

35. (a) Find the general solution of
$$y'' + \mu^2 y = a\sin bx$$
where a is an arbitrary constant and μ and b are positive constants such that $\mu \neq b$.

  (b) Use part (a) and Exercise 25 to find the general solution of
$$y'' + \mu^2 y = \sum_{k=1}^{n} a_k \sin k\pi x$$
where $\mu > 0$ and $\mu \neq k\pi$, $k = 1, 2, \ldots, n$.

36. Find the general solution of
$$y'' + \lambda^2 y = \sum_{k=1}^{n} a_k \cos k\pi x$$
where $\lambda > 0$ and $\lambda \neq k\pi$, $k = 1, 2, \ldots, n$. [*Hint:* See Exercise 35.]

37. Find all solutions of the equation
$$y'' - y' = 4 - 4x$$
(if any) with the property that $y'(x_0) = y''(x_0) = 0$ at some point x_0.

■ 19.3 VARIATION OF PARAMETERS

> *In this section* we shall consider an alternative method for finding a particular solution of a second-order linear nonhomogeneous equation that often works when the method of undetermined coefficients cannot be applied.*

As in the preceding section, we shall be concerned with equations of the form

$$y'' + py' + qy = r(x) \tag{1}$$

where p and q are constants and $r(x)$ is a continuous function of x. The method we shall discuss, called **variation of parameters**, is based on the (not very obvious) fact that if

$$y_c(x) = c_1 y_1(x) + c_2 y_2(x) \tag{2}$$

is the general solution of the complementary equation

$$y'' + py' + qy = 0$$

then it is possible to find functions $u(x)$ and $v(x)$ such that

$$y_p(x) = u(x)y_1(x) + v(x)y_2(x) \tag{3}$$

is a particular solution of (1).

To see how the functions $u(x)$ and $v(x)$ can be found, consider the derivative of (3):

$$y_p' = (uy_1' + vy_2') + (u'y_1 + v'y_2) \tag{4}$$

If we now differentiate y_p', we shall introduce second derivatives of the unknown functions $u(x)$ and $v(x)$. To avoid this complication we shall require that $u(x)$ and $v(x)$ satisfy the condition

$$u'y_1 + v'y_2 = 0 \tag{5}$$

so (4) simplifies to

$$y_p' = uy_1' + vy_2' \tag{6}$$

Thus,

$$y_p'' = uy_1'' + u'y_1' + vy_2'' + v'y_2' \tag{7}$$

On substituting (3), (6), and (7) in (1) and rearranging terms we obtain

$$u(y_1'' + py_1' + qy_1) + v(y_2'' + py_2' + qy_2) + u'y_1' + v'y_2' = r(x) \tag{8}$$

Since y_1 and y_2 are solutions of the complementary equation, we have

$$y_1'' + py_1' + qy_1 = 0 \quad \text{and} \quad y_2'' + py_2' + qy_2 = 0$$

so (8) simplifies to

$$u'y_1' + v'y_2' = r(x)$$

This equation together with (5) yields two equations in the two unknowns $u'(x)$ and $v'(x)$:

$$\begin{aligned} u'y_1 + v'y_2 &= 0 \\ u'y_1' + v'y_2' &= r(x) \end{aligned} \tag{9}$$

---

*It is assumed in this section that the reader is familiar with Cramer's rule, which is discussed in Appendix D.

It can be shown (see Boyce and DiPrima, *Elementary Differential Equations and Boundary Value Problems*, John Wiley & Sons, New York, 1992) that this system has a unique solution for u' and v'. Once this solution is found, u and v can be obtained by integration.

Example 1 Find the general solution of

$$y'' + y = \sec x \qquad (-\pi/2 < x < \pi/2) \tag{10}$$

Solution. We begin by finding the general solution of the complementary equation

$$y'' + y = 0$$

The auxiliary equation for this homogeneous equation is

$$m^2 + 1 = 0$$

which has roots $m_1 = i$ and $m_2 = -i$. Thus, the general solution of the complementary equation is

$$y_c(x) = c_1 \cos x + c_2 \sin x$$

Comparing this with (2) yields $y_1(x) = \cos x$ and $y_2(x) = \sin x$, so we shall look for a particular solution of the form

$$y_p(x) = u(x) \cos x + v(x) \sin x \tag{11}$$

Substituting $y_1 = \cos x$, $y_2 = \sin x$, and $r(x) = \sec x$ in (9) yields

$$u' \cos x + v' \sin x = 0$$
$$-u' \sin x + v' \cos x = \sec x \tag{12}$$

Solving this system for u' and v' by Cramer's rule (or any suitable method) we obtain

$$u' = \frac{\begin{vmatrix} 0 & \sin x \\ \sec x & \cos x \end{vmatrix}}{\begin{vmatrix} \cos x & \sin x \\ -\sin x & \cos x \end{vmatrix}} = \frac{-\sec x \sin x}{\cos^2 x + \sin^2 x} = \frac{-\tan x}{1} = -\tan x$$

$$v' = \frac{\begin{vmatrix} \cos x & 0 \\ -\sin x & \sec x \end{vmatrix}}{\begin{vmatrix} \cos x & \sin x \\ -\sin x & \cos x \end{vmatrix}} = \frac{\cos x \sec x}{\cos^2 x + \sin^2 x} = \frac{1}{1} = 1$$

Integrating u' and v' yields

$$u(x) = \int -\tan x \, dx = \ln |\cos x| \quad \text{and} \quad v(x) = \int 1 \, dx = x$$

[We set the constants of integration equal to zero because any functions $u(x)$ and $v(x)$ satisfying (12) will suffice.] Substituting these functions in (11) yields

$$y_p(x) = (\ln |\cos x|) \cos x + x \sin x$$

Thus, the general solution of (10) is

$$y(x) = y_c(x) + y_p(x) = c_1 \cos x + c_2 \sin x + (\ln |\cos x|) \cos x + x \sin x \qquad \blacktriangleleft$$

▶ **Exercise Set 19.3**

In Exercises 1–6, use variation of parameters to find the general solution of the differential equation and verify that the same solution results using undetermined coefficients.

1. $y'' + y = x^2$.

2. $y'' + 9y = 3x$.

3. $y'' + y' - 2y = 2e^x$.

4. $y'' + 5y' + 6y = e^{-x}$.

5. $y'' + 4y = \sin 2x$.

6. $y'' + 9y = \cos 3x$.

In Exercises 7–28, use variation of parameters to find the general solution of the differential equation.

7. $y'' + y = \tan x$.

8. $y'' + y = \cot x$.

9. $y'' - 2y' + y = e^x/x$.

10. $y'' - 4y' + 4y = e^{2x}/x$.

11. $y'' + y = 3\sin^2 x$.

12. $y'' + y = 6\cos^2 x$.

13. $y'' + y = \csc x$.

14. $y'' + 9y = 6\sec 3x$.

15. $y'' + y = \sec x \tan x$.

16. $y'' + y = \csc x \cot x$.

17. $y'' + 2y' + y = e^{-x}/x^2$.

18. $y'' - y = x^2 e^x$.

19. $y'' + 4y' + 4y = xe^{-x}$.

20. $y'' + 4y' + 4y = xe^{2x}$.

21. $y'' + y = \sec^2 x$.

22. $y'' + y = \sec^3 x$.

23. $y'' - 2y' + y = e^x/x^2$.

24. $y'' - 2y' + y = x^3 e^x$.

25. $y'' - y = e^x \cos x$.

26. $y'' - 2y' + 2y = e^{2x} \sin x$.

27. $y'' + 2y' + y = e^{-x} \ln|x|$.

28. $y'' - 3y' + 2y = \dfrac{e^x}{1 + e^x}$.

29. Use the method of variation of parameters to show that the general solution of $y'' + y = r(x)$ is
$$y = c_1 \cos x + c_2 \sin x + g(x) \cos x + h(x) \sin x$$
where
$$g(x) = -\int r(x) \sin x \, dx$$
$$h(x) = \int r(x) \cos x \, dx$$

30. Use the method of variation of parameters to show that if y_1 and y_2 are linearly independent solutions of the equation $y'' + py' + qy = 0$, then a particular solution of $y'' + py' + qy = r(x)$ is given by
$$y_p(x) = -y_1(x) \int \frac{y_2(x)r(x)}{W(x)} \, dx + y_2(x) \int \frac{y_1(x)r(x)}{W(x)} \, dx$$
where $W(x) = y_1(x)y_2'(x) - y_1'(x)y_2(x)$.

■ **19.4** VIBRATION OF A SPRING

> *In this section we shall use second-order linear differential equations with constant coefficients to study the vibration of a spring.*

We shall be interested in solving the following problem:

> **19.4.1**   THE VIBRATING SPRING PROBLEM.   *As shown in* Figure 19.4.1, *let a mass attached to a vertical spring be allowed to settle into an equilibrium position. Then, let the spring be stretched (or compressed) by pulling (or pushing) the mass, and finally, let the mass be released, thereby causing it to undergo a vibratory motion. We shall be interested in finding a formula for the position of the mass at any time t.*

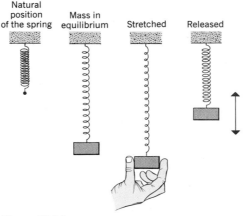

Natural position of the spring   Mass in equilibrium   Stretched   Released

Figure 19.4.1

To solve this problem we shall need three results from physics:

19.4.2   HOOKE'S LAW.   If a spring is stretched (or compressed) l units beyond its natural position, then it pulls back (or pushes) with a force of magnitude

$$F = kl$$

where k is a positive constant, called the **spring constant**. This constant depends on such factors as the thickness of the spring, the material from which it is made, and the units of force and length; it is measured in units of force per unit of length.

19.4.3   NEWTON'S SECOND LAW OF MOTION.   If an object with mass M is subjected to a force **F**, then it undergoes an acceleration **a** satisfying

$$\mathbf{F} = M\mathbf{a}$$

19.4.4   WEIGHT.   The gravitational force exerted by the earth on an object is called the **weight** (more precisely, **earth weight**) of the object. It follows from Newton's Second Law of Motion that an object with mass M has a weight whose magnitude w is given by

$$w = Mg \tag{1}$$

where g is a constant, called the **acceleration due to gravity**. Near the surface of the earth an approximate value of g is $g \approx 32$ ft/sec^2 if length is measured in feet and time in seconds, or $g \approx 9.8$ m/sec^2 if length is measured in meters and time in seconds.

In any problem involving length, mass, time, and force it is important that the units of measurement be consistent. The most important systems of measurement are summarized in Table 19.4.1.

Table 19.4.1

SYSTEM OF MEASUREMENT	LENGTH	TIME	MASS	FORCE
British system	foot (ft)	second (sec)	slug	pound (lb)
SI system	meter (m)	second (sec)	kilogram (kg)	newton (N)
CGS system	centimeter (cm)	second (sec)	gram (g)	dyne

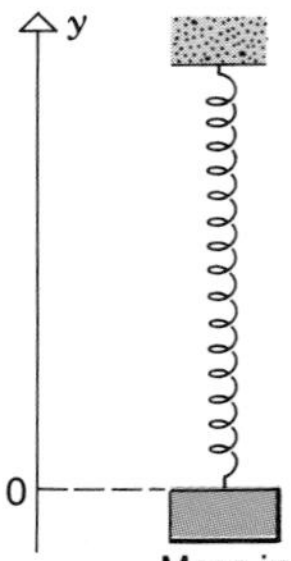

Figure 19.4.2

To solve the vibrating spring problem posed above, we introduce a coordinate axis (a y-axis) with the positive direction up and the origin at the bottom end of the spring when the mass is in its equilibrium position (Figure 19.4.2). If we take $t = 0$ to be the time at which the mass is released, then at each subsequent time t, the end of the spring has a position $y(t)$, a velocity $y'(t)$, and an acceleration $y''(t)$.* Because the positive direction of the y-axis is up, force, velocity, and acceleration are positive when directed up and negative when directed down.

Before we can solve the spring problem, it will be necessary to derive a preliminary result from the equilibrium conditions for the mass. Let us assume that the spring constant is k, the mass of the object is M, and the spring is stretched l units beyond its natural length when the mass is in equilibrium (Figure 19.4.3). When the mass is in its equilibrium

---

*For convenience of terminology we shall refer to $y(t)$, $y'(t)$, and $y''(t)$ as the position, velocity, and acceleration of the mass, respectively.

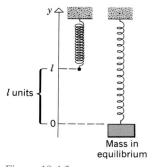

Figure 19.4.3

position its downward weight, $-Mg$, is balanced exactly by the upward force, kl, of the spring, so the sum of these forces must be zero;* that is,

$$kl - Mg = 0$$

or

$$kl = Mg \qquad (2)$$

The basic strategy for solving the spring problem is to find the total force $F(t)$ that acts on the mass at time t. Then, since the acceleration of the mass at time t is $y''(t)$, it will follow from Newton's Second Law of Motion that

$$My''(t) = F(t)$$

or

$$y''(t) = F(t)/M \qquad (3)$$

which is a second-order linear differential equation that can be solved for the position function $y(t)$.

At each instant, the force $F(t)$ in (3) consists of four possible components:

$F_g(t) =$ the force of gravity (weight of the mass)

$F_s(t) =$ the force of the spring

$F_d(t) =$ the **damping force** or frictional force exerted on the mass by the surrounding medium (air, water, oil, etc.)

$F_e(t) =$ external forces due to such factors as movement in the spring support, magnetic forces acting on the mass, and so on

☐ **UNDAMPED FREE VIBRATIONS**

The simplest case occurs when there is no damping ($F_d = 0$) and the mass is *free* of external forces ($F_e = 0$). In this case the only forces acting on the mass are the force of gravity, F_g, and the spring force, F_s. Let us try to calculate these forces when the mass is at an arbitrary point $y(t)$.

When the mass is at the point $y(t)$, the spring is stretched (or compressed) $l - y(t)$ units from its natural length (Figure 19.4.4), so by Hooke's law the spring exerts a force of

$$F_s(t) = k(l - y(t))$$

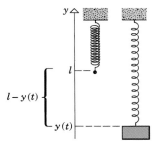

Figure 19.4.4

on the mass. Adding this to the force of gravity,

$$F_g(t) = -Mg$$

acting on the mass yields the total force acting on the mass:

$$F_s(t) + F_g(t) = k(l - y(t)) - Mg$$

But $kl = Mg$ from (2), so

$$F_s(t) + F_g(t) = -ky(t) \qquad (4)$$

Substituting this expression for $F(t)$ in (3) yields

$$y''(t) = -\frac{k}{M} y(t)$$

or

$$y''(t) + \frac{k}{M} y(t) = 0 \qquad (5)$$

which is a second-order linear differential equation with constant coefficients.

---

*We shall assume that the weight of the spring is small relative to the weight of the mass and can be neglected.

Because the mass is *released* (i.e., has zero initial velocity) at time $t = 0$, we have $y'(0) = 0$. If we assume, in addition, that the position of the mass at time $t = 0$ is $y(0) = y_0$, then we have two initial conditions that can be combined with (5) to yield an initial-value problem for $y(t)$:

$$y'' + \frac{k}{M}y = 0$$

$$y(0) = y_0, \quad y'(0) = 0$$

(6)

The auxiliary equation for the differential equation in (6) is

$$m^2 + \frac{k}{M} = 0$$

which has roots $m_1 = \sqrt{k/M}\,i$, $m_2 = -\sqrt{k/M}\,i$ (since k and M are positive), so the general solution of the differential equation is

$$y(t) = c_1 \cos\left(\sqrt{k/M}\,t\right) + c_2 \sin\left(\sqrt{k/M}\,t\right)$$

From the initial conditions $y(0) = y_0$ and $y'(0) = 0$ it follows that $c_1 = y_0$, $c_2 = 0$ (verify), so the solution of (6) is

$$y(t) = y_0 \cos\left(\sqrt{k/M}\,t\right)$$

(7)

This formula describes a periodic vibration with an **amplitude** of $|y_0|$ and a **period T** given by

$$T = \frac{2\pi}{\sqrt{k/M}} = 2\pi\sqrt{M/k}$$

(8)

Physically, the period is the time in seconds required for one complete oscillation (Figure 19.4.5). The reciprocal of the period is called the **frequency** of the oscillation. In the SI system the unit of frequency is the **hertz** (Hz); that is, 1 hertz = 1 Hz = 1 oscillation per second. Thus the frequency f is given by

$$f = \frac{1}{T} = \frac{1}{2\pi}\sqrt{k/M}$$

(9)

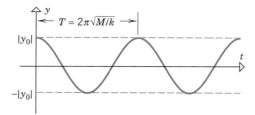

Figure 19.4.5

REMARK.   In the preceding problem, the spring oscillates indefinitely with constant amplitude because there are no frictional forces to dissipate the energy in the mass–spring system.

Example 1   Suppose that the top of a spring is fixed to a ceiling and a mass attached to the bottom end stretches the spring 0.2 m. If the mass is pulled 0.5 m below its equilibrium position and released, find

(a)   a formula for the position of the mass at any time t;

(b)   the amplitude, period, and frequency of the motion.

Solution (a).   The appropriate formula is (7). Although we are not given the mass M or the spring constant k, it does not matter because we are given that the mass stretches the spring $l = 0.2$ m and we know that $g = 9.8$ m/sec^2, so (2) implies that

$$\frac{k}{M} = \frac{g}{l} = \frac{9.8}{0.2} = 49 \text{ sec}^{-2}$$

Also, the mass is initially 0.5 m *below* its equilibrium position, so $y_0 = -0.5$. Therefore, from (7) the formula for the position of the mass at time t is

$$y(t) = -0.5 \cos 7t$$

Solution (b).   From (8) and (9)

$$\text{amplitude} = |y_0| = |-0.5| = 0.5 \text{ m}$$
$$\text{period} = T = 2\pi\sqrt{M/k} = 2\pi\sqrt{1/49} = 2\pi/7 \text{ sec}$$
$$\text{frequency} = f = 1/T = 7/(2\pi) \text{ Hz} \quad \blacktriangleleft$$

□ **DAMPED FREE VIBRATIONS**

We shall now consider the solution of the spring problem in the case where damping cannot be neglected; that is, $F_d \neq 0$.

Physicists have shown that under appropriate conditions the damping force F_d is opposite to the direction of motion of the mass (i.e., tends to slow the mass down) and has a magnitude that is proportional to the speed of the mass (the greater the speed, the greater the effect of friction). This type of damping force is described by an equation of the form

$$F_d(t) = -cy'(t) \tag{10}$$

where c is a positive constant, called the **damping constant**. The damping constant depends on the viscosity of the surrounding medium; it is measured in units of force per unit of velocity (lb/ft/sec or, equivalently, lb · sec/ft, for example).

It follows from (4) and (10) that

$$F_s(t) + F_g(t) + F_d(t) = -ky(t) - cy'(t)$$

If we substitute this expression in (3) for $F(t)$, we obtain

$$y''(t) = -\frac{k}{M}y(t) - \frac{c}{M}y'(t)$$

or

$$y''(t) + \frac{c}{M}y'(t) + \frac{k}{M}y(t) = 0$$

Combining this equation with the initial conditions $y(0) = y_0$, $y'(0) = 0$ yields the following initial-value problem whose solution describes the motion of the mass subject to damping:

$$y'' + \frac{c}{M}y' + \frac{k}{M}y = 0$$
$$y(0) = y_0, \quad y'(0) = 0 \tag{11}$$

The form of the solution to (11) will depend on whether the auxiliary equation

$$m^2 + \frac{c}{M}m + \frac{k}{M} = 0 \tag{12}$$

has distinct real roots, equal real roots, or complex roots. We leave it for the reader to show that the roots are distinct and real if $c^2 > 4kM$, equal and real if $c^2 = 4kM$, and complex if $c^2 < 4kM$ (Exercise 22).

☐ **UNDERDAMPED**
VIBRATIONS

The cases $c^2 > 4kM$ and $c^2 = 4kM$ are called ***overdamped*** and ***critically damped***, respectively. In these cases vibration in the usual sense does not occur. We shall leave the analysis of these cases for the exercises and concentrate on the case $c^2 < 4kM$ in which true vibratory motion occurs. This is called the ***underdamped case***.

In the underdamped case the roots of the auxiliary equation in (12) are

$$m_1 = -\frac{c}{2M} + \frac{\sqrt{4Mk - c^2}}{2M} i, \quad m_2 = -\frac{c}{2M} - \frac{\sqrt{4Mk - c^2}}{2M} i$$

(Verify.) For convenience, let

$$\alpha = \frac{c}{2M}, \quad \beta = \frac{\sqrt{4Mk - c^2}}{2M} \tag{13}$$

so the solution of the differential equation in (11) is

$$y(t) = e^{-\alpha t}(c_1 \cos \beta t + c_2 \sin \beta t) \tag{14}$$

It follows that

$$y'(t) = -\alpha e^{-\alpha t}(c_1 \cos \beta t + c_2 \sin \beta t) + \beta e^{-\alpha t}(-c_1 \sin \beta t + c_2 \cos \beta t)$$

so the initial conditions $y(0) = y_0$, $y'(0) = 0$ yield the equations

$$c_1 = y_0$$
$$-\alpha c_1 + \beta c_2 = 0$$

which can be solved to obtain

$$c_1 = y_0, \quad c_2 = \alpha y_0/\beta$$

Substituting these values in (14) yields the solution of (11):

$$y(t) = \frac{y_0}{\beta} e^{-\alpha t}(\beta \cos \beta t + \alpha \sin \beta t) \tag{15}$$

In the exercises (Exercise 23) we ask the reader to use the cosine addition formula to show that this solution can be rewritten in the alternative form

$$y(t) = \frac{y_0\sqrt{\alpha^2 + \beta^2}}{\beta} e^{-\alpha t} \cos (\beta t - \omega) \tag{16a}$$

where

$$\omega = \tan^{-1}\left(\frac{\alpha}{\beta}\right), \quad 0 < \omega < \pi/2 \tag{16b}$$

Since $\cos(\beta t - \omega)$ has values between $+1$ and -1, it follows from (16a) and (16b) that the graph of $y(t)$ oscillates between the curves

$$y = \frac{y_0\sqrt{\alpha^2 + \beta^2}}{\beta} e^{-\alpha t} \quad \text{and} \quad y = -\frac{y_0\sqrt{\alpha^2 + \beta^2}}{\beta} e^{-\alpha t}$$

Thus, the graph of $y(t)$ resembles a cosine curve, but with decreasing amplitude (Figure 19.4.6). Strictly speaking, the function $y(t)$ is not periodic. However, $\cos(\beta t - \omega)$ has a period of $2\pi/\beta$, so the displacement $y(t)$ reaches a relative maximum at times spaced $2\pi/\beta$ units apart. Thus, we shall *define* the ***period*** T and ***frequency*** f of the motion to be (Figure 19.4.6)

$$T = \frac{2\pi}{\beta} = \frac{4M\pi}{\sqrt{4Mk - c^2}} \quad \text{and} \quad f = \frac{1}{T} = \frac{\sqrt{4Mk - c^2}}{4M\pi} \tag{17–18}$$

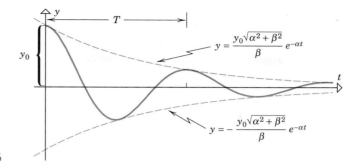

Figure 19.4.6

If we think of a relative maximum as marking the end of one cycle of motion and the start of the next, then the period T is the time required for the completion of one cycle, and the frequency is the number of cycles that occur per unit time. The physically significant fact is that the frequency and period of the mass–spring system do not change as the amplitudes of the vibrations get smaller (why?). Galileo, who first observed this fact, used it to help design clocks.

Example 2 A spring with a spring constant of $k = 4$ lb/ft is attached to a ceiling and a 64-lb weight is attached to the bottom end. The weight is pushed 2 ft above its equilibrium position and released with an initial velocity of 0. Assuming that the damping constant due to air resistance is $c = 4$ lb · sec/ft, find

(a) a formula for the position of the weight at any time t;

(b) the period and frequency of the vibration.

Solution (a). We shall first use (13) to find α and β, after which we can use either (15) or (16a) and (16b). The attached weight is 64 lb, so from (1) its mass is

$$M = \frac{w}{g} = \frac{64}{32} = 2 \text{ slugs}$$

Thus,

$$\alpha = \frac{c}{2M} = \frac{4}{4} = 1, \quad \beta = \frac{\sqrt{4Mk - c^2}}{2M} = \frac{4}{4} = 1$$

Since the weight is initially 2 ft above its equilibrium position, we have $y_0 = 2$, so from (15) the position function of the weight is

$$y(t) = 2e^{-t}(\cos t + \sin t)$$

Alternatively, we can apply (16a) and (16b), but we must first calculate ω. Since

$$\omega = \tan^{-1}\left(\frac{\alpha}{\beta}\right) = \tan^{-1}(1) = \frac{\pi}{4}$$

(16a) and (16b) yield the alternative formula

$$y(t) = 2\sqrt{2}\,e^{-t} \cos\left(t - \frac{\pi}{4}\right)$$

Solution (b). From (17) and (18)

$$T = \frac{2\pi}{\beta} = 2\pi \text{ sec}$$

$$f = \frac{1}{T} = \frac{1}{2\pi} \text{ Hz} \qquad \blacktriangleleft$$

☐ **FORCED VIBRATIONS**   If there are external forces such as movement of the spring support or magnetic forces acting on the mass, then $F_e \neq 0$ and the vibrations are said to be *forced*. The study of forced vibrations leads to important phenomena such as *resonance* and *beats*. These ideas are touched on in the Technology Exercises. Readers interested in this topic are referred to *Elementary Differential Equations and Boundary Value Problems*, John Wiley & Sons, New York, 1992, by William E. Boyce and Richard C. DiPrima.

▶ Exercise Set 19.4

In this exercise set assume that the y-axis is oriented as in Figure 19.4.2.

1.   A mass of 1 kg is attached to a vertical spring with spring constant $k = 0.25$ N/m. The mass is pushed 0.3 m above its equilibrium position and released.

   (a)   Find an initial-value problem whose solution $y(t)$ is the position function of the mass, assuming that there is no damping.

   (b)   Solve the initial-value problem.

   (c)   Check the solution in part (b) using Formula (7).

2.   A weight of 64 lb is attached to a vertical spring with spring constant $k = 8$ lb/ft. The weight is pushed 1 ft above its equilibrium position and released.

   (a)   Find an initial-value problem whose solution $y(t)$ is the position function of the weight, assuming that there is no damping.

   (b)   Solve the initial-value problem.

   (c)   Check the solution in part (b) using Formula (7).

3.   A mass attached to a vertical spring stretches the spring 0.05 m. The mass is pulled 0.12 m below its equilibrium position and released.

   (a)   Find an initial-value problem whose solution $y(t)$ is the position function of the mass, assuming that there is no damping.

   (b)   Solve the initial-value problem.

   (c)   Check the solution in part (b) using Formula (7).

4.   A mass attached to a vertical spring stretches the spring 0.2 m. The mass is pulled 0.1 m below its equilibrium position and released.

   (a)   Find an initial-value problem whose solution $y(t)$ is the position function of the mass, assuming that there is no damping.

   (b)   Solve the initial-value problem.

   (c)   Check the solution in part (b) using Formula (7).

5.   A mass of 0.5 kg is attached to a vertical spring with spring constant $k = 32$ N/m. The mass is pushed 0.2 m above its equilibrium position and released. Use Formulas (7), (8), and (9) to find

   (a)   the position function of the weight

   (b)   the amplitude of the vibration

   (c)   the period of the vibration

   (d)   the frequency of the vibration.

6.   A mass of 2 slugs is attached to a vertical spring with spring constant $k = 4$ lb/ft. The mass is pushed 1 ft above its equilibrium position and released. Use Formulas (7), (8), and (9) to find

   (a)   the position function of the mass

   (b)   the amplitude of the vibration

   (c)   the period of the vibration

   (d)   the frequency of the vibration.

7.   A mass attached to a vertical spring stretches the spring 1 in. The mass is pulled 3 in. below its equilibrium position and released. Use Formulas (7), (8), and (9) to find

   (a)   the position function of the mass

   (b)   the amplitude of the vibration

   (c)   the period of the vibration

   (d)   the frequency of the vibration.

8.   A mass attached to a vertical spring stretches the spring 0.8 m. The mass is pulled 0.2 m below its equilibrium position and released. Use Formulas (7), (8), and (9) to find

   (a)   the position function of the mass

   (b)   the amplitude of the vibration

   (c)   the period of the vibration

   (d)   the frequency of the vibration.

9.   A weight of 32 lb is attached to a vertical spring with spring constant $k = 8$ lb/ft. The surrounding medium has a damping constant of $c = 4$ lb · sec/ft. The weight is pulled 3 ft below its equilibrium position and released.

   (a)   Find an initial-value problem whose solution $y(t)$ is the position function of the weight.

   (b)   Solve the initial-value problem.

   (c)   Check the solution in part (b) using Formula (15).

   (d)   Express the solution in the form of (16a).

   (e)   Find the period of the vibration.

   (f)   Find the frequency of the vibration.

10. A mass of 3 slugs is attached to a vertical spring with spring constant $k = 9$ lb/ft. The surrounding medium has a damping constant of $c = 6$ lb · sec/ft. The mass is pulled 1 ft below its equilibrium position and released.

(a) Find an initial-value problem whose solution $y(t)$ is the position function of the mass.

(b) Solve the initial-value problem.

(c) Check the solution in part (b) using Formula (15).

(d) Express the solution in the form of (16a).

(e) Find the period of the vibration.

(f) Find the frequency of the vibration.

11. A mass of 25 g is attached to a vertical spring with spring constant $k = 3$ dynes/cm. The surrounding medium has a damping constant of $c = 10$ dynes · sec/cm. The mass is pushed 5 cm above its equilibrium position and released.

(a) Use Formula (16a) to find the position function of the mass.

(b) Find the period of the vibration.

(c) Find the frequency of the vibration.

12. A mass of 490 g is attached to a vertical spring with spring constant $k = 40$ dynes/cm. The surrounding medium has a damping constant of $c = 140$ dynes · sec/cm. The mass is pushed 20 cm above its equilibrium position and released.

(a) Use Formula (16a) to find the position function of the mass.

(b) Find the period of the vibration.

(c) Find the frequency of the vibration.

If the object in Figure 19.4.1 is given an initial velocity v_0 rather than being released with initial velocity 0, then in Formulas (6) and (11) the initial conditions become $y(0) = y_0$, $y'(0) = v_0$. Use this fact in Exercises 13–16.

---

13. A weight of 3 lb attached to a vertical spring stretches the spring $\frac{1}{2}$ ft. While in its equilibrium position, the weight is struck to give it a downward initial velocity of 2 ft/sec. Assuming that there is no damping, find

(a) the position function of the weight

(b) the amplitude of the vibration

(c) the period of the vibration

(d) the frequency of the vibration.

14. A mass of 64 slugs attached to a vertical spring stretches the spring $1\frac{1}{2}$ in. The mass is pulled 4 in. below its equilibrium position and struck to give it a downward initial velocity of 8 ft/sec. Assuming that there is no damping, find

(a) the position function of the weight

(b) the amplitude of the vibration

(c) the period of the vibration

(d) the frequency of the vibration.

15. A weight of 4 lb is attached to a vertical spring with spring constant $k = 6\frac{1}{4}$ lb/ft. The weight is pushed $\frac{1}{3}$ ft above its equilibrium position and struck to give it a downward initial velocity of 5 ft/sec. Assuming that the damping constant is $c = \frac{1}{4}$ lb · sec/ft, find the position function of the weight.

16. A weight of 49 dynes is attached to a vertical spring with spring constant $k = \frac{1}{4}$ dyne/cm. The weight is pushed 1 cm above its equilibrium position and struck to give it an upward initial velocity of 2 cm/sec. Assuming that the damping constant is $c = \frac{1}{5}$ dyne · sec/cm, find the position function of the weight.

17. A weight of w pounds is attached to a spring, then pulled below its equilibrium position and released, thereby causing it to vibrate with a period of 3 sec. When 4 additional pounds are added, the period becomes 5 sec. Assuming there is no damping, find

(a) the spring constant k (b) the weight w.

18. As illustrated in the following figure, let a toy cart of mass M be attached to a wall by a spring with spring constant k, and let an x-axis be introduced as shown with its origin at the point where the cart is in equilibrium. Suppose that the cart is pulled or pushed horizontally, then released. Find a differential equation for the position function of the cart if

(a) there is no damping

(b) there is a damping constant of c.

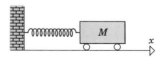

19. A cylindrical buoy of height h and radius r floats in the water with its axis vertical. By **Archimedes' principle** the water exerts an upward force on the buoy (the buoyancy force) with magnitude equal to the weight of the water displaced. Neglecting all forces except the buoyancy force and the force of gravity, determine the period with which the buoy will vibrate vertically if it is depressed slightly from its equilibrium position and released. Let h and r be measured in inches and take ρ lb/in^3 to be the density of water and δ lb/in^3 to be the density of the buoy material (density = weight per unit of volume).

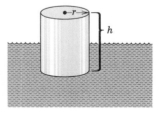

20. Suppose that the mass of the object in Figure 19.4.1 is M and the spring constant is k. Show that if the mass is displaced y_0 units from its equilibrium position and given an initial velocity of v_0 rather than being released with velocity 0, then the solution of (5) is given by

$$y(t) = y_0 \cos\left(\sqrt{\frac{k}{M}}\, t\right) + v_0 \sqrt{\frac{M}{k}} \sin\left(\sqrt{\frac{k}{M}}\, t\right)$$

21. A mass attached to a vertical spring is displaced from its equilibrium position and released, thereby causing it to vibrate with amplitude $|y_0|$ and period T (no damping).

(a) Show that the velocity of the mass has maximum magnitude $2\pi |y_0|/T$ and that the maximum occurs when the mass is at its equilibrium position.

(b) Show that the acceleration of the mass has maximum magnitude $4\pi^2 |y_0|/T^2$ and that the maximum occurs when the mass is at a top or bottom point of its motion.

22. Prove that the roots of Equation (12) are distinct and real, equal and real, or complex according to whether $c^2 > 4kM$, $c^2 = 4kM$, or $c^2 < 4kM$.

23. Use the addition formula for cosine to show that Formula (15) can be rewritten as (16a).

24. Overdamped Motion

(a) Prove that in the case of overdamped motion $(c^2 > 4kM)$ the solution of (11) is

$$y(t) = \frac{y_0}{m_2 - m_1}(m_2 e^{m_1 t} - m_1 e^{m_2 t})$$

where m_1 and m_2 are the distinct real roots of (12).

(b) Prove that $\lim\limits_{t \to +\infty} y(t) = 0$. [*Hint:* Show that m_1 and m_2 are negative.]

25. Critically Damped Motion

(a) Prove that in the case of critically damped motion $(c^2 = 4kM)$ the solution of (11) is

$$y(t) = y_0 e^{-\alpha t}(1 + \alpha t)$$

(b) Prove that $\lim\limits_{t \to +\infty} y(t) = 0$.

(c) Prove that $y(t) \neq 0$ for any positive value of t (assuming that the initial displacement y_0 is not zero).

◆ TECHNOLOGY EXERCISES   Chapter 19

Most of these exercises require access to a graphing calculator or a computer algebra system (CAS) such as *Mathematica*, *Maple*, or *Derive*. When you are asked to *find* an answer or to *solve* an equation, you may choose to find an exact result or a numerical approximation, depending on the particular technology you are using and on your own imagination. The form of your answers may differ from those of other students or from those in the answer section of the text, depending on how you solve the problems and the accuracy you use in your numerical approximations. Those exercises that are more appropriate for a CAS than a graphing calculator are labeled with the icon ◆.

The accompanying figure shows a mass–spring system in which an object of mass M is suspended by a spring and linked to a piston that moves in a *dashpot* containing a viscous fluid. If there are no external forces acting on the system, then the object is said to have **free motion** and the motion of the object is completely determined by the displacement and velocity of the object at time $t = 0$, the stiffness of the spring as measured by the spring constant k, and the viscosity of the fluid in the dashpot as measured by the damping constant c. Mathematically, the displacement $y = y(t)$ of the object from its equilibrium position is the solution of an initial-value problem of the

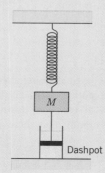

Dashpot

form

$$y'' + Ay' + By = 0, \quad y(0) = y_0, \quad y'(0) = v_0$$

where y_0 is the initial displacement of the object from equilibrium and v_0 is its initial velocity. We showed in Section 19.4 that the coefficient A is determined by the mass and the damping constant, and the coefficient B is determined by the mass and the spring constant. If an external force $F(t)$ is applied to the object, then the object is said to have **forced motion** and its displacement from equilibrium is the solution of an initial-value problem of the form

$$y'' + Ay' + By = f(t), \quad y(0) = y_0, \quad y'(0) = v_0$$

where $f(t) = F(t)/M$. In Section 19.4 we considered only motion in which the object is released from rest, that is, $v_0 = 0$. In the following exercises we shall allow the object to have a nonzero initial velocity, and we shall explore phenomena associated with forced motion. In all of the exercises assume that distance is measured in centimeters and time in seconds.

1. **Critically damped free motion**

   (a) Solve the initial-value problem $y'' + 2.4y' + 1.44y = 0$, $y(0) = 1$, $y'(0) = 2$ and graph $y = y(t)$ on the interval $[0, 5]$.

   (b) Find the maximum distance above the equilibrium position attained by the object.

   (c) The graph of $y(t)$ suggests that the object does not pass through the equilibrium position. Show that this is so.

◆ 2. **Overdamped free motion**

   (a) Solve the initial-value problem $y'' + 5y' + 2y = 0$, $y(0) = 1/2$, $y'(0) = -4$ and graph $y = y(t)$ on the interval $[0, 5]$.

   (b) Find the maximum distance below the equilibrium position attained by the object.

   (c) The graph of $y(t)$ suggests that the object passes through the equilibrium position exactly once. With what speed does the object pass through the equilibrium position?

◆ 3. **Underdamped free oscillation**

   (a) Solve the initial-value problem $y'' + y' + 5y = 0$, $y(0) = 1$, $y'(0) = -3.5$ and graph $y = y(t)$ on the interval $[0, 8]$.

   (b) Find the maximum distance below the equilibrium position attained by the object.

   (c) Find the velocity of the object when it passes through the equilibrium position the first time.

   (d) Find, by inspection, the acceleration of the object when it passes through the equilibrium position the first time. [*Hint:* Examine the differential equation and use the result in part (c).]

◆ 4. **Underdamped free oscillation**

   (a) Solve the initial-value problem $y'' + y' + 3y = 0$, $y(0) = -2$, $y'(0) = v_0$.

   (b) Find the largest positive value of v_0 for which the object will rise no higher than 1 cm above the equilibrium position. [*Hint:* Use a trial-and-error strategy. Estimate v_0 to the nearest hundredth of a cm/sec.]

   (c) Graph the solution of the initial-value problem on the interval $[0, 8]$ using the value of v_0 obtained in part (b).

In the absence of an external force, the frequency of oscillation of an undamped mass–spring system is called the **natural frequency** of the system. This terminology is used in Exercises 5 and 6.

◆ 5. **Undamped forced oscillation—beats:** Beats occur when there is a periodic external force with a frequency that is close to the natural frequency of the system. Physically, this results in a periodic variation of the amplitude of the oscillations. To illustrate this idea, consider a mass–spring system in which the motion is governed by the initial-value problem $y'' + 36y = \sin 5t$, $y(0) = 0$, $y'(0) = 0$.

   (a) Find the natural frequency of the system and the frequency of the external force.

   (b) Solve the initial-value problem and graph $y = y(t)$ on the interval $[0, 4\pi]$.

   (c) Solve the initial-value problems

   $$y'' + 36y = \sin 5.5t, \quad y(0) = 0, \quad y'(0) = 0$$
   $$y'' + 36y = \sin 5.8t, \quad y(0) = 0, \quad y'(0) = 0$$

   and graph their solutions. Make a conjecture about the effect on the oscillations as the frequency of the external force is made closer to the natural frequency of the system.

6. **Undamped forced oscillation—resonance:** Resonance occurs when there is a periodic external force with a frequency that is equal to the natural frequency of the system.

In this case the oscillations of the object increase without bound. To illustrate this idea, consider a mass–spring system in which the motion is governed by the initial-value problem $y'' + 36y = \sin 6t$, $y(0) = 0$, $y'(0) = 0$.

(a) Show that the natural frequency of the system is the same as the frequency of the external force.

(b) Solve the initial-value problem and graph $y = y(t)$ on the interval $[0, 4\pi]$.

(c) As a practical matter, the displacement of the object will be limited by factors such as the natural length of the spring, the breaking point of the spring, or the presence of obstructions above or below the object. Suppose that the bottom of the object encounters an obstruction when it is 1.3 cm below the equilibrium position. When and with what speed will the object collide with the obstruction?

7. Effect of initial velocity on overdamped free motion

(a) Solve the initial-value problem $y'' + 3.5y' + 3y = 0$, $y(0) = 1$, $y'(0) = v_0$.

(b) Use the result in part (a) to find the solutions for $v_0 = 2$, $v_0 = -1$, and $v_0 = -4$ and graph all three solutions on the interval $[0, 4]$ in the same coordinate system.

(c) Discuss the effect of the initial velocity on the motion of the object.

◆ **8. Effect of spring stiffness on forced motion**

(a) Solve the initial-value problem $y'' + 3y' + y = 3 \sin 2t$, $y(0) = 2$, $y'(0) = -1$ and graph the solution on the interval $[0, 20]$.

(b) Suppose that the spring in part (a) is replaced by a stiffer one. Mathematically, this increases the coefficient of y in the differential equation. Make a conjecture about the effect that an increase in spring stiffness has on the motion of the object. Test your conjecture by solving the initial-value problem $y'' + 3y' + 2y = 3 \sin 2t$, $y(0) = 2$, $y'(0) = -1$ and graphing the solution on the interval $[0, 20]$.

SUPPLEMENTARY MATERIAL ▬▬▬▬

Brook Taylor

Brook Taylor (1685–1731)

11 INFINITE SERIES

11.1 SEQUENCES

This chapter is concerned with the study of "infinite series," which, loosely speaking, are sums with infinitely many terms. The material in this chapter has far-reaching applications in engineering and science and is the cornerstone for many branches of mathematics. In this initial section we shall develop some preliminary results that are important in their own right.

□ **DEFINITION OF A SEQUENCE**

In everyday language, we use the term "sequence" to suggest a succession of objects or events given in a specified order. *Informally* speaking, the term "sequence" in mathematics is used to describe an unending succession of numbers. Some possibilities are

$$1, 2, 3, 4, \ldots, \qquad 1, \tfrac{1}{2}, \tfrac{1}{3}, \tfrac{1}{4}, \ldots,$$

$$2, 4, 6, 8, \ldots, \qquad 1, -1, 1, -1, \ldots$$

In each case, the three dots are used to suggest that the sequence continues indefinitely, following the obvious pattern. The numbers in a sequence are called the ***terms*** of the sequence. The terms may be described according to the positions they occupy. Thus, a sequence has a *first term* a_1, a *second term* a_2, a *third term* a_3, and so forth. Because a sequence continues indefinitely, there is no last term.

The most common way to specify a sequence is to give a formula that relates the terms in the sequence to their term numbers. For example, in the sequence

$$2, 4, 6, 8, \ldots$$

each term is twice the term number; that is, the nth term in the sequence is given by the formula $2n$. We denote this by writing the sequence as

$$2, 4, 6, 8, \ldots, 2n, \ldots$$

or more compactly in ***bracket notation*** as

$$\{2n\}_{n=1}^{+\infty}$$

which conveys that the sequence can be generated by successively substituting the integer values $n = 1, 2, 3, \ldots$ into the formula $2n$.

Example 1   List the first five terms of the sequence $\{2^n\}_{n=1}^{+\infty}$.

Solution.   Substituting $n = 1, 2, 3, 4, 5$ into the formula 2^n yields

$$2^1, 2^2, 2^3, 2^4, 2^5, \ldots$$

or, equivalently,

$$2, 4, 8, 16, 32, \ldots \quad \blacktriangleleft$$

Example 2   Express the following sequences in bracket notation.

(a) $\dfrac{1}{2}, \dfrac{2}{3}, \dfrac{3}{4}, \dfrac{4}{5}, \ldots$ (b) $\dfrac{1}{2}, \dfrac{1}{4}, \dfrac{1}{8}, \dfrac{1}{16}, \ldots$

(c) $\dfrac{1}{2}, -\dfrac{2}{3}, \dfrac{3}{4}, -\dfrac{4}{5}, \ldots$ (d) $1, 3, 5, 7, \ldots$

Solution (a).   The nth term in the following table is obtained by observing that for each term in the sequence the numerator is the same as the term number, and the denominator is one greater than the term number.

TERM NUMBER	1	2	3	4	$\cdots$	n	$\cdots$
TERM	$\dfrac{1}{2}$	$\dfrac{2}{3}$	$\dfrac{3}{4}$	$\dfrac{4}{5}$	$\cdots$	$\dfrac{n}{n+1}$	$\cdots$

Thus, the sequence can be written as $\left\{\dfrac{n}{n+1}\right\}_{n=1}^{+\infty}$.

Solution (b).   The nth term in the following table is obtained by rewriting the denominators in the sequence as powers of 2 and observing that for each term the exponent of the denominator is the same as the term number.

TERM NUMBER	1	2	3	4	$\cdots$	n	$\cdots$
TERM	$\dfrac{1}{2}$	$\dfrac{1}{2^2}$	$\dfrac{1}{2^3}$	$\dfrac{1}{2^4}$	$\cdots$	$\dfrac{1}{2^n}$	$\cdots$

Thus, the sequence can be written as $\left\{\dfrac{1}{2^n}\right\}_{n=1}^{+\infty}$.

Solution (c).   This sequence is identical to that in part (a), except for the alternating signs. Thus, the nth term in the sequence can be obtained by multiplying the nth term in part (a) by $(-1)^{n+1}$. This factor produces the correct alternating signs, since its successive values, starting with $n = 1$, are $1, -1, 1, -1, \ldots$. Thus, the sequence can be written as

$$\left\{(-1)^{n+1}\frac{n}{n+1}\right\}_{n=1}^{+\infty}$$

Solution (d).   The nth term in the following table is obtained by observing that each term is one less than twice the term number.

TERM NUMBER	1	2	3	4	$\cdots$	n	$\cdots$
TERM	1	3	5	7	$\ldots$	$2n - 1$	$\ldots$

Thus, the sequence can be written as $\{2n - 1\}_{n=1}^{+\infty}$.   ◀

Frequently we shall want to write down a sequence without specifying the numerical values of the terms. We do this by writing

$$a_1, a_2, \ldots, a_n, \ldots$$

or in bracket notation

$$\{a_n\}_{n=1}^{+\infty}$$

REMARK.   Sometimes it will be convenient to omit the limits in the bracket notation; thus, the preceding sequence can also be written as $\{a_n\}$. Moreover, there is nothing special about the letters a and n; any other letters may be used.

At the start of this section we described a sequence as an unending succession of numbers. However, this is not a satisfactory mathematical definition, since the word "succession" is itself an undefined term. To motivate an appropriate definition, consider the sequence of even integers

$$2, 4, 6, 8, \ldots, 2n, \ldots$$

We can think of the nth term as a formula for the function

$$f(n) = 2n, \quad n = 1, 2, 3, \ldots$$

whose domain is the set of positive integers, and we can think of the sequence as a listing of the function values

$$f(1), f(2), f(3), \ldots, f(n), \ldots$$

This suggests the following definition.

11.1.1   DEFINITION.   A *sequence* or *infinite sequence* is a function whose domain is the set of positive integers; that is, $\{a_n\}_{n=1}^{+\infty}$ is an alternative notation for the function $f(n) = a_n$, $n = 1, 2, 3, \ldots$.

□ **GRAPHS OF SEQUENCES**

Because sequences are functions, we may inquire about the graph of a sequence. For example, the graph of the sequence $\{1/n\}_{n=1}^{+\infty}$ is the graph of the equation

$$y = \frac{1}{n}, \quad n = 1, 2, 3, \ldots$$

Because the right side of this equation is defined only for positive integer values of n, the graph consists of a succession of isolated points (Figure 11.1.1a). This is in marked distinction to the graph of

$$y = \frac{1}{x}, \quad x \geq 1$$

which is a continuous curve (Figure 11.1.1b).

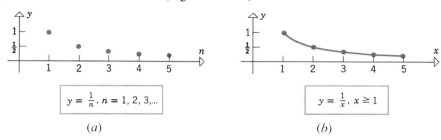

Figure 11.1.1

(a) (b)

☐ **LIMIT OF A SEQUENCE**

In Figure 11.1.2 we have sketched the graphs of four sequences, each of which behaves differently as n increases:

- The terms in the sequence $\{n + 1\}$ increase without bound.
- The terms in the sequence $\{(-1)^{n+1}\}$ oscillate between -1 and 1.
- The terms in the sequence $\{n/(n + 1)\}$ increase toward a ''limiting value'' of 1.
- The terms in the sequence $\{1 + (-\frac{1}{2})^n\}$ also tend toward a ''limiting value'' of 1, but do so in an oscillatory fashion.

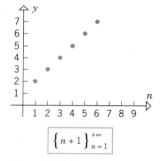

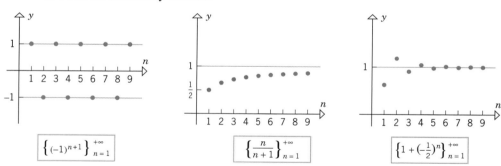

Figure 11.1.2

Our next goal is to make the concept of the ''limit'' of a sequence precise. Figure 11.1.3 conveys the basic idea: A sequence $\{a_n\}$ converges to a limit L if for any positive number ϵ there is a point in the sequence after which all terms lie between the lines $y = L + \epsilon$ and $y = L - \epsilon$; that is, eventually the terms in the sequence lie within ϵ units of L. Phrased another way, *a sequence $\{a_n\}$ converges to a limit L if the terms in the sequence eventually become arbitrarily close to L.* The following definition makes these ideas precise.

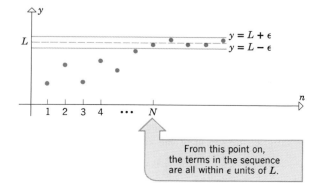

Figure 11.1.3

11.1.2 DEFINITION. A sequence $\{a_n\}$ is said to **converge** to the **limit** L if given any $\epsilon > 0$, there is a positive integer N such that $|a_n - L| < \epsilon$ for $n \geq N$. In this case we write

$$\lim_{n \to +\infty} a_n = L$$

A sequence that does not converge to some finite limit is said to **diverge**.

Example 3 The first two sequences in Figure 11.1.2 diverge, while the second two sequences converge to 1; that is,

$$\lim_{n \to +\infty} \frac{n}{n+1} = 1 \quad \text{and} \quad \lim_{n \to +\infty} \left(1 + (-\tfrac{1}{2})^n\right) = 1 \quad \blacktriangleleft$$

The following theorem, which we state without proof, shows that the familiar properties of limits apply to sequences.

11.1.3 THEOREM. *Suppose that the sequences $\{a_n\}$ and $\{b_n\}$ converge to limits L_1 and L_2, respectively, and c is a constant. Then*

(a) $\displaystyle \lim_{n \to +\infty} c = c$

(b) $\displaystyle \lim_{n \to +\infty} ca_n = c \lim_{n \to +\infty} a_n = cL_1$

(c) $\displaystyle \lim_{n \to +\infty} (a_n + b_n) = \lim_{n \to +\infty} a_n + \lim_{n \to +\infty} b_n = L_1 + L_2$

(d) $\displaystyle \lim_{n \to +\infty} (a_n - b_n) = \lim_{n \to +\infty} a_n - \lim_{n \to +\infty} b_n = L_1 - L_2$

(e) $\displaystyle \lim_{n \to +\infty} (a_n b_n) = \lim_{n \to +\infty} a_n \cdot \lim_{n \to +\infty} b_n = L_1 L_2$

(f) $\displaystyle \lim_{n \to +\infty} \left(\frac{a_n}{b_n}\right) = \frac{\displaystyle\lim_{n \to +\infty} a_n}{\displaystyle\lim_{n \to +\infty} b_n} = \frac{L_1}{L_2} \quad (\text{if } L_2 \neq 0)$

REMARK. It follows from this theorem that the algebraic techniques used to find limits of the form $\displaystyle\lim_{x \to +\infty}$ can also be used for limits of the form $\displaystyle\lim_{n \to +\infty}$.

Example 4 In each part, determine whether the sequence converges or diverges. If it converges, find the limit.

(a) $\left\{\dfrac{n}{2n+1}\right\}_{n=1}^{+\infty}$ (b) $\left\{(-1)^{n+1} \dfrac{n}{2n+1}\right\}_{n=1}^{+\infty}$

(c) $\left\{(-1)^{n+1} \dfrac{1}{n}\right\}_{n=1}^{+\infty}$ (d) $\{8 - 2n\}_{n=1}^{+\infty}$ (e) $\left\{\dfrac{n}{e^n}\right\}_{n=1}^{+\infty}$

Solution (a). Dividing numerator and denominator by n yields

$$\lim_{n \to +\infty} \frac{n}{2n+1} = \lim_{n \to +\infty} \frac{1}{2 + 1/n} = \frac{\displaystyle\lim_{n \to +\infty} 1}{\displaystyle\lim_{n \to +\infty} (2 + 1/n)} = \frac{\displaystyle\lim_{n \to +\infty} 1}{\displaystyle\lim_{n \to +\infty} 2 + \lim_{n \to +\infty} 1/n}$$

$$= \frac{1}{2+0} = \frac{1}{2}$$

Thus, the sequence converges to $\frac{1}{2}$.

Solution (b). This sequence is the same as that in part (a), except for the factor of $(-1)^{n+1}$, which oscillates between $+1$ and -1. Thus, the terms in this sequence oscillate between positive and negative values, with the odd-numbered terms being identical to those in part (a) and the even-numbered terms being the negatives of those in part (a). Since the sequence in part (a) has a limit of $\frac{1}{2}$, it follows that the odd-numbered terms in this sequence approach $\frac{1}{2}$, while the even-numbered terms approach $-\frac{1}{2}$. Therefore, this sequence has no limit—it diverges.

Solution (c). Since $\lim\limits_{n \to +\infty} 1/n = 0$, the product $(-1)^{n+1}(1/n)$ oscillates between positive and negative values, with the odd-numbered terms approaching 0 through positive values and the even-numbered terms approaching 0 through negative values. Thus,

$$\lim_{n \to +\infty} (-1)^{n+1}\frac{1}{n} = 0$$

so the sequence converges to 0.

Solution (d). $\lim\limits_{n \to +\infty} (8 - 2n) = -\infty$, so the sequence $\{8 - 2n\}_{n=1}^{+\infty}$ diverges.

Solution (e). We want to find $\lim\limits_{n \to +\infty} n/e^n$, which is an indeterminate form of type ∞/∞. Unfortunately, we cannot apply L'Hôpital's rule directly since e^n and n are not differentiable functions (n assumes only integer values). However, we can apply L'Hôpital's rule to the related problem $\lim\limits_{x \to +\infty} x/e^x$ to obtain

$$\lim_{x \to +\infty} \frac{x}{e^x} = \lim_{x \to +\infty} \frac{1}{e^x} = 0$$

We conclude from this that $\lim\limits_{n \to +\infty} n/e^n = 0$ since the values of n/e^n and x/e^x are the same when x is a positive integer. ◀

Example 5 Show that $\lim\limits_{n \to +\infty} \sqrt[n]{n} = 1$.

Solution.

$$\lim_{n \to +\infty} \sqrt[n]{n} = \lim_{n \to +\infty} n^{1/n} = \lim_{n \to +\infty} e^{(1/n)\ln n} = e^0 = 1 \qquad ◀$$

> By L'Hôpital's rule $\lim\limits_{n \to +\infty} (1/n)\ln n = 0$

Sometimes the even-numbered and odd-numbered terms of a sequence behave sufficiently differently that it is desirable to investigate their convergence separately. The following theorem, whose proof is omitted, is helpful for that purpose.

11.1.4 THEOREM. *A sequence converges to a limit L if and only if the sequences of even-numbered terms and odd-numbered terms both converge to L.*

Example 6 The sequence

$$\frac{1}{2}, \frac{1}{3}, \frac{1}{2^2}, \frac{1}{3^2}, \frac{1}{2^3}, \frac{1}{3^3}, \dots$$

converges to 0, since the even-numbered terms and the odd-numbered terms both converge to 0, and the sequence

$$1, \tfrac{1}{2}, 1, \tfrac{1}{3}, 1, \tfrac{1}{4}, \dots$$

diverges, since the odd-numbered terms converge to 1 and the even-numbered terms converge to 0. ◀

☐ **SEQUENCES DEFINED RECURSIVELY**

Sometimes sequences are defined by specifying one or more initial terms and giving a formula that relates each subsequent term to terms that precede it. Such sequences are said to be defined **recursively**, and the defining formula is called a **recursion formula**.

Example 7 Let $\{a_n\}$ be the sequence defined by $a_1 = 1$ and the recursion formula

$$a_{n+1} = \tfrac{1}{2}(a_n + 3/a_n) \tag{1}$$

for $n \geq 1$. The values of $a_2, a_3, a_4, \ldots$ can be obtained by successively substituting $n = 1, 2, 3, \ldots$ in (1):

$$a_2 = \tfrac{1}{2}(a_1 + 3/a_1) = \tfrac{1}{2}(1 + 3) = 2 \qquad \boxed{n = 1}$$

$$a_3 = \tfrac{1}{2}(a_2 + 3/a_2) = \tfrac{1}{2}(2 + \tfrac{3}{2}) = \tfrac{7}{4} \qquad \boxed{n = 2}$$

$$a_4 = \tfrac{1}{2}(a_3 + 3/a_3) = \tfrac{1}{2}(\tfrac{7}{4} + \tfrac{12}{7}) = \tfrac{97}{56} \qquad \boxed{n = 3}$$

$\vdots$ ◀

REMARK. The recursion formula in the preceding example results from applying Newton's Method to the equation $x^2 - 3 = 0$ [verify using Formula (4) of Section 4.8]. It will be shown in the exercises of the next section that the resulting sequence converges to $\sqrt{3}$.

▶ Exercise Set 11.1

In Exercises 1–18, list the first five terms of the sequence, determine whether the sequence converges, and if so find the limit. (When writing out the terms of the sequence, you need not find numerical values; leave the terms in the first form you obtain.)

1. $\left\{\dfrac{n}{n+2}\right\}_{n=1}^{+\infty}.$

2. $\left\{\dfrac{n^2}{2n+1}\right\}_{n=1}^{+\infty}.$

3. $\{2\}_{n=1}^{+\infty}.$

4. $\left\{\ln\left(\dfrac{1}{n}\right)\right\}_{n=1}^{+\infty}.$

5. $\left\{\dfrac{\ln n}{n}\right\}_{n=1}^{+\infty}.$

6. $\left\{n\sin\dfrac{\pi}{n}\right\}_{n=1}^{+\infty}.$

7. $\{1 + (-1)^n\}_{n=1}^{+\infty}.$

8. $\left\{\dfrac{(-1)^{n+1}}{n^2}\right\}_{n=1}^{+\infty}.$

9. $\left\{(-1)^n\dfrac{2n^3}{n^3+1}\right\}_{n=1}^{+\infty}.$

10. $\left\{\dfrac{n}{2^n}\right\}_{n=1}^{+\infty}.$

11. $\left\{\dfrac{(n+1)(n+2)}{2n^2}\right\}_{n=1}^{+\infty}.$

12. $\left\{\dfrac{\pi^n}{4^n}\right\}_{n=1}^{+\infty}.$

13. $\{\cos(3/n)\}_{n=1}^{+\infty}.$

14. $\left\{\cos\dfrac{\pi n}{2}\right\}_{n=1}^{+\infty}.$

15. $\{n^2 e^{-n}\}_{n=1}^{+\infty}.$

16. $\{\sqrt{n^2 + 3n} - n\}_{n=1}^{+\infty}.$

17. $\left\{\left(\dfrac{n+3}{n+1}\right)^n\right\}_{n=1}^{+\infty}.$

18. $\left\{\left(1 - \dfrac{2}{n}\right)^n\right\}_{n=1}^{+\infty}.$

In Exercises 19–26, express the sequence in the notation $\{a_n\}_{n=1}^{+\infty}$, determine whether the sequence converges, and if so find its limit.

19. $\dfrac{1}{2}, \dfrac{3}{4}, \dfrac{5}{6}, \dfrac{7}{8}, \ldots$

20. $0, \dfrac{1}{2^2}, \dfrac{2}{3^2}, \dfrac{3}{4^2}, \ldots$

21. $\dfrac{1}{3}, \dfrac{1}{9}, \dfrac{1}{27}, \dfrac{1}{81}, \ldots$

22. $-1, 2, -3, 4, -5, \ldots$

23. $\left(1 - \dfrac{1}{2}\right), \left(\dfrac{1}{2} - \dfrac{1}{3}\right), \left(\dfrac{1}{3} - \dfrac{1}{4}\right), \left(\dfrac{1}{4} - \dfrac{1}{5}\right), \ldots$

24. $3, \dfrac{3}{2}, \dfrac{3}{2^2}, \dfrac{3}{2^3}, \ldots$

25. $(\sqrt{2} - \sqrt{3}), (\sqrt{3} - \sqrt{4}), (\sqrt{4} - \sqrt{5}), \ldots$

26. $\dfrac{1}{3^5}, -\dfrac{1}{3^6}, \dfrac{1}{3^7}, -\dfrac{1}{3^8}, \ldots$

27. Let $\{a_n\}$ be the sequence defined recursively by $a_1 = \sqrt{6}$ and $a_{n+1} = \sqrt{6 + a_n}$ for $n \geq 1$.

   (a) List the first three terms of the sequence.

   (b) It can be shown that the sequence $\{a_n\}$ converges. Assuming this to be so, find its limit L. [*Hint:*
   $$\lim_{n \to +\infty} a_n = \lim_{n \to +\infty} a_{n+1} = L.]$$

28. Let a_1 and k be any positive real numbers and let $\{a_n\}$ be the sequence defined recursively by $a_{n+1} = \tfrac{1}{2}(a_n + k/a_n)$ for $n \geq 1$. Assuming that this sequence converges, find its limit using the hint in Exercise 27(b).

29. The *Fibonacci sequence* is defined by $a_{n+2} = a_n + a_{n+1}$ for $n \geq 1$, where $a_1 = a_2 = 1$.

(a) List the first eight terms of the sequence.

(b) Find $\lim\limits_{n \to +\infty} (a_{n+1}/a_n)$ assuming that it exists.

[*Hint:* $\lim\limits_{n \to +\infty} (a_{n+1}/a_n) = \lim\limits_{n \to +\infty} (a_{n+2}/a_{n+1})$.]

30. The nth term a_n of the sequence $1, 2, 1, 4, 1, 6, \ldots$ is best written in the form

$$a_n = \begin{cases} 1, & \text{if } n \text{ is odd} \\ n, & \text{if } n \text{ is even} \end{cases}$$

since a single formula applicable to all terms would be too complicated to be useful. By considering even and odd terms separately, find a similar expression for the nth term of the sequence

(a) $1, \dfrac{1}{2^2}, 3, \dfrac{1}{2^4}, 5, \dfrac{1}{2^6}, \ldots$

(b) $1, \dfrac{1}{3}, \dfrac{1}{3}, \dfrac{1}{5}, \dfrac{1}{5}, \dfrac{1}{7}, \dfrac{1}{7}, \dfrac{1}{9}, \dfrac{1}{9}, \ldots$

31. Consider the sequence $\{a_n\}_{n=1}^{+\infty}$, where

$$a_n = \frac{1}{n^2} + \frac{2}{n^2} + \cdots + \frac{n}{n^2}$$

(a) Write out the first four terms of the sequence.

(b) Find the limit of the sequence. [*Hint:* Sum up the terms in the formula for a_n.]

32. Follow the directions in Exercise 31 with

$$a_n = \frac{1^2}{n^3} + \frac{2^2}{n^3} + \cdots + \frac{n^2}{n^3}$$

33. If we accept the fact that the sequence $\{1/n\}_{n=1}^{+\infty}$ converges to the limit $L = 0$, then according to Definition 11.1.2, for every $\epsilon > 0$, there exists an integer N such that $|a_n - L| = |(1/n) - 0| < \epsilon$ when $n \geq N$. In each part, find

the smallest possible value of N for the given value of ϵ.

(a) $\epsilon = 0.5$ (b) $\epsilon = 0.1$ (c) $\epsilon = 0.001$.

34. If we accept the fact that the sequence

$$\left\{ \frac{n}{n+1} \right\}_{n=1}^{+\infty}$$

converges to the limit $L = 1$, then according to Definition 11.1.2, for every $\epsilon > 0$ there exists an integer N such that

$$|a_n - L| = \left| \frac{n}{n+1} - 1 \right| < \epsilon$$

when $n \geq N$. In each part, find the smallest value of N for the given value of ϵ.

(a) $\epsilon = 0.25$ (b) $\epsilon = 0.1$ (c) $\epsilon = 0.001$.

35. Use Definition 11.1.2 to prove that

(a) the sequence $\{1/n\}_{n=1}^{+\infty}$ converges to 0

(b) the sequence $\left\{ \dfrac{n}{n+1} \right\}_{n=1}^{+\infty}$ converges to 1.

36. Consider the sequence $\{a_n\}_{n=1}^{+\infty}$ whose nth term is

$$a_n = \sum_{k=1}^{n} \frac{1}{1 + k/n} \cdot \frac{1}{n}$$

Show that $\lim\limits_{n \to +\infty} a_n = \ln 2$. [*Hint:* Interpret $\lim\limits_{n \to +\infty} a_n$ as a definite integral.]

37. (a) Show that a polygon with n equal sides inscribed in a circle of radius r has perimeter $p_n = 2rn \sin(\pi/n)$.

(b) By finding the limit of the sequence $\{p_n\}_{n=1}^{+\infty}$, derive the formula for the circumference of the circle.

38. Find $\lim\limits_{n \to +\infty} r^n$, where r is a real number. [*Hint:* Consider the cases $|r| < 1$, $|r| > 1$, $r = 1$, and $r = -1$ separately.]

39. Find the limit of the sequence $\{(2^n + 3^n)^{1/n}\}_{n=1}^{+\infty}$.

11.2 MONOTONE SEQUENCES

Sometimes the critical information about a sequence is whether it converges or not, with the limit being of lesser interest. In this section we shall discuss results that are used to study convergence of sequences.

□ TERMINOLOGY

We begin with some terminology.

11.2.1 DEFINITION. A sequence $\{a_n\}$ is called

increasing if $a_1 < a_2 < a_3 < \cdots < a_n < \cdots$

nondecreasing if $a_1 \leq a_2 \leq a_3 \leq \cdots \leq a_n \leq \cdots$

decreasing if $a_1 > a_2 > a_3 > \cdots > a_n > \cdots$

nonincreasing if $a_1 \geq a_2 \geq a_3 \geq \cdots \geq a_n \geq \cdots$

A sequence that is either nondecreasing or nonincreasing is called **monotone**, and a sequence that is increasing or decreasing is called **strictly monotone**. Observe that a strictly monotone sequence is monotone, but not conversely.

Example 1

$$\frac{1}{2}, \frac{2}{3}, \frac{3}{4}, \ldots, \frac{n}{n+1}, \ldots \quad \text{is increasing}$$

$$1, \frac{1}{2}, \frac{1}{3}, \ldots, \frac{1}{n}, \ldots \quad \text{is decreasing}$$

$$1, 1, 2, 2, 3, 3, \ldots \quad \text{is nondecreasing}$$

$$1, 1, \frac{1}{2}, \frac{1}{2}, \frac{1}{3}, \frac{1}{3}, \ldots \quad \text{is nonincreasing}$$

All four of these sequences are monotone, but the sequence

$$1, -\frac{1}{2}, \frac{1}{3}, -\frac{1}{4}, \ldots, (-1)^{n+1}\frac{1}{n}, \ldots$$

is not. The first and second sequences are strictly monotone. ◀

TESTING FOR MONOTONICITY

In order for a sequence to be increasing, *all* pairs of successive terms, a_n and a_{n+1}, must satisfy $a_n < a_{n+1}$, or equivalently, $a_{n+1} - a_n > 0$. More generally, monotone sequences can be classified as follows:

DIFFERENCE BETWEEN SUCCESSIVE TERMS	CLASSIFICATION
$a_{n+1} - a_n > 0$	Increasing
$a_{n+1} - a_n < 0$	Decreasing
$a_{n+1} - a_n \geq 0$	Nondecreasing
$a_{n+1} - a_n \leq 0$	Nonincreasing

Frequently, one can *guess* whether a sequence is increasing, decreasing, nondecreasing, or nonincreasing after writing out some of the initial terms. However, to be certain that the guess is correct, a precise mathematical proof is needed. The following example illustrates a method for doing this.

Example 2 Show that

$$\frac{1}{2}, \frac{2}{3}, \frac{3}{4}, \ldots, \frac{n}{n+1}, \ldots$$

is an increasing sequence.

Solution. It is intuitively clear that the sequence is increasing. To prove that this is so, let

$$a_n = \frac{n}{n+1}$$

We can obtain a_{n+1} by replacing n by $n+1$ in this formula. This yields

$$a_{n+1} = \frac{n+1}{(n+1)+1} = \frac{n+1}{n+2}$$

Thus, for $n \geq 1$

$$a_{n+1} - a_n = \frac{n+1}{n+2} - \frac{n}{n+1} = \frac{n^2 + 2n + 1 - n^2 - 2n}{(n+1)(n+2)}$$

$$= \frac{1}{(n+1)(n+2)} > 0$$

which proves that the sequence is increasing. ◀

If a_n and a_{n+1} are any successive terms in an increasing sequence, then $a_n < a_{n+1}$. If the terms in the sequence are all positive, then we can divide both sides of this inequality by a_n to obtain $1 < a_{n+1}/a_n$ or equivalently, $a_{n+1}/a_n > 1$. More generally, monotone sequences with *positive* terms can be classified as follows:

RATIO OF SUCCESSIVE TERMS	CLASSIFICATION
$a_{n+1}/a_n > 1$	Increasing
$a_{n+1}/a_n < 1$	Decreasing
$a_{n+1}/a_n \geq 1$	Nondecreasing
$a_{n+1}/a_n \leq 1$	Nonincreasing

Example 3   Show that the sequence in Example 2 is increasing by examining the ratio of successive terms.

Solution.   As shown in the solution of Example 2,

$$a_n = \frac{n}{n+1} \quad \text{and} \quad a_{n+1} = \frac{n+1}{n+2}$$

Thus,

$$\frac{a_{n+1}}{a_n} = \frac{(n+1)/(n+2)}{n/(n+1)} = \frac{n+1}{n+2} \cdot \frac{n+1}{n} = \frac{n^2 + 2n + 1}{n^2 + 2n} \tag{1}$$

Since the numerator in (1) exceeds the denominator, the ratio exceeds 1, that is, $a_{n+1}/a_n > 1$ for $n \geq 1$. This proves that the sequence is increasing.   ◄

The following example illustrates still a third technique for determining whether a sequence is increasing or decreasing.

Example 4   In Examples 2 and 3 we proved that the sequence

$$\frac{1}{2}, \frac{2}{3}, \frac{3}{4}, \ldots, \frac{n}{n+1}, \ldots$$

is increasing by considering the difference and ratio of successive terms. Alternatively, we can proceed as follows. Let

$$f(x) = \frac{x}{x+1}$$

so the nth term in the given sequence is $a_n = f(n)$. The function f is increasing for $x \geq 1$ since

$$f'(x) = \frac{(x+1)(1) - x(1)}{(x+1)^2} = \frac{1}{(x+1)^2} > 0$$

Thus,

$$a_n = f(n) < f(n+1) = a_{n+1}$$

which proves that the given sequence is increasing.   ◄

In general, if $f(n) = a_n$ is the nth term of a sequence, and if f is differentiable for $x \geq 1$, then we have the following results:

DERIVATIVE OF f FOR $x \geq 1$	CLASSIFICATION OF THE SEQUENCE WITH $a_n = f(n)$
$f'(x) > 0$	Increasing
$f'(x) < 0$	Decreasing
$f'(x) \geq 0$	Nondecreasing
$f'(x) \leq 0$	Nonincreasing

☐ **EVENTUALLY MONOTONE SEQUENCES**

If the terms of a sequence fail to have a certain property from the start, but the terms have that property from some point on, then we say that the sequence has the property *eventually*. For example, a sequence $\{a_n\}$ would be called *eventually monotone* if there is some integer N such that the sequence is *monotone* for $n \geq N$.

We shall illustrate this idea with a sequence involving *factorials*. For this purpose recall that if n is a positive integer, then $n!$ (read "n factorial") denotes the product of the first n positive integers, that is,

$$n! = 1 \cdot 2 \cdot 3 \cdots n \quad \text{or equivalently,} \quad n! = n(n-1)(n-2)\cdots 1$$

Furthermore, it is agreed by convention that $0! = 1$.

Example 5   Show that the sequence $\left\{ \dfrac{10^n}{n!} \right\}_{n=1}^{+\infty}$ is eventually decreasing.

Solution.   We have

$$a_n = \frac{10^n}{n!} \quad \text{and} \quad a_{n+1} = \frac{10^{n+1}}{(n+1)!}$$

so

$$\frac{a_{n+1}}{a_n} = \frac{10^{n+1}/(n+1)!}{10^n/n!} = \frac{10^{n+1}\, n!}{10^n\,(n+1)!} = 10 \frac{n!}{(n+1)n!} = \frac{10}{n+1} \tag{2}$$

From (2), $a_{n+1}/a_n < 1$ for all $n \geq 10$, so the sequence is eventually decreasing.   ◀

☐ **CONVERGENCE OF MONOTONE SEQUENCES**

The following two theorems, whose proofs are discussed at the end of this section, show that a monotone sequence either converges or becomes infinite—divergence by oscillation cannot occur.

11.2.2   THEOREM.   *If $a_1 \leq a_2 \leq a_3 \leq \cdots \leq a_n \leq \cdots$ is a nondecreasing sequence, then there are two possibilities:*

(a)   *There is a constant M, called an **upper bound** for the sequence, such that $a_n \leq M$ for all n, in which case the sequence converges to a limit L satisfying $L \leq M$.*

(b)   *No upper bound exists, in which case $\lim\limits_{n \to +\infty} a_n = +\infty$.*

11.2.3   THEOREM.   *If $a_1 \geq a_2 \geq a_3 \geq \cdots \geq a_n \geq \cdots$ is a nonincreasing sequence, then there are two possibilities:*

(a)   *There is a constant M, called a **lower bound** for the sequence, such that $a_n \geq M$ for all n, in which case the sequence converges to a limit L satisfying $L \geq M$.*

(b)   *No lower bound exists, in which case $\lim\limits_{n \to +\infty} a_n = -\infty$.*

It should be noted that these results do not give a method for obtaining limits; they tell us only whether a limit exists.

Example 6   Use Theorems 11.2.2 and 11.2.3 to show that the sequence $\left\{ \dfrac{n}{n+1} \right\}_{n=1}^{+\infty}$ converges.

Solution.   We showed in Examples 2 and 3 that the given sequence is increasing (hence nondecreasing). It is evident that the number $M = 1$ is an upper bound for the sequence since

$$a_n = \frac{n}{n+1} < 1, \quad n = 1, 2, \ldots$$

Thus, by Theorem 11.2.2 the sequence converges to some limit L such that $L \leq M = 1$. This is indeed the case since

$$\lim_{n \to +\infty} \frac{n}{n+1} = \lim_{n \to +\infty} \frac{1}{1 + 1/n} = 1 \quad \blacktriangleleft$$

☐ **AN INTUITIVE VIEW OF CONVERGENCE**

Informally stated, the convergence or divergence of a sequence does not depend on the behavior of the "initial terms" of the sequence, but rather on the behavior of the "tail end." Thus, for a sequence $\{a_n\}$ to converge to a limit L, it does not matter if the initial terms are far from L, just so the terms in the sequence are eventually arbitrarily close to L. This being the case, one can add, delete, or alter *finitely* many terms without affecting the convergence, divergence, or the limit (if it exists).

As one would expect from the preceding discussion, it can be proved that Theorems 11.2.2 and 11.2.3 hold for sequences that are eventually nondecreasing and eventually nonincreasing, respectively.

Example 7   Show that the sequence $\left\{ \dfrac{10^n}{n!} \right\}_{n=1}^{+\infty}$ converges and find its limit.

Solution.   We showed in Example 5 that the sequence is eventually decreasing. Since all terms in the sequence are positive, it is bounded below by $M = 0$, and hence Theorem 11.2.3 guarantees that it converges to a limit L such that $L \geq 0$. However, the limit is not evident directly from the formula $10^n/n!$ for the nth term, so we will need some ingenuity to obtain it.

Recall from Formula (2) of Example 5 that successive terms in the given sequence are related by the recursion formula

$$a_{n+1} = \frac{10}{n+1} a_n \tag{3}$$

where $a_n = 10^n/n!$. We shall take the limit as $n \to +\infty$ of both sides of (3) and use the fact that

$$\lim_{n \to +\infty} a_{n+1} = \lim_{n \to +\infty} a_n = L$$

We obtain

$$L = \lim_{n \to +\infty} a_{n+1} = \lim_{n \to +\infty} \left(\frac{10}{n+1} a_n \right) = \lim_{n \to +\infty} \frac{10}{n+1} \lim_{n \to +\infty} a_n = 0 \cdot L = 0$$

so that

$$L = \lim_{n \to +\infty} \frac{10^n}{n!} = 0 \quad \blacktriangleleft$$

REMARK.   In the exercises we will show that the technique illustrated in the preceding example can be adapted to obtain the following limit

$$\lim_{n \to +\infty} \frac{x^n}{n!} = 0 \tag{4}$$

for any real value of x (Exercise 25). This result, which shows that $n!$ eventually increases more rapidly than any positive integer power of x, will be useful in our later work.

■ SOME PROOFS

In this text we have not been concerned with a rigorous development of the real number system; we have simply accepted the familiar properties of real numbers without proof, and indeed, we have not even attempted to define the term "real number." Although this is sufficient for many purposes, it was recognized by the late nineteenth century that the study of limits and functions in calculus requires a precise axiomatic formulation of the real numbers analogous to the axiomatic development of Euclidean geometry. Although we will not attempt to pursue this development, we will need to discuss one of the axioms about real numbers in order to prove Theorems 11.2.2 and 11.2.3. But first we shall introduce some terminology.

If S is a nonempty set of real numbers, then we call u an ***upper bound*** for S if u is greater than or equal to every number in S, and we call l a ***lower bound*** for S if l is smaller than or equal to every number in S. For example, if S is the set of numbers in the interval $(1, 3)$, then $u = 4$, 10, and 100 are upper bounds for S and $l = -10$, 0, and $\frac{1}{2}$ are lower bounds for S. Observe also that $u = 3$ is the smallest of all upper bounds and $l = 1$ is the largest of all lower bounds. The existence of a smallest upper bound and a greatest lower bound for S is not accidental; it is a consequence of the following axiom.

> **11.2.4**   AXIOM (***The Completeness Axiom***).   *If a nonempty set S of real numbers has an upper bound, then it has a smallest upper bound (called the **least upper bound**), and if a nonempty set S of real numbers has a lower bound, then it has a largest lower bound (called the **greatest lower bound**).*

Proof of Theorem 11.2.2.

(a)   Assume there exists a number M such that $a_n \leq M$ for $n = 1, 2, \ldots$. Then M is an upper bound for the set of terms in the sequence. By the Completeness Axiom there is a least upper bound for the terms, call it L. Now let ϵ be any positive number. Since L is the least upper bound for the terms, $L - \epsilon$ is not an upper bound for the terms, which means that there is at least one term a_N such that

$$a_N > L - \epsilon$$

Moreover, since $\{a_n\}$ is a nondecreasing sequence, we must have

$$a_n \geq a_N > L - \epsilon \tag{5}$$

when $n \geq N$. But a_n cannot exceed L since L is an upper bound for the terms. This observation together with (5) tells us that $L \geq a_n > L - \epsilon$ for $n \geq N$, so all terms from the Nth on are within ϵ units of L. This is exactly the requirement to have

$$\lim_{n \to +\infty} a_n = L$$

Finally, $L \leq M$ since M is an upper bound for the terms and L is the least upper bound. This proves part (a).

(b) If there is no number M such that $a_n \leq M$ for $n = 1, 2, \ldots$, then no matter how large we choose M, there is a term a_N such that

$$a_N > M$$

and, since the sequence is nondecreasing,

$$a_n \geq a_N > M$$

when $n \geq N$. Thus, the terms in the sequence become arbitrarily large as n increases. That is,

$$\lim_{n \to +\infty} a_n = +\infty \quad \blacksquare$$

The proof of Theorem 11.2.3 will be omitted since it is similar to the proof of 11.2.2.

▶ Exercise Set 11.2

In Exercises 1–6, use $a_{n+1} - a_n$ to show that the given sequence $\{a_n\}$ is strictly monotone and classify it as increasing or decreasing.

1. $\left\{\dfrac{1}{n}\right\}_{n=1}^{+\infty}$.

2. $\left\{1 - \dfrac{1}{n}\right\}_{n=1}^{+\infty}$.

3. $\left\{\dfrac{n}{2n+1}\right\}_{n=1}^{+\infty}$.

4. $\left\{\dfrac{n}{4n-1}\right\}_{n=1}^{+\infty}$.

5. $\{n - 2^n\}_{n=1}^{+\infty}$.

6. $\{n - n^2\}_{n=1}^{+\infty}$.

In Exercises 7–12, use a_{n+1}/a_n to show that the given sequence $\{a_n\}$ is strictly monotone and classify it as increasing or decreasing.

7. $\left\{\dfrac{n}{2n+1}\right\}_{n=1}^{+\infty}$.

8. $\left\{\dfrac{2^n}{1+2^n}\right\}_{n=1}^{+\infty}$.

9. $\{ne^{-n}\}_{n=1}^{+\infty}$.

10. $\left\{\dfrac{10^n}{(2n)!}\right\}_{n=1}^{+\infty}$.

11. $\left\{\dfrac{n^n}{n!}\right\}_{n=1}^{+\infty}$.

12. $\left\{\dfrac{5^n}{2^{(n^2)}}\right\}_{n=1}^{+\infty}$.

In Exercises 13–18, use differentiation to show that the sequence is strictly monotone and classify it as increasing or decreasing.

13. $\left\{\dfrac{n}{2n+1}\right\}_{n=1}^{+\infty}$.

14. $\left\{3 - \dfrac{1}{n}\right\}_{n=1}^{+\infty}$.

15. $\left\{\dfrac{1}{n + \ln n}\right\}_{n=1}^{+\infty}$.

16. $\{ne^{-2n}\}_{n=1}^{+\infty}$.

17. $\left\{\dfrac{\ln(n+2)}{n+2}\right\}_{n=1}^{+\infty}$.

18. $\{\tan^{-1} n\}_{n=1}^{+\infty}$.

In Exercises 19–24, use any method to show that the sequence is eventually increasing or eventually decreasing.

19. $\{2n^2 - 7n\}_{n=1}^{+\infty}$.

20. $\{n^3 - 4n^2\}_{n=1}^{+\infty}$.

21. $\left\{\dfrac{n}{n^2 + 10}\right\}_{n=1}^{+\infty}$.

22. $\left\{n + \dfrac{17}{n}\right\}_{n=1}^{+\infty}$.

23. $\left\{\dfrac{n!}{3^n}\right\}_{n=1}^{+\infty}$.

24. $\{n^5 e^{-n}\}_{n=1}^{+\infty}$.

25. Prove:

$$\lim_{n \to +\infty} \frac{x^n}{n!} = 0$$

for any real value of x. [*Hint:* Consider the cases $x = 0$ and $x \neq 0$. For $x \neq 0$, show that $\lim\limits_{n \to +\infty} |x|^n/n! = 0$ and use the Squeezing Theorem (Theorem 2.8.2).]

26. Show that

$$\lim_{n \to +\infty} \frac{n!}{n^n} = 0$$

27. Let $\{a_n\}$ be the sequence defined recursively by $a_1 = \sqrt{2}$ and $a_{n+1} = \sqrt{2 + a_n}$ for $n \geq 1$.

(a) List the first three terms of the sequence.

(b) Show that $a_n < 2$ for $n \geq 1$.

(c) Show that $a_{n+1}^2 - a_n^2 = (2 - a_n)(1 + a_n)$ for $n \geq 1$.

(d) Use the results in parts (b) and (c) to show that $\{a_n\}$ is an increasing sequence. [*Hint:* If x and y are positive real numbers such that $x^2 - y^2 > 0$, then it follows by factoring that $x - y > 0$.]

(e) Show that $\{a_n\}$ converges and find its limit L.

28. Let $\{a_n\}$ be the sequence defined recursively by $a_1 = 1$ and $a_{n+1} = \frac{1}{2}(a_n + 3/a_n)$ for $n \geq 1$ (see Example 7, Section 11.1).

(a) Show that $a_n \geq \sqrt{3}$ for $n \geq 2$. [*Hint:* What is the minimum value of $\frac{1}{2}(x + 3/x)$ for $x > 0$?]

(b) Show that $\{a_n\}$ is eventually nonincreasing. [*Hint:* Examine $a_{n+1} - a_n$ or a_{n+1}/a_n and use the result in part (a).]

(c) Show that $\{a_n\}$ converges and find its limit L.

29. (a) Show that if $\{a_n\}_{n=1}^{+\infty}$ is a nonincreasing sequence, then $\{-a_n\}_{n=1}^{+\infty}$ is a nondecreasing sequence.

(b) Use part (a) and Theorem 11.2.2 to help prove Theorem 11.2.3.

30. (a) Deduce the inequalities

$$\int_1^n \ln x \, dx < \ln n! < \int_1^{n+1} \ln x \, dx$$

from Figure 11.2.1 for $n \geq 2$ by comparing appropriate areas.

(b) Use the result in part (a) to show that

$$\frac{n^n}{e^{n-1}} < n! < \frac{(n+1)^{n+1}}{e^n}, \quad n > 1$$

31. Use the Squeezing Theorem (Theorem 2.8.2) and the result in Exercise 30(b) to show that

$$\lim_{n \to +\infty} \frac{\sqrt[n]{n!}}{n} = \frac{1}{e}$$

32. Use the left inequality in Exercise 30(b) to show that $\lim_{n \to +\infty} \sqrt[n]{n!} = +\infty$.

33. (a) Show that $\left\{ \dfrac{n^n}{n! e^n} \right\}_{n=1}^{+\infty}$ is a decreasing sequence.
[*Hint:* From Exercise 20(d) of Section 7.5 we have $(1 + 1/x)^x < e$ for $x > 0$.]

(b) Does the sequence converge? Explain.

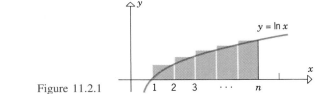
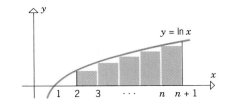

Figure 11.2.1

11.3 INFINITE SERIES

The purpose of this section is to discuss sums that contain infinitely many terms. The most familiar examples of such sums occur in the decimal representation of real numbers. For example, when we write $\frac{1}{3}$ in the decimal form $\frac{1}{3} = 0.3333\ldots$ we mean

$$\frac{1}{3} = 0.3 + 0.03 + 0.003 + 0.0003 + \cdots$$

which suggests that the decimal representation of $\frac{1}{3}$ can be viewed as a sum of infinitely many real numbers.

□ **SUMS OF INFINITE SERIES**

Our first objective is to define what is meant by the "sum" of infinitely many real numbers. We begin with some terminology.

11.3.1 DEFINITION. An *infinite series* is an expression that can be written in the form

$$\sum_{k=1}^{\infty} u_k = u_1 + u_2 + u_3 + \cdots + u_k + \cdots$$

The numbers $u_1, u_2, u_3, \ldots$ are called the *terms* of the series.

Since it is physically impossible to add infinitely many numbers together, sums of infinite series are defined and computed by an indirect limiting process. To motivate the basic idea, consider the infinite decimal

$$0.3333\ldots \tag{1}$$

which can be viewed as the infinite series

$$0.3 + 0.03 + 0.003 + 0.0003 + \cdots$$

or equivalently,

$$\frac{3}{10} + \frac{3}{10^2} + \frac{3}{10^3} + \frac{3}{10^4} + \cdots \tag{2}$$

Since (1) is the decimal expansion of $\frac{1}{3}$, any reasonable definition for the sum of an infinite series should yield $\frac{1}{3}$ for the sum of (2). To obtain such a definition, consider the following sequence of (finite) sums:

$$s_1 = \frac{3}{10} = 0.3$$

$$s_2 = \frac{3}{10} + \frac{3}{10^2} = 0.33$$

$$s_3 = \frac{3}{10} + \frac{3}{10^2} + \frac{3}{10^3} = 0.333$$

$$s_4 = \frac{3}{10} + \frac{3}{10^2} + \frac{3}{10^3} + \frac{3}{10^4} = 0.3333$$

$$\vdots$$

The sequence of numbers $s_1, s_2, s_3, s_4, \ldots$ can be viewed as a succession of approximations to the "sum" of the infinite series, which we want to be $\frac{1}{3}$. As we progress through the sequence, more and more terms of the infinite series are used, and the approximations get better and better, suggesting that the desired sum of $\frac{1}{3}$ might be the *limit* of this sequence of approximations. To see that this is so, we must calculate the limit of the general term in the sequence of approximations, namely

$$s_n = \frac{3}{10} + \frac{3}{10^2} + \cdots + \frac{3}{10^n} \tag{3}$$

The problem of calculating

$$\lim_{n \to +\infty} s_n = \lim_{n \to +\infty} \left(\frac{3}{10} + \frac{3}{10^2} + \cdots + \frac{3}{10^n} \right)$$

is complicated by the fact that both the last term and the number of terms in the sum change with n. It is best to rewrite such limits in a closed form in which the number of terms does not vary, if possible. (See the remark following Example 3 in Section 5.4.) To do this, we multiply both sides of (3) by $\frac{1}{10}$ to obtain

$$\frac{1}{10} s_n = \frac{3}{10^2} + \frac{3}{10^3} + \cdots + \frac{3}{10^n} + \frac{3}{10^{n+1}} \tag{4}$$

and then subtract (4) from (3) to obtain

$$s_n - \frac{1}{10} s_n = \frac{3}{10} - \frac{3}{10^{n+1}}$$

$$\frac{9}{10} s_n = \frac{3}{10} \left(1 - \frac{1}{10^n} \right)$$

$$s_n = \frac{1}{3} \left(1 - \frac{1}{10^n} \right)$$

Since $1/10^n \to 0$ as $n \to +\infty$, it follows that

$$\lim_{n \to +\infty} s_n = \lim_{n \to +\infty} \frac{1}{3}\left(1 - \frac{1}{10^n}\right) = \frac{1}{3}$$

which we denote by writing

$$\frac{1}{3} = \frac{3}{10} + \frac{3}{10^2} + \frac{3}{10^3} + \cdots + \frac{3}{10^n} + \cdots$$

Motivated by the preceding example, we are now ready to define the general concept of the "sum" of an infinite series

$$u_1 + u_2 + u_3 + \cdots + u_k + \cdots$$

We begin with some terminology: Let s_n denote the sum of the first n terms of the series. Thus,

$$s_1 = u_1$$

$$s_2 = u_1 + u_2$$

$$s_3 = u_1 + u_2 + u_3$$

$$\vdots$$

$$s_n = u_1 + u_2 + u_3 + \cdots + u_n = \sum_{k=1}^{n} u_k$$

The number s_n is called the ***nth partial sum*** of the series and the sequence $\{s_n\}_{n=1}^{+\infty}$ is called the ***sequence of partial sums***.

WARNING.   In everyday language the words "sequence" and "series" are often used interchangeably. However, this is not so in mathematics—mathematically, a sequence is a *succession* and a series is a *sum*. It is essential that you keep this distinction in mind.

As n increases, the partial sum $s_n = u_1 + u_2 + \cdots + u_n$ includes more and more terms of the series. Thus, if s_n tends toward a limit as $n \to +\infty$, it is reasonable to view this limit as the sum of *all* the terms in the series. This suggests the following definition.

11.3.2   DEFINITION.   Let $\{s_n\}$ be the sequence of partial sums of the series $u_1 + u_2 + u_3 + \cdots + u_k + \cdots$. If the sequence $\{s_n\}$ converges to a limit S, then the series is said to ***converge***, and S is called the ***sum*** of the series. We denote this by writing

$$S = \sum_{k=1}^{\infty} u_k$$

If the sequence of partial sums diverges, then the series is said to ***diverge***. A divergent series has no sum.

Example 1   Determine whether the series

$$1 - 1 + 1 - 1 + 1 - 1 + \cdots$$

converges or diverges. If it converges, find the sum.

Solution.   It is tempting to conclude that the sum of the series is zero by arguing that the positive and negative terms cancel one another. However, this is *not correct*; the problem is that algebraic operations that hold for finite sums do not carry over to infinite series in all cases. Later, we shall discuss conditions under which familiar algebraic operations can be applied to infinite series, but for this example we turn directly to Definition 11.3.2. The partial sums are

$$s_1 = 1$$

$$s_2 = 1 - 1 = 0$$

$$s_3 = 1 - 1 + 1 = 1$$

$$s_4 = 1 - 1 + 1 - 1 = 0$$

and so forth. Thus, the sequence of partial sums is

$$1, 0, 1, 0, 1, 0, \ldots$$

Since this is a divergent sequence, the given series diverges and consequently has no sum. ◄

☐ **GEOMETRIC SERIES**

The series encountered thus far are examples of *geometric series*. A geometric series is one of the form

$$a + ar + ar^2 + ar^3 + \cdots + ar^{k-1} + \cdots \quad (a \neq 0)$$

where each term is obtained by multiplying the preceding one by a constant r. The multiplier r is called the ***ratio*** for the series. Some examples of geometric series are

$$1 + 2 + 4 + 8 + \cdots + 2^{k-1} + \cdots \qquad \boxed{a = 1,\ r = 2}$$

$$\frac{3}{10} + \frac{3}{10^2} + \frac{3}{10^3} + \cdots + \frac{3}{10^k} + \cdots \qquad \boxed{a = \tfrac{3}{10},\ r = \tfrac{1}{10}}$$

$$\frac{1}{2} - \frac{1}{4} + \frac{1}{8} - \frac{1}{16} + \cdots + (-1)^{k+1} \frac{1}{2^k} + \cdots \qquad \boxed{a = \tfrac{1}{2},\ r = -\tfrac{1}{2}}$$

$$1 + 1 + 1 + \cdots + 1 + \cdots \qquad \boxed{a = 1,\ r = 1}$$

$$1 - 1 + 1 - 1 + \cdots + (-1)^{k+1} + \cdots \qquad \boxed{a = 1,\ r = -1}$$

The following theorem is the fundamental result on convergence of geometric series.

11.3.3 THEOREM. *A geometric series*

$$a + ar + ar^2 + \cdots + ar^{k-1} + \cdots \quad (a \neq 0)$$

converges if $|r| < 1$ and diverges if $|r| \geq 1$. If the series converges, then the sum is

$$\frac{a}{1-r} = a + ar + ar^2 + \cdots + ar^{k-1} + \cdots$$

Proof. Let us treat the case $|r| = 1$ first. If $r = 1$, then the series is

$$a + a + a + \cdots + a + \cdots$$

so the nth partial sum is $s_n = na$ and $\lim\limits_{n \to +\infty} s_n = \lim\limits_{n \to +\infty} na = \pm \infty$ (the sign depending on whether a is positive or negative). This proves divergence. If $r = -1$, the series is

$$a - a + a - a + \cdots$$

so the sequence of partial sums is

$$a, 0, a, 0, a, 0, \ldots$$

which diverges.

Now let us consider the case where $|r| \neq 1$. The nth partial sum of the series is

$$s_n = a + ar + ar^2 + \cdots + ar^{n-1} \tag{5}$$

Multiplying both sides of (5) by r yields

$$rs_n = ar + ar^2 + \cdots + ar^{n-1} + ar^n \tag{6}$$

and subtracting (6) from (5) gives

$$s_n - rs_n = a - ar^n$$

or

$$(1 - r)s_n = a - ar^n \tag{7}$$

Since $r \neq 1$ in the case we are considering, this can be rewritten as

$$s_n = \frac{a - ar^n}{1 - r} = \frac{a}{1 - r} - \frac{ar^n}{1 - r} \tag{8}$$

If $|r| < 1$, then $\lim\limits_{n \to +\infty} r^n = 0$ (can you see why?), so $\{s_n\}$ converges. From (8)

$$\lim_{n \to +\infty} s_n = \frac{a}{1 - r}$$

If $|r| > 1$, then either $r > 1$ or $r < -1$. In the case $r > 1$, $\lim\limits_{n \to +\infty} r^n = +\infty$, and in the case $r < -1$, r^n oscillates between positive and negative values that grow in magnitude, so $\{s_n\}$ diverges in both cases.   ∎

Example 2   The series

$$5 + \frac{5}{4} + \frac{5}{4^2} + \cdots + \frac{5}{4^{k-1}} + \cdots$$

is a geometric series with $a = 5$ and $r = \frac{1}{4}$. Since $|r| = \frac{1}{4} < 1$, the series converges and the sum is

$$\frac{a}{1 - r} = \frac{5}{1 - \frac{1}{4}} = \frac{20}{3} \quad ◄$$

Example 3   Find the rational number represented by the repeating decimal

$$0.784784784\ldots$$

Solution.   We can write

$$0.784784784\ldots = 0.784 + 0.000784 + 0.000000784 + \cdots$$

so the given decimal is the sum of a geometric series with $a = 0.784$ and $r = 0.001$. Thus,

$$0.784784784\ldots = \frac{a}{1 - r} = \frac{0.784}{1 - 0.001} = \frac{0.784}{0.999} = \frac{784}{999} \quad ◄$$

Example 4   Determine whether the series

$$\sum_{k=1}^{\infty} \frac{1}{k(k+1)} = \frac{1}{1 \cdot 2} + \frac{1}{2 \cdot 3} + \frac{1}{3 \cdot 4} + \frac{1}{4 \cdot 5} + \cdots$$

converges or diverges. If it converges, find the sum.

Solution.   The nth partial sum of the series is

$$s_n = \sum_{k=1}^{n} \frac{1}{k(k+1)} = \frac{1}{1 \cdot 2} + \frac{1}{2 \cdot 3} + \frac{1}{3 \cdot 4} + \cdots + \frac{1}{n(n+1)}$$

To calculate $\lim\limits_{n \to +\infty} s_n$ we shall rewrite s_n in closed form. This can be accomplished by using the method of partial fractions to obtain (verify)

$$\frac{1}{k(k+1)} = \frac{1}{k} - \frac{1}{k+1}$$

from which we obtain the telescoping sum

$$
\begin{aligned}
s_n &= \sum_{k=1}^{n} \left(\frac{1}{k} - \frac{1}{k+1} \right) \\
&= \left(1 - \frac{1}{2}\right) + \left(\frac{1}{2} - \frac{1}{3}\right) + \left(\frac{1}{3} - \frac{1}{4}\right) + \cdots + \left(\frac{1}{n} - \frac{1}{n+1}\right) \\
&= 1 + \left(-\frac{1}{2} + \frac{1}{2}\right) + \left(-\frac{1}{3} + \frac{1}{3}\right) + \cdots + \left(-\frac{1}{n} + \frac{1}{n}\right) - \frac{1}{n+1} \\
&= 1 - \frac{1}{n+1}
\end{aligned}
$$

so

$$\sum_{k=1}^{\infty} \frac{1}{k(k+1)} = \lim_{n \to +\infty} s_n = \lim_{n \to +\infty} \left(1 - \frac{1}{n+1}\right) = 1 \quad \blacktriangleleft$$

□ **HARMONIC SERIES**

One of the most important of all diverging series is the ***harmonic series***,

$$\sum_{k=1}^{\infty} \frac{1}{k} = 1 + \frac{1}{2} + \frac{1}{3} + \frac{1}{4} + \frac{1}{5} + \cdots$$

which arises in connection with the overtones produced by a vibrating musical string. It is not immediately evident that this series diverges. However, the divergence will become apparent when we examine the partial sums in detail. Because the terms in the series are all positive, the partial sums

$$s_1 = 1, \ s_2 = 1 + \tfrac{1}{2}, \ s_3 = 1 + \tfrac{1}{2} + \tfrac{1}{3}, \ s_4 = 1 + \tfrac{1}{2} + \tfrac{1}{3} + \tfrac{1}{4}, \ldots$$

form an increasing sequence

$$s_1 < s_2 < s_3 < \cdots < s_n < \cdots$$

Thus, by Theorem 11.2.2 we can prove divergence by demonstrating that there is no constant M that is greater than or equal to *every* partial sum. To this end, we shall consider some selected partial sums, namely $s_2, s_4, s_8, s_{16}, s_{32}, \ldots$. Note that the subscripts are successive powers of 2, so that these are the partial sums of the form s_{2^n}. These partial sums satisfy the inequalities

$$s_2 = 1 + \tfrac{1}{2} > \tfrac{1}{2} + \tfrac{1}{2} = \tfrac{2}{2}$$

$$s_4 = s_2 + \tfrac{1}{3} + \tfrac{1}{4} > s_2 + \left(\tfrac{1}{4} + \tfrac{1}{4}\right) = s_2 + \tfrac{1}{2} > \tfrac{3}{2}$$

$$s_8 = s_4 + \tfrac{1}{5} + \tfrac{1}{6} + \tfrac{1}{7} + \tfrac{1}{8} > s_4 + \left(\tfrac{1}{8} + \tfrac{1}{8} + \tfrac{1}{8} + \tfrac{1}{8}\right) = s_4 + \tfrac{1}{2} > \tfrac{4}{2}$$

$$s_{16} = s_8 + \tfrac{1}{9} + \tfrac{1}{10} + \tfrac{1}{11} + \tfrac{1}{12} + \tfrac{1}{13} + \tfrac{1}{14} + \tfrac{1}{15} + \tfrac{1}{16}$$
$$> s_8 + \left(\tfrac{1}{16} + \tfrac{1}{16} + \tfrac{1}{16} + \tfrac{1}{16} + \tfrac{1}{16} + \tfrac{1}{16} + \tfrac{1}{16} + \tfrac{1}{16}\right) = s_8 + \tfrac{1}{2} > \tfrac{5}{2}$$

$$\vdots$$

$$s_{2^n} > \frac{n+1}{2}$$

If M is any constant, we can find a positive integer n such that $(n + 1)/2 > M$. But for this n

$$s_{2^n} > \frac{n + 1}{2} > M$$

so that no constant M is greater than or equal to *every* partial sum of the harmonic series. This proves divergence.

▶ Exercise Set 11.3

1. In each part, find the first four partial sums; find a closed form for the nth partial sum; and determine whether the series converges (if so, give the sum).

   (a) $\displaystyle\sum_{k=1}^{\infty} \frac{2}{5^{k-1}}$ (b) $\displaystyle\sum_{k=1}^{\infty} \frac{1}{(k + 1)(k + 2)}$

   (c) $\displaystyle\sum_{k=1}^{\infty} \frac{2^{k-1}}{4}$.

In Exercises 2–16, determine whether the series converges or diverges. If it converges, find the sum.

2. $\displaystyle\sum_{k=1}^{\infty} \frac{1}{5^k}$.   3. $\displaystyle\sum_{k=1}^{\infty} \left(-\frac{3}{4}\right)^{k-1}$.

4. $\displaystyle\sum_{k=1}^{\infty} \left(\frac{2}{3}\right)^{k+2}$.   5. $\displaystyle\sum_{k=1}^{\infty} (-1)^{k-1} \frac{7}{6^{k-1}}$.

6. $\displaystyle\sum_{k=1}^{\infty} 4^{k-1}$.   7. $\displaystyle\sum_{k=1}^{\infty} \left(-\frac{3}{2}\right)^{k+1}$.

8. $\displaystyle\sum_{k=1}^{\infty} \left(\frac{1}{k + 3} - \frac{1}{k + 4}\right)$.

9. $\displaystyle\sum_{k=1}^{\infty} \frac{1}{(k + 2)(k + 3)}$.   10. $\displaystyle\sum_{k=1}^{\infty} \left(\frac{1}{2^k} - \frac{1}{2^{k+1}}\right)$.

11. $\displaystyle\sum_{k=1}^{\infty} \frac{1}{9k^2 + 3k - 2}$.   12. $\displaystyle\sum_{k=2}^{\infty} \frac{1}{k^2 - 1}$.

13. $\displaystyle\sum_{k=1}^{\infty} \frac{4^{k+2}}{7^{k-1}}$.   14. $\displaystyle\sum_{k=1}^{\infty} (e/\pi)^{k-1}$.

15. $\displaystyle\sum_{k=1}^{\infty} (-1/2)^k$.   16. $\displaystyle\sum_{k=3}^{\infty} \frac{5}{k - 2}$.

In Exercises 17–22, express the repeating decimal as a fraction.

17. $0.4444\ldots$ 18. $0.9999\ldots$

19. $5.373737\ldots$ 20. $0.159159159\ldots$

21. $0.782178217821\ldots$ 22. $0.451141414\ldots$

23. Find a closed form for the nth partial sum of the series

    $$\ln\frac{1}{2} + \ln\frac{2}{3} + \ln\frac{3}{4} + \cdots + \ln\frac{n}{n + 1} + \cdots$$

    and determine whether the series converges.

24. A ball is dropped from a height of 10 m. Each time it strikes the ground it bounces vertically to a height that is $\frac{3}{4}$ of the preceding height. Find the total distance the ball will travel if it is allowed to bounce indefinitely.

25. Show: $\displaystyle\sum_{k=2}^{\infty} \ln(1 - 1/k^2) = -\ln 2$.

26. Show: $\displaystyle\sum_{k=1}^{\infty} \frac{\sqrt{k + 1} - \sqrt{k}}{\sqrt{k^2 + k}} = 1$.

27. Show: $\displaystyle\sum_{k=1}^{\infty} \left(\frac{1}{k} - \frac{1}{k + 2}\right) = \frac{3}{2}$.

28. (a) Find A and B such that

    $$\frac{6^k}{(3^{k+1} - 2^{k+1})(3^k - 2^k)} = \frac{2^k A}{3^k - 2^k} + \frac{2^k B}{3^{k+1} - 2^{k+1}}$$

    (b) Use the result in part (a) to help show that

    $$\sum_{k=1}^{\infty} \frac{6^k}{(3^{k+1} - 2^{k+1})(3^k - 2^k)} = 2$$

    [This problem appeared in the Forty-Fifth Annual William Lowell Putnam Mathematical Competition.]

29. Show: $\displaystyle\frac{1}{1 \cdot 3} + \frac{1}{3 \cdot 5} + \frac{1}{5 \cdot 7} + \cdots = \frac{1}{2}$.

30. Show: $\displaystyle\frac{1}{1 \cdot 3} + \frac{1}{2 \cdot 4} + \frac{1}{3 \cdot 5} + \cdots = \frac{3}{4}$.

31. Use geometric series to show that

    (a) $\displaystyle\sum_{k=0}^{\infty} (-1)^k x^k = \frac{1}{1 + x}$ if $-1 < x < 1$

    (b) $\displaystyle\sum_{k=0}^{\infty} (x - 3)^k = \frac{1}{4 - x}$ if $2 < x < 4$

    (c) $\displaystyle\sum_{k=0}^{\infty} (-1)^k x^{2k} = \frac{1}{1 + x^2}$ if $-1 < x < 1$.

In Exercises 32–35, find all values of x for which the series converges, and for these values find its sum.

32. $x - x^3 + x^5 - x^7 + x^9 - \cdots$.

33. $\displaystyle\frac{1}{x^2} + \frac{2}{x^3} + \frac{4}{x^4} + \frac{8}{x^5} + \frac{16}{x^6} + \cdots$.

34. $e^{-x} + e^{-2x} + e^{-3x} + e^{-4x} + e^{-5x} + \cdots$.

35. $\sin x - \frac{1}{2}\sin^2 x + \frac{1}{4}\sin^3 x - \frac{1}{8}\sin^4 x + \cdots.$

36. Prove the following decimal equality assuming that $a_n \neq 9$:

$$0.a_1 a_2 \ldots a_n 9999\ldots = 0.a_1 a_2 \ldots (a_n + 1)0000\ldots$$

37. Let a_1 be any real number and define

$$a_{n+1} = \frac{1}{2}(a_n + 1) \text{ for } n = 1, 2, 3, \ldots$$

Show that the sequence $\{a_n\}_{n=1}^{+\infty}$ converges and find its limit. [*Hint:* Express a_n in terms of a_1.]

38. Lines L_1 and L_2 form an angle θ, $0 < \theta < \pi/2$, at their point of intersection P (Figure 11.3.1). A point P_0 is chosen that is on L_1 and a units from P. Starting from P_0 a zig-zag path is constructed by successively going back and forth between L_1 and L_2 along a perpendicular from one line to the other. Find the following sums in terms of θ:

(a) $P_0 P_1 + P_1 P_2 + P_2 P_3 + \cdots$

(b) $P_0 P_1 + P_2 P_3 + P_4 P_5 + \cdots$

(c) $P_1 P_2 + P_3 P_4 + P_5 P_6 + \cdots.$

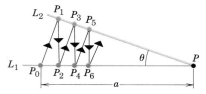

Figure 11.3.1

39. As shown in Figure 11.3.2, suppose that an angle θ is bisected using a straightedge and compass to produce ray R_1, then the angle between R_1 and the initial side is bisected to produce ray R_2. Thereafter, rays $R_3, R_4, R_5, \ldots$ are constructed in succession by bisecting the angle between the preceding two rays. Show that the sequence of angles that these rays make with the initial side has a limit of $\theta/3$. [This problem is based on *Trisection of an Angle in an Infinite Number of Steps* by Eric Kincannon, which appeared in *The College Mathematics Journal*, Vol. 21, No. 5, November 1990.]

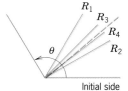

Initial side   Figure 11.3.2

40. In his *Treatise on the Configurations of Qualities and Motions* (written in the 1350s), the French Bishop of Lisieux, Nicole Oresme, used a geometric method to find the sum of the series

$$\sum_{k=1}^{\infty} \frac{k}{2^k} = \frac{1}{2} + \frac{2}{4} + \frac{3}{8} + \frac{4}{16} + \cdots$$

In Figure 11.3.3a each term in the series is represented by the area of a rectangle. In Figure 11.3.3b the configuration in part (a) has been divided into rectangles with areas $A_1, A_2, A_3, \ldots$. Find the sum $A_1 + A_2 + A_3 + \cdots$. [For convenience, the horizontal and vertical scales in Figure 11.3.3 are different.]

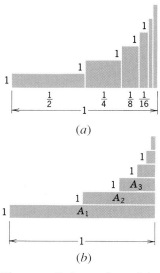

(a)

(b)   Figure 11.3.3

41. The great Swiss mathematician Leonhard Euler (biography on p. 52) made occasional errors in his pioneering work on infinite series. For example, Euler deduced that

$$\frac{1}{2} = 1 - 1 + 1 - 1 + \cdots$$

and

$$-1 = 1 + 2 + 4 + 8 + \cdots$$

by substituting $x = -1$ and $x = 2$ in the formula

$$\frac{1}{1-x} = 1 + x + x^2 + x^3 + \cdots$$

What was the error in his reasoning?

11.4 CONVERGENCE TESTS

In the preceding section we found sums of series and investigated convergence by first writing the nth partial sum s_n in closed form and then examining its limit. However, it is relatively rare that the nth partial sum of a series can be written in closed form; for most series, convergence or divergence is determined by using convergence tests, some of which we shall introduce in this section. Once it is established that a series converges, the sum of the series can generally be approximated to any degree of accuracy by a partial sum with sufficiently many terms.

☐ **THE DIVERGENCE TEST**

The kth term in an infinite series Σu_k is sometimes called the **general term** of the series. The following theorem establishes a relationship between the limit of the general term and the convergence properties of a series.

11.4.1 THEOREM (*The Divergence Test*).

(a) *If* $\lim\limits_{k \to +\infty} u_k \neq 0$, *then the series* Σu_k *diverges.*

(b) *If* $\lim\limits_{k \to +\infty} u_k = 0$, *then the series* Σu_k *may either converge or diverge.*

Proof (a). To prove this result, it suffices to show that if the series converges, then $\lim\limits_{k \to +\infty} u_k = 0$ (why?). We shall prove this alternative form of (a).

Let us assume that the series converges. The general term u_k can be written as

$$u_k = s_k - s_{k-1} \tag{1}$$

where s_k is the sum of the first k terms and s_{k-1} is the sum of the first $k - 1$ terms. If S denotes the sum of the series, then $\lim\limits_{k \to +\infty} s_k = S$, and since $(k - 1) \to +\infty$ as $k \to +\infty$, we also have $\lim\limits_{k \to +\infty} s_{k-1} = S$. Thus, from (1)

$$\lim\limits_{k \to +\infty} u_k = \lim\limits_{k \to +\infty} (s_k - s_{k-1}) = S - S = 0$$

Proof (b). To prove this result, it suffices to produce both a convergent series and a divergent series for which $\lim\limits_{k \to +\infty} u_k = 0$. The following series both have this property:

$$1/2 + 1/2^2 + \cdots + 1/2^k + \cdots \quad \text{and} \quad 1 + 1/2 + 1/3 + \cdots + 1/k + \cdots$$

The first is a convergent geometric series and the second is the divergent harmonic series. ▌

The alternative form of part (a) given in the preceding proof is sufficiently important that we state it separately for future reference.

11.4.2 THEOREM. *If the series Σu_k converges, then $\lim\limits_{k \to +\infty} u_k = 0$.*

Example 1   The series

$$\sum_{k=1}^{\infty} \frac{k}{k+1} = \frac{1}{2} + \frac{2}{3} + \frac{3}{4} + \cdots + \frac{k}{k+1} + \cdots$$

diverges since

$$\lim_{k \to +\infty} \frac{k}{k+1} = \lim_{k \to +\infty} \frac{1}{1 + 1/k} = 1 \neq 0 \quad \blacktriangleleft$$

WARNING.   The converse of Theorem 11.4.2 is false. To prove that a series converges it does not suffice to show that $\lim_{k \to +\infty} u_k = 0$, since this property may hold for divergent as well as convergent series. For example, the kth term of the divergent harmonic series $1 + 1/2 + 1/3 + \cdots + 1/k + \cdots$ approaches zero as $k \to +\infty$, and the kth term of the convergent geometric series $1/2 + 1/2^2 + \cdots + 1/2^k + \cdots$ tends to zero as $k \to +\infty$.

☐ **ALGEBRAIC PROPERTIES OF INFINITE SERIES**

For brevity, the proof of the following result is omitted.

11.4.3   THEOREM.

(a)   *If Σu_k and Σv_k are convergent series, then $\Sigma(u_k + v_k)$ and $\Sigma(u_k - v_k)$ are convergent series and the sums of these series are related by*

$$\sum_{k=1}^{\infty} (u_k + v_k) = \sum_{k=1}^{\infty} u_k + \sum_{k=1}^{\infty} v_k$$

$$\sum_{k=1}^{\infty} (u_k - v_k) = \sum_{k=1}^{\infty} u_k - \sum_{k=1}^{\infty} v_k$$

(b)   *If c is a nonzero constant, then the series Σu_k and $\Sigma c u_k$ both converge or both diverge. In the case of convergence, the sums are related by*

$$\sum_{k=1}^{\infty} c u_k = c \sum_{k=1}^{\infty} u_k$$

(c)   *Convergence or divergence is unaffected by deleting a finite number of terms from a series; in particular, for any positive integer K, the series*

$$\sum_{k=1}^{\infty} u_k = u_1 + u_2 + u_3 + \cdots$$

$$\sum_{k=K}^{\infty} u_k = u_K + u_{K+1} + u_{K+2} + \cdots$$

both converge or both diverge.

REMARK.   Do not read too much into part (*c*) of this theorem. Although the convergence is not affected when a finite number of terms is deleted from the beginning of a convergent series, the *sum* of a convergent series is changed by the removal of these terms.

Example 2   Find the sum of the series

$$\sum_{k=1}^{\infty} \left(\frac{3}{4^k} - \frac{2}{5^{k-1}} \right)$$

Solution.   The series

$$\sum_{k=1}^{\infty} \frac{3}{4^k} = \frac{3}{4} + \frac{3}{4^2} + \frac{3}{4^3} + \cdots$$

is a convergent geometric series ($a = \frac{3}{4}$, $r = \frac{1}{4}$), and the series

$$\sum_{k=1}^{\infty} \frac{2}{5^{k-1}} = 2 + \frac{2}{5} + \frac{2}{5^2} + \frac{2}{5^3} + \cdots$$

is also a convergent geometric series ($a = 2$, $r = \frac{1}{5}$). Thus, from Theorems 11.4.3(a) and 11.3.3 the given series converges and

$$\sum_{k=1}^{\infty} \left(\frac{3}{4^k} - \frac{2}{5^{k-1}} \right) = \sum_{k=1}^{\infty} \frac{3}{4^k} - \sum_{k=1}^{\infty} \frac{2}{5^{k-1}} = \frac{\frac{3}{4}}{1 - \frac{1}{4}} - \frac{2}{1 - \frac{1}{5}} = -\frac{3}{2} \quad \blacktriangleleft$$

Example 3 Determine whether the following series converge or diverge.

(a) $\displaystyle\sum_{k=1}^{\infty} \frac{5}{k} = 5 + \frac{5}{2} + \frac{5}{3} + \cdots + \frac{5}{k} + \cdots$

(b) $\displaystyle\sum_{k=10}^{\infty} \frac{1}{k} = \frac{1}{10} + \frac{1}{11} + \frac{1}{12} + \cdots$

Solution. The first series is a constant times the divergent harmonic series, and hence diverges by part (b) of Theorem 11.4.3. The second series results by deleting the first nine terms from the divergent harmonic series, and hence diverges by part (c) of Theorem 11.4.3. $\blacktriangleleft$

☐ **THE INTEGRAL TEST**

The expressions

$$\sum_{k=1}^{\infty} \frac{1}{k^2} \quad \text{and} \quad \int_{1}^{+\infty} \frac{1}{x^2} \, dx$$

are related in that the integrand in the improper integral results when the index k in the general term of the series is replaced by x and the limits of summation in the series are replaced by the corresponding limits of integration. The following theorem, which is proved at the end of this section, shows that there is a relationship between the convergence of the series and the integral.

11.4.4 THEOREM (***The Integral Test***). *Let Σu_k be a series with positive terms, and let $f(x)$ be the function that results when k is replaced by x in the formula for u_k. If f is decreasing and continuous on the interval $[a, +\infty)$, then*

$$\sum_{k=1}^{\infty} u_k \quad \text{and} \quad \int_{a}^{+\infty} f(x) \, dx$$

both converge or both diverge.

Example 4 Use the integral test to determine whether the following series converge or diverge.

(a) $\displaystyle\sum_{k=1}^{\infty} \frac{1}{k}$ (b) $\displaystyle\sum_{k=1}^{\infty} \frac{1}{k^2}$

Solution (a). We already know that this is the divergent harmonic series, so the integral test will simply provide another way of establishing the divergence. If we replace k by x in the general term $1/k$, we obtain the function $f(x) = 1/x$, which is decreasing and continuous for $x \geq 1$ (as required to apply the integral test with $a = 1$). Since

$$\int_{1}^{+\infty} \frac{1}{x} \, dx = \lim_{l \to +\infty} \int_{1}^{l} \frac{1}{x} \, dx = \lim_{l \to +\infty} [\ln l - \ln 1] = +\infty$$

the integral diverges and consequently so does the series.

Solution (b). If we replace k by x in the general term $1/k^2$, we obtain the function $f(x) = 1/x^2$, which is decreasing and continuous for $x \geq 1$. Since

$$\int_1^{+\infty} \frac{1}{x^2}\,dx = \lim_{l \to +\infty} \int_1^l \frac{dx}{x^2} = \lim_{l \to +\infty} \left[-\frac{1}{x} \right]_1^l = \lim_{l \to +\infty} \left[1 - \frac{1}{l} \right] = 1$$

the integral converges and consequently the series converges by the integral test with $a = 1$. ◄

REMARK. In part (b) of the preceding example, do *not* erroneously conclude that the sum of the series is 1 because the value of the corresponding integral is 1. It can be proved that the sum of the series is actually $\pi^2/6$ and, indeed, the sum of the first two terms alone exceeds 1.

☐ *p*-SERIES

The series in Example 4 are special cases of a class of series called *p-series* or **hyper-harmonic series**. A *p*-series is an infinite series of the form

$$\sum_{k=1}^{\infty} \frac{1}{k^p} = 1 + \frac{1}{2^p} + \frac{1}{3^p} + \cdots + \frac{1}{k^p} + \cdots$$

where $p > 0$. Examples of *p*-series are

$$\sum_{k=1}^{\infty} \frac{1}{k} = 1 + \frac{1}{2} + \frac{1}{3} + \cdots + \frac{1}{k} + \cdots \qquad \boxed{p = 1}$$

$$\sum_{k=1}^{\infty} \frac{1}{k^2} = 1 + \frac{1}{2^2} + \frac{1}{3^2} + \cdots + \frac{1}{k^2} + \cdots \qquad \boxed{p = 2}$$

$$\sum_{k=1}^{\infty} \frac{1}{\sqrt{k}} = 1 + \frac{1}{\sqrt{2}} + \frac{1}{\sqrt{3}} + \cdots + \frac{1}{\sqrt{k}} + \cdots \qquad \boxed{p = \tfrac{1}{2}}$$

The following theorem tells when a *p*-series converges.

11.4.5 THEOREM (*Convergence of p-Series*).

$$\sum_{k=1}^{\infty} \frac{1}{k^p} = 1 + \frac{1}{2^p} + \frac{1}{3^p} + \cdots + \frac{1}{k^p} + \cdots$$

converges if $p > 1$ and diverges if $0 < p \leq 1$.

Proof. To establish this result when $p \neq 1$, we shall use the integral test.

$$\int_1^{+\infty} \frac{1}{x^p}\,dx = \lim_{l \to +\infty} \int_1^l x^{-p}\,dx = \lim_{l \to +\infty} \frac{x^{1-p}}{1-p} \bigg]_1^l = \lim_{l \to +\infty} \left[\frac{l^{1-p}}{1-p} - \frac{1}{1-p} \right]$$

If $p > 1$, then $1 - p < 0$, so $l^{1-p} \to 0$ as $l \to +\infty$. Thus, the integral converges [its value is $-1/(1-p)$] and consequently the series also converges. For $0 < p < 1$, it follows that $1 - p > 0$ and $l^{1-p} \to +\infty$ as $l \to +\infty$, so the integral and the series diverge. The case $p = 1$ is the harmonic series, which was previously shown to diverge. ∎

Example 5

$$1 + \frac{1}{\sqrt[3]{2}} + \frac{1}{\sqrt[3]{3}} + \cdots + \frac{1}{\sqrt[3]{k}} + \cdots$$

diverges since it is a *p*-series with $p = \tfrac{1}{3} < 1$. ◄

☐ **PROOF OF THE INTEGRAL TEST**

Before we can prove the integral test, we need a basic result about convergence of series with *nonnegative* terms. If $u_1 + u_2 + u_3 + \cdots + u_k + \cdots$ is such a series, then its sequence of partial sums is nondecreasing, that is,

$$s_1 \le s_2 \le s_3 \le \cdots \le s_n \le \cdots$$

Thus, from Theorem 11.2.2 the sequence of partial sums converges to a limit S if and only if it has some upper bound M, in which case $S \le M$. If no upper bound exists, then the sequence of partial sums diverges. Since convergence of the sequence of partial sums corresponds to convergence of the series, we have the following theorem.

11.4.6 THEOREM. *If Σu_k is a series with nonnegative terms, and if there is a constant M such that*

$$s_n = u_1 + u_2 + \cdots + u_n \le M$$

for every n, then the series converges and the sum S satisfies $S \le M$. If no such M exists, then the series diverges.

In words, this theorem implies that *a series with nonnegative terms converges if and only if its sequence of partial sums is bounded above.*

Proof of Theorem 11.4.4.   We need only show that the series converges when the integral converges and that the series diverges when the integral diverges. The remaining cases are logical implications of these. For simplicity, we will limit the proof to the case where $a = 1$. Assume that $f(x)$ satisfies the hypotheses of the theorem for $x \ge 1$. Since

$$f(1) = u_1, f(2) = u_2, \ldots, f(n) = u_n, \ldots$$

the values of $u_1, u_2, \ldots, u_n, \ldots$ can be interpreted as the areas of the rectangles shown in Figure 11.4.1.

The following inequalities (for $n > 1$) result by comparing the areas under the curve $y = f(x)$ to the areas of the rectangles in Figure 11.4.1:

$$\int_1^{n+1} f(x)\, dx < u_1 + u_2 + \cdots + u_n = s_n \qquad \boxed{\text{Figure 11.4.1}a}$$

$$s_n - u_1 = u_2 + u_3 + \cdots + u_n < \int_1^n f(x)\, dx \qquad \boxed{\text{Figure 11.4.1}b}$$

These inequalities can be combined as

$$\int_1^{n+1} f(x)\, dx < s_n < u_1 + \int_1^n f(x)\, dx \qquad (2)$$

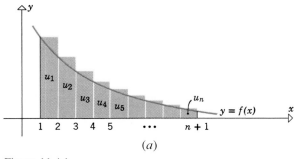

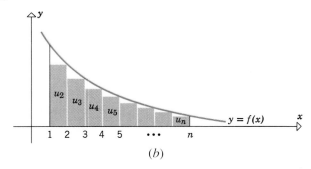

Figure 11.4.1

(a)   (b)

If the integral $\int_1^{+\infty} f(x)\, dx$ converges to a finite value L, then from the right-hand inequality in (2)

$$s_n < u_1 + \int_1^n f(x)\, dx < u_1 + \int_1^{+\infty} f(x)\, dx = u_1 + L$$

Thus, each partial sum is less than the finite constant $u_1 + L$, and the series converges by Theorem 11.4.6. On the other hand, if the integral $\int_1^{+\infty} f(x)\,dx$ diverges, then

$$\lim_{n \to +\infty} \int_1^{n+1} f(x)\,dx = +\infty$$

so that from the left-hand inequality in (2), $\lim_{n \to +\infty} s_n = +\infty$. This implies that the series also diverges. ∎

▶ **Exercise Set 11.4**   Ⓒ 37–40

In Exercises 1–4, use Theorem 11.4.3 to find the sum of the series.

1. $\displaystyle\sum_{k=1}^{\infty} \left[\frac{1}{2^k} + \frac{1}{4^k} \right].$

2. $\displaystyle\sum_{k=1}^{\infty} \left[\frac{1}{5^k} - \frac{1}{k(k+1)} \right].$

3. $\displaystyle\sum_{k=2}^{\infty} \left[\frac{1}{k^2 - 1} - \frac{7}{10^{k-1}} \right].$

4. $\displaystyle\sum_{k=1}^{\infty} \left[\frac{7}{3^k} + \frac{6}{(k+3)(k+4)} \right].$

5. In each part, determine whether the given p-series converges or diverges.

(a) $\displaystyle\sum_{k=1}^{\infty} \frac{1}{k^3}$

(b) $\displaystyle\sum_{k=1}^{\infty} \frac{1}{\sqrt{k}}$

(c) $\displaystyle\sum_{k=1}^{\infty} k^{-1}$

(d) $\displaystyle\sum_{k=1}^{\infty} k^{-2/3}$

(e) $\displaystyle\sum_{k=1}^{\infty} k^{-4/3}$

(f) $\displaystyle\sum_{k=1}^{\infty} \frac{1}{\sqrt[4]{k}}$

(g) $\displaystyle\sum_{k=1}^{\infty} \frac{1}{\sqrt[3]{k^5}}$

(h) $\displaystyle\sum_{k=1}^{\infty} \frac{1}{k^\pi}.$

In Exercises 6–8, use the divergence test to show that the series diverges.

6. (a) $\displaystyle\sum_{k=1}^{\infty} \frac{k+1}{k+2}$

(b) $\displaystyle\sum_{k=1}^{\infty} \ln k.$

7. (a) $\displaystyle\sum_{k=1}^{\infty} \frac{k^2 + k + 3}{2k^2 + 1}$

(b) $\displaystyle\sum_{k=1}^{\infty} \left(1 + \frac{1}{k} \right)^k.$

8. (a) $\displaystyle\sum_{k=1}^{\infty} \cos k\pi$

(b) $\displaystyle\sum_{k=1}^{\infty} \frac{e^k}{k}.$

In Exercises 9–30, determine whether the series converges or diverges.

9. $\displaystyle\sum_{k=1}^{\infty} \frac{1}{k+6}.$

10. $\displaystyle\sum_{k=1}^{\infty} \frac{3}{5k}.$

11. $\displaystyle\sum_{k=1}^{\infty} \frac{1}{5k+2}.$

12. $\displaystyle\sum_{k=1}^{\infty} \frac{k}{1+k^2}.$

13. $\displaystyle\sum_{k=1}^{\infty} \frac{1}{1+9k^2}.$

14. $\displaystyle\sum_{k=1}^{\infty} \frac{1}{(4+2k)^{3/2}}.$

15. $\displaystyle\sum_{k=1}^{\infty} \frac{1}{\sqrt{k+5}}.$

16. $\displaystyle\sum_{k=1}^{\infty} \frac{1}{\sqrt[k]{e}}.$

17. $\displaystyle\sum_{k=1}^{\infty} \frac{1}{\sqrt[3]{2k-1}}.$

18. $\displaystyle\sum_{k=3}^{\infty} \frac{\ln k}{k}.$

19. $\displaystyle\sum_{k=1}^{\infty} \frac{k}{\ln(k+1)}.$

20. $\displaystyle\sum_{k=1}^{\infty} k e^{-k^2}.$

21. $\displaystyle\sum_{k=1}^{\infty} \frac{1}{(k+1)[\ln(k+1)]^2}.$

22. $\displaystyle\sum_{k=1}^{\infty} \frac{k^2 + 1}{k^2 + 3}.$

23. $\displaystyle\sum_{k=1}^{\infty} \left(1 + \frac{1}{k} \right)^{-k}.$

24. $\displaystyle\sum_{k=1}^{\infty} \frac{1}{\sqrt{k^2 + 1}}.$

25. $\displaystyle\sum_{k=1}^{\infty} \frac{\tan^{-1} k}{1 + k^2}.$

26. $\displaystyle\sum_{k=1}^{\infty} \operatorname{sech}^2 k.$

27. $\displaystyle\sum_{k=5}^{\infty} 7k^{-p} \quad (p > 1).$

28. $\displaystyle\sum_{k=1}^{\infty} 7(k+5)^{-p} \quad (p \le 1).$

29. $\displaystyle\sum_{k=1}^{\infty} k^2 \sin^2 \left(\frac{1}{k} \right).$

30. $\displaystyle\sum_{k=1}^{\infty} k^2 e^{-k^3}.$

31. Prove: $\displaystyle\sum_{k=2}^{\infty} \frac{1}{k(\ln k)^p}$ converges if $p > 1$ and diverges if $p \le 1$.

32. Prove: $\displaystyle\sum_{k=3}^{\infty} \frac{1}{k(\ln k)[\ln(\ln k)]^p}$ converges if $p > 1$ and diverges if $p \le 1$.

33. Prove: If Σu_k converges and Σv_k diverges, then $\Sigma(u_k + v_k)$ diverges and $\Sigma(u_k - v_k)$ diverges. [*Hint:* Assume that $\Sigma(u_k + v_k)$ converges and use Theorem 11.4.3 to obtain a contradiction. Similarly, for $\Sigma(u_k - v_k)$.]

34. Find examples to show that $\Sigma(u_k + v_k)$ and $\Sigma(u_k - v_k)$ may converge or may diverge if Σu_k and Σv_k both diverge.

35. With the help of Exercise 33, determine whether the given series in parts (a)–(d) converge or diverge.

(a) $\displaystyle\sum_{k=1}^{\infty} \left[\left(\frac{2}{3} \right)^{k-1} + \frac{1}{k} \right]$

(b) $\displaystyle\sum_{k=1}^{\infty}\left[\dfrac{k^2}{1+k^2}+\dfrac{1}{k(k+1)}\right]$

(c) $\displaystyle\sum_{k=1}^{\infty}\left[\dfrac{1}{3k+2}+\dfrac{1}{k^{3/2}}\right]$

(d) $\displaystyle\sum_{k=2}^{\infty}\left[\dfrac{1}{k(\ln k)^2}-\dfrac{1}{k^2}\right]$.

36. If the sum S of a convergent series $\sum_{k=1}^{\infty}u_k$ of positive terms is approximated by its nth partial sum s_n, then the error in the approximation is $S-s_n$. Let $f(x)$ be the function that results when k is replaced by x in the formula for u_k. Show that if f is decreasing for $x\geq n$, then

$$\int_{n+1}^{+\infty} f(x)\,dx < S - s_n < \int_{n}^{+\infty} f(x)\,dx$$

The result in Exercise 36 can be written as

$$s_n + \int_{n+1}^{+\infty} f(x)\,dx < S < s_n + \int_{n}^{+\infty} f(x)\,dx$$

which provides an upper and lower bound on the sum S of the series. Use this result in part (a) of Exercises 37 and 38.

---

37. (a) Find the partial sum s_{10} of the series $\sum_{k=1}^{\infty}1/k^3$, and use it to obtain an upper and lower bound on the sum S of the series. Express your results to four decimal places.

(b) Use the right-hand inequality in Exercise 36 to find a value of n to ensure that the error in approximating S by s_n is less than 10^{-3}.

38. (a) Find the partial sum s_6 of the series $\sum_{k=1}^{\infty}1/k^4$, and use it to obtain an upper and lower bound on the sum S

of the series. Express your results to five decimal places. [*Note:* Compare your bounds with the exact value of S, which is $\pi^4/90$.]

(b) Use the right-hand inequality in Exercise 36 to find a value of n to ensure that the error in approximating S by s_n is less than 10^{-5}.

39. Let s_n be the nth partial sum of the divergent series $\sum_{k=1}^{\infty}1/k$.

(a) Use inequality (2) to show that for $n\geq 2$

$$\ln(n+1) < s_n < 1 + \ln n$$

Then find integer upper and lower bounds for $s_{1,000,000}$.

(b) Find a value of n to ensure that $s_n > 100$.

40. Let s_n be the nth partial sum of the divergent series $\sum_{k=1}^{\infty}1/\sqrt{k}$.

(a) Use inequality (2) to show that for $n\geq 2$

$$2\sqrt{n+1}-2 < s_n < 2\sqrt{n}-1$$

Then find integer upper and lower bounds for $s_{10,000}$.

(b) Find a value of n to ensure that $s_n > 100$.

41. Let $\sum u_k$ be a series with positive terms and $f(x)$ the function that results when k is replaced by x in the formula for u_k. Suppose that f is decreasing and continuous on the interval $[1,+\infty)$. Let $\{a_n\}_{n=1}^{+\infty}$ be the sequence defined by

$$a_n = \sum_{k=1}^{n} u_k - \int_{1}^{n} f(x)\,dx$$

(a) Use inequality (2) to show that $0 < a_n < u_1$.

(b) Show that $\{a_n\}$ is a decreasing sequence by considering $a_{n+1}-a_n$.

(c) Prove that the sequence $\{a_n\}$ converges.

■ 11.5 ADDITIONAL CONVERGENCE TESTS

In this section we shall develop some additional convergence tests for series with positive terms.

☐ **THE COMPARISON TEST**

The following result is a convergence test in its own right and will also serve as the foundation for other convergence tests that we will develop.

11.5.1 THEOREM (*The Comparison Test*). *Let $\sum a_k$ and $\sum b_k$ be series with nonnegative terms and suppose that*

$$a_1 \leq b_1,\ a_2 \leq b_2,\ a_3 \leq b_3,\ \ldots,\ a_k \leq b_k,\ \ldots$$

(a) *If the "bigger series" $\sum b_k$ converges, then the "smaller series" $\sum a_k$ also converges.*

(b) *If the "smaller series" $\sum a_k$ diverges, then the "bigger series" $\sum b_k$ also diverges.*

Proof (a). Suppose that the series Σb_k converges and its sum is B. Then for all n

$$b_1 + b_2 + \cdots + b_n \le \sum_{k=1}^{\infty} b_k = B$$

From our hypothesis it follows that

$$a_1 + a_2 + \cdots + a_n \le b_1 + b_2 + \cdots + b_n \le B$$

Thus, each partial sum of the series Σa_k is less than or equal to B, which implies that Σa_k converges by Theorem 11.4.6.

Proof (b). This part is really just an alternative phrasing of part (a). If Σa_k diverges, then Σb_k must diverge since convergence of Σb_k would imply convergence of Σa_k, contrary to the hypothesis. ∎

REMARK. As one would expect, it is not essential in Theorem 11.5.1 that the condition $a_k \le b_k$ hold for all k, as stated; the conclusions of the theorem remain true if this condition is eventually true.

☐ **THE RATIO TEST**

Since the comparison test requires a little ingenuity to use, we shall wait until the next section before applying it. For now, we shall use the comparison test to develop some other tests that are easier to apply.

11.5.2 THEOREM (*The Ratio Test*). *Let Σu_k be a series with positive terms and suppose that*

$$\rho = \lim_{k \to +\infty} \frac{u_{k+1}}{u_k}$$

(a) *If $\rho < 1$, the series converges.*
(b) *If $\rho > 1$ or $\rho = +\infty$, the series diverges.*
(c) *If $\rho = 1$, the series may converge or diverge, so that another test must be tried.*

Proof (a). The number ρ must be nonnegative since it is the limit of u_{k+1}/u_k, which is positive for all k. In this part of the proof we assume that $\rho < 1$, so that $0 \le \rho < 1$.

   We will prove convergence by showing that the terms of the given series are eventually less than the terms of a convergent geometric series. For this purpose, choose any real number r such that $0 < \rho < r < 1$. Since the limit of u_{k+1}/u_k is ρ, and $\rho < r$, the terms of the sequence $\{u_{k+1}/u_k\}$ must eventually be less than r. Thus, there is a positive integer K such that for $k \ge K$ we have

$$\frac{u_{k+1}}{u_k} < r \quad \text{or} \quad u_{k+1} < r u_k$$

This yields the inequalities

$$u_{K+1} < r u_K$$
$$u_{K+2} < r u_{K+1} < r^2 u_K$$
$$u_{K+3} < r u_{K+2} < r^3 u_K \tag{1}$$
$$u_{K+4} < r u_{K+3} < r^4 u_K$$
$$\vdots$$

But $0 < r < 1$, so

$$r u_K + r^2 u_K + r^3 u_K + \cdots$$

is a convergent geometric series. From the inequalities in (1) and the comparison test it follows that

$$u_{K+1} + u_{K+2} + u_{K+3} + \cdots$$

must also be a convergent series. Thus, $u_1 + u_2 + u_3 + \cdots + u_k + \cdots$ converges by Theorem 11.4.3(c).

Proof (b).   In this part we will prove divergence by showing that the limit of the general term is not zero. Since the limit of u_{k+1}/u_k is ρ and $\rho > 1$, the terms in the sequence $\{u_{k+1}/u_k\}$ must eventually be greater than 1. Thus, there is a positive integer K such that for $k \geq K$ we have

$$\frac{u_{k+1}}{u_k} > 1 \quad \text{or} \quad u_{k+1} > u_k$$

This yields the inequalities

$$u_{K+1} > u_K$$

$$u_{K+2} > u_{K+1} > u_K \tag{2}$$

$$u_{K+3} > u_{K+2} > u_K$$

$$u_{K+4} > u_{K+3} > u_K$$

$$\vdots$$

Since $u_K > 0$, it follows from the inequalities in (2) that $\lim\limits_{k \to +\infty} u_k \neq 0$, and thus $u_1 + u_2 + \cdots + u_k + \cdots$ diverges by part (a) of Theorem 11.4.1. The proof in the case where $\rho = +\infty$ is omitted.

Proof (c).   The divergent harmonic series and the convergent p-series with $p = 2$ both have $\rho = 1$ (verify), so the ratio test does not distinguish between convergence and divergence when $\rho = 1$.   ∎

Example 1   Use the ratio test to determine whether the following series converge or diverge.

$$\text{(a)} \ \sum_{k=1}^{\infty} \frac{1}{k!} \quad \text{(b)} \ \sum_{k=1}^{\infty} \frac{k}{2^k} \quad \text{(c)} \ \sum_{k=1}^{\infty} \frac{k^k}{k!} \quad \text{(d)} \ \sum_{k=1}^{\infty} \frac{(2k)!}{4^k}$$

Solution (a).   The series converges, since

$$\rho = \lim_{k \to +\infty} \frac{u_{k+1}}{u_k} = \lim_{k \to +\infty} \frac{1/(k+1)!}{1/k!} = \lim_{k \to +\infty} \frac{k!}{(k+1)!} = \lim_{k \to +\infty} \frac{1}{k+1} = 0 < 1$$

Solution (b).   The series converges, since

$$\rho = \lim_{k \to +\infty} \frac{u_{k+1}}{u_k} = \lim_{k \to +\infty} \frac{k+1}{2^{k+1}} \cdot \frac{2^k}{k} = \frac{1}{2} \lim_{k \to +\infty} \frac{k+1}{k} = \frac{1}{2} < 1$$

Solution (c).   The series diverges, since

$$\rho = \lim_{k \to +\infty} \frac{u_{k+1}}{u_k} = \lim_{k \to +\infty} \frac{(k+1)^{k+1}}{(k+1)!} \cdot \frac{k!}{k^k} = \lim_{k \to +\infty} \frac{(k+1)^k}{k^k}$$

See Theorem 7.5.15(b)

$$= \lim_{k \to +\infty} \left(1 + \frac{1}{k}\right)^k = e > 1$$

Solution (d).   The series diverges, since

$$\rho = \lim_{k \to +\infty} \frac{u_{k+1}}{u_k} = \lim_{k \to +\infty} \frac{[2(k+1)]!}{4^{k+1}} \cdot \frac{4^k}{(2k)!} = \lim_{k \to +\infty} \left(\frac{(2k+2)!}{(2k)!} \cdot \frac{1}{4} \right)$$

$$= \frac{1}{4} \lim_{k \to +\infty} (2k+2)(2k+1) = +\infty \quad \blacktriangleleft$$

Example 2   Determine whether the series

$$1 + \frac{1}{3} + \frac{1}{5} + \frac{1}{7} + \cdots + \frac{1}{2k-1} + \cdots$$

converges or diverges.

Solution.   The ratio test is of no help since

$$\rho = \lim_{k \to +\infty} \frac{u_{k+1}}{u_k} = \lim_{k \to +\infty} \frac{1}{2(k+1)-1} \cdot \frac{2k-1}{1} = \lim_{k \to +\infty} \frac{2k-1}{2k+1} = 1$$

However, the integral test proves that the series diverges since

$$\int_1^{+\infty} \frac{dx}{2x-1} = \lim_{l \to +\infty} \int_1^l \frac{dx}{2x-1} = \lim_{l \to +\infty} \frac{1}{2} \ln(2x-1) \Big]_1^l = +\infty \quad \blacktriangleleft$$

□ **THE ROOT TEST**

Sometimes the following result is easier to apply than the ratio test.

11.5.3   THEOREM (*The Root Test*).   *Let Σu_k be a series with positive terms and suppose that*

$$\rho = \lim_{k \to +\infty} \sqrt[k]{u_k} = \lim_{k \to +\infty} (u_k)^{1/k}$$

(a)   *If $\rho < 1$, the series converges.*
(b)   *If $\rho > 1$, or $\rho = +\infty$, the series diverges.*
(c)   *If $\rho = 1$, the series may converge or diverge, so that another test must be tried.*

Since the proof of the root test is similar to the proof of the ratio test, we shall omit it.

Example 3   Use the root test to determine whether the following series converge or diverge.

$$\text{(a)} \quad \sum_{k=2}^{\infty} \left(\frac{4k-5}{2k+1} \right)^k \qquad \text{(b)} \quad \sum_{k=1}^{\infty} \frac{1}{(\ln(k+1))^k}$$

Solution (a).   The series diverges, since

$$\rho = \lim_{k \to +\infty} (u_k)^{1/k} = \lim_{k \to +\infty} \frac{4k-5}{2k+1} = 2 > 1$$

Solution (b).   The series converges, since

$$\rho = \lim_{k \to +\infty} (u_k)^{1/k} = \lim_{k \to +\infty} \frac{1}{\ln(k+1)} = 0 < 1 \quad \blacktriangleleft$$

□ **COMMENTS ON NOTATION**

Until now we have written most of our infinite series with the summation index beginning at 1. If the summation index begins at some other integer, it is always possible to rewrite the series so that the summation index starts at 1. For example, the series

$$\sum_{k=0}^{\infty} \frac{2^k}{k!} = 1 + 2 + \frac{2^2}{2!} + \frac{2^3}{3!} + \cdots \tag{3}$$

can be written as

$$\sum_{k=1}^{\infty} \frac{2^{k-1}}{(k-1)!} = 1 + 2 + \frac{2^2}{2!} + \frac{2^3}{3!} + \cdots \tag{4}$$

However, the index of summation need not start at 1 to apply the convergence tests. For example, we can apply the ratio test to (3) without converting to the more complicated form (4). Doing so yields

$$\rho = \lim_{k \to +\infty} \frac{u_{k+1}}{u_k} = \lim_{k \to +\infty} \frac{2^{k+1}}{(k+1)!} \cdot \frac{k!}{2^k} = \lim_{k \to +\infty} \frac{2}{k+1} = 0$$

which shows that the series converges since $\rho < 1$.

▶ Exercise Set 11.5   [C] 39–42

In Exercises 1–6, apply the ratio test. According to the test, does the series converge, does the series diverge, or are the results inconclusive?

1. $\displaystyle\sum_{k=1}^{\infty} \frac{3^k}{k!}$.

2. $\displaystyle\sum_{k=1}^{\infty} \frac{4^k}{k^2}$.

3. $\displaystyle\sum_{k=2}^{\infty} \frac{1}{5k}$.

4. $\displaystyle\sum_{k=1}^{\infty} k\left(\frac{1}{2}\right)^k$.

5. $\displaystyle\sum_{k=1}^{\infty} \frac{k!}{k^3}$.

6. $\displaystyle\sum_{k=1}^{\infty} \frac{k}{k^2+1}$.

In Exercises 7–10, apply the root test. According to the test, does the series converge, does the series diverge, or are the results inconclusive?

7. $\displaystyle\sum_{k=1}^{\infty} \left(\frac{3k+2}{2k-1}\right)^k$.

8. $\displaystyle\sum_{k=1}^{\infty} \left(\frac{k}{100}\right)^k$.

9. $\displaystyle\sum_{k=1}^{\infty} \frac{k}{5^k}$.

10. $\displaystyle\sum_{k=1}^{\infty} (1 - e^{-k})^k$.

In Exercises 11–32, use any appropriate test to determine whether the series converges.

11. $\displaystyle\sum_{k=1}^{\infty} \frac{2^k}{k^3}$.

12. $\displaystyle\sum_{k=1}^{\infty} \frac{1}{k^2}$.

13. $\displaystyle\sum_{k=0}^{\infty} \frac{7^k}{k!}$.

14. $\displaystyle\sum_{k=1}^{\infty} \frac{1}{2k+1}$.

15. $\displaystyle\sum_{k=1}^{\infty} \frac{k^2}{5^k}$.

16. $\displaystyle\sum_{k=1}^{\infty} \frac{k! \, 10^k}{3^k}$.

17. $\displaystyle\sum_{k=1}^{\infty} k^{50} e^{-k}$.

18. $\displaystyle\sum_{k=1}^{\infty} \frac{k^2}{k^3+1}$.

19. $\displaystyle\sum_{k=1}^{\infty} k\left(\frac{2}{3}\right)^k$.

20. $\displaystyle\sum_{k=1}^{\infty} k^k$.

21. $\displaystyle\sum_{k=2}^{\infty} \frac{1}{k \ln k}$.

22. $\displaystyle\sum_{k=1}^{\infty} \frac{2^k}{k^3+1}$.

23. $\displaystyle\sum_{k=1}^{\infty} \left(\frac{4}{7k-1}\right)^k$.

24. $\displaystyle\sum_{k=1}^{\infty} \frac{(k!)^2 2^k}{(2k+2)!}$.

25. $\displaystyle\sum_{k=0}^{\infty} \frac{(k!)^2}{(2k)!}$.

26. $\displaystyle\sum_{k=1}^{\infty} \frac{1}{k^2+25}$.

27. $\displaystyle\sum_{k=1}^{\infty} \frac{1}{1+\sqrt{k}}$.

28. $\displaystyle\sum_{k=1}^{\infty} \frac{k!}{k^k}$.

29. $\displaystyle\sum_{k=1}^{\infty} \frac{\ln k}{e^k}$.

30. $\displaystyle\sum_{k=1}^{\infty} \frac{k!}{e^{k^2}}$.

31. $\displaystyle\sum_{k=0}^{\infty} \frac{(k+4)!}{4! \, k! \, 4^k}$.

32. $\displaystyle\sum_{k=1}^{\infty} \left(\frac{k}{k+1}\right)^{k^2}$.

In Exercises 33–35, show that the series converges.

33. $1 + \dfrac{1 \cdot 2}{1 \cdot 3} + \dfrac{1 \cdot 2 \cdot 3}{1 \cdot 3 \cdot 5} + \dfrac{1 \cdot 2 \cdot 3 \cdot 4}{1 \cdot 3 \cdot 5 \cdot 7} + \cdots$.

34. $1 + \dfrac{1 \cdot 3}{3!} + \dfrac{1 \cdot 3 \cdot 5}{5!} + \dfrac{1 \cdot 3 \cdot 5 \cdot 7}{7!} + \cdots$.

35. $\dfrac{2!}{1} + \dfrac{3!}{1 \cdot 4} + \dfrac{4!}{1 \cdot 4 \cdot 7} + \dfrac{5!}{1 \cdot 4 \cdot 7 \cdot 10} + \cdots$.

36. For which positive values of α does $\sum_{k=1}^{\infty} \alpha^k / k^\alpha$ converge?

37. (a) Show: $\displaystyle\lim_{k \to +\infty} (\ln k)^{1/k} = 1$. [*Hint:* Let $y = (\ln x)^{1/x}$

and find $\displaystyle\lim_{x \to +\infty} \ln y$.]

(b) Use the result in part (a) and the root test to show that $\sum_{k=1}^{\infty} (\ln k)/3^k$ converges.

(c) Show that the series converges using the ratio test.

38. If the sum S of a convergent series $\sum_{k=1}^{\infty} u_k$ of positive terms is approximated by the nth partial sum s_n, then the error in the approximation is defined as

$$S - s_n = \sum_{k=n+1}^{\infty} u_k = u_{n+1} + u_{n+2} + u_{n+3} + \cdots$$

Let $r_k = u_{k+1}/u_k$.

(a) Prove: If r_k is *decreasing* for all $k \geq n + 1$ and if $r_{n+1} < 1$, then

$$S - s_n < \frac{u_{n+1}}{1 - r_{n+1}}$$

[*Hint:* Show that the sum of the series $\sum_{k=n+1}^{\infty} u_k$ is less than the sum of the convergent geometric series

$$u_{n+1} + r_{n+1}u_{n+1} + r_{n+1}^2 u_{n+1} + \cdots]$$

(b) Prove: If r_k is *increasing* for all $k \geq n + 1$ and if $\lim\limits_{k \to +\infty} r_k = \rho < 1$, then

$$S - s_n < \frac{u_{n+1}}{1 - \rho}$$

[*Hint:* Show that the sum of the series $\sum_{k=n+1}^{\infty} u_k$ is less than the sum of the convergent geometric series

$$u_{n+1} + \rho u_{n+1} + \rho^2 u_{n+1} + \cdots]$$

In Exercises 39–42, use the results in Exercise 38. For each exercise do the following:

(a) Compute the stated partial sum and find an upper bound on the error in approximating S by the partial sum. Express your results to five decimal places.

(b) Find a value of n to ensure that s_n will approximate S with an error that is less than 10^{-5}.

39. $S = \sum\limits_{k=1}^{\infty} \dfrac{1}{k!}$; s_5.

40. $S = \sum\limits_{k=1}^{\infty} \dfrac{k}{3^k}$; s_8.

41. $S = \sum\limits_{k=1}^{\infty} \dfrac{1}{k2^k}$; s_7.

42. $S = \sum\limits_{k=1}^{\infty} \dfrac{1}{\sqrt{k}3^k}$; s_4.

11.6   THE LIMIT COMPARISON TEST

In this section we shall discuss procedures for applying the comparison test and we shall state an alternative version of this test that is easier to work with. Before starting, we remind the reader that the comparison test applies only to series with positive terms.

☐ **SOME USEFUL INFORMAL PRINCIPLES**

The convergence tests that we will discuss in this section require a little more ingenuity than those in the preceding sections in that they will require us to make an initial *guess* at whether the series in question is likely to converge or diverge. The subsequent steps in the test will depend on that initial guess.

To help with the guessing process in the first step we have formulated some principles that sometimes *suggest* whether a series is likely to converge or diverge. We have called these ''informal principles'' because they are not intended as formal theorems. In fact, we shall not guarantee that they *always* work. However, they work often enough to be useful as a starting point for the comparison test.

11.6.1   INFORMAL PRINCIPLE.   *Constant terms in the denominator of u_k can usually be deleted without affecting the convergence or divergence of the series.*

Example 1   Use the above principle to help guess whether the following series converge or diverge.

(a) $\sum\limits_{k=1}^{\infty} \dfrac{1}{2^k + 1}$ (b) $\sum\limits_{k=5}^{\infty} \dfrac{1}{\sqrt{k} - 2}$ (c) $\sum\limits_{k=1}^{\infty} \dfrac{1}{(k + \frac{1}{2})^3}$

Solution (a).   Deleting the constant 1 suggests that

$$\sum_{k=1}^{\infty} \frac{1}{2^k + 1} \quad \text{behaves like} \quad \sum_{k=1}^{\infty} \frac{1}{2^k}$$

The modified series is a convergent geometric series, so the given series is likely to converge.

Solution (b).   Deleting the -2 suggests that

$$\sum_{k=5}^{\infty} \frac{1}{\sqrt{k-2}} \quad \text{behaves like} \quad \sum_{k=5}^{\infty} \frac{1}{\sqrt{k}}$$

The modified series is a portion of a divergent p-series $(p = \frac{1}{2})$, so the given series is likely to diverge.

Solution (c).   Deleting the $\frac{1}{2}$ suggests that

$$\sum_{k=1}^{\infty} \frac{1}{(k + \frac{1}{2})^3} \quad \text{behaves like} \quad \sum_{k=1}^{\infty} \frac{1}{k^3}$$

The modified series is a convergent p-series $(p = 3)$, so the given series is likely to converge.   ◀

11.6.2   INFORMAL PRINCIPLE.   *If a polynomial in k appears as a factor in the numerator or denominator of u_k, all but the highest power of k in the polynomial may usually be deleted without affecting the convergence or divergence of the series.*

Example 2   Use the above principle to help guess whether the following series converge or diverge.

$$\text{(a)} \quad \sum_{k=1}^{\infty} \frac{1}{\sqrt{k^3 + 2k}} \qquad \text{(b)} \quad \sum_{k=1}^{\infty} \frac{6k^4 - 2k^3 + 1}{k^5 + k^2 - 2k}$$

Solution (a).   Deleting the term $2k$ suggests that

$$\sum_{k=1}^{\infty} \frac{1}{\sqrt{k^3 + 2k}} \quad \text{behaves like} \quad \sum_{k=1}^{\infty} \frac{1}{\sqrt{k^3}} = \sum_{k=1}^{\infty} \frac{1}{k^{3/2}}$$

Since the modified series is a convergent p-series $(p = \frac{3}{2})$, the given series is likely to converge.

Solution (b).   Deleting all but the highest powers of k in the numerator and also in the denominator suggests that

$$\sum_{k=1}^{\infty} \frac{6k^4 - 2k^3 + 1}{k^5 + k^2 - 2k} \quad \text{behaves like} \quad \sum_{k=1}^{\infty} \frac{6k^4}{k^5} = \sum_{k=1}^{\infty} 6 \left(\frac{1}{k} \right)$$

Since each term in the modified series is a constant times the corresponding term in the divergent harmonic series, the given series is likely to diverge.   ◀

☐ **THE LIMIT COMPARISON TEST**

The following result, which is proved at the end of the section, can be used to establish convergence or divergence by examining the limit of the ratio of the general term of the series in question with the general term of a series whose convergence properties are known.

11.6.3   THEOREM (***The Limit Comparison Test***).   *Let Σa_k and Σb_k be series with positive terms and suppose that*

$$\rho = \lim_{k \to +\infty} \frac{a_k}{b_k}$$

If ρ is finite and $\rho > 0$, then the series both converge or both diverge.

The cases where $\rho = 0$ or $\rho = +\infty$ are discussed in the exercises (Exercise 44).

Example 3   Use the limit comparison test to determine whether the following series converge or diverge.

(a) $\displaystyle\sum_{k=1}^{\infty} \frac{1}{2k^2 - k}$ (b) $\displaystyle\sum_{k=1}^{\infty} \frac{1}{k - \frac{1}{4}}$ (c) $\displaystyle\sum_{k=1}^{\infty} \frac{3k^3 - 2k^2 + 4}{k^5 - k^3 + 2}$

Solution (a).   Using Principle 11.6.2, the given series behaves like the series

$$\sum_{k=1}^{\infty} \frac{1}{2k^2} = \frac{1}{2} \sum_{k=1}^{\infty} \frac{1}{k^2} \tag{1}$$

which is a constant times a convergent p-series. Thus, the given series is likely to converge. To prove this, we apply Theorem 11.6.3 with

$$a_k = \frac{1}{2k^2 - k} \quad \text{and} \quad b_k = \frac{1}{2k^2}$$

We obtain

$$\rho = \lim_{k \to +\infty} \frac{a_k}{b_k} = \lim_{k \to +\infty} \frac{2k^2}{2k^2 - k} = \lim_{k \to +\infty} \frac{2}{2 - 1/k} = 1$$

Since ρ is finite and positive, it follows from Theorem 11.6.3 that the given series converges, since (1) converges.

Solution (b).   Using Principle 11.6.1, the series behaves like the divergent harmonic series

$$\sum_{k=1}^{\infty} \frac{1}{k} \tag{2}$$

Thus, the given series is likely to diverge. To prove this, we apply Theorem 11.6.3 with

$$a_k = \frac{1}{k - \frac{1}{4}} \quad \text{and} \quad b_k = \frac{1}{k}$$

We obtain

$$\rho = \lim_{k \to +\infty} \frac{a_k}{b_k} = \lim_{k \to +\infty} \frac{k}{k - \frac{1}{4}} = \lim_{k \to +\infty} \frac{1}{1 - \dfrac{1}{4k}} = 1$$

Since ρ is finite and positive, it follows from Theorem 11.6.3 that the given series diverges, since (2) diverges.

Solution (c).   From Principle 11.6.2, the series behaves like

$$\sum_{k=1}^{\infty} \frac{3k^3}{k^5} = \sum_{k=1}^{\infty} \frac{3}{k^2} \tag{3}$$

which converges since it is a constant times a convergent p-series. Thus, the given series is likely to converge. To prove this, we apply the limit comparison test to series (3) and the given series. We obtain

$$\rho = \lim_{k \to +\infty} \frac{\dfrac{3k^3 - 2k^2 + 4}{k^5 - k^3 + 2}}{\dfrac{3}{k^2}} = \lim_{k \to +\infty} \frac{3k^5 - 2k^4 + 4k^2}{3k^5 - 3k^3 + 6} = 1$$

Since ρ is finite and nonzero, it follows from Theorem 11.6.3 that the given series converges, since (3) converges.   ◀

☐ **THE COMPARISON TEST**

We shall now discuss some techniques for applying the comparison test (Theorem 11.5.1). However, as a practical matter, you should try the limit comparison test before attempting the comparison test, since it is usually easier to apply; the comparison test should be viewed as a last resort.

There are two basic steps required to apply the comparison test to a series Σu_k of positive terms:

- Guess at whether the series Σu_k converges or diverges.
- Find a series that proves the guess to be correct. Thus, if the guess is divergence, we must find a divergent series whose terms are "smaller" than the corresponding terms of Σu_k, and if the guess is convergence, we must find a convergent series whose terms are "bigger" than the corresponding terms of Σu_k.

Example 4   Use the comparison test to determine whether the following series converge or diverge.

(a) $\displaystyle\sum_{k=1}^{\infty} \frac{1}{k - \frac{1}{4}}$ (b) $\displaystyle\sum_{k=1}^{\infty} \frac{1}{\sqrt[3]{k} + 5}$ (c) $\displaystyle\sum_{k=1}^{\infty} \frac{1}{2k^2 + k}$

Solution (a).   In part (b) of Example 3 we used the limit comparison test to show that this series diverges. To reach the same conclusion by the comparison test, we note (as in Example 3) that the series behaves like the divergent harmonic series, and hence is likely to diverge. Thus, our goal is to find a divergent series that is "smaller" than the given series. We can do this by dropping the constant $-\frac{1}{4}$ in the denominator, thereby *decreasing* the size of the general term:

$$\frac{1}{k - \frac{1}{4}} > \frac{1}{k} \quad \text{for } k = 1, 2, \ldots$$

Thus, the given series diverges by the comparison test, since $\displaystyle\sum_{k=1}^{\infty} \frac{1}{k}$ diverges.

Solution (b).   Using Principle 11.6.1, the series behaves like the divergent *p*-series

$$\sum_{k=1}^{\infty} \frac{1}{\sqrt[3]{k}} \tag{4}$$

$(p = \frac{1}{3})$, and hence is likely to diverge. As in the preceding example, our goal is to find a divergent series that is "smaller" than the given series; but here we cannot achieve this by dropping the constant in the denominator, since that would decrease the denominator, thereby increasing the size of the general term, which is contrary to our objective of finding a smaller series. Instead, we will increase the denominator by replacing the constant 5 with a quantity that is eventually greater than 5. A convenient choice for this quantity is $\sqrt[3]{k}$, which is greater than 5 for $k > 125$. Thus,

$$\frac{1}{\sqrt[3]{k} + 5} \geq \frac{1}{\sqrt[3]{k} + \sqrt[3]{k}} = \frac{1}{2\sqrt[3]{k}} \quad \text{for } k = 125, 126, \ldots$$

But the series $\displaystyle\sum_{k=125}^{\infty} \frac{1}{2\sqrt[3]{k}}$ diverges, since (4) diverges (why?), and hence the given series diverges by the comparison test.

Solution (c).   Using Principle 11.6.2, the series behaves like the convergent series

$$\sum_{k=1}^{\infty} \frac{1}{2k^2} = \frac{1}{2} \sum_{k=1}^{\infty} \frac{1}{k^2} \tag{5}$$

and hence is likely to converge. Thus, our goal is to find a "bigger" convergent series. We can do this by dropping the k from the denominator, since that decreases the denominator and *increases* the size of the general term:

$$\frac{1}{2k^2 + k} < \frac{1}{2k^2} \quad \text{for } k = 1, 2, \ldots$$

Thus, the given series converges, since (5) converges. ◄

■ PROOF OF THE LIMIT COMPARISON TEST

We conclude this section with a proof of the limit comparison test.

Proof of Theorem 11.6.3. We need only show that Σb_k converges when Σa_k converges and that Σb_k diverges when Σa_k diverges, since the remaining cases are logical implications of these (why?). The idea of the proof is to apply the comparison test to Σa_k and suitable multiples of Σb_k. For this purpose let ϵ be any positive number. Since

$$\rho = \lim_{k \to +\infty} \frac{a_k}{b_k}$$

it follows that eventually the terms in the sequence $\{a_k/b_k\}$ must be within ϵ units of ρ; that is, there is a positive integer K such that for $k \geq K$ we have

$$\rho - \epsilon < \frac{a_k}{b_k} < \rho + \epsilon$$

In particular, if we take $\epsilon = \rho/2$, then for $k \geq K$ we have

$$\frac{1}{2}\rho < \frac{a_k}{b_k} < \frac{3}{2}\rho \quad \text{or} \quad \frac{1}{2}\rho b_k < a_k < \frac{3}{2}\rho b_k$$

Thus, by the comparison test we can conclude that

$$\sum_{k=K}^{\infty} \frac{1}{2}\rho b_k \quad \text{converges if} \quad \sum_{k=K}^{\infty} a_k \quad \text{converges} \tag{6}$$

$$\sum_{k=K}^{\infty} \frac{3}{2}\rho b_k \quad \text{diverges if} \quad \sum_{k=K}^{\infty} a_k \quad \text{diverges} \tag{7}$$

But the convergence or divergence of a series is not affected by deleting finitely many terms or by multiplying the general term by a nonzero constant, so (6) and (7) imply that

$$\sum_{k=1}^{\infty} b_k \quad \text{converges if} \quad \sum_{k=1}^{\infty} a_k \quad \text{converges}$$

$$\sum_{k=1}^{\infty} b_k \quad \text{diverges if} \quad \sum_{k=1}^{\infty} a_k \quad \text{diverges} \quad ■$$

► Exercise Set 11.6

In Exercises 1–6, use the limit comparison test to determine whether the series converges or diverges.

1. $\displaystyle\sum_{k=1}^{\infty} \frac{4k^2 - 2k + 6}{8k^7 + k - 8}$.

2. $\displaystyle\sum_{k=1}^{\infty} \frac{1}{9k + 6}$.

3. $\displaystyle\sum_{k=1}^{\infty} \frac{5}{3^k + 1}$.

4. $\displaystyle\sum_{k=1}^{\infty} \frac{k(k + 3)}{(k + 1)(k + 2)(k + 5)}$.

5. $\displaystyle\sum_{k=1}^{\infty} \frac{1}{\sqrt[3]{8k^2 - 3k}}$.

6. $\displaystyle\sum_{k=1}^{\infty} \frac{1}{(2k + 3)^{17}}$.

In Exercises 7–12, prove that the series converges by the comparison test.

7. $\displaystyle\sum_{k=1}^{\infty} \frac{1}{3^k + 5}$.

8. $\displaystyle\sum_{k=1}^{\infty} \frac{2}{k^4 + k}$.

9. $\displaystyle\sum_{k=1}^{\infty} \frac{1}{5k^2 - k}$.

10. $\displaystyle\sum_{k=1}^{\infty} \frac{k}{8k^3 + 2k^2 - 1}$.

11. $\displaystyle\sum_{k=1}^{\infty} \frac{2^k - 1}{3^k + 2k}$.

12. $\displaystyle\sum_{k=1}^{\infty} \frac{5 \sin^2 k}{k!}$.

In Exercises 13–18, prove that the series diverges by the comparison test.

13. $\displaystyle\sum_{k=1}^{\infty} \frac{3}{k - \frac{1}{4}}$.

14. $\displaystyle\sum_{k=1}^{\infty} \frac{1}{\sqrt{k + 8}}$.

15. $\displaystyle\sum_{k=1}^{\infty} \frac{9}{\sqrt{k} + 1}$.

16. $\displaystyle\sum_{k=2}^{\infty} \frac{k + 1}{k^2 - k}$.

17. $\displaystyle\sum_{k=1}^{\infty} \frac{k^{4/3}}{8k^2 + 5k + 1}$.

18. $\displaystyle\sum_{k=1}^{\infty} \frac{k^{-1/2}}{2 + \sin^2 k}$.

In Exercises 19–34, use any method to determine whether the series converges or diverges. In some cases, you may have to use tests from earlier sections.

19. $\displaystyle\sum_{k=1}^{\infty} \frac{1}{k^3 + 2k + 1}$.

20. $\displaystyle\sum_{k=1}^{\infty} \frac{1}{(3 + k)^{2/5}}$.

21. $\displaystyle\sum_{k=1}^{\infty} \frac{1}{9k - 2}$.

22. $\displaystyle\sum_{k=1}^{\infty} \frac{\ln k}{k}$.

23. $\displaystyle\sum_{k=1}^{\infty} \frac{\sqrt{k}}{k^3 + 1}$.

24. $\displaystyle\sum_{k=1}^{\infty} \frac{4}{2 + 3^k k}$.

25. $\displaystyle\sum_{k=1}^{\infty} \frac{1}{\sqrt{k(k + 1)}}$.

26. $\displaystyle\sum_{k=1}^{\infty} \frac{2 + (-1)^k}{5^k}$.

27. $\displaystyle\sum_{k=1}^{\infty} \frac{2 + \sqrt{k}}{(k + 1)^3 - 1}$.

28. $\displaystyle\sum_{k=1}^{\infty} \frac{4 + |\cos k|}{k^3}$.

29. $\displaystyle\sum_{k=1}^{\infty} \frac{1}{4 + 2^{-k}}$.

30. $\displaystyle\sum_{k=1}^{\infty} \frac{\sqrt{k} \ln k}{k^3 + 1}$.

31. $\displaystyle\sum_{k=1}^{\infty} \frac{\tan^{-1} k}{k^2}$.

32. $\displaystyle\sum_{k=1}^{\infty} \frac{5^k + k}{k! + 3}$.

33. $\displaystyle\sum_{k=1}^{\infty} \frac{\ln k}{k\sqrt{k}}$.

34. $\displaystyle\sum_{k=1}^{\infty} \frac{\cos (1/k)}{k^2}$.

35. Use the limit comparison test to show that the series
$$\sum_{k=1}^{\infty} (1 - \cos (1/k)) \text{ converges.}$$
$\left[\text{\textit{Hint:} Compare with the series } \displaystyle\sum_{k=1}^{\infty} 1/k^2.\right]$

36. Use the limit comparison test to show that the series
$$\sum_{k=1}^{\infty} \sin (\pi/k) \text{ diverges.}$$
$\left[\text{\textit{Hint:} Compare with the series } \displaystyle\sum_{k=1}^{\infty} \pi/k.\right]$

37. Use the comparison test to determine whether $\displaystyle\sum_{k=1}^{\infty} \frac{\ln k}{k^2}$ converges or diverges. [*Hint:* $\ln x < \sqrt{x}$ by Exercise 54 of Section 7.2.]

38. Determine whether $\displaystyle\sum_{k=2}^{\infty} \frac{1}{(\ln k)^2}$ converges or diverges. [*Hint:* See the hint in Exercise 37.]

39. Let a, b, and p be positive constants. For which values of p does the series $\displaystyle\sum_{k=1}^{\infty} \frac{1}{(a + bk)^p}$ converge?

40. (a) Show that $k^k \geq k!$ and use this result to prove that the series $\displaystyle\sum_{k=1}^{\infty} k^{-k}$ converges by the comparison test.

  (b) Prove convergence using the root test.

41. Use the limit comparison test to investigate convergence of $\displaystyle\sum_{k=1}^{\infty} \frac{(k + 1)^2}{(k + 2)!}$.

42. Use the limit comparison test to investigate convergence of the series $1 + \frac{1}{3} + \frac{1}{5} + \frac{1}{7} + \cdots$.

43. Prove that $\sum_{k=1}^{\infty} 1/k!$ converges by comparison with a suitable geometric series.

44. Let Σa_k and Σb_k be series with positive terms. Prove:

  (a) If $\displaystyle\lim_{k \to +\infty} (a_k/b_k) = 0$ and Σb_k converges, then Σa_k converges.

  (b) If $\displaystyle\lim_{k \to +\infty} (a_k/b_k) = +\infty$ and Σb_k diverges, then Σa_k diverges.

■ 11.7 ALTERNATING SERIES; CONDITIONAL CONVERGENCE

So far our emphasis has been on series with positive terms. In this section we shall discuss series containing both positive and negative terms.

□ **ALTERNATING SERIES**

Of special importance are series whose terms are alternately positive and negative. These are called ***alternating series***. Some examples are

$$1 - 1 + 1 - 1 + \cdots + (-1)^{k+1} + \cdots$$

$$1 - \frac{1}{2} + \frac{1}{3} - \frac{1}{4} + \cdots + (-1)^{k+1}\frac{1}{k} + \cdots$$

$$-1 + \frac{1}{2!} - \frac{1}{3!} + \frac{1}{4!} - \cdots + (-1)^k\frac{1}{k!} + \cdots$$

In general, an alternating series has one of the following two forms:

$$\sum_{k=1}^{\infty} (-1)^{k+1} a_k = a_1 - a_2 + a_3 - a_4 + \cdots \tag{1}$$

$$\sum_{k=1}^{\infty} (-1)^k a_k = -a_1 + a_2 - a_3 + a_4 - \cdots \tag{2}$$

where the a_k's are assumed to be positive in both cases.

The following theorem is the key result on convergence of alternating series.

11.7.1 THEOREM (*Alternating Series Test*). *An alternating series of either form* (1) *or form* (2) *converges if the following two conditions are satisfied:*

(*a*) $a_1 > a_2 > a_3 > \cdots > a_k > \cdots$

(*b*) $\displaystyle\lim_{k \to +\infty} a_k = 0$

Proof. We will consider only alternating series of form (1). The idea of the proof is to show that if conditions (*a*) and (*b*) hold, then the sequences of even-numbered and odd-numbered partial sums converge to a common limit S. It will then follow from Theorem 11.1.4 that the entire sequence of partial sums converges to S.

Figure 11.7.1 shows how successive partial sums satisfying conditions (*a*) and (*b*) appear when plotted on a horizontal axis. The even-numbered partial sums

$$s_2, s_4, s_6, s_8, \ldots, s_{2n}, \ldots$$

form an increasing sequence bounded above by a_1, and the odd-numbered partial sums

$$s_1, s_3, s_5, \ldots, s_{2n-1}, \ldots$$

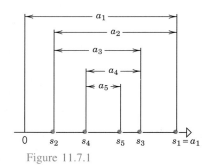

Figure 11.7.1

form a decreasing sequence bounded below by 0. Thus, by Theorems 11.2.2 and 11.2.3, the even-numbered partial sums converge to some limit S_E and the odd-numbered partial sums converge to some limit S_O. To complete the proof we must show that $S_E = S_O$. But the $(2n)$-th term in the series is $-a_{2n}$, so that $s_{2n} - s_{2n-1} = -a_{2n}$, which can be written as

$$s_{2n-1} = s_{2n} + a_{2n}$$

However, $2n \to +\infty$ and $2n - 1 \to +\infty$ as $n \to +\infty$, so that

$$S_O = \lim_{n \to +\infty} s_{2n-1} = \lim_{n \to +\infty} (s_{2n} + a_{2n}) = S_E + 0 = S_E$$

which completes the proof. ∎

REMARK.   As might be expected, it is not essential for condition (*a*) in the alternating series test to hold for all terms; an alternating series will converge if condition (*b*) is true and condition (*a*) holds eventually.

Example 1   Use the alternating series test to show that the following series converge.

(a) $\displaystyle\sum_{k=1}^{\infty} (-1)^{k+1} \frac{1}{k}$ (b) $\displaystyle\sum_{k=1}^{\infty} (-1)^{k+1} \frac{k+3}{k(k+1)}$

Solution (*a*).   The two conditions in the alternating series test are satisfied since

$$a_k = \frac{1}{k} > \frac{1}{k+1} = a_{k+1} \quad \text{and} \quad \lim_{k \to +\infty} a_k = \lim_{k \to +\infty} \frac{1}{k} = 0$$

Solution (*b*).   The two conditions in the alternating series test are satisfied, since

$$\frac{a_{k+1}}{a_k} = \frac{k+4}{(k+1)(k+2)} \cdot \frac{k(k+1)}{k+3} = \frac{k^2 + 4k}{k^2 + 5k + 6} = \frac{k^2 + 4k}{(k^2 + 4k) + (k+6)} < 1$$

so

$$a_k > a_{k+1}$$

and

$$\lim_{k \to +\infty} a_k = \lim_{k \to +\infty} \frac{k+3}{k(k+1)} = \lim_{k \to +\infty} \frac{\dfrac{1}{k} + \dfrac{3}{k^2}}{1 + \dfrac{1}{k}} = 0 \quad \blacktriangleleft$$

REMARK.   The series in part (a) of the preceding example is called the ***alternating harmonic series***. It is important to keep in mind that this series converges, whereas the harmonic series diverges.

REMARK.   If an alternating series violates condition (*b*) of the alternating series test, then the series must diverge by the divergence test (11.4.1). However, if condition (*b*) is satisfied, but condition (*a*) is not, the series can either converge or diverge.*

□ APPROXIMATING SUMS
OF ALTERNATING SERIES

The following theorem is concerned with the error that results when the sum of an alternating series is approximated by a partial sum.

11.7.2   THEOREM.   *If an alternating series satisfies the conditions of the alternating series test, and if the sum S of the series is approximated by the nth partial sum s_n, thereby resulting in an error of $S - s_n$, then*

$$|S - s_n| < a_{n+1}$$

Moreover, the sign of the error is the same as that of the coefficient of a_{n+1} in the series.

---

*The interested reader will find some nice examples in an article by R. Lariviere, "On a Convergence Test for Alternating Series," *Mathematics Magazine*, Vol. 29, 1956, p. 88.

550

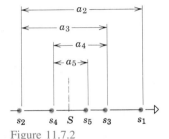

Figure 11.7.2

Proof. We shall prove the theorem for series of form (1). Referring to Figure 11.7.2 and keeping in mind our observation in the proof of Theorem 11.7.1 that the odd-numbered partial sums form a decreasing sequence converging to S and the even-numbered partial sums form an increasing sequence converging to S, we see that successive partial sums oscillate from one side of S to the other in smaller and smaller steps with the odd-numbered partial sums being greater than S and the even-numbered partial sums being less than S. Thus, the sum S falls between any two successive partial sums; that is, for any positive integer n

$$s_n < S < s_{n+1} \quad \text{or} \quad s_{n+1} < S < s_n$$

depending on whether n is even or odd. In either case,

$$|S - s_n| < |s_{n+1} - s_n| \tag{3}$$

But $s_{n+1} - s_n = \pm a_{n+1}$ (the sign depending on whether n is even or odd), so (3) yields $|S - s_n| < a_{n+1}$, which completes the proof. ■

REMARK. In words, the preceding theorem states that for series satisfying the conditions of the alternating series test, the magnitude of the error that results from approximating the sum of the series by the nth partial sum is less than the magnitude of the first term of the series *after* the partial sum.

Example 2 As shown in Example 1, the alternating harmonic series

$$1 - \frac{1}{2} + \frac{1}{3} - \frac{1}{4} + \cdots + (-1)^{k+1}\frac{1}{k} + \cdots$$

satisfies the conditions of the alternating series test; hence, the series has a sum S, which we know must lie between any two successive partial sums. In particular, it must lie between

$$s_7 = 1 - \frac{1}{2} + \frac{1}{3} - \frac{1}{4} + \frac{1}{5} - \frac{1}{6} + \frac{1}{7} = \frac{319}{420}$$

and

$$s_8 = 1 - \frac{1}{2} + \frac{1}{3} - \frac{1}{4} + \frac{1}{5} - \frac{1}{6} + \frac{1}{7} - \frac{1}{8} = \frac{533}{840}$$

so

$$\frac{533}{840} < S < \frac{319}{420} \tag{4}$$

Later in this chapter we shall show that the sum S of the alternating harmonic series is $\ln 2$. If we accept this to be so for now, it follows from (4) that

$$\frac{533}{840} < \ln 2 < \frac{319}{420}$$

or with the help of a calculator,

$$0.6345 < \ln 2 < 0.7596$$

The value of $\ln 2$, rounded to four decimal places, is 0.6931, which is consistent with these inequalities. It follows from Theorem 11.7.2 that

$$\left|\ln 2 - s_7\right| = \left|\ln 2 - \frac{319}{420}\right| < a_8 = \frac{1}{8}$$

and

$$\left|\ln 2 - s_8\right| = \left|\ln 2 - \frac{533}{840}\right| < a_9 = \frac{1}{9} \quad \blacktriangleleft$$

□ **ABSOLUTE AND CONDITIONAL CONVERGENCE**

The series

$$1 - \frac{1}{2} - \frac{1}{2^2} + \frac{1}{2^3} + \frac{1}{2^4} - \frac{1}{2^5} - \frac{1}{2^6} + \cdots$$

does not fit in any of the categories studied so far—it has mixed signs, but is not alternating. We shall now develop some convergence tests that can be applied to such series.

11.7.3 DEFINITION. A series

$$\sum_{k=1}^{\infty} u_k = u_1 + u_2 + \cdots + u_k + \cdots$$

is said to **converge absolutely** if the series of absolute values

$$\sum_{k=1}^{\infty} |u_k| = |u_1| + |u_2| + \cdots + |u_k| + \cdots$$

converges.

Example 3 The series

$$1 - \frac{1}{2} - \frac{1}{2^2} + \frac{1}{2^3} + \frac{1}{2^4} - \frac{1}{2^5} - \frac{1}{2^6} + \cdots$$

converges absolutely since the series of absolute values

$$1 + \frac{1}{2} + \frac{1}{2^2} + \frac{1}{2^3} + \frac{1}{2^4} + \frac{1}{2^5} + \frac{1}{2^6} + \cdots$$

is a convergent geometric series. On the other hand, the alternating harmonic series

$$1 - \frac{1}{2} + \frac{1}{3} - \frac{1}{4} + \frac{1}{5} - \cdots$$

does not converge absolutely since the series of absolute values

$$1 + \frac{1}{2} + \frac{1}{3} + \frac{1}{4} + \frac{1}{5} + \cdots$$

diverges. ◄

Absolute convergence is of importance because of the following theorem.

11.7.4 THEOREM. *If the series*

$$\sum_{k=1}^{\infty} |u_k| = |u_1| + |u_2| + \cdots + |u_k| + \cdots$$

converges, then so does the series

$$\sum_{k=1}^{\infty} u_k = u_1 + u_2 + \cdots + u_k + \cdots$$

In other words, if a series converges absolutely, then it converges.

Proof. Our proof is based on a trick. We shall show that the series

$$\sum_{k=1}^{\infty} (u_k + |u_k|) \tag{5}$$

converges. Since $\Sigma|u_k|$ is assumed to converge, it will then follow from Theorem 11.4.3(a) that Σu_k converges, since

$$\sum_{k=1}^{\infty} u_k = \sum_{k=1}^{\infty} [(u_k + |u_k|) - |u_k|]$$

For all k, the value of $u_k + |u_k|$ is either 0 or $2|u_k|$, depending on whether u_k is negative or not. Thus, for all values of k

$$0 \le u_k + |u_k| \le 2|u_k| \tag{6}$$

But $\Sigma 2|u_k|$ is a convergent series since it is a constant times the convergent series $\Sigma|u_k|$. Thus, from (6), series (5) converges by the comparison test. ∎

Example 4 In Example 3 we showed that

$$1 - \frac{1}{2} - \frac{1}{2^2} + \frac{1}{2^3} + \frac{1}{2^4} - \frac{1}{2^5} - \frac{1}{2^6} + \cdots$$

converges absolutely. It follows from Theorem 11.7.4 that the series converges. ◀

Example 5 Show that the series $\displaystyle\sum_{k=1}^{\infty} \frac{\cos k}{k^2}$ converges.

Solution. Since $|\cos k| \le 1$ for all k,

$$\left|\frac{\cos k}{k^2}\right| \le \frac{1}{k^2}$$

Thus,

$$\sum_{k=1}^{\infty} \left|\frac{\cos k}{k^2}\right|$$

converges by the comparison test, and consequently

$$\sum_{k=1}^{\infty} \frac{\cos k}{k^2}$$

converges. ◀

If $\Sigma|u_k|$ *diverges*, no conclusion can be drawn about the convergence or divergence of Σu_k. For example, consider the two series

$$1 - \frac{1}{2} + \frac{1}{3} - \frac{1}{4} + \cdots + (-1)^{k+1}\frac{1}{k} + \cdots \tag{7}$$

$$-1 - \frac{1}{2} - \frac{1}{3} - \frac{1}{4} - \cdots - \frac{1}{k} - \cdots \tag{8}$$

Series (7), the alternating harmonic series, converges, whereas series (8), being a constant times the harmonic series, diverges. Yet in each case the series of absolute values is

$$1 + \frac{1}{2} + \frac{1}{3} + \cdots + \frac{1}{k} + \cdots$$

which diverges. A series such as (7), which is convergent, but not absolutely convergent, is called ***conditionally convergent***.

☐ **THE RATIO TEST FOR ABSOLUTE CONVERGENCE**

The following version of the ratio test is useful for investigating absolute convergence.

> **11.7.5**   THEOREM(***Ratio Test for Absolute Convergence***).   *Let Σu_k be a series with nonzero terms and suppose that*
>
> $$\rho = \lim_{k \to +\infty} \frac{|u_{k+1}|}{|u_k|}$$
>
> *(a)   If $\rho < 1$, the series Σu_k converges absolutely and therefore converges.*
> *(b)   If $\rho > 1$ or if $\rho = +\infty$, then the series Σu_k diverges.*
> *(c)   If $\rho = 1$, no conclusion about convergence or absolute convergence can be drawn from this test.*

The proof is discussed in the exercises.

Example 6   The series

$$\sum_{k=1}^{\infty} (-1)^k \frac{2^k}{k!}$$

converges absolutely since

$$\rho = \lim_{k \to +\infty} \frac{|u_{k+1}|}{|u_k|} = \lim_{k \to +\infty} \frac{2^{k+1}}{(k+1)!} \cdot \frac{k!}{2^k} = \lim_{k \to +\infty} \frac{2}{k+1} = 0 < 1 \qquad \blacktriangleleft$$

The following review is included as a ready reference to convergence tests.

Review of Convergence Tests

NAME	STATEMENT	COMMENTS
Divergence Test (11.4.1)	If $\lim_{k \to +\infty} u_k \neq 0$, then Σu_k diverges.	If $\lim_{k \to +\infty} u_k = 0$, Σu_k may or may not converge.
Integral Test (11.4.4)	Let Σu_k be a series with positive terms, and let $f(x)$ be the function that results when k is replaced by x in the formula for u_k. If f is decreasing and continuous for $x \geq 1$, then $$\sum_{k=1}^{\infty} u_k \quad \text{and} \quad \int_1^{+\infty} f(x)\, dx$$ both converge or both diverge.	Use this test when $f(x)$ is easy to integrate. This test only applies to series that have positive terms.
Comparison Test (11.5.1)	Let Σa_k and Σb_k be series with nonnegative terms such that $$a_1 \leq b_1,\ a_2 \leq b_2, \ldots, a_k \leq b_k, \ldots$$ If Σb_k converges, then Σa_k converges, and if Σa_k diverges, then Σb_k diverges.	Use this test as a last resort; other tests are often easier to apply. This test only applies to series with nonnegative terms.

(continued on p. 554)

Review of Convergence Tests (Continued)

NAME	STATEMENT	COMMENTS
Ratio Test (11.5.2)	Let Σu_k be a series with positive terms and suppose that $$\rho = \lim_{k \to +\infty} \frac{u_{k+1}}{u_k}$$ (a) Series converges if $\rho < 1$. (b) Series diverges if $\rho > 1$ or $\rho = +\infty$. (c) No conclusion if $\rho = 1$.	Try this test when u_k involves factorials or kth powers.
Root Test (11.5.3)	Let Σu_k be a series with positive terms such that $$\rho = \lim_{k \to +\infty} \sqrt[k]{u_k}$$ (a) Series converges if $\rho < 1$. (b) Series diverges if $\rho > 1$ or $\rho = +\infty$. (c) No conclusion if $\rho = 1$.	Try this test when u_k involves kth powers.
Limit Comparison Test (11.6.3)	Let Σa_k and Σb_k be series with positive terms such that $$\rho = \lim_{k \to +\infty} \frac{a_k}{b_k}$$ If $0 < \rho < +\infty$, then both series converge or both diverge.	This is easier to apply than the comparison test, but still requires some skill in choosing the series Σb_k for comparison.
Alternating Series Test (11.7.1)	The series $$a_1 - a_2 + a_3 - a_4 + \cdots$$ $$-a_1 + a_2 - a_3 + a_4 - \cdots$$ converge if (a) $a_1 > a_2 > a_3 > \cdots$ (b) $\lim_{k \to +\infty} a_k = 0$	This test applies only to alternating series. It is assumed that $a_k > 0$ for all k.
Ratio Test for Absolute Convergence (11.7.5)	Let Σu_k be a series with nonzero terms such that $$\rho = \lim_{k \to +\infty} \frac{|u_{k+1}|}{|u_k|}$$ (a) Series converges absolutely if $\rho < 1$. (b) Series diverges if $\rho > 1$ or $\rho = +\infty$. (c) No conclusion if $\rho = 1$.	The series need not have positive terms and need not be alternating to use this test.

▶ Exercise Set 11.7   Ⓒ *38–46, 56*

In Exercises 1–6, determine whether the given alternating series converges or diverges.

1. $\displaystyle\sum_{k=1}^{\infty} \frac{(-1)^{k+1}}{2k + 1}$.

2. $\displaystyle\sum_{k=1}^{\infty} (-1)^{k+1} \frac{k}{3^k}$.

3. $\displaystyle\sum_{k=1}^{\infty} (-1)^{k+1} \frac{k + 1}{3k + 1}$.

4. $\displaystyle\sum_{k=1}^{\infty} (-1)^{k+1} \frac{k + 4}{k^2 + k}$.

5. $\displaystyle\sum_{k=1}^{\infty} (-1)^{k+1} e^{-k}$.

6. $\displaystyle\sum_{k=3}^{\infty} (-1)^k \frac{\ln k}{k}$.

In Exercises 7–12, use the ratio test for absolute convergence to determine whether the series converges absolutely or diverges.

7. $\displaystyle\sum_{k=1}^{\infty} \left(-\frac{3}{5}\right)^k$.

8. $\displaystyle\sum_{k=1}^{\infty} (-1)^{k+1} \frac{2^k}{k!}$.

9. $\displaystyle\sum_{k=1}^{\infty} (-1)^{k+1} \frac{3^k}{k^2}$.

10. $\displaystyle\sum_{k=1}^{\infty} (-1)^k \left(\frac{k}{5^k}\right)$.

11. $\displaystyle\sum_{k=1}^{\infty} (-1)^k \left(\frac{k^3}{e^k}\right)$.

12. $\displaystyle\sum_{k=1}^{\infty} (-1)^{k+1} \frac{k^k}{k!}$.

In Exercises 13–30, classify the series as absolutely convergent, conditionally convergent, or divergent.

13. $\displaystyle\sum_{k=1}^{\infty} \frac{(-1)^{k+1}}{3k}$.

14. $\displaystyle\sum_{k=1}^{\infty} \frac{(-1)^{k+1}}{k^{4/3}}$.

15. $\displaystyle\sum_{k=1}^{\infty} \frac{(-4)^k}{k^2}$.

16. $\displaystyle\sum_{k=1}^{\infty} \frac{(-1)^{k+1}}{k!}$.

17. $\displaystyle\sum_{k=1}^{\infty} \frac{\cos k\pi}{k}$.

18. $\displaystyle\sum_{k=3}^{\infty} \frac{(-1)^k \ln k}{k}$.

19. $\displaystyle\sum_{k=1}^{\infty} (-1)^{k+1} \left(\frac{k+2}{3k-1}\right)^k$.

20. $\displaystyle\sum_{k=1}^{\infty} \frac{(-1)^{k+1}}{k^2+1}$.

21. $\displaystyle\sum_{k=1}^{\infty} (-1)^{k+1} \frac{k+2}{k(k+3)}$.

22. $\displaystyle\sum_{k=1}^{\infty} \frac{(-1)^{k+1} k^2}{k^3+1}$.

23. $\displaystyle\sum_{k=1}^{\infty} \sin \frac{k\pi}{2}$.

24. $\displaystyle\sum_{k=1}^{\infty} \frac{\sin k}{k^3}$.

25. $\displaystyle\sum_{k=2}^{\infty} \frac{(-1)^k}{k \ln k}$.

26. $\displaystyle\sum_{k=1}^{\infty} \frac{(-1)^k}{\sqrt{k(k+1)}}$.

27. $\displaystyle\sum_{k=2}^{\infty} \left(-\frac{1}{\ln k}\right)^k$.

28. $\displaystyle\sum_{k=1}^{\infty} \frac{(-1)^{k+1}}{\sqrt{k+1}+\sqrt{k}}$.

29. $\displaystyle\sum_{k=2}^{\infty} \frac{(-1)^k (k^2+1)}{k^3+2}$.

30. $\displaystyle\sum_{k=1}^{\infty} \frac{k \cos k\pi}{k^2+1}$.

In Exercises 31–34, the series satisfies the conditions of the alternating series test. For the stated value of n, use Theorem 11.7.2 to find an upper bound on the magnitude of the error that results if the sum of the series is approximated by the nth partial sum.

31. $\displaystyle\sum_{k=1}^{\infty} \frac{(-1)^{k+1}}{k}$; $n = 7$.

32. $\displaystyle\sum_{k=1}^{\infty} \frac{(-1)^{k+1}}{k!}$; $n = 5$.

33. $\displaystyle\sum_{k=1}^{\infty} \frac{(-1)^{k+1}}{\sqrt{k}}$; $n = 99$.

34. $\displaystyle\sum_{k=1}^{\infty} \frac{(-1)^{k+1}}{(k+1) \ln (k+1)}$; $n = 3$.

In Exercises 35–38, the series satisfies the conditions of the alternating series test. Use Theorem 11.7.2 to find a value of n for which the nth partial sum is ensured to approximate the sum of the series to the stated accuracy.

35. $\displaystyle\sum_{k=1}^{\infty} \frac{(-1)^{k+1}}{k}$; $|\text{error}| < 0.0001$.

36. $\displaystyle\sum_{k=1}^{\infty} \frac{(-1)^{k+1}}{k!}$; $|\text{error}| < 0.00001$.

37. $\displaystyle\sum_{k=1}^{\infty} \frac{(-1)^{k+1}}{\sqrt{k}}$; $|\text{error}| < 0.005$.

38. $\displaystyle\sum_{k=1}^{\infty} \frac{(-1)^{k+1}}{(k+1) \ln (k+1)}$; $|\text{error}| < 0.1$.

In Exercises 39 and 40, use Theorem 11.7.2 to find an upper bound on the magnitude of the error that results if s_{10} is used to approximate the sum of the given *geometric* series. Compute s_{10} rounded to four decimal places and compare this value with the exact sum of the series.

39. $\dfrac{3}{4} - \dfrac{3}{8} + \dfrac{3}{16} - \dfrac{3}{32} + \cdots$.

40. $1 - \dfrac{2}{3} + \dfrac{4}{9} - \dfrac{8}{27} + \cdots$.

In Exercises 41–44, the series satisfies the conditions of the alternating series test. Use Theorem 11.7.2 to find a value of n for which s_n is ensured to approximate the sum of the series with an error that is less than 10^{-4} in magnitude. Compute s_n rounded to five decimal places and compare this value with the sum of the series.

41. $\sin 1 = 1 - \dfrac{1}{3!} + \dfrac{1}{5!} - \dfrac{1}{7!} + \cdots$.

42. $\cos 1 = 1 - \dfrac{1}{2!} + \dfrac{1}{4!} - \dfrac{1}{6!} + \cdots$.

43. $\ln \dfrac{3}{2} = \dfrac{1}{1 \cdot 2} - \dfrac{1}{2 \cdot 2^2} + \dfrac{1}{3 \cdot 2^3} - \dfrac{1}{4 \cdot 2^4} + \cdots$.

44. $\dfrac{\pi}{16} = \dfrac{1}{1^5 + 4 \cdot 1} - \dfrac{1}{3^5 + 4 \cdot 3}$
$\qquad\qquad + \dfrac{1}{5^5 + 4 \cdot 5} - \dfrac{1}{7^5 + 4 \cdot 7} + \cdots$.

45. For the series $\dfrac{\pi^2}{12} = 1 - \dfrac{1}{2^2} + \dfrac{1}{3^2} - \dfrac{1}{4^2} + \cdots$

   (a) use Theorem 11.7.2 to find a value of n for which s_n is ensured to approximate the sum of the series with an error that is less than 5×10^{-3} in magnitude

   (b) compute s_{10} and show that the magnitude of the error is less than 5×10^{-3}, thus showing that the value of n obtained from Theorem 11.7.2 is a conservative estimate.

46. For the series $\dfrac{\pi}{4} = 1 - \tfrac{1}{3} + \tfrac{1}{5} - \tfrac{1}{7} + \cdots$

   (a) use Theorem 11.7.2 to find a value of n for which s_n is ensured to approximate the sum of the series with an error that is less than 10^{-2} in magnitude

   (b) compute s_{26} and show that the magnitude of the error is less than 10^{-2}, thus showing that the value of n obtained from Theorem 11.7.2 is a conservative estimate.

47. Prove: If Σa_k converges absolutely, then Σa_k^2 converges.

48. Show that the converse of the result in Exercise 47 is false by finding a series for which Σa_k^2 converges, but $\Sigma |a_k|$ diverges.

49. Prove Theorem 11.7.1 for series of the form
$$-a_1 + a_2 - a_3 + a_4 - \cdots + (-1)^k a_k + \cdots$$

50. Prove Theorem 11.7.5. [*Hint:* Theorem 11.7.4 will help in part (*a*). For part (*b*), it may help to review the proof of Theorem 11.5.2.]

51. The sum of an absolutely convergent series is independent of the order in which the terms are added, but the terms of a conditionally convergent series can be rearranged to converge to any given value, or even diverge. For example, let S be the sum of the conditionally convergent alternating harmonic series,
$$S = 1 - \frac{1}{2} + \frac{1}{3} - \frac{1}{4} + \frac{1}{5} - \frac{1}{6} + \cdots$$

Rearrange the terms in this series to get
$$\left(1 - \frac{1}{2} - \frac{1}{4}\right) + \left(\frac{1}{3} - \frac{1}{6} - \frac{1}{8}\right) + \left(\frac{1}{5} - \frac{1}{10} - \frac{1}{12}\right) + \cdots$$

Show that this rearrangement results in a series that converges to $S/2$. [*Hint:* Add the first two terms within each pair of parentheses.]

52. Based on the discussion in Exercise 51, rearrange the terms in the convergent series
$$1 - \frac{1}{\sqrt{2}} + \frac{1}{\sqrt{3}} - \frac{1}{\sqrt{4}} + \frac{1}{\sqrt{5}} - \frac{1}{\sqrt{6}} + \cdots$$
as
$$\left(1 + \frac{1}{\sqrt{3}} - \frac{1}{\sqrt{2}}\right) + \left(\frac{1}{\sqrt{5}} + \frac{1}{\sqrt{7}} - \frac{1}{\sqrt{4}}\right) + \cdots$$
$$= \sum_{k=1}^{\infty} \left(\frac{1}{\sqrt{4k-3}} + \frac{1}{\sqrt{4k-1}} - \frac{1}{\sqrt{2k}}\right)$$

Show that this rearrangement results in a series that diverges to $+\infty$. [*Hint:* Note that

$$1/\sqrt{4k-3} > 1/\sqrt{4k} \quad \text{and} \quad 1/\sqrt{4k-1} > 1/\sqrt{4k}.$$

Show that the kth term in the rearranged series is greater than $(1 - 1/\sqrt{2})/\sqrt{k}$.]

> In Exercises 53–55, use parts (*a*) and (*b*) of Theorem 11.4.3 and the fact that the sum of an absolutely convergent series is independent of the order in which the terms are added.

53. Given: $\dfrac{\pi^2}{6} = 1 + \dfrac{1}{2^2} + \dfrac{1}{3^2} + \dfrac{1}{4^2} + \cdots$.

Show: $\dfrac{\pi^2}{8} = 1 + \dfrac{1}{3^2} + \dfrac{1}{5^2} + \dfrac{1}{7^2} + \cdots$.

54. Given: $\dfrac{\pi^4}{90} = 1 + \dfrac{1}{2^4} + \dfrac{1}{3^4} + \dfrac{1}{4^4} + \cdots$.

Show: $\dfrac{\pi^4}{96} = 1 + \dfrac{1}{3^4} + \dfrac{1}{5^4} + \dfrac{1}{7^4} + \cdots$.

55. Given: $\dfrac{\pi^2}{6} = 1 + \dfrac{1}{2^2} + \dfrac{1}{3^2} + \dfrac{1}{4^2} + \cdots$.

Show: $\dfrac{\pi^2}{12} = 1 - \dfrac{1}{2^2} + \dfrac{1}{3^2} - \dfrac{1}{4^2} + \cdots$.

56. A small bug moves back and forth along a straight line as follows: it walks D units, stops and reverses direction, walks $D/2$ units, stops and reverses direction, walks $D/3$ units, stops and reverses direction, walks $D/4$ units, stops and reverses direction, and so forth, until it stops for the 1000th time. Given that $D = 180$ cm, find upper and lower bounds for

(a) the final distance between the bug and its starting point; [*Hint:* Use Theorem 11.7.2 and the fact that $1 - \frac{1}{2} + \frac{1}{3} - \frac{1}{4} + \cdots = \ln 2$.]

(b) the total distance traveled by the bug. [*Hint:* Use inequality (2) in Section 11.4.]

■ **11.8** POWER SERIES

> *In previous sections we studied series with constant terms. In this section we shall consider series whose terms involve variables. Such series are of fundamental importance in many branches of mathematics and the physical sciences.*

□ **POWER SERIES IN** x

If $c_0, c_1, c_2, \ldots$ are constants and x is a variable, then a series of the form

$$\sum_{k=0}^{\infty} c_k x^k = c_0 + c_1 x + c_2 x^2 + \cdots + c_k x^k + \cdots \qquad (1)$$

is called a *power series in x*. Some examples are

$$\sum_{k=0}^{\infty} x^k = 1 + x + x^2 + x^3 + \cdots$$

$$\sum_{k=0}^{\infty} \frac{x^k}{k!} = 1 + x + \frac{x^2}{2!} + \frac{x^3}{3!} + \cdots$$

$$\sum_{k=0}^{\infty} (-1)^k \frac{x^{2k}}{(2k)!} = 1 - \frac{x^2}{2!} + \frac{x^4}{4!} - \frac{x^6}{6!} + \cdots$$

If a numerical value is substituted for x in a power series $\Sigma c_k x^k$, then the resulting series of constants may either converge or diverge. The problem of determining those values of x for which a given power series converges is addressed by the following theorem whose proof is omitted.

11.8.1 THEOREM. *For any power series in x, exactly one of the following is true*:

(a) *The series converges only for $x = 0$.*
(b) *The series converges absolutely (and hence converges) for all real values of x.*
(c) *The series converges absolutely (and hence converges) for all x in some finite open interval $(-R, R)$, and diverges if $x < -R$ or $x > R$. At either of the points $x = R$ or $x = -R$ the series may converge absolutely, converge conditionally, or diverge, depending on the particular series.*

☐ **RADIUS AND INTERVAL OF CONVERGENCE**

Theorem 11.8.1 states that the set of values for which a power series in x converges is always an interval centered at 0; we call this the ***interval of convergence*** (Figure 11.8.1). In part (*a*) of Theorem 11.8.1 the interval of convergence reduces to a single point (the origin), in which case we say that the series has ***radius of convergence $R = 0$***; in part (*b*) the interval of convergence is infinite (the entire real line), in which case we say that the series has ***radius of convergence $R = +\infty$***; and in part (*c*) the interval extends between $-R$ and R, in which case we say that the series has ***radius of convergence R***.

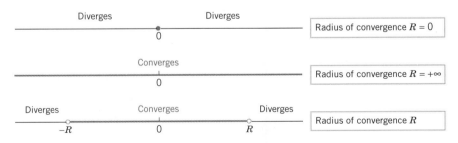

Figure 11.8.1

As illustrated in the following example, the main tool for finding the interval of convergence of a power series is the ratio test for absolute convergence (Theorem 11.7.5).

Example 1 Find the interval of convergence and radius of convergence of the following power series.

(a) $\displaystyle\sum_{k=0}^{\infty} x^k$ (b) $\displaystyle\sum_{k=0}^{\infty} \frac{x^k}{k!}$ (c) $\displaystyle\sum_{k=0}^{\infty} k! x^k$ (d) $\displaystyle\sum_{k=0}^{\infty} \frac{(-1)^k x^k}{3^k(k+1)}$

Solution (a). We shall apply the ratio test for absolute convergence. We have

$$\rho = \lim_{k \to +\infty} \left| \frac{u_{k+1}}{u_k} \right| = \lim_{k \to +\infty} \left| \frac{x^{k+1}}{x^k} \right| = \lim_{k \to +\infty} |x| = |x|$$

so the ratio test for absolute convergence implies that the series converges absolutely if $\rho = |x| < 1$ and diverges if $\rho = |x| > 1$. The test is inconclusive if $|x| = 1$ (i.e., if $x = 1$ or $x = -1$), so convergence at these points must be investigated separately. At these points the series becomes

$$\sum_{k=0}^{\infty} 1^k = 1 + 1 + 1 + 1 + \cdots \qquad \boxed{x = 1}$$

$$\sum_{k=0}^{\infty} (-1)^k = 1 - 1 + 1 - 1 + \cdots \qquad \boxed{x = -1}$$

both of which diverge; thus, the interval of convergence for the given power series is $(-1, 1)$, and the radius of convergence is $R = 1$.

Solution (b). Applying the ratio test for absolute convergence, we obtain

$$\rho = \lim_{k \to +\infty} \left| \frac{u_{k+1}}{u_k} \right| = \lim_{k \to +\infty} \left| \frac{x^{k+1}}{(k+1)!} \cdot \frac{k!}{x^k} \right| = \lim_{k \to +\infty} \left| \frac{x}{k+1} \right| = 0$$

Since $\rho < 1$ for all x, the series converges absolutely for all x. Thus, the interval of convergence is $(-\infty, +\infty)$ and the radius of convergence is $R = +\infty$.

Solution (c). If $x \neq 0$, then the ratio test for absolute convergence yields

$$\rho = \lim_{k \to +\infty} \left| \frac{u_{k+1}}{u_k} \right| = \lim_{k \to +\infty} \left| \frac{(k+1)! x^{k+1}}{k! x^k} \right| = \lim_{k \to +\infty} |(k+1)x| = +\infty$$

Therefore, the series diverges for all nonzero values of x. Consequently, the interval of convergence is the single point $x = 0$ and the radius of convergence is $R = 0$.

Solution (d). Since $|(-1)^k| = |(-1)^{k+1}| = 1$, we obtain

$$\rho = \lim_{k \to +\infty} \left| \frac{u_{k+1}}{u_k} \right| = \lim_{k \to +\infty} \left| \frac{x^{k+1}}{3^{k+1}(k+2)} \cdot \frac{3^k(k+1)}{x^k} \right|$$

$$= \lim_{k \to +\infty} \left[\frac{|x|}{3} \cdot \left(\frac{k+1}{k+2} \right) \right]$$

$$= \frac{|x|}{3} \lim_{k \to +\infty} \left(\frac{1 + 1/k}{1 + 2/k} \right) = \frac{|x|}{3}$$

The ratio test for absolute convergence implies that the series converges absolutely if $|x| < 3$ and diverges if $|x| > 3$. The ratio test fails to provide any information when $|x| = 3$, so the cases $x = -3$ and $x = 3$ need separate analyses. Substituting $x = -3$ in the given series yields

$$\sum_{k=0}^{\infty} \frac{(-1)^k(-3)^k}{3^k(k+1)} = \sum_{k=0}^{\infty} \frac{(-1)^k(-1)^k 3^k}{3^k(k+1)} = \sum_{k=0}^{\infty} \frac{1}{k+1}$$

which is the divergent harmonic series $1 + \frac{1}{2} + \frac{1}{3} + \frac{1}{4} + \cdots$. Substituting $x = 3$ in the given series yields

$$\sum_{k=0}^{\infty} \frac{(-1)^k 3^k}{3^k(k+1)} = \sum_{k=0}^{\infty} \frac{(-1)^k}{k+1} = 1 - \frac{1}{2} + \frac{1}{3} - \frac{1}{4} + \cdots$$

which is the conditionally convergent alternating harmonic series. Thus, the interval of convergence for the given series is $(-3, 3]$ and the radius of convergence is $R = 3$.   ◄

☐ **POWER SERIES IN** $x - a$

If a is a constant and $x - a$ is substituted for x in (1), then the resulting series,

$$\sum_{k=0}^{\infty} c_k(x - a)^k = c_0 + c_1(x - a) + c_2(x - a)^2 + \cdots + c_k(x - a)^k + \cdots$$

is called a *power series in x − a*. Some examples are

$$\sum_{k=0}^{\infty} \frac{(x-1)^k}{k+1} = 1 + \frac{(x-1)}{2} + \frac{(x-1)^2}{3} + \frac{(x-1)^3}{4} + \cdots \qquad \boxed{a = 1}$$

$$\sum_{k=0}^{\infty} \frac{(-1)^k(x+3)^k}{k!} = 1 - (x+3) + \frac{(x+3)^2}{2!} - \frac{(x+3)^3}{3!} + \cdots \qquad \boxed{a = -3}$$

The first of these is a power series in $x - 1$ and the second is a power series in $x + 3$. Note that a power series in x is the special case of a power series in $x - a$ in which $a = 0$.

The conditions for convergence of a power series in $x - a$ can be obtained by substituting $x - a$ for x in Theorem 11.8.1. This leads to the following theorem.

11.8.2   THEOREM.   *For a power series $\Sigma c_k(x-a)^k$, exactly one of the following is true:*

(a)   The series converges only for $x = a$.

(b)   The series converges absolutely (and hence converges) for all real values of x.

(c)   The series converges absolutely (and hence converges) for all x in some finite open interval $(a - R, a + R)$ and diverges if $x < a - R$ or $x > a + R$. At either of the points $x = a - R$ or $x = a + R$, the series may converge absolutely, converge conditionally, or diverge, depending on the particular series.

It follows from the preceding theorem that the set of values for which a power series in $x - a$ converges is always an interval centered at $x = a$; we call this the *interval of convergence* (Figure 11.8.2). In part (*a*) of Theorem 11.8.2 the interval of convergence reduces to the single point $x = a$, in which case we say that the series has *radius of convergence R = 0*; and in part (*b*) the interval of convergence is infinite (the entire real line), in which case we say that the series has *radius of convergence R = +∞*; and in part (*c*) the interval extends between $a - R$ and $a + R$, in which case we say that the series has *radius of convergence R*.

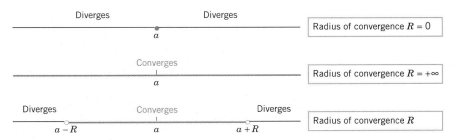

Figure 11.8.2

Example 2   Find the interval of convergence and radius of convergence of the series

$$\sum_{k=1}^{\infty} \frac{(x-5)^k}{k^2}$$

Solution.   We apply the ratio test for absolute convergence.

$$\rho = \lim_{k \to +\infty} \left| \frac{u_{k+1}}{u_k} \right| = \lim_{k \to +\infty} \left| \frac{(x-5)^{k+1}}{(k+1)^2} \cdot \frac{k^2}{(x-5)^k} \right|$$

$$= \lim_{k \to +\infty} \left[|x-5| \left(\frac{k}{k+1} \right)^2 \right]$$

$$= |x-5| \lim_{k \to +\infty} \left(\frac{1}{1+1/k} \right)^2 = |x-5|$$

Thus, the series converges absolutely if $|x - 5| < 1$, or $-1 < x - 5 < 1$, or $4 < x < 6$. The series diverges if $x < 4$ or $x > 6$.

To determine the convergence behavior at the endpoints $x = 4$ and $x = 6$, we substitute these values in the given series. If $x = 6$, the series becomes

$$\sum_{k=1}^{\infty} \frac{1^k}{k^2} = \sum_{k=1}^{\infty} \frac{1}{k^2} = 1 + \frac{1}{2^2} + \frac{1}{3^2} + \frac{1}{4^2} + \cdots$$

which is a convergent p-series ($p = 2$). If $x = 4$, the series becomes

$$\sum_{k=1}^{\infty} \frac{(-1)^k}{k^2} = -1 + \frac{1}{2^2} - \frac{1}{3^2} + \frac{1}{4^2} - \cdots$$

Since this series converges absolutely, the interval of convergence for the given series is $[4, 6]$. The radius of convergence is $R = 1$ (Figure 11.8.3). ◀

REMARK.   The ratio test should never be used for testing for convergence at the endpoints of the interval of convergence, since the ratio ρ is always 1 at those points (why?).

Figure 11.8.3

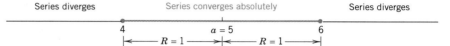

| Series diverges | Series converges absolutely | Series diverges |

$a = 5$

$\longleftarrow R = 1 \longrightarrow \longleftarrow R = 1 \longrightarrow$

4   6

▶ Exercise Set 11.8

In Exercises 1–24, find the radius of convergence and the interval of convergence.

1. $\displaystyle\sum_{k=0}^{\infty} \frac{x^k}{k + 1}$.

2. $\displaystyle\sum_{k=0}^{\infty} 3^k x^k$.

3. $\displaystyle\sum_{k=0}^{\infty} \frac{(-1)^k x^k}{k!}$.

4. $\displaystyle\sum_{k=0}^{\infty} \frac{k!}{2^k} x^k$.

5. $\displaystyle\sum_{k=1}^{\infty} \frac{5^k}{k^2} x^k$.

6. $\displaystyle\sum_{k=2}^{\infty} \frac{x^k}{\ln k}$.

7. $\displaystyle\sum_{k=1}^{\infty} \frac{x^k}{k(k + 1)}$.

8. $\displaystyle\sum_{k=0}^{\infty} \frac{(-2)^k x^{k+1}}{k + 1}$.

9. $\displaystyle\sum_{k=1}^{\infty} (-1)^{k-1} \frac{x^k}{\sqrt{k}}$.

10. $\displaystyle\sum_{k=0}^{\infty} \frac{(-1)^k x^{2k}}{(2k)!}$.

11. $\displaystyle\sum_{k=0}^{\infty} (-1)^k \frac{x^{2k+1}}{(2k + 1)!}$.

12. $\displaystyle\sum_{k=1}^{\infty} (-1)^k \frac{x^{3k}}{k^{3/2}}$.

13. $\displaystyle\sum_{k=0}^{\infty} \frac{3^k}{k!} x^k$.

14. $\displaystyle\sum_{k=2}^{\infty} (-1)^{k+1} \frac{x^k}{k(\ln k)^2}$.

15. $\displaystyle\sum_{k=0}^{\infty} \frac{x^k}{1 + k^2}$.

16. $\displaystyle\sum_{k=0}^{\infty} \frac{(x - 3)^k}{2^k}$.

17. $\displaystyle\sum_{k=1}^{\infty} (-1)^{k+1} \frac{(x + 1)^k}{k}$.

18. $\displaystyle\sum_{k=0}^{\infty} (-1)^k \frac{(x - 4)^k}{(k + 1)^2}$.

19. $\displaystyle\sum_{k=0}^{\infty} \left(\frac{3}{4}\right)^k (x + 5)^k$.

20. $\displaystyle\sum_{k=1}^{\infty} \frac{(2k + 1)!}{k^3} (x - 2)^k$.

21. $\displaystyle\sum_{k=1}^{\infty} (-1)^k \frac{(x + 1)^{2k+1}}{k^2 + 4}$.

22. $\displaystyle\sum_{k=1}^{\infty} \frac{(\ln k)(x - 3)^k}{k}$.

23. $\displaystyle\sum_{k=0}^{\infty} \frac{\pi^k (x - 1)^{2k}}{(2k + 1)!}$.

24. $\displaystyle\sum_{k=0}^{\infty} \frac{(2x - 3)^k}{4^{2k}}$.

In Exercises 25–27, show the first four terms of the series, and find the radius of convergence.

25. $\displaystyle\sum_{k=1}^{\infty} \frac{1 \cdot 2 \cdot 3 \cdots k}{1 \cdot 4 \cdot 7 \cdots (3k - 2)} x^k$.

26. $\displaystyle\sum_{k=1}^{\infty} (-1)^k \frac{1 \cdot 2 \cdot 3 \cdots k}{1 \cdot 3 \cdot 5 \cdots (2k - 1)} x^{2k+1}$.

27. $\displaystyle\sum_{k=1}^{\infty} \frac{1 \cdot 3 \cdot 5 \cdots (2k - 1)}{(2k - 2)!} x^k$.

28. Use the root test to find the interval of convergence of

$$\sum_{k=2}^{\infty} \frac{x^k}{(\ln k)^k}.$$

29. Find the interval of convergence of $\displaystyle\sum_{k=0}^{\infty} \frac{(x - a)^k}{b^k}$, where $b > 0$.

30. Find the radius of convergence of the power series

$$\sum_{k=0}^{\infty} \frac{(pk)!}{(k!)^p} x^k,$$ where p is a positive integer.

31. Find the radius of convergence of the power series

$$\sum_{k=0}^{\infty} \frac{(k + p)!}{k!(k + q)!} x^k,$$ where p and q are positive integers.

32. Prove: If $\displaystyle\lim_{k \to +\infty} |c_k|^{1/k} = L$, where $L \neq 0$, then $1/L$ is the radius of convergence of the power series $\sum_{k=0}^{\infty} c_k x^k$.

33. Prove: If the power series $\sum_{k=0}^{\infty} c_k x^k$ has radius of convergence R, then the series $\sum_{k=0}^{\infty} c_k x^{2k}$ has radius of convergence $\sqrt{R}$.

34. Prove: If the interval of convergence of the series $\sum_{k=0}^{\infty} c_k (x-a)^k$ is $(a-R, a+R]$, then the series converges conditionally at $a + R$.

■ **11.9** TAYLOR AND MACLAURIN SERIES

One of the early applications of calculus was the computation of approximate numerical values for functions such as $\sin x$, $\ln x$, *and* e^x. *One common method for obtaining such values is to approximate the function by a polynomial, then use that polynomial to compute the desired numerical values. In this section we shall discuss procedures for approximating functions by polynomials, and in the next section we shall investigate the errors in these approximations. In this section we shall also see how polynomial approximations lead naturally to the important problem of finding a power series that converges to a specified function.*

☐ APPROXIMATING
FUNCTIONS BY
POLYNOMIALS

The problem of primary interest in this section can be phrased informally as follows:

PROBLEM. *Given a function* f *and a point* a *on the x-axis, find a polynomial of specified degree that best approximates the function* f *in the "vicinity" of the point* a.

As stated, the problem is somewhat vague in that we have not specified any requirements on f such as continuity, differentiability, and so forth, and it is not evident what we mean by the "best approximation in the vicinity of a point." However, the problem is suggestive enough to get us started, and we shall resolve the ambiguities as we progress.

Suppose that we are interested in approximating a function f in the vicinity of the point $a = 0$ by a polynomial

$$p(x) = c_0 + c_1 x + \cdots + c_n x^n \tag{1}$$

Because $p(x)$ has $n + 1$ coefficients, it seems reasonable that we should be able to impose $n + 1$ conditions on this polynomial to achieve a good approximation to $f(x)$. Because the point $a = 0$ is the center of interest, our strategy will be to choose the coefficients of $p(x)$ so that the value of p and its first n derivatives are the same as the value of f and its first n derivatives at $a = 0$. By forcing this high degree of "match" at $a = 0$, it is reasonable to hope that $f(x)$ and $p(x)$ will remain close over some interval (possibly quite small) centered at $a = 0$. Thus, we shall assume that f can be differentiated n times at 0, and we shall try to find the coefficients in (1) such that

$$f(0) = p(0), \quad f'(0) = p'(0), \quad f''(0) = p''(0), \quad \ldots, \quad f^{(n)}(0) = p^{(n)}(0) \tag{2}$$

We have

$$p(x) \quad = c_0 + c_1 x + c_2 x^2 + c_3 x^3 + \cdots + c_n x^n$$

$$p'(x) \quad = c_1 + 2c_2 x + 3c_3 x^2 + \cdots + nc_n x^{n-1}$$

$$p''(x) \quad = 2c_2 + 3 \cdot 2c_3 x + \cdots + n(n-1)c_n x^{n-2}$$

$$p'''(x) \quad = 3 \cdot 2c_3 + \cdots + n(n-1)(n-2)c_n x^{n-3}$$

$$\vdots$$

$$p^{(n)}(x) = n(n-1)(n-2)\cdots(1)c_n$$

Thus, to satisfy (2) we must have

$$f(0) \quad = p(0) \quad = c_0$$

$$f'(0) \quad = p'(0) \quad = c_1$$

$$f''(0) \quad = p''(0) \quad = 2c_2 = 2!c_2$$

$$f'''(0) \quad = p'''(0) \quad = 3 \cdot 2c_3 = 3!c_3$$

$$\vdots$$

$$f^{(n)}(0) = p^{(n)}(0) = n(n-1)(n-2)\cdots(1)c_n = n!c_n$$

which yields the following values for the coefficients of $p(x)$:

$$c_0 = f(0), \quad c_1 = f'(0), \quad c_2 = \frac{f''(0)}{2!}, \quad c_3 = \frac{f'''(0)}{3!}, \quad \ldots, \quad c_n = \frac{f^{(n)}(0)}{n!}$$

□ **MACLAURIN POLYNOMIALS**

The polynomial that results by using these coefficients in (1) is called the *nth Maclaurin* polynomial for f*. In summary, we have the following definition.

> **11.9.1** DEFINITION. If f can be differentiated n times at 0, then we define the ***nth Maclaurin polynomial for f*** to be
>
> $$p_n(x) = f(0) + f'(0)x + \frac{f''(0)}{2!}x^2 + \frac{f'''(0)}{3!}x^3 + \cdots + \frac{f^{(n)}(0)}{n!}x^n \qquad (3)$$
>
> This polynomial has the property that its value and the values of its first n derivatives match the value of $f(x)$ and its first n derivatives when $x = 0$.

Example 1 Find the Maclaurin polynomials p_0, p_1, p_2, p_3, and p_n for e^x.

Solution. Let $f(x) = e^x$. Thus,

$$f'(x) = f''(x) = f'''(x) = \cdots = f^{(n)}(x) = e^x$$

and

$$f(0) = f'(0) = f''(0) = f'''(0) = \cdots = f^{(n)}(0) = e^0 = 1$$

---

*COLIN MACLAURIN (1698–1746). Scottish mathematician. Maclaurin's father, a minister, died when the boy was only six months old, and his mother when he was nine years old. He was then raised by an uncle who was also a minister. Maclaurin entered Glasgow University as a divinity student, but transferred to mathematics after one year. He received his Master's degree at age 17 and, in spite of his youth, began teaching at Marischal College in Aberdeen, Scotland. He met Isaac Newton during a visit to London in 1719 and from that time on became Newton's disciple. During that era, some of Newton's analytic methods were bitterly attacked by major mathematicians and much of Maclaurin's important mathematical work resulted from his efforts to defend Newton's ideas geometrically. Maclaurin's work, *A Treatise of Fluxions* (1742), was the first systematic formulation of Newton's methods. The treatise was so carefully done that it was a standard of mathematical rigor in calculus until the work of Cauchy in 1821.

Maclaurin was an outstanding experimentalist. He devised numerous ingenious mechanical devices, made important astronomical observations, performed actuarial computations for insurance societies, and helped to improve maps of the islands around Scotland.

Therefore,

$$p_0(x) = f(0) = 1$$

$$p_1(x) = f(0) + f'(0)x = 1 + x$$

$$p_2(x) = f(0) + f'(0)x + \frac{f''(0)}{2!}x^2 = 1 + x + \frac{x^2}{2!} = 1 + x + \frac{1}{2}x^2$$

$$p_3(x) = f(0) + f'(0)x + \frac{f''(0)}{2!}x^2 + \frac{f'''(0)}{3!}x^3$$

$$= 1 + x + \frac{x^2}{2!} + \frac{x^3}{3!} = 1 + x + \frac{1}{2}x^2 + \frac{1}{6}x^3$$

$$p_n(x) = f(0) + f'(0)x + \frac{f''(0)}{2!}x^2 + \cdots + \frac{f^{(n)}(0)}{n!}x^n$$

$$= 1 + x + \frac{x^2}{2!} + \cdots + \frac{x^n}{n!} \quad \blacktriangleleft$$

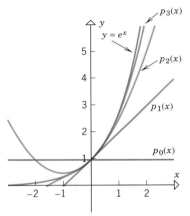

Figure 11.9.1

Figure 11.9.1 shows the graphs of e^x (in blue) and the graphs of the first four Maclaurin polynomials. Note that the graphs of $p_1(x)$, $p_2(x)$, and $p_3(x)$ are virtually indistinguishable from the graph of e^x near the origin, so that these polynomials are good approximations of e^x for x near 0. However, the farther x is from 0, the poorer these approximations become. This is typical of the Maclaurin polynomials for a function $f(x)$; they provide good approximations of $f(x)$ near 0, but the accuracy diminishes as x progresses away from 0. (In the next section we shall investigate the accuracy of such approximations.)

To obtain polynomial approximations of $f(x)$ that have their best accuracy near a general point $x = a$, it will be convenient to express the polynomials in powers of $x - a$, so that they have the form

$$p(x) = c_0 + c_1(x - a) + c_2(x - a)^2 + \cdots + c_n(x - a)^n \tag{4}$$

Since the point $x = a$ is the center of interest, we want to choose the coefficients so that the values of $p(x)$ and its first n derivatives match the values of $f(x)$ and its first n derivatives at a. We leave it as an exercise for the reader to show that if f can be differentiated n times at a, then the desired coefficients are

$$c_0 = f(a), \quad c_1 = f'(a), \quad c_2 = \frac{f''(a)}{2!}, \quad c_3 = \frac{f'''(a)}{3!}, \quad \ldots, \quad c_n = \frac{f^{(n)}(a)}{n!}$$

Substituting these values in (4) we obtain a polynomial called the *nth Taylor* polynomial about $x = a$ for f.

---

*BROOK TAYLOR (1685–1731). English mathematician. Taylor was born of well-to-do parents. Musicians and artists were entertained frequently in the Taylor home, which undoubtedly had a lasting influence on young Brook. In later years, Taylor published a definitive work on the mathematical theory of perspective and obtained major mathematical results about the vibrations of strings. There also exists an unpublished work, *On Musick*, that was intended to be part of a joint paper with Isaac Newton. Taylor's life was scarred with unhappiness, illness, and tragedy. Because his first wife was not rich enough to suit his father, the two men argued bitterly and parted ways. Subsequently, his wife died in childbirth. Then, after he remarried, his second wife also died in childbirth, though his daughter survived. Taylor's most productive period was from 1714 to 1719, during which time he wrote on a wide range of subjects—magnetism, capillary action, thermometers, perspective, and calculus. In his final years, Taylor devoted his writing efforts to religion and philosophy. According to Taylor, the results that bear his name were motivated by coffeehouse conversations about works of Newton on planetary motion and works of Halley ("Halley's comet") on roots of polynomials.

Taylor's writing style was so terse and hard to understand that he never received credit for many of his innovations.

11.9.2   DEFINITION.   If f can be differentiated n times at a, then we define the **nth Taylor polynomial for f about $x = a$** to be

$$p_n(x) = f(a) + f'(a)(x - a) + \frac{f''(a)}{2!}(x - a)^2$$

$$+ \frac{f'''(a)}{3!}(x - a)^3 + \cdots + \frac{f^{(n)}(a)}{n!}(x - a)^n \quad (5)$$

REMARK.   Observe that with $a = 0$, the nth Taylor polynomial for f is the nth Maclaurin polynomial for f; that is, the Maclaurin polynomials are special cases of the Taylor polynomials.

Example 2   Find the first four Taylor polynomials for $\ln x$ about $x = 2$.

Solution.   Let $f(x) = \ln x$. Thus,

$$f(x) \;= \ln x \qquad f(2) \;= \ln 2$$
$$f'(x) = 1/x \qquad f'(2) = 1/2$$
$$f''(x) = -1/x^2 \qquad f''(2) = -1/4$$
$$f'''(x) = 2/x^3 \qquad f'''(2) = 1/4$$

Substituting in (5) with $a = 2$ yields

$$p_0(x) = f(2) = \ln 2$$

$$p_1(x) = f(2) + f'(2)(x - 2) = \ln 2 + \tfrac{1}{2}(x - 2)$$

$$p_2(x) = f(2) + f'(2)(x - 2) + \frac{f''(2)}{2!}(x - 2)^2 = \ln 2 + \tfrac{1}{2}(x - 2) - \tfrac{1}{8}(x - 2)^2$$

$$p_3(x) = f(2) + f'(2)(x - 2) + \frac{f''(2)}{2!}(x - 2)^2 + \frac{f'''(2)}{3!}(x - 2)^3$$

$$= \ln 2 + \tfrac{1}{2}(x - 2) - \tfrac{1}{8}(x - 2)^2 + \tfrac{1}{24}(x - 2)^3$$

The graph of $\ln x$ (in blue) and its first four Taylor polynomials about $x = 2$ are shown in Figure 11.9.2. As expected, these polynomials produce their best approximations of $\ln x$ near 2.   ◄

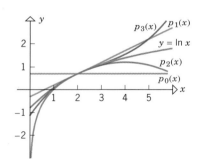

Figure 11.9.2

□ **SIGMA NOTATION FOR TAYLOR AND MACLAURIN POLYNOMIALS**

Frequently, it is convenient to express the defining formula for the Taylor polynomial in sigma notation. To do this, we use the notation $f^{(k)}(a)$ to denote the kth derivative of f at $x = a$, and we make the added convention that $f^{(0)}(a)$ denotes $f(a)$. This enables us to write

$$\sum_{k=0}^{n} \frac{f^{(k)}(a)}{k!}(x - a)^k = f(a) + f'(a)(x - a) + \frac{f''(a)}{2!}(x - a)^2 + \cdots + \frac{f^{(n)}(a)}{n!}(x - a)^n$$

In particular, the nth Maclaurin polynomial for $f(x)$ can be written as

$$\sum_{k=0}^{n} \frac{f^{(k)}(0)}{k!}x^k = f(0) + f'(0)x + \frac{f''(0)}{2!}x^2 + \cdots + \frac{f^{(n)}(0)}{n!}x^n \quad (6)$$

□ **TAYLOR AND MACLAURIN SERIES**

For a fixed value of x near a, one would expect that the approximation of $f(x)$ by its Taylor polynomial $p_n(x)$ about $x = a$ should improve as n increases, since increasing n has the effect of matching higher and higher derivatives of $f(x)$ with those of $p_n(x)$ at $x = a$. Indeed, its seems plausible that one might be able to achieve any desired degree of accuracy by choosing n sufficiently large; that is, the values of $p_n(x)$ might actually converge to $f(x)$ as $n \rightarrow +\infty$. Should this happen, we would have

$$f(x) = \lim_{n \rightarrow +\infty} \sum_{k=0}^{n} \frac{f^{(k)}(a)}{k!}(x-a)^k = \sum_{k=0}^{\infty} \frac{f^{(k)}(a)}{k!}(x-a)^k$$

In the next section we shall study conditions under which the series on the right actually converges to $f(x)$. For the remainder of this section, we will focus on the computational aspects of finding these series. We make the following definition.

> **11.9.3** DEFINITION. If f has derivatives of all orders at a, then we define the **Taylor series for f about $x = a$** to be
>
> $$\sum_{k=0}^{\infty} \frac{f^{(k)}(a)}{k!}(x-a)^k = f(a) + f'(a)(x-a)$$
> $$+ \frac{f''(a)}{2!}(x-a)^2 + \cdots + \frac{f^{(k)}(a)}{k!}(x-a)^k + \cdots \quad (7)$$

In the special case where $a = 0$, the Taylor series for f is called the **Maclaurin series for f**. In this case the series has the form

$$\sum_{k=0}^{\infty} \frac{f^{(k)}(0)}{k!} x^k = f(0) + f'(0)x + \frac{f''(0)}{2!}x^2 + \cdots + \frac{f^{(k)}(0)}{k!}x^k + \cdots \quad (8)$$

REMARK. Because the summation index in the Taylor series for f about $x = a$ starts at 0, it is convenient to think of the initial term in the series as the zeroth term. With this convention the nth term in the series is the term involving $(x-a)^n$. It then follows that the nth partial sum of the Taylor series is the nth Taylor polynomial.

Example 3 Find the Maclaurin series for
  (a) e^x (b) $\sin x$ (c) $\cos x$

Solution (a). In Example 1 we found the nth Maclaurin polynomial for the function e^x to be

$$\sum_{k=0}^{n} \frac{x^k}{k!} = 1 + x + \frac{x^2}{2!} + \cdots + \frac{x^n}{n!}$$

Thus, the Maclaurin series for e^x is

$$\sum_{k=0}^{\infty} \frac{x^k}{k!} = 1 + x + \frac{x^2}{2!} + \frac{x^3}{3!} + \cdots + \frac{x^k}{k!} + \cdots$$

Solution (b). In the Maclaurin polynomials for $\sin x$, only the odd powers of x appear explicitly. To see this, let $f(x) = \sin x$; thus,

$$f(x) \ = \sin x \qquad f(0) \ = 0$$
$$f'(x) = \cos x \qquad f'(0) = 1$$
$$f''(x) = -\sin x \qquad f''(0) = 0$$
$$f'''(x) = -\cos x \qquad f'''(0) = -1$$

Since $f^{(4)}(x) = \sin x = f(x)$, the pattern $0, 1, 0, -1$ will repeat over and over as we evaluate successive derivatives at 0. Therefore, the successive Maclaurin polynomials for $\sin x$ are

$$p_0(x) = 0$$

$$p_1(x) = 0 + x$$

$$p_2(x) = 0 + x + 0$$

$$p_3(x) = 0 + x + 0 - \frac{x^3}{3!}$$

$$p_4(x) = 0 + x + 0 - \frac{x^3}{3!} + 0$$

$$p_5(x) = 0 + x + 0 - \frac{x^3}{3!} + 0 + \frac{x^5}{5!}$$

$$p_6(x) = 0 + x + 0 - \frac{x^3}{3!} + 0 + \frac{x^5}{5!} + 0$$

$$p_7(x) = 0 + x + 0 - \frac{x^3}{3!} + 0 + \frac{x^5}{5!} + 0 - \frac{x^7}{7!}$$

$$\vdots$$

Because of the zero terms, each even-numbered Maclaurin polynomial [after $p_0(x)$] is the same as the preceding odd-numbered Maclaurin polynomial; that is,

$$p_{2n+1}(x) = p_{2n+2}(x) = x - \frac{x^3}{3!} + \frac{x^5}{5!} - \frac{x^7}{7!} + \cdots + (-1)^n \frac{x^{2n+1}}{(2n+1)!} \quad (n = 0, 1, 2, \ldots)$$

Thus, the Maclaurin series for $\sin x$ is

$$\sum_{k=0}^{\infty} (-1)^k \frac{x^{2k+1}}{(2k+1)!} = x - \frac{x^3}{3!} + \frac{x^5}{5!} - \frac{x^7}{7!} + \cdots + (-1)^k \frac{x^{2k+1}}{(2k+1)!} + \cdots$$

The graphs of $\sin x$, $p_1(x)$, $p_3(x)$, $p_5(x)$, and $p_7(x)$ are shown in Figure 11.9.3.

Solution (c). In the Maclaurin polynomials for $\cos x$, only the even powers of x appear explicitly; the computations are similar to those in part (b). The reader should be able to show that

$$p_0(x) = p_1(x) = 1$$

$$p_2(x) = p_3(x) = 1 - \frac{x^2}{2!}$$

$$p_4(x) = p_5(x) = 1 - \frac{x^2}{2!} + \frac{x^4}{4!}$$

$$p_6(x) = p_7(x) = 1 - \frac{x^2}{2!} + \frac{x^4}{4!} - \frac{x^6}{6!}$$

In general, the Maclaurin polynomials for $\cos x$ are

$$p_{2n}(x) = p_{2n+1}(x) = 1 - \frac{x^2}{2!} + \frac{x^4}{4!} - \cdots + (-1)^n \frac{x^{2n}}{(2n)!} \quad (n = 0, 1, 2, \ldots)$$

from which it follows that the Maclaurin series for $\cos x$ is

$$\sum_{k=0}^{\infty} (-1)^k \frac{x^{2k}}{(2k)!} = 1 - \frac{x^2}{2!} + \frac{x^4}{4!} - \frac{x^6}{6!} + \cdots + (-1)^k \frac{x^{2k}}{(2k)!} + \cdots$$

The graphs of $\cos x$, $p_0(x)$, $p_2(x)$, $p_4(x)$, and $p_6(x)$ are shown in Figure 11.9.4. ◄

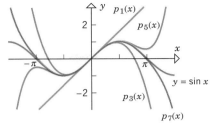

Figure 11.9.3

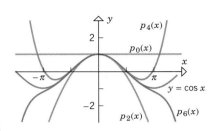

Figure 11.9.4

Example 4   Find the Taylor series about $x = 1$ for $1/x$.

Solution.   Let $f(x) = 1/x$ so that

$$f(x) = \frac{1}{x} \qquad\qquad f(1) = 1$$

$$f'(x) = -\frac{1}{x^2} \qquad\qquad f'(1) = -1$$

$$f''(x) = \frac{2}{x^3} \qquad\qquad f''(1) = 2!$$

$$f'''(x) = -\frac{3 \cdot 2}{x^4} \qquad\qquad f'''(1) = -3!$$

$$f^{(4)}(x) = \frac{4 \cdot 3 \cdot 2}{x^5} \qquad\qquad f^{(4)}(1) = 4!$$

$$\vdots \qquad\qquad\qquad \vdots$$

$$f^{(k)}(x) = (-1)^k \frac{k!}{x^{k+1}} \qquad f^{(k)}(1) = (-1)^k k!$$

$$\vdots \qquad\qquad\qquad \vdots$$

Thus, substituting in (7) with $a = 1$ yields

$$\sum_{k=0}^{\infty} \frac{(-1)^k k!}{k!} (x-1)^k = \sum_{k=0}^{\infty} (-1)^k (x-1)^k$$

$$= 1 - (x-1) + (x-1)^2 - (x-1)^3 + \cdots \qquad \blacktriangleleft$$

We conclude this section by emphasizing again that we have no guarantee that the Maclaurin and Taylor series of a function f converge, or if they do converge, that the sum is $f(x)$. Convergence questions will be addressed in the next section.

▶ Exercise Set 11.9

In Exercises 1–12, find the fourth Maclaurin polynomial ($n = 4$) for the given function.

1. e^{-2x}.

2. $\dfrac{1}{1+x}$.

3. $\sin 2x$.

4. $e^x \cos x$.

5. $\tan x$.

6. $x^3 - x^2 + 2x + 1$.

7. xe^x.

8. $\tan^{-1} x$.

9. $\sec x$.

10. $\sqrt{1+x}$.

11. $\ln(3 + 2x)$.

12. $\sinh x$.

In Exercises 13–22, find the third Taylor polynomial ($n = 3$) about $x = a$ for the given function.

13. e^x; $a = 1$.

14. $\ln x$; $a = 1$.

15. $\sqrt{x}$; $a = 4$.

16. $x^4 + x - 3$; $a = -2$.

17. $\cos x$; $a = \dfrac{\pi}{4}$.

18. $\tan x$; $a = \dfrac{\pi}{3}$.

19. $\sin \pi x$; $a = -\dfrac{1}{3}$.

20. $\csc x$; $a = \dfrac{\pi}{2}$.

21. $\tan^{-1} x$; $a = 1$.

22. $\cosh x$; $a = \ln 2$.

In Exercises 23–31, find the Maclaurin series for the given function. Express your answer in sigma notation.

23. e^{-x}.

24. e^{ax}.

25. $\dfrac{1}{1 + x}$.

26. xe^x.

27. $\ln (1 + x)$.

28. $\sin \pi x$.

29. $\cos \left(\dfrac{x}{2}\right)$.

30. $\sinh x$.

31. $\cosh x$.

In Exercises 32–39, find the Taylor series about $x = a$ for the given function. Express your answer in sigma notation.

32. $\dfrac{1}{x}$; $a = 3$.

33. $\dfrac{1}{x}$; $a = -1$.

34. e^x; $a = 2$.

35. $\ln x$; $a = 1$.

36. $\cos x$; $a = \pi/2$.

37. $\sin \pi x$; $a = 1/2$.

38. $\dfrac{1}{x + 2}$; $a = 3$.

39. $\sinh x$; $a = \ln 4$.

40. Prove: The value of $p_n(x) = \displaystyle\sum_{k=0}^{n} \dfrac{f^{(k)}(a)}{k!} (x - a)^k$ and its first n derivatives match the value of $f(x)$ and its first n derivatives at $x = a$.

■ 11.10 TAYLOR FORMULA WITH REMAINDER; CONVERGENCE OF TAYLOR SERIES

In the preceding section we anticipated the possibility that under appropriate conditions the Taylor series about $x = a$ for a function f might converge to $f(x)$ for values of x near a. In this section we shall establish conditions under which convergence occurs.

☐ **TAYLOR'S THEOREM**

If p_n denotes the nth Taylor polynomial about $x = a$ for a function f, and if we approximate $f(x)$ by $p_n(x)$ at a point x, then the difference $f(x) - p_n(x)$ is denoted by

$$R_n(x) = f(x) - p_n(x) \tag{1}$$

and is called the **nth remainder for f** about $x = a$. This equation can be rewritten as

$$f(x) = p_n(x) + R_n(x)$$

or more explicitly as

$$f(x) = \sum_{k=0}^{n} \frac{f^{(k)}(a)}{k!} (x - a)^k + R_n(x) \tag{2}$$

It is evident from (1) that $p_n(x) \to f(x)$ as $n \to +\infty$ if and only if $R_n(x) \to 0$ as $n \to +\infty$. But $p_n(x)$ is the nth partial sum of the Taylor series for f about $x = a$, so we have the following theorem.

11.10.1 THEOREM. *The equality*

$$f(x) = \sum_{k=0}^{\infty} \frac{f^{(k)}(a)}{k!} (x - a)^k$$

holds if and only if $\displaystyle\lim_{n \to +\infty} R_n(x) = 0$.

Thus, the problem of determining whether the Taylor series for f about $x = a$ converges to $f(x)$ at a point x reduces to determining whether the nth remainder for f has a limit of zero at the point. The following theorem, which is proved at the end of this section, provides a formula for the nth remainder that will be useful for investigating this convergence question.

11.10.2   THEOREM (*Taylor's Theorem*).   *Suppose that a function f can be differentiated $n + 1$ times at each point in an interval containing the point a, and let $R_n(x)$ be the nth remainder for f about $x = a$. Then for each x in that interval there is at least one point c between a and x such that*

$$R_n(x) = \frac{f^{(n+1)}(c)}{(n+1)!}(x - a)^{n+1} \tag{3}$$

REMARK.   In this theorem the statement that c is ''between'' a and x should be interpreted to mean that c is in the interval (a, x) if $a < x$, or in the interval (x, a) if $x < a$, or $c = a$ if $x = a$.

□ **LAGRANGE'S FORM OF THE REMAINDER**

Historically, (3) was not discovered by Taylor but rather by Joseph Louis Lagrange.* For this reason this formula is commonly called *Lagrange's form of the remainder*. In this formula values for c are unspecified; all that can be said about them is that they depend on a, x, and n and that they lie between a and x. Other formulas for the remainder can be found in advanced calculus texts.

Substituting (3) in (2) yields

$$f(x) = \sum_{k=0}^{n} \frac{f^{(k)}(a)}{k!}(x - a)^k + \frac{f^{(n+1)}(c)}{(n+1)!}(x - a)^{n+1}$$

---

*JOSEPH LOUIS LAGRANGE (1736–1813). French–Italian mathematician and astronomer. Lagrange, the son of a public official, was born in Turin, Italy. (Baptismal records list his name as Giuseppe Lodovico Lagrangia.) Although his father wanted him to be a lawyer, Lagrange was attracted to mathematics and astronomy after reading a memoir by the astronomer Halley. At age 16 he began to study mathematics on his own and by age 19 was appointed to a professorship at the Royal Artillery School in Turin. The following year Lagrange sent Euler solutions to some famous problems using new methods that eventually blossomed into a branch of mathematics called calculus of variations. These methods and Lagrange's applications of them to problems in celestial mechanics were so monumental that by age 25 he was regarded by many of his contemporaries as the greatest living mathematician.

In 1776, on the recommendation of Euler, he was chosen to succeed Euler as the director of the Berlin Academy. During his stay in Berlin, Lagrange distinguished himself not only in celestial mechanics, but also in algebraic equations and the theory of numbers. After twenty years in Berlin, he moved to Paris at the invitation of Louis XVI. He was given apartments in the Louvre and treated with great honor, even during the revolution.

Napoleon was a great admirer of Lagrange and showered him with honors—count, senator, and Legion of Honor. The years Lagrange spent in Paris were devoted primarily to didactic treatises summarizing his mathematical conceptions. One of Lagrange's most famous works is a memoir, *Mécanique Analytique*, in which he reduced the theory of mechanics to a few general formulas from which all other necessary equations could be derived.

It is an interesting historical fact that Lagrange's father speculated unsuccessfully in several financial ventures, so his family was forced to live quite modestly. Lagrange himself stated that if his family had money, he would not have made mathematics his vocation. In spite of his fame, Lagrange was always a shy and modest man. On his death, he was buried with honor in the Pantheon.

or equivalently,

$$f(x) = f(a) + f'(a)(x - a) + \frac{f''(a)}{2!}(x - a)^2 + \cdots$$

$$+ \frac{f^{(n)}(a)}{n!}(x - a)^n + \frac{f^{(n+1)}(c)}{(n + 1)!}(x - a)^{n+1} \qquad (4)$$

which is called **Taylor's formula with remainder**. In the special case where $a = 0$ this formula becomes

$$f(x) = f(0) + f'(0)x + \frac{f''(0)}{2!}x^2 + \cdots + \frac{f^{(n)}(0)}{n!}x^n + \frac{f^{(n+1)}(c)}{(n + 1)!}x^{n+1} \qquad (5)$$

In (4) the number c is between a and x and in (5) it is between 0 and x.

Example 1   From Example 1 of the preceding section, the nth Maclaurin polynomial for $f(x) = e^x$ is

$$1 + x + \frac{x^2}{2!} + \cdots + \frac{x^n}{n!}$$

Since $f^{(n+1)}(x) = e^x$, it follows that

$$f^{(n+1)}(c) = e^c$$

Thus, from (5)

$$e^x = 1 + x + \frac{x^2}{2!} + \cdots + \frac{x^n}{n!} + \frac{e^c}{(n + 1)!}x^{n+1} \qquad (6)$$

where c is between 0 and x. Note that (6) is valid for all real values of x since the hypotheses of Taylor's Theorem are satisfied on the interval $(-\infty, +\infty)$ (verify). ◄

Example 2   Show that the Maclaurin series for e^x converges to e^x for all x; that is,

$$e^x = \sum_{k=0}^{\infty} \frac{x^k}{k!} = 1 + x + \frac{x^2}{2!} + \frac{x^3}{3!} + \frac{x^4}{4!} + \cdots \qquad -\infty < x < +\infty$$

Solution.   We must show that $R_n(x) \to 0$ for all x as $n \to +\infty$. Thus, using Lagrange's form of the remainder given in (6) we must show that for all x

$$\lim_{n \to +\infty} R_n(x) = \lim_{n \to +\infty} \frac{e^c}{(n + 1)!}x^{n+1} = 0 \qquad (7)$$

In Lagrange's remainder formula the value of x does not depend on n, so for purposes of proving the preceding limit, we can treat x as a constant. However, c does depend on n, so it must be treated as a function of n. This creates a complication, since we have no explicit formula for this function, and hence we cannot obtain the limit directly. To circumvent this difficulty, we will resort to an indirect approach based on the Squeezing Theorem (2.8.2). For this purpose, let r be the right endpoint of the closed interval whose endpoints are 0 and x. Thus, $r = 0$ or $r = x$, depending on whether $x \le 0$ or $x > 0$. Since the exponential function is an increasing function, and since c lies between 0 and x, we are guaranteed that $e^c \le e^r$. Therefore,

$$0 \le |R_n(x)| = \left| \frac{e^c}{(n + 1)!}x^{n+1} \right| = e^c \frac{|x|^{n+1}}{(n + 1)!} \le e^r \frac{|x|^{n+1}}{(n + 1)!}$$

Since r is constant, it follows from Formula (4) of Section 11.2, with $n + 1$ in place of n and $|x|$ in place of x, that

$$\lim_{n \to +\infty} e^r \frac{|x|^{n+1}}{(n+1)!} = e^r \lim_{n \to +\infty} \frac{|x|^{n+1}}{(n+1)!} = e^r \cdot 0 = 0$$

From this limit and the Squeezing Theorem, the preceding inequalities imply that for all x

$$\lim_{n \to +\infty} |R_n(x)| = 0 \quad \text{and consequently} \quad \lim_{n \to +\infty} R_n(x) = 0 \quad \blacktriangleleft$$

Example 3   Show that the Maclaurin series for $\cos x$ converges to $\cos x$ for all x; that is,

$$\cos x = \sum_{k=0}^{\infty} (-1)^k \frac{x^{2k}}{(2k)!} = 1 - \frac{x^2}{2!} + \frac{x^4}{4!} - \frac{x^6}{6!} + \cdots \quad -\infty < x < +\infty$$

Solution.   As in the preceding example, we must show that $R_n(x) \to 0$ for all x as $n \to +\infty$. For this purpose let $f(x) = \cos x$, so that for all x we have

$$f^{(n+1)}(x) = \pm \cos x \quad \text{or} \quad f^{(n+1)}(x) = \pm \sin x$$

In all of these cases $|f^{(n+1)}(x)| \leq 1$, so that for all possible values of c

$$|f^{(n+1)}(c)| \leq 1$$

Hence, using Lagrange's form of the remainder we have

$$0 \leq |R_n(x)| = \left| \frac{f^{(n+1)}(c)}{(n+1)!} x^{n+1} \right| \leq \frac{|x|^{n+1}}{(n+1)!}$$

But

$$\lim_{n \to +\infty} \frac{|x|^{n+1}}{(n+1)!} = 0$$

so the Squeezing Theorem implies that $\lim_{n \to +\infty} R_n(x) = 0$ for all x. $\blacktriangleleft$

Example 4   Find the Taylor series for $\sin x$ about $x = \pi/2$ and show that the series converges to $\sin x$ for all x.

Solution.   Let $f(x) = \sin x$. Thus,

$$f(x) = \sin x \qquad f\left(\frac{\pi}{2}\right) = \sin \frac{\pi}{2} = 1$$

$$f'(x) = \cos x \qquad f'\left(\frac{\pi}{2}\right) = \cos \frac{\pi}{2} = 0$$

$$f''(x) = -\sin x \qquad f''\left(\frac{\pi}{2}\right) = -\sin \frac{\pi}{2} = -1$$

$$f'''(x) = -\cos x \qquad f'''\left(\frac{\pi}{2}\right) = -\cos \frac{\pi}{2} = 0$$

Since $f^{(4)}(x) = \sin x$, the pattern $1, 0, -1, 0$ will repeat over and over as we evaluate successive derivatives at $\pi/2$. Thus, the Taylor series representation about $x = \pi/2$ for $\sin x$ is

$$\sin x = 1 - \frac{1}{2!}\left(x - \frac{\pi}{2}\right)^2 + \frac{1}{4!}\left(x - \frac{\pi}{2}\right)^4 - \frac{1}{6!}\left(x - \frac{\pi}{2}\right)^6 + \cdots \tag{8}$$

which we are trying to show is valid for all x. As shown in Example 3, $\left|f^{(n+1)}(c)\right| \leq 1$, so

$$0 \leq |R_n(x)| = \left|\frac{f^{(n+1)}(c)}{(n+1)!}\left(x - \frac{\pi}{2}\right)^{n+1}\right| \leq \frac{\left|x - \frac{\pi}{2}\right|^{n+1}}{(n+1)!}$$

From Formula (4) of Section 11.2, with $|x - \pi/2|$ replacing x and $(n+1)!$ replacing $n!$, it follows that

$$\lim_{n \to +\infty} \frac{\left|x - \frac{\pi}{2}\right|^{n+1}}{(n+1)!} = 0$$

so that by the same argument given in Example 3 we have $\lim\limits_{n \to +\infty} R_n(x) = 0$ for all x. This shows that (8) is valid for all x. ◄

REMARK.   It can be shown that the Taylor series for e^x, $\sin x$, and $\cos x$ about any point $x = a$ converges to these functions for all x.

Sometimes Maclaurin series can be obtained by substituting in other Maclaurin series.

Example 5   Using the Maclaurin series

$$e^x = 1 + x + \frac{x^2}{2!} + \frac{x^3}{3!} + \frac{x^4}{4!} + \cdots \quad -\infty < x < +\infty$$

we can derive the Maclaurin series for e^{-x} by substituting $-x$ for x to obtain

$$e^{-x} = 1 + (-x) + \frac{(-x)^2}{2!} + \frac{(-x)^3}{3!} + \frac{(-x)^4}{4!} + \cdots \quad -\infty < -x < +\infty$$

or

$$e^{-x} = 1 - x + \frac{x^2}{2!} - \frac{x^3}{3!} + \frac{x^4}{4!} - \cdots \quad -\infty < x < +\infty$$

From the Maclaurin series for e^x and e^{-x} we can obtain the Maclaurin series for $\cosh x$ by writing

$$\cosh x = \frac{1}{2}(e^x + e^{-x}) = \frac{1}{2}\left(\left[1 + x + \frac{x^2}{2!} + \frac{x^3}{3!} + \frac{x^4}{4!} + \cdots\right]\right.$$

$$\left. + \left[1 - x + \frac{x^2}{2!} - \frac{x^3}{3!} + \frac{x^4}{4!} + \cdots\right]\right)$$

or

$$\cosh x = 1 + \frac{x^2}{2!} + \frac{x^4}{4!} + \cdots \quad -\infty < x < +\infty \quad ◄$$

REMARK.   Although we could have derived the preceding Maclaurin series directly, indirect methods are useful when it is messy to calculate the higher derivatives required for a Maclaurin series. There is, however, a loose thread in the logic of Example 5. We have produced a power series in x that converges to $\cosh x$ for all x. But isn't it conceivable that we have produced a power series in x *different* from the Maclaurin series? In Section 11.12 we shall show that if a power series in $x - a$ converges to $f(x)$ on some interval containing a, then the series must be the Taylor series for f about $x = a$. Thus, we are assured the series obtained for $\cosh x$ is, in fact, the Maclaurin series.

Example 6   Using the Maclaurin series

$$\frac{1}{1-x} = 1 + x + x^2 + x^3 + \cdots \quad -1 < x < 1$$

we can derive the Maclaurin series for $1/(1 - 2x^2)$ by substituting $2x^2$ for x to obtain

$$\frac{1}{1 - 2x^2} = 1 + (2x^2) + (2x^2)^2 + (2x^2)^3 + \cdots \qquad -1 < 2x^2 < 1$$

or

$$\frac{1}{1 - 2x^2} = 1 + 2x^2 + 4x^4 + 8x^6 + \cdots = \sum_{k=0}^{\infty} 2^k x^{2k} \qquad -1 < 2x^2 < 1$$

Since $2x^2 \geq 0$ for all x, the convergence condition $-1 < 2x^2 < 1$ can be written in the equivalent form $0 \leq 2x^2 < 1$ or $0 \leq x^2 < 1/2$ or $-1/\sqrt{2} < x < 1/\sqrt{2}$. ◀

☐ **BINOMIAL SERIES**

If m is a real number, then the Maclaurin series for $(1 + x)^m$ is called the **binomial series**; it is given by (verify)

$$1 + mx + \frac{m(m-1)}{2!}x^2 + \frac{m(m-1)(m-2)}{3!}x^3 + \cdots$$

REMARK.   If m is a nonnegative integer, then $f(x) = (1 + x)^m$ is a polynomial of degree m, so

$$f^{(m+1)}(0) = f^{(m+2)}(0) = f^{(m+3)}(0) = \cdots = 0$$

and the binomial series reduces to the familiar binomial expansion

$$(1 + x)^m = 1 + mx + \frac{m(m-1)}{2!}x^2 + \frac{m(m-1)(m-2)}{3!}x^3 + \cdots + x^m$$

which is valid for $-\infty < x < +\infty$.

It can be proved that if m is not a nonnegative integer, then the binomial series converges to $(1 + x)^m$ if $|x| < 1$. Thus, for such values of x

$$(1 + x)^m = 1 + mx + \frac{m(m-1)}{2!}x^2 + \cdots$$
$$+ \frac{m(m-1)(m-2)\cdots(m-k+1)}{k!}x^k + \cdots \qquad (9)$$

or in sigma notation

$$(1 + x)^m = 1 + \sum_{k=1}^{\infty} \frac{m(m-1)\cdots(m-k+1)}{k!}x^k \qquad \text{if } |x| < 1 \qquad (10)$$

Example 7   Express $1/\sqrt{1 + x}$ as a binomial series.

Solution.   Substituting $m = -\frac{1}{2}$ in (9) yields

$$\frac{1}{\sqrt{1 + x}} = 1 - \frac{1}{2}x + \frac{(-\frac{1}{2})(-\frac{1}{2} - 1)}{2!}x^2 + \frac{(-\frac{1}{2})(-\frac{1}{2} - 1)(-\frac{1}{2} - 2)}{3!}x^3$$
$$+ \cdots + \frac{(-\frac{1}{2})(-\frac{3}{2})(-\frac{5}{2})\cdots(-\frac{1}{2} - k + 1)}{k!}x^k + \cdots$$

$$= 1 - \frac{1}{2}x + \frac{1 \cdot 3}{2^2 \cdot 2!}x^2 - \frac{1 \cdot 3 \cdot 5}{2^3 \cdot 3!}x^3 + \cdots$$

$$+ (-1)^k \frac{1 \cdot 3 \cdot 5 \cdots (2k - 1)}{2^k k!}x^k + \cdots \qquad ◀$$

For reference, we have listed in Table 11.10.1 the Maclaurin series for a number of important functions, and we have indicated the interval over which the series converges to the function. It should be noted that the intervals of convergence stated for $\ln(1+x)$ and $\tan^{-1}x$ are somewhat difficult to obtain directly. However, these intervals can be obtained by indirect methods that we shall study in the last section of this chapter.

Table 11.10.1

MACLAURIN SERIES	INTERVAL OF CONVERGENCE
$\dfrac{1}{1-x} = \displaystyle\sum_{k=0}^{\infty} x^k = 1 + x + x^2 + x^3 + \cdots$	$-1 < x < 1$
$e^x = \displaystyle\sum_{k=0}^{\infty} \dfrac{x^k}{k!} = 1 + x + \dfrac{x^2}{2!} + \dfrac{x^3}{3!} + \dfrac{x^4}{4!} + \cdots$	$-\infty < x < +\infty$
$\sin x = \displaystyle\sum_{k=0}^{\infty} (-1)^k \dfrac{x^{2k+1}}{(2k+1)!} = x - \dfrac{x^3}{3!} + \dfrac{x^5}{5!} - \dfrac{x^7}{7!} + \cdots$	$-\infty < x < +\infty$
$\cos x = \displaystyle\sum_{k=0}^{\infty} (-1)^k \dfrac{x^{2k}}{(2k)!} = 1 - \dfrac{x^2}{2!} + \dfrac{x^4}{4!} - \dfrac{x^6}{6!} + \cdots$	$-\infty < x < +\infty$
$\ln(1+x) = \displaystyle\sum_{k=0}^{\infty} (-1)^k \dfrac{x^{k+1}}{k+1} = x - \dfrac{x^2}{2} + \dfrac{x^3}{3} - \dfrac{x^4}{4} + \cdots$	$-1 < x \le 1$
$\tan^{-1}x = \displaystyle\sum_{k=0}^{\infty} (-1)^k \dfrac{x^{2k+1}}{2k+1} = x - \dfrac{x^3}{3} + \dfrac{x^5}{5} - \dfrac{x^7}{7} + \cdots$	$-1 \le x \le 1$
$\sinh x = \displaystyle\sum_{k=0}^{\infty} \dfrac{x^{2k+1}}{(2k+1)!} = x + \dfrac{x^3}{3!} + \dfrac{x^5}{5!} + \dfrac{x^7}{7!} + \cdots$	$-\infty < x < +\infty$
$\cosh x = \displaystyle\sum_{k=0}^{\infty} \dfrac{x^{2k}}{(2k)!} = 1 + \dfrac{x^2}{2!} + \dfrac{x^4}{4!} + \dfrac{x^6}{6!} + \cdots$	$-\infty < x < +\infty$
$(1+x)^m = 1 + \displaystyle\sum_{k=1}^{\infty} \dfrac{m(m-1)\cdots(m-k+1)}{k!} x^k$	$-1 < x < 1*$ $(m \ne 0, 1, 2, \ldots)$

*The behavior at the endpoints depends on m: For $m \ge 0$ the series converges absolutely at both endpoints; for $m \le -1$ the series diverges at both endpoints; and for $-1 < m < 0$ the series converges conditionally at $x = 1$ and diverges at $x = -1$.

■ PROOF OF TAYLOR'S THEOREM

Proof of Theorem 11.10.2. By hypothesis, f can be differentiated $n + 1$ times at each point in an interval containing the point a. Choose any point b in this interval. To be specific, we shall assume that $b > a$. (The cases $b < a$ and $b = a$ are left to the reader.) Let $p_n(x)$ be the nth Taylor polynomial for $f(x)$ about $x = a$ and define

$$h(x) = f(x) - p_n(x) \tag{11}$$

$$g(x) = (x - a)^{n+1} \tag{12}$$

Because $f(x)$ and $p_n(x)$ have the same value and the same first n derivatives at $x = a$, it follows that

$$h(a) = h'(a) = h''(a) = \cdots = h^{(n)}(a) = 0 \tag{13}$$

Also, we leave it for the reader to show that

$$g(a) = g'(a) = g''(a) = \cdots = g^{(n)}(a) = 0 \tag{14}$$

and that $g(x)$ and its first n derivatives are nonzero when $x \neq a$.

It is straightforward to check that h and g satisfy the hypotheses of the Extended Mean-Value Theorem (10.2.2) on the interval $[a, b]$, so that there is a point c_1 with $a < c_1 < b$ such that

$$\frac{h(b) - h(a)}{g(b) - g(a)} = \frac{h'(c_1)}{g'(c_1)} \tag{15}$$

or, from (13) and (14),

$$\frac{h(b)}{g(b)} = \frac{h'(c_1)}{g'(c_1)} \tag{16}$$

If we now apply the Extended Mean-Value Theorem to h' and g' over the interval $[a, c_1]$, we may deduce that there is a point c_2 with $a < c_2 < c_1 < b$ such that

$$\frac{h'(c_1) - h'(a)}{g'(c_1) - g'(a)} = \frac{h''(c_2)}{g''(c_2)}$$

or, from (13) and (14),

$$\frac{h'(c_1)}{g'(c_1)} = \frac{h''(c_2)}{g''(c_2)}$$

which, when combined with (16), yields

$$\frac{h(b)}{g(b)} = \frac{h''(c_2)}{g''(c_2)}$$

It should now be clear that if we continue in this way, applying the Extended Mean-Value Theorem to the successive derivatives of h and g, we shall eventually obtain a relationship of the form

$$\frac{h(b)}{g(b)} = \frac{h^{(n+1)}(c_{n+1})}{g^{(n+1)}(c_{n+1})} \tag{17}$$

where $a < c_{n+1} < b$. However, $p_n(x)$ is a polynomial of degree n, so that its $(n + 1)$-st derivative is zero. Thus, from (11)

$$h^{(n+1)}(c_{n+1}) = f^{(n+1)}(c_{n+1}) \tag{18}$$

Also, from (12), the $(n + 1)$-st derivative of $g(x)$ is the constant $(n + 1)!$, so that

$$g^{(n+1)}(c_{n+1}) = (n + 1)! \tag{19}$$

Substituting (18) and (19) into (17) yields

$$\frac{h(b)}{g(b)} = \frac{f^{(n+1)}(c_{n+1})}{(n + 1)!}$$

Letting $c = c_{n+1}$, and using (11) and (12), it follows that

$$f(b) - p_n(b) = \frac{f^{(n+1)}(c)}{(n + 1)!}(b - a)^{n+1}$$

But this is precisely Formula (3) in Taylor's Theorem (Theorem 11.10.2), with the exception that the variable here is b rather than x. Thus, to finish we need only replace b by x.   ∎

▶ Exercise Set 11.10

For the functions in Exercises 1–14, find Lagrange's form of the remainder for the given values of a and n.

1. e^{2x}; $a = 0$; $n = 5$.

2. $\cos x$; $a = 0$; $n = 8$.

3. $\dfrac{1}{x + 1}$; $a = 0$; $n = 4$.

4. $\tan x$; $a = 0$; $n = 2$.

5. xe^x; $a = 0$; $n = 3$.

6. $\ln(1 + x)$; $a = 0$; $n = 5$.

7. $\tan^{-1} x$; $a = 0$; $n = 2$.

8. $\sinh x$; $a = 0$; $n = 6$.

9. $\sqrt{x}$; $a = 4$; $n = 3$.

10. $\dfrac{1}{x}$; $a = 1$; $n = 5$.

11. $\sin x$; $a = \dfrac{\pi}{6}$; $n = 4$.

12. $\cos \pi x$; $a = \dfrac{1}{2}$; $n = 2$.

13. $\dfrac{1}{(1+x)^2}$; $a = -2$; $n = 5$.

14. $\csc x$; $a = \dfrac{\pi}{2}$; $n = 1$.

> In Exercises 15–18, find Lagrange's form of the remainder $R_n(x)$ when $a = 0$.

15. $f(x) = \dfrac{1}{1-x}$.

16. $f(x) = e^{-x}$.

17. $f(x) = e^{2x}$.

18. $f(x) = \ln(1+x)$.

19. Prove: The Maclaurin series for $\sin x$ converges to $\sin x$ for all x.

20. Prove: The Taylor series for $\sin x$ about $x = \pi/4$ converges to $\sin x$ for all x.

21. Prove: The Taylor series for e^x about $x = 1$ converges to e^x for all x.

22. (a) Prove: The Maclaurin series for $\ln(1+x)$ converges to $\ln(1+x)$ if $0 \le x \le 1$. [*Remark:* The Maclaurin series actually converges to $\ln(1+x)$ on the interval $(-1, 1]$, but the Lagrange form of the remainder is not strong enough to establish this fact.]

    (b) Use $x = 1$ in the Maclaurin series for $\ln(1+x)$, Table 11.10.1, to show that
$$\ln 2 = 1 - \frac{1}{2} + \frac{1}{3} - \frac{1}{4} + \cdots$$

23. Prove: The Taylor series for e^x about any point $x = a$ converges to e^x for all x.

24. Prove: The Taylor series for $\sin x$ about any point $x = a$ converges to $\sin x$ for all x.

25. Prove: The Taylor series for $\cos x$ about any point $x = a$ converges to $\cos x$ for all x.

26. Derive (9). (Do not try to prove convergence.)

> In Exercises 27–50, use the Maclaurin series in Table 11.10.1 to obtain the Maclaurin series for the given function. In each case give the first four terms of the series and specify the interval on which the series converges to the function.

27. e^{-2x}.

28. $x^2 e^x$.

29. xe^{-x}.

30. e^{x^2}.

31. $\sin 2x$.

32. $\cos 2x$.

33. $x^2 \cos x$.

34. $\sin(x^2)$.

35. $\sin^2 x$. [*Hint:* $\sin^2 x = \frac{1}{2}(1 - \cos 2x)$.]

36. $\cos^2 x$. [*Hint:* $\cos^2 x = \frac{1}{2}(1 + \cos 2x)$.]

37. $\ln(1 - x^2)$.

38. $\ln(1 + 2x)$.

39. $\dfrac{1}{1 - 4x^2}$.

40. $\dfrac{x}{1 - x}$.

41. $\dfrac{x^2}{1 + 3x}$.

42. $\dfrac{x}{1 + x^2}$.

43. $x \sinh 2x$.

44. $\cosh(x^2)$.

45. $\sqrt{1 + 3x}$.

46. $\sqrt{1 + x^2}$.

47. $\dfrac{1}{(1 - 2x)^2}$.

48. $\dfrac{x}{(1 + 2x)^3}$.

49. $\dfrac{x}{\sqrt{1 - x^2}}$.

50. $x(1 - x^2)^{3/2}$.

51. Use the Maclaurin series for $1/(1 - x)$ to express $1/x$ in powers of $x - 1$. Find the interval of convergence.
$$\left[\text{Hint: } \frac{1}{x} = \frac{1}{1 + (x - 1)}. \right]$$

52. Show that the series
$$1 - \frac{x}{2!} + \frac{x^2}{4!} - \frac{x^3}{6!} + \cdots$$
converges to the function
$$f(x) = \begin{cases} \cos \sqrt{x}, & x \ge 0 \\ \cosh \sqrt{-x}, & x < 0 \end{cases}$$
[*Hint:* Use the Maclaurin series for $\cos x$ and $\cosh x$ to obtain series for $\cos \sqrt{x}$ where $x \ge 0$, and $\cosh \sqrt{-x}$ where $x \le 0$.]

53. If m is any real number, and k is a nonnegative integer, then we define the ***binomial coefficients*** $\dbinom{m}{k}$ by the formulas $\dbinom{m}{0} = 1$ and
$$\binom{m}{k} = \frac{m(m - 1)(m - 2) \cdots (m - k + 1)}{k!}$$
for $k \ge 1$.

Express Formula (10) in terms of binomial coefficients.

> In Exercises 54–57, use the known Maclaurin series for e^x, $\sin x$, and $\cos x$ to help find the sum of the given series.

54. $2 + \dfrac{4}{2!} + \dfrac{8}{3!} + \dfrac{16}{4!} + \cdots$.

55. $\pi - \dfrac{\pi^3}{3!} + \dfrac{\pi^5}{5!} - \dfrac{\pi^7}{7!} + \cdots$.

56. $1 - \dfrac{e^2}{2!} + \dfrac{e^4}{4!} - \dfrac{e^6}{6!} + \cdots$.

57. $1 - \ln 3 + \dfrac{(\ln 3)^2}{2!} - \dfrac{(\ln 3)^3}{3!} + \cdots$.

58. (a) Use the Maclaurin series for $\dfrac{1}{1 - x}$ to help find the Maclaurin series for the function
$$f(x) = \frac{x}{1 - x^2}$$

    (b) Use the result in part (a) to help find $f^{(5)}(0)$ and $f^{(6)}(0)$.

59. (a) Use the Maclaurin series for $\cos x$ to find the Maclaurin series for $f(x) = x^2 \cos 2x$.

(b) Use the result in part (a) to help find $f^{(5)}(0)$.

60. The purpose of this exercise is to show that the Taylor series of a function f may possibly converge to a value different from $f(x)$ for certain x. Let

$$f(x) = \begin{cases} e^{-1/x^2}, & x \neq 0 \\ 0, & x = 0 \end{cases}$$

(a) Use the definition of a derivative to show that $f'(0) = 0$.

(b) With some difficulty it can be shown that $f^{(n)}(0) = 0$ for $n \geq 2$. Accepting this fact, show that the Maclaurin series of f converges for all x, but converges to $f(x)$ only at the point $x = 0$.

11.11 COMPUTATIONS USING TAYLOR SERIES

In this section we shall show how Taylor and Maclaurin series can be used to obtain approximate values for trigonometric functions and logarithms.

☐ ACCURACY OF APPROXIMATIONS

In practical applications where a function f is approximated by a Taylor polynomial, it is important to have some way of estimating the error in the approximation. For this purpose we introduce the following terminology: An approximation is said to be **accurate to n decimal places** if the magnitude of the error is less than 0.5×10^{-n}. For example,

DESCRIPTION	MAGNITUDE OF THE ERROR IS LESS THAN	
1 decimal-place accuracy	0.05	$= 0.5 \times 10^{-1}$
2 decimal-place accuracy	0.005	$= 0.5 \times 10^{-2}$
3 decimal-place accuracy	0.0005	$= 0.5 \times 10^{-3}$
4 decimal-place accuracy	0.00005	$= 0.5 \times 10^{-4}$

☐ APPROXIMATING e

In the preceding section we showed that for all x

$$e^x = 1 + x + \frac{x^2}{2!} + \frac{x^3}{3!} + \frac{x^4}{4!} + \cdots + \frac{x^k}{k!} + \cdots$$

In particular, if we let $x = 1$, we obtain the following expression for e as the sum of an infinite series

$$e = 1 + 1 + \frac{1}{2!} + \frac{1}{3!} + \frac{1}{4!} + \cdots + \frac{1}{k!} + \cdots \tag{1}$$

Thus, we can approximate e to any degree of accuracy using an appropriate partial sum

$$e \approx 1 + 1 + \frac{1}{2!} + \frac{1}{3!} + \cdots + \frac{1}{n!} \tag{2}$$

The following example shows how Lagrange's remainder formula can be used to investigate the accuracy of such an approximation.

Example 1 Use (2) to approximate e to four decimal-place accuracy.

Solution. It follows from Taylor's formula with remainder that

$$e^x = 1 + x + \frac{x^2}{2!} + \cdots + \frac{x^n}{n!} + \frac{e^c}{(n+1)!} x^{n+1}$$

where c is between 0 and x (see Example 1, Section 11.10). Thus, in the case $x = 1$ we obtain

$$e = 1 + 1 + \frac{1}{2!} + \cdots + \frac{1}{n!} + \frac{e^c}{(n+1)!}$$

where c is between 0 and 1. This tells us that the magnitude of the error in approximation (2) is

$$|R_n(1)| = \left| \frac{e^c}{(n+1)!} \right| = \frac{e^c}{(n+1)!} \qquad (3)$$

where $0 < c < 1$. Since $c < 1$, it follows that

$$e^c < e^1 = e$$

so that from (3)

$$|R_n(1)| < \frac{e}{(n+1)!} \qquad (4)$$

This inequality provides an upper bound on the magnitude of the error R_n. Unfortunately, this inequality is not very useful since the right side involves the quantity e, which we are trying to estimate. However, if we use the fact that $e < 3$, then we can replace (4) with the following less precise but more useful result:

$$|R_n(1)| < \frac{3}{(n+1)!} \qquad (4a)$$

It follows that if we choose n so that

$$|R_n(1)| < \frac{3}{(n+1)!} < 0.5 \times 10^{-4} = 0.00005 \qquad (5)$$

then approximation (2) will be accurate to four decimal places. An appropriate value for n may be found by trial and error. For example, using a calculator one can evaluate $3/(n+1)!$ for $n = 0, 1, 2, \ldots$ until a value of n satisfying (5) is obtained. We leave it for the reader to show that $n = 8$ is the first positive integer satisfying (5). Thus, to four decimal-place accuracy

$$e \approx 1 + 1 + \frac{1}{2!} + \frac{1}{3!} + \frac{1}{4!} + \frac{1}{5!} + \frac{1}{6!} + \frac{1}{7!} + \frac{1}{8!} \approx 2.7183 \qquad (6)$$

◀

REMARK. It should be noted that there are two types of errors that result when computing with series. The first, called *truncation error*, is the error that results when the entire series is approximated by a partial sum. The second kind of error, called *roundoff error*, results when decimal approximations are used. For example, (6) involves a truncation error of at most 0.5×10^{-4} (four decimal-place accuracy). However, to obtain the numerical value in (6) we used a calculator to evaluate the left side, resulting in roundoff error due to the limitations of the calculator. The problem of controlling roundoff error is surprisingly difficult and is studied in courses in a branch of mathematics called *numerical analysis*. As a rule of thumb, to achieve n decimal-place accuracy in a final result, one should use more than $n + 1$ decimal places in each intermediate computation, then round off to n decimal places at the end. However, even this procedure may occasionally not produce n decimal-place accuracy. For purposes of this text, we recommend that you perform all intermediate calculations with the maximum number of decimal places that your calculator will allow, then round off the final result.

☐ AN UPPER BOUND ON THE REMAINDER

The technique illustrated in Example 1 can be applied to a variety of problems involving approximations by Taylor polynomials. To see how, recall from Taylor's Theorem that the absolute value of the error that results when $f(x)$ is approximated by its nth Taylor polynomial $p_n(x)$ about $x = a$ is

$$|R_n(x)| = |f(x) - p_n(x)| = \left| \frac{f^{(n+1)}(c)}{(n+1)!} (x-a)^{n+1} \right| \tag{7}$$

In this formula, c is an unknown number between a and x, so that the value of $f^{(n+1)}(c)$ usually cannot be determined. However, it is frequently possible to determine an upper bound on the size of $|f^{(n+1)}(c)|$; that is, one can often find a constant M such that $|f^{(n+1)}(c)| \leq M$. For such an M, it follows from (7) that

$$|R_n(x)| = |f(x) - p_n(x)| \leq \frac{M}{(n+1)!} |x-a|^{n+1} \tag{8}$$

which gives an upper bound on the magnitude of the error $R_n(x)$. The example that follows uses this result.

☐ APPROXIMATING TRIGONOMETRIC FUNCTIONS

To approximate the value of a function f at a point x_0 using a Taylor series, two factors enter into the selection of the point $x = a$ for the series. First, the point $x = a$ must be selected so that f and its derivatives can be evaluated at a, since those values are needed to find Taylor series. Second, it is desirable to choose a as close as possible to x_0 since the "rate of convergence" of a Taylor series is usually most rapid close to a; that is, fewer terms are required in a partial sum to achieve a given level of accuracy. For example, to approximate $\sin 3°$ ($= \pi/60$ radians), it would be reasonable to take $a = 0$, since $\pi/60$ is close to 0, and the successive derivatives of $\sin x$ are easy to evaluate at 0.

Example 2   Use the Maclaurin series for $\sin x$ to approximate $\sin 3°$ to five decimal-place accuracy.

Solution.   In the Maclaurin series

$$\sin x = x - \frac{x^3}{3!} + \frac{x^5}{5!} - \frac{x^7}{7!} + \cdots \tag{9}$$

the angle x is assumed to be in radians (because the differentiation formulas for the trigonometric functions were derived with this assumption). Since $3° = \pi/60$ radians, it follows from (9) that

$$\sin 3° = \sin \frac{\pi}{60} = \left(\frac{\pi}{60} \right) - \frac{\left(\frac{\pi}{60} \right)^3}{3!} + \frac{\left(\frac{\pi}{60} \right)^5}{5!} - \frac{\left(\frac{\pi}{60} \right)^7}{7!} + \cdots \tag{10}$$

We must now decide how many terms in this series must be kept in order to obtain five decimal-place accuracy. We shall consider two possible approaches, one using Lagrange's remainder formula, and the other exploiting the fact that (10) satisfies the conditions of the alternating series test.

If we let $f(x) = \sin x$, then the magnitude of the error that results when $\sin x$ is approximated by its nth Maclaurin polynomial is

$$|R_n(x)| = \left| \frac{f^{(n+1)}(c)}{(n+1)!} x^{n+1} \right|$$

where c is between 0 and x. Since $f^{(n+1)}(c)$ is either $\pm \sin c$ or $\pm \cos c$, it follows that $|f^{(n+1)}(c)| \leq 1$, so from (8) with $a = 0$ and $M = 1$ we have

$$|R_n(x)| \leq \frac{|x|^{n+1}}{(n+1)!}$$

In particular, if $x = \pi/60$, then

$$|R_n(\pi/60)| \leq \frac{\left(\dfrac{\pi}{60}\right)^{n+1}}{(n+1)!}$$

Thus, for five decimal-place accuracy, we must choose n so that

$$\frac{\left(\dfrac{\pi}{60}\right)^{n+1}}{(n+1)!} < 0.5 \times 10^{-5} = 0.000005$$

By trial and error with the help of a calculator or computer, the reader can check that $n = 3$ is the smallest n that works. Thus, in (10) we need only keep terms up to the third power for five decimal-place accuracy, that is,

$$\sin 3° \approx \left(\frac{\pi}{60}\right) - \frac{\left(\dfrac{\pi}{60}\right)^3}{3!} \approx 0.05234 \tag{11}$$

An alternative approach to determining n uses the fact that (10) satisfies the conditions of the alternating series test, Theorem 11.7.1. (Verify.) Thus, by Theorem 11.7.2, if we use only those terms up to and including

$$\pm \frac{\left(\dfrac{\pi}{60}\right)^m}{m!} \qquad \boxed{m \text{ is an odd positive integer.}}$$

then the magnitude of the error will be at most

$$\frac{\left(\dfrac{\pi}{60}\right)^{m+2}}{(m+2)!}$$

Thus, for five decimal-place accuracy, we look for the first positive odd integer m such that

$$\frac{\left(\dfrac{\pi}{60}\right)^{m+2}}{(m+2)!} < 0.5 \times 10^{-5} = 0.000005$$

By trial and error, $m = 3$ is the first such integer. Thus, to five decimal-place accuracy

$$\sin 3° \approx \left(\frac{\pi}{60}\right) - \frac{\left(\dfrac{\pi}{60}\right)^3}{3!} \approx 0.05234 \tag{12}$$

which is the same as our earlier result.   ◄

□ APPROXIMATING
LOGARITHMS

The Maclaurin series

$$\ln(1 + x) = x - \frac{x^2}{2} + \frac{x^3}{3} - \frac{x^4}{4} + \cdots \qquad -1 < x \leq 1 \tag{13}$$

is the starting point for the approximation of natural logarithms. Unfortunately, the usefulness of this series is limited because of its slow convergence and the restriction $-1 < x \leq 1$. However, if we replace x by $-x$ in this series, we obtain

$$\ln(1 - x) = -x - \frac{x^2}{2} - \frac{x^3}{3} - \frac{x^4}{4} - \cdots \qquad -1 \leq x < 1 \tag{14}$$

and on subtracting (14) from (13) we obtain

$$\ln\left(\frac{1 + x}{1 - x}\right) = 2\left(x + \frac{x^3}{3} + \frac{x^5}{5} + \frac{x^7}{7} + \cdots\right) \qquad -1 < x < 1 \tag{15}$$

Series (15), first obtained by James Gregory* in 1668, can be used to compute the natural logarithm of any positive number y by letting

$$y = \frac{1+x}{1-x}$$

or equivalently,

$$x = \frac{y-1}{y+1} \tag{16}$$

and noting that $-1 < x < 1$. For example, to compute $\ln 2$ we let $y = 2$ in (16), which yields $x = 1/3$. Substituting this value in (15) gives

$$\ln 2 = 2\left[(1/3) + \frac{(1/3)^3}{3} + \frac{(1/3)^5}{5} + \frac{(1/3)^7}{7} + \cdots \right]$$

Adding the four terms shown and then rounding to four decimal places at the end yields

$$\ln 2 \approx 0.6931$$

It is of interest to note that in the case $x = 1$, series (13) yields

$$\ln 2 = 1 - \frac{1}{2} + \frac{1}{3} - \frac{1}{4} + \frac{1}{5} - \cdots$$

Although this result gives the sum of the alternating harmonic series, the series converges too slowly to be of computational value.

□ APPROXIMATING π

If we let $x = 1$ in the Maclaurin series

$$\tan^{-1} x = x - \frac{x^3}{3} + \frac{x^5}{5} - \frac{x^7}{7} + \cdots \qquad -1 \le x \le 1 \tag{17}$$

we obtain

$$\frac{\pi}{4} = \tan^{-1} 1 = 1 - \frac{1}{3} + \frac{1}{5} - \frac{1}{7} + \cdots$$

or

$$\pi = 4\left[1 - \frac{1}{3} + \frac{1}{5} - \frac{1}{7} + \cdots \right]$$

This famous series, obtained by Leibniz in 1674, converges too slowly to be of computational importance. A more practical procedure for approximating π uses the identity

$$\frac{\pi}{4} = \tan^{-1}\frac{1}{2} + \tan^{-1}\frac{1}{3} \tag{18}$$

By using this identity and series (17) to approximate $\tan^{-1}\frac{1}{2}$ and $\tan^{-1}\frac{1}{3}$, the value of π can be effectively approximated to any degree of accuracy.

---

*JAMES GREGORY (1638–1675). Scottish mathematician and astronomer. Gregory, the son of a minister, was famous in his time as the inventor of the Gregorian reflecting telescope, so named in his honor. Although he is not generally ranked with the great mathematicians, much of his work relating to calculus was studied by Leibniz and Newton and undoubtedly influenced some of their discoveries. There is a manuscript, discovered posthumously, which shows that Gregory had anticipated Taylor series well before Taylor.

► Exercise Set 11.11 C 1–17, 19

1. Use inequality (4a) to find a value of n to ensure that (2) will approximate e to
   (a) five decimal-place accuracy
   (b) ten decimal-place accuracy.

In Exercises 2–8, apply Lagrange's form of the remainder.

2. Use $x = -1$ in the Maclaurin series for e^x to approximate $1/e$ to three decimal-place accuracy.

3. Use $x = \frac{1}{2}$ in the Maclaurin series for e^x to approximate $\sqrt{e}$ to four decimal-place accuracy.

4. Use the Maclaurin series for $\sin x$ to approximate $\sin 4°$ to five decimal-place accuracy.

5. Use the Maclaurin series for $\cos x$ to approximate $\cos (\pi/20)$ to four decimal-place accuracy.

6. Use an appropriate Taylor series for $\sin x$ to approximate $\sin 85°$ to four decimal-place accuracy.

7. Use the first two terms of series (15) to approximate $\ln 1.25$. Round your answer to three decimal places.

8. Use the first six terms of series (15) to approximate $\ln 3$. Round your answer to four decimal places.

9. Use the Maclaurin series for $\tan^{-1} x$ to approximate $\tan^{-1} 0.1$ to three decimal-place accuracy. [*Hint:* Use the fact that (17) is an alternating series.]

10. Use the Maclaurin series for $\sinh x$ to approximate $\sinh 0.5$ to three decimal-place accuracy.

11. Use the Maclaurin series for $\cosh x$ to approximate $\cosh 0.1$ to four decimal-place accuracy.

12. Use an appropriate Taylor series for $\sqrt[3]{x}$ to approximate $\sqrt[3]{28}$ to three decimal-place accuracy.

13. Find an interval of values for x containing $x = 0$ over which $\sin x$ can be approximated by $x - x^3/3!$ with three decimal-place accuracy ensured.

14. Find an interval of values for x over which e^x can be approximated by $1 + x + x^2/2!$ with three decimal-place accuracy ensured.

15. Find an upper bound on the magnitude of the error in the approximation

$$\cos x \approx 1 - x^2/2! + x^4/4!$$

if $-0.2 \le x \le 0.2$.

16. Find an upper bound on the magnitude of the error in the approximation $\ln (1 + x) \approx x$ if $|x| < 0.01$.

17. In each part find the number of terms of the stated series that are required to ensure that the sum of the terms approximates $\ln 2$ to six decimal-place accuracy.

   (a) The alternating harmonic series

   (b) Series (15).

18. Prove identity (18).

19. Approximate $\tan^{-1} \frac{1}{2}$ and $\tan^{-1} \frac{1}{3}$ to three decimal-place accuracy, and then use identity (18) to approximate π.

■ 11.12 DIFFERENTIATION AND INTEGRATION OF POWER SERIES

In this section we shall show that if a power series in $x - a$ converges in some interval and has a sum of $f(x)$ in that interval, then the power series must be the Taylor series about $x = a$ for the function f. This result is important for many reasons, but it is important for computational purposes because it will enable us to find Taylor series without using the defining formula for such series. This is often necessary in cases where successive derivatives of f are prohibitively complicated to compute. In this section we shall also consider conditions under which a Taylor series can be differentiated or integrated term by term.

☐ **POWER SERIES REPRESENTATIONS OF FUNCTIONS**

If a function f is expressed as the sum of a power series for all x in some interval, then we shall say that the power series *represents* f on the interval or that the series is a ***power series representation of f*** on the interval. For example, we know from our study of geometric series that

$$\frac{1}{1 - x} = 1 + x + x^2 + x^3 + \cdots \quad -1 < x < 1$$

Thus, the series $1 + x + x^2 + x^3 + \cdots$ represents the function

$$f(x) = \frac{1}{1 - x}$$

on the interval $(-1, 1)$.

The following theorems show that if a function f is represented by a power series on an interval, then differentiating the series term by term produces a power series representation of f' on the interval, and integrating the series term by term produces a power series representation of the integral of f on the interval.

11.12.1 THEOREM (*Differentiation of Power Series*).   *If a function f is represented by a power series, say*

$$f(x) = \sum_{k=0}^{\infty} c_k(x-a)^k$$

where the series has a nonzero radius of convergence R, then:

(a) *The series of differentiated terms*

$$\sum_{k=0}^{\infty} \frac{d}{dx}[c_k(x-a)^k] = \sum_{k=1}^{\infty} kc_k(x-a)^{k-1}$$

   has radius of convergence R.

(b) *The function f is differentiable on the interval $(a - R, a + R)$, and for every x in this interval*

$$f'(x) = \sum_{k=0}^{\infty} \frac{d}{dx}[c_k(x-a)^k]$$

To paraphrase this theorem informally, *a power series representation of a function can be differentiated term by term on any open interval within the interval of convergence.*

Example 1   To illustrate this theorem, we shall use the Maclaurin series

$$\sin x = x - \frac{x^3}{3!} + \frac{x^5}{5!} - \frac{x^7}{7!} + \cdots \qquad -\infty < x < +\infty$$

$$\cos x = 1 - \frac{x^2}{2!} + \frac{x^4}{4!} - \frac{x^6}{6!} + \cdots \qquad -\infty < x < +\infty$$

$$e^x = 1 + x + \frac{x^2}{2!} + \frac{x^3}{3!} + \frac{x^4}{4!} + \cdots \qquad -\infty < x < +\infty$$

to obtain the familiar derivative formulas

$$\frac{d}{dx}[\sin x] = \cos x \quad \text{and} \quad \frac{d}{dx}[e^x] = e^x$$

Differentiating the Maclaurin series for $\sin x$ and e^x term by term yields

$$\frac{d}{dx}[\sin x] = \frac{d}{dx}\left[x - \frac{x^3}{3!} + \frac{x^5}{5!} - \frac{x^7}{7!} + \cdots\right]$$

$$= 1 - 3\frac{x^2}{3!} + 5\frac{x^4}{5!} - 7\frac{x^6}{7!} + \cdots$$

$$= 1 - \frac{x^2}{2!} + \frac{x^4}{4!} - \frac{x^6}{6!} + \cdots = \cos x$$

$$\frac{d}{dx}[e^x] = \frac{d}{dx}\left[1 + x + \frac{x^2}{2!} + \frac{x^3}{3!} + \cdots\right]$$

$$= 1 + 2\frac{x}{2!} + 3\frac{x^2}{3!} + 4\frac{x^3}{4!} + \cdots$$

$$= 1 + x + \frac{x^2}{2!} + \frac{x^3}{3!} + \cdots = e^x \qquad \blacktriangleleft$$

11.12.2 THEOREM (*Integration of Power Series*). *If a function f is represented by a power series, say*

$$f(x) = \sum_{k=0}^{\infty} c_k(x - a)^k$$

where the series has a nonzero radius of convergence R, then:

(a) *The series of integrated terms*

$$\sum_{k=0}^{\infty} \left[\int c_k(x - a)^k \, dx \right] = \sum_{k=0}^{\infty} \frac{c_k}{k + 1} (x - a)^{k+1}$$

has radius of convergence R.

(b) *The function f is continuous on the interval $(a - R, a + R)$ and for all x in this interval*

$$\int f(x) \, dx = \sum_{k=0}^{\infty} \left[\int c_k(x - a)^k \, dx \right] + C$$

(c) *For all α and β in the interval $(a - R, a + R)$, the series*

$$\sum_{k=0}^{\infty} \left[\int_{\alpha}^{\beta} c_k(x - a)^k \, dx \right]$$

converges absolutely and

$$\int_{\alpha}^{\beta} f(x) \, dx = \sum_{k=0}^{\infty} \left[\int_{\alpha}^{\beta} c_k(x - a)^k \, dx \right]$$

To paraphrase this theorem informally, *a power series representation of a function can be integrated term by term on any interval within the interval of convergence.*

REMARK.   Note that in part (*b*) a separate constant of integration is not introduced for each term in the series; rather, a single constant C is added to the entire series.

Example 2   To illustrate part (*b*) of the preceding theorem, we shall use the Maclaurin series for $\sin x$ and $\cos x$ to obtain the familiar integration formula

$$\int \cos x \, dx = \sin x + C$$

Integrating the Maclaurin series for $\cos x$ term by term yields

$$\int \cos x \, dx = \int \left[1 - \frac{x^2}{2!} + \frac{x^4}{4!} - \frac{x^6}{6!} + \cdots \right] dx$$

$$= \left[x - \frac{x^3}{3(2!)} + \frac{x^5}{5(4!)} - \frac{x^7}{7(6!)} + \cdots \right] + C$$

$$= \left[x - \frac{x^3}{3!} + \frac{x^5}{5!} - \frac{x^7}{7!} + \cdots \right] + C = \sin x + C \qquad \blacktriangleleft$$

Example 3   The integral

$$\int_0^1 e^{-x^2} \, dx$$

cannot be evaluated directly because there is no elementary antiderivative of e^{-x^2}. However, it is possible to approximate the integral by some numerical technique such as

Simpson's rule. Still another possibility is to represent e^{-x^2} by its Maclaurin series and then integrate term by term in accordance with part (c) of Theorem 11.12.2. This produces a series that converges to the integral.

The simplest way to obtain the Maclaurin series for e^{-x^2} is to replace x by $-x^2$ in the Maclaurin series

$$e^x = 1 + x + \frac{x^2}{2!} + \frac{x^3}{3!} + \frac{x^4}{4!} + \cdots$$

to obtain

$$e^{-x^2} = 1 - x^2 + \frac{x^4}{2!} - \frac{x^6}{3!} + \frac{x^8}{4!} - \cdots$$

Therefore,

$$\int_0^1 e^{-x^2}\, dx = \int_0^1 \left[1 - x^2 + \frac{x^4}{2!} - \frac{x^6}{3!} + \frac{x^8}{4!} - \cdots \right] dx$$

$$= \left[x - \frac{x^3}{3} + \frac{x^5}{5(2!)} - \frac{x^7}{7(3!)} + \frac{x^9}{9(4!)} - \cdots \right]_0^1$$

$$= 1 - \frac{1}{3} + \frac{1}{5 \cdot 2!} - \frac{1}{7 \cdot 3!} + \frac{1}{9 \cdot 4!} - \cdots \tag{1}$$

Thus, we have found a series that converges to the value of the integral $\int_0^1 e^{-x^2}\, dx$. Using the first three terms in this series we obtain the approximation

$$\int_0^1 e^{-x^2}\, dx \approx 1 - \frac{1}{3} + \frac{1}{10} = \frac{23}{30} \approx 0.767$$

Since series (1) satisfies the conditions of the alternating series test, it follows from Theorem 11.7.2 that the magnitude of the error in this approximation is at most $1/(7 \cdot 3!) = 1/42 \approx 0.0238$. Greater accuracy can be obtained by using more terms in the series.   ◄

☐ **POWER SERIES REPRESENTATIONS MUST BE TAYLOR SERIES**

The following theorem shows that if a function f is represented by a power series in $x - a$ on an interval, then that series must be the Taylor series for f; thus, Taylor series are the only power series that can represent functions on an interval.

11.12.3   THEOREM.   *If*

$$f(x) = c_0 + c_1(x - a) + c_2(x - a)^2 + \cdots + c_n(x - a)^n + \cdots$$

for all x in some open interval containing a, then the series is the Taylor series for f about a.

Proof.   By repeated application of Theorem 11.12.1(b) we obtain

$$f(x) = c_0 + c_1(x - a) + c_2(x - a)^2 + c_3(x - a)^3 + c_4(x - a)^4 + \cdots$$

$$f'(x) = c_1 + 2c_2(x - a) + 3c_3(x - a)^2 + 4c_4(x - a)^3 + \cdots$$

$$f''(x) = 2!c_2 + (3 \cdot 2)c_3(x - a) + (4 \cdot 3)c_4(x - a)^2 + \cdots$$

$$f'''(x) = 3!c_3 + (4 \cdot 3 \cdot 2)c_4(x - a) + \cdots$$

$$\vdots$$

On substituting $x = a$, all the powers of $x - a$ drop out leaving

$$f(a) = c_0, \quad f'(a) = c_1, \quad f''(a) = 2!c_2, \quad f'''(a) = 3!c_3, \quad \ldots$$

from which we obtain

$$c_0 = f(a), \quad c_1 = f'(a), \quad c_2 = \frac{f''(a)}{2!}, \quad c_3 = \frac{f'''(a)}{3!}, \quad \ldots$$

which shows that the coefficients $c_0, c_1, c_2, c_3, \ldots$ are precisely the coefficients in the Taylor series about a for $f(x)$. ∎

REMARK.   The preceding theorem tells us that no matter how we arrive at a power series in $x - a$ converging to $f(x)$, be it by substitution, by integration, by differentiation, or by algebraic manipulation, the resulting series will be the Taylor series about a for $f(x)$.

Example 4   Find the Maclaurin series for $\tan^{-1} x$.

Solution.   We could calculate this Maclaurin series directly. However, we can also exploit Theorem 11.12.2 by first observing that

$$\int \frac{1}{1 + x^2}\, dx = \tan^{-1} x + C$$

and then integrating the Maclaurin series for $1/(1 + x^2)$ term by term. Since

$$\frac{1}{1 - x} = 1 + x + x^2 + x^3 + x^4 + \cdots \qquad -1 < x < 1$$

it follows on replacing x by $-x^2$ that

$$\frac{1}{1 + x^2} = 1 - x^2 + x^4 - x^6 + x^8 - \cdots \qquad -1 < x < 1 \tag{2}$$

Thus,

$$\tan^{-1} x + C = \int \frac{1}{1 + x^2}\, dx = \int [1 - x^2 + x^4 - x^6 + x^8 - \cdots]\, dx$$

or

$$\tan^{-1} x = \left[x - \frac{x^3}{3} + \frac{x^5}{5} - \frac{x^7}{7} + \frac{x^9}{9} - \cdots \right] - C$$

The constant of integration may be evaluated by substituting $x = 0$ and using the condition $\tan^{-1} 0 = 0$. This gives $C = 0$, so that

$$\tan^{-1} x = x - \frac{x^3}{3} + \frac{x^5}{5} - \frac{x^7}{7} + \frac{x^9}{9} - \cdots \tag{3}$$

We are guaranteed by Theorem 11.12.3 that the series we have produced is the Maclaurin series for $\tan^{-1} x$ and that it actually converges to $\tan^{-1} x$ for $-1 < x < 1$. ◀

REMARK.   Theorems 11.12.1 and 11.12.2 say nothing about the behavior of the differentiated and integrated series at the endpoints $a - R$ and $a + R$. Indeed, by differentiating termwise, convergence may be lost at one or both endpoints; and by integrating termwise, convergence may be gained at one or both endpoints. As an illustration, we derived series (3) for $\tan^{-1} x$ by integrating series (2) for $1/(1 + x^2)$. We omit the proof that the series for $\tan^{-1} x$ converges to $\tan^{-1} x$ on the interval $[-1, 1]$, while the series for $1/(1 + x^2)$ converges to $1/(1 + x^2)$ only on $(-1, 1)$. Thus, convergence was gained at both endpoints by integrating.

MISCELLANEOUS TECHNIQUES FOR OBTAINING TAYLOR SERIES

We conclude this section with some techniques for obtaining Taylor series that would be messy to obtain directly.

$$1 - x^2 + \frac{x^4}{2} - \cdots$$
$$\times \quad x - \frac{x^3}{3} + \frac{x^5}{5} - \cdots$$
$$\overline{\quad x - x^3 + \frac{x^5}{2} - \cdots}$$
$$\qquad -\frac{x^3}{3} + \frac{x^5}{3} - \frac{x^7}{6} + \cdots$$
$$\qquad\qquad \frac{x^5}{5} - \frac{x^7}{5} + \cdots$$
$$\overline{\quad x - \frac{4}{3}x^3 + \frac{31}{30}x^5 - \cdots}$$

Example 5 Find the first three terms that occur in the Maclaurin series for $e^{-x^2} \tan^{-1} x$.

Solution. Using the series for e^{-x^2} and $\tan^{-1} x$ obtained in Examples 3 and 4 gives

$$e^{-x^2} \tan^{-1} x = \left(1 - x^2 + \frac{x^4}{2} - \cdots \right)\left(x - \frac{x^3}{3} + \frac{x^5}{5} - \cdots \right)$$

Multiplying, as shown in the margin, we obtain

$$e^{-x^2} \tan^{-1} x = x - \frac{4}{3}x^3 + \frac{31}{30}x^5 - \cdots \qquad -1 < x < 1$$

More terms in the series can be obtained by including more terms in the factors. ◀

Example 6 Find the first three nonzero terms in the Maclaurin series for $\tan x$.

$$x + \frac{x^3}{3} + \frac{2x^5}{15} + \cdots$$
$$1 - \frac{x^2}{2} + \frac{x^4}{24} - \cdots \overline{\left)\, x - \frac{x^3}{6} + \frac{x^5}{120} - \cdots \right.}$$
$$\qquad x - \frac{x^3}{2} + \frac{x^5}{24} - \cdots$$
$$\overline{\qquad \frac{x^3}{3} - \frac{x^5}{30} + \cdots}$$
$$\qquad \frac{x^3}{3} - \frac{x^5}{6} + \cdots$$
$$\overline{\qquad\qquad \frac{2x^5}{15} + \cdots}$$

Solution. Instead of computing the series directly, we write

$$\tan x = \frac{\sin x}{\cos x} = \frac{x - \dfrac{x^3}{3!} + \dfrac{x^5}{5!} - \cdots}{1 - \dfrac{x^2}{2!} + \dfrac{x^4}{4!} - \cdots}, \qquad -\frac{\pi}{2} < x < \frac{\pi}{2}$$

Dividing, as shown in the margin, we obtain

$$\tan x = x + \frac{x^3}{3} + \frac{2x^5}{15} + \cdots \qquad -\frac{\pi}{2} < x < \frac{\pi}{2}$$ ◀

▶ **Exercise Set 11.12** Ⓒ *14–21*

In Exercises 1–4, obtain the stated results by differentiating or integrating Maclaurin series term by term.

1. (a) $\dfrac{d}{dx}[e^x] = e^x$ (b) $\displaystyle\int e^x \, dx = e^x + C.$

2. (a) $\dfrac{d}{dx}[\cos x] = -\sin x$

(b) $\displaystyle\int \sin x \, dx = -\cos x + C.$

3. (a) $\dfrac{d}{dx}[\sinh x] = \cosh x$

(b) $\displaystyle\int \sinh x \, dx = \cosh x + C.$

4. (a) $\dfrac{d}{dx}[\ln(1 + x)] = \dfrac{1}{1 + x}$

(b) $\displaystyle\int \frac{1}{1 + x} \, dx = \ln(1 + x) + C.$

5. Derive the Maclaurin series for $1/(1 + x)^2$ by differentiating an appropriate Maclaurin series term by term.

6. By differentiating an appropriate series, show that

$$\sum_{k=1}^{\infty} kx^k = \frac{x}{(1 - x)^2} \qquad \text{for } -1 < x < 1$$

$$\left[\text{Hint: Consider } x \frac{d}{dx}\left[\frac{1}{1 - x} \right]. \right]$$

7. By integrating an appropriate series, show that

$$\sum_{k=1}^{\infty} \frac{x^k}{k} = \ln\left(\frac{1}{1 - x} \right) \qquad \text{for } -1 < x < 1$$

$$\left[\text{Hint: } \ln\left(\frac{1}{1 - x} \right) = -\ln(1 - x). \right]$$

8. Use the result of Exercise 6 to find the sum of the series

$$\frac{1}{3} + \frac{2}{3^2} + \frac{3}{3^3} + \frac{4}{3^4} + \cdots$$

9. Use the result of Exercise 7 to find the sum of the series

$$\frac{1}{4} + \frac{1}{2(4^2)} + \frac{1}{3(4^3)} + \frac{1}{4(4^4)} + \cdots$$

10. Find the sum

$$\sum_{k=0}^{\infty} \frac{k+1}{k!} = 1 + 2 + \frac{3}{2!} + \frac{4}{3!} + \frac{5}{4!} + \cdots$$

[*Hint:* Differentiate the Maclaurin series for xe^x.]

11. Find the sum of the series

$$2 + 6x + 12x^2 + 20x^3 + \cdots$$

[*Hint:* Find the second derivative of the Maclaurin series for $1/(1-x)$.]

12. Find the sum $\sum_{k=1}^{\infty} \frac{k^2}{4^k}$. [*Hint:* Differentiate the Maclaurin series for $1/(1-x)$, multiply by x, differentiate, and multiply by x again.]

13. Let $f(x) = \sum_{k=0}^{\infty} (-1)^k \frac{x^{k+1}}{k+1}$

$$= x - \frac{x^2}{2} + \frac{x^3}{3} - \frac{x^4}{4} + \cdots$$

(a) Use the ratio test to show that the series converges for all x in the interval $(-1, 1)$.

(b) Use part (a) of Theorem 11.12.1 to find a power series for $f'(x)$. What is its interval of convergence?

(c) From the series obtained in part (b), deduce that $f'(x) = 1/(1+x)$ and hence that

$$f(x) = \ln(1+x) \quad \text{for } -1 < x < 1$$

[*Remark:* The Lagrange form of the remainder can be used to show that the Maclaurin series for $\ln(1+x)$ converges to $\ln(1+x)$ for $x = 1$ as well.]

In Exercises 14–21, use series to approximate the value of the integral to three decimal-place accuracy.

14. $\int_0^1 \sin x^2 \, dx.$

15. $\int_0^1 \cos \sqrt{x} \, dx.$

16. $\int_0^{0.1} \frac{\sin x}{x} \, dx.$

17. $\int_0^{1/2} \frac{dx}{1+x^4}.$

18. $\int_0^{1/2} \tan^{-1} 2x^2 \, dx.$

19. $\int_0^{0.1} e^{-x^3} \, dx.$

20. $\int_0^{0.2} \sqrt[3]{1+x^4} \, dx.$

21. $\int_0^{1/2} \frac{dx}{\sqrt[4]{x^2+1}}.$

In Exercises 22–29, use any method to find the first four nonzero terms in the Maclaurin series of the given function.

22. $x^4 e^x.$

23. $e^{-x^2} \cos x.$

24. $\frac{x^2}{1+x^4}.$

25. $\frac{\sin x}{e^x}.$

26. $\tanh x.$

27. $x \ln(1-x^2).$

28. $\frac{\ln(1+x)}{1-x}.$

29. $x^2 e^{4x} \sqrt{1+x}.$

30. Obtain the familiar result, $\lim_{x \to 0} (\sin x)/x = 1$, by finding a power series for $(\sin x)/x$ and taking the limit term by term.

31. Use the method of Exercise 30 to find the limits.

(a) $\lim_{x \to 0} \frac{1 - \cos x}{\sin x}$

(b) $\lim_{x \to 0} \frac{\ln \sqrt{1+x} - \sin 2x}{x}.$

32. (a) Use the relationship

$$\int \frac{1}{\sqrt{1-x^2}} \, dx = \sin^{-1} x + C$$

to find the first four nonzero terms in the Maclaurin series for $\sin^{-1} x$.

(b) Express the series in sigma notation.

(c) What is the radius of convergence?

33. (a) Use the relationship

$$\int \frac{1}{\sqrt{1+x^2}} \, dx = \sinh^{-1} x + C$$

to find the first four nonzero terms in the Maclaurin series for $\sinh^{-1} x$.

(b) Express the series in sigma notation.

(c) What is the radius of convergence?

34. Prove: If the power series $\sum_{k=0}^{\infty} a_k x^k$ and $\sum_{k=0}^{\infty} b_k x^k$ have the same sum on an interval $(-r, r)$, then $a_k = b_k$ for all values of k.

◆ TECHNOLOGY EXERCISES   Chapter 11

Most of these exercises require access to a graphing calculator or a computer algebra system (CAS) such as *Mathematica*, *Maple*, or *Derive*. When you are asked to *find* an answer or to *solve* an equation, you may choose to find an exact result or a numerical approximation, depending on the particular technology you are using and on your own imagination. The form of your answers may differ from those of other students or from those in the answer section of the text, depending on how you solve the problems and the accuracy you use in your numerical approximations. Those exercises that are more appropriate for a CAS than a graphing calculator are labeled with the icon ◆.

◆ **1. An approximation for π:** The Maclaurin series for $\tan^{-1} x$ converges on the interval $[-1, 1]$ and is given by

$$\tan^{-1} x = x - \frac{x^3}{3} + \frac{x^5}{5} - \frac{x^7}{7} + \cdots \qquad (1)$$

If $x = 1$, then we obtain

$$\frac{\pi}{4} = 1 - \frac{1}{3} + \frac{1}{5} - \frac{1}{7} + \cdots$$

(a) Use Theorem 11.7.2 to find a value for n to ensure that the nth partial sum s_n for the series for $\pi/4$ will approximate $\pi/4$ with an error that is less than 10^{-3} in magnitude.

(b) Using the value of n obtained in part (a), find an approximation for π and a bound on the magnitude of the error in this approximation.

(c) Verify that the bound obtained in part (b) is correct by comparing your approximation with the value for π produced by a calculator or CAS.

2. A better approximation for π: A formula for $\pi/4$ discovered by the British astronomer and mathematician John Machin in 1706 is

$$\frac{\pi}{4} = 4 \tan^{-1} \frac{1}{5} - \tan^{-1} \frac{1}{239}$$

(a) Let s_m be the mth partial sum for (1) in Exercise 1 with $x = 1/5$, and let s_n' be the nth partial sum for (1) with $x = 1/239$. Use Theorem 11.7.2 to find values of m and n to ensure that s_m and s_n' will approximate $\tan^{-1}(1/5)$ and $\tan^{-1}(1/239)$, respectively, with errors less than 10^{-7}.

(b) Using the values of m and n obtained in part (a), find an approximation for π from Machin's formula.

(c) Show that the magnitude of the error in the approximation to π obtained in part (b) is less than 2×10^{-6},

and verify that this bound is correct by comparing your approximation with the value for π produced by a calculator or CAS. [*Hint:* Use the triangle inequality (Theorem 1.2.5) to find the bound.]

◆ **3. An even better approximation for π:** In 1914, the brilliant Indian mathematician Ramanujan showed that

$$\frac{1}{\pi} = \frac{\sqrt{8}}{9801} \sum_{k=0}^{\infty} \frac{(4k)! \, (1103 + 26{,}390k)}{(k!)^4 \, 396^{4k}}$$

(a) Use the first two terms in Ramanujan's formula to obtain an approximation for π. [*Note:* Because of the high degree of accuracy in this formula, we suggest that you display 20 digits.]

(b) Compare your approximation for π in part (a) with the value for π in Figure 1.1.3.

◆ **4. Approximating tan (1/2):** Let $f(x) = \tan x$, and let $p_5(x)$ be the Maclaurin polynomial of degree 5 for $f(x)$.

(a) Use Lagrange's form of the remainder to find a bound on the magnitude of the error in using $p_5(1/2)$ to approximate $\tan (1/2)$.

(b) Find $p_5(1/2)$ and verify that the bound found in part (a) is correct by comparing $p_5(1/2)$ with the value of $\tan (1/2)$ obtained from a calculator or computer.

◆ **5. Maclaurin polynomials for tan x:** Graph $f(x) = \tan x$ together with the Maclaurin polynomials $p_1(x)$, $p_3(x)$, $p_5(x)$, and $p_7(x)$ in the same coordinate system over the interval $[-1.5, 1.5]$.

◆ **6. Maclaurin series for rational functions:** Let

$$f(x) = \frac{2x^3 + 5x - 9}{x^4 - x^3 - 3x^2 - 3x - 18}$$

(a) Find the partial fraction decomposition of $f(x)$.

(b) Use the result of part (a) to find the radius of convergence of the Maclaurin series for $f(x)$.

7. **The Riemann zeta function:** For real values of $x > 1$, the function defined by the convergent p-series

$$\zeta(x) = \sum_{k=1}^{\infty} \frac{1}{k^x}$$

is called the *Riemann zeta function*.

(a) Let s_n be the nth partial sum for the series for $\zeta(3.7)$. Find a value for n to ensure that s_n will approximate $\zeta(3.7)$ with an error that is less than 10^{-5}. [*Hint:* From Exercise 36 in Section 11.4, if S is the sum of the series, then

$$S - s_n < \int_n^{+\infty} f(x)\, dx$$

where $f(x) = 1/x^{3.7}$.]

(b) Approximate $\zeta(3.7)$ by s_n using the value of n obtained in part (a).

8. **Euler's constant:** The number γ defined by

$$\gamma = \lim_{n \to +\infty} \left(\sum_{k=1}^{n} \frac{1}{k} - \ln n \right)$$

$$= \lim_{n \to +\infty} \left(1 + \frac{1}{2} + \frac{1}{3} + \cdots + \frac{1}{n} - \ln n \right)$$

is called *Euler's constant*. Approximate Euler's constant by finding the value of

$$\sum_{k=1}^{n} \frac{1}{k} - \ln n$$

for $n = 10,000$. [*Note:* For comparison, the value of γ to ten decimal places is 0.5772156649.]

9. **The Bernoulli numbers:** Let $f(x) = x/(e^x - 1)$. Note that f and its derivatives are defined everywhere except at $x = 0$; however, if we define $f(0) = 1$, then it can be shown that $f^{(k)}(x)$ is continuous at $x = 0$ for $k = 0, 1, 2, 3, \ldots$, and thus f has a Maclaurin series of the form

$$\frac{x}{e^x - 1} = \sum_{k=0}^{\infty} \frac{B_k}{k!} x^k$$

where $B_k = f^{(k)}(0)$, $k = 0, 1, 2, 3, \ldots$. The number B_k is called the kth **Bernoulli number**. Use the continuity of $f^{(k)}(x)$ at $x = 0$ to find B_0 through B_6 from the formula

$$B_k = \lim_{x \to 0} f^{(k)}(x)$$

Approximation of functions: The nth Taylor polynomial for a function $f(x)$ about $x = a$ is generally a good approximation for $f(x)$ when x is near a, but the accuracy usually deteriorates as the distance between x and a increases. In Exercises 10 and 11 we will consider alternative approximations that may produce better accuracy than a Taylor polynomial.

10. **Polynomial approximations**

(a) Let $E_1(x) = \cos x - p_4(x)$, where

$$p_4(x) = 1 - \frac{x^2}{2!} + \frac{x^4}{4!}$$

is the fourth-degree Maclaurin polynomial approximation for $\cos x$. Graph $E_1(x)$ for $-1 \le x \le 1$. Find the maximum value of $|E_1(x)|$ on the interval $[-1, 1]$.

(b) Using methods from numerical analysis, it can be shown that the polynomial

$$p(x) = 0.99995795 - 0.49924045 x^2 \\ + 0.03962674 x^4$$

is a good approximation for $\cos x$ on the interval $[-1, 1]$. Let $E_2(x) = \cos x - p(x)$. Graph $E_2(x)$ for $-1 \le x \le 1$. Find the maximum value of $|E_2(x)|$ on the interval $[-1, 1]$.

(c) Discuss the advantages and disadvantages of the approximations $p_4(x)$ and $p(x)$ for $\cos x$ in parts (a) and (b).

11. **Rational approximations**

(a) Let $E_1(x) = e^x - p_5(x)$, where

$$p_5(x) = 1 + x + \frac{x^2}{2!} + \frac{x^3}{3!} + \frac{x^4}{4!} + \frac{x^5}{5!}$$

is the fifth-degree Maclaurin polynomial approximation for e^x. Graph $E_1(x)$ for $-1 \le x \le 1$. Find the maximum value of $|E_1(x)|$ on the interval $[-1, 1]$.

(b) Using methods from numerical analysis, it can be shown that the rational function

$$r(x) = \frac{1 + \frac{3}{5}x + \frac{3}{20}x^2 + \frac{1}{60}x^3}{1 - \frac{2}{5}x + \frac{1}{20}x^2}$$

is a good approximation for e^x on the interval $[-1, 1]$. Let $E_2(x) = e^x - r(x)$. Graph $E_2(x)$ for $-1 \le x \le 1$. Find the maximum value of $|E_2(x)|$ on the interval $[-1, 1]$.

(c) Discuss the advantages and disadvantages of the approximations $p_5(x)$ and $r(x)$ for e^x in parts (a) and (b).

■ 7.7 FIRST-ORDER DIFFERENTIAL EQUATIONS AND APPLICATIONS

Many of the principles in science and engineering concern relationships between changing quantities. Since rates of change are represented mathematically by derivatives, such principles often lead to equations involving unknown functions and their derivatives. In this section we shall study some basic equations of this type, and in Chapter 19 we go a little further into this topic.

□ TERMINOLOGY

A **differential equation** is an equation involving one or more derivatives of an unknown function. The **order** of a differential equation is defined to be the order of the highest derivative it contains. Here are some examples:

DIFFERENTIAL EQUATION	ORDER
$\dfrac{dy}{dx} = 3y$	1
$\dfrac{d^2y}{dx^2} - 6\dfrac{dy}{dx} + 8y = 0$	2
$y' - y = e^{2x}$	1
$\dfrac{d^3y}{dt^3} - t\dfrac{dy}{dt} + (t^2 - 1)y = e^t$	3

In the first three equations, $y = y(x)$ is an unknown function of x, and in the last equation $y = y(t)$ is an unknown function of t. The variable t is generally used in applied problems involving time. Except in such problems, we shall use x as the independent variable.

A function $y = y(x)$ is a **solution** of a differential equation if the equation is satisfied for all x in some interval when y and its derivatives are substituted. For example, $y = e^{2x}$ is a solution of the differential equation

$$\frac{dy}{dx} - y = e^{2x} \tag{1}$$

since substituting y and its derivative on the left side of this equation yields

$$\frac{dy}{dx} - y = 2e^{2x} - e^{2x} = e^{2x}$$

This is not the only solution; for any choice of the constant C the function

$$y = Ce^x + e^{2x} \tag{2}$$

is also a solution, since

$$\frac{dy}{dx} - y = (Ce^x + 2e^{2x}) - (Ce^x + e^{2x}) = e^{2x}$$

One can prove that *all* solutions of (1) can be obtained by substituting values for the constant C. A solution of a differential equation from which all solutions can be derived by substituting values for arbitrary constants is called the **general solution** of the equation. Usually, the general solution of an nth-order differential equation contains n arbitrary constants. For example, (2), which is the general solution of the first-order equation (1), contains one arbitrary constant.

□ INITIAL-VALUE
   PROBLEMS

When a physical problem leads to a differential equation, there are usually conditions in the problem that determine specific values for the arbitrary constants in the general solution of the equation. For a first-order equation, a condition that specifies the value of the unknown function $y(x)$ at some point $x = x_0$ is called an **initial condition**. A first-order differential equation together with one initial condition constitutes a **first-order initial-value problem**.

Example 1   The solution of the initial-value problem

$$\frac{dy}{dx} - y = e^{2x}, \quad y(0) = 3$$

can be obtained by substituting the initial condition $x = 0$, $y = 3$ in the general solution (2) to find C. We obtain

$$3 = Ce^0 + e^0 = C + 1$$

Thus, $C = 2$, and the solution of the initial-value problem, which is obtained by substituting this value of C in (2), is

$$y = 2e^x + e^{2x} \qquad \blacktriangleleft$$

☐ **FIRST-ORDER EQUATIONS**

The simplest first-order differential equations are of the form

$$\frac{dy}{dx} = f(x) \tag{3}$$

Such equations can sometimes be solved by integration. For example, if

$$\frac{dy}{dx} = x^3 \tag{4}$$

then

$$y = \int x^3 \, dx = \frac{x^4}{4} + C$$

is the general solution of (4).

☐ **FIRST-ORDER SEPARABLE EQUATIONS**

First-order differential equations in which the right side of (3) involves both x and y rather than x alone can be complicated to solve exactly, and one must often settle for numerical approximations of solutions. However, if the equation can be expressed in the form

$$\frac{dy}{dx} = \frac{g(x)}{h(y)} \tag{5}$$

then the general solution can sometimes be found by rewriting the equation in the differential form

$$h(y) \, dy = g(x) \, dx \tag{6}$$

and integrating both sides to obtain

$$\int h(y) \, dy = \int g(x) \, dx + C$$

A first-order differential equation of form (5) is said to be **separable**, and the process of rewriting (5) in form (6) is called **separating the variables**.

Example 2   Solve the initial-value problem:

$$\frac{dy}{dx} = -4xy^2, \quad y(0) = 1$$

Solution. We first solve the differential equation by separating variables:

$$\frac{dy}{y^2} = -4x\,dx$$

$$\int \frac{dy}{y^2} = \int -4x\,dx$$

$$-\frac{1}{y} = -2x^2 + C$$

or on multiplying by -1, taking reciprocals, and writing K in place of $-C$,

$$y = \frac{1}{2x^2 + K} \tag{7}$$

The initial condition, $y(0) = 1$, requires that $y = 1$ when $x = 0$. Substituting these values in (7) yields $K = 1$. Thus, the solution of the initial-value problem is

$$y = \frac{1}{2x^2 + 1} \qquad \blacktriangleleft$$

The graph of a solution of a differential equation is sometimes called an ***integral curve***. Because the general solution of a first-order linear equation involves an arbitrary constant, it produces a family of integral curves. Some integral curves for the differential equation in the preceding example are shown in Figure 7.7.1. As illustrated in that figure, the initial condition ($y = 1$ if $x = 0$) isolates the unique integral curve in the family that passes through the point $(0, 1)$.

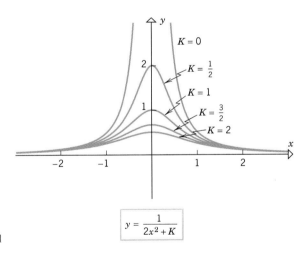

Figure 7.7.1

□ **IMPLICIT SOLUTIONS**

Sometimes solutions of differential equations are expressed as implicitly defined functions. For example,

$$\ln y = xy + C \tag{8}$$

defines a solution of

$$\frac{dy}{dx} = \frac{y^2}{1 - xy} \tag{9}$$

for any value of the constant C, since implicit differentiation of (8) yields

$$\frac{1}{y}\frac{dy}{dx} = x\frac{dy}{dx} + y \quad \text{or} \quad \frac{dy}{dx} - xy\frac{dy}{dx} = y^2$$

from which (9) follows.

Example 3 Solve the equation $x(y - 1)\dfrac{dy}{dx} = y$.

Solution. Separating the variables and integrating yields

$$x(y - 1)\,dy = y\,dx$$

$$\frac{y - 1}{y}\,dy = \frac{dx}{x}$$

$$\int \frac{y - 1}{y}\,dy = \int \frac{dx}{x}$$

$$\int \left(1 - \frac{1}{y}\right) dy = \int \frac{dx}{x}$$

$$y - \ln |y| = \ln |x| + C$$

or, using properties of logarithms,

$$y = \ln |xy| + C$$

This gives a solution implicitly as a function of x; in this case there is no simple formula for y explicitly as a function of x. ◀

☐ **FIRST-ORDER LINEAR EQUATIONS**

Not every first-order differential equation is separable. For example, it is impossible to separate the variables in the equation

$$\frac{dy}{dx} + 2xy = xe^{-x^2}$$

However, this equation can be solved by a different method that we shall now consider. A first-order differential equation is called **linear** if it is expressible in the form

$$\frac{dy}{dx} + p(x)y = q(x) \tag{10}$$

where the functions $p(x)$ and $q(x)$ may or may not be constant. Some examples are

$$\frac{dy}{dx} + x^2y = e^x, \qquad \frac{dy}{dx} + (\sin x)y + x^3 = 0, \qquad \frac{dy}{dx} + 5y = 2$$

$$\boxed{p(x) = x^2,\, q(x) = e^x} \qquad \boxed{p(x) = \sin x,\, q(x) = -x^3} \qquad \boxed{p(x) = 5,\, q(x) = 2}$$

One procedure for solving (10) is based on the observation that if we define $\mu = \mu(x)$ by

$$\mu = e^{\int p(x)\,dx}$$

then

$$\frac{d\mu}{dx} = e^{\int p(x)\,dx} \cdot \frac{d}{dx} \int p(x)\,dx = \mu p(x)$$

Thus,

$$\frac{d}{dx}(\mu y) = \mu \frac{dy}{dx} + \frac{d\mu}{dx} y = \mu \frac{dy}{dx} + \mu p(x)y \tag{11}$$

If (10) is multiplied through by μ, and then simplified using (11), it becomes

$$\mu \frac{dy}{dx} + \mu p(x)y = \mu q(x)$$

$$\frac{d}{dx}(\mu y) = \mu q(x)$$

This equation can be solved by integrating both sides to obtain

$$\mu y = \int \mu q(x)\, dx + C \quad \text{or} \quad y = \frac{1}{\mu}\left[\int \mu q(x)\, dx + C\right]$$

To summarize, (10) can be solved in three steps:

Step 1. Calculate

$$\mu = e^{\int p(x)\, dx}$$

This is called the ***integrating factor***. Since any μ will suffice, we can take the constant of integration to be zero in this step.

Step 2. Multiply both sides of (10) by μ and express the result as

$$\frac{d}{dx}(\mu y) = \mu q(x)$$

Step 3. Integrate both sides of the equation obtained in Step 2 and then solve for y. Be sure to include a constant of integration in this step.

Example 4 Solve the equation $\dfrac{dy}{dx} - 4xy = x$.

Solution. Since $p(x) = -4x$, the integrating factor is

$$\mu = e^{\int(-4x)\, dx} = e^{-2x^2}$$

If we multiply the given equation by μ, we obtain

$$\frac{d}{dx}(e^{-2x^2}y) = xe^{-2x^2}$$

Integrating both sides of this equation yields

$$e^{-2x^2}y = \int xe^{-2x^2}\, dx = -\tfrac{1}{4}e^{-2x^2} + C$$

and then multiplying both sides by e^{2x^2} yields

$$y = -\tfrac{1}{4} + Ce^{2x^2} \quad \blacktriangleleft$$

REMARK. The equation in Example 4 can also be solved by separating variables.

Example 5 Solve the equation $x\dfrac{dy}{dx} - y = x$, where $x > 0$.

Solution. To put the equation in form (10), we divide through by x to obtain

$$\frac{dy}{dx} - \frac{1}{x}y = 1 \tag{12}$$

Since $p(x) = -1/x$, the integrating factor is

$$\mu = e^{\int -(1/x)\, dx} = e^{-\ln|x|} = \frac{1}{|x|} = \frac{1}{x}$$

(The absolute value was dropped because of the assumption that $x > 0$.) If we multiply (12) by μ, we obtain

$$\frac{d}{dx}\left(\frac{1}{x}y\right) = \frac{1}{x}$$

Integrating both sides of this equation yields

$$\frac{1}{x}y = \int \frac{1}{x}\,dx = \ln x + C \quad \text{or} \quad y = x \ln x + Cx \qquad \blacktriangleleft$$

☐ **APPLICATIONS**

We conclude this section with some applications of first-order differential equations.

Example 6 (*Geometry*)   Find a curve in the xy-plane that passes through $(0, 3)$ and whose tangent line at a point (x, y) has slope $2x/y^2$.

Solution.   Since the slope of the tangent line is dy/dx, we have

$$\frac{dy}{dx} = \frac{2x}{y^2} \tag{13}$$

and, since the curve passes through $(0, 3)$, we have the initial condition

$$y(0) = 3 \tag{14}$$

Equation (13) is separable and can be written as

$$y^2\,dy = 2x\,dx$$

so

$$\int y^2\,dy = \int 2x\,dx \quad \text{or} \quad \tfrac{1}{3}y^3 = x^2 + C$$

From the initial condition, (14), it follows that $C = 9$, and so the curve has the equation

$$\tfrac{1}{3}y^3 = x^2 + 9 \quad \text{or} \quad y = (3x^2 + 27)^{1/3} \qquad \blacktriangleleft$$

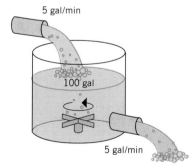

5 gal/min

100 gal

5 gal/min

Figure 7.7.2

Example 7 (*Mixing Problems*)   At time $t = 0$, a tank contains 4 lb of salt dissolved in 100 gal of water. Suppose that brine containing 2 lb of salt per gallon of water is allowed to enter the tank at a rate of 5 gal/min and that the mixed solution is drained from the tank at the same rate (Figure 7.7.2). Find the amount of salt in the tank after 10 min ($t = 10$).

Solution.   Let $y(t)$ be the amount of salt (in pounds) at time t. We are interested in finding $y(10)$, the amount of salt at time $t = 10$. We will begin by finding an expression for dy/dt, the rate of change of the amount of salt in the tank at time t. Clearly,

$$\frac{dy}{dt} = \text{rate in} - \text{rate out} \tag{15}$$

where *rate in* is the rate at which salt enters the tank and *rate out* is the rate at which salt leaves the tank. But

$$\text{rate in} = (2\text{ lb/gal}) \cdot (5\text{ gal/min}) = 10\text{ lb/min}$$

At time t, the mixture contains $y(t)$ lb of salt in 100 gal of water; thus, the concentration of salt at time t is $y(t)/100$ lb/gal and

$$\text{rate out} = \left(\frac{y(t)}{100}\text{ lb/gal}\right) \cdot (5\text{ gal/min}) = \frac{y(t)}{20}\text{ lb/min}$$

Therefore, (15) can be written as

$$\frac{dy}{dt} = 10 - \frac{y}{20}$$

or

$$\frac{dy}{dt} + \frac{y}{20} = 10 \tag{16}$$

which is a first-order linear differential equation. We also have the initial condition

$$y(0) = 4 \qquad (17)$$

since the tank contains 4 lb of salt at time $t = 0$.

Multiplying both sides of (16) by the integrating factor

$$\mu = e^{\int (1/20)\,dt} = e^{t/20}$$

yields

$$\frac{d}{dt}(e^{t/20}y) = 10e^{t/20}$$

$$e^{t/20}y = \int 10e^{t/20}\,dt = 200e^{t/20} + C$$

$$y(t) = 200 + Ce^{-t/20}$$

From the initial condition, (17), it follows that

$$4 = 200 + C \quad \text{or} \quad C = -196$$

so

$$y(t) = 200 - 196e^{-t/20}$$

Thus, at time $t = 10$ the amount of salt in the tank is

$$y(10) = 200 - 196e^{-0.5} \approx 81.1 \text{ lb} \qquad \blacktriangleleft$$

 **EXPONENTIAL GROWTH**

Many quantities increase or decrease with time in proportion to the amount of the quantity present. Some examples are human population, bacteria in a culture, drug concentration in the bloodstream, radioactivity, and the values of certain kinds of investments. We shall show how differential equations can be used to study the growth and decay of such quantities.

> **7.7.1** DEFINITION.   A quantity is said to have an ***exponential growth (decay) model*** if at each instant of time its rate of increase (decrease) is proportional to the amount of the quantity present.

Consider a quantity with an exponential growth or decay model and denote by $y(t)$ the amount of the quantity present at time t. We assume that $y(t) > 0$ for all t. Since the rate of change of $y(t)$ is proportional to the amount present, it follows that $y(t)$ satisfies

$$\frac{dy}{dt} = ky \qquad (18)$$

where k is a constant of proportionality.

The constant k in (18) is called the ***growth constant*** if $k > 0$ and the ***decay constant*** if $k < 0$. If $k > 0$, then $dy/dt > 0$, so that y is increasing with time (a growth model). If $k < 0$, then $dy/dt < 0$, so that y is decreasing with time (a decay model).

REMARK.   The constant k in (18) is sometimes called the ***growth rate*** (a negative growth rate meaning that y is decreasing). Strictly speaking, this is not correct since the growth rate of y is not k, but rather dy/dt ($= ky$). Although it is rarely done, it would be more accurate to call k the ***relative growth rate***, since it follows from (18) that

$$k = \frac{dy/dt}{y}$$

However, it is so common to call k the growth rate that we shall use this standard terminology in this section. The growth rate k is usually expressed as a percentage; thus, a growth rate of 3% means $k = 0.03$ and a growth rate of -500% means $k = -5$.

If we are given a quantity y with an exponential growth or decay model, and if we know the amount y_0 present at some initial time $t = 0$, then we can find the amount present at any time t by solving the initial-value problem

$$\frac{dy}{dt} = ky, \quad y(0) = y_0 \tag{19}$$

This differential equation is both separable and linear and thus can be solved by either of the methods we have studied in this section. We shall treat it as a linear equation. Rewriting the differential equation as

$$\frac{dy}{dt} - ky = 0$$

and multiplying through by the integrating factor

$$\mu = e^{\int -k\,dt} = e^{-kt}$$

yields

$$\frac{d}{dt}(e^{-kt}y) = 0$$

After integrating,

$$e^{-kt}y = C \quad \text{or} \quad y = Ce^{kt}$$

From the initial condition, $y(0) = y_0$, it follows that $C = y_0$; thus, the solution of (19) is

$$y(t) = y_0 e^{kt} \tag{20}$$

Exponential models have proved useful in studies of population growth. Although populations (e.g., people, bacteria, and flowers) grow in discrete steps, we can apply the results of this section if we are willing to approximate the population graph by a continuous curve $y = y(t)$ (Figure 7.7.3).

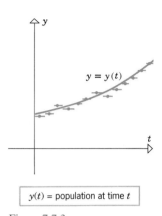

$y(t)$ = population at time t

Figure 7.7.3

Example 8 (*Population Growth*)   According to United Nations data, the world population at the beginning of 1990 was approximately 5.3 billion and growing at a rate of about 2% per year. Assuming an exponential growth model, estimate the world population at the beginning of the year 2015.

Solution.   Let

t = time elapsed from the beginning of 1990 (in years)
y = world population (in billions)

Since the beginning of 1990 corresponds to $t = 0$, it follows from the given data that

$$y_0 = y(0) = 5.3 \text{ (billion)}$$

Since the growth rate is 2% ($k = 0.02$), it follows from (20) that the world population at time t will be

$$y(t) = y_0 e^{kt} = 5.3 e^{0.02t} \tag{21}$$

Since the beginning of the year 2015 corresponds to an elapsed time of $t = 25$ years ($2015 - 1990 = 25$), it follows from (21) that the world population by the year 2015 will be

$$y(25) = 5.3 e^{0.02(25)} = 5.3 e^{0.5} \approx 5.3(1.6487) = 8.7381 \text{ (billion)}$$

which is a population of approximately 8.7 billion.   ◄

□ **DOUBLING AND HALVING TIME**

If a quantity y has an exponential growth model, then the time required for it to double in size is called the ***doubling time***. Similarly, if y has an exponential decay model, then the time required for it to reduce in value by half is called the ***halving time***. As it turns out, doubling and halving times depend only on the growth rate and not on the amount present initially. To see why, suppose y has an exponential growth model so that

$$y = y_0 e^{kt} \quad (k > 0)$$

At any fixed time t_1 let

$$y_1 = y_0 e^{kt_1} \tag{22}$$

be the value of y, and let T denote the amount of time required for y to double in size. Thus, at time $t_1 + T$ the value of y will be $2y_1$, so

$$2y_1 = y_0 e^{k(t_1+T)} = y_0 e^{kt_1} e^{kT}$$

or, from (22),

$$2y_1 = y_1 e^{kT}$$

Thus,

$$2 = e^{kT} \quad \text{and} \quad \ln 2 = kT$$

Therefore, the doubling time T is

$$T = \frac{1}{k} \ln 2 \tag{23}$$

which does not depend on y_0 or t_1. We leave it as an exercise to show that the halving time for a quantity with an exponential decay model ($k < 0$) is

$$T = -\frac{1}{k} \ln 2 \tag{24}$$

Example 9   It follows from (23) that at the current 2% annual growth rate, the doubling time for the world population is

$$T = \frac{1}{0.02} \ln 2 \approx \frac{1}{0.02} (0.6931) = 34.655$$

or approximately 35 years. Thus, with a continued 2% annual growth rate the population of 5.3 billion in 1990 will double to 10.6 billion by the year 2025 and will double again to 21.2 billion by 2060.   ◀

Radioactive elements continually undergo a process of disintegration called ***radioactive decay***. It is a physical fact that at each instant of time the rate of decay is proportional to the amount of the element present. Consequently, the amount of any radioactive element has an exponential decay model. For radioactive elements, halving time is called ***half-life***.

Example 10 (*Radioactive Decay*)   The radioactive element carbon-14 has a half-life of 5750 years. If 100 grams of this element are present initially, how much will be left after 1000 years?

Solution.   From (24) the decay constant is

$$k = -\frac{1}{T} \ln 2 \approx -\frac{1}{5750} (0.6931) \approx -0.00012$$

Thus, if we take $t = 0$ to be the present time, then $y_0 = y(0) = 100$, so that (20) implies that the amount of carbon-14 after 1000 years will be

$$y(1000) = 100 e^{-0.00012(1000)} = 100 e^{-0.12} \approx 100(0.88692) = 88.692$$

Thus, about 89 grams of carbon-14 will remain.   ◀

▶ Exercise Set 7.7   Ⓒ *27, 29–32, 34–41, 46*

In Exercises 1–6, solve the given separable differential equation. Where convenient, express the solution explicitly as a function of *x*.

1. $\dfrac{dy}{dx} = \dfrac{y}{x}$.

2. $\dfrac{dy}{dx} = \dfrac{x^3}{(1+x^4)y}$.

3. $\sqrt{1+x^2}\,y' + x(1+y) = 0$.

4. $3\tan y - \dfrac{dy}{dx}\sec x = 0$.

5. $e^{-y}\sin x - y'\cos^2 x = 0$.

6. $\dfrac{dy}{dx} = 1 - y + x^2 - yx^2$.

In Exercises 7–12, solve the given first-order linear differential equation. Where convenient, express the solution explicitly as a function of *x*.

7. $\dfrac{dy}{dx} + 3y = e^{-2x}$.

8. $\dfrac{dy}{dx} - \dfrac{5}{x}y = x$ $(x > 0)$.

9. $y' + y = \cos(e^x)$.

10. $2\dfrac{dy}{dx} + 4y = 1$.

11. $x^2 y' + 3xy + 2x^5 = 0$ $(x > 0)$.

12. $\dfrac{dy}{dx} + y - \dfrac{1}{1+e^x} = 0$.

In Exercises 13–18, solve the initial-value problems.

13. $\dfrac{dy}{dx} - xy = x$, $y(0) = 3$.

14. $2y\dfrac{dy}{dx} = 3x^2(x^3+1)^{-1/2}$, $y(2) = 1$.

15. $\dfrac{dy}{dt} + y = 2$, $y(0) = 1$.

16. $y' - xe^y = 2e^y$, $y(0) = 0$.

17. $y^2 t\dfrac{dy}{dt} - t + 1 = 0$, $y(1) = 3$ $(t > 0)$.

18. $y'\cosh x + y\sinh x = \cosh^2 x$, $y(0) = \frac{1}{4}$.

In Exercises 19–22, find an equation of the curve in the *xy*-plane that passes through the given point and whose tangent at (x, y) has the given slope.

19. $(1, -1)$; slope $= \dfrac{y^2}{3\sqrt{x}}$.

20. $(1, 1)$; slope $= \dfrac{3x^2}{2y}$.

21. $(2, 0)$; slope $= xe^y$.

22. $(1, -1)$; slope $= 2y + 3$, $y > -\frac{3}{2}$.

The following discussion is needed for Exercises 23 and 24. Suppose that a tank containing a liquid is vented to the air at the top and has an outlet at the bottom through which the liquid can drain. It follows from **Torricelli's law** in physics that if the outlet is opened at time $t = 0$, then at each instant the depth of the liquid $h(t)$ and the area $A(h)$ of the liquid's surface are related by

$$A(h)\frac{dh}{dt} = -k\sqrt{h}$$

where k is a positive constant that depends on such factors as the viscosity of the liquid and the cross-sectional area of the outlet. Use this result in Exercises 23 and 24, assuming that h is in feet, $A(h)$ is in square feet, and t is in seconds. A calculator will be useful.

23. Suppose that the cylindrical tank in Figure 7.7.4 is filled to a depth of 4 ft at time $t = 0$ and that the constant in Torricelli's law is $k = 0.025$.

(a) Find $h(t)$.

(b) How many minutes will it take for the tank to drain completely?

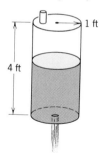

Figure 7.7.4

24. Follow the directions of Exercise 23 for the cylindrical tank in Figure 7.7.5, assuming that the tank is filled to a depth of 4 ft at time $t = 0$ and that the constant in Torricelli's law is $k = 0.025$.

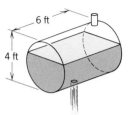

Figure 7.7.5

25. Suppose that a particle moving along the *x*-axis encounters a resisting force that results in an acceleration of $a = dv/dt = -0.04v^2$. Given that $x = 0$ cm and $v = 50$ cm/sec at time $t = 0$, find the velocity v and position x as a function of t for $t \geq 0$.

26. Suppose that a particle moving along the *x*-axis encounters a resisting force that results in an acceleration of

$a = dv/dt = -0.02\sqrt{v}$. Given that $x = 0$ cm and $v = 9$ cm/sec at time $t = 0$, find the velocity v and position x as a function of t for $t \geq 0$.

27. A rocket, fired upward from rest at time $t = 0$, has an initial mass of m_0 (including its fuel). Assuming that the fuel is consumed at a constant rate k, the mass m of the rocket, while fuel is being burned, will be given by $m = m_0 - kt$. It can be shown that if air resistance is neglected and the fuel gases are expelled at a constant speed c relative to the rocket, then the velocity v of the rocket will satisfy the equation

$$m\frac{dv}{dt} = ck - mg$$

where g is the acceleration due to gravity.

(a) Find $v(t)$ keeping in mind that the mass m is a function of t.

(b) Suppose that the fuel accounts for 80% of the initial mass of the rocket and that all of the fuel is consumed in 100 sec. Find the velocity of the rocket in meters/second at the instant the fuel is exhausted. (Use $g = 9.8$ m/sec^2 and $c = 2500$ m/sec.)

28. An object of mass m is dropped from rest at time $t = 0$. If we assume that the only forces acting on the object are the constant force due to gravity and a retarding force of air resistance, which is proportional to the velocity $v(t)$ of the object, then the velocity will satisfy the equation

$$m\frac{dv}{dt} = mg - kv$$

where k is a positive constant and g is the acceleration due to gravity.

(a) Find $v(t)$.

(b) Use the result in part (a) to find $\lim\limits_{t \to +\infty} v(t)$.

(c) Find the distance $x(t)$ that the object has fallen at time t given that $x = 0$ when $t = 0$.

29. An object of constant mass is projected upward from the surface of the earth. Neglecting air resistance, the gravitational force exerted by the earth on the object results in an acceleration of

$$a = dv/dt = -gR^2/x^2$$

where x is the distance of the object from the center of the earth, R is the radius of the earth, and g is the acceleration due to gravity at the surface of the earth. From the chain rule $dv/dt = v\,dv/dx$, so v and x must satisfy the equation

$$v\,dv/dx = -gR^2/x^2$$

(a) Find v^2 in terms of x given that $v = v_0$ when $x = R$.

(b) Use the result in part (a) to show that the velocity cannot reach zero if $v_0 \geq \sqrt{2gR}$.

(c) The minimum value of v_0 that is required for the object not to fall back to the earth is called the **escape velocity**. Assuming that $g = 32$ ft/sec^2 and

$R = 3960$ mi, show that the escape velocity is approximately 6.9 mi/sec.

30. A bullet of mass m, fired straight up with an initial velocity of v_0, is slowed by the force of gravity and a drag force of air resistance kv^2, where g is the constant acceleration due to gravity and k is a positive constant. As the bullet moves upward, its velocity v satisfies the equation

$$m\frac{dv}{dt} = -(kv^2 + mg)$$

(a) Show that if x is the position of the bullet at time t (Figure 7.7.6), then

$$mv\frac{dv}{dx} = -(kv^2 + mg)$$

Figure 7.7.6

(b) Express x in terms of v given that $x = 0$ when $v = v_0$.

(c) Assuming that $v_0 = 988$ m/sec, $g = 9.8$ m/sec^2, $m = 3.56 \times 10^{-3}$ kg, and $k = 7.3 \times 10^{-6}$, use the result in part (b) to find out how high the bullet rises. [*Hint:* Find the velocity of the bullet at its highest point.]

31. At time $t = 0$, a tank contains 25 lb of salt dissolved in 50 gal of water. Then brine containing 4 lb of salt per gallon of water is allowed to enter the tank at a rate of 2 gal/min and the mixed solution is drained from the tank at the same rate.

(a) How much salt is in the tank at an arbitrary time t?

(b) How much salt is in the tank after 25 min?

32. A tank initially contains 200 gal of pure water. Then at time $t = 0$ brine containing 5 lb of salt per gallon of water is allowed to enter the tank at a rate of 10 gal/min and the mixed solution is drained from the tank at the same rate.

(a) How much salt is in the tank at an arbitrary time t?

(b) How much salt is in the tank after 30 min?

33. A tank with a 1000-gal capacity initially contains 500 gal of brine containing 50 lb of salt. At time $t = 0$, pure water is added at a rate of 20 gal/min and the mixed solution is drained off at a rate of 10 gal/min. How much salt is in the tank when it reaches the point of overflowing?

34. The number of bacteria in a certain culture grows exponentially at a rate of 1% per hour. Assuming that 10,000 bacteria are present initially, find

   (a) the number of bacteria present at any time t

   (b) the number of bacteria present after 5 hr

   (c) the time required for the number of bacteria to reach 45,000.

35. Polonium-210 is a radioactive element with a half-life of 140 days. Assume that a sample has a mass of 10 mg initially.

   (a) Find a formula for the amount that will remain after t days.

   (b) How much will remain after 10 weeks?

36. In a certain chemical reaction a substance decomposes at a rate proportional to the amount present. Tests show that under appropriate conditions 15,000 grams will reduce to 5000 grams in 10 hours.

   (a) Find a formula for the amount that will remain from a 15,000-gram sample after t hours.

   (b) How long will it take for 50% of an initial sample of y_0 grams to decompose?

37. One hundred fruit flies are placed in a breeding container that can support a population of at most 5000 flies. If the population grows exponentially at a rate of 2% per day, how long will it take for the container to reach capacity?

38. In 1960 the American scientist W. F. Libby won the Nobel prize for his discovery of carbon dating, a method for determining the age of certain fossils. Carbon dating is based on the fact that nitrogen is converted to radioactive carbon-14 by cosmic radiation in the upper atmosphere. This radioactive carbon is absorbed by plant and animal tissue through the life processes while the plant or animal lives. However, when the plant or animal dies the absorption process stops and the amount of carbon-14 decreases through radioactive decay. Suppose that tests on a fossil show that 70% of its carbon-14 has decayed. Estimate the age of the fossil, assuming a half-life of 5750 years for carbon-14.

39. Forty percent of a radioactive substance decays in 5 years. Find the half-life of the substance.

40. Assume that if the temperature is constant, then the atmospheric pressure p varies with the altitude h (above sea level) in such a way that $dp/dh = kp$, where k is a constant.

   (a) Find a formula for p in terms of k, h, and the atmospheric pressure p_0 at sea level.

   (b) Given that p measures 15 lb/in^2 at sea level and 12 lb/in^2 at 5000 ft above sea level, find the pressure at 10,000 ft (assuming temperature is constant).

41. The town of Grayrock had a population of 10,000 in 1980 and 12,000 in 1990.

   (a) Assuming an exponential growth model, estimate the population in 2000.

   (b) What is the doubling time for the town's population?

42. Prove: If a quantity A has an exponential growth or decay model and A has values A_1 and A_2 at times t_1 and t_2, respectively, then the growth rate k is

$$k = \frac{1}{t_1 - t_2} \ln\left(\frac{A_1}{A_2}\right)$$

43. Newton's law of cooling states that the rate at which an object cools is proportional to the difference in temperature between the object and the surrounding medium. Show that if C is the constant temperature of a surrounding medium, then the temperature $T(t)$ at time t of a cooling object is given by

$$T(t) = (T_0 - C)e^{kt} + C$$

where T_0 is the temperature of the object at $t = 0$ and k is a negative constant.

44. A liquid with an initial temperature of 200° is enclosed in a metal container that is held at a constant temperature of 80°. If the liquid cools to 120° in 30 min, what will the temperature be after 1 hr? (Use the result of Exercise 43.)

45. Suppose that P dollars is invested at an annual interest rate of $r \times 100\%$. If the accumulated interest is credited to the account at the end of the year, then the interest is said to be *compounded annually*; if it is credited at the end of each six-month period, then it is said to be *compounded semi-annually*; and if it is credited at the end of each three-month period, then it is said to be *compounded quarterly*. The more frequently the interest is compounded, the better it is for the investor since more of the interest is itself earning interest.

   (a) Show that if interest is compounded n times a year at equally spaced intervals, then the value A of the investment after t years is

$$A = P\left(1 + \frac{r}{n}\right)^{nt}$$

   (b) One can imagine interest to be compounded each day, each hour, each minute, and so forth. Carried to the limit one can conceive of interest compounded at each instant of time; this is called **continuous compounding**. Thus, from part (a), the value A of P dollars after t years when invested at an annual rate of $r \times 100\%$, compounded continuously, is

$$A = \lim_{n \to +\infty} P\left(1 + \frac{r}{n}\right)^{nt}$$

   Use the fact that $\lim_{x \to 0}(1 + x)^{1/x} = e$ to prove that $A = Pe^{rt}$.

   (c) Use the result in part (b) to show that money invested at continuous compound interest increases at a rate proportional to the amount present.

46. (a) If $1000 is invested at 8% per year compounded continuously (Exercise 45), what will the investment be worth after 5 years?

(b) If it is desired that an investment at 8% per year compounded continuously should have a value of $10,000 after 10 years, how much should be invested now?

(c) How long does it take for an investment at 8% per year compounded continuously to double in value?

47. Derive Formula (24) for halving time.

48. Let a quantity have an exponential growth model with growth rate k. How long does it take for the quantity to triple in size?

APPENDIX D

Cramer's Rule

We are concerned here with solving systems of equations of the forms

$$a_1 x + b_1 y = k_1$$
$$a_2 x + b_2 y = k_2.$$

and

$$a_1 x + b_1 y + c_1 z = k_1$$
$$a_2 x + b_2 y + c_2 z = k_2$$
$$a_3 x + b_3 y + c_3 z = k_3$$

(1–2)

The first is a system of ***two linear equations in two unknowns***, x and y; and the second is a system of ***three linear equations in three unknowns***, x, y, and z. Values of the unknowns that satisfy *all* of the equations of a system are said to form a ***solution*** of the system.

The graph of each equation in (1) is a line in 2-space. Thus, if (x, y) is a point that lies on both lines, then the coordinates of this point satisfy both equations in (1), and x and y form a solution of the system. Similarly, the graph of each equation in (2) is a plane in 3-space, and the coordinates of any point (x, y, z) that lies on all three planes form a solution of the system. Figures D.1 and D.2 illustrate that in both cases there are only three possibilities—the system has no solutions, the system has exactly one solution, or the system has infinitely many solutions.

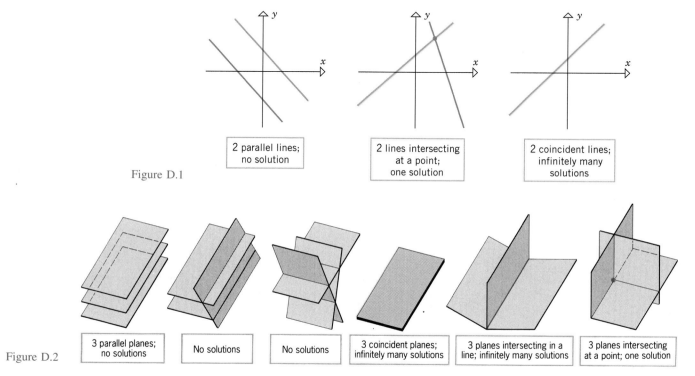

| 2 parallel lines; no solution | 2 lines intersecting at a point; one solution | 2 coincident lines; infinitely many solutions |

Figure D.1

| 3 parallel planes; no solution | No solutions | No solutions | 3 coincident planes; infinitely many solutions | 3 planes intersecting in a line; infinitely many solutions | 3 planes intersecting at a point; one solution |

Figure D.2

The determinants

$$\begin{vmatrix} a_1 & b_1 \\ a_2 & b_2 \end{vmatrix} \quad \text{and} \quad \begin{vmatrix} a_1 & b_1 & c_1 \\ a_2 & b_2 & c_2 \\ a_3 & b_3 & c_3 \end{vmatrix}.$$

are called the **coefficient determinants** for the systems in (1) and (2). It is proved in linear algebra courses that in both cases the system has a unique solution if and only if its coefficient determinant is *not* zero. In this event the solution can be expressed as a ratio of determinants, using one of the following theorems, which are special cases of a general result called **Cramer's rule**.*

THEOREM **1** (**Cramer's Rule for Two Unknowns**).   *The system of equations*

$$a_1 x + b_1 y = k_1$$

$$a_2 x + b_2 y = k_2$$

has a unique solution if and only if its coefficient determinant is not zero, in which case the solution can be expressed as

$$x = \frac{\begin{vmatrix} k_1 & b_1 \\ k_2 & b_2 \end{vmatrix}}{\begin{vmatrix} a_1 & b_1 \\ a_2 & b_2 \end{vmatrix}}, \quad y = \frac{\begin{vmatrix} a_1 & k_1 \\ a_2 & k_2 \end{vmatrix}}{\begin{vmatrix} a_1 & b_1 \\ a_2 & b_2 \end{vmatrix}}$$

THEOREM **2** (**Cramer's Rule for Three Unknowns**).   *The system of equations*

$$a_1 x + b_1 y + c_1 z = k_1$$

$$a_2 x + b_2 y + c_2 z = k_2$$

$$a_3 x + b_3 y + c_3 z = k_3$$

has a unique solution if and only if its coefficient determinant is not zero, in which case the solution can be expressed as

$$x = \frac{\begin{vmatrix} k_1 & b_1 & c_1 \\ k_2 & b_2 & c_2 \\ k_3 & b_3 & c_3 \end{vmatrix}}{\begin{vmatrix} a_1 & b_1 & c_1 \\ a_2 & b_2 & c_2 \\ a_3 & b_3 & c_3 \end{vmatrix}}, \quad y = \frac{\begin{vmatrix} a_1 & k_1 & c_1 \\ a_2 & k_2 & c_2 \\ a_3 & k_3 & c_3 \end{vmatrix}}{\begin{vmatrix} a_1 & b_1 & c_1 \\ a_2 & b_2 & c_2 \\ a_3 & b_3 & c_3 \end{vmatrix}}, \quad z = \frac{\begin{vmatrix} a_1 & b_1 & k_1 \\ a_2 & b_2 & k_2 \\ a_3 & b_3 & k_3 \end{vmatrix}}{\begin{vmatrix} a_1 & b_1 & c_1 \\ a_2 & b_2 & c_2 \\ a_3 & b_3 & c_3 \end{vmatrix}}$$

REMARK.   There is a pattern to the formulas in these theorems. In each formula the determinant in the denominator is formed from the coefficients of the unknowns. The determinant in the numerator differs from the determinant in the denominator in that the coefficients of the unknown being calculated are replaced by the k's. It is assumed in these theorems that the system is written so that like unknowns are aligned vertically and the constants appear by themselves on the right side of each equation.

---

*GABRIEL CRAMER (see p. A32 for biography).

Example 1   Use Cramer's rule to solve

(a)
$$5x - 2y = -1$$
$$2x + 3y = 3$$

(b)
$$x + 2z = 6$$
$$-3x + 4y + 6z = 30$$
$$-x - 2y + 3z = 8$$

Solution (a).

$$x = \frac{\begin{vmatrix} -1 & -2 \\ 3 & 3 \end{vmatrix}}{\begin{vmatrix} 5 & -2 \\ 2 & 3 \end{vmatrix}} = \frac{3}{19}; \quad y = \frac{\begin{vmatrix} 5 & -1 \\ 2 & 3 \end{vmatrix}}{\begin{vmatrix} 5 & -2 \\ 2 & 3 \end{vmatrix}} = \frac{17}{19}$$

Solution (b).

$$x = \frac{\begin{vmatrix} 6 & 0 & 2 \\ 30 & 4 & 6 \\ 8 & -2 & 3 \end{vmatrix}}{\begin{vmatrix} 1 & 0 & 2 \\ -3 & 4 & 6 \\ -1 & -2 & 3 \end{vmatrix}} = \frac{-10}{11}; \quad y = \frac{\begin{vmatrix} 1 & 6 & 2 \\ -3 & 30 & 6 \\ -1 & 8 & 3 \end{vmatrix}}{\begin{vmatrix} 1 & 0 & 2 \\ -3 & 4 & 6 \\ -1 & -2 & 3 \end{vmatrix}} = \frac{18}{11}; \quad z = \frac{\begin{vmatrix} 1 & 0 & 6 \\ -3 & 4 & 30 \\ -1 & -2 & 8 \end{vmatrix}}{\begin{vmatrix} 1 & 0 & 2 \\ -3 & 4 & 6 \\ -1 & -2 & 3 \end{vmatrix}} = \frac{38}{11}$$

◀

▶ Exercise Set D

In Exercises 1–8, solve the system using Cramer's rule.

1.
$$3x - 4y = -5$$
$$2x + y = 4.$$

2.
$$-x + 3y = 8$$
$$2x + 5y = 7.$$

3.
$$2x_1 - 5x_2 = -2$$
$$4x_1 + 6x_2 = 1.$$

4.
$$3a + 2b = 4$$
$$-a + b = 7.$$

5.
$$x + 2y + z = 3$$
$$2x + y - z = 0$$
$$x - y + z = 6.$$

6.
$$x - 3y + z = 4$$
$$2x - y = -2$$
$$4x - 3z = 0.$$

7.
$$x_1 + x_2 - 2x_3 = 1$$
$$2x_1 - x_2 + x_3 = 2$$
$$x_1 - 2x_2 - 4x_3 = -4.$$

8.
$$r + s + t = 2$$
$$r - s - 2t = 0$$
$$-r + 2s + t = 4.$$

GABRIEL CRAMER (1704–1752). Swiss mathematician. Although Cramer does not rank with the great mathematicians of his time, his contributions as a disseminator of mathematical ideas have earned him a well-deserved place in the history of mathematics. The son of a physician, Cramer was born and educated in Geneva, Switzerland. At age 20 he competed for, but failed to secure, the chair of philosophy at the Académie de Calvin at Geneva. However, the awarding magistrates were sufficiently impressed with Cramer and a fellow competitor to create a new chair of mathematics for both men to share. Alternately, each assumed the full responsibility and salary associated with the chair for two or three years while the other traveled. During his travels Cramer met many of the great mathematicians and scientists of his day: the Bernoullis, Euler, Halley, D'Alembert, and others. Many of these contacts and friendships led to extensive correspondence in which information about new mathematical discoveries was transmitted. Eventually, Cramer became sole possessor of the mathematics chair and the chair of philosophy as well.

Cramer's mathematical work was primarily in geometry and probability; he had relatively little knowledge of calculus and did not use it to any great extent in his work. In 1730 he finished second to Johann I Bernoulli in a competition for a prize offered by the Paris Academy to explain properties of planetary orbits.

Cramer's most widely known work, Introduction à l'analyse des lignes courbes algébriques (1750), was a study and classification of algebraic curves; Cramer's rule appeared in the appendix. Although the rule bears his name, variations of the basic idea were formulated earlier by Leibniz (and even earlier by Chinese mathematicians). However, Cramer's superior notation helped clarify and popularize the technique.

Perhaps Cramer's most important contributions stemmed from his work as an editor of the mathematical creations of others. He edited and published the works of Jacob I Bernoulli and Leibniz.

Overwork combined with a fall from a carriage eventually led to his death in 1752. Cramer was apparently a good-natured and pleasant person, though he never married. His interests were broad. He wrote on philosophy of law and government and the history of mathematics. He served in public office, participated in artillery and fortifications activities for the government, instructed workers on techniques of cathedral repair, and undertook excavations of cathedral archives. Cramer received numerous honors for his activities.

9. Use Cramer's rule to solve the rotation equations

$$x = x' \cos \theta - y' \sin \theta$$
$$y = x' \sin \theta + y' \cos \theta$$

for x' and y' in terms of x and y.

10. Solve the following system of equations for the unknown angles α, β, and γ, where $-\pi/2 \leq \alpha \leq \pi/2$, $0 \leq \beta \leq \pi$, and $-\pi/2 < \gamma < \pi/2$:

$$2 \sin \alpha - \cos \beta + 3 \tan \gamma = 3$$
$$4 \sin \alpha + 2 \cos \beta - 2 \tan \gamma = 2$$
$$6 \sin \alpha - 3 \cos \beta + \tan \gamma = 9$$

[*Hint:* First solve for $\sin \alpha$, $\cos \beta$, and $\tan \gamma$.]

APPENDIX E

Complex Numbers

■ E.I DEFINITIONS; ALGEBRA OF COMPLEX NUMBERS

□ DEFINITION OF A
COMPLEX NUMBER

Since $x^2 \geq 0$ for every real number x, the equation $x^2 = -1$ has no real solutions. To deal with this problem, mathematicians of the eighteenth century introduced the *imaginary number* $i = \sqrt{-1}$, which they assumed had the property

$$i^2 = (\sqrt{-1})^2 = -1$$

Expressions of the form $a + bi$ were called *complex numbers*, and these were manipulated according to the standard rules of algebra with the added property that $i^2 = -1$. By the beginning of the nineteenth century it was recognized that a complex number $a + bi$ could be viewed as an ordered pair of real numbers (a, b) or as a vector $\langle a, b \rangle$. This is the approach we will follow.

DEFINITION 1. A *complex number* is an ordered pair of real numbers, denoted either by (a, b) or $a + bi$. The number a is called the *real part* of the complex number, and the number b is called the *imaginary part*.

Sometimes it is convenient to use a single letter, such as z, to denote a complex number. Thus, we might write

$$z = a + bi$$

in which case the real and imaginary parts of z are denoted by $\text{Re}(z)$ and $\text{Im}(z)$, respectively; that is, $\text{Re}(z) = a$ and $\text{Im}(z) = b$. Real numbers can be regarded as complex numbers that have an imaginary part of zero; that is, $a = a + 0i$. In this sense the complex number system is an extension of the real number system.

Example 1

z (ORDERED PAIR NOTATION)	z (EQUIVALENT NOTATION)	$\text{Re}(z)$	$\text{Im}(z)$
$(3, 4)$	$3 + 4i$	3	4
$(-1, 2)$	$-1 + 2i$	-1	2
$(0, 1)$	$0 + i = i$	0	1
$(2, 0)$	$2 + 0i = 2$	2	0
$(4, -2)$	$4 + (-2)i = 4 - 2i$	4	-2

◀

☐ **THE COMPLEX PLANE**

A complex number $z = a + bi$ can be represented geometrically as a point or a vector in an xy-coordinate system, in which case we call the x-axis the **real axis**, the y-axis the **imaginary axis**, and the plane itself the **complex plane** (Figure E.1). Figure E.2 shows the complex numbers in Example 1 as points and vectors in the complex plane.

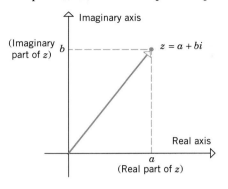

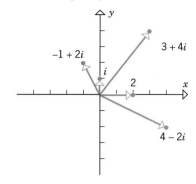

Figure E.1   Figure E.2

☐ **OPERATIONS ON COMPLEX NUMBERS**

The **sum** and **difference** of two complex numbers are defined by adding or subtracting corresponding real and imaginary parts; that is,

$$(a + bi) + (c + di) = (a + c) + (b + d)i \tag{1}$$

$$(a + bi) - (c + di) = (a - c) + (b - d)i \tag{2}$$

Example 2

$$(4 - 5i) + (-1 + 6i) = (4 - 1) + (-5 + 6)i = 3 + i$$

$$(4 - 5i) - (-1 + 6i) = (4 + 1) + (-5 - 6)i = 5 - 11i \quad \blacktriangleleft$$

The sum and difference of complex numbers z_1 and z_2 can be visualized geometrically by viewing the numbers as vectors (Figure E.3).

Multiplication of two complex numbers can be motivated by expanding the expression $(a + bi)(c + di)$ according to the usual laws of algebra, but treating i^2 as -1; that is,

$$(a + bi)(c + di) = ac + bdi^2 + adi + bci = (ac - bd) + (ad + bc)i$$

Thus, we define the **product** of complex numbers as

$$(a + bi)(c + di) = (ac - bd) + (ad + bc)i \tag{3}$$

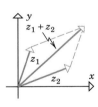

The sum of two complex numbers

For computational purposes, we suggest using the method that led to the formula in this definition, rather than substituting in the formula itself.

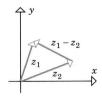

The difference of two complex numbers

Figure E.3

Example 3

$$5(2 + 4i) = 10 + 20i$$

$$(3 + 2i)(4 + i) = 12 + 3i + 8i + 2i^2 = 12 + 11i - 2 = 10 + 11i$$

$$i(1 + i)(1 - 2i) = i(1 - 2i + i - 2i^2) = i(3 - i) = 3i - i^2 = 1 + 3i \quad \blacktriangleleft$$

Division of complex numbers can be motivated by a process that is similar to "rationalizing" quotients in algebra. To obtain the real and imaginary parts of the quotient

$$\frac{a + bi}{c + di}$$

we multiply the numerator and denominator by $c - di$ and simplify:

$$\frac{a + bi}{c + di} = \frac{a + bi}{c + di} \cdot \frac{c - di}{c - di} = \frac{(ac + bd) + (bc - ad)i}{c^2 - d^2 i^2}$$

$$= \frac{(ac + bd) + (bc - ad)i}{c^2 + d^2}$$

which suggests that we define the **quotient** of complex numbers as

$$\frac{a + bi}{c + di} = \left(\frac{ac + bd}{c^2 + d^2}\right) + \left(\frac{bc - ad}{c^2 + d^2}\right)i \tag{4}$$

Again, we recommend that computations be performed using the procedure that led to the formula in this definition rather than substituting in the formula itself. Moreover, note that division of complex numbers is undefined if the denominator is zero, just as for real numbers.

Example 4

$$\frac{3 + 4i}{1 - 2i} = \frac{3 + 4i}{1 - 2i} \cdot \frac{1 + 2i}{1 + 2i} = \frac{-5 + 10i}{5} = -1 + 2i \qquad \blacktriangleleft$$

☐ **COMPLEX CONJUGATES**

If $z = a + bi$ is any complex number, then the **complex conjugate** (or **conjugate**) **of** z is denoted by $\bar{z}$ and is defined as

$$\bar{z} = a - bi$$

Example 5

z	$\bar{z}$
$3 + 4i$	$3 - 4i$
$-2 - 3i$	$-2 + 3i$
i	$-i$
5	5

$\blacktriangleleft$

We have already seen that complex conjugates arise in the division of complex numbers. They also arise in solving polynomial equations

$$a_0 x^n + a_1 x^{n-1} + \cdots + a_{n-1} x + a_n = 0$$

with real coefficients. It can be proved that such equations always have solutions in the complex number system and that if z is a solution, then so is $\bar{z}$. Thus, the solutions with nonzero imaginary parts occur in conjugate pairs. For example, from the quadratic formula the solutions of the equation $x^2 + x + 2 = 0$ are

$$x = \frac{-1 \pm \sqrt{-7}}{2} = \frac{-1 \pm \sqrt{7}i}{2}$$

so the solutions $x = -\frac{1}{2} + \frac{1}{2}\sqrt{7}i$ and $x = -\frac{1}{2} - \frac{1}{2}\sqrt{7}i$ are complex conjugates.

Geometrically, complex conjugates are reflections of one another about the real axis

(Figure E.4). Moreover, we leave it for the exercises to show that complex conjugates have the following properties:

$$\bar{\bar{z}} = z, \quad \overline{z_1 + z_2} = \bar{z}_1 + \bar{z}_2, \quad \overline{z_1 - z_2} = \bar{z}_1 - \bar{z}_2$$
$$\overline{z_1 z_2} = \bar{z}_1 \bar{z}_2, \quad \overline{(z_1/z_2)} = \bar{z}_1/\bar{z}_2 \tag{5}$$

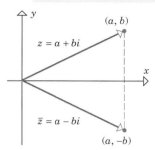

Figure E.4

☐ **MODULUS**

If a complex number z is viewed as a vector, then its length is called the **modulus** of z and is denoted by $|z|$. Thus, if $z = a + bi$, then

$$|z| = \sqrt{a^2 + b^2} \tag{6}$$

Figure E.5

(Figure E.5). For example, if $z = 3 - 4i$, then

$$|z| = |3 - 4i| = \sqrt{3^2 + (-4)^2} = 5$$

If z is a real number, say $z = a + 0i = a$, then

$$|z| = \sqrt{a^2 + 0^2} = \sqrt{a^2} = |a|$$

so the modulus of a real number is the same as the absolute value of that number. Observe that

$$z\bar{z} = (a + bi)(a - bi) = a^2 - abi + bai - b^2 i^2 = a^2 + b^2 = |z|^2$$

so the modulus and complex conjugate of z are related by

$$z\bar{z} = |z|^2 \tag{7}$$

Moreover, it can be shown that

$$|z_1 z_2| = |z_1||z_2| \quad \text{and} \quad |z_1/z_2| = |z_1|/|z_2| \tag{8}$$

▶ **Exercise Set E.I**

1. In each part plot the point and sketch the vector that corresponds to the given complex number.

   (a) $2 + 3i$ (b) -4

   (c) $-3 - 2i$ (d) $-5i$.

2. Express each complex number in Exercise 1 as an ordered pair of real numbers.

3. In each part use the given information to find the real numbers x and y.

   (a) $x - iy = -2 + 3i$

   (b) $(x + y) + (x - y)i = 3 + i$.

4. Given that $z_1 = 1 - 2i$ and $z_2 = 4 + 5i$, find

   (a) $z_1 + z_2$ (b) $z_1 - z_2$

   (c) $4z_1$ (d) $-z_2$

   (e) $3z_1 + 4z_2$ (f) $\frac{1}{2}z_1 - \frac{3}{2}z_2$.

5. In each part solve for z.

   (a) $z + (1 - i) = 3 + 2i$ (b) $-5z = 5 + 10i$

   (c) $(i - z) + (2z - 3i) = -2 + 7i$.

6. In each part sketch the vectors $z_1, z_2, z_1 + z_2$, and $z_1 - z_2$.

   (a) $z_1 = 3 + i, \ z_2 = 1 + 4i$

   (b) $z_1 = -2 + 2i, \ z_2 = 4 + 5i$.

7. In each part sketch the vectors z and kz.

   (a) $z = 1 + i, \ k = 2$

   (b) $z = -3 - 4i, \ k = -2$

   (c) $z = 4 + 6i, \ k = \frac{1}{2}$.

8. In each part find real numbers k_1 and k_2 that satisfy the equation.

    (a) $k_1 i + k_2(1 + i) = 3 - 2i$

    (b) $k_1(2 + 3i) + k_2(1 - 4i) = 7 + 5i$.

9. In each part find $z_1 z_2$, z_1^2, and z_2^2.

    (a) $z_1 = 3i,\ z_2 = 1 - i$

    (b) $z_1 = 4 + 6i,\ z_2 = 2 - 3i$

    (c) $z_1 = \frac{1}{3}(2 + 4i),\ z_2 = \frac{1}{2}(1 - 5i)$.

10. Given that $z_1 = 2 - 5i$ and $z_2 = -1 - i$, find

    (a) $z_1 - z_1 z_2$ (b) $(z_1 + 3z_2)^2$

    (c) $[z_1 + (1 + z_2)]^2$ (d) $iz_2 - z_1^2$.

In Exercises 11–18, perform the calculations and express the result in the form $a + bi$.

11. $(1 + 2i)(4 - 6i)^2$.   **12.** $(2 - i)(3 + i)(4 - 2i)$.

13. $(1 - 3i)^3$.   **14.** $i(1 + 7i) - 3i(4 + 2i)$.

15. $[(2 + i)(\frac{1}{2} + \frac{3}{4}i)]^2$.   **16.** $(\sqrt{2} + i) - \sqrt{2}i(1 + \sqrt{2}i)$.

17. $(1 + i + i^2 + i^3)^{100}$.   **18.** $(3 - 2i)^2 - (3 + 2i)^2$.

19. In each part find $\bar{z}$.

    (a) $z = 2 + 7i$ (b) $z = -3 - 5i$

    (c) $z = 5i$ (d) $z = -i$

    (e) $z = -9$ (f) $z = 0$.

20. In each part find $|z|$.

    (a) $z = i$ (b) $z = -7i$

    (c) $z = -3 - 4i$ (d) $z = 1 + i$

    (e) $z = -8$ (f) $z = 0$.

21. Verify that $z\bar{z} = |z|^2$ for

    (a) $z = 2 - 4i$ (b) $z = -3 + 5i$

    (c) $z = \sqrt{2} - \sqrt{2}i$.

22. Given that $z_1 = 1 - 5i$ and $z_2 = 3 + 4i$, find

    (a) z_1/z_2 (b) $\bar{z}_1/z_2$

    (c) $z_1/\bar{z}_2$ (d) $\overline{(z_1/z_2)}$

    (e) $z_1/|z_2|$ (f) $|z_1/z_2|$.

23. In each part find $1/z$.

    (a) $z = i$ (b) $z = 1 - 5i$

    (c) $z = \dfrac{-i}{7}$.

24. Given that $z_1 = 1 + i$ and $z_2 = 1 - 2i$, find

    (a) $z_1 - \dfrac{z_1}{z_2}$ (b) $\dfrac{z_1 - 1}{z_2}$

    (c) $z_1^2 - \dfrac{iz_1}{z_2}$ (d) $\dfrac{z_1}{iz_2}$.

In Exercises 25–32, perform the calculations and express the result in the form $a + bi$.

25. $\dfrac{i}{1 + i}$.   **26.** $\dfrac{2}{(1 - i)(3 + i)}$.

27. $\dfrac{1}{(3 + 4i)^2}$.   **28.** $\dfrac{2 + i}{i(-3 + 4i)}$.

29. $\dfrac{\sqrt{3} + i}{(1 - i)(\sqrt{3} - i)}$.   **30.** $\dfrac{1}{i(3 - 2i)(1 + i)}$.

31. $\dfrac{i}{(1 - i)(1 - 2i)(1 + 2i)}$.   **32.** $\dfrac{1 - 2i}{3 + 4i} - \dfrac{2 + i}{5i}$.

33. In each part solve for z.

    (a) $iz = 2 - i$ (b) $(4 - 3i)\bar{z} = i$.

34. Use the properties in (5) to show that

    (a) $\overline{\bar{z} + 5i} = z - 5i$ (b) $\overline{iz} = -i\bar{z}$

    (c) $\dfrac{\overline{i + z}}{i - z} = -1$.

35. In each part sketch the set of points in the complex plane that satisfies the equation.

    (a) $|z| = 2$ (b) $|z - (1 + i)| = 1$

    (c) $|z - i| = |z + i|$ (b) $\mathrm{Im}(\bar{z} + i) = 3$.

36. In each part sketch the set of points in the complex plane that satisfies the given condition(s).

    (a) $|z + i| \le 1$ (b) $1 < |z| < 2$

    (c) $|2z - 4i| < 1$ (d) $|z| \le |z + i|$.

37. Show that

    (a) $\mathrm{Im}(iz) = \mathrm{Re}(z)$ (b) $\mathrm{Re}(iz) = -\mathrm{Im}(z)$

    (c) $\mathrm{Re}(\bar{iz}) = -\mathrm{Im}(z)$ (d) $\mathrm{Im}(\bar{iz}) = -\mathrm{Re}(z)$

    (e) $\mathrm{Re}(i\bar{z}) = \mathrm{Im}(z)$ (f) $\mathrm{Im}(i\bar{z}) = \mathrm{Re}(z)$.

38. In each part solve the equation by the quadratic formula and check your results by substituting the solutions into the given equation.

    (a) $x^2 + 2x + 2 = 0$ (b) $x^2 - x + 1 = 0$.

39. (a) Show that if n is a positive integer, then the only possible values for i^n are 1, -1, i, and $-i$. [*Hint:* The value of i^n can be determined from the remainder when n is divided by 4.]

    (b) Find i^{2509}.

40. (a) Use the result in part (a) of Exercise 39 to show that if n is a positive integer, then the only possible values for $(1/i)^n$ are 1, -1, i, and $-i$.

    (b) Find $(1/i)^{2509}$.

41. Prove:

    (a) $\dfrac{1}{2}(z + \bar{z}) = \mathrm{Re}(z)$ (b) $\dfrac{1}{2i}(z - \bar{z}) = \mathrm{Im}(z)$.

42. Prove: $z = \bar{z}$ if and only if z is a real number.

43. Given that $z_1 = x_1 + iy_1$ and $z_2 = x_2 + iy_2$, find

    (a) $\mathrm{Re}\left(\dfrac{z_1}{z_2}\right)$ (b) $\mathrm{Im}\left(\dfrac{z_1}{z_2}\right)$.

44. Show: If $(\bar{z})^2 = z^2$, then z is either real or pure imaginary.

45. Show that $|z| = |\bar{z}|$.

46. Show:

(a) $\overline{z_1 - z_2} = \bar{z}_1 - \bar{z}_2$

(b) $\overline{z_1 z_2} = \bar{z}_1 \bar{z}_2$

(c) $\overline{(z_1/z_2)} = \bar{z}_1/\bar{z}_2$

(d) $\bar{\bar{z}} = z$.

47. (a) Use the properties in (5) to show that $\overline{z^2} = (\bar{z})^2$.

(b) Show that if n is a positive integer, then $\overline{z^n} = (\bar{z})^n$.

(c) Is the result in part (b) true if n is a negative integer? Explain.

48. Let $p(x) = a_0 + a_1 x + a_2 x^2 + \cdots + a_n x^n$ be a polynomial for which the coefficients $a_0, a_1, a_2, \ldots, a_n$ are real. Use (5) and the result in part (b) of Exercise 47 to show that if z is a solution of the equation $p(x) = 0$, then so is $\bar{z}$.

49. Show: For any complex number z, $|\text{Re}(z)| \le |z|$ and $|\text{Im}(z)| \le |z|$.

50. Show that

$$\frac{|\text{Re}(z)| + |\text{Im}(z)|}{\sqrt{2}} \le |z|$$

[*Hint:* Let $z = x + iy$ and use the fact that $(|x| - |y|)^2 \ge 0$.]

51. Show that if $z_1 z_2 = 0$, then $z_1 = 0$ or $z_2 = 0$.

■ E.II POLAR FORM; DEMOIVRE'S THEOREM

☐ **POLAR FORM OF A COMPLEX NUMBER**

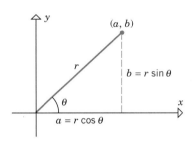

Figure E.6

If we view $z = a + bi$ as a point (a, b) in the complex plane, and if we let r and θ be polar coordinates of that point with $r \ge 0$, then it follows that

$$a = r \cos \theta, \quad b = r \sin \theta$$

(Figure E.6), so z can be expressed as

$$z = a + bi = (r \cos \theta) + (r \sin \theta)i = r(\cos \theta + i \sin \theta)$$

from which it follows that

$$z = r(\cos \theta + i \sin \theta) \tag{1}$$

We call this a **polar form of z**. Observe that the quantity r in this formula is the modulus of z and that r and θ are related to the real and imaginary parts of z by

$$r = |z| = \sqrt{a^2 + b^2} \quad \text{and} \quad \tan \theta = b/a \tag{2-3}$$

The angle θ is called an **argument of z** and is denoted by

$$\theta = \arg z$$

The argument of z is not uniquely determined because we can add or subtract any multiple of 2π from θ to produce another value of the argument. However, if $z \ne 0$, then there is only one value of the argument in radians that satisfies

$$-\pi < \theta \le \pi$$

This is called the **principal argument of z** and is denoted by

$$\theta = \text{Arg } z$$

Example 1 Express the following complex numbers in polar form using their principal arguments:

(a) $z = 1 + \sqrt{3}i$ (b) $z = -1 - i$

Solution (a). To find r and Arg(z), we start with Formulas (2) and (3). We have $a = 1$ and $b = \sqrt{3}$, so

$$r = |z| = \sqrt{1^2 + (\sqrt{3})^2} = \sqrt{4} = 2 \quad \text{and} \quad \tan \theta = \sqrt{3}$$

From these values and the fact that z corresponds to a point in the first quadrant (Figure E.7), we conclude that $\theta = \text{Arg}(z) = \pi/3$, and hence a polar form of z is

$$z = 2\left(\cos\frac{\pi}{3} + i\sin\frac{\pi}{3}\right)$$

Solution (b). In this case we have $a = -1$ and $b = -1$, so it follows from (2) and (3) that

$$r = |z| = \sqrt{(-1)^2 + (-1)^2} = \sqrt{2} \quad \text{and} \quad \tan\theta = 1$$

From these values and the fact that z corresponds to a point in the third quadrant (Figure E.8), we conclude that $\theta = \text{Arg}(z) = -3\pi/4$, and hence a polar form of z is

$$z = \sqrt{2}\left[\cos\left(\frac{-3\pi}{4}\right) + i\sin\left(\frac{-3\pi}{4}\right)\right] \quad \blacktriangleleft$$

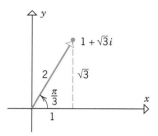

Figure E.7

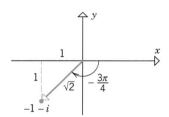

Figure E.8

□ **GEOMETRIC INTERPRETATION OF MULTIPLICATION AND DIVISION**

We now show how polar forms can be used to give geometric interpretations of multiplication and division of complex numbers. Let

$$z_1 = r_1(\cos\theta_1 + i\sin\theta_1) \quad \text{and} \quad z_2 = r_2(\cos\theta_2 + i\sin\theta_2)$$

Multiplying, we obtain

$$z_1 z_2 = r_1 r_2[(\cos\theta_1\cos\theta_2 - \sin\theta_1\sin\theta_2) + i(\sin\theta_1\cos\theta_2 + \cos\theta_1\sin\theta_2)]$$

Recall the trigonometric identities

$$\cos(\theta_1 + \theta_2) = \cos\theta_1\cos\theta_2 - \sin\theta_1\sin\theta_2$$

$$\sin(\theta_1 + \theta_2) = \sin\theta_1\cos\theta_2 + \cos\theta_1\sin\theta_2$$

We obtain the following polar form of $z_1 z_2$:

$$z_1 z_2 = r_1 r_2[\cos(\theta_1 + \theta_2) + i\sin(\theta_1 + \theta_2)] \tag{4}$$

In words, *the product of two complex numbers can be obtained by multiplying their moduli and adding their arguments* (Figure E.9).

We leave it as an exercise to show that if $z_2 \neq 0$, then

$$\frac{z_1}{z_2} = \frac{r_1}{r_2}[\cos(\theta_1 - \theta_2) + i\sin(\theta_1 - \theta_2)] \tag{5}$$

In words, *the quotient of two complex numbers can be obtained by dividing their moduli and subtracting their arguments.*

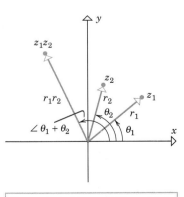

The product of two complex numbers

Figure E.9

Example 2 Use the polar forms obtained in Example 1 to express the following product and quotient in polar form.

(a) $(1 + \sqrt{3}i)(-1 - i)$ (b) $\dfrac{1 + \sqrt{3}i}{-1 - i}$

Solution. From Example 1 we have

$$1 + \sqrt{3}i = 2\left(\cos\frac{\pi}{3} + i\sin\frac{\pi}{3}\right)$$

$$-1 - i = \sqrt{2}\left[\cos\left(-\frac{3\pi}{4}\right) + i\sin\left(-\frac{3\pi}{4}\right)\right]$$

Thus, by multiplying the moduli and adding the arguments we obtain

$$(1 + \sqrt{3}i)(-1 - i) = 2\sqrt{2}\left[\cos\left(-\frac{5\pi}{12}\right) + i\sin\left(-\frac{5\pi}{12}\right)\right]$$

and by dividing the moduli and subtracting the arguments we obtain

$$\frac{1 + \sqrt{3}i}{-1 - i} = \frac{2}{\sqrt{2}}\left[\cos\left(\frac{13\pi}{12}\right) + i\sin\left(\frac{13\pi}{12}\right)\right]$$

$$= \sqrt{2}\left[\cos\left(\frac{13\pi}{12}\right) + i\sin\left(\frac{13\pi}{12}\right)\right] \quad \blacktriangleleft$$

REMARK.   Applying Formulas (4) and (5) may not result in polar forms with principal arguments, even if θ_1 and θ_2 are principal arguments. For example, in the preceding computations the argument that resulted for the product was the principal argument, but the argument for the quotient was not. If a principal argument is desired, then it may be necessary to add or subtract an appropriate multiple of 2π. For example, subtracting 2π from the argument for the quotient in part (b) yields $-11\pi/12$, which is the principal argument for the quotient.

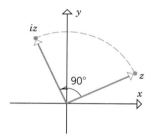

Figure E.10

Example 3   The complex number i has a modulus of 1 and an argument of $\pi/2$, so for any complex number z the product iz has the same modulus as z, but its argument is $\pi/2$ ($= 90°$) greater than that of z. Thus, multiplying z by i has the geometric effect of rotating z counterclockwise by $90°$ (Figure E.10). $\blacktriangleleft$

☐ **DEMOIVRE'S FORMULA**

If n is a positive integer and $z = r(\cos\theta + i\sin\theta)$, then from Formula (4),

$$z^n = \underbrace{z \cdot z \cdot z \cdots z}_{n\text{ factors}} = r^n[\cos\underbrace{(\theta + \theta + \cdots + \theta)}_{n\text{ terms}} + i\sin\underbrace{(\theta + \theta + \cdots + \theta)}_{n\text{ terms}}]$$

that is,

$$z^n = r^n(\cos n\theta + i\sin n\theta) \tag{6}$$

In words, *a complex number can be raised to the nth power by raising the modulus to the nth power and multiplying the argument by n.*

Although we derived (6) for positive integer n, this formula is valid for all integer values of n under appropriate conditions (Exercise 17). In the special case where $r = 1$, we have $z = \cos\theta + i\sin\theta$, so (6) becomes

$$(\cos\theta + i\sin\theta)^n = \cos n\theta + i\sin n\theta \tag{7}$$

which is called **DeMoivre's* Formula**.

---

* ABRAHAM DEMOIVRE (1667–1754) was a French mathematician who made important contributions to probability, statistics, and trigonometry. He developed the concept of statistically independent events, wrote a major influential treatise on probability, and helped transform trigonometry from a branch of geometry into a branch of analysis through his use of complex numbers. In spite of his important work, he barely managed to eke out a living as a tutor and a consultant on gambling and insurance.

Example 4   Find $(1 + \sqrt{3}i)^9$.

Solution.   We could use the binomial formula and simplify, but it is much easier to apply (6). Using the polar form obtained in Example 1 we have

$$(1 + \sqrt{3}i)^9 = \left[2 \left(\cos \frac{\pi}{3} + i \sin \frac{\pi}{3} \right) \right]^9 = 2^9 (\cos 3\pi + i \sin 3\pi) = -2^9 = -512 \quad \blacktriangleleft$$

☐ **FINDING *n*th ROOTS OF COMPLEX NUMBERS**

We define an ***n*th root** of a nonzero complex number z to be any complex number w that satisfies the equation

$$w^n = z \tag{8}$$

A formula for the nth roots of z can be obtained by expressing w and z in polar form, say

$$w = \rho(\cos \alpha + i \sin \alpha) \quad \text{and} \quad z = r(\cos \theta + i \sin \theta)$$

It now follows from (8) with (6) applied to w that

$$\rho^n (\cos n\alpha + i \sin n\alpha) = r(\cos \theta + i \sin \theta)$$

which implies that

$$\rho^n = r, \quad \cos n\alpha = \cos \theta, \quad \sin n\alpha = \sin \theta$$

Since ρ and r are nonnegative real numbers, it follows from the first of these equations that ρ is the nonnegative real nth root of r, that is, $\rho = \sqrt[n]{r}$; and it follows from the last two equations that

$$n\alpha = \theta + 2k\pi \quad \text{or} \quad \alpha = \frac{\theta}{n} + \frac{2k\pi}{n}, \quad k = 0, \pm 1, \pm 2, \ldots$$

Thus, the values of $w = \rho(\cos \alpha + i \sin \alpha)$ that satisfy (8) are given by

$$w = \sqrt[n]{r} \left[\cos \left(\frac{\theta}{n} + \frac{2k\pi}{n} \right) + i \sin \left(\frac{\theta}{n} + \frac{2k\pi}{n} \right) \right], \quad k = 0, \pm 1, \pm 2, \ldots$$

Although there are infinitely many values of k, it can be shown (Exercise 16) that $k = 0$, 1, 2, ..., $n - 1$ produce distinct values of w satisfying (8), but all other choices of k yield duplicates of these. Therefore, there are exactly n different nth roots of $z = r(\cos \theta + i \sin \theta)$, and these are given by

$$w = \sqrt[n]{r} \left[\cos \left(\frac{\theta}{n} + \frac{2k\pi}{n} \right) + i \sin \left(\frac{\theta}{n} + \frac{2k\pi}{n} \right) \right], \quad k = 0, 1, 2, \ldots, n - 1 \tag{9}$$

REMARK.   Observe that each of these nth roots has the same modulus, namely $\sqrt[n]{r}$ $(= \sqrt[n]{|z|})$, and that the nth roots corresponding to successive values of k have arguments that differ by $2\pi/n$ radians. This implies that the nth roots of z are equally spaced around a circle of radius $\sqrt[n]{|z|}$ and that if one nth root of z is found, then the remaining $n - 1$ roots can be obtained by rotating this root through successive increments of $2\pi/n$ radians.

Example 5   Find all cube roots of -8.

Solution.   Since -8 lies on the negative real axis, it can be expressed in polar form as

$$-8 = 8(\cos \pi + i \sin \pi)$$

Thus, from (9) with $n = 3$ the cube roots of -8 are given by

$$w = \sqrt[3]{8} \left[\cos \left(\frac{\pi}{3} + \frac{2k\pi}{3} \right) + i \sin \left(\frac{\pi}{3} + \frac{2k\pi}{3} \right) \right], \quad k = 0, 1, 2$$

If we denote these cube roots by w_0, w_1, and w_2, respectively, then it follows from this formula that

$$w_0 = 2\left(\cos\frac{\pi}{3} + i\sin\frac{\pi}{3}\right) = 2\left(\frac{1}{2} + \frac{\sqrt{3}}{2}i\right) = 1 + \sqrt{3}i$$

$$w_1 = 2(\cos\pi + i\sin\pi) = 2(-1) = -2$$

$$w_2 = 2\cos\left(\frac{5\pi}{3} + i\sin\frac{5\pi}{3}\right) = 2\left(\frac{1}{2} - \frac{\sqrt{3}}{2}i\right) = 1 - \sqrt{3}i$$

Alternative Solution. By inspection, -2 is a cube root of -8. If we represent this root as a point or a vector in the complex plane, then the remaining two roots can be obtained by rotating this root through two increments of $2\pi/3$ ($= 120°$), as shown in Figure E.11. ◀

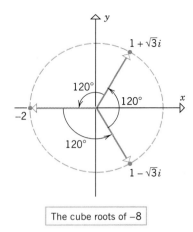

The cube roots of –8

Figure E.11

☐ **COMPLEX EXPONENTIALS**

To define the complex exponential function e^z ($= e^{a+ib}$), we start by defining e^{ib}. A motivation for the definition of this function can be obtained by formally substituting ib for x in the Maclaurin series for e^x and simplifying using the relationships

$$i^2 = -1, \quad i^3 = i\cdot i^2 = -i, \quad i^4 = i\cdot i^3 = 1, \quad i^5 = i\cdot i^4 = i, \quad i^6 = i\cdot i^5 = -1,\ldots$$

This yields

$$e^{ib} = 1 + ib + \frac{(ib)^2}{2!} + \frac{(ib)^3}{3!} + \frac{(ib)^4}{4!} + \frac{(ib)^5}{5!} + \frac{(ib)^6}{6!} + \cdots$$

$$= 1 + ib - \frac{b^2}{2!} - i\frac{b^3}{3!} + \frac{b^4}{4!} + i\frac{b^5}{5!} - \frac{b^6}{6!} - \cdots$$

$$= \left(1 - \frac{b^2}{2!} + \frac{b^4}{4!} - \frac{b^6}{6!} + \cdots\right) + i\left(b - \frac{b^3}{3!} + \frac{b^5}{5!} - \cdots\right)$$

$$= \cos b + i\sin b$$

where the last step follows from the Maclaurin series for $\cos b$ and $\sin b$. This suggests that e^{ib} be defined as

$$e^{ib} = \cos b + i\sin b \tag{10}$$

which is sometimes called the ***Euler formula***. The definition of e^z can now be motivated by writing $e^z = e^{a+ib} = e^a e^{ib}$, which suggests that we define

$$e^z = e^{a+ib} = e^a(\cos b + i\sin b) \tag{11}$$

Example 6

$$e^{i\pi} = \cos\pi + i\sin\pi = -1$$

$$e^{1+2\pi i} = e^1(\cos 2\pi + i\sin 2\pi) = e$$

$$e^{2+(\pi i/2)} = e^2(\cos\pi/2 + i\sin\pi/2) = e^2 i \qquad ◀$$

☐ **EXPONENTIAL NOTATION FOR POLAR FORMS**

It follows from (10) that the polar form of a complex number can be expressed in exponential notation as

$$z = r(\cos\theta + i\sin\theta) = re^{i\theta} \tag{12}$$

It can be proved that complex exponents follow the same algebraic laws as real exponents. For example, if $z_1 = r_1 e^{i\theta_1}$ and $z_2 = r_2 e^{i\theta_2}$, then

$$z_1 z_2 = r_1 r_2 e^{i\theta_1 + i\theta_2} = r_1 r_2 e^{i(\theta_1 + \theta_2)}$$

$$\frac{z_1}{z_2} = \frac{r_1}{r_2} e^{i\theta_1 - i\theta_2} = \frac{r_1}{r_2} e^{i(\theta_1 - \theta_2)}$$

which are precisely Formulas (4) and (5), but expressed in exponential notation. Moreover, it follows from (12) that

$$z^n = (re^{i\theta})^n = r^n e^{in\theta} \tag{13}$$

which is precisely Formula (6), but in exponential notation.

Finally, if z has modulus r and argument θ, then $\bar{z}$ has modulus r and argument $-\theta$, since z and $\bar{z}$ are symmetric about the real axis. Thus, if $z = re^{i\theta}$, then $\bar{z} = re^{-i\theta}$, that is,

$$\overline{re^{i\theta}} = re^{-i\theta} \tag{14}$$

▶ Exercise Set E.II

1. In each part find the principal argument of z.
   (a) $z = 1$
   (b) $z = i$
   (c) $z = -i$
   (d) $z = 1 + i$
   (e) $z = -1 + \sqrt{3}i$
   (f) $z = 1 - i$.

2. In each part find the value of $\theta = \arg(1 - \sqrt{3}i)$ that satisfies the given condition.
   (a) $0 \le \theta < 2\pi$
   (b) $-\pi < \theta \le \pi$
   (c) $-\frac{\pi}{6} \le \theta < \frac{11\pi}{6}$.

3. In each part express the complex number in polar form using its principal argument.
   (a) $2i$
   (b) -4
   (c) $5 + 5i$
   (d) $-6 + 6\sqrt{3}i$
   (e) $-3 - 3i$
   (f) $2\sqrt{3} - 2i$.

4. Given that $z_1 = 2(\cos \pi/4 + i \sin \pi/4)$ and $z_2 = 3(\cos \pi/6 + i \sin \pi/6)$, find a polar form of
   (a) $z_1 z_2$
   (b) z_1/z_2
   (c) z_2/z_1
   (d) z_1^5/z_2^2.

5. Express $z_1 = i$, $z_2 = 1 - \sqrt{3}i$, and $z_3 = \sqrt{3} + i$ in polar form, and use your results to find $z_1 z_2/z_3$. Check your results by performing the calculations without using polar forms.

6. Use Formula (6) to find
   (a) $(1 + i)^{12}$
   (b) $\left(\frac{1}{\sqrt{2}} - \frac{1}{\sqrt{2}}i\right)^{-6}$
   (c) $(\sqrt{3} + i)^7$
   (d) $(1 - \sqrt{3}i)^{-10}$.

7. In each part find all the roots and sketch them as vectors in the complex plane.
   (a) $(-i)^{1/2}$
   (b) $(1 + \sqrt{3}i)^{1/2}$
   (c) $(-27)^{1/3}$
   (d) $(i)^{1/3}$
   (e) $(-1)^{1/4}$
   (f) $(-8 + 8\sqrt{3}i)^{1/4}$.

8. Use the method in the alternative solution of Example 5 to find all cube roots of 1.

9. Use the method in the alternative solution of Example 5 to find all sixth roots of 1.

10. Find all square roots of $1 + i$ and express your results in polar form.

11. In each part find all solutions of the equation.
    (a) $z^4 - 16 = 0$
    (b) $z^4 + 64 = 0$.

12. Find the four solutions of the equation $z^4 + 8 = 0$ and use your results to factor $z^4 + 8$ into two quadratic factors with real coefficients.

13. It was shown in Example 3 that multiplying z by i rotates z counterclockwise by 90°. What is the geometric effect of dividing z by i?

14. In each part use (13) to calculate the given power.
    (a) $(1 + i)^8$
    (b) $(-2\sqrt{3} + 2i)^{-9}$.

15. In each part find $\text{Re}(z)$ and $\text{Im}(z)$.
    (a) $z = 3e^{i\pi}$
    (b) $z = 3e^{-i\pi}$
    (c) $\bar{z} = \sqrt{2}e^{\pi i/2}$
    (d) $\bar{z} = -3e^{-2\pi i}$.

16. (a) Show that if $z \neq 0$, then the values of w in Formula (9) are all different.
    (b) Show that integer values of k other than $k = 0, 1, 2, \ldots, n - 1$ produce values of w that are duplicates of those in Formula (9).

17. Assuming that $z \neq 0$ and $z^0 = 1$, show that Formula (6) is valid if $n = 0$ or n is a negative integer.

18. Derive Formula (5).

ANSWERS TO
ODD-NUMBERED EXERCISES

▶ **Exercise Set 14.1 (Page 662)**

1. (a) $\sqrt{14}$; $(1, \frac{1}{2}, \frac{3}{2})$ (b) $\sqrt{11}$; $(\frac{9}{2}, \frac{3}{2}, \frac{9}{2})$
   (c) $\sqrt{30}$; $(\frac{1}{2}, -\frac{1}{2}, 4)$ (d) $\sqrt{42}$; $(\frac{3}{2}, 1, -\frac{5}{2})$
3. $(-6, 2, 1)$, $(-6, 2, -2)$, $(-6, 1, -2)$, $(4, 2, 1)$,
   $(4, 1, 1)$, $(4, 1, -2)$, $(4, 2, -2)$, $(-6, 1, 1)$
5. (b) $(2, 1, 6)$ (c) 49
7. distance to x-axis is $\sqrt{y_0^2 + z_0^2}$
   distance to y-axis is $\sqrt{x_0^2 + z_0^2}$
9. $(x + 2)^2 + (y - 4)^2 + (z + 1)^2 = 36$
11. $x^2 + (y - 1)^2 + z^2 = 9$
13. (a) $(x - 2)^2 + (y + 1)^2 + (z + 3)^2 = 9$
    (b) $(x - 2)^2 + (y + 1)^2 + (z + 3)^2 = 1$
    (c) $(x - 2)^2 + (y + 1)^2 + (z + 3)^2 = 4$
15. $(x + \frac{1}{2})^2 + (y - 2)^2 + (z - 2)^2 = \frac{5}{4}$
17. $(x - 3)^2 + (y + 2)^2 + (z - 4)^2 = 41$
19. $x^2 + y^2 + z^2 = 30 \pm 2\sqrt{29}$
21. sphere, center $(-5, -2, -1)$, radius 7
23. sphere, center $(\frac{1}{2}, \frac{3}{4}, -\frac{5}{4})$, radius $\frac{3}{4}\sqrt{6}$ 25. no graph
27. largest $3 + \sqrt{6}$; smallest $3 - \sqrt{6}$
29. all points outside the circular cylinder
    $(y + 3)^2 + (z - 2)^2 = 16$
31. $(2 - \sqrt{3})R$
33. (a) 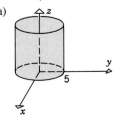 (b)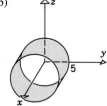

    (c)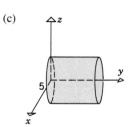

35. (a) 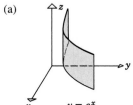 (b)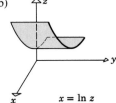
    $y = e^x$ $x = \ln z$

    (c)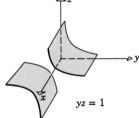
    $yz = 1$

37. (a) 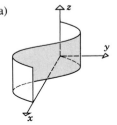 (b)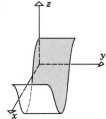

39. (a) 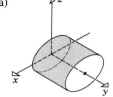 (b)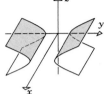

37. (a) 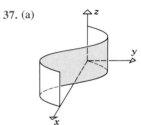 (b)

39. (a) 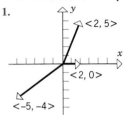 (b)

▶ **Exercise Set 14.2 (Page 670)**

1.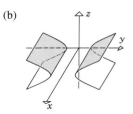

3.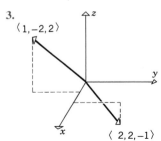

5.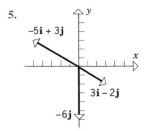

7.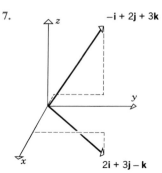

9. (a) $\langle -1, 3 \rangle$ (b) $\langle -7, 2 \rangle$ (c) $\langle 2, 1 \rangle$ (d) $\langle -8, 7 \rangle$
11. (a) $\langle -3, 6, 1 \rangle$ (b) $\langle 1, -3, -5 \rangle$ 13. $(4, -4)$
15. $(4, -4)$ 17. $(8, -1, -3)$
19. (a) $-5\mathbf{i} - 2\mathbf{j}$ (b) $8\mathbf{i} + 10\mathbf{j}$ (c) $-2\mathbf{i} + 4\mathbf{j}$
    (d) $40\mathbf{i} + 36\mathbf{j}$ (e) $-20\mathbf{i} + 22\mathbf{j}$ (f) $-8\mathbf{i} - 8\mathbf{j}$
21. (a) $-\mathbf{i} + 4\mathbf{j} - 2\mathbf{k}$ (b) $18\mathbf{i} + 12\mathbf{j} - 6\mathbf{k}$ (c) $-\mathbf{i} - 5\mathbf{j} - 2\mathbf{k}$
    (d) $40\mathbf{i} - 4\mathbf{j} - 4\mathbf{k}$ (e) $-2\mathbf{i} - 16\mathbf{j} - 18\mathbf{k}$
    (f) $-\mathbf{i} + 13\mathbf{j} - 2\mathbf{k}$
23. (a) $\sqrt{2}$ (b) 2 (c) 3 25. (a) $\sqrt{14}$ (b) 3
27. (a) $\sqrt{41}$ (b) 7 (c) $3\sqrt{29} + 8$ (d) $3\sqrt{10}$
    (e) $\langle \frac{3}{5}, \frac{4}{5} \rangle$ (f) 1
29. (a) $2\sqrt{3}$ (b) $\sqrt{14} + \sqrt{2}$ (c) $2\sqrt{14} + 2\sqrt{2}$
    (d) $2\sqrt{37}$ (e) $\dfrac{1}{\sqrt{6}}\mathbf{i} + \dfrac{1}{\sqrt{6}}\mathbf{j} - \dfrac{2}{\sqrt{6}}\mathbf{k}$ (f) 1
31. $\langle -\frac{2}{3}, 1 \rangle$ 33. $\mathbf{u} = \frac{5}{7}\mathbf{i} + \frac{2}{7}\mathbf{j} + \frac{1}{7}\mathbf{k}$, $\mathbf{v} = \frac{8}{7}\mathbf{i} - \frac{1}{7}\mathbf{j} - \frac{4}{7}\mathbf{k}$
43. $-\frac{3}{5}\mathbf{i} + \frac{4}{5}\mathbf{j}$ 45. $-\dfrac{1}{\sqrt{14}}(3\mathbf{i} - 2\mathbf{j} + \mathbf{k})$
47. $\dfrac{1}{3\sqrt{2}}(4\mathbf{i} + \mathbf{j} - \mathbf{k})$ 49. $\langle -\frac{3}{2}, 2 \rangle$ 51. $6\mathbf{i} - 8\mathbf{j} - 2\mathbf{k}$
53. circle; radius 1, center at (x_0, y_0)
55. (a) sphere; radius 2, center at $(0, 0, 0)$
    (b) sphere; radius 3, center at (x_0, y_0, z_0)
    (c) all points on and inside the sphere of radius 1, centered at (x_0, y_0, z_0)
57. (a) $\langle 1/\sqrt{2}, -1/\sqrt{2} \rangle$, $\langle -1/\sqrt{2}, 1/\sqrt{2} \rangle$
    (b) $\langle 1/\sqrt{2}, 1/\sqrt{2} \rangle$, $\langle -1/\sqrt{2}, -1/\sqrt{2} \rangle$
59. $\langle \frac{13}{4}, \frac{21}{4} \rangle$ 61. (a) $\langle 1/2, \sqrt{3}/2 \rangle$ (b) $\langle -2\sqrt{2}, 2\sqrt{2} \rangle$
63. $(1, 1), (-1, -1)$ 77. (a) $\langle \frac{3}{2}, \frac{1}{2} \rangle$

▶ **Exercise Set 14.3 (Page 677)**

1. (a) -10 (b) -3 (c) 0 (d) -20
3. (a) obtuse (b) acute (c) obtuse (d) orthogonal
5. (a) $\frac{14}{13}\mathbf{i} + \frac{21}{13}\mathbf{j}$ (b) $\langle 2, 6 \rangle$ (c) $-\frac{11}{13}\mathbf{i} + \mathbf{j} + \frac{55}{13}\mathbf{k}$
    (d) $\langle -\frac{32}{89}, -\frac{12}{89}, \frac{73}{89} \rangle$
9. (a) 6 (b) 36 (c) $24\sqrt{5}$ (d) $24\sqrt{5}$
11. $\dfrac{1}{5\sqrt{2}}, \dfrac{4}{\sqrt{65}}, \dfrac{9}{\sqrt{130}}$ 13. The right angle is at vertex B.
15. (a) the line through the origin and perpendicular to $\mathbf{r}_0$
    (b) the line through (x_0, y_0) and perpendicular to $\mathbf{r}_0$
    (c) circle; center at $(\frac{1}{2}x_0, \frac{1}{2}y_0)$, radius $\frac{1}{2}\|\mathbf{r}_0\|$
17. (a) $-\frac{3}{4}$ (b) $\frac{1}{7}$ (c) $\dfrac{48 \pm 25\sqrt{3}}{11}$ (d) $\frac{4}{3}$
21. (a) $\frac{9}{5}$ (b) $\frac{13}{5}$ 23. (a) $\frac{4}{3}$ (b) $\frac{1}{3}\sqrt{137}$
25. -12 ft · lb 31. $71°$

▶ **Exercise Set 14.4 (Page 685)**

1. $\langle 7, 10, 9 \rangle$ 3. $\langle -4, -6, -3 \rangle$
5. (a) $\langle -20, -67, -9 \rangle$ (b) $\langle -78, 52, -26 \rangle$
    (c) $\langle 24, 0, -16 \rangle$ (d) $\langle -12, -22, -8 \rangle$
    (e) $\langle 0, -56, -392 \rangle$ (f) $\langle 0, 56, 392 \rangle$
9. $\pm \dfrac{1}{\sqrt{5}}(2\mathbf{j} + \mathbf{k})$ 11. $2\mathbf{v} \times \mathbf{u}$
13. (a) $\frac{1}{2}\sqrt{374}$ (b) $9\sqrt{13}$ 15. (a) $\frac{1}{2}\sqrt{26}$ (b) $\frac{1}{3}\sqrt{26}$
17. (a) $2\sqrt{\dfrac{141}{29}}$ (b) $\frac{1}{3}\sqrt{137}$

19. ambiguous, needs parentheses   **21.** 80   **23.** 1
25. (a) 16   (b) 45
27. (a) 9   (b) $\sqrt{122}$ (c) $\sin^{-1}\left(\frac{9}{14}\right)$ **29.** $\mathbf{u} \times \mathbf{v}$
37. (a) $\frac{2}{3}$ (b) $\frac{1}{2}$

▶ **Exercise Set 14.5** (Page 690)

1. $x = 3 + 2t,\ y = -2 + 3t$ **3.** $x = 4,\ y = 1 + 2t$
5. $x = 5 - 3t,\ y = -2 + 6t,\ z = 1 + t$
7. $x = -t,\ y = 6t,\ z = t$
9. same as Exercise 1 with $0 \le t \le 1$
11. same as Exercise 3 with $0 \le t \le 1$
13. same as Exercise 5 with $0 \le t \le 1$
15. same as Exercise 7 with $0 \le t \le 1$
17. $x = -5 + 2t,\ y = 2 - 3t$ **19.** $x = 3 + 4t,\ y = -4 + 3t$
21. $x = -1 + 3t,\ y = 2 - 4t,\ z = 4 + t$
23. $x = -2 + 2t,\ y = -t,\ z = 5 + 2t$
25. $x = 3 + t,\ y = 7,\ z = 0$ **27.** $(-1, 1),\ (3, 9)$
29. (a) $(-2, 10, 0)$ (b) $(-2, 0, -5)$
    (c) does not intersect yz-plane
31. $(0, 4, -2),\ (4, 0, 6)$
33. $x = x_1 + at,\ y = y_1 + bt,\ z = z_1 + ct$ **35.** $(1, -1, 2)$
39. (a) no   (b) yes   **41.** $\left(1, \frac{14}{3}, -\frac{5}{3}\right)$
45. (a) $\langle x, y \rangle = \langle 2, -1 \rangle + t \langle -7, 4 \rangle$ (b) $\langle x, y \rangle = \langle 0, 3 \rangle + t \langle 4, 0 \rangle$
47. the line segment joining the points $(1, 0)$ and $(-3, 6)$
49. $2\sqrt{5}$ **51.** $\sqrt{35/6}$
53. (b) $84°$ (c) $x = 7 + t,\ y = -1,\ z = -2 + t$
55. $x = t,\ y = 2 + t,\ z = 1 - t$
57. (a) $\sqrt{17}$ cm   (b) $\frac{1}{2}\sqrt{14}$ cm

▶ **Exercise Set 14.6** (Page 696)

1. $x + 4y + 2z = 28$ **3.** $z = 0$
5. (a) $2y - z = 1$ (b) $x + 9y - 5z = 16$
7. (a) yes   (b) no   **9.** (a) yes   (b) no
11. (a) $35°$ (b) $79°$
13. (a) $z = 0$ (b) $y = 0$ (c) $x = 0$
15. $4x - 2y + 7z = 0$ **17.** $4x - 13y + 21z = -14$
19. $x = 5 - 2t,\ y = 5t,\ z = -2 + 11t$ **21.** $x + y - 3z = 6$
23. $7x + y + 9z = 25$ **25.** $2x + 4y + 8z = 29$
29. (a) $x = -\frac{11}{7} - 23t,\ y = -\frac{12}{7} + t,\ z = -7t$
    (b) $x = -5t,\ y = -3t,\ z = 0$
31. (b) $\frac{2}{5}$ **33.** $\frac{5}{3}$ **35.** $\frac{137}{21}$ **37.** $\frac{5}{3\sqrt{6}}$ **39.** $\frac{25}{\sqrt{126}}$
41. $\frac{95}{\sqrt{1817}}$ **45.** $(x - 2)^2 + (y - 1)^2 + (z + 3)^2 = \frac{121}{14}$

▶ **Exercise Set 14.7** (Page 706)

1. (a) $4x^2 + y^2 = 4$; ellipse   (b) $y^2 + z^2 = 3$; circle
    (c) $4x^2 + z^2 = 3$; ellipse
3. (a) $9x^2 - z^2 = 16$; hyperbola   (b) $y^2 + z^2 = 20$; circle
    (c) $9x^2 - y^2 = 20$; hyperbola
5. (a) $z = 4y^2$; parabola   (b) $z = 9x^2 + 16$; parabola
    (c) $9x^2 + 4y^2 = 4$; ellipse

7.
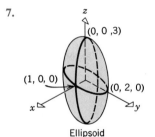
Ellipsoid

9.
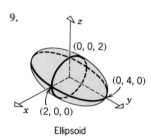
Ellipsoid

11.
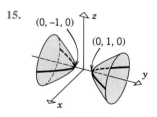
Hyperboloid
of one sheet

13.
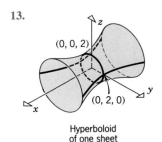
Hyperboloid
of one sheet

15.
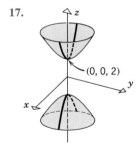
Hyperboloid
of two sheets

17.
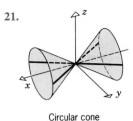
Hyperboloid
of two sheets

19.
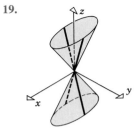
Elliptic cone

21.
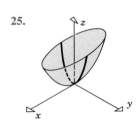
Circular cone

23.
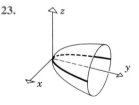

Circular paraboloid

25.
Elliptic paraboloid

27.

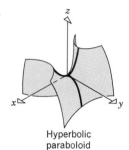

Hyperbolic
paraboloid

29. (a)

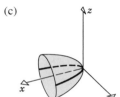

Hyperboloid
of one sheet

(b)

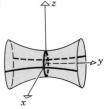

Hyperboloid
of two sheets

(c)

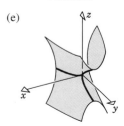

Paraboloid

(d)

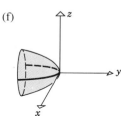

Cone

(e)

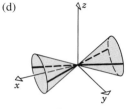

Hyperbolic
paraboloid

(f)

Paraboloid

31.

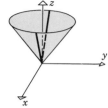

33.

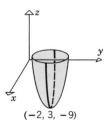

35.

$(-2, 3, -9)$
Circular
paraboloid

37.

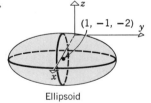

Ellipsoid

39. $(0, -1, 5)$

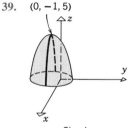

Circular
paraboloid

45. $x^2 + y^2 = 2$; circle   **47.** $x^2 + y^2 - 2x = 0$; circle

49. $y = \frac{1}{2}x^2 - \frac{1}{2}$; parabola

51. $x^2 + y^2 + 4x = 5$, $x \geq 0$; circular arc

53. (a) focus: $\left(k, 0, \frac{k^2}{9} + 1\right)$, vertex: $\left(k, 0, \frac{k^2}{9}\right)$

(b) foci: $(\sqrt{5k}, 0, k)$, $(-\sqrt{5k}, 0, k)$;
endpoints of major axis: $(3\sqrt{k}, 0, k)$, $(-3\sqrt{k}, 0, k)$;
endpoints of minor axis: $(0, 2\sqrt{k}, k)$, $(0, -2\sqrt{k}, k)$

55. $z = \frac{1}{4}(x^2 + y^2)$; paraboloid

▶ **Exercise Set 14.8 (Page 712)**

1. (a) $(8, \pi/6, -4)$ (b) $(5\sqrt{2}, 3\pi/4, 6)$ (c) $(2, \pi/2, 0)$
(d) $(8, 5\pi/3, 6)$ (e) $(2, 7\pi/4, 1)$ (f) $(0, 0, 1)$

3. (a) $(2\sqrt{2}, \pi/3, 3\pi/4)$ (b) $(2, 7\pi/4, \pi/4)$
(c) $(6, \pi/2, \pi/3)$ (d) $(10, 5\pi/6, \pi/2)$
(e) $(8\sqrt{2}, \pi/4, \pi/6)$ (f) $(2\sqrt{2}, 5\pi/3, 3\pi/4)$

5. (a) $(2\sqrt{3}, \pi/6, \pi/6)$ (b) $(\sqrt{2}, \pi/4, 3\pi/4)$
(c) $(2, 3\pi/4, \pi/2)$ (d) $(4\sqrt{3}, 1, 2\pi/3)$
(e) $(4\sqrt{2}, 5\pi/6, \pi/4)$ (f) $(2\sqrt{2}, 0, 3\pi/4)$

7.

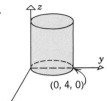

$(3, 0, 0)$
$x^2 + y^2 = 9$

9.

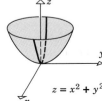

$z = x^2 + y^2$

11.

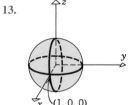

$(0, 4, 0)$
$x^2 + (y-2)^2 = 4$

13.

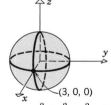

$(1, 0, 0)$
$x^2 + y^2 + z^2 = 1$

15.

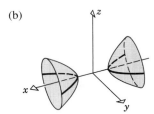

$(3, 0, 0)$
$x^2 + y^2 + z^2 = 9$

17.

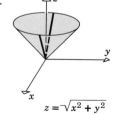

$z = \sqrt{x^2 + y^2}$

19.

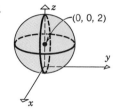

$x^2 + y^2 + (z-2)^2 = 4$

21.

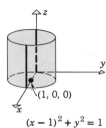

$(x-1)^2 + y^2 = 1$

23. (a) $z = 3$ (b) $\rho = 3 \sec \phi$
25. (a) $z = 3r^2$ (b) $\rho = \frac{1}{3} \csc \phi \cot \phi$
27. (a) $r = 2$ (b) $\rho = 2 \csc \phi$
29. (a) $r^2 + z^2 = 9$ (b) $\rho = 3$
31. (a) $r(2 \cos \theta + 3 \sin \theta) + 4z = 1$
(b) $\rho(2 \sin \phi \cos \theta + 3 \sin \phi \sin \theta + 4 \cos \phi) = 1$
33. (a) $r^2 \cos^2 \theta = 16 - z^2$ (b) $\rho^2(1 - \sin^2 \phi \sin^2 \theta) = 16$
35. all points on or above the paraboloid $z = x^2 + y^2$ that are also on or below the plane $z = 4$
37. all points on or between the concentric spheres $\rho = 1$ and $\rho = 3$
39. spherical: $(4000, \pi/6, \pi/6)$
rectangular: $(1000\sqrt{3}, 1000, 2000\sqrt{3})$
41.

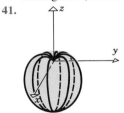

▶ **TECHNOLOGY EXERCISES, CHAPTER 14 (Page 713)**

1. (a) $40°$ (b) -0.682328
3. (b) $60°$ (c) $55°, 125°$
(d) maximum $140°$, minimum $40°$
5. 0.800444
7. $x = 3.932993$, $y = 0.644331$, $z = -0.355669$ **9.** 50.96

▶ **Exercise Set 15.1 (Page 719)**

1. $(-\infty, +\infty)$; $-\mathbf{i} - 3\pi\mathbf{j}$ **3.** $[2, +\infty)$; $-\mathbf{i} - \ln 3\mathbf{j} + \mathbf{k}$
5. $\mathbf{r} = 3 \cos t\mathbf{i} + (t + \sin t)\mathbf{j}$
7. $\mathbf{r} = 2t\mathbf{i} + 2 \sin 3t\mathbf{j} + 5 \cos 3t\mathbf{k}$
9. $x = 3t^2$, $y = -2$, $z = 0$
11. $x = 2t - 1$, $y = -3\sqrt{t}$, $z = \sin 3t$
13. line in the xy-plane through $(2, 0)$, parallel to $-3\mathbf{i} - 4\mathbf{j}$
15. line through $(0, -3, 1)$, parallel to $2\mathbf{i} + 3\mathbf{k}$
17. ellipse in the plane $z = -1$, center at $(0, 0, -1)$, major axis of length 6 parallel to x-axis, minor axis of length 4 parallel to y-axis
19. $-\frac{3}{2}$ **21.** $(\frac{5}{2}, 0, \frac{3}{2})$
23.

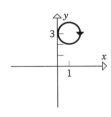

25.

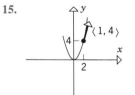

27.

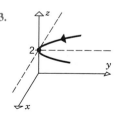

29.

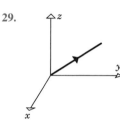

31.

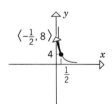

$(0, 2, \pi/2)$
$(2, 0, 0)$

33.

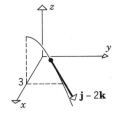

37. lies on the sphere $x^2 + y^2 + z^2 = 4$ and the plane $z = \sqrt{3}x$; center $(0, 0, 0)$, radius 2
39. $3/(2\pi)$ **41.** conical helix
43. $x = t$, $y = \frac{1}{4}t^2 - 1$, $z = \frac{1}{4}t^2 + 1$; parabola
45. $x = 3 \cos t$, $y = 3 \sin t$, $z = 9 \cos^2 t$
47. $x = 1 + \cos t$, $y = \frac{1}{2} \sin t$, $z = 2 + 2 \cos t$

▶ **Exercise Set 15.2 (Page 726)**

1. $9\mathbf{i} + 6\mathbf{j}$ **3.** $\mathbf{j}$ **5.** $2\mathbf{i} - 3\mathbf{j} + 4\mathbf{k}$ **7.** $\frac{1}{2}\pi\mathbf{i} + \mathbf{k}$
11. $5\mathbf{i} + (1 - 2t)\mathbf{j}$ **13.** $-\frac{1}{t^2}\mathbf{i} + \sec^2 t\mathbf{j} + 2e^{2t}\mathbf{k}$
15.

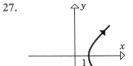

17.

$\langle -\frac{1}{2}, 8 \rangle$

19.

21.

$\mathbf{j} - 2\mathbf{k}$
$-2\mathbf{k}$

23. $x = 1 + 2t$, $y = 2 - t$, $z = 0$
25. $x = 1 - \sqrt{3}\pi t$, $y = \sqrt{3} + \pi t$, $z = 1 + 3t$
27. $\mathbf{r} = -\mathbf{i} + 2\mathbf{j} + t(2\mathbf{i} + \frac{3}{4}\mathbf{j})$
29. $\mathbf{r} = 4\mathbf{i} + \mathbf{j} + t(-4\mathbf{i} + \mathbf{j} + 4\mathbf{k})$ **31.** $2y + z = \pi/2$
33. (a) $(-2, 4, 6)$, $(1, 1, -3)$ (b) $76°, 71°$ **35.** $68°$
37. $3t\mathbf{i} + 2t^2\mathbf{j} + \mathbf{C}$ **39.** $\langle 0, -\frac{2}{3} \rangle$ **41.** $\frac{52}{3}\mathbf{i} + 4\mathbf{j}$
43. $\langle (t-1)e^t, t(\ln t - 1) \rangle + \mathbf{C}$ **45.** $\frac{1}{3}t^3\mathbf{i} - t^2\mathbf{j} + \ln |t|\mathbf{k} + \mathbf{C}$
47. $\frac{1}{2}(e^2 - 1)\mathbf{i} + (1 - e^{-1})\mathbf{j} + \frac{1}{2}\mathbf{k}$
49. (a) $\mathbf{F} = 3t\mathbf{i} - 2\mathbf{j} - 3t^2\mathbf{k}$ (b) 30
51. $(1 + \sin t)\mathbf{i} - (\cos t)\mathbf{j}$ **53.** $(t^4 + 2)\mathbf{i} - (t^2 + 4)\mathbf{j}$
55. $2(t - 1)\mathbf{i} + \frac{1}{2} \ln \frac{1}{2}(t^2 + 1)\mathbf{j} + \frac{1}{2}(t^2 - 1)\mathbf{k}$
57. (a) $7t^6$ (b) $12t \sec t \tan t + 12 \sec t - \dfrac{\sin t}{t} - (\cos t) \ln t$

▶ **Exercise Set 15.3 (Page 734)**

3. smooth   **5.** not smooth   **7.** $4\mathbf{i} + 8(4\tau + 1)\mathbf{j}$

9. $2\tau e^{\tau^2}\mathbf{i} - 8\tau e^{-\tau^2}\mathbf{j}$ **11.** $\sqrt{14}$ **13.** 28   **15.** $e - e^{-1}$

17. $t_0\sqrt{a^2 + c^2}$ **19.** $x = \frac{3}{5}s - 2, \; y = \frac{4}{5}s + 3$

21. $x = 3 + \cos s, \; y = 2 + \sin s \; (0 \le s \le 2\pi)$

23. $x = \frac{1}{3}[(3s + 1)^{2/3} - 1]^{3/2}, \; y = \frac{1}{2}[(3s + 1)^{2/3} - 1] \; (s \ge 0)$

25. $x = \left(\dfrac{s}{\sqrt{2}} + 1\right)\cos\left[\ln\left(\dfrac{s}{\sqrt{2}} + 1\right)\right],$

$y = \left(\dfrac{s}{\sqrt{2}} + 1\right)\sin\left[\ln\left(\dfrac{s}{\sqrt{2}} + 1\right)\right]$

where $0 \le s \le \sqrt{2}(e^{\pi/2} - 1)$

27. $x = (\sqrt{2s + 1} - 1)\cos(\sqrt{2s + 1} - 1),$

$y = (\sqrt{2s + 1} - 1)\sin(\sqrt{2s + 1} - 1),$

$z = \frac{2}{3}\sqrt{2}\,(\sqrt{2s + 1} - 1)^{3/2} \; (s \ge 0)$

31. (b) $x = a(\cos\sqrt{2s/a} + \sqrt{2s/a}\,\sin\sqrt{2s/a})$ $(s \ge 0)$
      $y = a(\sin\sqrt{2s/a} - \sqrt{2s/a}\,\cos\sqrt{2s/a})$

33. (a) $\frac{9}{2}$ (b) $9 - 2\sqrt{6}$ **35.** (a) $\sqrt{3}(1 - e^{-2})$ (b) $4\sqrt{5}$

37. (a) $2t + \dfrac{1}{t}$ (b) $2t + \dfrac{1}{t}$ (c) $8 + \ln 3$

▶ **Exercise Set 15.4 (Page 739)**

1.

$T = -\dfrac{\sqrt{3}}{2}\mathbf{i} + \dfrac{1}{2}\mathbf{j}$

$\left(\dfrac{5}{2}, \dfrac{5\sqrt{3}}{2}\right)$

$N = -\dfrac{1}{2}\mathbf{i} - \dfrac{\sqrt{3}}{2}\mathbf{j}$

3.

$T = \dfrac{2}{\sqrt{5}}\mathbf{i} + \dfrac{1}{\sqrt{5}}\mathbf{j}$

$N = \dfrac{1}{\sqrt{5}}\mathbf{i} - \dfrac{2}{\sqrt{5}}\mathbf{j}$

5.

$N = -\dfrac{1}{\sqrt{2}}\mathbf{i} + \dfrac{1}{\sqrt{2}}\mathbf{j}$

$T = \dfrac{1}{\sqrt{2}}\mathbf{i} + \dfrac{1}{\sqrt{2}}\mathbf{j}$

$\left(\dfrac{1}{2}, \dfrac{1}{3}\right)$

7.

$T = -\dfrac{4}{\sqrt{97}}\mathbf{i} + \dfrac{9}{\sqrt{97}}\mathbf{j}$

$(2\sqrt{2}, 9\sqrt{2}/2)$

$N = -\dfrac{9}{\sqrt{97}}\mathbf{i} - \dfrac{4}{\sqrt{97}}\mathbf{i}$

9. $T = -\dfrac{4}{\sqrt{17}}\mathbf{i} + \dfrac{1}{\sqrt{17}}\mathbf{k}, \; N = -\mathbf{j}$ **11.** $T = \mathbf{i}, \; N = \mathbf{j}$

13. $T = -\dfrac{3}{\sqrt{10}}\mathbf{i} + \dfrac{1}{\sqrt{10}}\mathbf{k}, \; N = -\mathbf{j}$

15. $T = \dfrac{1}{\sqrt{5}}\mathbf{j} + \dfrac{2}{\sqrt{5}}\mathbf{k}, \; N = -\dfrac{2}{\sqrt{5}}\mathbf{j} + \dfrac{1}{\sqrt{5}}\mathbf{k}$

17. (a) $T = \frac{3}{5}\mathbf{i} + \frac{4}{5}\mathbf{j}$

(b)

$(7, 6)$

$T(5)$

$(4, 2)$

$(1, -2)$

21. See Exercise 1.   **23.** See Exercise 3.

25. See Exercise 9.   **27.** See Exercise 11.

29. $(\frac{4}{5}\cos t)\mathbf{i} - (\frac{4}{5}\sin t)\mathbf{j} - \frac{3}{5}\mathbf{k}$

▶ **Exercise Set 15.5 (Page 745)**

1. $\dfrac{96}{125}$ **3.** $\dfrac{6}{5\sqrt{10}}$ **5.** $\dfrac{3}{2\sqrt{2}}$ **7.** $\dfrac{4}{17}$ **9.** 1   **11.** $\dfrac{2}{5}$

13. $\dfrac{2}{5\sqrt{5}}$ **17.** 1   **19.** $\dfrac{1}{\sqrt{2}}$ **21.** $\dfrac{4}{5\sqrt{5}}$

23. $\cos x$; maximum for $x = 0$ **25.** See Exercise 1.

27. See Exercise 3.   **29.** See Exercise 5.

31. $\kappa(0) = a/b^2, \; \kappa(\pi/2) = b/a^2$ **33.** 1   **35.** $\dfrac{3}{2\sqrt{2a}}$

37. $\kappa(t) = \dfrac{1}{4a}\csc\dfrac{t}{2}$

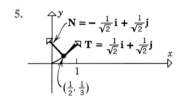

$\dfrac{1}{4a}$

2π

39.

$\rho = 1$

41.

$\rho = 2\sqrt{2}$

43.

$\rho = 2\sqrt{2}$ $(1, 0)$

45.

$\rho = 4$

47. $\rho(0) = \dfrac{1}{2}$

$\rho\left(\dfrac{\pi}{2}\right) = 4$

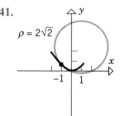

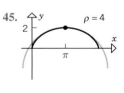

49. $2|p|$ **51.** $(3, 0), (-3, 0)$

53. maximum 2, minimum $1/\sqrt{2}$ **55.** $\kappa = \dfrac{1}{\sqrt{1 + a^2\, r}}$

57. $\sim 2.86°/\text{cm}$ **61.** $\dfrac{1}{2r}$ **69.** $\dfrac{2}{(t^2 + 2)^2}$

71. $-\dfrac{\sqrt{2}}{(e^t + e^{-t})^2}$

▶ **Exercise Set 15.6(Page 757)**

1.

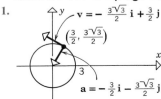

$v = -\dfrac{3\sqrt{3}}{2}\mathbf{i} + \dfrac{3}{2}\mathbf{j}$

$\left(\dfrac{3}{2}, \dfrac{3\sqrt{3}}{2}\right)$

$a = -\dfrac{3}{2}\mathbf{i} - \dfrac{3\sqrt{3}}{2}\mathbf{j}$

$\mathbf{v}(t) = -3\sin t\,\mathbf{i} + 3\cos t\,\mathbf{j}$
$\mathbf{a}(t) = -3\cos t\,\mathbf{i} - 3\sin t\,\mathbf{j}$
$\|\mathbf{v}(t)\| = 3$

3.

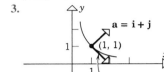

$\mathbf{a} = \mathbf{i} + \mathbf{j}$
$(1, 1)$
$\mathbf{v} = \mathbf{i} - \mathbf{j}$

$\mathbf{v}(t) = e^t\mathbf{i} - e^{-t}\mathbf{j}$
$\mathbf{a}(t) = e^t\mathbf{i} + e^{-t}\mathbf{j}$
$\|\mathbf{v}(t)\| = \sqrt{e^{2t} + e^{-2t}}$

5.

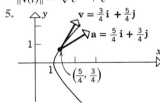

$\mathbf{v} = \dfrac{3}{4}\mathbf{i} + \dfrac{5}{4}\mathbf{j}$
$\mathbf{a} = \dfrac{5}{4}\mathbf{i} + \dfrac{3}{4}\mathbf{j}$
$\left(\dfrac{5}{4}, \dfrac{3}{4}\right)$

$\mathbf{v}(t) = (\sinh t + \cosh t)\,\mathbf{j}$
$\mathbf{a}(t) = \cosh t\,\mathbf{i} + \sinh t\,\mathbf{j}$
$\|\mathbf{v}(t)\| = \sqrt{\sinh^2 t + \cosh^2 t}$

7. $\mathbf{v} = \mathbf{i} + \mathbf{j} + \mathbf{k}, \|\mathbf{v}\| = \sqrt{3}, \mathbf{a} = \mathbf{j} + 2\mathbf{k}$

9. $\mathbf{v} = -\sqrt{2}\mathbf{i} + \sqrt{2}\mathbf{j} + \mathbf{k}, \|\mathbf{v}\| = \sqrt{5}, \mathbf{a} = -\sqrt{2}\mathbf{i} - \sqrt{2}\mathbf{j}$

11. $\mathbf{v} = e^{\pi/2}\mathbf{i} - e^{\pi/2}\mathbf{j} + \mathbf{k}, \|\mathbf{v}\| = \sqrt{2e^\pi + 1}, \mathbf{a} = -2e^{\pi/2}\mathbf{j}$

13. $3\sqrt{2}; \mathbf{r} = 24\mathbf{i} + 8\mathbf{j}$ **15.** maximum 6, minimum 3

17. $15°$ **19.** $\mathbf{r} = -\dfrac{19}{16}\mathbf{i} + \dfrac{3}{2}\mathbf{j} + \dfrac{3}{16}\mathbf{k}$

21. The particles move counterclockwise around the circle $x^2 + y^2 = 4$; $\|\mathbf{r}_1'\| = 6, \|\mathbf{r}_2'\| = 4t$.

23. $\mathbf{r}(t) = -16t^2\mathbf{j}, \mathbf{v}(t) = -32t\mathbf{j}$

25. $\mathbf{r}(t) = (t + \cos t - 1)\mathbf{i} + (\sin t - t + 1)\mathbf{j}$
$\mathbf{v}(t) = (1 - \sin t)\mathbf{i} + (\cos t - 1)\mathbf{j}$

27. $\mathbf{r}(t) = \dfrac{1}{2}t^2\mathbf{i} + \mathbf{j} + \dfrac{1}{6}t^3\mathbf{k}, \mathbf{v}(t) = t\mathbf{i} + \dfrac{1}{2}t^2\mathbf{k}$

29. $\mathbf{r}(t) = (t - \sin t - 1)\mathbf{i} + (1 - \cos t)\mathbf{j} + e^t\mathbf{k}$
$\mathbf{v}(t) = (1 - \cos t)\mathbf{i} + \sin t\,\mathbf{j} + e^t\mathbf{k}$

33. $8\mathbf{i} + \dfrac{26}{3}\mathbf{j}, \dfrac{1}{3}(13\sqrt{13} - 5\sqrt{5})$ **35.** $0, 12\pi$

---

37. $2\mathbf{i} - \dfrac{2}{3}\mathbf{j} + \sqrt{2}\ln 3\mathbf{k}, \dfrac{8}{3}$ **39.** $a_T = 0, a_N = 2$

41. $a_T = 0, a_N = \sqrt{2}$ **43.** $a_T = 2\sqrt{5}, a_N = 2\sqrt{5}$

45. $a_T = 22/\sqrt{14}, a_N = \sqrt{38/7}$ **47.** $a_T = 0, a_N = 3$

49. $a_T = -3, a_N = 2; \mathbf{T} = -\mathbf{j}, \mathbf{N} = \mathbf{i}$

51. $a_T = \dfrac{4}{3}, a_N = \dfrac{1}{3}\sqrt{29}; \mathbf{T} = \dfrac{2}{3}\mathbf{i} + \dfrac{2}{3}\mathbf{j} + \dfrac{1}{3}\mathbf{k},$

$\mathbf{N} = \dfrac{1}{3\sqrt{29}}(\mathbf{i} - 8\mathbf{j} + 14\mathbf{k})$

53. $\dfrac{3}{2}$ **55.** $-\pi/\sqrt{2}$ **57.** $\dfrac{1}{8}$ **59.** $\sqrt{29}/27$

61. 8.41×10^{10} km/sec^2 **63.** $\dfrac{18}{(1 + 4x^2)^{3/2}}$

65. (a) $x = 160t, y = 160\sqrt{3}t - 16t^2, t \ge 0$
  (b) 1200 ft   (c) $1600\sqrt{3}$ ft   (d) 320 ft/sec

67. $40\sqrt{3}$ ft   **69.** 800 ft/sec   **71.** $15°, 75°$

73. (b) $45°; v_0^2/g$ **75.** (a) 2.62 sec   (b) 181.5 ft

▶ **Exercise Set 15.7(Page 764)**

3. (b) 10.94 km/sec   **7.** (a) 22,278 mi   (b) 6869 mi/hr

9. 248 years

11. (a) 8864.5 km   (b) 0.208   (c) 138 min

13. (a) 17,293 mi/hr
  (b) $e = 0.07065$; apogee altitude $= 809$ mi

15. (b) $r = \dfrac{5.544 \times 10^7}{1 + 0.206\cos\theta}$

▶ **TECHNOLOGY EXERCISES, CHAPTER 15(Page 766)**

1. 1.338583   **3.** 1.494 m   **5.** 32.63

7. (b) $80°$ (c) $a_T = 0.43$ m/sec^2, $a_N = 2.46$ m/sec^2

9. (b) $17.8°$ (c) 2.34 hr

▶ **Exercise Set 16.1(Page 781)**

1. (a) 5   (b) 3   (c) 1   (d) -2 (e) $9a^3 + 1$
  (f) $a^3b^2 - a^2b^3 + 1$

3. (a) $x^2 - y^2 + 3$ (b) $3x^3y^4 + 3$ **5.** $x^3e^{x^3(3y+1)}$

7. (a) $t^2 + 3t^{10}$ (b) 0   (c) 3076

9. (a) 19   (b) -9 (c) 3   (d) $a^6 + 3$ (e) $-t^8 + 3$
  (f) $(a + b)(a - b)^2b^3 + 3$

11. $(y + 1)e^{x^2(y+1)z^2}$ **13.** (a) t^{14} (b) 0   (c) 16,384

15.

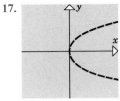

17.

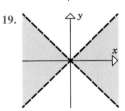

19.

21. all points above or on the line $y = -2$

23. all points above the line $y = 2x$

25. all points not on the plane $x + y + z = 0$

27. all points within the cylinder $x^2 + y^2 = 1$

29.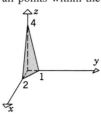

31.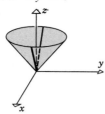

33.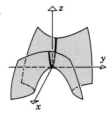

35.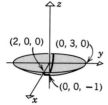

37.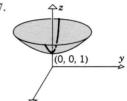

39.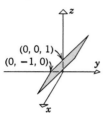

41.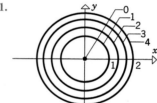

43.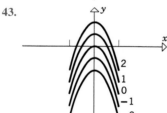

45.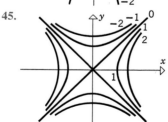

47.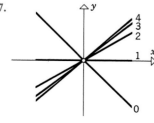

49.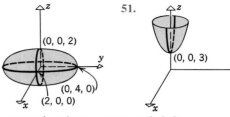

51.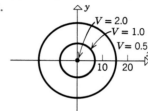

53. concentric spheres, center at $(2, 0, 0)$

55. concentric cylinders, common axis the y-axis

57. (a) $x^2 - 2x^3 + 3xy = 0$
(b) $x^2 - 2x^3 + 3xy = 0$
(c) $x^2 - 2x^3 + 3xy = -18$

59. (a) $x^2 + y^2 - z = 5$
(b) $x^2 + y^2 - z = -2$
(c) $x^2 + y^2 - z = 0$

61.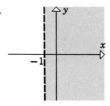

63. (a) A (b) B

65. (a) decrease (b) increase (c) increase (d) decrease

67. (a) open (b) neither (c) closed (d) closed

69. (a) bounded (b) unbounded (c) unbounded
(d) unbounded

71. $x = u$, $y = v$, $z = u^2$, where $-\infty < u < +\infty$, $-\infty < v < +\infty$

73. $x = u$, $y = v$, $z = \sqrt{4 - u^2}$, where $-2 \le u \le 2$, $1 \le v \le 3$

75. $x = r\cos\theta$, $y = r\sin\theta$, $z = 1/(1 + r^2)$, where $r \ge 0$,
$0 \le \theta < 2\pi$

77. $x = r\cos\theta$, $y = r\sin\theta$, $z = r^2\sin 2\theta$, where $r \ge 0$,
$0 \le \theta < 2\pi$

79. $x = r\cos\theta$, $y = r\sin\theta$, $z = \sqrt{9 - r^2}$, where $0 \le r \le \sqrt{5}$,
$0 \le \theta < 2\pi$

81. $x = r\cos\theta$, $y = r\sin\theta$, $z = 3 - 2r\sin\theta$, where $0 \le r \le 2$,
$0 \le \theta < 2\pi$

83. $x = \frac{1}{2}\rho\cos\theta$, $y = \frac{1}{2}\rho\sin\theta$, $z = \frac{1}{2}\sqrt{3}\rho$, where $\rho \ge 0$,
$0 \le \theta < 2\pi$

85. $x = 3\cos v$, $y = \frac{3}{2}\sin v$, $z = u$, where $0 \le u \le 5$, $0 \le v < 2\pi$

87. the plane $z = x - 2y$

89. the portion of the elliptic cylinder $x^2/9 + y^2/4 = 1$ for
$2 \le z \le 4$

91. the portion of the elliptic cone $z = \sqrt{x^2/9 + y^2/16}$ for
$0 \le z \le 1$

▶ **Exercise Set 16.2 (Page 789)**

1.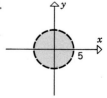

3.

5.

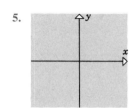

7.

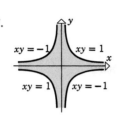

9. all of 3-space

11. all points not on the cylinder $x^2 + z^2 = 1$

13. 35   15. -8 17. 0   19. does not exist   21. 1

23. 0   25. does not exist   27. 0   29. 0   31. $\frac{8}{3}$

33. 0   35. does not exist   41. $-\pi/2$ 43. no

45. choose $\delta = \sqrt{\epsilon}$ 47. choose $\delta = \sqrt{\epsilon}$

▶ **Exercise Set 16.3 (Page 796)**

1. $\dfrac{\partial z}{\partial x} = 9x^2 y^2$, $\dfrac{\partial z}{\partial y} = 6x^3 y$

3. $\dfrac{\partial z}{\partial x} = 8xy^3 e^{x^2 y^3}$, $\dfrac{\partial z}{\partial y} = 12x^2 y^2 e^{x^2 y^3}$

5. $\dfrac{\partial z}{\partial x} = \dfrac{x^3}{y^{3/5} + x} + 3x^2 \ln(1 + xy^{-3/5})$, $\dfrac{\partial z}{\partial y} = -\dfrac{3x^4}{5(y^{8/5} + xy)}$

7. $f_x(x, y) = \frac{3}{2} x^2 y (5x^2 - 7)(3x^5 y - 7x^3 y)^{-1/2}$,
$f_y(x, y) = \frac{1}{2} x^3 (3x^2 - 7)(3x^5 y - 7x^3 y)^{-1/2}$

9. $f_x(x, y) = \dfrac{y^{-1/2}}{y^2 + x^2}$,

$f_y(x, y) = -\dfrac{xy^{-3/2}}{y^2 + x^2} - \dfrac{3}{2} y^{-5/2} \tan^{-1}\left(\dfrac{x}{y}\right)$

11. $f_x(x, y) = -\frac{4}{3} y^2 \sec^2 x (y^2 \tan x)^{-7/3}$,
$f_y(x, y) = -\frac{8}{3} y \tan x (y^2 \tan x)^{-7/3}$

13. (a) -6 (b) -21 15. (a) $1/\sqrt{17}$ (b) $8/\sqrt{17}$

17. $\dfrac{\partial z}{\partial x} = -\dfrac{x}{z}$, $\dfrac{\partial z}{\partial y} = -\dfrac{y}{z}$

19. $\dfrac{\partial z}{\partial x} = -\dfrac{2x + yz^2 \cos(xyz)}{xyz \cos(xyz) + \sin(xyz)}$

$\dfrac{\partial z}{\partial y} = -\dfrac{xz^2 \cos(xyz)}{xyz \cos(xyz) + \sin(xyz)}$

21. $f_{xx} = 8$, $f_{yy} = -96xy^2 + 140y^3$, $f_{xy} = f_{yx} = -32y^3$

23. $f_{xx} = e^x \cos y$, $f_{yy} = -e^x \cos y$, $f_{xy} = f_{yx} = -e^x \sin y$

25. $f_{xx} = -\dfrac{16}{(4x - 5y)^2}$, $f_{yy} = -\dfrac{25}{(4x - 5y)^2}$,

$f_{xy} = f_{yx} = \dfrac{20}{(4x - 5y)^2}$

27. (a) $30xy^4 - 4$ (b) $60x^2 y^3$ (c) $60x^3 y^2$

29. (a) -30 (b) -125 (c) 150

31. (a) $\dfrac{\partial^3 f}{\partial x^3}$ (b) $\dfrac{\partial^3 f}{\partial y^2 \partial x}$ (c) $\dfrac{\partial^4 f}{\partial x^2 \partial y^2}$ (d) $\dfrac{\partial^4 f}{\partial y^3 \partial x}$

33. $\dfrac{\partial w}{\partial x} = 2xy^4 z^3 + y$, $\dfrac{\partial w}{\partial y} = 4x^2 y^3 z^3 + x$, $\dfrac{\partial w}{\partial z} = 3x^2 y^4 z^2 + 2z$

35. $\dfrac{\partial w}{\partial x} = \dfrac{2x}{y^2 + z^2}$, $\dfrac{\partial w}{\partial y} = -\dfrac{2y(x^2 + z^2)}{(y^2 + z^2)^2}$, $\dfrac{\partial w}{\partial z} = -\dfrac{2z(y^2 - x^2)}{(y^2 + z^2)^2}$

37. $\dfrac{\partial w}{\partial x} = \dfrac{x}{\sqrt{x^2 + y^2 + z^2}}$, $\dfrac{\partial w}{\partial y} = \dfrac{y}{\sqrt{x^2 + y^2 + z^2}}$,

$\dfrac{\partial w}{\partial z} = \dfrac{z}{\sqrt{x^2 + y^2 + z^2}}$

39. $f_x = -\dfrac{y^2 z^3}{x^2 y^4 z^6 + 1}$, $f_y = -\dfrac{2xyz^3}{x^2 y^4 z^6 + 1}$,

$f_z = -\dfrac{3xy^2 z^2}{x^2 y^4 z^6 + 1}$

41. $f_x = 4xyz \cosh \sqrt{z} \sinh(x^2 yz) \cosh(x^2 yz)$,
$f_y = 2x^2 z \cosh \sqrt{z} \sinh(x^2 yz) \cosh(x^2 yz)$,
$f_z = 2x^2 y \cosh \sqrt{z} \sinh(x^2 yz) \cosh(x^2 yz)$

$+ \dfrac{\sinh \sqrt{z} \sinh^2(x^2 yz)}{2\sqrt{z}}$

43. (a) -80 (b) 40   (c) -60

45. (a) $2/\sqrt{7}$ (b) $4/\sqrt{7}$ (c) $1/\sqrt{7}$

47. $\dfrac{\partial w}{\partial x} = -\dfrac{x}{w}$, $\dfrac{\partial w}{\partial y} = -\dfrac{y}{w}$, $\dfrac{\partial w}{\partial z} = -\dfrac{z}{w}$

49. $\dfrac{\partial w}{\partial x} = -\dfrac{yzw \cos(xyz)}{2w + \sin(xyz)}$, $\dfrac{\partial w}{\partial y} = -\dfrac{xzw \cos(xyz)}{2w + \sin(xyz)}$,

$\dfrac{\partial w}{\partial z} = -\dfrac{xyw \cos(xyz)}{2w + \sin(xyz)}$

51. (a) $15x^2 y^4 z^7 + 2y$ (b) $35x^3 y^4 z^6 + 3y^2$ (c) $21x^2 y^5 z^6$
(d) $42x^3 y^5 z^5$ (e) $140x^3 y^3 z^6 + 6y$ (f) $30xy^4 z^7$
(g) $105x^2 y^4 z^6$ (h) $210xy^4 z^6$

55. 6   57. (a) 8   (b) -2

59. (a) $\dfrac{\partial V}{\partial r} = 2\pi rh$ (b) $\dfrac{\partial V}{\partial h} = \pi r^2$ (c) 48π (d) 64π

61. (a) $\frac{1}{5}$ (b) $-\frac{25}{8}$

63. $\dfrac{\partial z}{\partial x} = -1 - \dfrac{\cos(x - y)}{\cos(x + z)}$, $\dfrac{\partial z}{\partial y} = \dfrac{\cos(x - y)}{\cos(x + z)}$,

$\dfrac{\partial^2 z}{\partial x \partial y} = -\dfrac{\cos^2(x + z)\sin(x - y) + \cos^2(x - y)\sin(x + z)}{\cos^3(x + z)}$

65. (a) 4 degrees per centimeter   (b) 8 degrees per centimeter

69. (a) Both are positive; $\partial T/\partial x$.   (b) All are negative.

▶ **Exercise Set 16.4 (Page 807)**

1. 0.872   3. $\frac{7}{2}$ 5. $\epsilon_1 = 0$, $\epsilon_2 = \Delta x$

7. $\epsilon_1 = y\Delta x + \Delta x \Delta y$, $\epsilon_2 = 2x\Delta x$

13. $f_{xy} = f_{yx} = 12x^2 + 6x$

15. $f_{xy} = f_{yx} = -6xy^2 \sin(x^2 + y^3)$

17. (a) 4   (b) 5   19. $42t^{13}$ 21. $\dfrac{3 \sin(1/t)}{t^2}$

23. $-\frac{10}{3} t^{7/3} e^{1 - t^{10/3}}$

25. $\dfrac{\partial z}{\partial u} = 24u^2 v^2 - 16uv^3 - 2v + 3$,

$\dfrac{\partial z}{\partial v} = 16u^3 v - 24u^2 v^2 - 2u - 3$

27. $\dfrac{\partial z}{\partial u} = -\dfrac{2 \sin u}{3 \sin v}$, $\dfrac{\partial z}{\partial v} = -\dfrac{2 \cos u \cos v}{3 \sin^2 v}$

29. $\dfrac{\partial z}{\partial u} = e^u$, $\dfrac{\partial z}{\partial v} = 0$ 31. $\dfrac{\partial z}{\partial u} = \dfrac{2e^{2u}}{1 + e^{4u}}$, $\dfrac{\partial z}{\partial v} = 0$

33. $\dfrac{\partial T}{\partial r} = 3r^2 \sin \theta \cos^2 \theta - 4r^3 \sin^3 \theta \cos \theta$,

$\dfrac{\partial T}{\partial \theta} = -2r^3 \sin^2 \theta \cos \theta + r^4 \sin^4 \theta$

$+ r^3 \cos^3 \theta - 3r^4 \cos^2 \theta \sin^2 \theta$

35. $\dfrac{\partial t}{\partial x} = \dfrac{x^2 + y^2}{4x^2 y^3}$, $\dfrac{\partial t}{\partial y} = \dfrac{y^2 - 3x^2}{4xy^4}$ 37. $-\pi$

39. $\left.\dfrac{\partial z}{\partial r}\right|_{r=2,\,\theta=\pi/6} = \sqrt{3}e^{\sqrt{3}}, \left.\dfrac{\partial z}{\partial \theta}\right|_{r=2,\,\theta=\pi/6} = (2-4\sqrt{3})e^{\sqrt{3}}$

41. $\dfrac{x^2-y^2}{2xy-y^2}$ **43.** $\dfrac{2\sqrt{xy}-y}{x-6\sqrt{xy}}$ **45.** -39 mi/hr

47. $-\frac{7}{36}\sqrt{3}$ rad/sec   **49.** $\frac{4}{3}-16\ln 3$ °C/sec

55. (b) $z = C_1\ln r + C_2$

65. (a) $n=2$ (b) $n=1$ (c) $n=3$ (d) $n=-4$

67. (a) $\dfrac{\partial w}{\partial x} = \dfrac{\partial f}{\partial x} + \dfrac{\partial f}{\partial y}\dfrac{\partial y}{\partial x}$ (b) $\dfrac{\partial w}{\partial z} = \dfrac{\partial f}{\partial y}\dfrac{\partial y}{\partial z}$

▶ **Exercise Set 16.5 (Page 814)**

1. $48x - 14y - z = 64$; $x = 1 + 48t$, $y = -2 - 14t$, $z = 12 - t$

3. $x - y - z = 0$; $x = 1 + t$, $y = -t$, $z = 1 - t$

5. $3y - z = -1$; $x = \pi/6$, $y = 3t$, $z = 1 - t$

7. $3x - 4z = -25$; $x = -3 + \frac{3}{4}t$, $y = 0$, $z = 4 - t$

9. $7\,dx - 2\,dy$ **11.** $\dfrac{y}{1+x^2y^2}\,dx + \dfrac{x}{1+x^2y^2}\,dy$

13. 0.10 **15.** 0.03

17. (a) all points on the x-axis and y-axis   (b) $(0, -2, -4)$

19. $(\frac{1}{2}, -2, -\frac{3}{4})$ **23.** 0.088 cm   **25.** 8% **27.** $r\%$

29. 2% **31.** 0.004 rad   **33.** 0.3%

35. (a) $(r+s)\%$ (b) $(r+s)\%$ (c) $(2r+3s)\%$
(d) $(3r+\frac{1}{2}s)\%$

37. (a) $(-2, 1, 5), (0, 3, 9)$
(b) $4/(3\sqrt{14})$ at $(-2, 1, 5)$, $4/\sqrt{222}$ at $(0, 3, 9)$

▶ **Exercise Set 16.6 (Page 820)**

1. $4\mathbf{i} - 8\mathbf{j}$ **3.** $\dfrac{x}{x^2+y^2}\mathbf{i} + \dfrac{y}{x^2+y^2}\mathbf{j}$ **5.** $-36\mathbf{i} - 12\mathbf{j}$

7. $4\mathbf{i} + 4\mathbf{j}$ **9.** $6\sqrt{2}$ **11.** $-3/\sqrt{10}$ **13.** 0

15. $-8\sqrt{2}$ **17.** $\dfrac{1}{2\sqrt{2}}$ **19.** $\dfrac{1}{2} + \dfrac{\sqrt{3}}{8}$ **21.** $2\sqrt{2}$

23.   **25.**

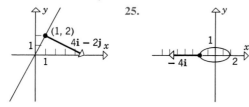

27. $\dfrac{3}{\sqrt{13}}\mathbf{i} - \dfrac{2}{\sqrt{13}}\mathbf{j}$, $4\sqrt{13}$ **29.** $\frac{4}{5}\mathbf{i} - \frac{3}{5}\mathbf{j}$, 1

31. $-\dfrac{1}{\sqrt{10}}\mathbf{i} - \dfrac{3}{\sqrt{10}}\mathbf{j}$, $-2\sqrt{10}$

33. $\dfrac{3}{\sqrt{10}}\mathbf{i} - \dfrac{1}{\sqrt{10}}\mathbf{j}$, $-\sqrt{5}$ **35.** $1/\sqrt{5}$ **37.** $-\frac{3}{2}e$

39. $-\dfrac{4}{\sqrt{17}}\mathbf{i} + \dfrac{1}{\sqrt{17}}\mathbf{j}$, $\dfrac{4}{\sqrt{17}}\mathbf{i} - \dfrac{1}{\sqrt{17}}\mathbf{j}$

41. (a) 5 (b) 10 (c) $-5\sqrt{5}$ **43.** $8/\sqrt{29}$

45. all points on the ellipse $9x^2 + y^2 = 9$

47. $36/\sqrt{17}$ **49.** (a) $2e^{-\pi/2}\mathbf{i}$

▶ **Exercise Set 16.7 (Page 827)**

1. $165t^{32}$ **3.** $-2t\cos(t^2)$ **5.** 3264 **7.** -320

9. $-\frac{314}{741}$ **11.** $72/\sqrt{14}$ **13.** $-\frac{8}{63}$

15. $\dfrac{1}{\sqrt{2}}\mathbf{i} - \dfrac{1}{\sqrt{2}}\mathbf{j}$, $3\sqrt{2}$ **17.** $-\dfrac{1}{\sqrt{2}}\mathbf{i} + \dfrac{1}{\sqrt{2}}\mathbf{j}$, $\dfrac{\sqrt{2}}{2}$

19. $\dfrac{1}{\sqrt{266}}\mathbf{i} - \dfrac{11}{\sqrt{266}}\mathbf{j} + \dfrac{12}{\sqrt{266}}\mathbf{k}$, $-\sqrt{266}$

21. $3/\sqrt{11}$ **23.** $-\frac{10}{3}\mathbf{i} + \frac{5}{3}\mathbf{j} + \frac{10}{3}\mathbf{k}$

25. $3x - 2y + 6z = -49$;
$x = -3 + 3t$, $y = 2 - 2t$, $z = -6 + 6t$

27. $x - y - 15z = -17$; $x = 3 + t$, $y = 5 - t$, $z = 1 - 15t$

31. $(1, \frac{2}{3}, \frac{2}{3})$, $(-1, -\frac{2}{3}, -\frac{2}{3})$ **33.** $8\,dx - 3\,dy + 4\,dz$

35. $\dfrac{yz}{1+(xyz)^2}\,dx + \dfrac{xz}{1+(xyz)^2}\,dy + \dfrac{xy}{1+(xyz)^2}\,dz$

37. 0.96 **39.** 2.35 cm^3 **41.** 39 ft^2 **43.** 14%

▶ **Exercise Set 16.8 (Page 831)**

1. $\dfrac{\partial f}{\partial v} = 8vw^3x^4y^5$, $\dfrac{\partial f}{\partial w} = 12v^2w^2x^4y^5$,

$\dfrac{\partial f}{\partial x} = 16v^2w^3x^3y^5$, $\dfrac{\partial f}{\partial y} = 20v^2w^3x^4y^4$

3. $\dfrac{\partial f}{\partial v_1} = \dfrac{2v_1}{v_3^2 + v_4^2}$, $\dfrac{\partial f}{\partial v_2} = -\dfrac{2v_2}{v_3^2 + v_4^2}$,

$\dfrac{\partial f}{\partial v_3} = -\dfrac{2v_3(v_1^2 - v_2^2)}{(v_3^2 + v_4^2)^2}$, $\dfrac{\partial f}{\partial v_4} = -\dfrac{2v_4(v_1^2 - v_2^2)}{(v_3^2 + v_4^2)^2}$

5. $f_v(1, -2, 4, 8) = 128$, $f_w(1, -2, 4, 8) = -512$,
$f_x(1, -2, 4, 8) = 32$, $f_y(1, -2, 4, 8) = \frac{64}{3}$

7. $210t^{29}$ **9.** $\dfrac{\partial z}{\partial r} = \dfrac{2r\cos^2\theta}{r^2\cos^2\theta + 1}$, $\dfrac{\partial z}{\partial\theta} = -\dfrac{2r^2\sin\theta\cos\theta}{r^2\cos^2\theta + 1}$

11. $\dfrac{\partial w}{\partial\rho} = 2\rho(4\sin^2\phi + \cos^2\phi)$,

$\dfrac{\partial w}{\partial\phi} = 6\rho^2\sin\phi\cos\phi$, $\dfrac{\partial w}{\partial\theta} = 0$

13. $\dfrac{-4\cos 2y\sin 2y + 2y + 1}{2\sqrt{\cos^2 2y + y^2 + y}}$

15. $\sqrt{3} + \pi/6$ cm/sec, increasing

17. $\dfrac{\partial z}{\partial x} = -\dfrac{2x + 4z}{4x + 2z - 3y}$, $\dfrac{\partial z}{\partial y} = \dfrac{3z}{4x + 2z - 3y}$

27. $\dfrac{\partial z}{\partial x} = \dfrac{ye^x}{15\cos 3z + 3}$, $\dfrac{\partial z}{\partial y} = \dfrac{e^x}{15\cos 3z + 3}$

33. (a) $\dfrac{dw}{dt} = \dfrac{\partial w}{\partial x_1}\dfrac{dx_1}{dt} + \dfrac{\partial w}{\partial x_2}\dfrac{dx_2}{dt} + \dfrac{\partial w}{\partial x_3}\dfrac{dx_3}{dt} + \dfrac{\partial w}{\partial x_4}\dfrac{dx_4}{dt}$

(b) $\dfrac{\partial w}{\partial v_1} = \dfrac{\partial w}{\partial x_1}\dfrac{\partial x_1}{\partial v_1} + \dfrac{\partial w}{\partial x_2}\dfrac{\partial x_2}{\partial v_1} + \dfrac{\partial w}{\partial x_3}\dfrac{\partial x_3}{\partial v_1} + \dfrac{\partial w}{\partial x_4}\dfrac{\partial x_4}{\partial v_1}$,

$\dfrac{\partial w}{\partial v_2} = \dfrac{\partial w}{\partial x_1}\dfrac{\partial x_1}{\partial v_2} + \dfrac{\partial w}{\partial x_2}\dfrac{\partial x_2}{\partial v_2} + \dfrac{\partial w}{\partial x_3}\dfrac{\partial x_3}{\partial v_2} + \dfrac{\partial w}{\partial x_4}\dfrac{\partial x_4}{\partial v_2}$,

$\dfrac{\partial w}{\partial v_3} = \dfrac{\partial w}{\partial x_1}\dfrac{\partial x_1}{\partial v_3} + \dfrac{\partial w}{\partial x_2}\dfrac{\partial x_2}{\partial v_3} + \dfrac{\partial w}{\partial x_3}\dfrac{\partial x_3}{\partial v_3} + \dfrac{\partial w}{\partial x_4}\dfrac{\partial x_4}{\partial v_3}$

▶ **Exercise Set 16.9 (Page 841)**

1. $(0, 0)$ relative min   **3.** $(1, -2)$ saddle

5. $(2, -1)$ relative min

7. $(2, 1)$, $(-2, 1)$ saddle; $(0, 0)$ relative min

9. $(-1, -1)$, $(1, 1)$ relative min   **11.** $(0, 0)$ saddle

13. none

15. $(0, 0)$, $(4, 0)$, $(0, -2)$ saddle; $(\frac{4}{3}, -\frac{2}{3})$ relative min

17. $(-1, 0)$ relative max   **19.** $(\pi/2, \pi/2)$ relative max

21. (b) $(0, 0)$ relative min   23. critical point $(1, 0)$

25. absolute maximum 0 at $(0, 0)$;
absolute minimum -12 at $(0, 4)$

27. absolute maximum 3 at $(0, 1)$, $(2, 1)$;
absolute minimum -1 at $(1, 0)$, $(1, 2)$

29. absolute maximum $\frac{33}{4}$ at $(-\frac{1}{2}, \pm\frac{1}{2}\sqrt{15})$;
absolute minimum $-\frac{1}{4}$ at $(\frac{1}{2}, 0)$

33. $9, 9, 9$ 35. $(\sqrt{5}, 0, 0)$, $(-\sqrt{5}, 0, 0)$

37. $\frac{1}{27}$ 39. length and width 2 ft, height 4 ft

41. $\frac{3}{2}\sqrt{6}$ 43. length and width $\sqrt[3]{2V}$, height $\frac{1}{2}\sqrt[3]{2V}$

47. $y = 0.5x + 0.8$

49. For $0 \le x \le 1$ let $f(x, y) = \begin{cases} y, & 0 < y < 1 \\ \frac{1}{2}, & y = 0 \text{ or } y = 1 \end{cases}$.
For $-\infty < x < +\infty$ and $y > 0$ let $f(x, y) = y$.

▶ Exercise Set 16.10 **(Page 849)**

1. max $\sqrt{2}$ at $(\sqrt{2}, 1)$ and $(-\sqrt{2}, -1)$,
min $-\sqrt{2}$ at $(-\sqrt{2}, 1)$ and $(\sqrt{2}, -1)$

3. max $\sqrt{2}$ at $(1/\sqrt{2}, 0)$, min $-\sqrt{2}$ at $(-1/\sqrt{2}, 0)$

5. max 6 at $(\frac{4}{3}, \frac{2}{3}, -\frac{4}{3})$, min -6 at $(-\frac{4}{3}, -\frac{2}{3}, \frac{4}{3})$

7. max $\frac{1}{3\sqrt{3}}$ at $\left(\frac{1}{\sqrt{3}}, \frac{1}{\sqrt{3}}, \frac{1}{\sqrt{3}}\right)$, $\left(\frac{1}{\sqrt{3}}, -\frac{1}{\sqrt{3}}, -\frac{1}{\sqrt{3}}\right)$,
$\left(-\frac{1}{\sqrt{3}}, \frac{1}{\sqrt{3}}, -\frac{1}{\sqrt{3}}\right)$, $\left(-\frac{1}{\sqrt{3}}, -\frac{1}{\sqrt{3}}, \frac{1}{\sqrt{3}}\right)$;
min $-\frac{1}{3\sqrt{3}}$ at $\left(\frac{1}{\sqrt{3}}, \frac{1}{\sqrt{3}}, -\frac{1}{\sqrt{3}}\right)$, $\left(\frac{1}{\sqrt{3}}, -\frac{1}{\sqrt{3}}, \frac{1}{\sqrt{3}}\right)$,
$\left(-\frac{1}{\sqrt{3}}, \frac{1}{\sqrt{3}}, \frac{1}{\sqrt{3}}\right)$, $\left(-\frac{1}{\sqrt{3}}, -\frac{1}{\sqrt{3}}, -\frac{1}{\sqrt{3}}\right)$

9. $(\frac{2}{5}, \frac{19}{5})$ 11. $(1, -1, 1)$ 13. $\frac{1}{2}$

15. $(3, 6)$ closest, $(-3, -6)$ farthest   17. $9, 9, 9$

19. $(\sqrt{5}, 0, 0)$, $(-\sqrt{5}, 0, 0)$

21. length and width 2 ft, height 4 ft

▶ TECHNOLOGY EXERCISES, CHAPTER 16 **(Page 849)**

3. $67.7°$ 5. $-0.805428\mathbf{i} - 0.592292\mathbf{j} - 0.021803\mathbf{k}$; $7.37°/m$

7. (a) $\frac{\sin(ex) - \sin x}{x}$ (b) 0.909026

9. length and width 1.37 ft, height 0.53 ft; $20.12

▶ Exercise Set 17.1 **(Page 857)**

1. 7   3. 2   5. 2   7. $\frac{2}{15}(31 - 9\sqrt{3})$ 9. 3

11. $1 - \ln 2$ 13. $\frac{1}{2}(1 - \ln 2)$ 15. 0   17. $\frac{1}{3}$ 19. 1

21.
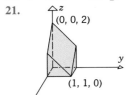
(0, 0, 2) ... (1, 1, 0)

23.
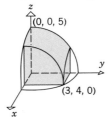
(0, 0, 5) ... (3, 4, 0)

25.
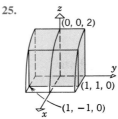
(0, 0, 2) ... (1, 1, 0), (1, -1, 0)

27. 172   29. 8

▶ Exercise Set 17.2 **(Page 864)**

1. $\frac{1}{40}$ 3. 9   5. $\pi/2$ 7. 1   9. $\frac{2}{3}a^3$ 11. $\frac{1}{12}$

13. 32   15. $-2/\pi$ 17. 576   19. $\frac{1}{2}(\sqrt{17} - 1)$

21. 0   23. $\pi/4 - \frac{1}{2}\ln 2$ 25. $\frac{50}{3}$ 27. $-\frac{1}{2}$ 29. $\frac{25}{2}$

31. $\sqrt{2} - 1$ 33. 32   35. $\frac{125}{12}$ 37. $\frac{56}{15}$ 39. 170

41. $\frac{27}{2}\pi$ 43. $\frac{20}{3}$ 45. $\pi/2$ 47. $\frac{2000}{3}$

49. $\displaystyle\int_0^{\sqrt{2}} \int_{y^2}^2 f(x, y)\, dx\, dy$ 51. $\displaystyle\int_1^{e^2} \int_{\ln x}^2 f(x, y)\, dy\, dx$

53. $\displaystyle\int_{-1}^1 \int_{-2\sqrt{1-y^2}}^{2\sqrt{1-y^2}} f(x, y)\, dx\, dy$

55. $\displaystyle\int_0^{\pi/2} \int_0^{\sin x} f(x, y)\, dy\, dx$ 57. $\frac{1}{8}(1 - e^{-16})$

59. $\frac{1}{3}(e^8 - 1)$ 61. $\frac{1}{8}\pi$ 63. $\frac{1}{3}(1 - \cos 8)$

65. (a) 0   (b) $-\frac{603}{40}$ 67. π 69. 0

▶ Exercise Set 17.3 **(Page 871)**

1. $\frac{1}{6}$ 3. 0   5. 0   7. $3\pi/2$ 9. $\pi/16$

11. $4\pi/3 + 2\sqrt{3}$ 13. $\frac{4}{3}(27 - 16\sqrt{2})\pi$ 15. $\frac{32}{9}$

17. $5\pi/32$ 19. $(1 - e^{-1})\pi$ 21. $\frac{\pi}{8}\ln 5$ 23. $\pi/8$

25. $\frac{16}{9}$ 27. $\frac{\pi}{2}\left(1 - \frac{1}{\sqrt{1+a^2}}\right)$ 29. $\frac{1}{4}\pi(\sqrt{5} - 1)$

31. $\frac{1}{9}(3\pi - 4)a^2 c$ 33. $\frac{4\pi}{3} + 2\sqrt{3} - 2$

35. (b) $\pi/4$ (c) $\sqrt{\pi}/2$

▶ Exercise Set 17.4 **(Page 877)**

1. 6π 3. $\sqrt{5}/6$ 5. $\frac{\pi}{6}(5\sqrt{5} - 1)$

7. $\frac{\pi}{18}(10\sqrt{10} - 1)$ 9. 8π 11. $2(\pi - 2)a^2$ 13. 128

17. $\frac{1}{6}(17^{3/2} - 5^{3/2})\pi$ 19. $4\pi a^2$ 21. $5\sqrt{5} - 1$

▶ Exercise Set 17.5 **(Page 883)**

1. 8   3. 7   5. $\frac{81}{5}$ 7. $\frac{128}{15}$ 9. $\frac{\pi}{2}(\pi - 3)$ 11. $\frac{1}{6}$

13. 4   15. $\frac{256}{15}$ 17. 9π 19. 2π

21. $\frac{\pi}{6}(8\sqrt{2} - 7)a^3$

23. (a)
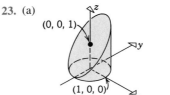
(0, 0, 1) ... (1, 0, 0)

(b)
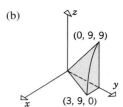
(0, 9, 9) ... (3, 9, 0)

25. (a) $\displaystyle\int_0^a \int_0^{b(1-x/a)} \int_0^{c(1-x/a-y/b)} dz\,dy\,dx,$

$\displaystyle\int_0^b \int_0^{a(1-y/b)} \int_0^{c(1-x/a-y/b)} dz\,dx\,dy,$

$\displaystyle\int_0^c \int_0^{a(1-z/c)} \int_0^{b(1-x/a-z/c)} dy\,dx\,dz,$

$\displaystyle\int_0^a \int_0^{c(1-x/a)} \int_0^{b(1-x/a-z/c)} dy\,dz\,dx,$

$\displaystyle\int_0^c \int_0^{b(1-z/c)} \int_0^{a(1-y/b-z/c)} dx\,dy\,dz,$

$\displaystyle\int_0^b \int_0^{c(1-y/b)} \int_0^{a(1-y/b-z/c)} dx\,dz\,dy$

27. $\frac{1}{6}a^3$ **29.** (a) 0   (b) $\frac{1}{2}(e^2-1)$ **31.** (a) 10   (b) 0

33. 0

▶ **Exercise Set 17.6 (Page 893)**

1. 10 feet to the right of m_1 **3.** $\frac{13}{20}, \left(\frac{190}{273}, \frac{6}{13}\right)$

5. $\frac{a^4}{8}, \left(\frac{8a}{15}, \frac{8a}{15}\right)$ **7.** $\left(\frac{2}{3}, \frac{1}{3}\right)$ **9.** $\left(-\frac{1}{2}, \frac{2}{5}\right)$

11. $\left(0, \dfrac{4(b^3-a^3)}{3\pi(b^2-a^2)}\right)$ **13.** $(0, \frac{8}{3})$ **15.** $\frac{a^4}{2}, \left(\frac{a}{3}, \frac{a}{2}, \frac{a}{2}\right)$

17. $\frac{1}{6}, (0, \frac{16}{35}, \frac{1}{2})$ **19.** $(\frac{1}{4}, \frac{1}{4}, \frac{1}{4})$ **21.** $(\frac{1}{2}, 0, \frac{3}{5})$

23. $\left(\frac{3a}{8}, \frac{3a}{8}, \frac{3a}{8}\right)$ **25.** $\frac{2}{3}\pi ka^3$ **29.** $\left(\dfrac{128}{105\pi}, \dfrac{128}{105\pi}\right)$

33. $2\pi^2 kab$ **35.** $\left(\dfrac{a}{3}, \dfrac{b}{3}\right)$ **37.** $\frac{1}{12}M(a^2+b^2)$

39. $\frac{1}{4}MR^2$ **41.** $\frac{3}{2}MR^2$ **43.** $\frac{2}{3}Ma^2$

▶ **Exercise Set 17.7 (Page 903)**

1. $\pi/4$ **3.** $\pi/16$ **5.** $81\pi/2$ **7.** $\frac{8}{3}(10\sqrt{5}-19)\pi$

9. $\frac{32}{9}(3\pi-4)$ **11.** $\frac{2}{3}(\sqrt{3}-1)\pi$ **13.** $9\sqrt{2}\pi$

15. $\frac{4}{3}\pi a^3$ **17.** $\frac{11}{6}\pi$ **19.** $\frac{1}{4}(2-\sqrt{2})\pi$ **21.** πka^4

23. $\left(0, 0, \dfrac{7}{16\sqrt{2}-14}\right)$ **25.** $\left(\dfrac{3a}{8}, \dfrac{3a}{8}, \dfrac{3a}{8}\right)$

27. $\left(0, \dfrac{195}{152}, 0\right)$ **29.** $\frac{1}{3}(1-e^{-1})\pi$ **31.** 81π

33. $\left(0, 0, \dfrac{2a}{5}\right)$ **35.** $(0, 0, \frac{11}{30})$ **37.** $\frac{63}{8}\pi$

41. $M(\frac{1}{4}R^2 + \frac{1}{3}h^2)$ **43.** $\frac{2}{5}MR^2$

45. (a) $2\pi k\delta(\sqrt{R^2+a^2} - \sqrt{R^2+(a+h)^2} + h)$

(b) Other components are zero because of symmetry.

47. $2\pi k\delta h\left(1 - \dfrac{h}{\sqrt{R^2+h^2}}\right)$

▶ **Exercise Set 17.8 (Page 915)**

1. -17 **3.** $\cos(u-v)$ **5.** $\frac{1}{9}$ **7.** $\dfrac{1}{4\sqrt{v^2-u^2}}$

9. 5   **11.** $1/v$ **13.** $\frac{3}{2}\ln 3$ **15.** $1 - \frac{1}{2}\sin 2$

17. 96π **23.** 21   **25.** $-\frac{2}{3}\ln 2$ **27.** $\frac{1}{4}\ln\frac{5}{2}$

29. $\frac{1}{2}[\ln(\sqrt{2}+1) - \pi/4]$ **31.** $\frac{35}{256}$ **33.** $2\ln 3$

35. $\frac{4}{15}\pi a^3 bc$ **37.** $\frac{105}{32}$

▶ **TECHNOLOGY EXERCISES, CHAPTER 17 (Page 917)**

1. 0.676089   **3.** 3.960251

5. (a) 2.859712   (b) 2.153777   (c) 38.699285

7. 1.980291

▶ **Exercise Set 18.1 (Page 926)**

1.

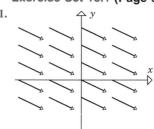

3.

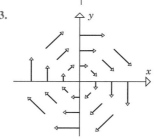

5. $\operatorname{div}\mathbf{F} = 2x + y$, $\operatorname{curl}\mathbf{F} = z\mathbf{i}$

7. $\operatorname{div}\mathbf{F} = 0$, $\operatorname{curl}\mathbf{F} = (40x^2z^4 - 12xy^3)\mathbf{i}$

$\qquad + (14y^3z + 3y^4)\mathbf{j} - (16xz^5 + 21y^2z^2)\mathbf{k}$

9. $\operatorname{div}\mathbf{F} = 2/\sqrt{x^2+y^2+z^2}$, $\operatorname{curl}\mathbf{F} = \mathbf{0}$ **21.** $f'(x)$

31. (b) $x^2 + y^2 = K$

33. $y = \ln x + K$

▶ **Exercise Set 18.2 (Page 936)**

1. $(17\sqrt{17}-1)/4$ **3.** 2   **5.** $\frac{13}{20}$

7. (a) $\frac{4}{3}$ (b) 0   (c) $\frac{4}{3}$ **9.** $-\frac{8}{3}$ **11.** $1 - \pi$ **13.** 3

15. $-1 - \pi/4$ **17.** 0   **19.** $1 - e^{-1}$ **21.** $1 - e^3$

23. $\frac{23}{6}$ **25.** $6k\sqrt{3}$ **27.** $5k\tan^{-1}3$ **29.** $\frac{3}{5}$

31. (a) $\frac{5}{4} - \pi/8 + \frac{1}{2}\tan^{-1}2$ (b) $\frac{1}{3}\tan^{-1}2 - \frac{2}{3}\tan^{-1}\frac{1}{2}$ (c) $\frac{3}{4}$

33. $-\frac{37}{2}$ **35.** -12

▶ **Exercise Set 18.3 (Page 945)**

1. $\frac{1}{2}x^2 + \frac{1}{2}y^2 + K$ **3.** not conservative

5. $x\cos y + y\sin x + K$ **7.** 13   **9.** -6 **11.** $9e^2$

13. 32   **15.** $-\frac{1}{2}$ **17.** $\pi/4$ **19.** $\ln 2 - 1$

21. (a) $-1/\sqrt{x^2+y^2}$ (b) $\frac{4}{5}$ (c) 0   **23.** Ce^x

29. $xz^2 - e^{-y}$ **31.** $xyz + z^3$ **33.** not conservative

35. (a) $1/\sqrt{6} - 1/\sqrt{14}$ (b) $1/\sqrt{6} - 1/\sqrt{14}$ (c) 0

▶ **Exercise Set 18.4 (Page 951)**

1. 0   **3.** 0   **5.** 0   **7.** 8π **9.** -4 **11.** -1

13. 0   **15.** πab **17.** $\frac{1}{2}ab$ **19.** $\frac{1}{2}abt_0$ **21.** $\frac{250}{3}$

23. (b) $\left(0, \dfrac{4a}{3\pi}\right)$

25. (c) $A = \frac{1}{2}[(x_1 y_2 - x_2 y_1) + (x_2 y_3 - x_3 y_2)$
$+ \cdots + (x_n y_1 - x_1 y_n)]$

(d) 8

27. 69

▶ **Exercise Set 18.5 (Page 957)**

1. $15\pi/\sqrt{2}$ **3.** $\pi/4$ **5.** $3/\sqrt{2}$ **7.** 9   **9.** $4\pi/3$

11. (a) $\dfrac{\sqrt{29}}{16} \displaystyle\int_0^6 \int_0^{(12-2x)/3} xy(12 - 2x - 3y)\, dy\, dx$

(b) $\dfrac{\sqrt{29}}{4} \displaystyle\int_0^3 \int_0^{(12-4z)/3} yz(12 - 3y - 4z)\, dy\, dz$

(c) $\dfrac{2\sqrt{29}}{9} \displaystyle\int_0^3 \int_0^{6-2z} xz(6 - x - 2z)\, dx\, dz$

13. $\displaystyle\int_0^4 \int_1^2 y^3 z\sqrt{1 + 4y^2}\, dy\, dz,\ \frac{1}{2}\int_0^4 \int_1^4 xz\sqrt{1 + 4x}\, dx\, dz$

15. $\dfrac{\pi \delta_0}{6}(5\sqrt{5} - 1)$ **17.** $\frac{4}{3}\pi \delta_0$ **19.** $\frac{1}{4}[37\sqrt{37} - 1]$

23. $(0, 0, \frac{149}{65})$ **25.** $93/\sqrt{10}$ **27.** $\pi/4$

▶ **Exercise Set 18.6 (Page 965)**

1. $\dfrac{2}{\sqrt{29}}\mathbf{i} + \dfrac{3}{\sqrt{29}}\mathbf{j} + \dfrac{4}{\sqrt{29}}\mathbf{k}$

3. (a) $-\dfrac{2}{\sqrt{21}}\mathbf{i} - \dfrac{4}{\sqrt{21}}\mathbf{j} + \dfrac{1}{\sqrt{21}}\mathbf{k}$

(b) $\dfrac{3}{5\sqrt{2}}\mathbf{i} - \dfrac{4}{5\sqrt{2}}\mathbf{j} + \dfrac{1}{\sqrt{2}}\mathbf{k}$ (c) $\frac{3}{5}\mathbf{i} - \frac{4}{5}\mathbf{k}$

5. 2π **7.** $\pi/2$ **9.** 54π **11.** $\frac{14}{3}\pi$ **13.** 0

15. $4\pi a^3$

19. (a) $\displaystyle\iint_R \mathbf{F} \cdot \left(\mathbf{i} - \frac{\partial x}{\partial y}\mathbf{j} - \frac{\partial x}{\partial z}\mathbf{k}\right) dA$

Front, R = projection on yz-plane

$\displaystyle\iint_R \mathbf{F} \cdot \left(-\mathbf{i} + \frac{\partial x}{\partial y}\mathbf{j} + \frac{\partial x}{\partial z}\mathbf{k}\right) dA$

Back, R = projection on yz-plane

(b) $\displaystyle\iint_R \mathbf{F} \cdot \left(-\frac{\partial y}{\partial x}\mathbf{i} + \mathbf{j} - \frac{\partial y}{\partial z}\mathbf{k}\right) dA$

Right, R = projection on xz-plane

$\displaystyle\iint_R \mathbf{F} \cdot \left(\frac{\partial y}{\partial x}\mathbf{i} - \mathbf{j} + \frac{\partial y}{\partial z}\mathbf{k}\right) dA$

Left, R = projection on xz-plane

21. 2π **23.** 18π **25.** $\frac{4}{9}$

▶ **Exercise Set 18.7 (Page 974)**

1. 8   **3.** 216π **5.** $3\pi a^2$ **7.** 180π **9.** $\frac{1}{24}$

11. $\pi/2$ **13.** $\frac{4608}{35}$ **17.** no sources or sinks

19. sources at all points except the origin; no sinks

▶ **Exercise Set 18.8 (Page 980)**

1. 2π **3.** 16π **5.** 0   **7.** πa^2 **9.** $\frac{3}{2}$ **11.** 0

▶ **TECHNOLOGY EXERCISES, CHAPTER 18 (Page 981)**

1. $-4.553124, 6.822229$ **3.** 12.4 m/sec

5. (a) 1   (b) $x^3 y^2 z^4$ **7.** 57.895751   **9.** $-5\pi/4$

▶ **Exercise Set 19.1 (Page 988)**

3. $y = c_1 e^x + c_2 e^{-4x}$ **5.** $y = c_1 e^x + c_2 x e^x$

7. $y = c_1 \cos \sqrt{5}x + c_2 \sin \sqrt{5}x$ **9.** $y = c_1 + c_2 e^x$

11. $y = c_1 e^{-2t} + c_2 t e^{-2t}$ **13.** $y = e^{2x}(c_1 \cos 3x + c_2 \sin 3x)$

15. $y = c_1 e^{-x/4} + c_2 e^{x/2}$ **17.** $y = 2e^x - e^{-3x}$

19. $y = (2 - 5x)e^{3x}$ **21.** $y = -e^{-2x}(3\cos x + 6\sin x)$

23. (a) $y'' - 3y' - 10y = 0$ (b) $y'' - 8y' + 16y = 0$
(c) $y'' + 2y' + 17y = 0$

25. (a) $k < 0$ or $k > 4$ (b) 0, 4   (c) $0 < k < 4$

27. (a) $y = (1/x)[c_1 \cos(\ln x) + c_2 \sin(\ln x)]$
(b) $y = c_1 x^{1+\sqrt{3}} + c_2 x^{1-\sqrt{3}}$

▶ **Exercise Set 19.2 (Page 994)**

1. $y = c_1 e^{-x} + c_2 e^{-5x} + \frac{1}{16} e^{3x}$

3. $y = c_1 e^{4x} + c_2 e^{5x} - 3x e^{5x}$

5. $y = (c_1 + c_2 x)e^{-x} + \frac{1}{2}x^2 e^{-x}$

7. $y = c_1 e^{3x} + c_2 e^{-4x} - \frac{13}{216} - \frac{1}{18}x - \frac{1}{3}x^2$

9. $y = c_1 + c_2 e^{6x} + \frac{5}{36}x - \frac{1}{12}x^2$

11. $y = c_1 + c_2 x - \frac{1}{2}x^2 + \frac{1}{20}x^5$

13. $y = c_1 e^{-x} + c_2 e^{2x} - 3\cos x - \sin x$

15. $y = c_1 e^{-2x} + c_2 e^{2x} - \frac{3}{8}\cos 2x - \frac{1}{4}\sin 2x$

17. $y = c_1 \cos x + c_2 \sin x - \frac{1}{2}x \cos x$

19. $y = c_1 e^x + c_2 e^{2x} + \frac{3}{4} + \frac{1}{2}x$

21. $y = e^{-2x}(c_1 \cos \sqrt{5}x + c_2 \sin \sqrt{5}x) - \frac{94}{729} + \frac{19}{81}x + \frac{1}{9}x^2$

23. $y = c_1 \cos 2x + c_2 \sin 2x - \frac{1}{8}x \cos 2x$

25. (b) $-\frac{3}{16} - \frac{1}{4}x + \frac{1}{5}x e^x$ **27.** $y = c_1 e^{-x} + c_2 e^x - 1 + \frac{1}{2}x e^x$

29. $y = c_1 \cos 2x + c_2 \sin 2x + \frac{1}{4} + \frac{1}{4}x + \frac{1}{3}\sin x$

31. $y = (c_1 + c_2 x)e^x + \frac{1}{4}x^2 e^x + \frac{1}{8}e^{-x}$

33. $y = c_1 \cos x + c_2 \sin x + 6 - 2\cos 2x$

35. (a) $y = c_1 \cos \mu x + c_2 \sin \mu x + \dfrac{a}{\mu^2 - b^2}\sin bx$

(b) $y = c_1 \cos \mu x + c_2 \sin \mu x + \displaystyle\sum_{k=1}^n \dfrac{a_k}{\mu^2 - k^2\pi^2}\sin k\pi x$

37. $x_0 = 1;\ y = 2x^2 - 4e^{x-1} + c,\ c$ arbitrary

▶ **Exercise Set 19.3 (Page 996)**

1. $y = c_1 \cos x + c_2 \sin x + x^2 - 2$

3. $y = c_1 e^x + c_2 e^{-2x} + \frac{2}{3}x e^x$

5. $y = c_1 \cos 2x + c_2 \sin 2x - \frac{1}{4}x \cos 2x$

7. $y = c_1 \cos x + c_2 \sin x - (\cos x)\ln|\sec x + \tan x|$

9. $y = c_1 e^x + c_2 x e^x + x e^x \ln|x|$

11. $y = c_1 \cos x + c_2 \sin x + \frac{3}{2} + \frac{1}{2}\cos 2x$

13. $y = c_1 \cos x + c_2 \sin x - x \cos x + (\sin x)\ln|\sin x|$

15. $y = c_1 \cos x + c_2 \sin x + x \cos x - (\sin x)\ln|\cos x|$

17. $y = c_1 e^{-x} + c_2 x e^{-x} - e^{-x}\ln|x|$

19. $y = c_1 e^{-2x} + c_2 x e^{-2x} + (x - 2)e^{-x}$

21. $y = c_1 \cos x + c_2 \sin x - 1 + (\sin x)\ln|\sec x + \tan x|$

23. $y = c_1 e^x + c_2 x e^x - e^x \ln|x|$

25. $y = c_1 e^x + c_2 e^{-x} + \frac{1}{5}e^x(2\sin x - \cos x)$

27. $y = c_1 e^{-x} + c_2 x e^{-x} + \frac{1}{2}x^2 e^{-x}\ln|x| - \frac{3}{4}x^2 e^{-x}$

▶ **Exercise Set 19.4 (Page 1004)**

1. (a) $y'' + 0.25y = 0$, $y(0) = 0.3$, $y'(0) = 0$
   (b) $y = 0.3 \cos 0.5t$

3. (a) $y'' + 196y = 0$, $y(0) = -0.12$, $y'(0) = 0$
   (b) $y = -0.12 \cos 14t$

5. (a) $y = 0.2 \cos 8t$ (b) 0.2 m   (c) $\pi/4$ sec   (d) $4/\pi$ Hz

7. (a) $y = -\frac{1}{4} \cos 8\sqrt{6}t$ (b) $\frac{1}{4}$ ft   (c) $\dfrac{\pi}{4\sqrt{6}}$ sec

   (d) $\dfrac{4\sqrt{6}}{\pi}$ Hz

9. (a) $y'' + 4y' + 8y = 0$, $y(0) = -3$, $y'(0) = 0$
   (b) $y = -3e^{-2t}(\cos 2t + \sin 2t)$
   (d) $y = -3\sqrt{2}e^{-2t}\cos(2t - \pi/4)$
   (e) π sec   (f) $1/\pi$ Hz

11. (a) $y = \frac{5}{2}\sqrt{6}e^{-t/5} \cos\left(\dfrac{\sqrt{2}}{5}t - \tan^{-1}\dfrac{1}{\sqrt{2}}\right)$

    (b) $\dfrac{10\pi}{\sqrt{2}}$ sec   (c) $\dfrac{\sqrt{2}}{10\pi}$ Hz

13. (a) $y = -\frac{1}{4} \sin 8t$ (b) $\frac{1}{4}$ ft   (c) $\pi/4$ sec   (d) $4/\pi$ Hz
15. $y = \frac{1}{3}e^{-t}(\cos 7t - 2 \sin 7t)$
17. (a) $\frac{1}{32}\pi^2$ lb/ft   (b) $\frac{9}{4}$ lb

19. $T = 2\pi \sqrt{\dfrac{\delta h}{\rho g}}$

▶ **TECHNOLOGY EXERCISES, CHAPTER 19 (Page 1006)**

1. (a) $y = e^{-1.2t} + 3.2te^{-1.2t}$ (b) 1.427364 cm
3. (a) $y = e^{-t/2}\cos(\sqrt{19}t/2) - \frac{6}{19}\sqrt{19}e^{-t/2}\sin(\sqrt{19}t/2)$
   (b) 1.054466 cm   (c) -3.210357 cm/sec
   (d) 3.210357 cm/sec^2
5. (a) $3/\pi$, $5/(2\pi)$ Hz   (b) $y = \frac{1}{11}\sin 5t - \frac{5}{66}\sin 6t$
   (c) $y = \frac{4}{23}\sin 5.5t - \frac{11}{69}\sin 6t$, $y = \frac{25}{59}\sin 5.8t - \frac{145}{354}\sin 6t$
7. (a) $y = (4 + 2v_0)e^{-3t/2} - (3 + 2v_0)e^{-2t}$

▶ **Exercise Set 11.1 (Page 515)**

1. $\frac{1}{3}, \frac{2}{4}, \frac{3}{5}, \frac{4}{6}, \frac{5}{7}$; converges to 1
3. 2, 2, 2, 2, 2; converges to 2
5. $\dfrac{\ln 1}{1}, \dfrac{\ln 2}{2}, \dfrac{\ln 3}{3}, \dfrac{\ln 4}{4}, \dfrac{\ln 5}{5}$; converges to 0
7. 0, 2, 0, 2, 0; diverges
9. $-1, \frac{16}{9}, -\frac{54}{28}, \frac{128}{65}, -\frac{250}{126}$; diverges
11. $\frac{6}{2}, \frac{12}{8}, \frac{20}{18}, \frac{30}{32}, \frac{42}{50}$; converges to $\frac{1}{2}$
13. $\cos 3, \cos \frac{3}{2}, \cos 1, \cos \frac{3}{4}, \cos \frac{3}{5}$; converges to 1
15. $e^{-1}, 4e^{-2}, 9e^{-3}, 16e^{-4}, 25e^{-5}$; converges to 0
17. 2, $(\frac{5}{3})^2$, $(\frac{6}{4})^3$, $(\frac{7}{5})^4$, $(\frac{8}{6})^5$; converges to e^2
19. $\left\{\dfrac{2n-1}{2n}\right\}_{n=1}^{+\infty}$; converges to 1
21. $\left\{\dfrac{1}{3^n}\right\}_{n=1}^{+\infty}$; converges to 0
23. $\left\{\dfrac{1}{n} - \dfrac{1}{n+1}\right\}_{n=1}^{+\infty}$; converges to 0
25. $\{\sqrt{n+1} - \sqrt{n+2}\}_{n=1}^{+\infty}$; converges to 0
27. (a) $\sqrt{6}, \sqrt{6+\sqrt{6}}, \sqrt{6+\sqrt{6+\sqrt{6}}}$ (b) 3
29. (a) 1, 1, 2, 3, 5, 8, 13, 21   (b) $(1 + \sqrt{5})/2$

31. (a) $1, \frac{3}{4}, \frac{2}{3}, \frac{5}{8}$ (b) $\frac{1}{2}$ 33. (a) 3   (b) 11   (c) 1001
39. 3

▶ **Exercise Set 11.2 (Page 522)**

1. decreasing   3. increasing   5. decreasing
7. increasing   9. decreasing   11. increasing
13. increasing   15. decreasing   17. decreasing
19. eventually increasing   21. eventually decreasing
23. eventually increasing
27. (a) $\sqrt{2}, \sqrt{2+\sqrt{2}}, \sqrt{2+\sqrt{2+\sqrt{2}}}$ (e) 2
33. (b) converges (decreasing and bounded below by 0)

▶ **Exercise Set 11.3 (Page 529)**

1. (a) converges to $\frac{5}{2}$ (b) converges to $\frac{1}{2}$ (c) diverges
3. $\frac{4}{7}$ 5. 6   7. diverges   9. $\frac{1}{3}$ 11. $\frac{1}{6}$ 13. $\frac{448}{3}$
15. $-\frac{1}{3}$ 17. $\frac{4}{9}$ 19. $\frac{532}{99}$ 21. $\frac{869}{1111}$ 23. diverges
33. $\dfrac{1}{x^2 - 2x}$, $|x| > 2$ 35. $\dfrac{2 \sin x}{2 + \sin x}$, $-\infty < x < +\infty$
37. 1
41. The series converges to $1/(1 - x)$ only for $-1 < x < 1$.

▶ **Exercise Set 11.4 (Page 536)**

1. $\frac{4}{3}$ 3. $-\frac{1}{36}$
5. (a) converges   (b) diverges   (c) diverges   (d) diverges
   (e) converges   (f) diverges   (g) converges
   (h) converges
9. diverges   11. diverges   13. converges
15. diverges   17. diverges   19. diverges
21. converges   23. diverges   25. converges
27. converges   29. diverges
35. (a) diverges   (b) diverges   (c) diverges   (d) converges
37. (a) 1.1975; $1.2016 < S < 1.2026$ (b) 23
39. (a) $13 < s_{1,000,000} < 15$ (b) 2.69×10^{43}

▶ **Exercise Set 11.5 (Page 541)**

1. converges   3. inconclusive   5. diverges
7. diverges   9. converges   11. diverges
13. converges   15. converges   17. converges
19. converges   21. diverges   23. converges
25. converges   27. diverges   29. converges
31. converges
39. (a) 1.71667; error < 0.00163 (b) 8
41. (a) 0.69226; error < 0.00098 (b) 13

▶ **Exercise Set 11.6 (Page 546)**

1. converges   3. converges   5. diverges
19. converges   21. diverges   23. converges
25. diverges   27. converges   29. diverges
31. converges   33. converges   37. converges
39. $p > 1$ 41. converges

▶ **Exercise Set 11.7 (Page 554)**

1. converges   3. diverges   5. converges
7. absolutely   9. diverges   11. absolutely
13. conditionally   15. divergent   17. conditionally
19. absolutely   21. conditionally   23. divergent

25. conditionally **27.** absolutely **29.** conditionally
31. 0.125 **33.** 0.1 **35.** 9999 **37.** 39,999
39. $|\text{error}| < 0.00074$; $s_{10} \approx 0.4995$; exact sum = 0.5
41. $n = 4$, $s_4 \approx 0.84147$; $\sin(1) \approx 0.841470985$
43. $n = 9$, $s_9 \approx 0.40553$; $\ln\frac{3}{2} \approx 0.405465108$
45. (a) 14 (b) 0.817962176; $|\text{error}| \approx 0.004504858$

▶ **Exercise Set 11.8 (Page 560)**

1. 1, $[-1, 1)$ **3.** $+\infty$, $(-\infty, +\infty)$ **5.** $\frac{1}{5}$, $[-\frac{1}{5}, \frac{1}{5}]$
7. 1, $[-1, 1]$ **9.** 1, $(-1, 1]$ **11.** $+\infty$, $(-\infty, +\infty)$
13. $+\infty$, $(-\infty, +\infty)$ **15.** 1, $[-1, 1]$ **17.** 1, $(-2, 0]$
19. $\frac{4}{3}$, $(-\frac{19}{3}, -\frac{11}{3})$ **21.** 1, $[-2, 0]$ **23.** $+\infty$, $(-\infty, +\infty)$
25. $x + \frac{1}{2}x^2 + \frac{3}{14}x^3 + \frac{3}{35}x^4 + \cdots$; $R = 3$
27. $x + \frac{3}{2}x^2 + \frac{5}{8}x^3 + \frac{7}{48}x^4 + \cdots$; $R = +\infty$
29. $(a - b, a + b)$ **31.** $+\infty$

▶ **Exercise Set 11.9 (Page 567)**

1. $1 - 2x + 2x^2 - \frac{4}{3}x^3 + \frac{2}{3}x^4$ **3.** $2x - \frac{4}{3}x^3$ **5.** $x + \frac{1}{3}x^3$
7. $x + x^2 + \frac{x^3}{2!} + \frac{x^4}{3!}$ **9.** $1 + \frac{1}{2}x^2 + \frac{5}{24}x^4$
11. $\ln 3 + \frac{2}{3}x - \frac{2}{9}x^2 + \frac{8}{81}x^3 - \frac{4}{81}x^4$
13. $e + e(x - 1) + \frac{e}{2!}(x - 1)^2 + \frac{e}{3!}(x - 1)^3$
15. $2 + \frac{1}{4}(x - 4) - \frac{1}{64}(x - 4)^2 + \frac{1}{512}(x - 4)^3$
17. $\frac{\sqrt{2}}{2} - \frac{\sqrt{2}}{2}\left(x - \frac{\pi}{4}\right) - \frac{\sqrt{2}}{4}\left(x - \frac{\pi}{4}\right)^2 + \frac{\sqrt{2}}{12}\left(x - \frac{\pi}{4}\right)^3$
19. $-\frac{\sqrt{3}}{2} + \frac{\pi}{2}\left(x + \frac{1}{3}\right) + \frac{\sqrt{3}\pi^2}{4}\left(x + \frac{1}{3}\right)^2 - \frac{\pi^3}{12}\left(x + \frac{1}{3}\right)^3$
21. $\frac{\pi}{4} + \frac{1}{2}(x - 1) - \frac{1}{4}(x - 1)^2 + \frac{1}{12}(x - 1)^3$
23. $\sum_{k=0}^{\infty} (-1)^k \frac{x^k}{k!}$ **25.** $\sum_{k=0}^{\infty} (-1)^k x^k$ **27.** $\sum_{k=1}^{\infty} (-1)^{k+1} \frac{x^k}{k}$
29. $\sum_{k=0}^{\infty} (-1)^k \frac{x^{2k}}{4^k(2k)!}$ **31.** $\sum_{k=0}^{\infty} \frac{x^{2k}}{(2k)!}$
33. $\sum_{k=0}^{\infty} (-1)(x + 1)^k$ **35.** $\sum_{k=1}^{\infty} (-1)^{k+1} \frac{(x - 1)^k}{k}$
37. $\sum_{k=0}^{\infty} (-1)^k \frac{\pi^{2k}}{(2k)!}\left(x - \frac{1}{2}\right)^{2k}$
39. $\sum_{k=0}^{\infty} \left(\frac{16 + (-1)^{k+1}}{8}\right) \frac{(x - \ln 4)^k}{k!}$

▶ **Exercise Set 11.10 (Page 575)**

1. $\frac{2^6 e^{2c}}{6!}x^6$ **3.** $-\frac{x^5}{(c + 1)^6}$ **5.** $\frac{(4 + c)e^c}{4!}x^4$
7. $-\frac{(1 - 3c^2)}{3(1 + c^2)^3}x^3$ **9.** $-\frac{5}{128c^{7/2}}(x - 4)^4$
11. $\frac{\cos c}{5!}\left(x - \frac{\pi}{6}\right)^5$ **13.** $\frac{7}{(1 + c)^8}(x + 2)^6$
15. $\frac{x^{n+1}}{(1 - c)^{n+2}}$ **17.** $\frac{2^{n+1}e^{2c}}{(n + 1)!}x^{n+1}$
27. $1 - 2x + 2x^2 - \frac{4}{3}x^3 + \cdots$; $(-\infty, +\infty)$
29. $x - x^2 + \frac{1}{2!}x^3 - \frac{1}{3!}x^4 + \cdots$; $(-\infty, +\infty)$

31. $2x - \frac{2^3}{3!}x^3 + \frac{2^5}{5!}x^5 - \frac{2^7}{7!}x^7 + \cdots$; $(-\infty, +\infty)$
33. $x^2 - \frac{1}{2!}x^4 + \frac{1}{4!}x^6 - \frac{1}{6!}x^8 + \cdots$; $(-\infty, +\infty)$
35. $x^2 - \frac{2^3}{4!}x^4 + \frac{2^5}{6!}x^6 - \frac{2^7}{8!}x^8 + \cdots$; $(-\infty, +\infty)$
37. $-x^2 - \frac{1}{2}x^4 - \frac{1}{3}x^6 - \frac{1}{4}x^8 - \cdots$; $(-1, 1)$
39. $1 + 4x^2 + 16x^4 + 64x^6 + \cdots$; $(-\frac{1}{2}, \frac{1}{2})$
41. $x^2 - 3x^3 + 9x^4 - 27x^5 + \cdots$; $(-\frac{1}{3}, \frac{1}{3})$
43. $2x^2 + \frac{2^3}{3!}x^4 + \frac{2^5}{5!}x^6 + \frac{2^7}{7!}x^8 + \cdots$; $(-\infty, +\infty)$
45. $1 + \frac{3}{2}x - \frac{9}{8}x^2 + \frac{27}{16}x^3 - \cdots$; $(-\frac{1}{3}, \frac{1}{3})$
47. $1 + 4x + 12x^2 + 32x^3 + \cdots$; $(-\frac{1}{2}, \frac{1}{2})$
49. $x + \frac{1}{2}x^3 + \frac{3}{8}x^5 + \frac{5}{16}x^7 + \cdots$; $(-1, 1)$
51. $\sum_{k=0}^{\infty} (-1)^k(x - 1)^k$; $(0, 2)$ **53.** $\sum_{k=0}^{\infty} \binom{m}{k} x^k$
55. $\sin \pi = 0$ **57.** $e^{-\ln 3} = \frac{1}{3}$
59. (a) $x^2 - 2x^4 + \frac{2}{3}x^6 - \frac{4}{45}x^8 + \cdots$ (b) 0

▶ **Exercise Set 11.11 (Page 581)**

1. (a) 9 (b) 13 **3.** 1.6487 **5.** 0.9877 **7.** 0.223
9. 0.100 **11.** 1.0050 **13.** $|x| < 0.569$ **15.** 9×10^{-8}
17. (a) 1,999,999 (b) 8 **19.** 3.140

▶ **Exercise Set 11.12 (Page 587)**

5. $\sum_{k=1}^{\infty} (-1)^{k+1} kx^{k-1}$ **9.** $\ln\frac{4}{3}$ **11.** $\frac{2}{(1 - x)^3}$
13. (b) $\sum_{k=0}^{\infty} (-1)^k x^k$; $(-1, 1)$ **15.** 0.764 **17.** 0.494
19. 0.100 **21.** 0.491 **23.** $1 - \frac{3}{2}x^2 + \frac{25}{24}x^4 - \frac{331}{720}x^6 + \cdots$
25. $x - x^2 + \frac{1}{3}x^3 - \frac{1}{30}x^5 + \cdots$
27. $-x^3 - \frac{1}{2}x^5 - \frac{1}{3}x^7 - \frac{1}{4}x^9 - \cdots$
29. $x^2 + \frac{9}{2}x^3 + \frac{79}{8}x^4 + \frac{683}{48}x^5 + \cdots$ **31.** (a) 0 (b) $-\frac{3}{2}$
33. (a) $x - \frac{1}{6}x^3 + \frac{3}{40}x^5 - \frac{5}{112}x^7$
(b) $x + \sum_{k=1}^{\infty} (-1)^k \frac{1 \cdot 3 \cdot 5 \cdots (2k - 1)}{2^k k!(2k + 1)} x^{2k+1}$ (c) 1

▶ **TECHNOLOGY EXERCISES, CHAPTER 11 (Page 589)**

1. (a) 500 (b) 3.139593; $|\text{error}| < 4 \times 10^{-3}$
(c) $|\text{error}| \approx 2 \times 10^{-3}$
3. (a) 3.141592653589793878 (b) $|\text{error}| \approx 6.4 \times 10^{-16}$
7. (a) 50 (b) 1.106279
9. $B_0 = 1$, $B_1 = -\frac{1}{2}$, $B_2 = \frac{1}{6}$, $B_3 = 0$, $B_4 = -\frac{1}{30}$, $B_5 = 0$, $B_6 = \frac{1}{42}$
11. (a) 0.001615 (b) 0.000333

▶ **Exercise Set 7.7 (Page 406)**

1. $y = Cx$ **3.** $y = Ce^{-\sqrt{1+x^2}} - 1$ **5.** $y = \ln(\sec x + C)$
7. $y = e^{-2x} + Ce^{-3x}$ **9.** $y = e^{-x}\sin(e^x) + Ce^{-x}$
11. $y = -\frac{2}{7}x^4 + Cx^{-3}$ **13.** $y = -1 + 4e^{x^2/2}$
15. $y = 2 - e^{-t}$ **17.** $y = \sqrt[3]{3t - 3\ln t + 24}$
19. $y = -3/(2\sqrt{x} + 1)$ **21.** $y = -\ln(3 - x^2/2)$
23. (a) $h(t) = (2 - 0.003979t)^2$ (b) about 8.4 min
25. $v = 50/(2t + 1)$, $x = 25\ln(2t + 1)$

27. (a) $v(t) = c \ln \dfrac{m_0}{m_0 - kt} - gt$ (b) 3044 m/sec

29. (a) $v^2 = 2gR^2/x + v_0{}^2 - 2gR$

31. (a) $y = 200 - 175e^{-t/25}$ (b) 136 lb

33. 25 lb   35. (a) $y = 10e^{-0.005t}$ (b) 7 mg

37. 196 days   39. 6.8 years

41. (a) 14,400   (b) 38 years

▶ **Exercise Set D (Page A2)**

1. $x = 1,\ y = 2$ 3. $x_1 = -\frac{7}{32},\ x_2 = \frac{5}{16}$

5. $x = 2,\ y = -1,\ z = 3$ 7. $x_1 = \frac{26}{21},\ x_2 = \frac{25}{21},\ x_3 = \frac{5}{7}$

9. $x' = x \cos\theta + y \sin\theta,\ y' = -x \sin\theta + y \cos\theta$

▶ **Exercise Set E.I (Page A7)**

1. (a–d)

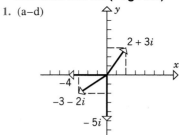

3. (a) $x = -2,\ y = -3$ (b) $x = 2,\ y = 1$

5. (a) $2 + 3i$ (b) $-1 - 2i$ (c) $-2 + 9i$

7. (a)

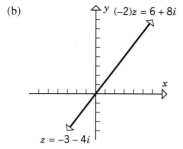

(b)

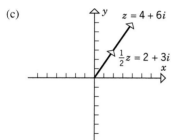

(c)

9. (a) $z_1 z_2 = 3 + 3i,\ z_1^2 = -9,\ z_2^2 = -2i$

(b) $z_1 z_2 = 26,\ z_1^2 = -20 + 48i,\ z_2^2 = -5 - 12i$

(c) $z_1 z_2 = \frac{11}{3} - i,\ z_1^2 = -\frac{4}{3} + \frac{16}{9}i,\ z_2^2 = -6 - \frac{5}{2}i$

11. $76 - 88i$ 13. $-26 + 18i$ 15. $-\frac{63}{16} + i$ 17. 0

19. (a) $2 - 7i$ (b) $-3 + 5i$ (c) $-5i$ (d) i (e) -9

(f) 0

23. (a) $-i$ (b) $\frac{1}{26} + \frac{5}{26}i$ (c) $7i$

25. $\frac{1}{2} + \frac{1}{2}i$ 27. $-\frac{7}{625} - \frac{24}{625}i$

29. $\frac{1}{4}(1 - \sqrt{3}) + \frac{1}{4}(1 + \sqrt{3})i$ 31. $-\frac{1}{10} + \frac{1}{10}i$

33. (a) $-1 - 2i$ (b) $-\frac{3}{25} - \frac{4}{25}i$

35. (a)   (b)

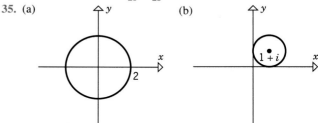

(c)   (d)

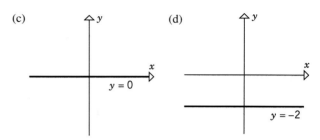

39. (b) i 43. (a) $\dfrac{x_1 x_2 + y_1 y_2}{x_2^2 + y_2^2}$ (b) $\dfrac{x_2 y_1 - x_1 y_2}{x_2^2 + y_2^2}$

47. (c) yes, if $z \ne 0$

▶ **Exercise Set E.II (Page A14)**

1. (a) 0   (b) $\pi/2$ (c) $-\pi/2$ (d) $\pi/4$ (e) $2\pi/3$

(f) $-\pi/4$

3. (a) $2[\cos(\pi/2) + i \sin(\pi/2)]$ (b) $4[\cos\pi + i \sin\pi]$

(c) $5\sqrt{2}[\cos(\pi/4) + i \sin(\pi/4)]$

(d) $12[\cos(2\pi/3) + i \sin(2\pi/3)]$

(e) $3\sqrt{2}[\cos(-3\pi/4) + i \sin(-3\pi/4)]$

(f) $4[\cos(-\pi/6) + i \sin(-\pi/6)]$

5. 1

7. (a)

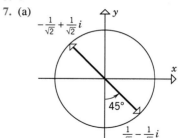

(b)

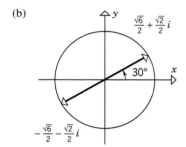

(c)

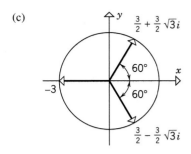

(d)

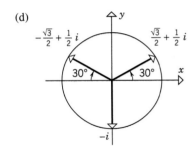

(e)

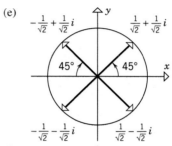

(f)

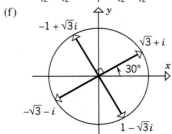

9.

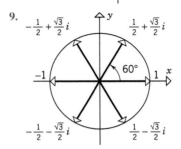

11. (a) $\pm 2, \pm 2i$ (b) $\pm(2 + 2i), \pm(2 - 2i)$

13. rotates z clockwise by 90°

15. (a) $\text{Re}(z) = -3$, $\text{Im}(z) = 0$ (b) $\text{Re}(z) = -3$, $\text{Im}(z) = 0$

   (c) $\text{Re}(z) = 0$, $\text{Im}(z) = -\sqrt{2}$ (d) $\text{Re}(z) = -3$, $\text{Im}(z) = 0$

Index

Photo Credits

☐ CHAPTER OPENERS

Preliminaries
Portrait: Courtesy of The David Eugene Smith Collection, Columbia University
Signature: Courtesy of The New York Public Library Picture Collection

Portrait: Courtesy of The New York Public Library Picture Collection
Signature: Courtesy of the Bettmann Archive

Chapter 11
Portrait: Courtesy of The David Eugene Smith Collection, Columbia University
Signature: Courtesy The British Library

Chapter 14
Portrait: Courtesy of The David Eugene Smith Collection, Columbia University
Signature: Courtesy Universitätsbibliothek Basel

Chapter 15
Portrait: Courtesy AIP Neils Bohr Library
Signature: Courtesy Smithsonian Institution

Chapter 16
Portrait and Signature: Courtesy Smithsonian Institution

Chapter 17
Portrait: Courtesy of The David Eugene Smith Collection, Columbia University
Signature: Courtesy Smithsonian Institution

Chapter 18
Portrait and Signature: Courtesy of the Bettmann Archive

Chapter 19
Portrait: Courtesy of The David Eugene Smith Collection, Columbia University
Signature: Courtesy The British Library

☐ CHAPTER OPENING
   PORTRAITS

The chapter opening portraits were created by HRS Electronic Text Management from black and white engravings. They were scanned into and manipulated by computer to create the color illustrations.

INTEGRALS CONTAINING $2au - u^2$

50. $\displaystyle\int \sqrt{2au - u^2}\, du = \frac{u - a}{2}\sqrt{2au - u^2} + \frac{a^2}{2}\sin^{-1}\left(\frac{u - a}{a}\right) + C$

51. $\displaystyle\int u\sqrt{2au - u^2}\, du = \frac{2u^2 - au - 3a^2}{6}\sqrt{2au - u^2} + \frac{a^3}{2}\sin^{-1}\left(\frac{u - a}{a}\right) + C$

52. $\displaystyle\int \frac{\sqrt{2au - u^2}\, du}{u} = \sqrt{2au - u^2} + a\sin^{-1}\left(\frac{u - a}{a}\right) + C$

53. $\displaystyle\int \frac{\sqrt{2au - u^2}\, du}{u^2} = -\frac{2\sqrt{2au - u^2}}{u} - \sin^{-1}\left(\frac{u - a}{a}\right) + C$

54. $\displaystyle\int \frac{du}{\sqrt{2au - u^2}} = \sin^{-1}\left(\frac{u - a}{a}\right) + C$

55. $\displaystyle\int \frac{u\, du}{\sqrt{2au - u^2}} = -\sqrt{2au - u^2} + a\sin^{-1}\left(\frac{u - a}{a}\right) + C$

56. $\displaystyle\int \frac{u^2\, du}{\sqrt{2au - u^2}} = -\frac{(u + 3a)}{2}\sqrt{2au - u^2} + \frac{3a^2}{2}\sin^{-1}\left(\frac{u - a}{a}\right) + C$

57. $\displaystyle\int \frac{du}{u\sqrt{2au - u^2}} = -\frac{\sqrt{2au - u^2}}{au} + C$

58. $\displaystyle\int \frac{du}{(2au - u^2)^{3/2}} = \frac{u - a}{a^2\sqrt{2au - u^2}} + C$

59. $\displaystyle\int \frac{u\, du}{(2au - u^2)^{3/2}} = \frac{u}{a\sqrt{2au - u^2}} + C$

INTEGRALS CONTAINING TRIGONOMETRIC FUNCTIONS

60. $\displaystyle\int \sin u\, du = -\cos u + C$

61. $\displaystyle\int \cos u\, du = \sin u + C$

62. $\displaystyle\int \tan u\, du = \ln|\sec u| + C$

63. $\displaystyle\int \cot u\, du = \ln|\sin u| + C$

64. $\displaystyle\int \sec u\, du = \ln|\sec u + \tan u| + C$
$\qquad\qquad = \ln|\tan(\tfrac{1}{4}\pi + \tfrac{1}{2}u)| + C$

65. $\displaystyle\int \csc u\, du = \ln|\csc u - \cot u| + C$
$\qquad\qquad = \ln|\tan\tfrac{1}{2}u| + C$

66. $\displaystyle\int \sec^2 u\, du = \tan u + C$

67. $\displaystyle\int \csc^2 u\, du = -\cot u + C$

68. $\displaystyle\int \sec u \tan u\, du = \sec u + C$

69. $\displaystyle\int \csc u \cot u\, du = -\csc u + C$

70. $\displaystyle\int \sin^2 u\, du = \tfrac{1}{2}u - \tfrac{1}{4}\sin 2u + C$

71. $\displaystyle\int \cos^2 u\, du = \tfrac{1}{2}u + \tfrac{1}{4}\sin 2u + C$

72. $\displaystyle\int \tan^2 u\, du = \tan u - u + C$

73. $\displaystyle\int \cot^2 u\, du = -\cot u - u + C$

74. $\displaystyle\int \sin^n u\, du = -\frac{1}{n}\sin^{n-1} u \cos u + \frac{n - 1}{n}\int \sin^{n-2} u\, du$

75. $\displaystyle\int \cos^n u\, du = \frac{1}{n}\cos^{n-1} u \sin u + \frac{n - 1}{n}\int \cos^{n-2} u\, du$

76. $\displaystyle\int \tan^n u\, du = \frac{1}{n - 1}\tan^{n-1} u - \int \tan^{n-2} u\, du$

77. $\displaystyle\int \cot^n u\, du = -\frac{1}{n - 1}\cot^{n-1} u - \int \cot^{n-2} u\, du$

78. $\displaystyle\int \sec^n u\, du = \frac{1}{n - 1}\sec^{n-2} u \tan u + \frac{n - 2}{n - 1}\int \sec^{n-2} u\, du$

79. $\displaystyle\int \csc^n u\, du = -\frac{1}{n - 1}\csc^{n-2} u \cot u + \frac{n - 2}{n - 1}\int \csc^{n-2} u\, du$

80. $\displaystyle\int \sin mu \sin nu\, du = -\frac{\sin(m + n)u}{2(m + n)} + \frac{\sin(m - n)u}{2(m - n)} + C$

81. $\displaystyle\int \cos mu \cos nu\, du = \frac{\sin(m + n)u}{2(m + n)} + \frac{\sin(m - n)u}{2(m - n)} + C$

82. $\displaystyle\int \sin mu \cos nu\, du = -\frac{\cos(m + n)u}{2(m + n)} - \frac{\cos(m - n)u}{2(m - n)} + C$

83. $\displaystyle\int u \sin u\, du = \sin u - u \cos u + C$

84. $\displaystyle\int u \cos u\, du = \cos u + u \sin u + C$

85. $\displaystyle\int u^2 \sin u\, du = 2u \sin u + (2 - u^2)\cos u + C$

86. $\displaystyle\int u^2 \cos u\, du = 2u \cos u + (u^2 - 2)\sin u + C$

87. $\displaystyle\int u^n \sin u\, du = -u^n \cos u + n\int u^{n-1}\cos u\, du$

88. $\displaystyle\int u^n \cos u\, du = u^n \sin u - n\int u^{n-1}\sin u\, du$

89. $\displaystyle\int \sin^m u \cos^n u\, du = -\frac{\sin^{m-1} u \cos^{n+1} u}{m + n} + \frac{m - 1}{m + n}\int \sin^{m-2} u \cos^n u\, du$
$\qquad\qquad\qquad = \frac{\sin^{m+1} u \cos^{n-1} u}{m + n} + \frac{n - 1}{m + n}\int \sin^m u \cos^{n-2} u\, du$